Physical and Chemical Mechanisms in Molecular Radiation Biology

BASIC LIFE SCIENCES

Ernest H. Y. Chu, Series Editor
The University of Michigan Medical School
Ann Arbor, Michigan

Alexander Hollaender, Founding Editor

Recent volumes in the series:

Physical and Chemical Mechanisms in Molecular Radiation Biology

Edited by

William A. Glass

Pacific Northwest Laboratories
Richland, Washington

and

Matesh N. Varma

U.S. Department of Energy
Germantown, Maryland

Plenum Press • New York and London

Library of Congress Cataloging in Publication Data

Physical and chemical mechanisms in molecular radiation biology / edited by William
A. Glass and Matesh N. Varma.
    p.   cm. — (Basic life sciences; v. 58)
Proceedings of a workshop held Sept. 3–6, 1990, at the National Academy of
Sciences Study Center, Woods Hole, Mass. and organized by the U.S. Dept. of Energy,
Office of Health and Environmental Research.
Includes bibliographical references and index.
ISBN 978-1-4684-7629-3   ISBN 978-1-4684-7627-9 (eBook)
DOI 10.1007/978-1-4684-7627-9
1. Molecular radiobiology — Congresses. I. Glass, William A. II. Varma, Matesh N.
III. United States. Dept. of Energy. Office of Health and Environmental Research. IV.
Series.
[DNLM: 1. Molecular Biology — congresses. 2. Radiobiology — congresses. W3
BA255 v. 58 / WN 160 P575 1990]
QP82.2.R3P53   1991
574.8′8 — dc20
DNLM/DLC   91-45566
for Library of Congress   CIP

Organized by the U.S. Department of Energy,
Office of Health and Environmental Research

Proceedings of a workshop on Physical and Chemical Mechanisms in
Molecular Radiation Biology, held September 3–6, 1990, at the
National Academy of Sciences Study Center, Woods Hole, Massachusetts

ISBN 978-1-4684-7629-3

© 1991 Plenum Press, New York
Softcover reprint of the hardcover 1st edition 1991
A Division of Plenum Publishing Corporation
233 Spring Street, New York, N.Y. 10013

In Appreciation

The body of knowledge represented in these proceedings is due, in large measure, to research made possible by the pioneering leadership and dedication of Dr. Robert W. Wood. For nearly thirty years, Dr. Wood has guided the programs of the Office of Health and Environmental Research and its predecessor organizations in this multidisciplinary field. He has successfully encouraged national and international cooperation in scientific exchange and among funding agencies to assure continuity of research aimed at understanding the physical and chemical mechanisms underlying the effects of ionizing radiation on human health. His vision and foresight in recognizing the fundamental value of this knowledge is greatly appreciated by the radiation research community.

Acknowledgments

The editors of these proceedings would like to offer special thanks to the superb production team from Pacific Northwest Laboratory: Suzanne Liebetrau and Sallie Ortiz, whose creativity and dedication to the Woods Hole project made possible the preparation of this manuscript; Mary Heid, whose computer expertise enhanced both quality and speed of production; Katherine Borgeson and Lisa Ballou, who contributed many hours of editing and proofing; and, last but certainly not least, Phyllis Parker, whose capability and patience greatly facilitated the production effort at all stages.

Preface

The fundamental understanding of the production of biological effects by ionizing radiation may well be one of the most important scientific objectives of mankind; such understanding could lead to the effective and safe utilization of the nuclear energy option. In addition, this knowledge will be of immense value in such diverse fields as radiation therapy and diagnosis and in the space program. To achieve the above stated objective, the U.S. Department of Energy (DOE) and its predecessors embarked upon a fundamental interdisciplinary research program some 35 years ago. A critical component of this program is the Radiological and Chemical Physics Program (RCPP).

When the RCPP was established, there was very little basic knowledge in the fields of physics, chemistry, and biology that could be directly applied to understanding the effects of radiation on biological systems. Progress of the RCPP program in its first 15 years was documented in the proceedings of a conference held at Airlie, Virginia, in 1972. At this conference, it was clear that considerable progress had been made in research on the physical and chemical processes in well-characterized systems that could be used to understand biological effects. During this period of time, most physical knowledge was obtained for the gas phase because the technology and instrumentation had not progressed to the point that measurements could be made in liquids more characteristic of biological materials. Furthermore, most biological experiments were in what we now call "Classical Radiobiology," such as studying dose-response curves at the animal, organ, or cellular level. The field of molecular biology was just being established. However, with the advent of new technology and instrumentation such as electrophoresis, polymer chain reaction (PCR), and flow cytometry, the identification of biological changes at a molecular level became possible. In the years following the Airlie House Conference, the RCPP recognized the potential of the new radiobiology and molecular biology in its program. In recent years, the Program has emphasized fundamental research in physical and chemical processes in the condensed phase and utilization of this research through studies to understand the molecular changes in biological systems.

To record the progress and advances that have been made in this effort since 1972, the conference documented in these proceedings was held in September 1990 at the National Academy of Sciences' Center at Woods Hole, Massachusetts. Leading scientists in the fields of physics, chemistry, and biology, many of whom have been or are presently involved in RCPP programs, were invited to this conference to review the advances in the last 20 years and to provide some direction for the future. The agenda was arranged to provide ample time for discussions. The scientific discussions following the formal presentations were recorded and made part of this conference proceedings.

These proceedings provide an important documentation of the scientific progress made in the Radiological and Chemical Physics Program. We hope that it will stimulate and attract new investigators to this program to tackle the challenge of understanding, from first principles, the biological effects produced by ionizing radiation.

Matesh N. Varma, Ph.D.
Office of Health and Environmental Research

Contents

Models of Radiation Effects

Molecular Radiation Biology

Introduction to the Problem

The Molecular Biology of Radiation Carcinogenesis

Eric J. Hall and Greg A. Freyer[a]

Abstract

Major new insights into carcinogenesis have come from recent advances in cellular and molecular biology. The concept of oncogenes provides a simple explanation for how agents as diverse as radiation, chemicals or retroviruses can induce tumors that are indistinguishable one from another. Oncogenes may be activated by a point mutation, by a chromosome translocation, or by amplification. Ionizing radiations are efficient at the first two mechanisms. While oncogenes are frequently associated with leukemias and lymphomas, they are associated with only 10 to 15% of human solid cancers. The importance of the loss of suppressor genes was suggested first from studies with human-hamster hybrid cells, but has since been shown to be of importance in an increasing number of human solid tumors, from rare tumors such as retinoblastoma to more common tumors such as small cell lung cancer and colorectal cancer. The mechanism of somatic homozygosity clearly involves several steps, some of which, such as a deletion, could be readily produced by ionizing radiation.

The multi-step nature of carcinogenesis can be demonstrated in the petri dish, where the transfection of multiple oncogenes is required to transform normal cells from short-term explants. It can be shown, too, in colorectal cancer in the human, where the activation of an oncogene and the loss of more than one suppressor gene may be involved in the progression from normal epithelium to a frank malignancy.

Introduction

The evolution of radiation biology over the past 75 years has been dictated by the development of techniques to study smaller and smaller biological entities, from whole organisms to cells to genes to base sequences in DNA.

In the 1930's and 40's, several separate and parallel studies took place: (i) the lethal effects of whole body irradiation of experimental animals; (ii) the genetic consequences of radiation from a gross study of offspring, but without an understanding of the mechanisms involved; (iii) chromosome changes and abberrations, the earliest visible signs of radiation damage in biological material; and (iv) the induction of cancer and leukemia in irradiated animals (and humans).

During the past decade, the development of techniques in molecular biology has enabled these diverse effects in radiation biology to be better understood and interpreted in terms of changes in the base sequences of the DNA in the relevant target cells.

(a) Center for Radiological Research, College of Physicians and Surgeons of Columbia University, New York

Physical and Chemical Mechanisms in Molecular Radiation Biology
Edited by W.A. Glass and M.N. Varma, Plenum Press, New York, 1991

As our level of sophistication has progressed from the whole organism to the tissue, to the cell, to the chromosome, and to the gene, we have arrived at the basic building block of biological material, the arrangement and sequence of nucleic acids in DNA. The diverse biological effects of ionizing radiations can all ultimately be understood and interpreted in terms of changes and/or disruptions in these base sequences.

Prospects for the immediate future, like progress in the recent past in radiation biology, reflect the development of molecular biology technology and the change in philosophy that it engenders.

Cancer as a Multi-Step Process

Cancer is by far the most important possible deleterious effect of low doses of ionizing radiation. Consequently, the processes underlying radiation-induced carcinogenesis must be central to the mission and interests of the Department of Energy.

The major new insights of the past decade have come from advances in cellular and molecular biology. A principal justification for studying cancer cells in vitro, abstracted from the entire organism, is that a neoplasm is usually considered as arising from a single cell that has undergone a critical change. Evidence for this includes the fact that some malignancies can be propagated by a single cell and many tumors have been shown to be clonal, in that every cell carries the same biochemical marker.

Despite its clonal origin, as the cells of a cancer grow and divide, progressive stages can be identified from preneoplasia to malignancy. These steps are usually described operationally as initiation, promotion, and progression. The progressive nature of cancer has been known for many years. It was first described in phenomenological terms for skin cancer in animals, but has more recently been described in exquisite molecular detail for colon cancer, as will be discussed later. Radiation carcinogenesis, in common with any other form of tumor production, is likely to be a complex multi-step process.

Oncogenes

Two decades ago, it was hard to understand how agents as diverse as retroviruses, x-rays, chemicals and ultraviolet light could all give rise to cancers that could not be distinguished one from another. It was known that all agents known to cause cancer also damage DNA, but that was as far as our understanding went. The discovery of oncogenes, illustrated in Fig. 1, provided a simple explanation for many experimental observations.

The relationship between viruses and cancer was first established by Peyton Rous in 1911. He found that a filterable extract from sarcomas (connective tissue tumors) could be injected to induce tumors in healthy chickens. The infective agent was later shown to be a virus whose genome is composed of RNA. The Rouse Sarcoma virus (RSV) thus belongs to the category of RNA viruses or retroviruses.[2-4]

A detailed understanding of the evolutionary origin of viral oncogenes came in the late 1970's from studies of the v-src gene.[5,6] J. Michael Bishop and Harold Varmus, of the University of California at San Francisco, used a DNA copy of v-src to probe the cellular DNA of various animals. A closely related homolog, cellular (c)-src, was found in the DNA from normal chickens, fish, mammals, humans, and even *Drosophila*. Subsequently, other viral oncogenes were shown to have related cellular forms that have been conserved from species to species during the course of evolution.[6-8]

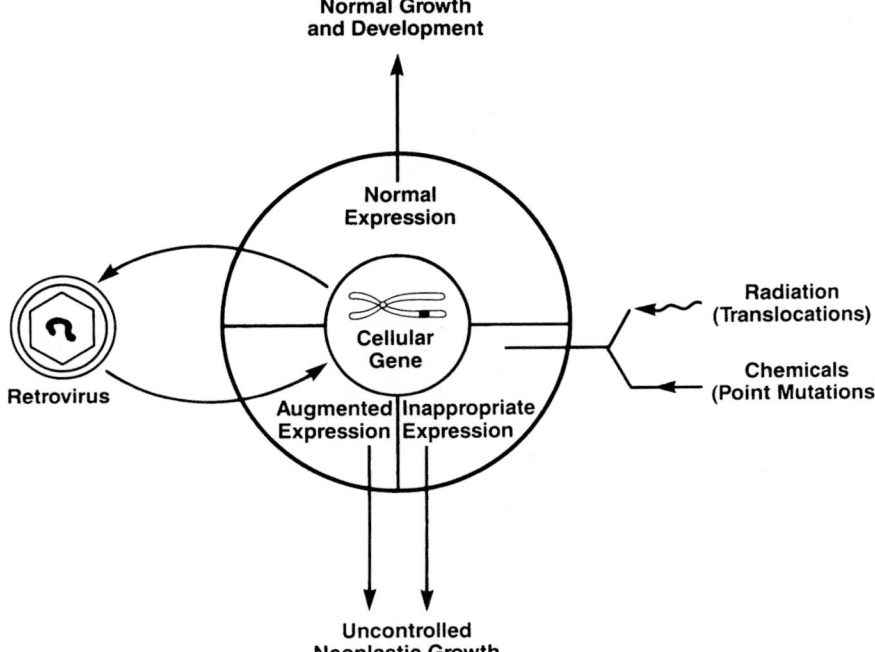

Normal Growth and Development

Normal Expression

Cellular Gene

Retrovirus

Augmented Expression | **Inappropriate Expression**

Radiation (Translocations)

Chemicals (Point Mutations)

Uncontrolled Neoplastic Growth

Figure 1.   A unified scheme to explain the involvement of an oncogene in the neoplastic transformation of cells by retroviruses, radiation, or chemical carcinogens. In the case of the retroviruses, host DNA sequences become reintegrated into recipient cells during virus infection and are then aberrantly expressed to produce the transformed phenotype. In the case of chemical carcinogens or radiation, damage to cellular DNA acts on endogenous oncogenes or long terminal repeat (LTR) sequences to produce aberrant expression of similar sequences, in the absence of a virus vector. Radiation, for example, is particularly efficient at producing translocations, while some chemicals are efficient at producing point mutations.[1]

In the early 1980's, Hidesaburo Hanafusa, at the Rockefeller University, found that retroviruses with an incomplete, defective src oncogene can still produce tumors in animals. Viruses subsequently isolated from the malignant tumors had reassembled a complete src gene, indicating that they had become oncogenic by "capturing" the missing portion of the src gene from the host-cell DNA.[9]

The work of Bishop, Varmus, and Hanafusa showed that retroviral oncogenes are modified cellular genes captured from the genomes of their vertebrate hosts. During provirus integration, all or part of the coding region of a cellular gene may be integrated within the sequences of the viral genome. Thus, the virus acts like a transducing phage, removing the cellular gene as part of its genome as it exits from the host DNA. The cellular gene is then packaged, along with viral sequences, into an infectious virus particle.[10]

So, in fact, cancer caused by retroviruses is mediated through cellular genes that have been mutated or have altered expression. Thus, non-altered oncogenes (proto-oncogenes) have normal cellular functions.[11-13] In most cases where their normal role is understood, proto-oncogenes function in regulating cell growth. It is the loss or alteration of this fine-tuned regulation that leads to overproliferation, which is one signature of cancer. The mechanism of transformation by agents such as chemicals and radiation must be through mutations in these same proto-oncogenes, producing activated oncogenes.

Detection of Activated Cellular Oncogenes

A critical feature of oncogenes is that they act in a dominant fashion. This means that the presence of a single copy of the oncogene in a cell is sufficient to produce the transformed phenotype, even in the presence of normal copies of the oncogene.[14]

Activated cellular oncogenes can be detected by transfection (Fig. 2).[15,16] High-molecular-weight DNA is precipitated by calcium phosphate and layered onto cells in tissue culture. This exogenous DNA is taken up and integrated into the genome of the host cells and expressed. Only a few cell types work well in this assay, most notably NIH 3T3 cells. Like normal cells, this cell line shows contact inhibition and does not produce tumors when injected into animals, but unlike normal cells, NIH 3T3 cells are "immortal," having an unlimited life span. When the fragment of DNA containing an activated cellular oncogene is integrated and expressed, transformation of that cell occurs, leading to loss of contact inhibition and unrestricted growth. In a petri dish, this leads to a piling up of cells to form a "focus." The identification of an oncogene by this process is called a "focus assay"; NIH 3T3 cells are the most common recipient cells, but C3H10T1/2 or Rat-3 cells may also be used.[17] Furthermore, these cells produce tumors when injected into animals.

This procedure has been utilized to identify and isolate several oncogenes from human tumors by the scheme illustrated in Fig. 3. High- molecular-weight DNA from cells that have been transformed by radiation or chemicals can transmit or propagate the malignant phenotype by transfection. This DNA can be probed by the technique of in situ hybridization to detect it in the presence of any of the known oncogenes.[18]

Human Oncogenes

To date, about 50 oncogenes have been identified.[19] Among them the most frequently found are members of the ras gene family: H-ras, K-ras, and N-ras.[20] H-ras and K-ras are the cellular homologues of the oncogenes present in the Harvey and Kirsten strains of murine sarcoma viruses. The N-ras locus has so far not been transduced by retroviruses. Ras oncogenes have been identified in almost every form of human cancer, with an overall incidence of 10% to 15%. In some types of tumors, this percentage is significantly lower (e.g., breast carcinoma), whereas in others (e.g., acute myelocytic leukemia) it could be as high as 25%.

Other human oncogenes identified by different experimental approaches were also found to represent cellular homologues of well-characterized retroviral oncogenes. For example, the oncogene involved in the chromosomal translocation characteristic of Burkitt's lymphoma, c-myc, was first identified as the oncogene of avian myelocyto-matosis virus.[21] Similarly, the oncogene implicated in the development of chronic myelogenous leukemia, c-abl, turned out to be the cellular homologue of v-abl.[22] The oncogene present in a human gastric carcinoma has recently been identified as c-raf-1 proto-oncogene.[23] This locus is the cellular homologue of v-raf, the oncogene of the 3611 strain of murine sarcoma virus.

These findings clearly illustrate the existence in human tumors of dominant oncogenes, some of which correspond to those previously characterized as responsible for the carcinogenic effect of acute transforming retroviruses.

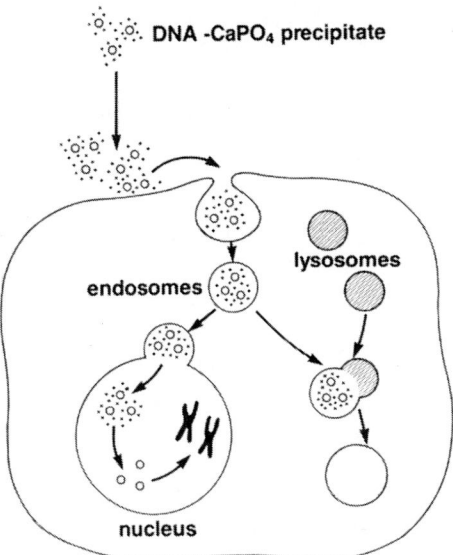

DNA -CaPO$_4$ precipitate

lysosomes

endosomes

nucleus

Figure 2. The technique of transfection. DNA from a mammalian cell, broken up into small fragments, can be introduced into another cell with the aid of calcium phosphate precipitates. The donor DNA is integrated into the genome of the recipient cell and is expressed at low frequency.

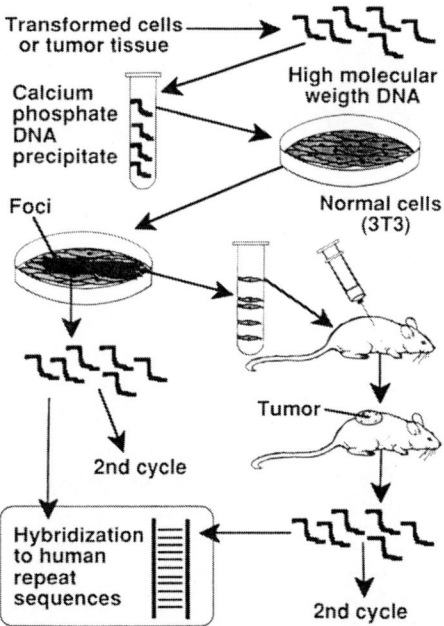

Transformed cells or tumor tissue

High molecular weight DNA

Calcium phosphate DNA precipitate

Foci

Normal cells (3T3)

2nd cycle

Tumor

Hybridization to human repeat sequences

2nd cycle

Figure 3. High-molecular-weight DNA from cells transformed by radiation or chemicals can transmit the malignant phenotype if transfected into non-transformed NIH 3T3 cells. DNA from the foci produced can be used to form tumors in animals, or used for a second round of transfections in vitro. The presence of known oncogenes can then be identified by "in situ" hybridization.

Activation of Oncogenes

There are several mechanisms by which proto-oncogenes can be activated to produce a malignant cell.

- A point mutation can occur, changing a single base pair which subsequently produces a protein with a single amino acid change. In the example of ras, single amino acid substitutions at amino acid positions 12, 13, 59, and 61 have been shown to activate this oncogene.[24] A point mutation in N-ras is found in the cancer cells of most patients suffering from acute leukemia.[25]

- A chromosomal rearrangement or translocation can occur, often placing a proto-oncogene next to a strong promoter and leading to its over expression, or producing a new fusion gene whose product acquires a new transforming activity. The process of translocation is illustrated in Fig. 4. Ionizing radiations are very effective at producing DNA breaks. These breaks may rejoin in such a way as to form a dicentric, which is readily recognizable and almost always lethal to the cell. An approximately equal number of translocations may occur. These are not lethal but are much more difficult to see except by the new techniques of chromosome painting. Translocations have been shown to be associated with several human cancer cells involving myc genes. A translocation between chromosomes 2 and 8 is responsible for myc activation in Burkitt's lymphoma.[21]

- Gene amplification, where many extra copies of a proto-oncogene exist in a cell, is associated with the activation of oncogenes in several cancers. The presence of multiple copies of a proto-oncogene leads to its overexpression. Gene amplification of N-myc is characteristic of many neuroblastomas.[26]

There are many known examples of these mechanisms of activation; one will be cited here for each. A point mutation can activate the N-ras oncogene and produce acute leukemia. A translocation between chromosomes 2 and 8 can move the myc oncogene to a position in which it is activated to result in Burkitt's lymphoma. Gene amplification involving the N-myc oncogene is responsible for neuroblastoma. Table I is a summary of known oncogenes, the chromosomal changes involved, and the human cancers associated with them. The next section describes in more detail the chromosomal changes involved in oncogene activation.

Chromosomal Changes and Oncogenes

Some of the most interesting evidence for the role of oncogenes comes from chromosomal changes in various human cancers. In chronic myelogenous leukemia (CML) the Philadelphia chromosome has been identified in 85% of cases.[27] This involves translocation of chromosome 9 to the long arm of chromosome 22. The incidence of leukemia in children with trisomy 21-23, or Down's syndrome, is 10 to 20 times greater than normal. In addition, syndromes that exhibit a high incidence of chromosomal breakage and rearrangements (e.g., Bloom's syndrome, Fanconi's anemia, and ataxia-telangiectasia) are all associated with an increased incidence of leukemia and solid cancers.[28]

Using the new banding techniques for staining chromosomes, consistent chromosomal abnormalities have been shown in various human cancers. The abnormalities are of two general types. The first is homogeneously staining extra chromosomes, known as a double minutes, or homogeneously staining regions (HSR) in specific chromosomes. These are believed to represent clusters of genes whose presence in cellular DNA has been amplified or increased manyfold by an increase in the number

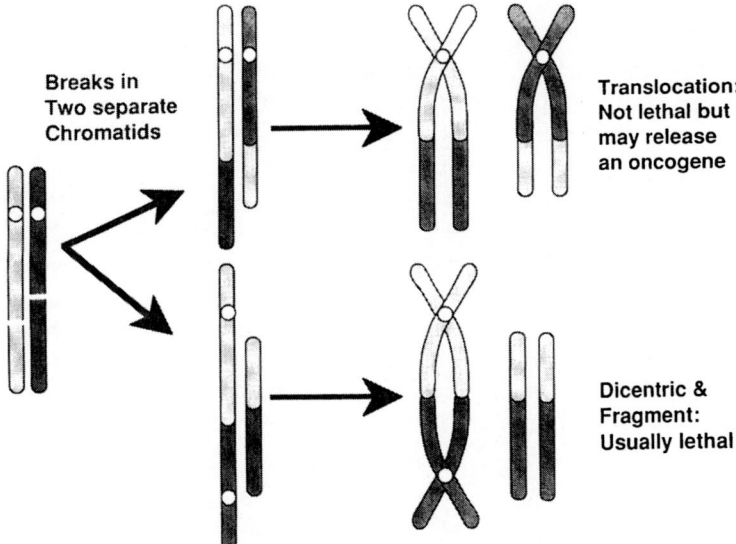

Breaks in Two separate Chromatids

Translocation: Not lethal but may release an oncogene

Dicentric & Fragment: Usually lethal

Figure 4. Diagram of the way in which oncogenes may be activated by a chromosomal translocation. When chromosomes broken by radiation rejoin, the most likely consequence is a rearrangement such as a dicentric, which is not compatible with viability and leads to cell death. On the other hand, a translocation is viable. If the rearrangement moves an oncogene from a place where it is usually dormant or suppressed to a location adjacent to an active region of another chromosome, the oncogene may be activated. This is the case in Burkitt's lymphoma, and several other human cancers.

Table I. Chromosomal Changes Leading to Oncogene Activation, and the Human Malignancies Associated with Them

Oncogene	Chromosomal Changes	Human Cancers
N-ras	deletion (1)	neuroblastoma
Blym	deletion (1)	neuroblastoma
fms	deletion (5)	acute nonlymphocytic leukemia
H-ras	deletion (11)	sarcoma
c-abl	translocation (9-12)	chronic myelogenous leukemia
c-myc	translocation (8-14)	B-cell lymphoma
	translocation (2-8)	Burkitt's lymphoma
N-myc	translocation (2-8)	Burkitt's lymphoma
raf	translocation (3-8)	parotid gland tumor
myb	translocation (6-14)	carcinoma
mos	translocation (3-8)	acute myelocytic leukemia
abl	translocation (9-22)	chronic myelogenous leukemia
sis	translocation (9-22)	chronic myelogenous leukemia
	translocation (8-22)	Burkitt's lymphoma
p-53	rearrangement	osteogenic sarcoma
N-myc	gene amplification	neuroblastoma
neu	gene amplication	breast carcinoma

of copies present.[29,30] For example, in a cell line derived from human neuroblastoma, the N-myc gene was found to be amplified.[31,32] There are many other examples, such as c-myb, and c-abl.

The second type includes specific chromosomal translocations, inversions or deletions. The most interesting example is Burkitt's lymphoma. One of these translocations can be demonstrated in this malignancy, involving the long arm of chromosome 8. The break point occurs in the region of the c-myc. In the most common rearrangements, the oncogene is translocated to the constant chain region of the immunoglobulin heavy chain gene, where its expression is altered.

These findings suggest a general model. Translocations (or inversions) bring together two genes that are normally far apart and under different regulatory controls. This could readily occur by a translocation as illustrated in Fig. 4. If one oncogene is related to control of growth or differentiation while the other codes for a protein that is actively transcribed by the cell, then activation of the growth control gene may occur, and cancer may develop. The real-life situation may well be much more complex, as is described below.

Radiation and Oncogenes

Numerous experimental studies have demonstrated, both in animals and in vitro, that radiation can cause cancer.[33-35] This has been demonstrated. That this occurs via direct or indirect damage to the DNA is a certainty and is born out by the studies of Borek in which DNA isolated from radiation-transformed C3H/10T1/2 cells was shown to transform recipient cells following transfection.[36] The actual molecular mechanism of radiation-induced transformation is unknown. Several studies have attempted to identify the oncogenes in radiation transformed cells by various indirect methods.[37,38] One method used is to search DNA isolated from radiation-transformed cells for mutations in known oncogenes. In this way K-ras and N-ras were shown to be activated in mouse lymphomas induced by gamma radiation.[39]

A second method is to determine whether any known oncogenes are overexpressed in transformed cells. This requires measuring the mRNA levels of known oncogenes by northern hybridization or by dot blot. Two studies used this method to examine gamma-irradiated C3H/10T1/2 cells.[36,40] Each study used several overexpressed, cloned oncogenes as probes and could not identify an oncogene. Both studies speculated on the strong possibility that gamma radiation could activate an as-yet-unidentified oncogene.

A more direct approach to this question is to isolate the oncogene(s) present in the transformed cells (Fig. 5). Recently, such an approach was used to isolate an oncogene from a gamma-irradiated C3H/10T1/2 cell line (Freyer et al. unpublished). While the gene has not yet been identified, 22 cloned oncogenes tested negative by hybridization. These data would support the speculation that a new or unique oncogene is involved in gamma-radiation transformation.

Carcinogenesis as a Multi-Stage Process

It is estimated that if one ignores induced mutations, such as those caused by radiation damage, and simply considers the mutation rate produced by mistakes made in replicating DNA that are missed by the DNA repair system, then an average-sized gene will be mutated once in every one million cells. If we consider that a human produces some 10^{17} cells during a lifetime, an interesting question arises: Why is cancer so uncommon? The answer was first demonstrated epidemiologically years ago. Today experiments with recombinant DNA techniques provide the evidence. Cancer is a multi-step process involving at least two—and likely more than two—oncogenes.

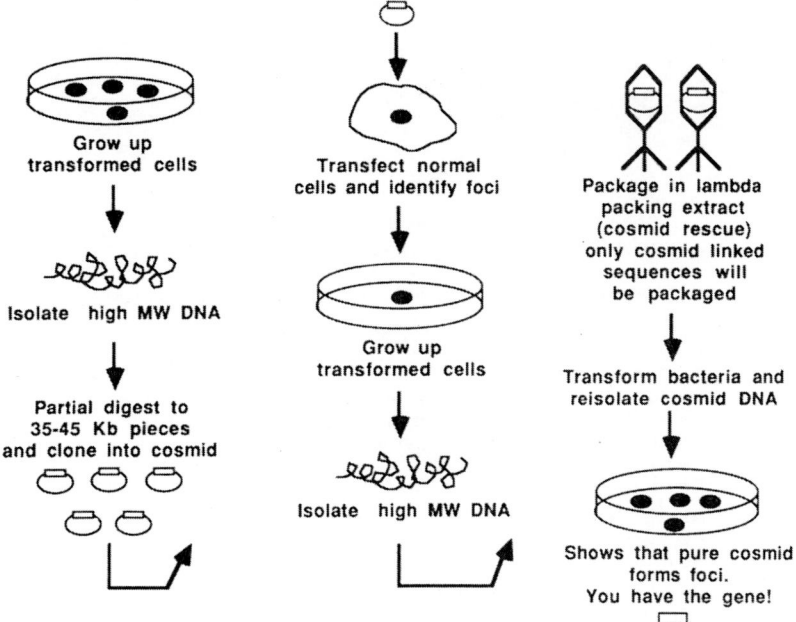

Figure 5. Proposed protocol used to isolate a new dominant transforming gene, different from any of the known oncogenes.

There is much evidence that carcinogenesis *in vivo*, in humans and other animals, is a multi-step process. At least two steps are involved, described operationally as initiation and promotion. Experimentally this was demonstrated eloquently with cloned oncogenes (Fig. 6). Initially it was shown that while the ras oncogene could transform established cell lines like NIH 3T3, it could not transform primary cells. This finding led to a series of studies in which combinations of oncogenes were co-transfected into primary rat embryo cells. The combination of ras plus the adenovirus oncogene E1A or myc led to transformation.[41,42] From these studies investigators recognized that oncogenes could be grouped into two categories; those that immortalized cells, and those that led to the loss-of-contact inhibition associated with the neoplastic state.

Recent evidence for the multi-step nature of cancer has come from clinical studies, most notably those from the laboratory of Volgelstein—studying the clinical progression of colo-rectal cancer.[44] These studies have beautifully demonstrated a correlation between 1) the clinical progression of the cancer from benign, through non-malignant adenomas, to full-blown cancer, and 2) the activation of oncogenes and the loss of anti-oncogenes. This progression is illustrated in Fig. 7.

Somatic Cell Hybrids and the Analysis of Malignancy

Somatic-cell-hybridization experiments, which preceded the discovery of oncogenes, showed that the cancerous phenotype could be suppressed by the introduction of normal genetic information via whole-cell fusion.[45,46] The fusion of a normal human fibroblast with a HeLa cell suppressed the malignant expression of the HeLa cell. This is illustrated in Fig. 8. Using the technique of single chromosome transfer[47,48] it is possible to show that transfer of chromosome 11 from normal fibroblasts into HeLa cells or HeLa-fibroblast hybrids results in the suppression of tumor-forming ability, as

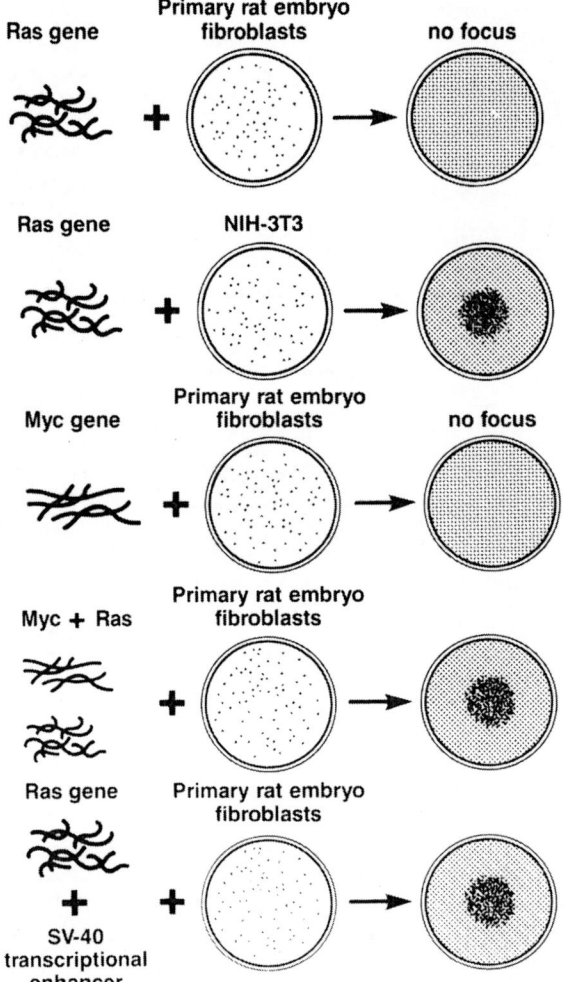

Figure 6. Schematic illustration of the cotransfection studies that showed that cooperation between both the ras and myc genes are required to transform primary nonimmortalized rat embryo fibroblasts, whereas only the ras gene is needed to transform the NIH/3T3 cell line, which is already immortal. The interpretation is that the myc gene confers immortality, while the ras gene is responsible for the morphological changes associated with oncogenic transformation.[41]

illustrated in Fig. 9.[49] In another study, transfer of chromosome 11 into a Wilms' tumor cell line also resulted in suppression of tumorigenicity, whereas transfer of other chromosomes had no effect.[50] These results indicate that human chromosome 11 carries one or more genes expressing strong tumor-suppressing abilities, loss of which leads to the expression of the malignant phenotype. These human fibroblast x HeLa hybrids can be used as the basis of a transformation assay for radiation. When exposed to X-rays or neutrons, a proportion of the hybrids expresses tumor-associated antigens in a dose-dependent manner.[51] The mechanism for the radiation-induced neoplastic transformation of these cells is the loss of the tumor suppressor gene on chromosome 11 of the normal fibroblast.

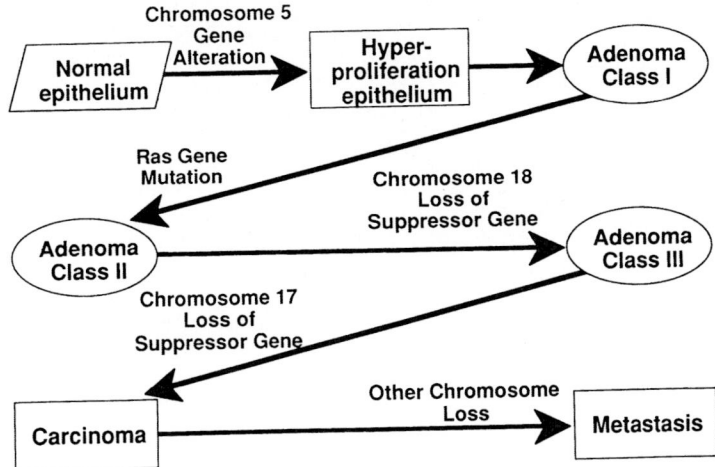

Figure 7. Cancer has long been thought to be a multi-step process, with operational terms such as initiation, promotion and progression. Colon cancer is an excellent example where this multi-step process can be characterized by a series of chromosomal and molecular events that accompany the evolution from hyperproliferation to a metastatic malignant tumor.[44]

Rb: A Recessive-Acting or Anti-Oncogene

The viral and cellular oncogenes discussed earlier can all be described as "dominant-acting." The presence of the gene, in an altered form or in increased copy number, contributes to malignancy. In 1987, a fundamentally different type of oncogene was found to be involved with the rare childhood tumor, retinoblastoma.[52,53] In this cancer, malignancy results from the absence of a functional copy of the retinoblastoma (Rb) gene, which is therefore said to be "recessive-acting." Rb is an anti-oncogene, because its presence (even in a single copy) inhibits formation of this particular cancer. Whereas the oncogenes described so far encourage cell proliferation, Rb suppresses the mitogenic response.

Retinoblastoma also illustrates the genetic hit model of oncogenesis. Children who develop the "familial" form of the disease inherit a normal copy of the Rb gene from one parent and a defective copy from the other parent (Fig. 10). Subsequently, a spontaneous mutation in a blast cell of the developing retina causes loss of the remaining normal copy and results in tumor cells all lacking functional Rb genes. Children who have the "sporadic" form of the disease inherit normal Rb genes from each parent and thus are born with two functional Rb genes per cell.[52,53] Two successive somatic mutations within a single cell of the developing retina must inactivate both copies of the Rb gene (Fig. 10).

Not surprisingly, retinoblastoma patients may show deletions and rearrangements at the Rb gene site on chromosome 13. Considering that chromosome abnormalities are widely observed in cancer cells, the loss of anti-oncogene function may turn out to be a common feature of cancers. Indeed, the Rb gene itself has now been implicated in several other human cancers, which indicates that it may play a generalized role in growth suppression in a variety of tissues. For example, patients who are cured of familial retinoblastoma are at increased risk of osteosarcoma, small cell lung cancer, and breast cancer; gene deletions or chromosome 13 abnormalities have been found in each of these tumors.

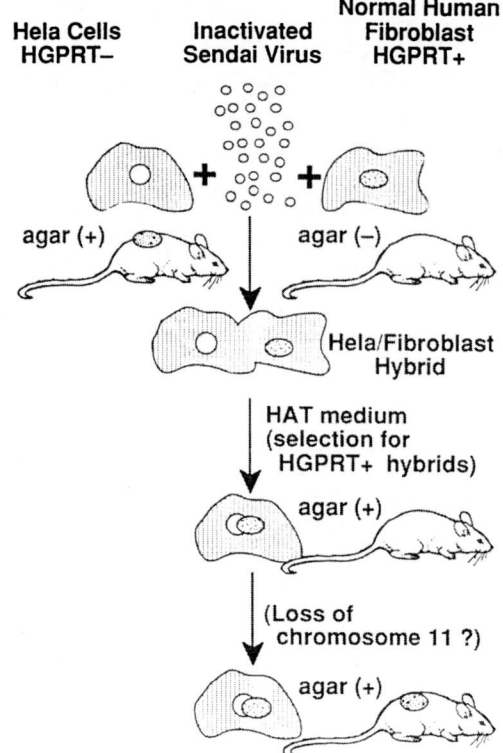

Hela Cells HGPRT−   **Inactivated Sendai Virus**   **Normal Human Fibroblast HGPRT+**

agar (+)   agar (−)

Hela/Fibroblast Hybrid

HAT medium (selection for HGPRT+ hybrids)

agar (+)

(Loss of chromosome 11 ?)

agar (+)

Figure 8.   Illustrating how hybrid cells can be used to demonstrate the way in which tumorigenicity may result from the loss of a suppressor gene. HeLa cells are tumorigenic while normal human fibroblasts are not. HeLa cells are transformed, as indicated by their ability to grow in soft agar, and tumorigenic, as evidenced by their ability to form tumors in immune suppressed animals. Normal human fibroblasts do not show either of these properties of malignancy. A hybrid formed by the fusion of a HeLa cell and a normal human fibroblast will grow in soft agar, but does not form tumors; the malignant phenotype of the HeLa cell is suppressed by the normal fibroblast. If however, chromosome 11 from the normal fibroblast is lost, tumorigenicity is restored, indicating that the repressed gene or oncogene is located on that chromosome.[46]

A dramatic gene addition experiment, in principle like those that demonstrated dominant-acting oncogenes, shows that Wilms' tumor of the kidney involves deletion of a possible growth-suppression gene on chromosome 11. If a normal copy of chromosome 11 is introduced into Wilms' tumor cells, they revert to normal and lose the ability to form tumors.[50]

The Impact of In Situ Hybridization on Radiation Cytogenetics

In an analogous manner to the explosion of detection techniques in molecular biology, procedures have developed for the specific detection of DNA sequences, RNA, or proteins at the level of the single cell or of a single chromosome in a mitotic spread.[54,55] Southern hybridizations involve the recognition of specific DNA sequences in DNA fragments separated on the basis of size by electrophoresis in a gel matrix. Northern hybridizations detect RNA derived from populations of cells and also separated electrophoretically in a gel matrix, while Western hybridizations detect proteins separated on the basis of both size and charge (two-dimensional).

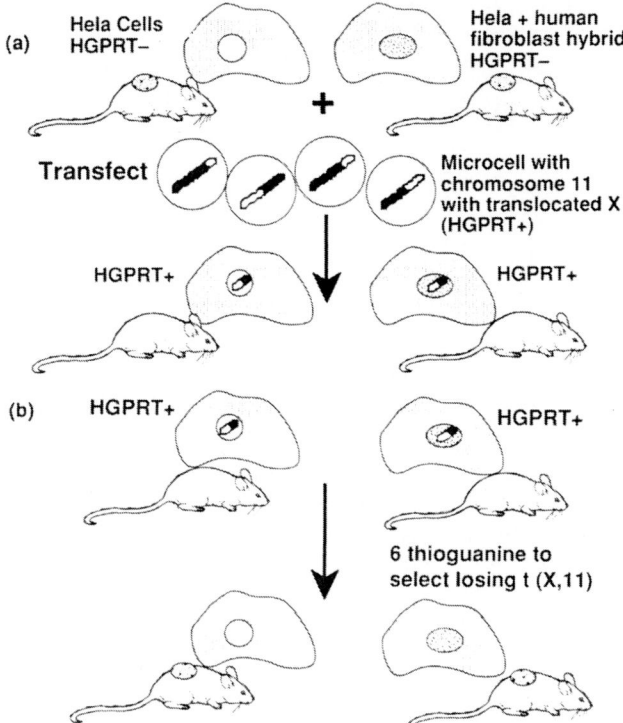

Figure 9. (a) Illustrating the suppression of malignancy by chromosome 11 from normal human fibroblasts. The tumorigenesis of HeLa cells or HeLa-fibroblast hybrids can be reversed by transfecting into the cells a microcell which contains only human chromosome 11. A section of the x chromosome is translocated into chromosome 11, which contains the HGPRT gene, so that cells can be selected that have taken up the microcell. (b) The reversion to tumorigenicity can be demonstrated in cells that lose chromosome 11.[48,89]

Immunohistochemistry has long been a powerful tool for detection of molecular entities in single cells, and the combination of histochemistry and molecular biological approaches has reached the point where unique genes or gene products are demonstrable in individual cells.

These in situ hybridization techniques have had a particularly profound effect in cytogenetics. The finding that non-coding regions of DNA are specifically different between individual chromosomes[54-56] and that regions of chromosomes (centromeres or telomeres) are recognizable by defined satellite sequences[56] has allowed for ready chromosome classification and detection of chromosome aberrations.

These major advances in the ability to discern between and within chromosomes has not merely transcended distinguishing changes based on various banding procedures in chromosomal spreads, it has also allowed (in some instances) the detection of numerical or structural chromosomal changes in interphase nuclei or micronuclei.[57-59] Further, individual chromosomes (or parts thereof, e.g., telomeres) can be three-dimensionally localized in interphase nuclei[60,61] and the potential is present for quantitating amounts of DNA target sequences in individual single cells.

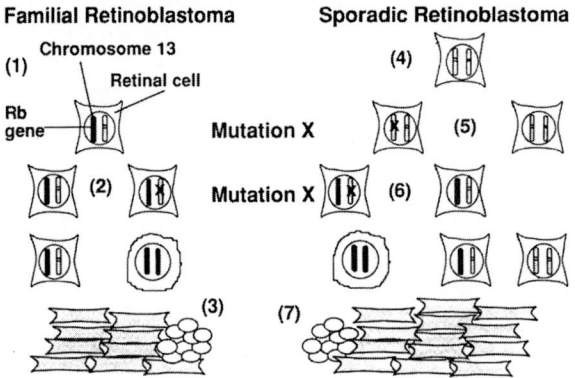

Figure 10. Rb Gene Mutations in Familial and Sporadic Retinoblastoma: In familial retinoblastoma, one normal and one mutated Rb gene are inherited (1). A subsequent mutation in any retinal cell inactivates the remaining Rb gene (2), leading to loss of growth control in a clone of tumor cells (3). In sporadic retinoblastoma, two normal Rb genes are inherited (4). The first mutation inactivates one copy of the Rb gene(5); a subsequent mutation within the same retinal cell inactivates the remaining copy of the Rb gene (6), leading to loss of growth control in a clone of tumor cells (7).

Hence, the detection of changes in individual cells allows closer attention to initiating events in such areas as radiation-induced carcinogenesis.

Historically the dicentric chromosome has been used as a monitor of radiation-induced change, with levels of dicentric chromosomes in mitotic spreads derived from human peripheral blood lymphocytes being accepted as indications of absorbed dose, i.e., a biological dosimeter. The complement to the dicentric chromosome, which involves an interchange between two chromosomes with one component having two centromeres and one none (and thereby a lethal event), is the translocation. This involves an interchange of chromosomal parts, is non-lethal, and has been shown to be specifically associated with human leukemias and lymphomas. Chromosome-specific probes allow for the ready and potentially automated detection of translocations, normally a difficult class of aberrations to detect, with profound implications for radiation cytogenetics.[54]

Detection of specific sequences in chromosomal spreads or interphase nuclei usually requires the probe (the hybridizing sequence) to be tagged with a molecule that directly or indirectly fluoresces, thereby allowing for micrometer-level localization. Fluorescent in situ hybridization (FISH) in appropriate circumstances therefore allows "fishing" for a given species in a "pool" (pool = potentially oncogenic oligonucleotide lengths) and will play an expanding role in evaluating radiation-induced genetic damage[54,55] in tumor cytogenetics[58,59] and in clinical cytogenetics.[55,57,60]

Different chromosome-specific sequences (probes) may also be tagged with molecules expressing different colors, which, when used in different combinations (and/or concentrations), allow for the detection of multi-colored chromosomal entities, i.e., chromosome painting. At present, more than two-thirds of the chromosomes of the human genome can be color highlighted in this manner.

A substantial and expanding array of cloned DNA probes representative of and specific for the majority of human chromosomes is available for an equally expanding variety of cytogenetic and molecular applications in the field of genomics. These applications will have an equally profound impact in the field of radiobiology.

Acknowledgment

This work was based on research supported by grants DE-F902-88ER 60631 and CA 12536.

References

1. J. M. Bishop. Oncogenes. *Sci. Am.* 246:80-92 (March 1982).
2. J. M. Bishop and H. E. Varmus. Functions and Origins of Retrovirial Transforming Genes. In *RNA Tumor Viruses. Molecular Biology of Tumor Viruses*, 2nd ed. R. Weiss, N. Teich, H. Varmus, and J. Coffin, pp. 990-1108. Cold Spring Harbor Laboratory, Cold Spring Harbor, New York (1982).
3. J. M. Bishop. Cellular Oncogenes and Retroviruses. *Ann. Rev. Biochem.* 52:301-354 (1983).
4. H. E. Varmus. The Molecular Genetics of Cellular Oncogenes. *Ann. Rev. Genet.* 18:553-612 (1984).
5. D. Stehelin, H. E. Varmus, J. M. Bishop, and P. K. Vogt. DNA Related to the Transforming Gene(s) of Avian Sarcoma Viruses Is Present in Normal Avian DNA. *Nature* 260:170-173 (1976).
6. D. H. Spector, H. E. Varmus, and J. M. Bishop. Nucleotide Sequences Related to the Transforming Gene of Avian Sarcoma Virus Are Present in DNA of Uninfected Vertebrates. *Proc. Natl. Acad. Sci. USA* 78:4102-6 (1978).
7. D. DeFoe-Jones, E. E. Scolnick, R. Koller, and R. Dhar. Ras-Related Gene Sequences Identified and Isolated from Saccharomyces Cerevisiae. *Nature* 306:707-709 (1983).
8. B. Z. Shilo and R. A. Weinberg. DNA Sequences Homologous to Vertebrate Oncogenes Are Conserved in Drosophila Melanogaster. *Proc. Natl. Acad. Sci. USA* 78:6789-6792 (1981).
9. T. Takeya and H. Hanafusa. Structure and Sequence of the Cellular Gene Homologous to the RSV src Gene and the Mechanism for Generating the Transforming Virus. *Cell* 32:881-890 (1983).
10. R. Weiss, N. Teich, H. Varmus, and J. Coffin, eds. *RNA Tumor Viruses*, 2nd ed., pp. 579-765. Cold Spring Harbor Laboratory, Cold Spring Harbor, New York (1985).
11. Y. Sugimoto, M. Whitman, L. C. Cantley, and R. L. Erickson. Evidence That the Rous Sarcoma Virus Transforming Gene Product Phosphorylates Phosphatidylinositol and Diacylglycerol. *Proc. Natl. Acad. Sci. USA* 81:2117-2122 (1984).
12. M. D. Waterfield, G. T. Scrace, N. Whittle, P. Stroobant, A. Johnsson, A. Wastesar, B. Westermark, C-H. Heldin, J. S. Huang, and T. F. Deuel. Platelet Derived Growth Factor is Structurally Related to the Putative Transforming Protein. Simian Sarcoma Virus, p. 28. *Nature* 304:35-39 (1983).
13. C. J. Sherr, G. W. Rettenmier, R. Sacca, M. F. Roussel, A. T. Look, and E. R. Stanley. The c-fms Proto-Oncogene Product Is Related to the Receptor for the Mononuclear Phagocyte Growth Factor, CSF-1. *Cell* 41:665-676 (1985).
14. G. M. Cooper. Cellular Transforming Genes. *Science* 217:801-806 (1982).
15. F. H. Graham and A. J. van der Eb. A New Technique for the Assay of Infectivity of Human Adenovirus 5 DNA. *Virology* 52:456-467 (1973).
16. A. Pellicer, D. Robins, B. Wold, R. Sweet, J. Jackson, I. Lowy, J. M. Roberts, G. K. Sim, S. Silverstein, and R. Axel. Altering Genotype and Phenotype by DNA-Mediated Transfer. *Science* 209:1414-1422.
17. C. Shih and R. A. Weinberg. Isolation of a Transforming Sequence from a Human Bladder Carcinoma Cell Line. *Cell* 29:161-169 (1982).
18. E. Southern. Detection of Specific Sequence Among DNA Fragments Separated by Gel Electrophoresis. *J. Mol. Biol.* 98:503-525 (1975).
19. B. Abberts, D. Broy, J. Lewis, M. Raff, K. Roberts, and J. D. Watson. *The Molecular Biology of the Cell,* 2nd ed., pp. 1187-1218. Garland Publishers Inc., New York (1989).
20. S. A. Aaronson and S. R. Tronick. The Role of Oncogenes in Human Neoplasin. In *Important Advances in Oncology*, ed. V. DeVita, S. Hellman, and S. Rosenberg. Philadelphia: Lippincott (1986).
21. R. S. Dalla-Favera, S. Martinotti, R. C. Gallo, J. Erikson, and C. M. Croce. Translocation and Rearrangements of the c-myc Oncogene Locus in Human Undifferentiated B-cell Lymphomas. *Science* 219:963-997.
22. E. Shtivelman, B. Lifshitz, R. P. Gale, and E. Canaani. Fused Transcript of abl and bcr Genes in Chronic Myelogenous Leukemia. *Nature* 315:550-554.
23. K. Shimizu, Y. Nakatsu, M. Sekiguchi, K. Hokamura, K. Tanaka, M. Terada, and T. Sugimura. Molecular Cloning of an Activated Human Oncogene, Homologous to v-raf from Primary Stomach Cancer. *Proc. Natl. Acad. Sci. USA* 82:5641-5645 (1985).

24. J. D. Watson, N. H. Hopkins, J. W. Roberts, J. A. Steitz, and A. M. Weiner. *Molecular Biology of the Gene*, 4th ed., pp. 1058-1096. Benjamin/Cummings, Menlo Park, California.
25. J. L. Bos. The ras Gene Family and Human Carcinogensis. *Mutat. Res.* 195:255-271, 1988.
26. G. M. Brodeur, R. C. Seeger, M. Schwab, H. E. Varmus, and J. M. Bishop. Amplification of N-myc in Untreated Human Neuroblastomas Correlates with Advanced Disease Stage. *Science* 224:1121-1124.
27. A. De Klein, A. G. van Kessel, G. Grosveld, C. R. Bartram, A. Hagemeijer, D. Bootsma, N. K. Spurr, N. Heisterkamp, J. Groffen, and J. R. Stephenson. A Cellular Oncogene is Translocated to the Philadelphia Chromosome in Chronic Myelocytic Leukemia. *Nature* 300:765-767.
28. E. Pimentel. Oncogenes and Human Cancer. *Cancer Genet. Cytogenet.* 14:347-368 (1985).
29. J. D. Rowley. Identification of the Constant Chromosome Regions Involved in Human Hematologic Malignant Disease. *Science* 216:749-751 (1982).
30. J. J. Yunis. Chromosomal Rearrangements, Genes, and Fragile Sites in Cancer: Clinical and Biological Implications. In *Important Advances in Oncology*, ed. V. DeVita, S. Hellman, and S. Rosenberg, pp. 93-128. Lippincott, Philadelphia, Pennsylvania (1986).
31. M. Schwab, K. H. Alitalo, K. Klempnauer, H. E. Varmus, J. M. Bishop, F. Gilbert, G. Brodeur, M. Goldstein, and J. Trent. Amplified DNA with Limited Homology to myc Cellular Oncogene Is Shared by Human Neuroblastoma Cell-lines and a Neuroblastoma Tumor. *Nature* 305:245-248.
32. G. M. Brodeur, R. C. Seeger, M. Schwab, H. E. Varmus, and J. M. Bishop. Amplification of N-myc in Untreated Human Neuroblastomas Correlates with Advanced Disease Stage. *Science* 224:1121-1124.
33. W. M. Court-Brown and R. Doll. Expectation of Life and Mortality from Cancer Among British Radiologists. *Br. Med. J.* 2:181-190 (1958).
34. H. S. Martland. Occurrence of Malignancy in Radioactive Persons. *Am. J. Cancer* 15:2435-2516 (1931).
35. G. W. Beebe, M. Ishida, and S. Jablon. Studies of the Mortality of A-bomb Survivors. I. Plan of Study and Mortality in the Medical Subsample (Selection 1), 1950-1958. *Radiat. Res.* 16:253-280 (1962).
36. C. Borek, A. Ong, and H. Mason. Distinctive Transforming Genes in X-ray - Transformed Mammalian Cells. *Proc. Natl. Acad. Sci. USA* 84:794-798 (1987).
37. J. Shuin, P. C. Billings, J. R. Lillehaug, S. R. Patierno, P. Roy-Burman, and J. R. Landolph. Enhanced Expression of c-myc and Decreased Expression of c-fos Protooncogenes in Chemically and Radiation Transformed C3H/10T/2C18 Mouse Embryo Cell Line. *Cancer Res.* 46:5302-5311 (1986).
38. I. Guerrero, A. Villasante, V. Corces, and A. Pellicer. Activation of a c-K-ras Oncogene by Somatic Mutation in Mouse Lymphomas Induced by Gamma Radiation. *Science* 225:1159-1162 (1984).
39. E. W. Newcomb, J. J. Steinberg, and A. Pellicer. Ras Oncogenes and Phenotypic Staging in N-methyl Nitrosourea and γ-Irradiation-Induced Thymic Lymphomas in C57BL/6J Mice. *Cancer Res.* 48:5514-5521 (1988).
40. B. Krolewski and J. B. Little. Molecular Analysis of DNA Isolated from the Different Stages of X-ray-Induced Transformation In Vitro. *Mol. Carcinog.* 2:27-33 (1989).
41. H. Land, L. F. Parada, and R. A. Weinberg. Tumorigenic Conversion of Primary Embryo Fibroblasts Requires at Least Two Cooperating Oncogenes. *Nature* 304:596-602 (1983).
42. H. E. Ruley. Adenovirus Early Region 1A Enables Virial and Cellular Transforming Genes to Transform Primary Cells in Culture. *Nature* 304:602-606 (1983).
43. E. J. Stanbridge. Identifying Tumor Suppressor Genes in Human Colorectal Cancer. *Science* 247:12-13 (1990).
44. E. R. Fearon, K. R. Cho, J. M. Nigro, S. E. Kern, J. W. Simons, J. M. Ruppert, S. R. Hamilton, A. C. Preisinger, G. Thomas, K. W. Kinzler, and B. Vogelstein. Identification of a Chromosome 189 Gene That Is Altered in Colorectal Cancers. *Science* 247:49-56 (1990).
45. H. Harris. Cell Fusion and the Analysis of Malignancy: The Croonian Lecture. *Proc. R. Soc. B* 179:1-20 (1971).
46. E. J. Stanbridge. Suppression of Malignancy in Human Cells. *Nature* 260:17-20 (1976).
47. R.E.K. Fournier and F. H. Ruddle. Microcell-Mediated Transfer of Murine Chromosomes into Mouse, Chinese Hamster, and Human Somatic Cells. *Proc. Natl. Acad. Sci. USA*, 74:319-323 (1977).
48. P. J. Saxon et al. Selective Transfer of Individual Human Chromosomes to Recipient Cells. *Mol. Cell Biol.* 5:140-146 (1985).
49. P. J. Saxon, E. S. Srivatsan, and E. J. Stanbridge. Introduction of Chromosome 11 Via Microcell Transfer Controls Tumorigenic Expression of HeLa Cells. *Embo J.* 5:3461-3466 (1986).

50. B. E. Weissman, P. J. Saxon, S. R. Pasquale, G. R. Jones, A. G. Geiser, E. J. Stanbridge. Introduction of a Normal Human Chromosome 11 into a Wilms' Tumor Cell Line Controls Its Tumorigenic Expression. *Science* 236:175-180 (1987).

51. J. L. Redpath, C. Sun, M. Colman, and E. J. Stanbridge. Neoplastic Transformation of Human Hybrid Cells by γ-Radiation: A Quantitative Assay. *Radiat. Res.* 110:468-472 (1987).

52. W. H. Lee, R. Bookstein, F. Hong, L. J. Young, J. Y. Shew, and EY-HP Lee. Human Retinablastoma Susceptibility Gene: Cloning, Identification, and Sequence. *Science* 235:1394-1399 (1987).

53. H.J.S. Huang, J. K. Yee, J. Y. Shew, P. L. Chen, R. Bookstein, T. Friedmann, EY-HP Lee, and W. H. Lee. Suppression of the Neoplastic Phenotype by Replacement of the RB Gene in Human Cancer Cells. *Science* 242:1563-1566 (1988).

54. D. Pinkel, T. Straume, and J. W. Gray. Cytogenetic Analysis Using Quantitative, High Sensitivity Fluorescence, Hybridization. *Proc. Natl. Acad. Sci. USA* 83:2934-2938 (1986).

55. D. Pinkel, J. Landegent, C. Collins, J. Fuscoe, R. Segraves, J. Lucas, and J. Gray. Fluorescence In Situ Hybridization with Human Chromosome Specific Libraries: Detection of Trisomy 21 and Translocations of Chromosome 4. *Proc. Natl. Acad. Sci. USA* 85:9138-9142 (1988).

56. J. S. Waye and H. F. Willard. Chromosome Specificity of Satellite DNA's: Short- and Long-Range Organization of a Diverged Dimeric Subset of Human Alpha Satellite from Chromosome 3. *Chromosoma* 97:475-480 (1989).

57. A. Jauch, C. Daumer, P. Lichter, J. Murken, T. Schroeder-Kurth, and T. Cremer. Chromosomal In Situ Suppression Hybridization of Human Gonosomes and Autosomes and Its Use in Clinical Cytogenetics. *Hum. Genet.* 85:145-150, 1990.

58. H. Van Dekken, J. G. Pizzolo, D. P. Kelsen, and M. R. Melamed. Targeted Cytogenic Analysis of Gastric Tumors by In Situ Hybridization with a Set of Chromosome Specific DNA Probes. *Cancer* 66:491-497 (1990).

59. T. Cremer, P. Lichter, J. Borden, D. C. Ward, and L. Manuelidis. Detection of Chromosome Aberrations in Metaphase and Interphase Tumor Cells by In Situ Hybridization Using Chromosome-Specific Library Probes. *Hum. Genet.* 80:235-246 (1988).

60. P. Lichter, T. Cremer, C. C. Tang, P. C. Watkins, L. Manuelidis, and D. C. Ward. Rapid Detection of Human Chromosome 21 Aberrations by In Situ Hybridization. *Proc. Natl. Acad. Sci. USA* 85:9664-9668 (1988).

61. J. C. Fuscoe, C. C. Collins, D. Pinkel, and J. W. Gray. An Efficient Method for Selecting Unique-Sequence Clones from DNA Libraries and Its Application to Fluorescent Staining of Human Chromosome 21 Using In Situ Hybridization. *Genomics* 5:100-109 (1989).

Discussion

Ward: Eric, you say that the people who look at these transformations don't generally look at them quantitatively. In the work of Pellicer, is it possible to back track the doses: he has seven specific changes at the same chromosomal point? Can you calculate whether these correspond to radiation events at that point?

Hall: I don't know. Marco Zaider recently asked a similar question. He asked the question concerning the transforming gene in the DNA library. We went through 300,000 clones to find it. Marco is asking "Is that consistent with the fact that with 400 rads you get essentially 100% transformation?" It is a good question, but I don't know the answer. We didn't think of it before, and Marco Zaider needs to do the relevant calculations.

Moolgavkar: One of the problems with Vogelstein's scheme for colon cancer, which you showed in the end, is that it is not clear to me that he has shown that any of those steps is necessary for malignant transformation. Some of them occur with high frequency, but none of them occurs in 100 percent of the tumors. Secondly, we know from gross evidence, i.e., epidemiologic and pathologic evidence, that not all colon cancers arise from visible polyps. You can get colon cancers apparently without any predisposing adenoma, so there is plenty of confusion here and I don't believe that the scheme is entirely correct. I think the data are consistent with many other schemes, not just the one proposed by Vogelstein.

Hall: I think you're right. The lesson that it tells you is that it is complicated, but it's not a single step.

Moolgavkar: No, it is not a single step, but I still contend that it is consistent with two rate-limiting events. A scenario that is equally consistent with the data is that you need homozygosity at one of two gene loci. Either the genes for familial polymers on Chromosome 5, or the as-yet-unmapped locus for nonpolymers familial colon cancer. The other mutations like the ras gene mutations simply serve to confer a selective group advantage on once-hit cells. So when people talk about multiple steps, I don't think that it is clear that multiple steps (or more than two) are necessary.

Hall: I think I said essentially the same thing; sorting out which are essential initiating events and which are the complications of growth and development is difficult. However, I am not as convinced as you are that it is only two steps; sometimes it may be more.

Ward: I know there has been a lot of discussion of this, but could you summarize for us why human cells cannot be used, either in the direct transformation or in the transfection experiments where you do not get transformation of human cells?

Hall: It is fortuitious historically that in rodent cell systems (particularly SHE cells in which in vitro oncogenic transformation by X-rays was first demonstrated by Carmia Borek), an early event in the expression of transformation is the morphological change. This allows transformed cells to be identified easily even if the frequency is 10^{-4} or 10^{-5}. Later developments include loss of anchorage dependence and tumorigenesis in immune-suppressed animals. For practical purposes, these endpoints are harder to score if they occur at low frequency.

In human cells, a morphological change is not an early event in the expression of the malignant phenotype. Coupled with the fact that the radiation-induced transformation incidence is much lower than in rodent cells, this makes the use of human cell systems impractical for quantitative studies. Transformed foci have been identified by irradiation of normal human cells, but the incidence is of the order of 10^{-15} for a dose of hundreds of rads. Various investigators have attempted to first immortalize human cells with agents as diverse as transfected oncogenes, viruses, chemical mutagens or radiation. In the case of gamma rays, a dose of over 4,000 rads, delivered over a period of a year, was found necessary, but this results in enormous, widespread chromosome damage. An ideal assay would score metastasizing tumors as an endpoint in human epithelial cells. In practice, the only practical highly quantitative systems score morphological changes in fibroblasts of rodent origin. Many laboratories are attempting to remedy this deficiency, but so far with little success.

Kasha: Eric, at the beginning you showed a slide on the general theme of oncogenes with chemical and radiation instigation, and near the end you showed a slide with 23 sub-classes of oncogenes. I have a general question about human cancer from the environmental effects, which are either of radiation background or chemical background origin. Now that so many different oncogenes are known, has anyone gone back in a systematic testing of these oncogene responses to chemical versus radiation perturbation?

Hall: I'm not sure I understand the question. Can I rephrase it? Are you asking if radiation has a signature in terms of the oncogene(s) activities?

Kasha: Yes, the question is, are some oncogenes more responsive to chemical induction and others to radiation induction?

Hall: The bulk of the oncogenes that have been identified are either from spontaneous human tumors or from chemically induced tumors or foci; not one of them is radiation-induced at the moment. No one has identified an activated oncogene in a radiation-induced transformed focus. That was my message.

Kasha: I have a few more definition questions. Does the term "point mutation" refer to radiation locus, cellular locus, or molecular locus?

Hall: That's not my field, particularly, but the point mutations associated with the ras activation are known at the base level, so I think it is molecular.

Ward: I think what they do is look for the gene--apparently the gene product is not there. They look for the gene, and if they can find it with a Southern blot--if the gene is still there, then they say that it point mutated to a level which they can't detect.

Hall: But they know the sequence change in the ras activation, don't they?

Ward: Sometimes, yes.

Kasha: Now I have an interdisciplinary question. At some point you made an interesting statement. You said that we went from the whole organism, to the cell, to the chromosome, to the base pair; and then you said, "We can't go lower than that." What is the limitation?

Hall: I said "presumably." I started life as a physicist, so I know better than to assume that we have ever reached the elementary unit.

Kasha: I was wondering, because from the physical sciences viewpoint, we do probe down to the sub-micron scale to molecular groups, to protons, electrons, and similar sub-base-pair components. When you say we can't go lower than that, do you mean in biological diagnosis?

Hall: At the present time, biologists regard the sequence in DNA as physicists once regarded protons and electrons, i.e., as the basic building bricks, but of course that may not be the case.

Varma: I follow on with what Mike said. You have pointed out advances in molecular biology, whereby now sophisticated biological end points such as transformation and transfection can be measured. Let's say you have an initial change that is produced by radiation, such as ionizational or excitation that follows through to chemical species, and we can determine the spatial and temporal position of these events. How can this information then be applied to the interpretation of the molecular end points that you mentioned?

Hall: You're assuming that radiation produces the initial event, but when you have a multi-step process, radiation may not be producing the initial event. It may be producing a second or a later event. In principle this could now be modeled. The information is probably available to start this.

Varma: Does the initial physical or chemical event have a relationship to the initial event that you identified as a point mutation or a transformation?

Hall: It seems to me, rightly or wrongly, that the so-called "new biology" is not involving any new *radiation* biology. You are still talking about DNA; that hasn't changed at all. In going from chromosome breakage to the loss of a suppressor gene or activation of an oncogene, the target is still DNA, so that is not a new question.

Weinstein: Given the detailed analysis that Eric presented, the question becomes more refined. It becomes possible, with what is known now in molecular biology, to identify the steps that could be affected at the molecular level to produce the observed lesions, those changes that Eric presented. And that is how one can begin to relate the physics of radiation biology with molecular mechanisms in biology. This is the basis of the question I asked during the talk. If in fact the gene is activated—and it doesn't matter at this point whether it is one of the now-classical oncogenes or a new gene that might be identified—then we know several mechanisms for gene activation. We can attempt to relate the type of lesions we know are produced by radiation to the type of mechanisms that we know can cause gene activation. One such proposal that I said I would present at the end of the week is that genes are activated when, say, repressors are removed; that is, the protein/DNA interaction is affected. If in fact radiation-induced lesions produce a change in the local structure of DNA, non-local change in the conformation of the DNA can be the result of this lesion. We know some of the rules of protein DNA interaction, and both the sequence and the conformation of the operator are important elements in the affirmity of the protein for the binding site. If these protein/ DNA interactions are affected, the proteins may not bind anymore, or new binding sites may be created. That can result either in repression of DNA translation, or in activation produced by removal of a repressor due to a local change which has induced a nonlocal conformational change, far from the point mutation. This mechanism has not been tested specifically, so I'm just presenting it as a hypothesis and I will try to show you why in my own presentation. The attractive part of this type of mechanism is that it relates specifically to the physics that occurs at a molecular level so that what we know from radiation physics can be related to the biology that we have now learned by looking at cellular processes.

von Sonntag: You show that when using restriction enzymes, some produced small DNA sections which still transformed, but some would cut in a way that no mutagenicity is introduced. This has been explained by assuming that the relevant gene has been cut somewhere in the middle by this type of restriction enzyme. There might be another possibility, and I'd just like to ask you whether that has been taken into account. Some of the restriction enzymes may produce blunt ends and some may produce sticky ends. When you introduce those with sticky ends into cells, they are digested rather rapidly and no longer persist, and those with the blunt ends have a better chance for survival.

Hall: I have no idea whether that was taken into account, but one could go back.

Paretzke: You were talking essentially about radiation effects on the DNA. Is there any other structure in a cell, like the cellular membrane or the nuclear membrane or other structures within the cell, where radiation-induced changes are of importance for carcinogenesis? Radiation physicists have to think in terms of sensitive targets.

Hall: With cell killing as an end point, the factors that can modulate the biological effect of a given radiation dose are those that occur *during* irradiation--dose, rate, dose fractionation, presence of oxygen or radical scavengers. The fraction of cells surviving a given dose cannot be modulated much by events after irradiation, except in terms of potentially lethal damage repair, which is a relatively small effect.

By contrast, the incidence of radiation-induced transformation can be modulated substantially by events that occur after irradiation. Changes made days, or even weeks, after the radiation exposure is completed can dramatically modify the expression of transformation. Tumor promoters and thyroid hormones *increase* transformation incidence, while retinoids, superoxide dismutase, selenium or 5-aminobenzamide *reduce* transformation. In some cases, agents that modify transformation do not enter the cell to affect DNA directly, but act via receptors on the cell membrane.

Zaider: I would like to point out that you can show in a very simple way, using microdosimetric arguments, that in order to have 4 Gray of low LET radiation pretransform practically all cells, you have to have at least 0.1 percent of the cellular nuclear volume sensitive to whatever causes transformation. This suggests that, if indeed the actual agent producing the transformation is such a rare oncogene (1 in

300,000), then one may argue that the immediate effect of radiation is not to activate this oncogene; but maybe there are a number of steps involved in between the primary event, called pretransformation, for which radiation may be responsible and the final end point.

Hall: So that's 0.1 percent, that's 1 in 10^3, and you never get above 10 percent transformation so that's 1 in 10^4. We are within a decade. With radiation, you never get above 1 percent transformation, at any dose—so we are within a decade.

Varma: The problem with that calculation is that you can get 0.1 percent sensitive volume of the cells, but I think that the more important thing is where the sensitive site is. DNA is not distributed homogeneously within the nucleus; is that where you have the probability difference?

Zaider: I think it is sufficiently homogeneous. My argument is based on an all-or-none approach, or, to put it differently, on the minimum amount of cellular material that needs to be traversed by ionizing radiation in order to have any effect at all.

Varma: Frankly, you've got to put in the spatial distribution of the sensitive sites. You might get only one since you may have only one specific critical site.

Weinstein: But the point, I think, is that the reason Marco thinks (and rightly so) that this is an answer to my question is that even if there is one site that is activated, it doesn't mean that there is only one means of activating it by actually affecting one base pair in that gene. There are many different ways that we know about by which you can activate.

Weinstein: With radiation, and in that respect just to answer Paretzke, not only membranes, but also the proteins that modulate gene expression, have to be considered as important.

Christophorou: Physics and chemistry contributed to biology in terms of basic knowledge, but also in terms of technique. My question is, what kind of technique(s) from physics and chemistry do you wish to see developed for radiation biology? Is there any particular need for measurement in radiation biology that physics can possibly develop techniques for?

Hall: We will have to leave that question for discussion on Wednesday. I can't answer that off the top of my head.

Ward: I would like to go back to what Marco was saying. If what you say is correct, Marco, then how do you explain the LET effect where, with increasing LET, the number of events per dose decreases but the transformation frequency increases?

Zaider: Although high-LET transformation data are not as clear as those available to lower LET, a small number of events per unit dose will only increase further the size of the sensitive volume.

Curtis: I would like to return to a point, Eric, that you mentioned earlier, and that is the fact that human cells have a lower transformation frequency than rodent cells; in fact, *in vitro* they are very difficult, or even impossible, to transform. Now, some people say that this might be due to the chromosomal stability of human cells. This brings up the idea that perhaps it is impossible to transform a normal human cell simply by radiation. Either before the radiation or after it, the interesting point is that there might have to be some event that occurs to sensitize it to make it transformable by radiation. Would you care to comment on this?

Hall: That is essentially what I said to Matt. In human cells, the frequency is so low that you can't make that into a routine quantitative assay. As I said, the estimate is around 10^{-15}, so you can't say, okay, let's do a dose response curve. But I don't think it's true to say that it can't happen. 10^{-15} is plenty high enough to account for all carcinogenesis in humans when you think how many cells there are.

Kasha: My question is in the spirit of Bill Glass's retrospective view. In the DOE Gainesville meeting, you gave what I thought was a very dramatic demonstration that when one extrapolates to zero radiation dose, the radiation cell lethalities do not extrapolate to zero, a non-linear extrapolation. Has that observation held up, and what is the current status of low-radiation-exposure research?

Hall: Are you referring to the idea that at the very low doses you still get some transformation? Well, on the advice of the various site visitors we've had, we have not pursued this problem because it was thought that the expenditure of a lot of effort and time was not justified. But interestingly enough, while we haven't pursued it, at the last microdosimetry meeting in Rome two other groups reported essentially what we found 10 or 15 years ago. The group from Italy (Madame Bettega) presented a curve, which was not linear and became shallower at low doses, with a significant transformation incidence still at doses of a few rads.

Kasha: That topic bears on a very important question about background radiation health effects.

Varma: On the question of the activation of oncogenes, we didn't go much into the repair, but does repair play any role in the eventual expression of malignancy? Does anyone here know from molecular biology whether the activation of an oncogene can be reversed once it is activated?

Hall: Once you have activated an oncogene by radiation, it is very difficult to see how that could be reversed.

Moolgavkar: I don't think that repair is really relevant once the cell has divided because these are all heritable changes. Once the cell has divided and the daughter cells have been formed, it is virtually impossible that the lesion will be repaired in all the cells.

Varma: Yes, but before the cell division, can you repair it? That was the question.

Hall: On that theme, for the title I was given you could have half a dozen or more totally different topics. I elected rightly or wrongly to say that carcinogenesis is the most important, so I'll do that. But last week I was talking to Martin Brown on the phone, and I felt inclined to rip the whole thing up and start again. I gathered that at Stanford, now, they have a mutant mouse, which is like a patient with AT, all the cells in the mouse are more sensitive by a factor of three. They all have D_o's of about 40 to 50 rads or less. The reason essentially is the same as in the genetic defect in AT, a repair deficiency. They have taken cells from various tissues, and they are all sensitive by about the same factor; presumably the animal would be wildly sensitive to carcinogenesis too. Obviously, I could have started again and talked about repair.

Kasha's Added Comment concerning Hall's System Analysis Diagram:

The experimental simulation of *the molecular environment of DNA in the cell* can, in part, be accomplished by the use of DNA preparations which have a higher nucleic acid density versus water than in the case of the dilute DNA aqueous solutions.

One nucleic acid preparation which is available is the solid film with equilibrium hydration. Natural DNAs from different biological species offer choices of different AT/GC ratios. Synthetic polynucleotides offer ranges of base-pair combinations. The nucleic acid or polynucleotide gels can be strongly oriented directionally by stroking between quartz plates.

Another condensed nucleic acid preparation is the liquid crystal DNA or chromatic system studied optically by Randolph Rill (Florida State University). This also offers dense nucleic acid material in a hydrated environment, with determined orientation.

Radiation studies on these systems may reveal responses intermediate to the behavior of dilute DNA solutions and the intact cell.

Radiological Physics

Atomic and Molecular Theory

Mitio Inokuti[a]

Abstract

The multifaceted role of theoretical physics in understanding the earliest stages of radiation action is discussed. Scientific topics chosen for the present discourse include photoabsorption, electron collisions, ionic collisions, and electron transport theory. Connections of atomic and molecular physics with condensed-matter physics are also discussed. The present article includes some historical perspective and an outlook for the future.

Introduction

Role of Theoretical Physics

In this paper, we discuss the role of theoretical physics in understanding the initial stages of radiation interactions with atoms and molecules in the context of radiation biology. At the outset, let us consider the role of physics in general in the same context. In three respects physics has contributed materially and will undoubtedly continue to do so.

First, physics provides instrumentation necessary for research in radiation biology. By "instrumentation," I mean a broad class of devices for the generation of radiations, for the measurement of radiations, and for the identification and analysis of radiation effects. This role of physics has long been well recognized; indeed, it is so obvious that we often fail to point it out.

Second, physics provides principles that are crucial for establishing a framework of sound reasoning. To cite an example in our context, it is the role of physics to reveal how different radiations are absorbed in different materials.

Third, physics provides "data" basic to the considerations of chemical and biological effects of radiation. Elementary examples in our context are the stopping powers of various materials for particles of various energies and related quantities such as the linear energy transfer (LET). Any serious discussion of the physical and chemical mechanisms of radiation effects requires knowledge of at least the stopping powers, and often more, e.g., of cross sections for individual energy transfer processes.

Theoretical physics in particular contributes in the same three respects: instrumentation, principles, and data. A recent example of theoretical physical contributions to instrumentation development is seen in computerized tomography, which originated in the mathematical problem[1,2] of determining electron density in matter from measurements of photon beam attenuation in various directions.

(a) Argonne National Laboratory, Argonne, Illinois

Physical and Chemical Mechanisms in Molecular Radiation Biology
Edited by W.A. Glass and M.N. Varma, Plenum Press, New York, 1991

(Although computerized tomography is primarily used in medical diagnosis and radiotherapy planning and is only indirectly related to radiation biology, I cite this example because of its great importance and to illustrate the wide-ranging significance of theoretical physics.) Contributions of theoretical physics in providing principles and data are plentiful and too well-known to be enumerated.

Relevant Subfields of Physics

Several subfields of physics play different roles in our context. They include particle and nuclear physics, atomic and molecular physics, condensed-matter physics, and statistical physics.

The role of particle and nuclear physics primarily concerns characterizing charged-particle spectra resulting from the incidence of baryons (e.g., pions, neutrons, and energetic heavy ions) with matter. Besides causing electromagnetic interactions, baryons strongly interact with nuclei at short distances, causing nuclear transformation and generating various charged particles. An example of excellent work on charged-particle spectra in materials of interest to radiation biology and radiological dosimetry is seen in the papers by Caswell and Coyne.[3-6]

In this paper, I discuss the role of atomic and molecular physics in some depth and also touch upon some areas of condensed-matter physics. As an introduction, I wish to make a single remark. Since radiation biology concerns the cell, which consists of condensed matter with many component molecules, condensed-matter physics is actually more relevant than atomic and molecular physics. However, atomic and molecular physics is basic and is more advanced in treating details. Thus, it serves as an underpinning for condensed-matter physics. In other words, any concept in condensed-matter physics must be consistent with the knowledge of atomic and molecular physics. Moreover, under certain circumstances, radiation interactions in condensed matter occur virtually in the same way as in individual atoms and molecules, the influence of atomic and molecular aggregation playing only a minor role. This circumstance occurs, for instance, for energy transfer from a fast charged particle far exceeding the binding energy of an electron in the relevant shell; more precisely, then, a secondary electron is ejected from individual atoms or molecules in condensed matter almost as if they were isolated. In contrast, an electron of low energy interacts with condensed matter in a notably different way.

The role of statistical physics in our context arises because radiation interactions with matter cause a multitude of collision processes. In order to evaluate the cumulative consequences of all these processes to the incident radiation and to matter, we must use some elements of the particle transport theory, either in the form of an analytic method or a Monte Carlo simulation. This topic belongs to statistical physics[7] and physical kinetics,[8] in the words of the Landau school.

Historical Perspective

In my view, the present conference belongs to a series that began at Oberlin, Ohio.[9] Subsequent meetings were held at Highland Park, Illinois,[10] and Airlie, Virginia.[11]

The Oberlin conference marked the beginning of the modern inquiry into our theme. The physics part of the discussion at Oberlin was presented by Morrison, Fano, Platzman, Evans, and others. Fano[12] discussed secondary electron spectra and the total ionization yield. Platzman presented two papers, one [13] on the earliest processes in radiation chemistry and biology and the other[14] on the stopping power with particular emphasis on the influence of chemical binding and molecular aggregation in condensed matter. The papers by Fano and by Platzman were most stimulating to me (I was then a graduate student at the University of Tokyo) and indeed posed many of the problems that have been pursued seriously by many investigators, myself included. The knowledge that has been obtained since then about some of the problems raised at Oberlin is extensive and solid. An example of such a problem is secondary-electron spectra, discussed here by Toburen.[15] Another example concerns the total ionization yield; progress in this topic up to the early 1970s is seen in the ICRU Report 31.[16]

A few points made in the Oberlin proceedings warrant mention as examples of Platzman's remarkable foresight. In the discussion following Fano's presentation, Platzman[17] pointed out the incorrectness of the then-standard notion that a heavy particle moving at speeds comparable to $e^2/\hbar = c/137$ does not ionize atoms or molecules with appreciable probability. This remark later led to the discovery of the electron promotion mechanism by Fano and Lichten,[18] which in turn led to extensive studies that continue even now. Another noteworthy point made by Platzman[14] concerns the importance of studying the spectra of water and other substances in the far-ultraviolet and soft X-ray regions, which through a forceful campaign by Fano[19] and Platzman eventually led to current research using synchrotron radiation in the U.S.

The Highland Park conference was most noteworthy because of Platzman's prediction of the hydrated electron, whose presence was firmly established through the absorption spectrum first reported by Hart and Boag.[20]

The Airlie conference represented the greatly enhanced richness of our knowledge by the early 1970's. The most notable achievements to me as a participant were the advent of molecular biology and the progress in radiation chemistry with the use of pulse radiolysis methods. By that time physics was showing steady advance, largely following up ideas put forth at Oberlin. These advances can be characterized as elaborate treatments of track structures, as exemplified by discussions of condensed-phase effects,[21] of detailed cross-section data and their applications,[22,23] and of microdosimetry.[24]

Here at Woods Hole, we face the challenge of keeping up the tradition of high-quality research. Progress in the two decades since the Airlie conference has been substantial indeed, as we shall discuss at some depth. Before undertaking a technical discussion, however, a few general remarks are in order. Because of the immense technical complexity of the theme, related research is necessarily of long range; in other words, there is little chance for anyone to clarify an entire mechanism in one stroke with a bright idea. The work is also obviously interdisciplinary and thus demands efforts by persons of various talents. Therefore, continued support of a substantial number of devoted scientists is essential for progress.

Another merit of this multidisciplinary effort concerns a variety of spinoffs in both basic science and applied technology. For example, the use of synchrotron radiation in spectroscopy in the far vacuum ultraviolet and soft X-ray regions was urged in the U.S. because of research needs in this area.[19] Now applications of synchrotron radiation have been extended to an immense variety of topics far beyond the original scope. At the same time, spectroscopic studies thus initiated have led to many important results in basic physics, e.g., the understanding of doubly excited and other autoionizing states of atoms and molecules or, more generally, of electron correlation effects.[25]

Atomic and Molecular Collisions

Preamble

As was recognized at the Oberlin conference, knowledge about individual collisions of energetic charged particles and photons with atoms and molecules is fundamental to the full understanding of radiation actions on matter. For considerations of radiation energy absorption in some depth, e.g., for analyzing track structure or calculating yields of excitation and ionization, we need cross sections for individual collisions of various kinds and for various particles.

Research on those individual collisions has a long history and continues to be pursued by many workers. However, not every result of this research is relevant to our theme. As I have been stressing for years,[26,27] cross-section data must fulfill what I call the trinity of requirements in order to be useful for the analysis of radiation actions and indeed for many other applications. In other words, they must be absolute, correct, and comprehensive in order to be useful for the analysis of radiation actions and indeed for other applications such as astrophysics, atmospheric physics, and fusion research. The meaning of the term "absolute" is plain; cross-section values must be given in cm^2 or any other absolute scale. The term "correct" is clear; cross-section values must be right, although the degree of required accuracy depends upon the specific application. The term "comprehensive" may call for explanation. In the analysis of radiation actions, we need cross-section values for a wide range of particle kinetic energy, energy transfer, and other variables characterizing a collision process such as scattering angle. This is true because particles of widely different energies exist in irradiated matter and because the outcome of a single collision is by no means unique. Therefore, we need cross-section values for a wide range of collision variables and for a variety of atomic and molecular species. However, the cross-section data in the literature are often relative as opposed to absolute, discordant as opposed to correct, or fragmentary as opposed to comprehensive. (This is because studies on atomic collision physics are largely carried out for purposes other than ours, most notably for advances in basic physics. To prove a point in physics, it is indeed sufficient and expedient to generate relative cross-section values over a crucial, limited range of variables, either in theory or experiment, although the requirement of correctness is crucial in all scientific contexts.)

In the following sections, I discuss several topics selected in part from the viewpoint of the trinity of requirements and in part because of their importance in other respects.

Photoabsorption

Let me begin with the photoabsorption cross section and related matters. The importance of the photoabsorption cross section in far vacuum ultraviolet and soft X-ray regions, i.e., for photon energies between about 10 eV and several keV, is twofold. First, photons in these regions interact strongly with materials of low and modest atomic numbers (including most of the materials in the biological cell) and therefore have a special role in the study of mechanisms in radiation biology. Second, there are close relations between the photoabsorption cross section and the cross section for glancing collisions of charged particles, which occur frequently when the particles are fast, as first pointed out by Bethe[28] and discussed fully in the literature as exemplified by Refs. 29-31. It is precisely because of these relations that Platzman[17] at Oberlin called for studies on oscillator strength spectra (which are the same as photoabsorption cross sections as functions of photon energy, apart from a universal constant) of water and other materials basic to radiation biology. In 1966, Platzman[32] also surveyed fragmentary data then available and gave his "educated guesses."

Today, our knowledge of the oscillator strength spectra is incomparably more extensive and reliable, as illustrated by reviews of data by Samson[33] on atoms and by Gallagher et al.[34] on molecules. These reviews focus on experimental data, but concomitant progress in theory is crucial. The 1968 review by Fano and Cooper[35] was a landmark in the modern theory of photoabsorption cross sections of atoms. Progress since then is seen in the reviews by Starace[36] and Amusia.[37]

The theory of photoabsorption cross sections of molecules has seen considerable development, although it has not yet reached the level of rigor that has been attained by the corresponding theory for atoms. Among other methods, the multiple-scattering method, pioneered by Dehmer and Dill,[38] is particularly notable for its general applicability to polyatomic molecules. Major accomplishments are 1) the demonstration of the possibility of realistic calculations fully taking into account the crucial effects of nonspherical molecular fields experienced by outgoing electrons, and 2) the elucidation of commonly occurring phenomena such as shape resonances and barrier effects. Full treatment of these topics is beyond the scope of the present discussion because a great deal of physics is involved in the transition of an electron from a bound orbit into an unbound orbit, especially from a molecule, which has a geometric structure as well as internal degrees of freedom.

Indeed, a great deal of physics and chemistry concerning excited and ionized states of polyatomic molecules awaits discovery. As we see in the review of data by Gallagher et al.,[34] the result of photoabsorption or energy transfer from any other agent in excess of the first ionization threshold energy may or may not be ionization; molecular dissociation into fragments is a competing process. Full elucidation of the competition between ionization, dissociation, and other relaxation processes of polyatomic molecules along the lines outlined by Henriksen[39] among others, will be an important task left for the future.

Electron Collisions

The importance to our theme of electron collisions with molecules is clear, because the absorption of any ionizing radiation in any matter produces many electrons whose kinetic energies are widely distributed.

Considered from the point of view of the trinity of requirements, the most significant development in the last two decades is the advent of cross-section compilations[40-42] for simpler molecules such as N_2, O_2, and H_2O. The compiled and recommended data are based in part on experiment and in part on theory.

The role of theory in cross-section determination, not only for electron collisions but also for all atomic and molecular collisions, is manifold. First, theory provides principles, i.e., a general framework of understanding applicable to all atoms and molecules. An example of this role is seen in the Bethe theory[28-31] and its major conclusions such as the relations between the photoabsorption and glancing collisions of charged particles. In my view, this role is the most important of all. Second, the principles found through theoretical study often point to the general systematics that correct cross sections should obey. For example, the principles tell us in general how cross sections for different atoms or molecules should depend on the electronic structure. Third, in certain exceptionally simple cases, theory leads to numerical calculations of sufficient reliability to be taken seriously. Finally, in rarer cases, theory opens up a novel perspective.

The theory of electron-molecule collisions has seen notable progress in the past two decades, as seen in recent monographs.[43-45] Here I choose to point out one development that seems to be especially important, i.e., the application of the multiple-scattering method.[38] For instance, Sato et al.[46] used this method and obtained cross sections in excellent agreement with experiment for elastic scattering by H_2O of electrons at 2-200 eV.

Ionic Collisions

Progress in the understanding of ion-atom and ion-molecule collisions pertinent to our theme is likewise noteworthy.[47-49] This progress started with the Fano-Lichten discovery of what we now call the electron-promotion mechanism, which accounts for the inelasticity of close collisions between particles carrying electrons.

For convenience, let me discuss separately collisions in different energy regions. First, at very high energies, i.e., at energies of many MeV/u and higher, the first Born approximation is adequate for most of the collisions, as has long been recognized.[28-30] Next, at energies between a few MeV/u and many keV/u, various perturbative methods[47] were developed in the past two decades, with partial success. Yet classical-trajectory Monte Carlo simulations[48] have turned out to be most successful for explaining many experimental results. None of the methods is fully justifiable for a wide range of energies and for a wide variety of collisions. Among the challenging problems are spectra of convoy electrons (i.e., secondary electrons that are ejected in the forward direction at speeds comparable to the speed of an incident ion), the recoil of target atoms, multiple electron ejection and multiple electron capture by an incident ion, and fragmentation of molecules. All these matters have been studied experimentally and await full theoretical treatment.

For collisions at intermediate energies, i.e., at energies of several keV/u, progress in theory has been considerable.[49] Electron capture by an incident ion and the excitation of a target atom are reasonably well understood within present theories. However, details of ionization processes such as ejected-electron spectra need further theoretical study. Ionic collisions with molecules have been treated only for simpler molecules such as H_2, N_2, and O_2. Furthermore, theoretical work on ionic collisions with polyatomic molecules, molecular clusters,[50] or solid surfaces[51] is in an early stage of exploration. These topics are evidently more relevant to radiation effect studies. In particular, some understanding seems most urgently needed of the general characteristics and systematics of collisional ionization and dissociation of polyatomic molecules such as hydrocarbons.

Collisions of ions at low energies, i.e., at energies of several eV, are relevant to radiation chemistry and hot-atom chemistry. Knowledge in this area is mainly experimental, and theoretical studies are highly desirable. With the advent of supercomputers, reasonably realistic determination of adiabatic-potential surfaces is becoming feasible for collision systems involving a considerable number of atoms, and this determination will provide a sound starting point for analysis. The next decade will see developments in theories of low-energy ionic collisions with polyatomic molecules, molecular clusters,[50] and solid surfaces.[51]

Efforts toward data compilation to satisfy the trinity of requirements are underway also for ionic collisions, as Toburen[15] discusses. An excellent example is seen in the recent work of Phelps,[52] who studied the collisions of a proton with molecular hydrogen with exceptional thoroughness. Inspection of the work gives a good impression of the level of current understanding and an appreciation of the rich physics involved.

General Remarks on Atomic Collision Theory

I wish to make a few points that seem to be essential for eventual theories of atomic and molecular collisions in general. My first point is the need for what I call the Wigner policy.[53] The idea is to distinguish at least two regions of space: one in which colliding partners are well separated and the other in which they temporarily form a combined dynamical system that allows full exchanges of energy, angular momentum, and other quantities describing the internal motion. Explicitly distinct treatment of the two regions is necessary because different physics is involved. Figures 1 and 2 illustrate the idea. Moreover, distinction of more than two spatial regions may be necessary.

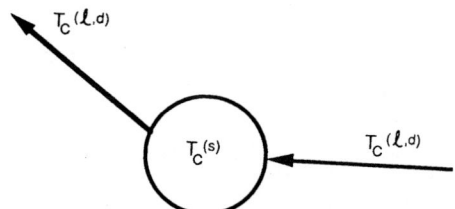

Figure 1. Schematic view of a collision of a particle with an atom or molecule. The particle, presumed for simplicity as structureless (i.e., as carrying no electrons), approaches from the right. The contribution of the approach at large distances to the scattering amplitude A is represented by a transition matrix $T_c^{(\ell,a)}$. In the superscript, "ℓ" signifies long range and "a" approach. The subscript c represents a set of quantum numbers that specify a channel. The contribution of the close encounter, in which the particle in effect merges with the target, is represented by another transition matrix $T_c^{(s)}$, where the superscript "s" signifies short range. The contribution of the departure to large distances is represented by still another transition matrix $T_c^{(\ell,d)}$, where "d" stands for departure. Consequently, we express the scattering amplitude as

$$A = \Sigma_c T_c^{(\ell,d)} T_c^{(s)} T_c^{(\ell,a)}.$$

The long-range transition matrices $T_c^{(\ell,d)}$ and $T_c^{(\ell,a)}$ are modified in condensed mattter, while the short-range transition matrix $T_c^{(s)}$ is the same in gas and in condensed matter.

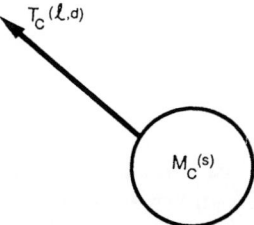

Figure 2. Schematic view of photoabsorption. The immediate result of photoabsorption is the transition of an electron to a state of higher energy while it remains confined to the spatial domain of the initial state. The contribution of this stage to the dipole matrix element may be written as $M_c^{(s)}$, where superscript "s" signifies short range and subscript "c" a set of quantum numbers that specify a channel. The energetic electron later departs toward large distances. The contribution of this stage may be represented by a transition matrix $T_c^{(\ell,d)}$. Consequently, we express the total dipole matrix element D as

$$D = \Sigma_c T_c^{(\ell,d)} M_c^{(s)}.$$

The long-range transition matrix $T_c^{(\ell,d)}$ is modified in condensed matter, while the short-range matrix element $M_c^{(s)}$ is the same in gas and in condensed matter.

This idea of distinguishing regions of space has been implemented in several formulations including the R-matrix theory, the quantum defect theory, and the frame transformation theory, as seen in the monograph by Fano and Rau.[54] However, the general importance of the idea, apart from the technicalities necessary for implementation in specific applications, does not seem to be fully appreciated. Indeed, the multiple-scattering method, to which I referred in connection with photoabsorption and electron collisions with molecules, may be viewed as a minimal version of the Wigner policy applied to electron problems. I believe that the Wigner policy needs to be also incorporated into bound-state calculations of molecules in quantum chemistry.

Much too often, I find that theoretical calculations, formulated from the point of view of analytical expediency, use the same wave function form for the entire space. It is true that under certain simple circumstances a single-form wave function is sufficient to yield a basically correct result. However, this simplification should be clearly justified in advance from full consideration of the physics. In general, we need to be prepared to start with the Wigner idea.

As a second point, I stress the need for theoretical work to establish methods for data representation and systematization. This kind of work is extremely important from the point of view of the trinity of requirements. An example is seen in the Bethe theory,[28-30] which gives a general analytic form for a cross section, and hence the stopping power, as a function of the charged-particle speed (though it is limited to high speeds). A more elementary example is the quantum-defect theory,[54] which gives the Rydberg formula for energy levels of highly excited states of simpler atoms and molecules and its generalizations for more complicated cases. According to this theory, individual energy levels as observed in experiments need not be tabulated; only a few key indices (i.e., the ionization threshold and the quantum defect in the simplest case) summarize all the information and thus are subjects of theoretical study.

Returning to the cross-section study, I point out another example of work in the same spirit. This example concerns the analytic representation of oscillator strength spectra and secondary-electron spectra,[55-59] based on studies of these spectra from the point of view of the theory of functions of a complex variable. I strongly suspect that work in this direction will be much more fruitful than we have so far appreciated.

Atoms and Molecules Versus Condensed Matter

The connection of atomic and molecular physics with condensed-matter physics is an important topic in our context. Indeed, in other chapters, Christophorou[60] discusses high-pressure gases as a prelude to condensed matter, and Ritchie[61] focuses on problems specific to condensed matter. Therefore, I intend to present a few introductory remarks. A more extensive account[62] of my views on this subject has been published.

Let me focus on the complex dielectric-response function $\epsilon(E)$, viz., the electric displacement generated in matter by an external electric field of unit strength that is spatially uniform and oscillates at angular frequency $\omega = E/\hbar$. Basic properties of $\epsilon(E)$ are fully discussed in textbooks such as that of Landau and Lifshitz.[63] The real part $\epsilon_1(E)$ describes the dispersion, and the imaginary part $\epsilon_2(E)$ describes the absorption, both of light at angular frequency ω. The probability that a glancing collision of a fast charged particle results in transfer of energy E to matter is proportional to

$$\Pi(E) = \text{Im}\left[\frac{-1}{\epsilon(E)}\right] = \frac{\epsilon_2(E)}{\epsilon_1^2(E) + \epsilon_2^2(E)}$$

In a low-density material, e.g., in a dilute gas, $\epsilon_1(E)$ is close to unity, and $\epsilon_2(E)$ is much smaller than unity at all E. Then, $\Pi(E)$ is virtually the same as $\epsilon_2(E)$, which means that the spectrum of energy transfer in a glancing collision of a fast-charged particle is the same as that for photoabsorption. In a high-density material, e.g., in condensed matter, $\Pi(E)$ may differ appreciably from $\epsilon_2(E)$. An extreme case occurs if $\epsilon_1(E) = 0$ at a value of E; then $\Pi(E) = 1/\epsilon_2(E)$ has a high value. This case is plasma excitation, well known in metals.

There are many other kinds of condensed-matter effects. A general criterion for the occurrence of molecular-aggregation effects has been given by Fano.[64] [For fuller discussions, see his other related papers.[29,65,66]] Application of the Fano criterion led to the conclusion[64] that the influence of molecular aggregation in water and organic materials of interest to radiation biology is certainly non-negligible at $E \gtrsim 25$ eV (corresponding to valence shell excitation) but is less spectacular than in metals, in agreement with an earlier assessment.[32]

Various effects have been recognized in the past two decades. A full discussion of these effects is given in Ref. 62. Here, I present in Table I a summary of our understanding. Although the knowledge summarized there is extensive, much remains to be done in the future. I strongly suspect that the coming decade will see the recognition of a few more crucial items to be included in Table I.

Table I. Condensed-Phase Effects on the Oscillator Strength Spectrum

Classification	Example	Remarks on Characteristics
Shift of the strength toward higher energies	Water, hydrocarbons, and organics	Occurs over wide ranges of excitation energies
Excitation of special modes of motion	Plasmons in metals	Occurs at specific energies and carries considerable strength
	Excitons in molecular crystals and ionic crystals	Occurs at specific energies and at minor strengths
Interaction of ejected electrons with other atoms and molecules	EXAFS (extended X-ray absorption fine structure)	Occurs at energies above 100 eV, slightly above the pertinent threshold, and at weak strengths, chiefly due to elastic scattering
Resonances	N_2, C_2H_4, C_2H_2, and other unsaturated hydrocarbons	Occurs at energies slightly above the threshold or even lower; influences on the strength are appreciable and lead to inelasticity

An extremely important development in the past two decades is the work of Sanche and co-workers,[67-74] who conducted a series of experiments to study collisions of electrons between tens of eV and about 1 eV with molecules of interest in the solid phase. The work has led to theoretical studies on the interactions of slow electrons in condensed matter, as exemplified in Refs. 75-80.

A major conclusion we have thus drawn may be summarized as follows. Recall the distinction between the long-range interactions and the short-range interactions of an electron with a molecule in condensed matter (see Figs. 1 and 2). Molecular aggregation certainly affects the long-range interactions but leaves the short-range interactions intact. In technical terms, the long-range transition matrix $T^{(\ell)}$ incorporates all molecular-aggregation effects. In contrast, the short-range transition matrix $T^{(s)}$ for electron interactions with an isolated molecule remains applicable to the analysis of the electron behavior in condensed matter. This example illustrates the general strategy of approach, to be fully implemented in the coming decade.

Electron Transport Theory

Preamble

Absorption of any ionizing radiation by matter invariably results in the production of many electrons having a wide range of kinetic energy T. These electrons in turn collide with molecules and thereby degrade in energy and possibly produce further electrons by ionization. The description of the cumulative consequences of many collisions for the electrons and the medium is the goal of the electron transport theory. This topic is important because it provides a link between the knowledge of individual collision processes and the earliest events in radiation biology.

The electron transport theory in a broad sense belongs to statistical physics[7] and physical kinetics,[8] and it has been studied for many years in various contexts such as the electrical conductivity of ordinary matter as well as of partially ionized gases, i.e., plasmas. In the context of radiation actions, modern work began with a paper by Spencer and Fano.[81]

To present an overview of this topic, it is convenient to use a graph, shown in Fig. 3, in which the horizontal axis represents the electron kinetic energy T and the vertical axis the energy transfer E upon a single collision. The figure is meant to apply to water, organics, or other ordinary molecular substances (as opposed to metals or semiconductors).

The vertical broken line indicates the first electronic excitation threshold E_1, which is several eV. Electrons with $T > E_1$ are capable of exciting electronic levels of molecules. The cross section for electronic excitation (or for ionization if T exceeds the ionization threshold I) is appreciable at T within a few multiples of E_1 and declines gradually at higher T, as indicated by the shades in the figure. More importantly, energy transfer E is greater than E_1 for excitation and greater than I for ionization. Consequently, the moderation of an electron in the electronic excitation domain ($T > E_1$) is extremely rapid.

At $T < E_1$, cross sections are usually somewhat smaller, as indicated by lighter shades in the figure. More importantly, energy transfer E upon a single collision is much smaller because electrons in the subexcitation domain can give energies only in smaller quanta to the vibrational, rotational, and translational degrees of molecules. Thus, this domain warrants a treatment separate from that of the electronic excitation domain.

The subexcitation domain is delimited at lower T. At T comparable to thermal energy E_{th} or below, electrons not only lose energy to molecules but also may gain energy from a molecule upon a collision. Therefore, the treatment of electrons in the thermal domain is still another subject.

Electronic Excitation Domain

The past two decades saw notable progress in several respects. The advent of modern computers made it practical to carry out Monte Carlo simulations of electron transport first at high energies as pioneered by Berger.[82] Developments since then have been extensive and show that Monte Carlo simulations are highly effective in

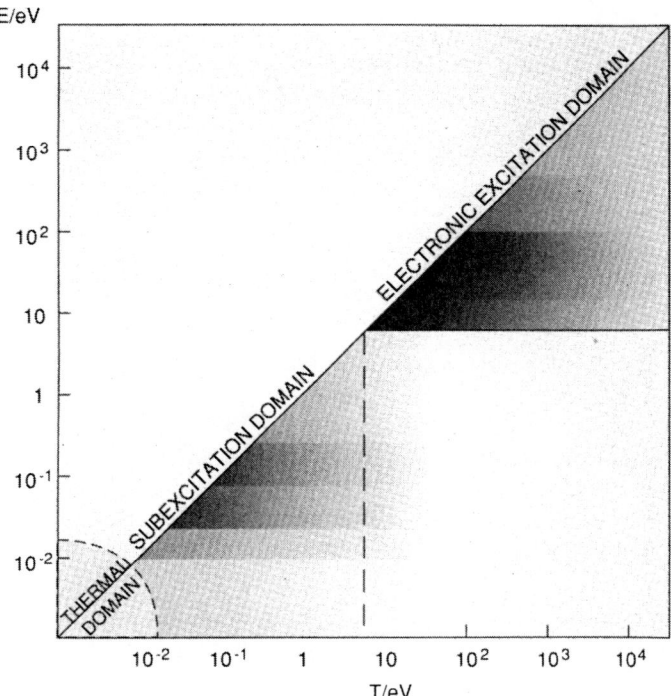

Figure 3. Schematic diagram for showing three distinct domains of electron transport in a molecular substance. The horizontal axis represents the kinetic energy T of an electron. The vertical axis represents the energy transfer E upon a single collision with a molecule. The vertical broken line indicates the first electronic excitation threshold E_1. The shade and fade roughly show the magnitudes of cross sections.

treating problems involving complicated geometry such as dose distributions in non-uniform media, as seen in a recent workshop proceedings.[83] Applications have been extended to lower electron energies with the use of realistic cross sections, as seen for instance in the work by Paretzke.[84]

Progress in analytic treatments has been also considerable. The Spencer-Fano theory has been extended in several directions, i.e., for treating time-dependent cases, media consisting of mixtures, and stochastic aspects.[85-91] A major merit of the analytic theory is to elucidate principles governing electron degradation phenomena, i.e., to answer questions such as how radiation energy absorbed in a mixture might be apportioned to each component of the mixture and how stochastic aspects, e.g., yield fluctuations, can be precisely evaluated. All our experience shows that the Monte Carlo simulation and the analytic theory complement each other in deepening our understanding.

Subexcitation Domain

More recently the behavior of electrons in the subexcitation domain has been studied extensively.[92-101] A general message resulting from the study is that subexcitation electron behavior depends upon the electronic structure of the individual medium and permits no simple scaling, for instance, in terms of the atomic number (which is valid to some extent in the electronic excitation domain).

Thermal Domain

Recent advances in the thermal domain are exemplified by the work by Shizgal and co-workers.[102-104] Among other achievements, they have interpreted the thermalization times of electrons in simpler gases and thus have shown principles governing the thermalization process in unprecedented clarity.

Additional Remarks and Outlook

Although progress in electron transport theory in the past two decades is notable, as even the above cursory summary indicates, much remains to be done in the future.

In order to treat fully electron transport in condensed matter, we should incorporate quantum-mechanical effects. Even in the electronic excitation domain, quantum-mechanical effects may be appreciable for electron energies below about 100 eV. Recall that the de Broglie wavelength of an electron at 150 eV is about 10^{-8} cm, i.e., comparable to the atomic size. The de Broglie wavelength of an electron in the subexcitation domain is longer, and that of an electron in the thermal domain is even longer.

Indeed, fully quantum-mechanical treatments have been carried out on the eventual stage of a thermal electron, i.e., the solvation in water clusters and liquids, as exemplified by work of Berne and co-workers,[105-107] Jonah et al.,[108] and Barnett et al.[109,110] It is extremely desirable to extend work of this kind to higher energies, i.e., energies of a few eV, although methods used in the studies of Refs. 105-110 require prohibitive resources even with a supercomputer. A successful resolution of this difficulty will establish a link between the traditional electron transport theory and the earliest stage of radiation chemistry.

Another related topic in radiation chemistry is the germinate-ion recombination in organic liquids, which leads to excited states of molecules that are important in the kinetics. This is an old subject discussed since the times of the conference at Oberlin. Traditionally, it has been treated with the use of the classical trajectory and of the diffusion equation. I find it questionable to approach this problem in this way. For electron-ion recombination, there is no doubt that some quantum-mechanical effects are decisive. I have long held an idea that a more realistic approach to the electron-ion recombination will be to consider the process from the point of view of the Fermi theory[111] of perturbed Rydberg states. A similar view has recently been expressed by Schiller.[112] The modern theory of Rydberg states is the quantum-defect theory as fully discussed by Fano and Rau.[53] A prototype of the approach I envision is seen in the recent work of Du and Greene,[113] who treated Rydberg states of clusters in which the competition between autoionization (i.e., electron escape) and dissociation is especially crucial. This topic is also closely related to the relaxation phenomena, such as the photofragmentation of molecules,[39] as I earlier mentioned, and as more generally treated by Dattagupta.[114]

Acknowledgments

I thank Ugo Fano, M. A. Dillon, and Mineo Kimura for generously providing ideas and information of great value during the writing of the present article. Work was supported in part by the U.S. Department of Energy, Assistant Secretary for Energy Research, Office of Health and Environmental Research, under Contract W-31-109-Eng-38.

References

1. A. M. Cormack. Representation of a Function by Its Line Integrals, with Some Radiological Applications. *J. Appl. Phys.* 34:2722-2727 (1963).

2. A. M. Cormack. Representation of a Function by Its Line Integrals, with Some Radiological Applications: II. *J. Appl. Phys.* 35:2908-2913 (1964).

3. R. S. Caswell. Deposition of Energy by Neutrons in Spherical Cavities. *Radiat. Res.* 27:92-107 (1966).

4. R. S. Caswell, J. J. Coyne, H. M. Gerstenberg, and E. J. Axton. Basic Data Necessary for Neutron Dosimetry. *Radiat. Prot. Dosim.* 23:41-44 (1988).

5. R. S. Caswell and J. J. Coyne. Effects of Track Structure on Neutron Microdosimetry and Nanodosimetry. *Nucl. Tracks Radiat. Meas.* 16:187-195 (1989).

6. J. J. Coyne, R. S. Caswell, J. Zoetelief, and B. R. L. Siebert. Improved Calculations of Microdosimetric Spectra for Low-Energy Neutrons. *Radiat. Prot. Dosim.*, in press.

7. L. D. Landau and E. M. Lifshitz. *Statistical Physics,* translated by J. B. Sykes and M. J. Kearsley, Pergamon Press, Oxford (1969).

8. E. M. Lifshitz and L. P. Pitaevskii. *Physical Kinetics*, translated by J. B. Sykes and R. N. Franklin, Pergamon Press, Oxford (1981).

9. J. J. Nickson (ed.). *Symposium on Radiobiology. The Basic Aspects of Radiation Effects on Living Cells,* Oberlin College, June 14-18, 1950. John Wiley and Sons, Inc., New York (1952).

10. J. L. Magee, M. D. Kamen, and R. L. Platzman (eds.). Physical and Chemical Aspects of Basic Mechanisms in Radiobiology. *Proceedings of an Informal Conference,* Highland Park, Illinois. Publication No. 305, National Academy of Sciences - National Research Council, Washington, D.C. (1953).

11. R. D. Cooper and R. W. Wood (eds.). *Physical Mechanisms in Radiation Biology.* Proceedings of a Conference held at Airlie, Virginia, Oct. 11-14, 1972. USAEC Technical Information Center, Oak Ridge, Tennessee (1974).

12. U. Fano. Secondary Electrons: Average Energy Loss Per Ionization. *Symposium on Radiobiology*, Oberlin College, June 14-18, 1950, pp. 13-24. John Wiley and Sons, Inc., New York (1952).

13. R. L. Platzman. On the Primary Process in Radiation Chemistry and Biology. *Symposium on Radiobology,* Oberlin College, June 14-18, 1950, pp. 97-116. John Wiley and Sons, Inc., New York (1952).

14. R. L. Platzman. Influence of Details of Electronic Binding on Penetration Phenomena, and the Penetration of Energetic Charged Particles Through Liquid Water. *Symposium on Radiobology,* Oberlin College, June 14-18, 1950, pp. 139-176. John Wiley and Sons, Inc., New York (1952).

15. L. H. Toburen. The Proceedings of the Present Conference.

16. International Commission on Radiation Units and Measurements. *Average Energy Required to Form an Ion Pair,* ICRU Report 31, Washington, D.C. (1974).

17. R. L. Platzman. *Symposium on Radiobology,* Oberlin College, June 14-18, 1950, pp. 20-21. John Wiley and Sons, Inc., New York (1952).

18. U. Fano and W. Lichten. Interpretation of Ar^+-Ar Collisions at 50 keV. *Phys. Rev. Lett.* 14:627-629 (1965).

19. U. Fano. Platzman's Analysis of the Delivery of Radiation Energy to Molecules. *Radiat. Res.* 64:217-232 (1975).

20. E. J. Hart and J. W. Boag. Absorption Spectrum of the Hydrated Electron in Water and in Aqueous Solutions. *J. Am. Chem. Soc.* 84:4090-4095 (1962).

21. W. Brandt and R. H. Ritchie. Primary Processes in the Physical Stage. Proceedings of a Conference held at Airlie, Virginia, October 11-14, 1972, pp. 20-46. USAEC Technical Information Center, Oak Ridge, Tennessee (1974).

22. M. Inokuti. Critique of Cross-Section Data Governing the Physical Stage of Radiation Action. Proceedings of a Conference held at Airlie, Virginia, October 11-14, 1972, pp. 51-67. USAEC Technical Information Center, Oak Ridge, Tennessee (1974).

23. A. E. S. Green and J. H. Miller. Atomic and Molecular Effects in the Physical Stage. Proceedings of a Conference held at Airlie, Virginia, October 11-14, 1972, pp. 68-111. USAEC Technical Information Center, Oak Ridge, Tennessee (1974).

24. H. H. Rossi and A. M. Kellerer. Effects of Spatial-Temporal Distribution of Primary Events. Proceedings of a Conference held at Airlie, Virginia, October 11-14, 1972, pp. 224-243. USAEC Technical Information Center, Oak Ridge, Tennessee (1974).

25. U. Fano. Correlations of Two Excited Electrons. *Rep. Prog. Phys.* 46:97 (1983).

26. M. Inokuti. Foreword to the Proceedings of the Workshop on Electronic and Ionic Collision Cross Sections Needed in the Modeling of Radiation Interactions with Matter, held on December 6-8, 1983, at Argonne National Laboratory, pp. iii-iv, Report ANL-84-28 (1984).

27. M. Inokuti. Cross Sections for Inelastic Collisions of Fast Charged Particles with Atoms and Molecules. *Proceedings of an Advisory Group Meeting on Nuclear and Atomic Data for Radiotherapy and Related Radiobiology,* Rijswijk, The Netherlands, September 16-20, 1987, pp. 57-365. International Atomic Energy Agency, Vienna (1987).

28. H. Bethe. Zur Theorie des Durchgangs schneller Korpuskularstrahlen durch Materie. *Ann. Physik* 5:325-400 (1930).

29. U. Fano. Penetration of Protons, Alpha Particles, and Mesons. *Ann. Rev. Nucl. Sci.* 13:1-66 (1963).

30. M. Inokuti. Inelastic Collisions of Fast Charged Particles with Atoms and Molecules—The Bethe Theory Revisited. *Rev. Mod. Phys.* 43:297-347 (1971).

31. M. Inokuti. VUV Absorption and Its Relation to the Effects of Ionizing Corpuscular Radiation. *Photochem. Photobiol.* 44:297-285 (1985).

32. R. L. Platzman. Energy Spectrum of Primary Activations in the Action of Ionizing Radiation. *Proceedings of the Fourth International Congress of Radiation Research,* Cortina d'Ampezzo, June-July 1966, G. Silini (ed.), pp. 20-42. North-Holland, Amsterdam (1967).

33. J. A. R. Samson. Atomic Photoionization. *Handbuch der Physik,* edited by S. Flügge (ed.), 31:123-213. Springer Verlag, Berlin (1982).

34. J. W. Gallagher, C. E. Brion, J. A. R. Samson, and P. W. Langhoff. Absolute Cross Sections for Molecular Photoabsorption, Partial Photoionization, and Ionic Photofragmentation Processes. *J. Phys. Chem. Ref. Data* 17:9-153 (1988).

35. U. Fano and J. W. Cooper. Spectral Distribution of Atomic Oscillator Strengths. *Rev. Mod. Phys.* 40:441-507 (1968); 41:724-725 (1969).

36. A. F. Starace. Theory of Atomic Photoionization. *Handbuch der Physik,* S. Flügge (ed.), 31:1-121. Springer Verlag, Berlin (1982).

37. M. Y. Amusia. *Atomic Photoeffect.* Plenum Press, New York (1990).

38. J. L. Dehmer and D. Dill. The Continuum Multiple-Scattering Approach to Electron-Molecule Scattering and Molecular Photoionization. *Electron-Molecule and Photon-Molecule Collisions,* T. Rescigno, V. McKoy, and B. Schneider (eds.), pp. 225-265. Plenum Press, New York (1979).

39. N. E. Henrikson. On the Evaluation of Branching Ratios in Molecular Photofragmentation. *Chem. Phys. Lett.* 169:229-235 (1990).

40. Y. Itikawa, M. Hayashi, A. Ichimura, K. Onda, K. Sakimoto, and K. Takayanagi. Cross Sections for Collisions of Electrons and Photons with Nitrogen Molecules. *J. Phys. Chem. Ref. Data* 15:985-1010 (1986).

41. Y. Itikawa, K. Ichimura, K. Onda, K. Sakimoto, K. Takayanagi, Y. Hatano, M. Hayash, H. Nishimura, and S. Tsurubuchi. Cross Sections for Collisions of Electrons and Photons with Oxygen Molecules. *J. Phys. Chem. Ref. Data* 18:23-42 (1989).

42. M. Hayashi. Electron Collision Cross Sections for Atoms and Molecules Determined from Beam and Swarm Data, in Atomic and Molecular Data for Radiotherapy. *Proceedings of an Advisory Group Meeting Organized by the International Atomic Energy Agency,* Vienna, June 13-16, 1988, IAEA-TECDOC-506, pp. 193-199. International Atomic Energy Agency, Vienna (1989).

43. I. Shimamura and K. Takayanagi (eds.). *Electron-Molecule Collisions.* Plenum Press, New York (1984).

44. L. G. Christophorou (ed.). *Electron-Molecule Interactions and Their Applications,* Vols. 1 and 2 (Academic Press, New York (1983-1984).

45. E. W. McDaniel. *Atomic Collisions: Electron and Photon Projectiles.* John Wiley & Sons, New York (1989).

46. H. Sato, M. Kimura, and K. Fujima. Elastic and Momentum Transfer Cross Sections in Electron Scattering by Water Molecules. *Chem. Phys. Lett.* 145:21-25 (1988).

47. R. H. Bransden. Charge Transfer and Ionization in Fast Collisions. *Electronic and Atomic Collisions: Invited Papers of the XV International Conference on the Physics of Electronic and Atomic Collisions,* Brighton, UK, July 1987, H. B. Gilbody, W. R. Newell, F. H. Read, and A.C.H. Smith (eds.), pp. 255-269. North-Holland, Amsterdam (1988).

48. R. E. Olson. Multiple Electron Capture and Ionization in Ion-Atom Collisions. *Electronic and Atomic Collisions: Invited papers of the XV International Conference on the Physics of Electronic and Atomic Collisions,* Brighton, UK, July 1987, H. B. Gilbody, W. R. Newell, F. H. Read, and A. C. H. Smith (eds.), pp. 271-285. North-Holland, Amsterdam (1988).

49. M. Kimura and N. F. Lane. The Low-Energy, Heavy-Particle Collisions—A Close-Coupling Treatment. *Advances in Atomic, Molecular, and Optical Physics,* Vol. 26, D. Bates and B. Bederson (eds.), pp. 80-160. Academic Press, Boston (1990).

50. E. R. Bernstein (ed.). *Atomic and Molecular Clusters.* Elsevier, Amsterdam (1990).

51. J. Los and J. J. C. Geerlings. Charge Exchange in Atom-Surface Collisions. *Phys. Rep.* 190:133-190 (1990).

52. A. V. Phelps. Cross Sections and Swarm Coefficients for H^+, H_2^+, H_3^+, H, H_2, and H^- in H_2 for Energies from 0.1 eV to 10 keV. *J. Phys. Chem. Ref. Data* 19:653-675 (1990).

53. M. Inokuti. The Future of Atomic Collision Theory. *Comments At. Mol. Phys.* 10:99-106 (1981).

54. U. Fano and A.R.P. Rau. *Atomic Collisions and Spectra.* Academic Press, Orlando, Florida (1986).

55. M. A. Dillon and M. Inokuti. Analytic Representation of the Dipole Oscillator-Strength Distribution. *J. Chem. Phys.* 74:6271-6277 (1981).

56. M. A. Dillon and M. Inokuti. Analytic Representation of the Dipole Oscillator-Strength Distribution: II. The Normalization Factor for Electron Continuum States in Atomic Fields. *J. Chem. Phys.* 82:4415-4424 (1985).

57. M. A. Dillon, M. Inokuti, and Z.-W. Wang. Analytic Representation of the Generalized Oscillator Strength for Ionization. *Radiat. Res.* 102:151-164 (1985).

58. M. Inokuti and M. A. Dillon. What Formulas Are Good for Representing Dipole and Generalized Oscillator-Strength Spectra? *Physics of Ionized Gases: Proceedings of the Twelfth Yugoslav Summer School and International Symposium on the Physics of Ionized Gases,* Sibenik, September 3-7, 1984, M. M. Popvic and P. Krstic (eds.), pp. 3-22. World Scientific, Singapore (1986).

59. M. Inokuti, M. A. Dillon, J. H. Miller, and K. Omidvar. Analytic Representation of Secondary-Electron Spectra. *J. Chem. Phys.* 87:6967-6972 (1987).

60. L. G. Christophorou. The Proceedings of the Present Conference.

61. R. H. Ritchie. The Proceedings of the Present Conference.

62. M. Inokuti. How Is Radiation Energy Absorption Different in the Condensed Phase and in the Gas? *Hoshasen Kagaku (Radiation Chemistry)* 49:2-14 (1990) [In Japanese]. Radiat. Effects and Defects in Solid, in press [in English].

63. L. D. Landau and E. M. Lifshitz. *Electrodynamics of Continuous Media,* translated by J. B. Sykes and J. S. Bell, Chapter IX. Pergamon Press, Oxford (1960).

64. U. Fano. Normal Modes of a Lattice of Oscillators with Many Resonances and Dipolar Coupling. *Phys. Rev.* 118:451-455 (1960).

65. U. Fano. Atomic Interactions in Dense Materials. *Phys. Rev.* 103:1202-1218 (1956).

66. U. Fano. *A Mechanism of Collective Phenomena.* A preprint of an article to be published.

67. L. Sanche. Transmission of 0-15 eV Monoenergetic Electrons Through Thin-Film Molecular Solids. *J. Chem. Phys.* 71:4860-4882 (1979).

68. G. Bader, G. Perluzzo, L. G. Caron, and L. Sanche. Structural-Order Effects in Low-Energy Electron Transmission Spectra of Condensed Ar, Kr, Xe, N_2, CO, and CO_2. *Phys. Rev. B* 30:78-84 (1984).

69. G. Perluzzo, L. Sanche, C. Gaubert, and R. Baudoing. Thickness-Dependent Interference Structure in the 0-15-eV Electron Transmission Spectra of Rare-Gas Films. *Phys. Rev. B* 30:4292-4296 (1984).

70. M. Michaud and L. Sanche. Interactions of Low-Energy Electrons (1-30 eV) with Condensed Molecules: I. Multiple Scattering Theory. *Phys. Rev. B* 30:6067-6077 (1984).

71. L. Sanche and M. Michaud. Interactions of Low-Energy Electrons (1-30 eV) with Condensed Molecules: Vibrational-Librational Excitation and Shape Resonances in N_2 and CO films. *Phys. Rev. B* 30:6078-6092 (1984).

72. M. Michaud and L. Sanche. Total Cross Sections for Slow-Electron (1-20 eV) Scattering in Solid H_2O. *Phys. Rev. A* 36:4672-4683 (1987).

73. M. Michaud and L. Sanche. Absolute Vibrational Excitation Cross Sections for Slow-Electron (1-18 eV) Scattering in Solid H_2O. *Phys. Rev. A* 36:4684-4699 (1987).

74. L. Sanche. Primary Interactions of Low-Energy Electrons in Condensed Matter. *Excess Electrons in Dielectric Media,* C. Ferrandini and J.-P. Jay-Gerin (eds.), to appear as a CRC Uniscience Book. (This review article summarizes the work of the author's group. Refs. 66-72 are representative of dozens of papers published by the group.)

75. U. Fano, J. A. Stephens, and M. Inokuti. Absence of Resonances in the Elastic Scattering of Electrons in Molecular Solids. *J. Chem. Phys.* 85:6239-6240 (1986).

76. U. Fano and J. A. Stephens. Slow Electrons in Condensed Matter. *Phys. Rev. B* 34:438-441 (1986).

77. U. Fano. Studies of Slow Electron Action on Condensed Media. *Radiat. Phys. Chem.* 32:95-97 (1988).

78. J. A. Stephens and U. Fano. Slow Electrons in Condensed Matter. The Large Polaron. *Phys. Rev. A* 38:3372-3376 (1988).

79. P. Knipp. Interaction of Slow Electrons with Density Fluctuations in Condensed Materials: Calculation of Stopping Power. *Phys. Rev. B* 37:12-17 (1988).

80. U. Fano and N.-Y. Du. Dissipative Polarization by Slow Electrons. *Appl. Radiat. Isot.* (in press).

81. L. V. Spencer and U. Fano. Energy Spectrum Resulting from Electron Slowing Down. *Phys. Rev.* 93:1172-1181 (1954).

82. M. J. Berger. Monte Carlo Calculation of the Penetration and Diffusion of Fast Charged Particles. *Methods in Computational Physics,* Vol. 1, B. Adler, S. Fernbach, and M. Rotenberg (eds.), pp. 135-215. Academic Press, New York (1963).

83. T. M. Jenkins, W. R. Nelson, and A. Rindi (eds.). *Monte Carlo Transport of Electrons and Photons.* Plenum Press, New York (1988).

84. H. G. Paretzke. Radiation Track Structure Theory. *Kinetics of Non-Homogeneous Processes,* G. R. Freeman (ed.), pp. 89-170. John Wiley & Sons, New York (1987).

85. A.R.P. Rau, M. Inokuti, and D. A. Douthat. Variational Treatment of Electron Degradation and Yields of Initial Molecular Species. *Phys. Rev. A* 18:971-988 (1978).

86. M. Inokuti, D. A. Douthat, and A. R. P. Rau. Statistical Fluctuations in the Ionization Yield and Their Relation to the Degradation Spectrum. *Phys. Rev. A* 22:445-453 (1983).

87. M. Inokuti and E. Eggarter. Theory of Initial Yields of Ions Generated by Electrons in Binary Mixtures: II. *J. Chem. Phys.* 86:3870-3875 (1987).

88. M. Inokuti, M. A. Dillon, and M. Kimura. Theory of Electron Degradation and Yields of Initial Molecular Species. *Int. J. Quantum Chem. Symp. Series* 21:251-266 (1987).

89. M. Inokuti, M. Kimura, and M. A. Dillon. Time-Dependent Aspects of Electron Degradation: II. General Theory. *Phys. Rev. A* 38:1217-1224 (1988).

90. K. Kowari, M. Kimura, and M. Inokuti. Electron Degradation and Yields of Initial Products: V. Degradation Spectra, the Ionization Yield, and the Fano Factor for Argon Under Electron Degradation. *Phys. Rev. A* 39:5545-5553 (1989).

91. K. Kowari, M. Inokuti, and M. Kimura. Time-Dependent Aspects of Electron Degradation: V. Ar - H_2 Mixtures. *Phys. Rev. A* 42:795-802 (1990).

92. M. A. Dillon, M. Inokuti, and M. Kimura. Time-Dependent Aspects of Electron Degradation: I. Subexcitation Electrons in Helium or Neon Admixed with Nitrogen. *Radiat. Phys. Chem.* 32:43-48 (1988).

93. A. Pagnamenta, M. Kimura, M. Inokuti, and K. Kowari. Electron Degradation and Yields of Initial Products: III. Dissociative Attachment in Carbon Dioxide. *J. Chem. Phys.* 89:6220-6225 (1988).

94. K. Kowari, M. Kimura, and M. Inokuti. Electron Degradation and Yields of Initial Products: II. Subexcitation Electrons in Molecular Nitrogen. *J. Chem. Phys.* 89:7229-7237 (1988).

95. M. Kimura, M. Inokuti, K. Kowari, M. A. Dillon, and A. Pagnamenta. Time-Dependent Aspects of Electron Degradation: IV. Subexcitation Electrons in Nitrogen and Carbon Dioxide. *Radiat. Phys. Chem.* 34:481-485 (1989).

96. M. A. Ishii, Mineo Kimura, Mitio Inokuti, and Ken-ichi Kowari. Electron Degradation and Yields of Initial Products: IV. Subexcitation Electrons in Molecular Oxygen. *J. Chem. Phys.* 90:3081-3086 (1989).

97. M. Kimura, M. Inokuti, and K. Kowari. Electron Degradation and Yields of Initial Products: VI. Energy Spectra of Subexcitation Electrons in Argon and Molecular Hydrogen. *Phys. Rev. A* 40:2316-2320 (1989).

98. M. Inokuti. Subexcitation Electrons in Gases, in Molecular Processes in Space, T. Watanabe, I. Shimamura, M. Shimizu, and Y. Itikawa (eds.), pp. 65-86. Plenum, London (1990).

99. M. Kimura and M. Inokuti. Subexcitation Electrons in Molecular Gases. Comments At. Mol. Phys. 24:269-286 (1990).

100. M. A. Ishii, M. Kimura, and M. Inokuti. Electron Degradation and Yields of Initial Products: VII. Subexcitation Electrons in Gaseous and Solid H_2O. *Phys. Rev. A* 42:6486-6496 (1990).

101. M. Inokuti. Subexcitation Electrons: An Appraisal of Our Understanding. *Appl. Radiat. Iso.* (in press).

102. B. Shizgal and D.R.A. McMahon. Electric Field Dependence of Transient Electron Transport in Rare-Gas Moderators. *Phys. Rev. A* 32:3669-3680 (1985).

103. B. Shizgal, D.R.A. McMahon, and L. A. Viehland. Thermalization of Electrons in Gases. *Radiat. Phys. Chem.* 34:35-50 (1989).

104. K. Kowari and B. Shizgal. Time Dependent Electron Energy Distribution Functions and Degradation Spectra: A Comparison of the Spencer-Fano Equation and the Boltzmann Equation. *Appl. Radiat. Iso.* (In press)

105. A. Wallqvist, D. Thirumalai, and B. J. Berne. Localization of an Excess Electron in Water Clusters. *J. Chem. Phys.* 85:1583-1591 (1986).

106. D. F. Coker, B. J. Berne, and D. Thirumalai. Path Integral Monte Carlo Studies of the Behavior of Excess Electrons in Simple Fluids. *J. Chem. Phys.* 86:5689-5702 (1987).

107. A. Wallqvist, D. Thirumalai, and B. J. Berne. Path Integral Monte Carlo Study of the Hydrated Electron. *J. Chem. Phys.* 86:6404-6418 (1987).

108. C. D. Jonah, C. Romero, and A. Rahman. Hydrated Electron Revisited Via the Feynman Path Integral Route. *Chem. Phys. Lett.* 123:209-214 (1986).

109. R. N. Barnett, U. Landman, C. L. Cleveland, and J. Jortner. Electron Localization in Water Clusters: I. Electron-Water Pseudopotential. *J. Chem. Phys.* 88:4421-4428 (1988).

110. R. N. Barnett, U. Landman, C. L. Cleveland, and J. Jortner. Electron Localization in Water Clusters: II. Surface and Internal States. *J. Chem. Phys.* 88:4429-4447 (1988).

111. E. Fermi. Sopra lo Spostamento per Pressione delle Righe Elevate delle Serie Spettrali. [On the Pressure Shift of the Higher-Order Lines of the Spectral Series] Nuovo Cimento 11:157-166 (1934). English translation is available as NTC-TRANS-II-580.

112. R. Schiller. Ion-Electron Pairs in Condensed Polar Media Treated as H-Like Atoms. *J. Chem. Phys.* 92:5527-5532 (1990).

113. N.-Y. Du and C. H. Greene. Multichannel Rydberg Spectra of Rare Gas Dimers. *J. Chem. Phys.* 90:6347-6360 (1989).

114. S. Dattagupta. Relaxation Phenomena in Condensed Matter Physics. Academic Press, Orlando, Florida (1987).

Discussion

Kopelman: What I've been really missing are the differences between gas phase and condensed phase. The secondary effects and the differences in energy degradation, and now you touched on that at the end with the branching ratios for dissociation, but there are those glaring differences. Just as an example, in molecular crystals you have exciton fission and then you have exciton fusion; and you have phonon creation and those phonons interact, so there is this whole gamut of phenomena, which is one of the big differences between the gas phase and the condensed phase. Especially important is that we don't only have energy dissipation in various new channels and branching, but we may even have "up" conversion phenomena, which will be very rare in a gas phase, though it could happen.

Inokuti: I totally agree with what you said. You see, I deliberately left out any discussion on the secondary effects, because I thought that belonged to Mike Kasha.

Kasha: In high-pressure gases, you begin to see phenomena such as the transition, or up conversion of excited states that you don't see in low-pressure gases. That means simultaneous excitation of two different molecules by one photon. That occurs through the gas phase at a hundred atmospheres but not at one atmosphere at all. And in the condensed phase, that's a common thing.

Kopelman: That's correct. So that at high pressures the difference becomes negligible.

Kasha: The thing that I see is that you and I are interested in very similar things, but in very different energy ranges. Our interests overlap, and where they overlap we still use different languages, so I have a few questions to ask you, and they are somewhere in this category. For example, in the water spectrum, which you show, there is a sharp peak and then a continuum that is a discrete sort of a peak. Now, what is that peak? Is that simply ionization, or could that be a resonance which is excitation of bonding electrons?

Inokuti: I think it's a well-known excitation, and the peak is sharp, generally, when you excite electrons that are not bonded strongly.

Kasha: I understand. You understand that in the same way. Now, I heard a lecture by Ugo Fano once on longitudinal excitons. That means longitudinal with the momentum and the propagation collinear, except instead of transversing excitons which are electromagnetic with the propagation direction of the perturbation at right angles. What I remember is that he showed that there is a unique cross section for absorption, which appears as a resonance owing to excitons. Do you remember that work? Now, that's another kind of resonance which is really coherent excitation of a line of atoms which are transferring momentum. I just wondered if that was part of your review.

Inokuti: Oh yes, that is very much a part of the discussion, you see, I think what Ugo Fano discussed is just to show origins of collective excitation. In the beginning he used the word "collective excitation," and the basic idea there is still good. There have been many applications of the idea.

Kasha: Okay, now, another way in which our interests overlap but sort of disagree is where you draw the boundary between molecular excitation and sub-excitation electrons. Your boundary came at 7 eV.

Inokuti: I had in mind water as a medium. I know that in water electronic excitation begins at 6.8 eV. Similarly, for most materials of our interest, first electronic excitation begins at 8 eV or even higher.

Kasha: Of course, in the kind of things that I study, we go much lower in energy and excitation. It's just a question of arbitrary definition. What I'd like to ask about is some excitation delayed on your part. Now, I don't think I really understand, I'm sure I don't understand what sub-excitation means, for example,....

Inokuti: You have lower electronic excitation energies in aromatic and unsaturated molecules, which you often study. The meaning of the term "subexcitation," you see, is "subelectronic excitation." This is so cumbersome that I just don't say it, but it is subelectronic excitation.

Kasha: I was going to ask you the question, where do metastable states of molecules come in?

Inokuti: That is a very nice, technical question. For example, let's consider an oxygen molecule. Of course, the oxygen molecule has a very low excited state of 1.0 eV and another at 1.6 eV, which are metastable. Those metastable states have such small cross-sections for excitation by electrons that, in practice, if you ask, "What is a subexcitation electron in molecular oxygen?" I would say it is an electron below about 9 eV at which strong electronic excitation begins. Practically speaking, the metastable production is a phenomenon in the subexcitation domain, because the total loss is really comparable to that of vibrational excitation.

Kasha: But, you know, the experience in electron scattering spectroscopy is that what are metastable states for electromagnetic radiation are not metastable states for electron scattering, and you get high cross-sections through some of the highly forbidden states for light absorption but which appear in the electron-scattering spectrum. That's a whole virtue in the way of using electron scattering for studying the molecule.

Inokuti: Yes, of course, the difference between electron scattering and photo-absorption is well known. Yet, the total excitation cross section for metastable states is often small, as in molecular oxygen.

Christophorou: Since the issue of definitions has come up, maybe one could suggest that instead of the term "subexcitation electrons," the term "subionization electrons" is used to describe electrons whose kinetic energy is below the first ionization onset energy of the medium.

To what accuracy do you really want, or need, the various cross sections to be measured? If we insist on very accurate data, obviously that would mean fewer data since accuracy goes along with two other things: cost and time.

Inokuti: I think the accuracy would depend on the purpose. It is meaningless to discuss the accuracy requirement without specifying the purpose.

Christophorou: In general, is, say, 10 percent accuracy sufficient?

Inokuti: That is an achievable limit, and for most purposes that is a typical requirement. I think my own preference is comprehensiveness of data over a wide range of variables. I will not press the need for accuracy; even 50 percent accuracy may suffice for some purposes. At the expense of accuracy, I would press on comprehensive coverage and a variety of data.

Kasha: In the condensed phase, it is found that the energy to ionize the molecule depends on the medium and is greatly lowered compared to the gas-phase value.

Inokuti: Yes, that is well known.

Varma: You have summarized in a very eloquent way where we are in 20 years from the Airlie House Conference, but I really would like to focus on the objective of this meeting. For example, the slide that you presented, which showed differences in gas phase and the condensed phase, specifically the shifting of oscillator strength to high energies. I would like to ask you how this affects the initial events in biology in your view.

Inokuti: So you want to know what comes further?

Varma: How are these things related to a particular biological endpoint? This is the important thing, and since we have spent 25 years studying these from the point of view of biology, I think that we must ask the question: "Why can't we make the connection to biological endpoint?" What is preventing us from doing it?

Inokuti: What do you mean by preventing the connection between physics and biology? Could you elaborate?

Varma: Yes, and finding out whether, for example, the shift in oscillator strength to higher energy is more important, or the fission or fusion of the DNA that Harel talked about, or whether what Mike is talking about in terms of the metastable states is the important mode.

Inokuti: I don't know that. I'm just trying to identify as many facts as possible and as many principles as possible involved in the radiation action. Most likely, there is no single dominant thing, but there are many facets of understanding needed for a complete picture of radiation biology. Clearly, chemistry plays an important role here.

Weinstein: That would not come from physics. What you need is a mechanistic hypothesis.

Inokuti: Yes, it doesn't come out of physics only.

Weinstein: It doesn't come out of physics, but you have to make the connection somewhere. This connection has to come from a mechanistic hypothesis for which physics will provide the details one needs to know in order to test the hypothesis, and to see how to make it go from a model system, where we always start, to the actual system for which we want to understand the actual behavior. But the connection cannot come from the measurement. The connection can come from a mechanism that connects the effect of irradiation to the molecular mechanism underlying the biological process.

Varma: I'll come back to my question; I don't believe that it has been answered. Maybe Mike has the answer.

Kasha: I will be the devil's advocate by saying that what Inokuti says he does know, let's say that I know perfectly and here's what it is. It is wrong to say that you can just narrow down to a specific thing. You've got to start with the high oscillator-strength end where the high energies are. That's the fundamental physics of getting the first energy down to the final place which is going to be a chemical trap. So all the things we are discussing are, how is it going to get down to the critical trap? What we have are many different mechanisms all over the place. Stepwise, degradation of energy, and finally there's got to be some radiation chemistry at a critical point, which then triggers the biological events that Eric was talking about. In other words, the chemical origin of the point mutation is one we have to work toward, and that goes between fundamental radiation-produced radical reactions and things which are low key to the studies, and then the chemists who know how to deal with the specific hetero-cyclic or peptide or other local chemistry. I have a hint or two of that in my talk, but that is the critical issue here. There is no one single thing that is the clue; it is the whole succession of steps, and what we ought to be doing is locating the biggest gap in what we know.

Varma: I think that you want to describe the initial physical and chemical processes and follow these up to the point when initial biological damage takes place. My question is, when you have these shifts due to the differences in the gas phase and the condensed phase, how do these shifts affect the probability of point mutation induction? I realize that this might be a tall order, but nevertheless, I am interested in your ideas on how this happens or ways in which it can be tested.

Kasha: I think that the missing data is the kind that Leroy Oakenstein was most interested in, the fast-time events before deducible biology takes place, and we didn't know what times they were. In fact, someone had on a slide, I think it was you, Matesh, a time scale for chemistry and for physics. In fact you had 10^{-19} seconds for physics and 10^{-12} for chemistry. Well, we think there is fempto-second chemistry going on, and now we can measure it. We have laboratories which are doing that. We are finding new things that were transients which we never suspected before. In fact, we know now that states which are produced in a picosecond or a hundred fempto-seconds are the critical chemical steps before you observe the photochemistry. We didn't know that before, and the very kinds of photochemistry we need in this kind of field, no one has sponsored research for that yet. There are very elementary steps that are missing. And I think when Harel, Loucas, and I, when all of us try to say, "What is the critical step?" to me it's the fast-time chemical trap that leads to more stable, slower steps, which are the usual kind that have been measured before.

Varma: But those fast chemical steps, do we know how fast they are and what physical mechanisms affect them?

Kasha: Yes. I think that at the forefront of modern photochemistry and radiation chemistry are very fast initial transients which are now being probed. There are conferences on this regularly now; there are two or three a year, and I think we are patching up that missing gap.

Christophorou: Although progress is being made, we still don't know much about the behavior and interactions of electrons in dense matter, i.e., high-pressure gases and liquids, in the subionization region. What we really need is a theory of slow-electron interaction in condensed phase, especially liquid matter. The existing theory is very poor.

Inokuti: Yes, it is very poor. I'm glad you pointed that out. For instance, if we go to a conference on radiation chemistry, we hear talks about geminate-ion recombination, and there they often speak about electron trajectories of low-energy electrons. These things don't seem to have any justification whatsoever.

Christophorou: It seems to me that someone has to make these things really understandable.

Inokuti: They are used to the trajectory notion. It's really no good for electrons of low energies for which wave-mechanical effects are crucial. There are a great many things to be done.

Varma: But the point is, if you talk to a biologist they are going to say that the excitation doesn't matter at all. That's the biologist's point of concern. You have to relate ionization to the biological effect.

Inokuti: That is clearly too limited an idea.

Kopelman: This comment may emphasize an obvious point here, but I'll comment for the record. We have too much emphasis on ionization, which comes from atomic physics. The key word I haven't heard yet, even though I know Mike knows it well, is "dissociation," because we have bonds breaking, we have radicals forming. The radicals may not be ionic and may not have any charges, but it is very certain, from what I may know and others may know better, that those radicals may be involved in the biological havoc. So, I'm trying to emphasize, it's not just subionization. We have to talk about subdissociation energy, and so on. It is the word "dissociation" I want to hear more.

Christophorou: I find it hard to make the connection here. I'm glad that you mentioned this because the most effective way to produce radicals is in the subexcitation or subionization region. When an electron comes as close as possible to these molecules, you lead to dissociation of the molecules. And that is really the essence, relevant to electron trapping and electron migration and radical formation. Do you agree with that?

Inokuti: Yes, I agree. The production of excited states and dissociation, of course, they don't come from fast particles alone. The direct excitation to the dissociation states, I think, is a small part in general. We know it for sure. It is not a negligible part, but I would say it is something like 20 to 30 percent. You see, a larger part essentially comes as a secondary process of slow electrons, that is, electrons doing something, most importantly, electron-ion recombination.

Varma: I just want to say a word on whether the ionizations are important or whether the excitations are important. There are a lot of biological experiments done as a function of frequency or energy of ionizing radiation. These experiments indicate that biological damage is predominantly produced by ionizations and not excitations. This, however, doesn't preclude the possibility that in some cases excitations may lead to dissociation or a radical and produce a biological change. But the majority of the biological damage is produced by ionization and not from excitations.

Inokuti: That interpretation is too simplistic. The slow electrons I talked about originate in ionization. Such electrons are, of course, absent in matter under low-frequency electromagnetic waves that are non-ionizing.

von Sonntag: We have no information whatsoever whether by the direct effect considerable fragmentations occur in DNA, and I think that this is a very interesting question. We know from photochemistry that the lower excited states will not produce such reactions. But it's now the question whether highly excited states are sufficiently long-lived to produce these kinds of reactions. We must study these questions. In, say, quasi-aromatic molecules, energy dissipation might, however, be extremely rapid.

Christophorou: We need to keep in mind that a molecule can dissociate even with very little energy. For example, in an exothermic dissociative electron attachment reaction, a "zero-energy" electron can cause molecular dissociation into neutral and negatively charged species.

von Sonntag: This doesn't happen in DNA, as far as we know; e.g., strand breakage is not induced by the solvated electron by splitting the carbon phosphate bond.

Chatterjee: Michael, maybe you can help me to understand. With DNA we know that this is dimer formation through photochemistry. But with ionizing radiation, the yield is very, very small, and I never understood this phenomena. Do we understand this?

Kasha: How much excitation occurs with ionizing radiation acting on a system? Is there some fraction of distribution that you can cite?

Chatterjee: It's large.

John Ward: The problem here is that the production of the dimer is wavelength dependent. If the energy is lower than that, it splits the dimer, and it is dependent on the energy which is imparted. It has to be around 260 nanometers.

von Sonntag: So the cross-section of forming these dimers is equivalent to that of photoabsorption so far as we know. It closely parallels the absorption spectrum. So even at the low end it is lower, a high energy range with low wavelength range; the quantum yield of dimer formation is rather uniform over the absorption spectrum.

Atomic and Molecular Physics in the Gas Phase

L. H. Toburen[a]

Abstract

The spatial and temporal distributions of energy deposition by high-linear-energy-transfer radiation play an important role in the subsequent chemical and biological processes leading to radiation damage. Because the spatial structures of energy deposition events are of the same dimensions as molecular structures in the mammalian cell, direct measurements of energy deposition distributions appropriate to radiation biology are infeasible. This circumstance has led to the development of models of energy transport based on a knowledge of atomic and molecular interactions that enable one to simulate energy transfer on an atomic scale. Such models require a detailed understanding of the interactions of ions and electrons with biologically relevant material. During the past 20 years, there has been a great deal of progress in our understanding of these interactions, much of it coming from studies in the gas phase. These studies provide information on the systematics of interaction cross sections, and lead to knowledge of the regions of energy deposition where molecular and phase effects are important—knowledge that guides development in appropriate theory. In this report, studies of the doubly differential cross sections, which are crucial to the development of stochastic energy deposition calculations and track structure simulation, are reviewed. We discuss areas of understanding and address directions for future work. Particular attention is given to experimental and theoretical findings that have changed the traditional view of secondary electron production for charged-particle interactions with atomic and molecular targets.

Introduction

The importance of the spatial and temporal distributions of ionization in determining the subsequent chemical and biological damage induced by ionizing radiation has long been recognized [see, for example, Lea[1]]. During the past 25 years, we have seen a continuing evolution in the need for understanding the details of these distributions in increasingly smaller volumes in order to interpret results obtained in studies of radiation biology. This need spawned the field of microdosimetry and has led to the development of computational tools in charged-particle track simulation to investigate energy deposition in volumes smaller than can be reached by experimental microdosimetric techniques. Computational techniques also provide flexibility to incorporate the target heterogeneity that is important to modeling biological media.

For high-linear-energy-transfer (high-LET) radiation, the traditional concept of dose, that of average energy imparted per unit mass, is inappropriate because of the highly localized nature of the energy deposition. For a dose of a few rads delivered by alpha particles, for example, only about one cell in ten may be traversed by an alpha particle, and the cells that do get hit may receive ten times the average dose estimate. Thus, assessing the biological effects from alpha particles or other high-LET

(a) Pacific Northwest Laboratory, Richland, Washington

Physical and Chemical Mechanisms in Molecular Radiation Biology
Edited by W.A. Glass and M.N. Varma, Plenum Press, New York, 1991

radiation requires a knowledge of the energy deposition, or dose, at the microscopic level. Studies of DNA strand breaks induced by ^{125}I, for example, indicate that an energy of 17.5 eV deposited in the sugar-phosphate is adequate to cause a DNA strand break.[2] These estimates are made by comparing energy deposition distributions determined from track simulation calculations with measurements of the frequency of strand breaks in cells irradiated with known amounts of ^{125}I incorporated into the cellular DNA. An experimental measurement of the energy deposited in volumes this small is, of course, infeasible; one must rely on accurate models of energy deposition and transport to make such comparisons.

A description of the interaction of high-LET radiation with cellular DNA requires a knowledge of the structures of both the particle track and the DNA target. An example of the relative size of the structural features of a 2-MeV alpha particle track and representative target structures of a DNA molecule is shown in Fig. 1. This illustration was prepared by Walt Wilson in our laboratory using the Monte Carlo track structure code MOCA15 that he has developed in collaboration with Herwig Paretzke.[3] This code scores both excitations and ionizations for interactions of charged particles in a water vapor medium scaled to the density of liquid water. For this example secondary electrons were followed until degraded in energy to 25 eV, where they are considered locally absorbed. The example given in Fig. 1 illustrates several features that can be derived from stochastic, atom-by-atom descriptions of a charged-particle track. First, the structures of the track resulting from energy transport by delta rays are of the same order of magnitude in size as the nucleosome structures of DNA. One may expect multiple sites of damage in the DNA if one of these track features should correspond in space to that of the nucleosome. The example also illustrates the effect of inner-shell ionization of the oxygen atom of the constituent water vapor molecule. The energetic delta ray shown moving up through the nucleo-some in Fig. 1a is the result of the ejection of an electron from the K-shell of oxygen; the second electron emanating from that point is the Auger electron following relaxa-tion of the inner-shell vacancy.

The end-on view of the track in Fig. 1b illustrates what some authors have described as the "core" and the "penumbra" regions of charged-particle tracks. It can be shown, however, that in this case this effect is simply the result of the projection displayed and has little physical meaning. To illustrate this we have expanded the scale in Fig. 1c to look at the individual energy deposition events on the same scale as the atomic positions of the DNA molecule. Note that on this scale there is no evidence of a track core. Certainly it is possible that ionizations may occur on adjacent atoms, but the probability is low for even a particle with LET this high (approximately 165 keV/μ). It should also be noted that this illustration is a two-dimensional projection of a three-dimensional structure; thus, individual ionizations/excitations are actually distributed further apart, on the average, than they appear in the figure.

It is obvious that a comprehensive knowledge of the interactions of charged particles with biological material must be known if one is to assess accurately the spatial patterns of excitations and ionizations for charged particles in the hetero-geneous environment of the biological cell. A detailed set of quantitative information is required for the production and subsequent slowing down of secondary electrons ejected in ionizing collisions between the moving ion and the atomic and molecular constituents of the media. Such information must be in the form of absolute cross sections and must incorporate knowledge of the atomic, molecular, and phase natures of the target. The full extent of the cross sections needed in track structure simulation depend on the intended application and the mechanism for subsequent chemical and biological damage assumed. Most applications of track structure simulation have

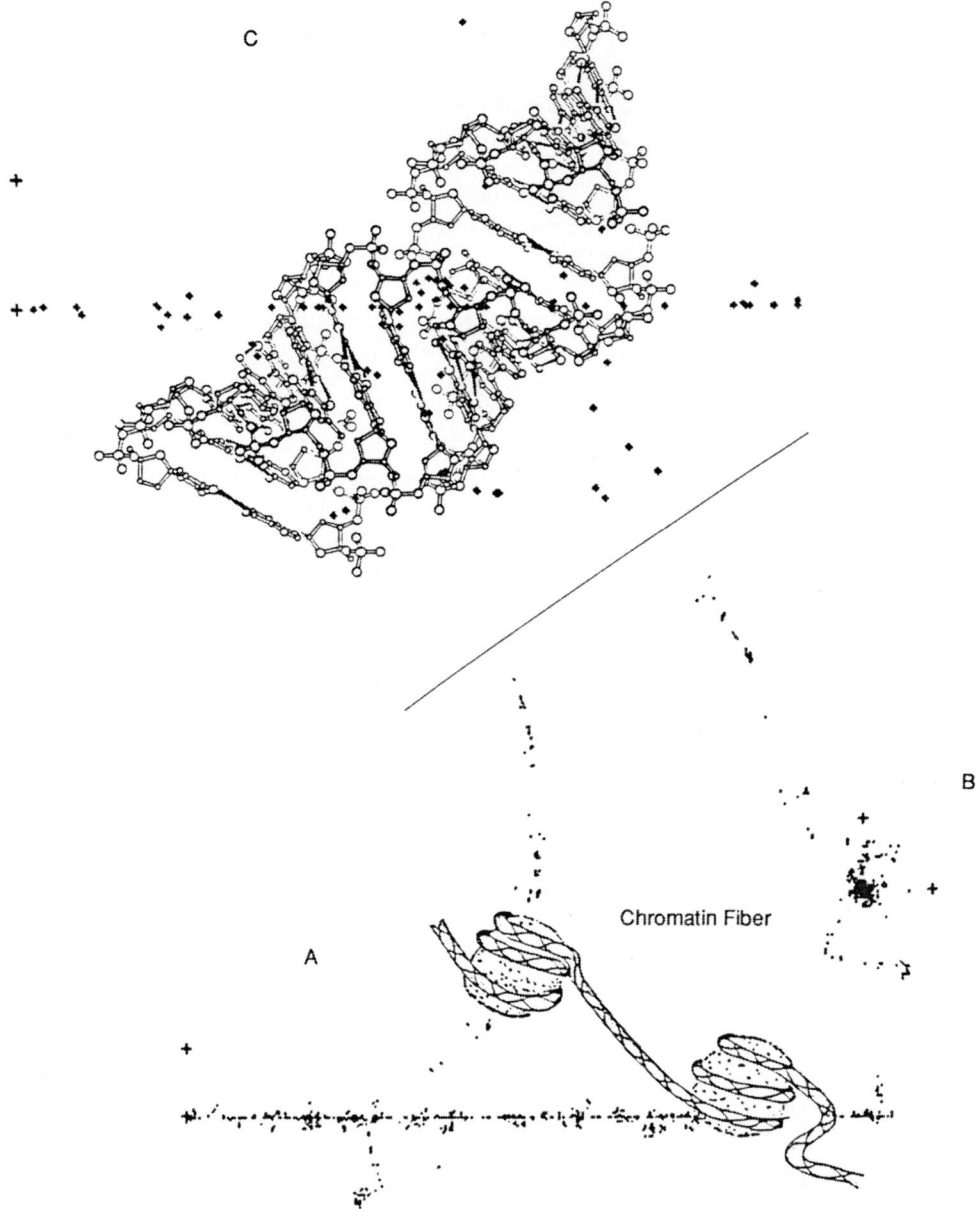

C

B

Chromatin Fiber

A

Figure 1. Simulated 2 MeV alpha particle track compared to spatial patterns of DNA structures, A: comparison to linker and nucleosomes structures, the distance between the two (+) markers is 10 nm; B: an end-on view of the same track segment as in A; C: a portion of the same track magnified to the dimensions comparable to the atomic positions of DNA, the distance between the two (+) markers is 10 Å.

involved the investigation of microdosimetric distributions of energy deposition by secondary electrons for different types of radiation of importance to the field of radiation dosimetry or have provided the initial pattern of energy deposition for investigation of the time sequence of chemical reactions that follow degradation of secondary electrons. Such calculations rely primarily on an accurate knowledge of the production and transport of secondary electrons that form the basic structure of charged-particle tracks.

Knowledge of the physics of electron production and degradation has been the key to development of reliable stochastic track structure models. However, there are other processes of energy deposition in the track of a charged particle that are less common but may produce the unusual events that the biological system is incapable of handling. Such events as multiple ionizations of constituents of the DNA may produce irreparable chemical/biological damage, or the correlated electrons emitted in such interactions may induce subsequent chemical damage unique to the biological repair system. Temporally and spatially correlated events may also be stimulated by inner-shell ionization or by simultaneous electron loss and target ionization involving projectiles that carry bound electrons, such as He^+ ions formed in the slowing down of alpha particles. Many of these processes are only beginning to be understood from a physical point of view, and their biological implications have yet to be investigated. The principal source of data needed as input for calculations that simulate the stochastic processes that form charged-particle tracks, such as those shown in Fig. 1, are the cross sections for the production and transport of secondary electrons. These cross sections must be absolute in magnitude and differential in ejected electron energy and emission angle. The considerable progress in measurements of these cross sections and in our ability to model their systematics during the past 20 years has been a key to performing reliable track structure simulations. Prior to these advances, homogeneous track structure models[4,5] were based on the early collision physics theory of Bohr.[6] In those models the probability of ionization was obtained from the free electron Rutherford cross section, electrons were assumed to be emitted perpendicular to the particle path, and straight line trajectories were assumed in order to calculate electron ranges. In addition, Chatterjee and Shaefer[5] assumed that half the energy lost in collisions between the incident particle and constituents of the medium went into excitation, thus contributing to a high energy density in the core of the particle track. As we shall see, many of these assumptions have proven to be inaccurate as experimental data have become available to test our understanding.

In this paper we look briefly at the extent of our knowledge of differential ionization cross sections for energy loss by charged particles. As Inokuti has stated, useful cross sections must be "right, absolute, and comprehensive".[7] In this review, these virtues of the experimental data are addressed. The discussion is organized by incident particle; we start by discussing electron impact, then proton impact, followed by heavier ion impact. The final section looks briefly at the relative importance of other processes, such as charge transfer and multiple ionization. Although an effort is made to be comprehensive in this review, the field is sufficiently large that some of the pertinent data are sure to have been inadvertently left out. We apologize to those investigators, and to the reader, for such emissions.

Doubly Differential Ionization Cross Sections

Electron Impact

The primary components of any track structure simulation are the production and slowing down of secondary electrons. Therefore it is important that one has a detailed knowledge of the interaction cross sections for electrons with the stopping media of interest. A recent review by Paretzke[3] provides an excellent guide to the literature of electron interactions of interest to radiobiology and to radiation chemistry. Table I provides a listing of the measured doubly differential electron emission cross sections, differential with respect to "ejected" electron energy and emission angle, obtained from a search of the literature. Since electrons are indistinguishable, the slower of the two electrons leaving a collision is defined as the secondary electron.

Table I. Published Double Differential Cross Sections - Electron Impact

| Target | Ion Energy Range (keV) | Ejected Electron | | Investigators |
		Energy (eV)	Angle (Degrees)	
He	500 - 300	$1-1/2(E_p-I)$	6 - 156	Shyn and Sharp[79]
He	100, 200	$3-(E_p-I)$	10 - 150	Rudd and DuBois[80]
He	50 - 2000	4 - 2000	30 - 150	Opal, Beaty, and Peterson[81,82]
He	200 - 2000	$2 - E_p$	30 - 150	Goruganthu and Bonham[83]
He	500, 1000	25 - 45	10 - 130	Oda, Nishimura, and Tahira[84]
Ne	100 - 500	$4-(E_p-I)$	10 - 150	DuBois and Rudd[85]
Ne	500	4 - 200	30 - 150	Opal, Beaty, and Peterson[81,82]
Ar	100 - 500	$4-(E_p-I)$	10 - 150	DuBois and Rudd[85]
Ar	500	4 - 200	30 - 150	Opal, Beaty, and Peterson[81,82]
Ar	1000	4 - 500	90	Mathis and Vroom[86]
Kr	500	4 - 200	30 - 150	Opal, Beaty, and Peterson[81,82]
Kr	1000	21 - 52	10 - 130	Oda, Nishimura, and Tahira[84]
Xe	500	4 - 200	30 - 150	Opal, Beaty, and Peterson[81,82]
H_2	100 - 500	$4-(E_p-I)$	10 - 150	DuBois and Rudd[85]
H_2	500	4 - 200	30 - 150	Opal, Beaty, and Peterson[81,82]
H_2	25 - 250	$1-1/2(E_p-I)$	12 - 156	Shyn, Sharp, and Kim[87]
N_2	100 - 500	$4-(E_p-I)$	10 - 150	DuBois and Rudd[85]
N_2	50 - 2000	4 - 200	30 - 150	Opal, Beaty, and Peterson[81,82]
N_2	200 - 2000	$2-E_p$	30 - 150	Goruganthu, Wilson, and Bonham[88]
N_2	1000	4 - 500	90	Mathis and Vroom[86]
N_2	50 - 400	$1-1/2(E_p-I)$	12 - 156	Shyn[89]
O_2	50 -2000	4 - 200	30 - 150	Opal, Beaty, and Peterson[81,82]
CH_4	200	4 - 200	30 - 150	Opal, Beaty, and Peterson[81,82]
CH_4	500, 1000	5 - 1000	15 - 148	Oda[11]
NH_3	200	4 - 200	30 - 150	Opal, Beaty, and Peterson[81,82]
H_2O	50 - 2000	$2-(E_p-I_p)$	15 - 150	Bolorizadeh and Rudd[12]
H_2O	500	4 - 200	30 - 150	Opal, Beaty, and Peterson[81,82]

(continued)

Table I. (continued)

| Target | Ion Energy Range (keV) | Ejected Electron | | Investigators |
		Energy (eV)	Angle (Degrees)	
H_2O	1000	4 - 500	90	Mathis and Vroom[86]
H_2O (clusters)	1000	4 - 500	90	Mathis and Vroom[86]
H_2O	500, 1000	5 - 1000	15 - 148	Oda[11]
C_2H_2	500	4 - 200	30 - 150	Opal, Beaty, and Peterson[81,82]
CO	800	0.9 - 393	30 - 150	Ma and Bonham[90]
CO	500	4 - 200	30 - 150	Opal, Beaty, and Peterson[81,82]
NO	500	4 - 200	30 - 150	Opal, Beaty, and Peterson[81,82]
CO_2	500	4 - 200	30 - 150	Opal, Beaty, and Peterson[81,82]
CO_2	50 - 400	$1-1/2(E_p-I)$	12 - 156	Shyn and Sharp[91]
CO_2	500, 1000	5 - 1000	15 - 148	Oda[11]

To completely define the collision for electron impact, one would need to measure triply differential cross sections, i.e., to also detect the scattering angle of the primary electron. A limited number of triply differential cross sections have been measured for simple gas targets such as helium and argon [see, for example Beaty et al.[8] and Hong and Beaty[9]]; however, such data have been of little practical use in track structure calculations and are considered out of the scope of the present review.

Although Table I illustrates that there are a relatively large amount of data available on the doubly differential cross sections for ionization by incident electrons, only a limited subset of this data is directly appropriate to targets of interest to radiological physics. In addition, where data have been obtained by different groups, such as the cross sections for ionization of water vapor shown in Fig. 2, there is considerable scatter in the data for different investigators. In general the agreement between the data of Opal et al.,[10] Oda,[11] and Bolorizadeh and Rudd[12] is quite good for intermediate angles. However, at both, large and small emission angles, the cross sections of Opal et al. tend to be smaller than the other two measurements. These differences result from different methods of accounting for the finite size of the target as one views it from different angles. The true cross sections are probably somewhere between the extremes represented by the data of Bolorizadeh and Rudd and of Opal et al.

Because of the scatter in experimental data from different sources, a good deal of effort has gone into theoretical techniques to evaluate the accuracy of measured cross sections. Following the lead of Platzman, Kim has explored the consistency of experimental data for electron and proton collisions using well-established theory.[13-17] The Mott cross section is used to test the behavior of fast electrons ejected by fast

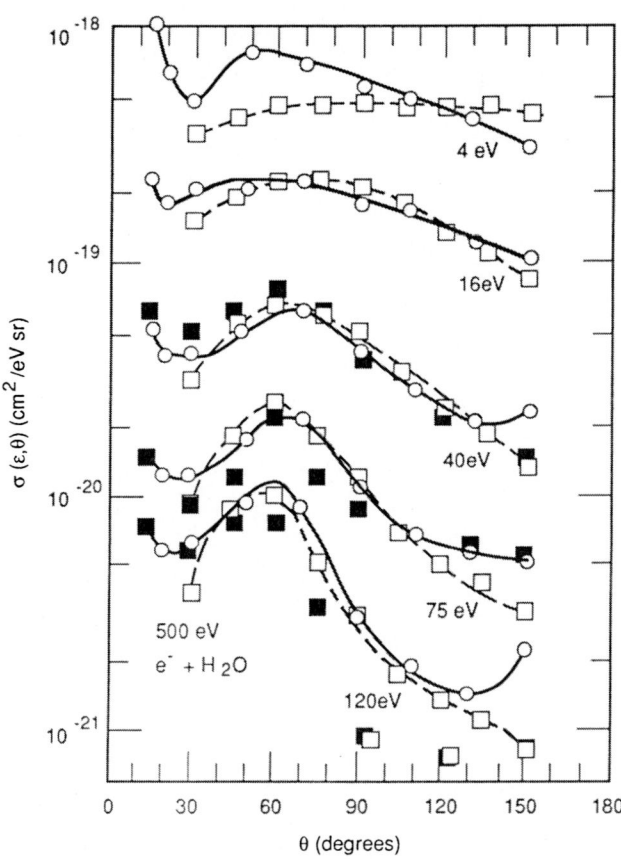

Figure 2. Angular distributions of electrons ejected from water vapor by 500 eV electron impact. The data are from (□) Opal et al.,[10] (■) Oda,[11] and (o) Bolorizadeh and Rudd.[12]

primary electrons, whereas the slow electrons are analyzed in terms of the dipole oscillator strengths as prescribed by the Born approximation. An example of the utility of this method is shown in Fig. 3, taken from Kim,[14] where the ratios of the experimental cross sections to the corresponding Rutherford cross sections are plotted as a function of the reciprocal of the energy loss. Plotted in this way, the area under the curve is proportional to the total ionization cross section; and the shape of the low-energy portion of the curve is representative of the dipole oscillator strength. The fraction of the electrons ejected with energies ϵ between the shaded vertical lines i.e., between ϵ=o and 15.6 eV, represents those electrons that are unable to produce further ionization as they slow down in the target medium. In the example shown in Fig. 3, the only experimental data used to establish the family of curves were singly differential cross sections for electron emission by 500 eV incident electrons. Those data were used, along with the dipole oscillator strengths, to define the overall shape of the curve for 500 eV primaries. The magnitude of the cross sections was then established by normalization of the area under the curve to the total ionization cross section. Curves for other primary energies could then be drawn by extrapolation based on maintaining (i) a curve shape consistent with the optical oscillator strengths, (ii) the proper integrated area consistent with total ionization cross sections, and (iii) the proper kinematic limit to the secondary electron energy consistent with the maximum energy transfer. Models of this type provide means to evaluate experimental consistency and to extrapolate data to regions where data are unavailable. They provide convenient methods of introducing data into computer codes for track structure

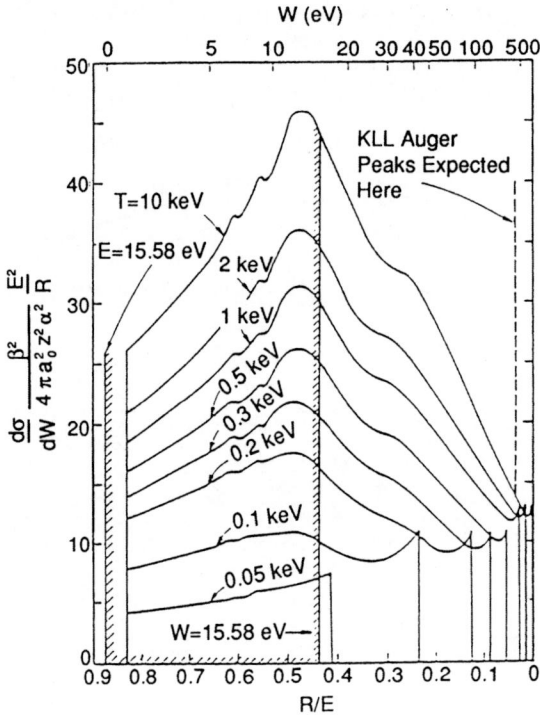

Figure 3. Secondary electron spectra for electron impact ionization of N_2. The areas under the curves have been normalized to the respective total ionization cross sections. The area between the shaded lines corresponds to the fraction of the electrons ejected with insufficient energies to produce further ionization in the target medium. These data of Kim[14] were reproduced with permission of the author.

calculations. This technique of data analysis and extrapolation takes advantage of the availability of a wide range of experimental data on total ionization cross sections [see, for example, reviews by Schram et al.[18] and Shimamura[19]] and oscillator strengths [see reviews by Berkowitz and by Gallagher et al.[20]]. In addition, the use of spectra based on oscillator strength distributions has the potential for application to both gas and condensed phase targets by simply using the proper oscillator strengths. These techniques will be described in more detail in the following sections with regard to their application to proton-induced ionization. For greater detail in the application of this method to electron-impact ionization, the interested reader is directed to the work of Kim referenced above, as well as to the review of Paretzke[3] and to studies by Miller and Manson[21] and Miller et al.[22]

One of the large gaps in our knowledge of interactions of electrons with biologically relevant material is the lack of direct measurements of these processes in the condensed phase. Presently the basis of condensed-phase electron transport used in track structure simulation is deduced from the theory of charged-particle interactions in condensed phase, and from oscillator strengths for photoabsorption.[20] During the past few years, however, significant advances in our understanding of electron interactions in the condensed phase have been brought about by the pioneering work of Leon Sanche and his coworkers.[23-26] Measurements that they have conducted on the scattering of low-energy electrons in thin films have provided detailed information on the energy loss mechanisms associated with the slowing down of slow electrons. A particularly interesting feature in the preliminary studies has been the similarity of elastic and inelastic electron scattering processes in the solid to those observed in the gas phase. For example, resonant processes such as transient negative-ion formation[24] are strong features in the energy-loss spectra for very low-energy electrons.

Proton Impact

There has been a wide range of experimental and theoretical studies of doubly differential cross sections for proton impact ionization of atomic and molecular targets. Much of this work was funded by the Radiological and Chemical Physics Program of the U.S. Department of Energy and was directed toward understanding the effects of molecular structure and to developing models applicable to track structure calculations. Reviews of doubly differential cross sections have been published by Toburen[27,28] and Rudd.[29] An updated listing, first published by Toburen,[27] of the doubly differential cross sections available for proton impact is presented as Table II. This list focuses on studies that report absolute cross sections and those that provide a broad spectrum of energies and angles. The table does not include studies of "convoy electrons" [see, for example, Breinig, et al.[30]] or studies that focus on a narrow angular range, e.g., electrons ejected at zero degrees; such studies are not considered highly relevant to Radiological Physics and would require a full review on their own. From Table II we see that data are available that span the regions of low (5 to 50 keV), intermediate (50 to 300 keV), and high (greater than about 300 keV) proton energies; these energy ranges reflect regions requiring different theoretical approaches. Only a few target molecules, e.g., hydrogen, nitrogen, oxygen, and water vapor, have been studied through all the energy ranges. Those molecules, however, provide a good representation of the constituents of tissue. The majority of the data base for investigating molecular effects has been developed in the region of high-energy protons.

An indication of the precision of the various measurements can be addressed by an evaluation of the uncertainties contributing to the individual measurements and by comparison of measurements of different investigators where they overlap. A comparison of doubly differential cross sections for ejection of electrons from nitrogen by 0.3 Mev protons measured by three different research groups is shown in Fig. 4. For ejected electron energies greater than about 15 eV, the agreement is well within the stated 20% uncertainties in the individual measurements. For lower energy ejected electrons, the data diverge due to the effects of stray electrostatic and magnetic fields on the transmission of the electrostatic energy analyzers used in the cross section measurements. To resolve the uncertainties at low energies, a time-of-flight (TOF) technique was developed that could measure relative cross sections for ejected electron energies in the range from 1 to 200 eV.[34] The solid lines in Fig. 4 were derived from TOF measurements normalized to the electrostatic results at 100 eV. This combination of electrostatic and TOF measurements provides reliable cross sections for the ejected energy range from 1 to 5000 eV, thus providing a wide range of data for analysis of cross sections systematics and for development of theoretical models.

To test the "correctness", as defined by Inokuti,[7] of the measured cross sections we can make use of simple theoretical arguments for the asymptotic behavior of cross sections as has been advocated by Kim and Inokuti.[16] For example, the Rutherford formula, given as

$$\frac{d\sigma}{dE} = \frac{4\pi a_o^2 z^2}{T} \frac{R^2}{E^2} \tag{1}$$

where a_o is the Bohr radius, z is the projectile charge, R is the Rydberg energy, $T=mv^2/2$ (m is the electron mass), and $E=\epsilon+I$ (ϵ is the ejected electron energy and I the ionization potential), should provide an accurate estimate of the cross section

Table II. Published Double Differential Cross Sections - Proton Impact

Target	Ion Energy Range (keV)	Ejected Electron Energy (eV)	Angle (Degrees)	Investigators
H_2	50 - 100	4 - 300	23 - 152	Kuyatt and Jorgensen[92]
H_2	50 - 100	1 - 500	10 - 160	Rudd and Jorgensen[93]
H_2	100 - 300	2 - 1000	10 - 160	Rudd, Sautter and Bailey[94]
H_2	300	2 - 800	20 - 130	Toburen[33]
H_2	300 - 1500	2 - 3500	20 - 130	Toburen and Wilson[95]
H_2	1000	100 - 1000	20 - 130	Toburen[96]
H_2	5 - 100	1.5 - 300	10 - 160	Rudd[50]
He	75 - 150	1 - 550	10 - 160	Cheng, Rudd, and Hsu[49]
He	50 - 150	1 - 500	10 - 160	Rudd and Jorgensen[93]
He	100 - 300	2 - 1000	10 - 160	Rudd, Sautter, and Bailey[94]
He	2 - 100	5 - 100	0 - 100	Gibson and Reid[97]
He	2000	30 - 1500	20 - 130	Toburen[96]
He	300	1 - 1030	20 - 150	Stolterfoht[98]
He	300 - 5000	1 - 8577	15 - 160	Manson, Toburen, Madison and Stolterfoht[99]
He	5 - 5000	1 - 8577	10 - 160	Rudd, Toburen and Stolterfoht[99]
He	5 - 100	10 - 200	10 - 160	Rudd, Webster, Blocker, and Madison[100]
He	300 - 1500	1 - 3500	15 - 125	Toburen, Manson, and Kim[35]
He	5 - 100	10 - 200	10 - 160	Rudd and Madison[101]
He	100 - 300	40 - 180	0	Crooks and Rudd[102]
Ne	7.5 - 150	1 - 550	10 - 160	Cheng, Rudd, and Hsu[49]
Ne	50 - 300	1.5 - 1057	10 - 160	Crooks and Rudd[32]
Ne	1000	1 - 2000	15 - 125	Toburen and Manson[103]
Ne	300 - 1500	1 - 3500	15 - 125	Toburen, Manson, and Kim[35]
Ar	50 - 300	1.5 - 1057	10 - 160	Crooks and Rudd[32]
Ar	5 - 1500	1 - 3500	15 - 160	Criswell, Toburen, and Rudd[104]
Ar	300 - 5000	1.1 - 10000	25 - 150	Gabler[105]
Ar	5 - 2000	1 - 3500	15 - 125	Criswell, Wilson, and Toburen[106]
Ar	1000	1 - 360	15 - 125	Manson and Toburen[107]
Ar	300 - 1500	1 - 3500	15 - 125	Toburen, Manson, and Kim[35]
Ar	100	3 - 250	160	Rudd, Jorgensen, and Volz[54]
Ar	5 - 20	1 - 26	30 - 140	Sataka, Okuno, Urakawa and Oda[108]
Ar	5 - 5000	1 - 10000	10 - 160	Rudd, Toburen, and Stolterfoht[109]

Table II. (continued)

Target	Ion Energy Range (keV)	Ejected Electron Energy (eV)	Ejected Electron Angle (Degrees)	Investigators
Kr	7.5 - 150	1 - 550	10 - 1601	Cheng, Rudd, and Hsu[49]
Kr	1000	1 - 3000	15 and 90	Manson and Toburen[110]
Kr	1000 - 4200	30, 136	15 - 90	Toburen and Manson[111]
Xe	300 - 2000	2 - 4620	20 - 130	Toburen[112]
N_2	300 - 1700	2 - 4000	20 - 130	Toburen[33]
N_2	50 - 300	1.5 - 1057	10 - 160	Crooks and Rudd[32]
N_2	200 - 500	1 - 1300	20 - 150	Stolterfoht[98]
N_2	5 - 70	1.5 - 300	10 - 160	Rudd[50]
N_2	200 - 500	1 - 1300	20 - 150	Stolterfoht[113]
O_2	50 - 300	1.5 - 1057	10 - 160	Crooks and Rudd[32]
O_2	300 - 1500	1 - 3500	15 - 125	Toburen and Wilson[114]
H_2O	15 - 150	1 - 3000	10 - 160	Bolorizadeh and Rudd[115]
H_2O	300 - 1500	1 - 3500	15 - 125	Toburen and Wilson[114]
CH_4	200 - 400	1 - 1270	20 - 150	Stolterfoht[98]
CH_4	300 - 1000	4 - 5000	20 - 130	Wilson and Toburen[40]
CH_4	250 - 2000	1 - 5000	20 - 130	Lynch, Toburen, and Wilson[36]
C_2H_2	300 - 1000	4 - 5000	20 - 130	Wilson and Toburen[40]
C_2H_4	300 - 1000	4 - 5000	20 - 130	Wilson and Toburen[40]
C_2H_6	300 - 1000	4 - 5000	20 - 130	Wilson and Toburen[40]
C_6H_6	300 - 2000	4 - 5000	20 - 130	Wilson and Toburen[40]
NH_3	250 - 2000	1 - 5000	20 - 130	Lynch, Toburen, and Wilson[36]
CH_3NH_2	250 - 2000	1 - 5000	20 - 130	Lynch, Toburen, and Wilson[36]
$(CH_3)_2NH$	250 - 2000	1 - 5000	20 - 130	Lynch, Toburen, and Wilson[36]
TeF_6	300 - 1800	1 - 5000	20 - 130	Toburen, Wilson, and Porter[116]
SF_6	300 - 1800	1 - 5000	20 - 130	Toburen, Wilson, and Porter[116]

when the energy loss is large compared to the binding energy of the ejected electron, but smaller than the kinematic limit of energy transfer in a binary collision. Thus if we plot the ratio of the measured cross section to the Rutherford cross section the ratio should approach a constant value for high energies of the ejected electron and, since the Rutherford formula gives the cross section per target electron, the magnitude of that constant should be equal to the number of electrons in the atomic or molecular target. In Fig. 5 this ratio, $Y(E,T)$, is plotted versus E for an atomic helium target. In this illustration the ratios approach a value of approximately 2.2, which indicates

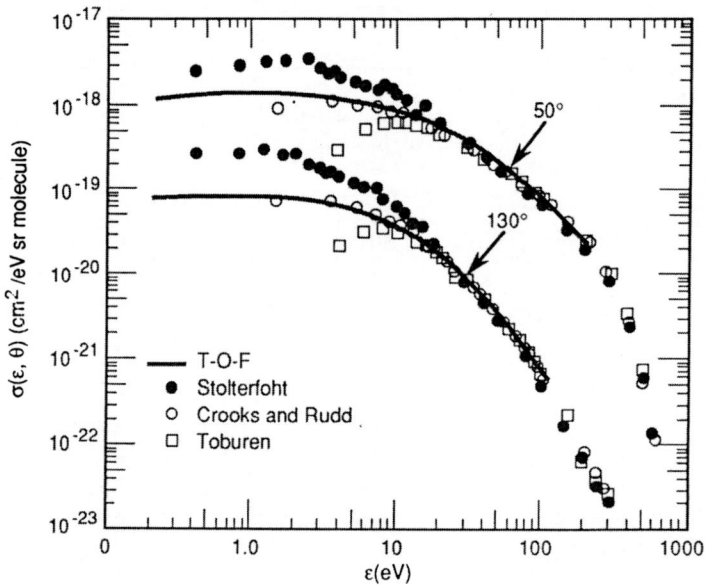

Figure 4. Comparison of absolute cross sections for ionization of N_2 by 0.3 MeV protons measured by (•) Stolterfoht,[31] (o) Crooks and Rudd,[32] and (□) Toburen,[33] and the relative line shape for low-energy ejected electrons measured by TOF techniques.[34]

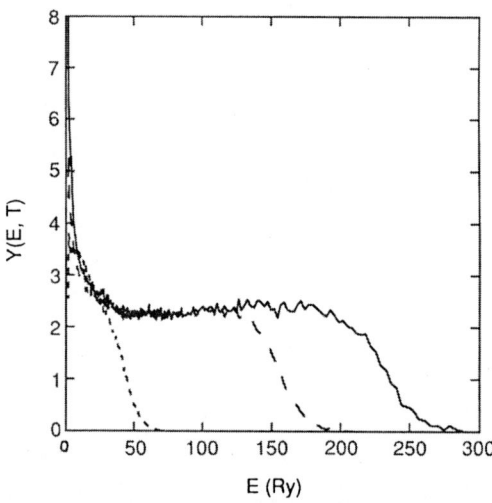

Figure 5. The ratio of the measured singly differential cross section for proton ionization of helium[35] to the Rutherford cross section plotted as a function of the energy loss E.

that the measured cross sections may be systematically 10% too large. Similar ratios are plotted in Fig. 6 for a wider range of proton energies and an atomic neon target. In principle, for an atom such as neon, where the electron can be ejected from inner shells or sub-shells, the ionization potential used in the Rutherford formula should reflect the origin of the electron. However, the experiments are not able to determine the origin of the detected electron; therefore, the ratio is calculated assuming that all electrons originate from the valance shell.[35] For neon the ratio $Y(E,T)$ approaches a value of approximately 10, indicative of the number of bound electrons in the atom;

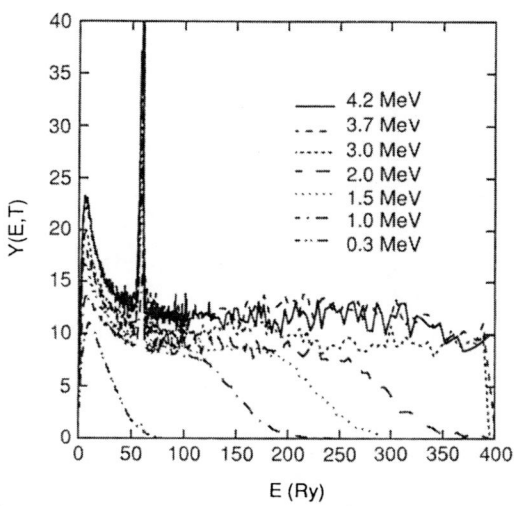

Figure 6. The ratio of the measured singly differential cross section to the Rutherford cross sections for ionization of neon by protons.

the peak observed at approximately 800 eV results from Auger electron emission following K-shell vacancy production. For the highest energy protons, $Y(E,T)$ approaches a value somewhat greater than 10, the total number of electrons in neon, again indicative that the experimental values may be systematically about 10% higher than would be expected. It should be noted, however, that the ratio for 1.5 MeV protons reaches a plateau at approximately 8, which is in good agreement with the Rutherford prediction if only the outer shell electrons participate; the inner-shell electrons are tightly bound and contribute little to the Rutherford cross section at this proton energy. An increased scatter in the data is observed for the higher energy ejected electrons shown in Fig. 6. This scatter occurs because the cross sections are becoming significantly smaller and statistical uncertainties are greater; the $1/E^2$ factor in the Rutherford cross section masks the absolute value of these cross sections. The Rutherford analysis generally confirms that the differential cross sections obtained in our work at the Pacific Northwest Laboratory (PNL) are accurate to within the stated 20% absolute uncertainties derived from the experimental parameters; data from other laboratories are generally at lower proton energies and not amenable to this theoretical test, but where data from different laboratories overlap, agreement is good.

Data have been obtained for a wide range of molecular targets for investigating the effects of molecular structure on electron emission cross sections. In Fig. 7 are displayed the singly differential electron emission cross sections for a number of carbon-containing molecules plotted as the ratio to the respective Rutherford cross sections. The horizonal lines with the number to the right each give the Rutherford estimate of the asymptotic value those cross sections should attain. This agreement between experimental data and the predictions is simply a confirmation of the Bragg rule for scaling cross sections for emission of fast electrons, or a statement of the independent particle model for ionization. One must be careful to stress that this scaling feature of collisions involving large energy loss does not apply to soft collisions; this will be discussed in detail later. It does, however, provide justification for application of simplified theory and scaling techniques that are very useful in track structure calculations.

The limits of applicability and reliability of various theoretical calculations can be assessed by comparison of their predictions to experimental data. In Fig. 8 the results

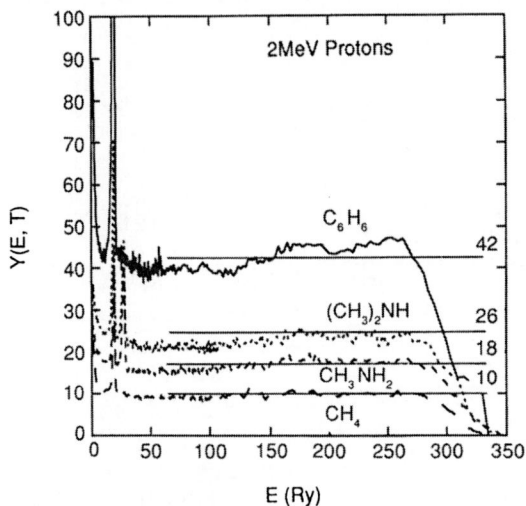

Figure 7. The ratio of the measured to Rutherford singly differential cross sections for ionization of a number of molecules by protons, data of Lynch et al.[36]

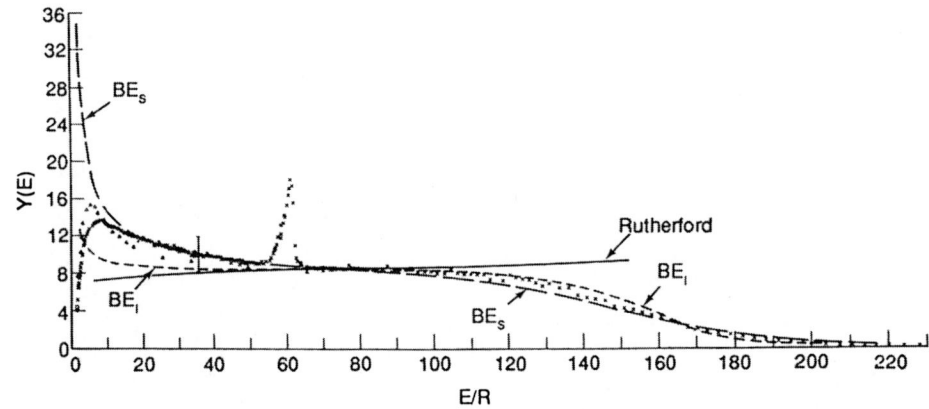

Figure 8. Comparison of measured and calculated singly differential cross sections for ionization of neon by 1.0 MeV protons; cross sections are presented as the ratio to the Rutherford cross section for outer-shell ionization of neon. The measured singly differential cross sections (x) are from Toburen et al.[35] The cross section for zero energy ejected electrons (□) is from Grissom.[37] The binary encounter calculations (o) are from a program of Rudd[29] and include results assuming the average kinetic energy of the bound electron is equal to the binding energy (BE_I) or given by Slater's rules (BE_S). The Rutherford cross section is calculated including inner-shell contributions from Kim[14] and the plane wave Born calculation (Δ) is described by Toburen et al.[35]

of binary encounter theory, the Born approximation, and Rutherford theory are plotted as the ratio to the corresponding Rutherford cross section for ionization of the outermost shell of neon. The primary reason for comparing results by dividing by the Rutherford cross section is that the principal dependence on energy loss, $1/E^2$, is removed and one can compare data on a linear, rather than a logarithmic scale, thus accentuating spectral features. The gradual increase in the plotted Rutherford cross section plotted in this way occurs because it was calculated by including electrons from all shells of neon, with their respective binding energies.[14] At larger values of the energy loss, the more tightly bound electrons contribute more to the total; the Rutherford cross section in the denominator of the calculated ratio is taken as the valance shell cross section (as discussed above).

For a 1 MeV proton interacting with a free electron, the Rutherford theory would predict an abrupt decrease in the cross section at approximately 160 Ry, as that is the kinematic maximum in the energy that can be transferred in the classical proton-electron collision. The measured cross sections show a gradual decrease in magnitude between approximately 100 and 160 Ry, reflecting the momentum distribution of the bound electrons. The increasing Rutherford cross section (owing to inner shell contributions) combined with the decrease in the experimental values near the maximum energy transfer (owing to binding effects) renders Rutherford theory inappropriate as a definitive test of the accuracy of measured cross sections in this proton energy range. At higher proton energies, as shown in Fig. 6, a plateau value can be expected over a broader ejected electron energy range.

Binary encounter theory [reviewed by Rudd[29]] extends the classical Rutherford-like approach to collisions with electrons that have an initial velocity distribution as a consequence of being bound to the atom. When integrated over a quantum mechanical distribution of bound electron velocities, the high-energy portion of the ejected electron spectra is well represented by this semi-classical approach. The only parameter that is not well defined in this computational technique is the mean of the initial distribution of kinetic energies exhibited by the orbital electrons. Calculations that assume that the initial kinetic energy is equal to the binding energy, BE_I, and that use the kinetic energy derived from Slater's rules [see Robinson[38]], BE_S, are shown in Fig. 8; the use of Slater's rules seems to provide a slightly better agreement at high energies than does the binding-energy approach. For low-energy electron emission, neither approach is very good, although the use of Slater's rules extends the agreement with experimental data to somewhat lower energies.

The results of calculations based on the plane wave Born approximation[35] are also shown in Fig. 8. This calculation is in good agreement with the measured differential cross sections for ejection of low-energy electrons and with the independent measurement of Grissom et al.[37] for electrons ejected with zero kinetic energy; the calculation was not carried out to ejected electron energies greater than 64 Ry (870 eV) because it is based on a partial wave analysis, and the number of partial waves necessary to describe the higher energy processes makes the calculation unwieldy. Similar calculations are in good agreement for helium targets.[39] The use of this technique for molecules has not been attempted, however, due to a lack of adequate wave functions to describe the molecular systems.

The data shown in Fig. 8 clearly illustrate the limitations of classical and semi-classical theory for predicting the cross sections for ejection of low-energy electrons. The relative importance of this region of the spectra, however, is illustrated in Fig. 9, where ejected electron distributions are displayed for ionization of several different molecules by 1 MeV protons. The hydrocarbons referenced in Fig. 9 are the same as those of Fig. 7, now scaled on a per "effective" electron basis. Here the effective number of electrons in a molecule is taken as the total minus the number of K-shell electrons;[36,40] from the data for neon shown in Fig. 6, one would not expect the K-shell electrons to contribute significantly to the emission cross sections for ionization of first row elements by 1 MeV protons. Note the large differences in the electron yields associated with different molecules as the ejected electron energies decrease below approximately 20 eV. This is also the portion of the emission spectrum that is the major contributor to the total yield of electrons and therefore directly influences the total ionization cross section.

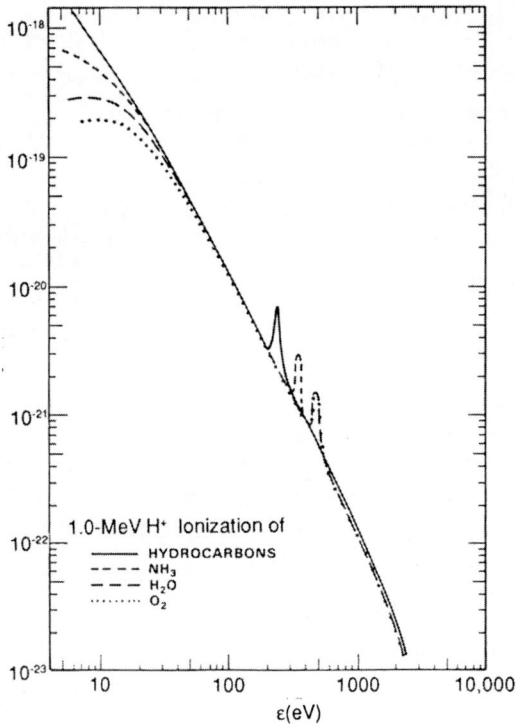

Figure 9. Singly differential cross sections for ionization of several molecules by 1 MeV protons.

As was discussed above for electron impact, one can make use of the analysis developed by Kim to focus on the accuracy and consistency of the low-energy portion of the ejected electron spectra. As pointed out by Kim and Noguchi,[17] the area under the curve in a plot of $Y(E,T)$ versus $1/E$ can easily be shown to be proportional to the total ionization cross section:

$$\sigma_{ion} = \int_{B}^{E_o} \frac{d\sigma}{dE}\, dE = \int_{1/E_o}^{1/B} \frac{d\sigma}{dE}\, E^2\, d\left(\frac{1}{E}\right) \tag{2}$$

where B is the electron binding energy and E_o is the incident ion energy. Such a plot is useful for testing absolute normalization of the differential cross sections and for determining the importance of specific features of the spectra as contributors to the overall yield of ionization. Total ionization cross sections are available with accuracies of 5% to 10% for a wide range of atoms and molecules.[41]

The Bethe-Born expansion of the differential cross section provides a convenient framework to investigate the features of the singly differential cross sections and their dependence on projectile parameters. The Bethe-Born formula can be written[22] as

$$\frac{d\sigma}{dE} = \frac{4\pi a_o^2 z^2}{T}\left[\frac{R^2}{E}\frac{df}{dE}\ln\left(\frac{4T}{R}\right) + B(E) + 0\left(\frac{E}{T}\right)\right] \tag{3}$$

where most symbols are defined as in Eq. 1 and df/dE is the optical oscillator strength distribution. Expressed in this way, $B(E)$ includes contributions independent of T, and

0(E/T) contains contributions of higher order in E/T. Because of the logarithmic dependence of the term involving the oscillator strength, the spectra should become increasingly optical in nature as the ion energy increases. This is seen in Fig. 10, where data for ionization of helium by protons of different energies are shown. Also shown are data for ionization by 500 eV electrons;[10] these are of comparable velocity to the 1 MeV protons. The data displayed in Fig. 10 show the importance of knowing the shape of the low-energy portion of the spectra if one is to be able to gain an accurate knowledge of the total yield of electrons. For proton impact on a helium target, a major fraction of the electrons have energies less than 25 eV ($1/E=0.27$ Ry^{-1}). This presents no difficulties where cross sections have been measured using both electro-static and TOF techniques, as was done for the helium data in Fig. 10. However, where TOF is not available, there may be large uncertainties in the low-energy portion of the spectra; see, for example, the data in Fig. 4.

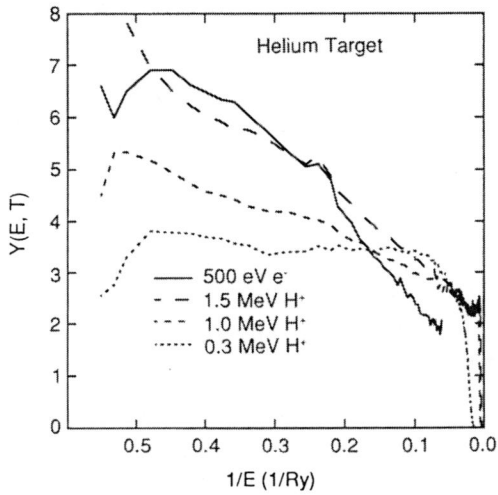

Figure 10. Ratio of the measured singly differential cross sections to the corresponding Rutherford values plotted as a function of 1/energy loss for ionization of neon by protons and electrons. The electron results are from Opal et al.[10] and the proton data are from Toburen et al.[35]

To overcome the uncertainties in low-energy cross section measurements, Miller et al.[42] applied the Bethe-Born approximation from Kim and Inokuti[16] to the analysis of proton impact data. Inspection of Eq. (3) indicates that if we have experimental data at any proton energy for which the Born approximation is valid, those data can be used with optical oscillator strengths to evaluate what has been called the "hard collision component" of the interaction B(E); this assumes that terms of higher order in E/T are negligible. Since B(E) is independent of the incident proton energy, once it is determined, that spectrum can be used to obtain cross sections at other energies. In practice experimental data are used to obtain B(E) for low-energy ejected electrons, and the results are than merged with binary encounter theory to obtain an estimate of B(E)+0(E/T) for high ejected electron energies to give the full spectrum; binary encounter has been shown to describe fast ejected electrons quite well. The hard collisions contribution, B(E), found by Wilson et al.[43] for ionization of water vapor by protons, is shown in Fig. 11. Although B(E) is theoretically independent of ion energy, there was considerable scatter among the data for B(E) derived from different proton energies. To determine the spectrum of B(E) from

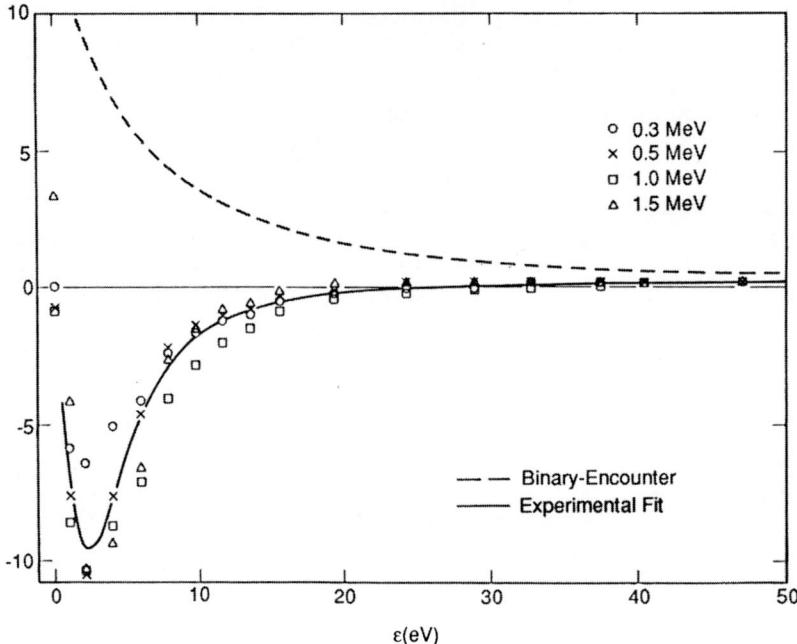

Figure 11. The hard collision component of the Bethe-Born cross section, B(E), for ionization of water vapor by protons.

these data for use in calculating emission cross sections, a simple average of the experimental values at different ion energies was performed (the solid line in Fig. 11). Cross sections derived from this determination of B(E) are shown in Fig. 12. This model of the differential cross sections is in good agreement with the 0.5- and 1.5-MeV data, as it must be, since these data were used in the determination of B(E). There is also good agreement with 3.0- and 4.2-MeV proton results obtained with a different experimental system and not included in the fitting process. An important asset of this technique is the high degree of accuracy that can be obtained for the low-energy portion of the spectra owing to the use of optical oscillator strengths that dominate in this region.

Models of the ejected electron spectra based on Eq. (3) rely on two primary sources of experimental data; differential electron emission cross sections for at least one ion energy and a source of optical oscillator strengths. We have discussed the electron spectra at length, but have said little regarding the availability of oscillator strengths. For the water vapor data discussed above, oscillator strengths were derived from the photo absorption cross sections compiled by Berkowitz,[20] using the expression

$$R\frac{df}{dE} = \frac{\sigma\,(Mb)}{8.07} \tag{4}$$

where σ is the photoabsorption cross section in units of megabarns and R=13.6. A good review of photoabsorption cross sections is also given in a technical report by McDaniel et al.[44] Oscillator strengths are also available for a number of molecules of biological interest, such as DNA[45,46] and DNA bases (Ref. 47, and M. Dillon,

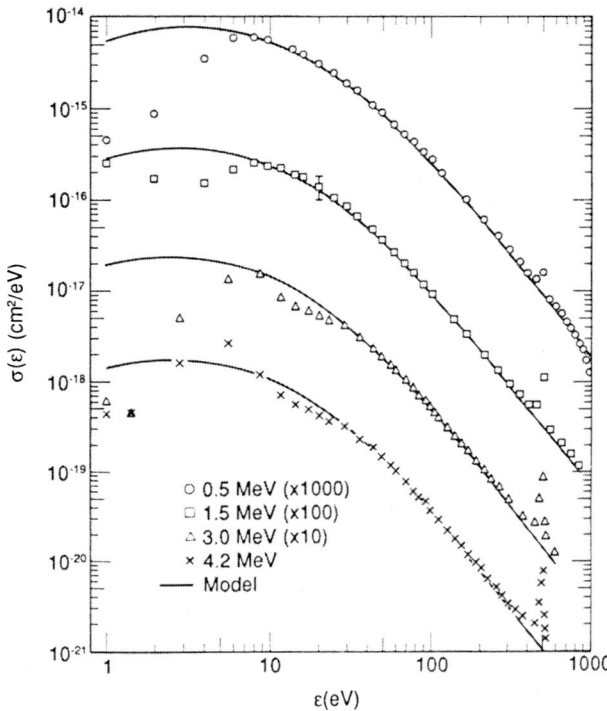

Figure 12. Singly differential cross sections from the model of Miller[42] compared to measurements for ionization of water vapor by protons.

private communication, 1990). We can also expect considerable progress in the measurement of photoabsorption cross sections as synchrotron light sources become more widely used.

One of the shortcomings of models of secondary electron emission cross sections based on Bethe-Born theory, as expressed in Eq. (3), is that the application is limited to ion energies that are sufficiently large for the Born approximation to be valid. A model developed by Rudd[48] overcomes this difficulty by incorporating aspects of molecular promotion theory to enable extensions of his model to low-energy ions. Rudd's model is not as versatile as the Bethe-Born, however, with respect to changing target parameters, and it requires a much more extensive set of data to determine the full range of model parameters. Rudd's model is based on the molecular promotion model at low energies and on the classical binary encounter approximation, modified to agree with Bethe-Born theory at higher energies. In total, Rudd's model requires 10 basic fitting parameters for each electronic shell of each target. These parameters have been published for proton impact ionization of H_2, He, and Ar[48] and He, Ne, Ar, and Kr.[49] Model parameters for N_2, CO_2, H_2O, and O_2 are available from Rudd by private communication. An example of the results of Rudd's model fit to molecular nitrogen data is shown in Fig. 13 for proton ionization of molecular nitrogen. This example illustrates the wide range in proton energy attainable by this model. The experimental data shown in Fig. 13 also illustrate the excellent agreement among the different experimental groups; Rudd,[50] Crooks and Rudd,[32] Toburen,[33] and Stolterfoht.[31] The arrows in the figure point to the electron energy where one would expect to see enhancement of the cross sections by the process of continuum-charge-transfer (CCT). This mechanism can be described as an electron being "dragged" out of the collision by the proton, owing to the coulomb attraction, but failing to be captured into a bound state of the projectile.[29] This mechanism should enhance the cross section for electron energies where the velocity of the out-going electron and proton are comparable; the arrows in Fig. 13 are placed at those electron energies. The lack of observable enhancement in the singly differential cross sections at the appropriate energy is evidence that this mechanism does not contribute markedly to the total electron yield; the model does not include any theoretical mechanism for this process.

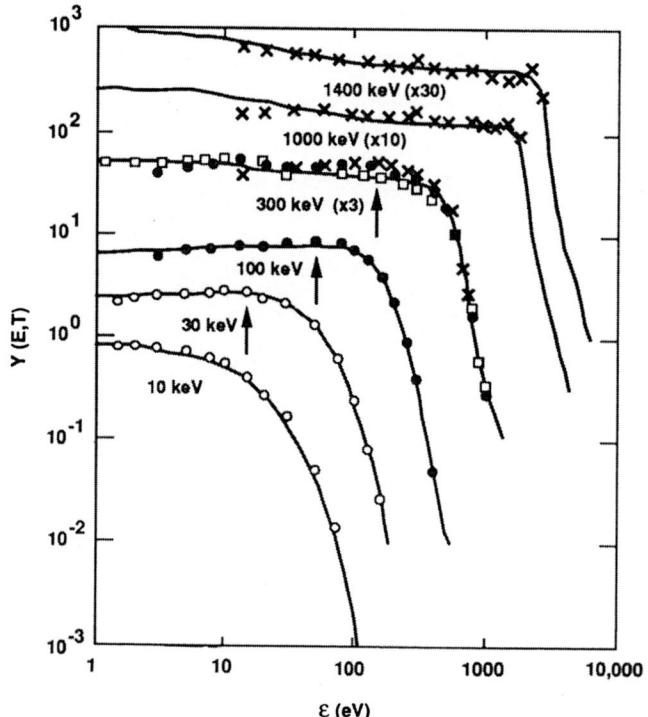

Figure 13. The calculated singly differential cross sections (solid lines) for ionization of N_2 by protons from the model of Rudd[48] compared to the experimental data of (o) Rudd,[50] (•) Crooks and Rudd,[32] (x) Toburen,[33] and (□) Stolterfoht.[31]

To this point the discussion has focused on singly differential cross sections for electron ejection. However, a crucial source of data for the determination of the spatial distributions of energy deposition around charged-particle tracks is the angular distribution of ejected electrons. In Fig. 14, angular distributions are shown for ejection of electrons of several energies from collisions of 2 MeV protons with a helium target. Also shown are the results of a binary encounter calculation of Bonsen and Vriens[51] and those of a plane wave Born calculation of Madison.[52] This illustration emphasizes that the electrons are not, as assumed in early radiobiological models,[4,5] ejected at 90 degrees to the proton path. There is a sizeable component of the cross section for electron ejection at both large and small angles. Also note that binary encounter theory underestimates the cross sections at large and small angles by as much as an order of magnitude. The use of the plane wave Born approximation improves the estimates at the large and small angles considerably; however, there are still discrepancies at small angles for intermediate electron energies. These remaining discrepancies are the result of an enhancement of the cross sections by the process of CCT.[29] This ionization mechanism, though not significant in the singly differential cross sections (see Fig. 13), plays a sizeable role in the doubly differential cross

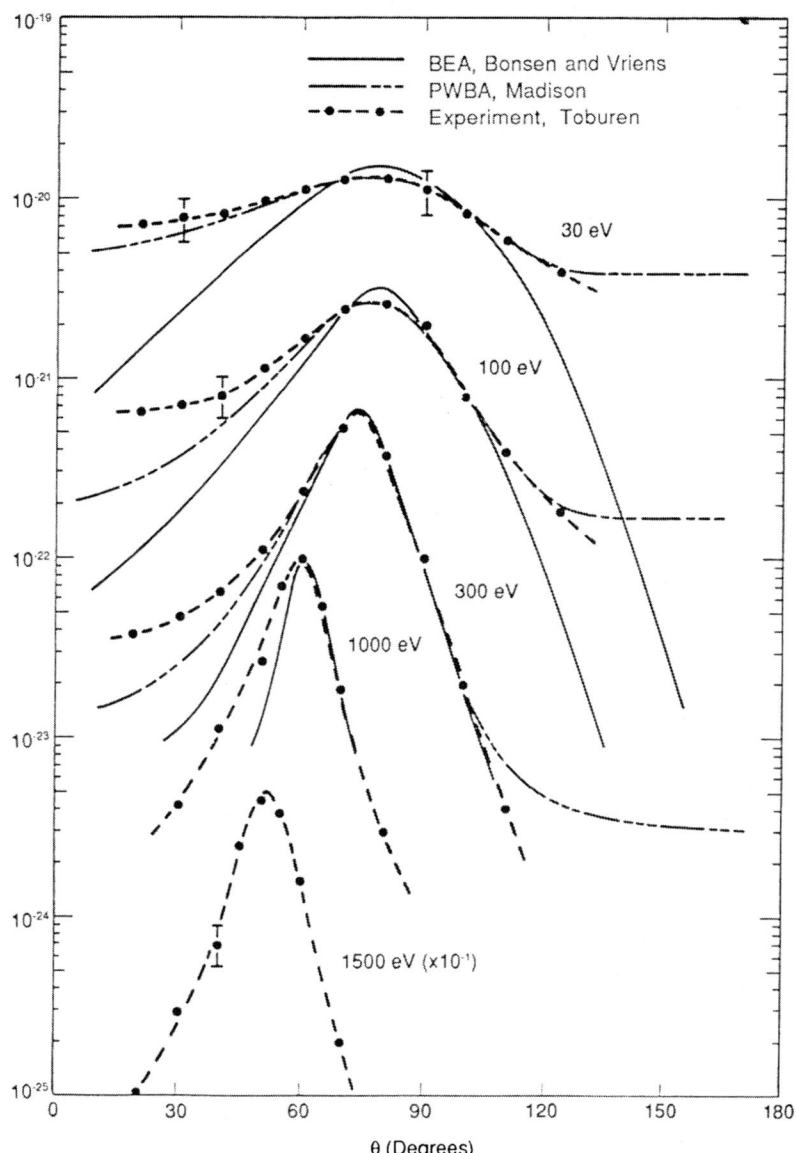

θ (Degrees)

Figure 14. Doubly differential cross sections for ionization of helium by 2 MeV protons. The binary encounter theory calculation (BEA) is from Bonson and Vriens[51] and the plane wave Born approximation is from the work of Madison.[52]

sections for small emission angles; it has not been included in the plane wave Born calculations shown in Fig. 14. The theory of the CCT process was first carried out by Macek.[53] Macek's results are compared in Fig. 15 to Born results without CCT and to measurements from our laboratory for electrons ejected with velocities near that of the incident proton (the equivalent electron velocity for a 1 MeV proton is 544 eV) where the maximum contribution from CCT is expected. The calculation of Macek is shown to be in excellent agreement with the measurements. It should again be emphasized that although the CCT contributions may enhance the doubly differential

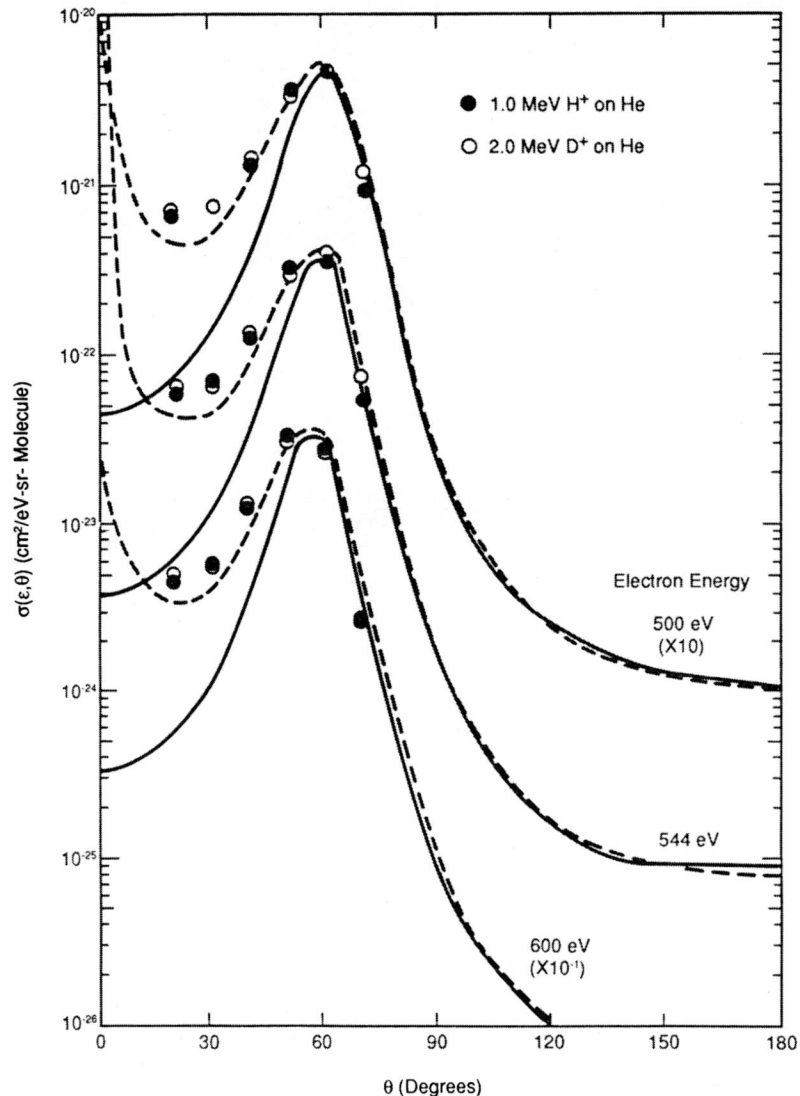

Figure 15. Doubly differential cross sections for emission of electrons of velocity comparable to the projectile compared to calculations of Macek[53] using the Fadeev method (---) and the plane wave Born approximation (——).

cross sections by as much as an order of magnitude in certain regions of the spectra, the contribution to the total yield of electrons is small.

For molecular targets, there is a great deal of similarity in the angular distributions of electrons ejected by protons for all the molecules we have studied except hydrogen. Data for a number of simple molecular targets are shown in Fig. 16 for 1.5 MeV proton impact. As in Fig. 9, we have scaled the cross sections compared in Fig. 16 by the number of weakly bound electrons. Scaled in this way, data for all molecules except hydrogen agree within experimental uncertainties. Had data for hydrocarbon molecules been included in the comparison [see for example, Wilson and Toburen[40]] they would have shown slightly higher values at the peaks in the angular distributions,

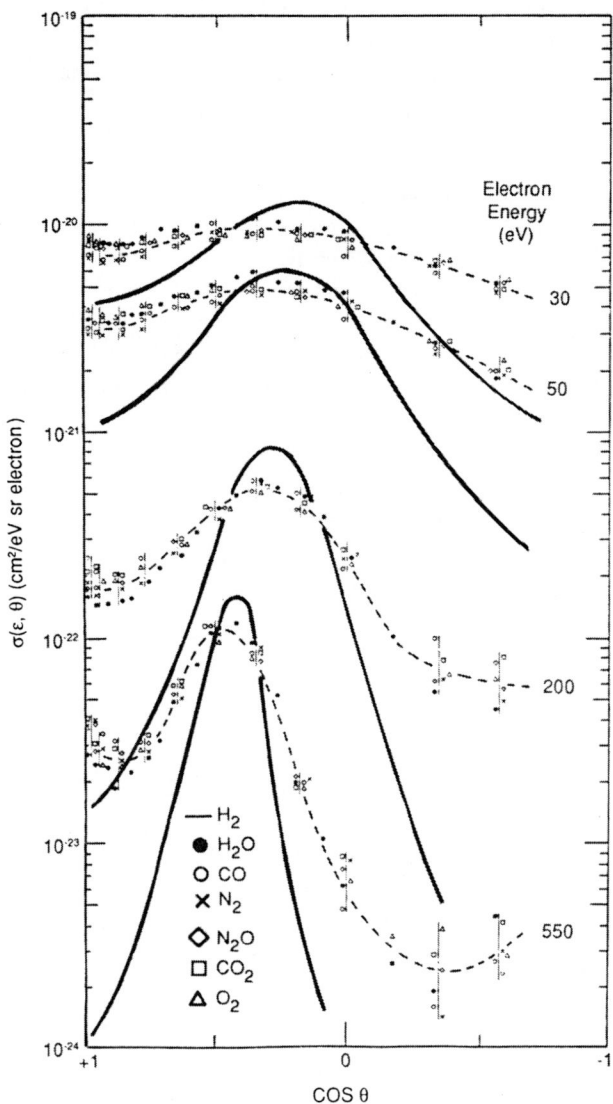

Figure 16. Doubly differential cross sections measured at PNL for ionization of a number of molecules by 1.5 MeV protons all scaled according to the number of loosely bound electrons.

i.e., somewhat more hydrogen-like; but when scaled in the same manner they would agree well with the molecules shown here at both large and small emission angles. If cross sections for higher ejected electron energies were plotted, the peaks in the angular distributions would move to smaller angles in agreement with classical kinematics, i.e., proportional to $\cos^2\theta$. Evidence of ionization via the CCT mechanism is seen as the increase in the cross sections for the smallest angles of ejection and the highest energy electrons shown in Fig. 16. We would expect enhancement due to CCT to be most evident in distributions for ejected electron energies near 817 eV, an electron energy with equivalent velocity to the 1.5 MeV proton.

The most dramatic differences observed among the data shown in Fig. 16 is the large difference between the scaled molecular hydrogen cross sections and those of all the other molecules. Differences of more than an order of magnitude occur at both small and large emission angles. The implication is even more important when we recall that semi-classical and hydrogenic Born calculations yield results that mimic the hydrogen measurements; e.g., see the binary encounter results shown in Fig. 14. It is only when realistic wave functions are used for both bound and continuum states that the Born approximation gives adequate agreement with measured doubly differential cross sections.[39] Unfortunately techniques have not been developed for application of Born theory to molecular targets, primarily because of the unavailability of adequate wave functions. The general trends of the data shown are, of course, representative for fast collisions. An examination of the angular distributions for low-energy proton impact shown in Fig. 17 indicates that the distributions peak at zero degrees for all ejected electron energies.[49]

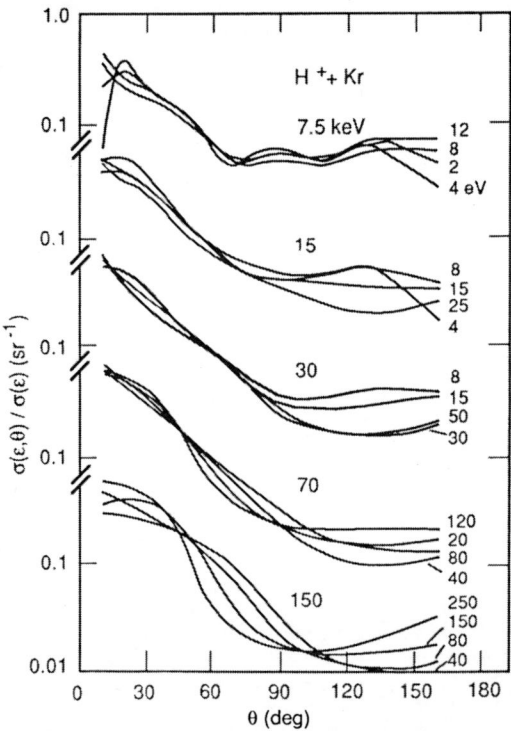

Figure 17. Angular distributions of electrons of several energies ejected from krypton by low-energy protons; this figure was reproduced from Cheng et al.[49]

From a review of the doubly differential cross sections for electron emission by protons, it would appear that the mechanisms responsible for ionization are well understood; this is particularly true for fast ions where the Born approximation in expected to be valid. Singly differential cross sections for large energy losses can be described well by binary encounter theory, and the low-energy portion of the spectra is accurately evaluated using Bethe-Born theory combined with optical oscillator strengths and measured spectra to evaluate the hard collisions component of the cross sections. Where sufficient experimental data exist, Rudd's model can be used to extrapolate cross sections over the complete range of proton energies. The angular distributions are not as fully understood. Binary encounter theory and hydrogenic Born calculations both underestimate the cross sections for emission of electrons into large and small angles. The similarity of angular distributions for a wide range of molecular targets, however, is conducive to the development of molecular models for use in track structure calculations. Charge-transfer-to-continuum states contribute to an enhancement of the cross sections for small emission angles, but do not contribute significantly to the total yield of electrons.

Structured Ion Impact

Studies of ionization of atomic and molecular targets by structured ion, ions that carry bound electrons and are sometimes referred to as "clothed" or "dressed" ions, have been under way for more than 20 years [see for example, Rudd et al.,[54] Cacak and Jorgensen,[55] Wilson and Toburen[56]] and have been discussed in reviews by Toburen,[57,27] Stolterfoht,[58] and Rudd.[29] Publications addressing doubly differential cross sections for a broad range of collision partners are presented in Table III; in preparing this table an attempt was made to limit the publications included to those that involve measurement of absolute cross sections and that cover a reasonably wide range of ejected electron energies and angles.

The primary differences between the spectra of electrons ejected by ions that carry bound electrons and those ejected by bare ions can be seen in Fig. 18, where spectra are shown for ejection of electrons from water vapor by 0.3 MeV/amu H^+, He^{++}, and He^+ ions;[59] the proton data have been multiplied by a factor of 4 for comparison to the helium ion data. Note that the scaled proton data are in excellent agreement with the He^{++} results over most of the energy and angular range. This is representative of the accuracy of Z^2 projectile-charge scaling of collision cross sections for bare ions. The greatest differences between the scaled proton and bare helium ion cross sections occur at the smallest angles and for electron energies near 160 eV; this is the electron energy at which the electron and ion have comparable velocity. These differences are attributed to the CCT mechanism of ionization, which has been predicted to have a Z^3 dependence on projectile charge.[60] In contrast to the excellent agreement between scaled H^+ and He^{++} cross sections, the emission cross sections for He^+ impact exhibit marked differences from the bare ion results. Most evident is the reduction in cross section for ejection of low-energy electrons. These electrons are ejected in distant "soft" collisions in which the bound electron provides an effective electrostatic shield of the helium ion nucleus.[61] Higher energy electrons are ejected, with increasingly close collisions that penetrate the shielding radius of the He^+ bound electron and are subject to the coulomb potential of the full nuclear charge; thus, cross sections for high-energy electrons ejected in He^+ collisions are similar to those for He^{++} impact. In principle, one would expect a gradual change in the nature of the cross sections, from low-energy electrons that are ejected in large-impact parameter collisions by what would appear to be a "heavy" proton, charge +1, to fast electrons ejected in close collisions by an effectively bare alpha particle, charge +2. Unfortunately the functional relationship that allows one to scale the charge as a function of energy loss, or impact parameter, has not been determined for different ions and molecular targets. The most obvious implication of the energy dependence of the "effective" nuclear charge is that the effective charge of stopping power theory, i.e., an effective charge that is only a function of the nuclear charge and particle velocity, is totally inadequate for use in any theory of differential energy loss by dressed ions.

Table III. Published Double Differential Cross Sections - Structured-Ion Impact

Reaction	Ion Energy Range (keV)	Ejected Electron		Investigators
		Energy Range (kev)	Angle (Degrees)	
$H_2^+ + H_2$	600 - 1500	2 - 2000	20 - 125	Wilson and Toburen[56]
$Ne^+ + Ne$	50 - 300	1.5 - 1000	10 - 160	Cacak and Jorgensen[55]
$Ne^{n+} + Ne$ (n=1-4)	25 - 800	1.6 - 1100	45 - 135	Woerlee, Gordeev, de Waard, and Saris[68]
$Ar^+ + Ar$	50 - 300	1.5 - 1000	10 - 160	Cacak and Jorgensen[55]
$Ar^+ + Ar$	100	3 - 250	160	Rudd, Jorgensen, and Volz[54]
$O^{n+} = O_2$ (n=4-8)	30000	10 - 4000	25 - 90	Stolterfoht et al.[117]
$\left.\begin{array}{l}H_2^+ \\ He^+\end{array}\right\} + \left\{\begin{array}{l}H_2 \\ He\end{array}\right.$	1000, 2000	20 - 1000	30 - 140	Oda and Nishimura[118]
$\left.\begin{array}{l}H_2^+ \\ He^o\end{array}\right\} + Ar$	5 - 20	1 - 26	30 - 140	Sataka, Okuno, Urakawa, and Oda[108]
$\left.\begin{array}{l}H^o + He \\ H_2^o + He \\ {}^3He^o + He \\ {}^4He^o + He\end{array}\right\}$	15 - 150	1.5 - 300	10 - 160	Fryar, Rudd, and Risley[119]
$H^o = He$	15 - 150	1.5 - 300	15 - 150	Rudd, Risley, Friar, and Rolfes[121]
$\left.\begin{array}{l}He^+ \\ \\ He^{++}\end{array}\right\} + \left\{\begin{array}{l}He \\ Ne \\ Ar\end{array}\right.$	1200	1 - 3500	15 - 125	Toburen and Wilson[122]
$\left.\begin{array}{l}He^+ \\ He^{++}\end{array}\right\} + Ar$	300 - 2000	1 - 4000	15 - 125	Toburen and Wilson[123]
$\left.\begin{array}{l}He^+ \\ He^{2+}\end{array}\right\} + H_2O$	300 - 2000	1 - 4000	15 - 125	Toburen, Wilson, and Popowich[59]
$H^o + H_2O$	20 - 150	1 - 300	10 - 160	Bolorizadeh and Rudd[124]
$\left.\begin{array}{l}H_2^+ \\ H_3^+\end{array}\right\} He$	5 - 20	2 - 50	30 - 120	Urakawa, Tokoro, and Oda[125]
$\left.\begin{array}{l}H^+ \\ H_2^+ \\ He^+\end{array}\right\} Ar$	5 - 20	2 - 26	30, 90	Sataka, Urakawa, and Oda[126]
$C^+ \left\{\begin{array}{l}He \\ Ne \\ Ar \\ CH_4\end{array}\right.$	1200	1 - 800	30, 90	Toburen[127]
$C^+ + He$	800 - 4200	10 - 1500	15 - 130	Reinhold, Schultz, Olson, Toburen, and DuBois[66]

Table III. (continued)

Reaction	Ion Energy Range (keV)	Ejected Electron		Investigators
		Energy Range (kev)	Angle (Degrees)	
C^{+n} + Ar (n=1-3)	1300 - 3000	1 - 4000	15 - 130	Toburen[128]
O^+ N^+ } Ar	50 - 500	5 - 500	16 - 160	Stolterfoht and Schneider[129]
Kr^{n+} + Kr (n=2-5)	50 - 1000	80 - 1000	45 - 135	Gordeev, Woerlee, de Waard, and Saris[130]
Kr^{n+} + Kr (n=2-5)	25 - 800	16 - 1100	45 - 135	Gordeev, Woerlee, De Waard, and Saris[131]
H^+ H_2^+ } + Ar He^+	5 - 20 keV	2 - 26	30 and 90	Sataka, Urakawa, and Oda[126]
U^{38+} } He Th^{38+} } Ar	6000/amu	5 - 5000	20 - 150	Schneider et al.[132]
Mo^{44+} - He	2500/amu	2 - 5000	20 - 160	Stolterfoht et al.[133]
U^{33+} - Ne	1400/amu	1 - 4000	20 - 90	Kelbch, Olson, Schmidt, Schmidt-Böcking, and Hagmann[134]
U^{33+} - Ar	1400/amu	1 - 4000	20 - 90	Kelbch, Olson, Schmidt, Schmidt-Böcking, and Hagmann[134]

The second feature of electron spectra for structured ions that is different from bare ions is the presence of a peak in the spectra from electrons that are stripped from the incident ion. These electrons are found predominantly in the forward directions, small emission angles in the laboratory reference frame, and at electron energies that correspond to electrons of the same velocity as the ion. Such a peak is visible in Fig. 18 in the 15- and 30-degree spectra at approximately 160 eV. In the 15-degree spectrum the contribution from projectile electron loss enhances the He^+ spectrum over that for He^{++}. The contribution of electron loss from the projectile to the total yield of electrons can be demonstrated by plotting the ratio of the measured cross section to the Rutherford cross section as a function of 1/E as was described earlier; for comparison with heavier ions, the Z^2 dependence of the Rutherford cross section must be implicitly included. Data for ionization of helium by 0.3 MeV/amu helium ions and protons are shown plotted in this way in Fig. 19. Excellent agreement between theory and experiment is observed between the H^+ and He^{++} induced cross sections; the only differences are at electron energies less than about 18 eV ($1/E=0.3$ Ry^{-1}) where one can expect larger uncertainties in the helium ion data because no TOF data are available to improve the accuracy of the low-energy data. In these spectra the electron loss peak is clearly seen in the He^+ data; it enhances the cross sections well over those for bare ions in the region near electron energies of

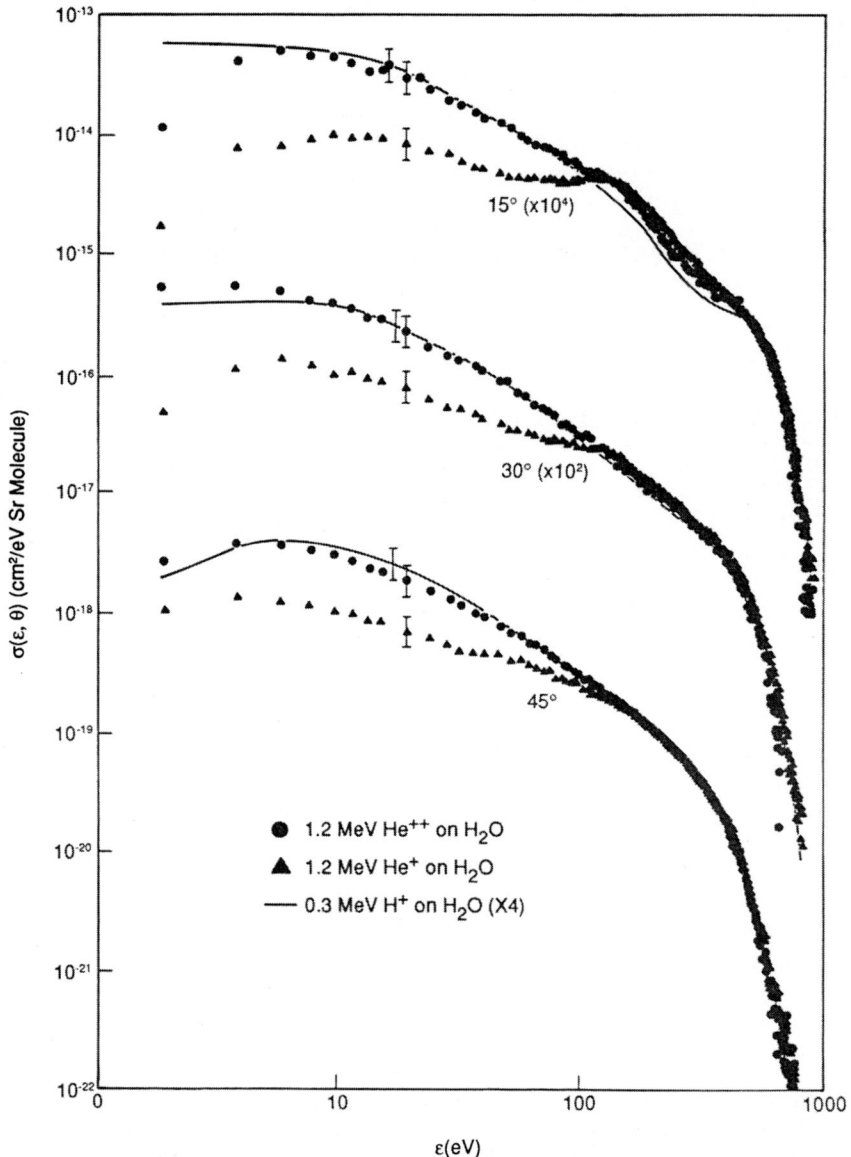

Figure 18. Doubly differential cross sections for electron emission from water vapor by 0.3 MeV/ amu protons and helium ions. The proton data have been multiplied by 4 (as implied by Z^2 scaling) for comparison to the helium ion results; the data are from Toburen et al.[59]

160 eV ($1/E=0.07$ Ry^{-1}). The actual contribution of this process to the total ionization is, however, hard to determine because the effects of screening make identification of the portion of the curve due to target ionization difficult. One can see from this illustration, recalling that equal areas under the curve contribute equally to the total ionization cross section, that the mean energy of the ejected electrons will certainly be greater for He$^+$ ions than the bare ions.

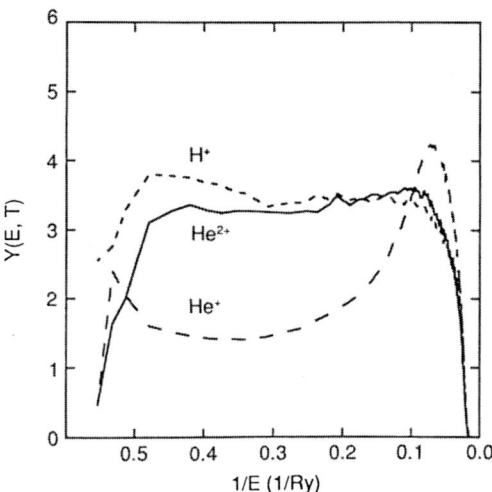

Figure 19. Ratio of the measured singly differential cross sections for ionization of helium by 0.3 MeV/amu protons and helium ions to the corresponding Rutherford cross sections.

Theoretical studies of structured ions have, until recently, been limited primarily to simple systems such as He^+-He [see DuBois and Manson[62] and references therein], although Stolterfoht and his colleagues have made a systematic study of the energy loss distributions in high-energy neon ion collisions.[63] A comparison of the doubly differential cross sections calculated with the Born approximation for the He^+-He collision system to spectra measured at 15 and 60 degrees with respect to the outgoing He^+ ion described in the work of Manson and Toburen[64] is shown in Fig. 20. Excellent agreement is observed for electrons ejected at 60 degrees, but differences of approximately a factor of two are found for the 15-degree spectra. Since the electrons are indistinguishable in these measurements, it could not be determined whether the discrepancy resulted from calculation of target or projectile ionization. More recent measurements were made in which electrons were detected in coincidence with either the transmitted He^+ or stripped He^{++} ion.[62] Those measurements demonstrated the inadequacy of the theoretical treatment to address simultaneous ionization processes, eg., ionization of both the projectile and target in a single collision. It is still not clear, however, whether the wave functions for the system are inadequate or whether the discrepancies were a result of a breakdown in the Born approximation itself. Recent measurements have been undertaken for the H^0-He collision system[65] that now indicate that the Born approximation is adequate to describe these few electron systems if adequate wave functions are used for discrete and continuum states.

An example of the spectrum of electrons ejected in He^+ collisions with water vapor in which electrons are detected in coincidence with the stripped He^{2+} ion is shown in Fig. 21. One would expect the coincidence spectrum to be dominated by electrons lost by the projectile, a spectrum that peaks at approximately 400 eV for the ion energy considered. The expected spectrum of electrons stripped from the projectiles, based on the transformation of an ejected electron spectrum from the projectile frame of reference to the laboratory frame, is shown as the dashed line in Fig. 21. The strong contribution of electrons at energies less than 400 eV is attributed to simultaneous projectile and target ionization. This simultaneous projectile and target ionization implies that there will be a significant amount of correlation between the two

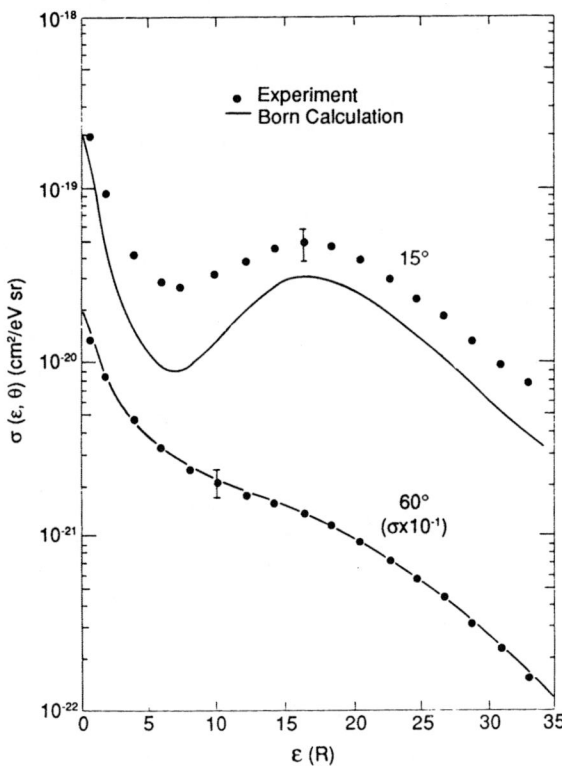

Figure 20. Comparison of the doubly differential cross sections for ionization of helium by 2 MeV He$^+$ to calculations based on the Born approximation.[64]

ejected electrons as they slow down in a biological medium. This could have impact on the subsequent chemical reactions that follow energy deposition.

Although a considerable amount of data has been generated for light structured ions such as H^0, H$_2^+$, H$_3^+$, He$^+$, etc., and the features are relatively well understood in the high-energy range, there are only scattered sets of data for heavier ions, and the theory of such collisions is only beginning to be developed. Because of their importance in neutron dosimetry, we have initiated studies of electron emission for collisions involving carbon and oxygen ions. The relative contributions of the spectra of electrons from ionization of helium by H$^+$, He$^+$, and C$^+$ ions are shown in Fig. 22. Since these spectra are all scaled by Z^2, only the highest energy cross sections, produced by very close collisions, may be expected to scale to the same values; low-energy ejected electron cross sections reflect the extent of screening by the bound electrons. There is evidence of a small peak at the equal velocity condition, $v_e = v_i$, that occurs for R/E=0.07 and indicates a small contribution of electron loss from the C$^+$ ion. A comparison of spectra for different C$^+$ ion energies is shown in Fig. 23. This illustration shows that the electron loss contribution grows as the ion velocity increases, reflecting the increasing electron loss cross section. The most obvious characteristic of energy loss in collisions of C$^+$ ions with atomic targets seen in Figs. 22 and 23 is the much lower fraction of low-energy electrons resulting from dressed ion collisions compared to bare ions. This leads to a much higher mean energy of the ejected electron in such collisions. Also note that it is not possible to scale proton cross sections to structured ion impact in any simple manner.

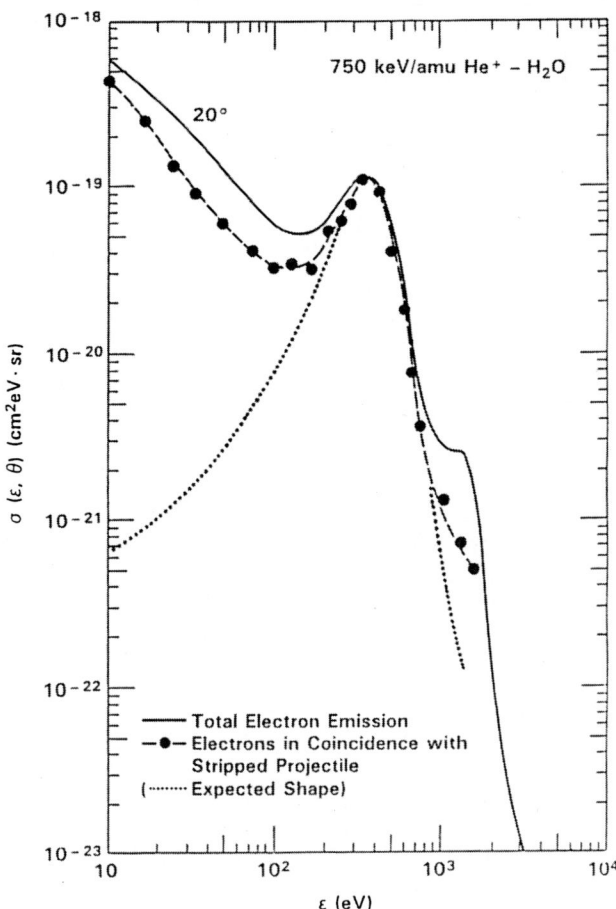

Figure 21. Comparison of the spectrum of electrons detected in coincidence with ionization of the He$^+$ projectile to that of all electrons ejected at 20 degrees with respect to the exiting ion.

In the energy range where we have studied differential cross sections for carbon and oxygen ion impact, 66 to 350 keV/u, the analysis advocated by Kim and discussed for protons in Figs. 5 to 8 does not enable a test of the absolute cross sections; there is no high-energy plateau in the plot of Y(E,T) versus E. This is illustrated in Fig. 24. Absolute cross sections must, therefore, be evaluated from experimental techniques and from recently measured total ionization cross sections.[66,67] Because of a combination of the effects of screening by projectile electrons and electron loss from the projectile, it is difficult to identify the origin of the spectral features observed in Fig. 24. The peak, or shoulder for lower energy ions, at the low-energy end of the spectra is a result of electron loss from the projectile; the maximum contribution from this process for 4.2 MeV ions should be at about 190 eV (E/R=16 Ry). The electron loss peaks appear at a somewhat higher energy than predicted by kinematics because they are on a rapidly increasing background of electrons from target ionization. The binary encounter peak should be about four times higher in energy, or at approximately 58 Rydbergs for the 4.2 MeV spectrum. At this particle velocity, one cannot expect a well-defined binary encounter peak (see for example the proton impact data of Fig. 6).

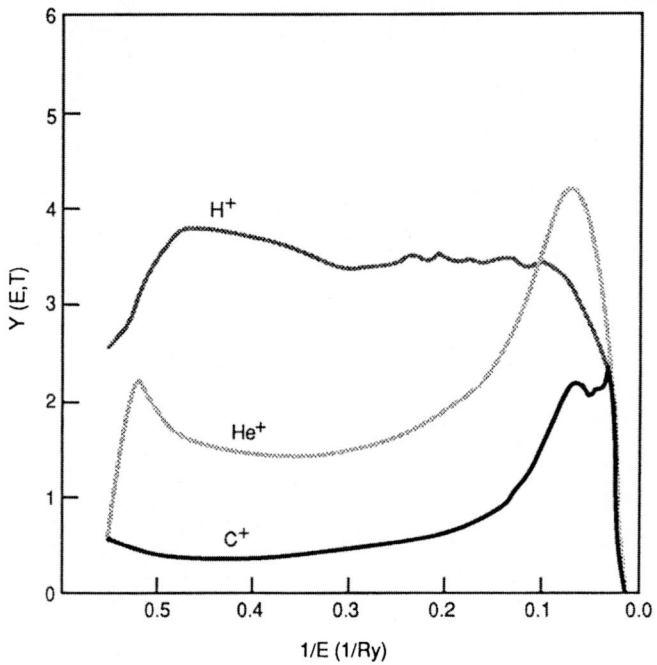

Figure 22. Comparison of the singly differential cross sections for ionization of helium by 0.3 MeV/amu H^+, He^+, and C^+ ions plotted as the ratio to the corresponding Rutherford cross sections.

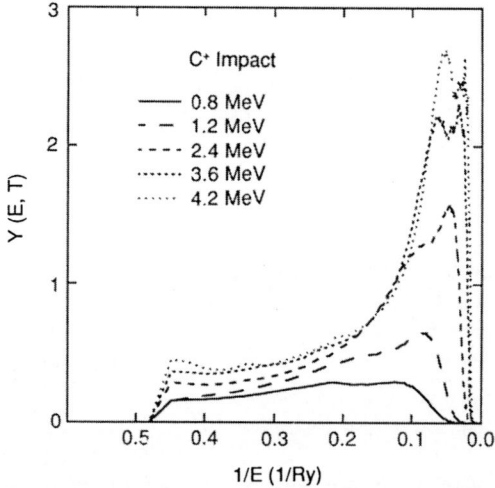

Figure 23. The ratio of measured and Rutherford cross sections for ionization of helium by singly charged carbon ions plotted as a function of the inverse of the energy loss.

The small peaks observed superimposed on the spectra in Fig. 24 for the two highest energy ions result from Auger transitions following inner-shell vacancy production in the carbon projectile. These transitions are observed at Doppler shifted energies in the laboratory determined by the kinematics of the collision. Since these spectra were obtained from integration of doubly differential electron energy distributions measured at discrete angles, these peaks carry forward as discrete peaks in the

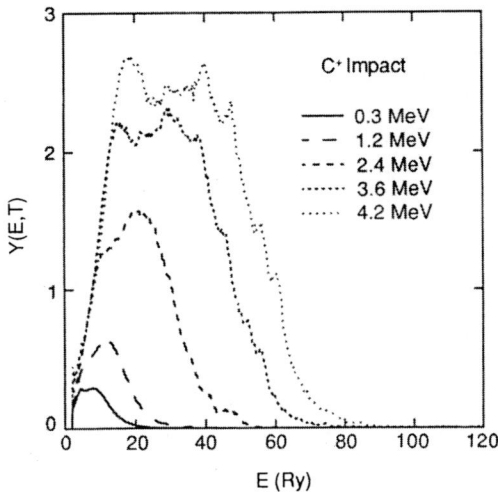

Figure 24. The ratio of measured and Rutherford cross sections for ionization of helium by singly-charged carbon ions plotted as a function of the energy loss.

integral spectrum. The intensity of these transitions in the doubly differential cross sections also provides a means to determine the consistency of the measured absolute cross sections and the energy scales by comparison with other measurements on inner-shell ionization.

There is presently very little information on the effects of projectile charge on the systematics of the doubly differential ionization cross sections, with the exception of studies undertaken for helium ions discussed earlier in this section. Measurements of doubly differential cross sections for collisions of 25 to 800 keV neon ions with neon by Woerlee et al.[68] showed little effect from a change in projectile charge from +1 to +3; the cross sections were reduced by about 10% per charge state independent of ejected electron energy. For these low energies, the molecular promotion model is the primary mechanism of ionization; it would predict that the more electrons there are bound to the collision partners, the more likely it is to promote a bound electron to a continuum state. For direct coulomb ionization, however, one would expect a more highly stripped ion to be more efficient at ejecting electrons from an atomic or molecular target. Cross sections for ejection of electrons from water vapor at 15 degrees with respect to outgoing oxygen ions of charge +1 to +3 are shown in Fig. 25. These data from our laboratory show the expected increase in cross section with projectile charge for ejection of low-energy electrons. They do not, however, increase as the square of the net projectile charge, even for the lowest ejected electrons energies; those ejected in the most distant collisions. Thus, even at the largest impact parameters encountered in these measurements, the ions do not appear as point charges. The data shown in Fig. 25 again illustrate the contribution to the spectra from electron loss by the projectile. This is observed in the peak at $v_e = v_i$ that seems to decrease in intensity as the charge state increases, reflecting the smaller number of projectile electrons available to be stripped. The binary encounter peak from direct interactions between the incident oxygen nucleus and a target electron is also visible at about 400 eV. At the extreme high-energy end of the spectra is the Doppler-shifted Auger electron spectra resulting from K-shell ionization of the oxygen projectile. Transitions of this type were responsible for the structure superimposed on the integral spectra shown in Fig. 24.

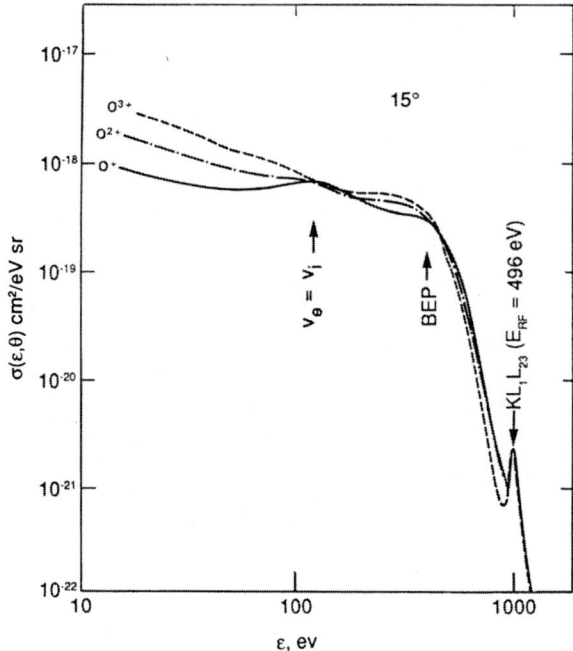

Figure 25. Doubly differential cross sections for ionization of water vapor by oxygen ions of different charge states. Arrows point out the position of the KLL-Auger spectrum resulting from relaxation of K-shell vacancies produced in the oxygen projectile, the binary peak (BEP) for electrons originating from the target, and the electron loss peak ($v_e = v_i$) of electrons originating from the projectile.

At the present time, there is little theory that can help in the analysis and prediction of differential cross sections for structured ions. The Born approximation appears to adequately describe the collision for simple systems such as ionization by H^0 or He^+ ions if reliable wave functions are available for discrete and continuum states. This theory, however, appears a long way from application to more complex molecular systems of interest in Radiological Physics; and, even where wave functions are available, the theory is usable only for high-energy ions. At the low-energy extreme, the use of a quasi-molecular description of the collision with electron promotion to the continuum via radial coupling has proven useful for modeling cross sections.[49,68] During the past few years, classical trajectory Monte Carlo (CTMC) techniques have been developed for use in calculation of doubly differential cross sections for emission of electrons by intermediate and high-energy ions.[66,67,69] This technique is attractive in that it provides, ab initio, absolute cross sections and includes continuum charge transfer and multiple ionization processes, as well as being appropriate to the intermediate-ion energy range. The primary disadvantage of this technique is the extraordinarily long computational time required. In Fig. 26 is shown a recent calculation of the singly differential cross sections for ejection of electrons in C^+ + He collisions at 200 keV/amu. There is good agreement between theory and experiment for the high-energy portion of the spectra, with the experimental data being about 50% larger than theory at the low energies. A strength of theory, as contrast to experiment, is that one is able to estimate the contribution, and the spectral shape, of electrons coming from either the target or the projectile. The dashed line in Fig. 26 is the calculated contribution of electrons ejected from the projectile. An example of the doubly differential cross sections derived for this collision system is shown in

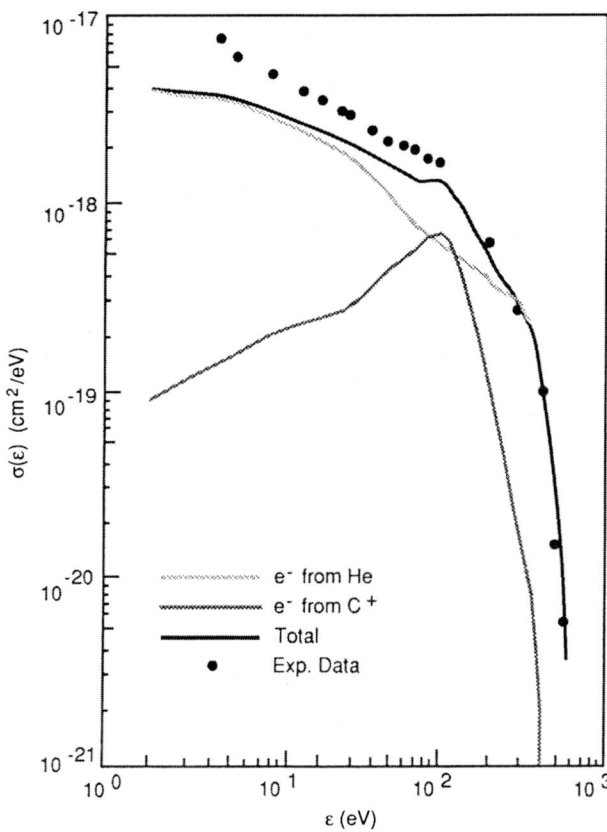

Figure 26. Comparison of measured singly differential cross sections for ionization of helium by C^+ to the classical trajectory Monte Carlo calculation of Reinhold et al.[66]

Fig. 27. For the emission angle shown, the experimental data are about 50% larger than the calculation; comparisons at smaller angles exhibit better agreement. Although there are discrepancies between the CTMC calculations and experiment, the agreement is comparable to, or better than, that seen earlier for Born calculations for simple collision systems, and the CTMC calculations can be applied to essentially any system because they do not rely on special system wave functions.

One could summarize our knowledge of differential ionization cross sections for structured ions as fragmentary. We have a reasonable understanding of collisions for light projectiles such as H^0 and He^+. Unfortunately, there is no theoretical means of calculating doubly differential cross sections for molecular targets at the present time, nor have models been developed for fitting or extrapolating such cross sections. Data that exist, however, show dramatically that scaling from bare ions to dressed ions is not possible with a single parameter, such as the effective charge of stopping power theory, and that the mean energy of electrons emitted in collisions involving structured ions is much higher than for bare ions. There is a long way to go before cross sections for structured ions will be understood with the same detail as bare ions, but recent advances in the development of theory and in experimental techniques give optimism to a vastly improved understanding in the near future.

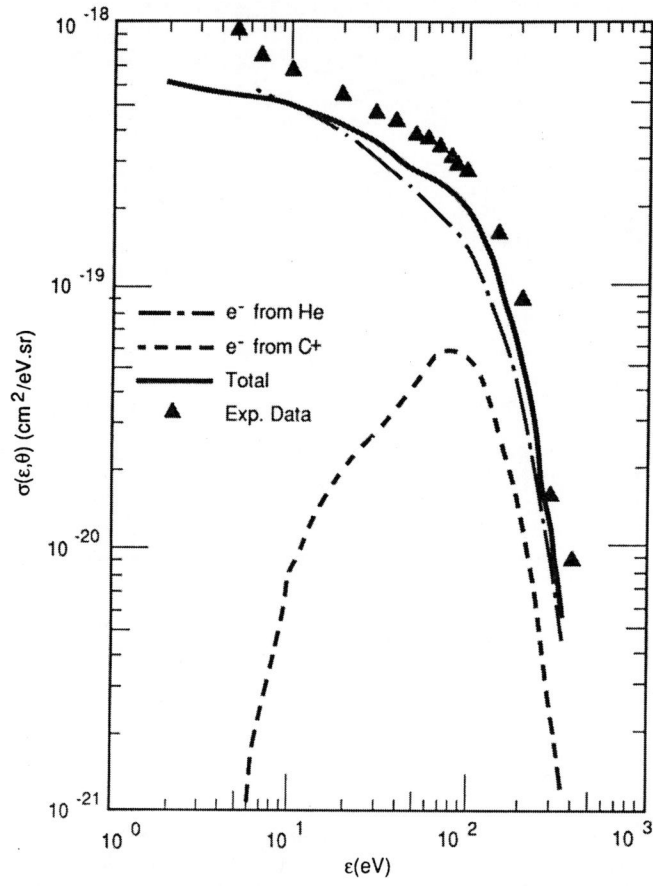

Figure 27. Comparison of measured doubly differential cross sections for emission of electrons at 50 degrees by C^+ ions to classical trajectory Monte Carlo calculations of Reinhold et al.[66]

Multiple Ionization and Charge Transfer

There is no question that the doubly differential cross sections for electron emission from charged-particle impact are of primary importance in the development of track structure descriptions of energy deposition by high-LET radiation. However, the studies of electron emission do not provide information regarding the fate of the target, nor do they provide information on the number of electrons that may be emitted in a single collision. The latter may be biologically significant since the multiply emitted electrons would slow down in a spatially and temporally correlated way. To describe fully the interactions appropriate to a charged-particle track, one must have information on the charge transfer cross sections, the change in the spectrum of electrons emitted following charge transfer, the number of electrons emitted per interaction and their individual energy and angular distributions, and the excitation and dissociation states of the target molecules. Obtaining detailed information on all of these processes, and the effect of target and projectile structure on them, is a tall order. As a first step, an assessment of the relative importance of the different processes can be very useful. For the past few years, a part of our effort at PNL has been devoted to measuring the spectrum of charge states of atomic and molecular products formed during ionizing collisions with light ions. These studies enable one to gain insight into the relative importance of different ionization channels appropriate to energy loss by ions and neutrals (see, for example, Refs. 70-78).

The principal interaction mechanisms leading to the release of free electrons involving collisions of H^+ and He^+ ions with a neon target are illustrated in Fig. 28 for ion energies from a few keV/μ to a few hundred keV/μ. As expected, direct ionization is the primary source of electrons for proton collisions throughout the energy range. Electron capture, which requires simultaneous target ionization to release a free electron, is at most a 5% contributor to free electron production. For helium ion impact, however, electron capture and loss can make a sizeable contribution to the free electron production. In the case of electron capture, this implies a high probability of simultaneous capture-plus-target ionization, leaving the target multiply ionized. It is also found that target ionization is produced with a high probability in collisions resulting in projectile ionization, although that is not obvious from this illustration. The message to be derived from this analysis is that, as projectiles heavier than protons are considered, processes other than direct ionization must be included if electron production along the track is to be fully described.

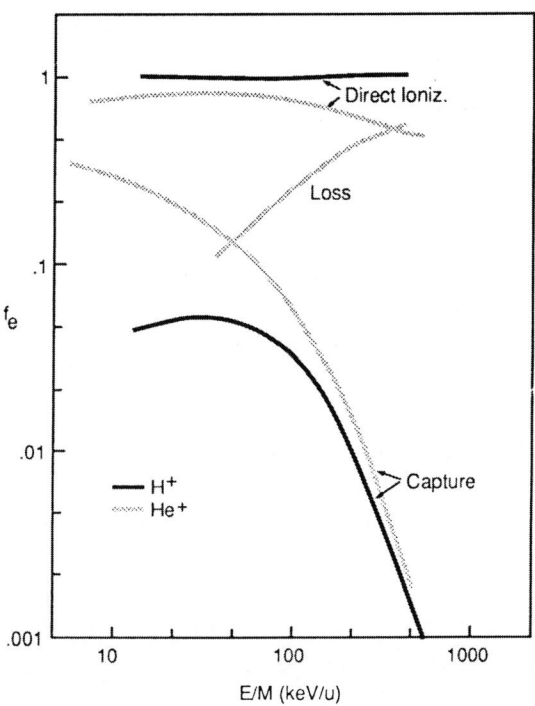

Figure 28. Comparison of the relative importance of the principal mechanisms for production of free electrons in collisions of protons and helium ions with neon; f_e represents the fraction of electrons produced by each mechanism.

In addition to electron capture-plus-projectile ionization that leaves the target atom doubly ionized, it is also possible to produce a doubly ionized target by direct ionization. The fraction of free electrons that results from multiple ionization of neon by protons and helium ions is shown in Fig. 29. This fraction never goes above 10% for proton impact; however, for alpha particles the fraction approaches 100% at very low energies and it is nearly 40% for He^+ ions over a large energy range. These data should remind us that one cannot simply scale data from protons to heavier ions. The data should also encourage us to investigate the biological consequences of multiple

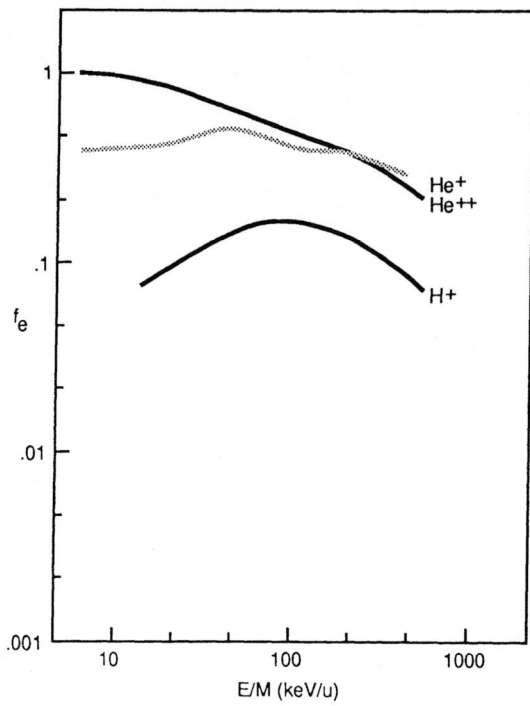

Figure 29. Comparison of the fractions of free electrons produced from multiple ionization events for protons and helium ions colliding with neon atoms.

ionization—What are the effects of a transient build-up of localized coulomb charge? It is no doubt true that multiple ionization is less frequent in molecular and condensed phase targets; however, the difference in multiple ionization cross sections for different ions may be manifest as different molecular fragmentation patterns and yields that may be equally important in leading to biologically important damage. In the near future, we should see an enhanced understanding of the relationship between the initial products of radiation and the subsequent chemistry and biology that is initiated. With the advances that are taking place in molecular biology, the molecular view of radiation damage from energy deposition to biological expression is within our grasp.

Acknowledgment

This work was supported by the Office of Health and Environmental Research (OHER), U.S. Department of Energy, under Contract DE-AC06-76RLO 1830.

References

1. D. E. Lea. *Actions of Radiation on Living Cells.* The Macmillan Company, New York (1947).
2. D. E. Charlton. Calculation of Single and Double Strand DNA Breakage from Incorporated [125]I. In *DNA Damage by Auger Emitters*, K. F. Baverstock and D. E. Charlton, eds., pp. 89-100. Taylor and Francis, New York (1988).
3. H. G. Paretzke. Radiation Track Structure Theory. In *Kinetics of Nonhomogeneous Processes*, G. R. Freeman, ed., pp. 89-170. John Wiley and Sons, New York (1987).
4. J. J. Butts and R. Katz. Theory of RBE for Heavy Ion Bombardment of Dry Enzymes and Viruses. *Radiat. Res.* 30:855-871 (1967).
5. A. Chatterjee and H. J. Shaefer. Microdosimetric Structure of Heavy Ion Tracks in Tissue. *Rad. and Environm. Biophys.* 13:215-227 (1976).
6. N. Bohr. The Penetration of Charged Particles through Matter. Det Kgl. Danske Videnskabernes Selskab. *Mathematisk-fysiske Meddelelser XVIII* 8:1-143 (1947).

7. M. Inokuti. *Introductory Remarks, Atomic and Molecular Data for Radiotherapy*, pp. 7-12. IAEA-TECDOC-506 (International Atomic Energy Agency, Vienna (1989).

8. E. C. Beaty, K. H. Hesselbacher, S. P. Hong, and J. H. Moore. Measurements of the Triple-Differential Cross Sections for Low-Energy Electron-Impact Ionization of Helium. *Phys. Rev. A* 17:1592-1599 (1978).

9. S. P. Hong and E. C. Beaty. Measurements of the Triple-Differential Cross Sections for Low-Energy Electron-Impact Ionization of Argon. *Phys. Rev A* 17:1829-1836 (1978).

10. C. E. Opal, W. K. Peterson, and E. C. Beaty. Measurements of Secondary-Electron Spectra Produced by Electron Impact Ionization of a Number of Simple Gases. *J. Chem. Phys.* 55:4100-4106 (1971).

11. N. Oda. Energy and Angular Distributions of Electrons from Atoms and Molecules by Electron Impact. *Radiat. Res.* 64:80-95 (1975).

12. M. A. Bolorizadeh and M. E. Rudd. Angular and Energy Dependence of Cross Sections for Ejection of Electrons from Water Vapor. I. 50-2000-eV Electron Impact. *Phys. Rev. A* 33:882-887 (1986).

13. Y.-K. Kim. Angular Distribution of Secondary Electrons in the Dipole Approximation. *Phys. Rev. A* 6:666-670 (1972).

14. Y.-K. Kim. *Secondary Electron Spectra, in Radiation Research, Biomedical, Chemical, and Physical Prospectives,* pp. 219-226. Academic Press, Inc., New York (1975).

15. Y.-K. Kim. Energy Distributions of Secondary Electrons I. Consistency of Experimental Data. *Radiat. Res.* 61:21-35 (1975).

16. Y.-K. Kim and M. Inokuti. Slow Electrons Ejected by Fast Charged Particles. *Phys. Rev. A* 7:1257-1260 (1973).

17. Y.-K. Kim and T. Noguchi. Secondary Electrons Ejected from He by Protons and Electrons. *Int. J. Radiat. Phys. Chem.* 7:77-82 (1975).

18. B. L. Schram, F. J. de Heer, M. J. van de Wiel, and J. Kistemaker. Ionization Cross Sections for Electrons (0.6-20 keV) in Noble and Diatomic Gases. *Physica* 31:94-112 (1965).

19. I. Shimamura. Cross Sections for Collisions of Electrons with Atoms and Molecules. *Scientific Papers of the Institute of Physical and Chemical Research (Japan)* 82:1-50 (1989).

20. J. Berkowitz. *Photoabsorption, Photoionization and Photoelectron Spectroscopy.* Academic Press, New York (1979).
    J. W. Gallagher, C. E. Brian, J.A.R. Samson, and P. W. Langhoff. Absolute Cross Sections for Molecular Photoabsorption, Partial Photoionization, and Ionic Photofragmentation Process. *J. Phys. Chem. Ref. Data* 17:9-153 (1988).

21. J. H. Miller and S. T. Manson. Differential Cross Sections for Ionization of Helium, Neon, and Argon by Fast Electrons. *Phys. Rev. A* 29:2435-2439 (1984).

22. J. H. Miller, W. E. Wilson, S. T. Manson, and M. E. Rudd. Differential Cross Sections for Ionization of Water Vapor by High-Velocity Bare Ions and Electrons. *J. Chem. Phys.* 86:157-162 (1987).

23. M. Michaud and L. Sanche. Total Cross Sections for Slow-Electron (1-20 eV) Scattering in Solid H_2O. *Phys. Rev. A* 36:4672-4683 (1987).

24. L. Sanche. Investigation of Ultra-Fast Events in Radiation Chemistry with Low-Energy Electrons. *Radiat. Phys. Chem.* 34:15-33 (1989).

25. P. Cloutier and L. Sanche. A Trochoidal Spectrometer for the Analysis of Low-Energy Inelastically Backscattered Electrons. *Rev. Sci. Instrum.* 60:1054-1060 (1989).

26. R. M. Marsolais and L. Sanche. Mechanisms Producing Inelastic Structures in Low-Energy Electron Transmission Spectra. *Phys. Rev. B* 38:11 118-11 130 (1988).

27. L. H. Toburen. Continuum Electron Emission in Heavy-Ion Collisions. In *Nuclear Methods 2; High-Energy Ion-Atom Collisions,* D. Brenyi and G. Hock, eds., pp. 53-82. Elsevier, New York (1982).

28. L. H. Toburen and W. E. Wilson. Secondary Electron Emission in Ion-Atom Collisions. In *Radiation Research,* S. Okada, M. Imamura, T. Terashima, and H. Yamaguchi, eds., pp. 80-88. Toppan Printing Co., Tokyo (1979).

29. M. E. Rudd. Mechanisms of Electron Production in Ion-Atom Collisions. *Radiat. Res.* 64:153-180 (1975).

30. M. Breinig, S. B. Elston, S. Huldt, L. Liljeby, C. R. Vane, S. D. Berry, G. A. Glass, M. Schauer, I. A. Sellin, G. D. Alton, S. Datz, S. Overbury, R. Laubert, and M. Suter. Experiments Concerning Electron Capture and Loss to the Continuum and Convoy Electron Production by Highly Ionized Projectiles in the 0.7-8.5 MeV/amu Range Traversing the Rare Gases, Polycrystalline Solids and Axial Channels in Gold. *Phys. Rev. A* 25:3015-3048 (1982).

31. N. Stolterfoht. Angular and Energy Distributions of Electrons Produced by 200-500 keV Protons in Gases: I. Experimental Arrangement. *Z. Phys.* 248:81-93 (1971).

32. J. B. Crooks and M. E. Rudd. Angular and Energy Distributions of Cross Sections for Electron Production by 50-300-keV Proton Impact on N_2, O_2, Ne, and Ar. *Phys. Rev. A* 3:1628-1634 (1971).

33. L. H. Toburen. Distributions in Energy and Angle of Electrons Ejected from Molecular Nitrogen by 0.3-1.7 MeV Protons. *Phys. Rev. A* 3:216-227 (1971).

34. L. H. Toburen and W. E. Wilson. Time-of-Flight Measurements of Low-Energy Electron Energy Distributions from Ion-Atom Collisions. *Rev. Sci. Instrum.* 46:851-854 (1975).

35. L. H. Toburen, S. T. Manson, and Y.-K. Kim. Energy Distributions of Secondary Electrons. III. Projectile Energy Dependence for Ionization of He, Ne, and Ar by Protons. *Phys. Rev. A* 17:148-159 (1978).

36. D. J. Lynch, L. H. Toburen, and W. E. Wilson. Electron Emission from Methane, Ammonia, Monomethylamine and Dimethylamine by 0.25 to 2.0 MeV Protons. *J. Chem. Phys.* 64:2616-2622 (1976).

37. J. T. Grissom, R. N. Compton, and W. R. Garrett. Slow Electrons from Electron Impact Ionization of He, Ne, and Ar. *Phys. Rev. A* 6:977-987 (1972).

38. B. B. Robinson. Modifications of the Impulse Approximation for Ionization and Detachment Cross Sections. *Phys. Rev.* 140:A764-768 (1965).

39. S. T. Manson, L. H. Toburen, D. H. Madison, and N. Stolterfoht. Energy and Angular Distributions of Electrons Ejected from Helium by Fast Protons and Electrons: Theory and Experiment. *Phys. Rev. A* 12:60-78 (1975).

40. W. E. Wilson and L. H. Toburen. Electron Emission from Proton-Hydrocarbon-Molecule Collisions. *Phys. Rev. A* 11:1303-1308 (1975).

41. M. E. Rudd, Y.-K. Kim, D. H. Madison, and J. W. Gallagher. Electron Production in Proton Collisions: Total Cross Sections. *Rev. Mod. Phys.* 57:965-994 (1985).

42. J. H. Miller, L. H. Toburen, and S. T. Manson. Differential Cross Sections for Ionization of Helium, Neon, and Argon by High-Velocity Ions. *Phys. Rev. A* 27:1337-1344 (1983).

43. W. E. Wilson, J. H. Miller, L. H. Toburen, and S. T. Manson. Differential Cross Sections for Ionization of Methane, Ammonia, and Water Vapor by High Velocity Ions. *J. Chem. Phys.* 80:5631-5638 (1984).

44. E. W. McDaniel, M. R. Flannery, E. W. Thomas, H. W. Ellis, K. J. McCann, S. T. Manson, J. W. Gallagher, J. R. Rumble, E. C. Beaty, and T. G. Roberts. *Compilation of Data Relevant to Nuclear Pumped Lasers,* Vol. V, pp. 1917-2011. Technical Report H-78-1. U.S. Army Missile Research and Development Command, Redstone Arsenal, Alabama (1979).

45. T. Inagaki, R. N. Hamm, and E. T. Arakawa. Optical and Dielectric Properties of DNA in the Extreme Ultraviolet. *J. Chem. Phys.* 61:4246-4250 (1974).

46. W. Sontag and F. Weibezahn. Absorption of DNA in the Region of Vacuum-uv. *Rad. and Environ. Biophys.* 12:169-174 (1975).

47. M. Fujii, T. Tamura, N. Mikami, and M. Ito. Electronic Spectra of Uracil in a Supersonic Jet. *Chem. Phys. Lett.* 126:583-587 (1986).

48. M. E. Rudd. Differential Cross Sections for Secondary Electron Production by Proton Impact. *Phys. Rev. A* 38:6129-6137 (1988).

49. W. Cheng, M. E. Rudd, and Y. Hsu. Differential Cross Sections for Ejection of Electrons from Rare Gases by 75-140 keV Protons. *Phys. Rev. A* 39:2359-2366 (1989).

50. M. E. Rudd. Energy and Angular Distributions of Electrons from 5-100-keV-Proton Collisions with Hydrogen and Nitrogen Molecules. *Phys. Rev. A* 20:787-796 (1979).

51. T. F. M. Bonson and L. Vriens. Angular Distribution of Electrons Ejected by Charged Particles I. Ionization of He and H_2 by Protons. *Physica* 47:307-319 (1970).

52. D. H. Madison. Angular Distribution of Electrons Ejected from Helium by Proton Impact. *Phys. Rev. A* 8:2449-2456 (1973).

53. J. Macek. Theory of the Forward Peak in the Angular Distribution of Electrons Ejected by Fast Protons. *Phys. Rev. A* 1:235-241 (1970).

54. M. E. Rudd, T. Jorgensen, Jr. and D. J. Volz. Electron Energy Spectra from Ar^+-Ar and H^+-Ar Collisions. *Phys. Rev.* 151:28-31 (1966).

55. R. K. Cacak and T. Jorgensen, Jr. Absolute Doubly Differential Cross Sections for Production of Electrons in Ne^+-Ne and Ar^+-Ar Collisions. *Phys. Rev. A* 4 2:1322-1327 (1970).

56. W. E. Wilson and L. H. Toburen. Electron Emission in H_2^+-H_2 Collisions from 0.6 to 1.5 MeV. *Phys. Rev. A* 7:1535-1544 (1973).

57. L. H. Toburen. *Atomic and Molecular Processes of Energy Loss by Energetic Charged Particles.* IAEA-TECDOC-506, International Atomic Energy Agency, pp. 160-168. Vienna (1989).

58. N. Stolterfoht. Excitation in Energetic Ion-Atom Collisions Accompanied by Electron Emission. In *Topics in Current Physics V.: Structure and Collisions of Ions and Atoms,* ed. I. A. Sellin, pp. 155-199. Springer-Verlag, Berlin (1978).

59. L. H. Toburen, W. E. Wilson, and R. J. Popowich. Secondary Electron Emission from Ionization of Water Vapor by 0.3- to 2.0-MeV He^+ and He^{++} Ions. *Radiat. Res.* 82:27-44 (1980).

60. K. Dettman, K. G. Harrison, and M. W. Lucas. Charge Exchange to the Continuum for Light Ions in Solids. *J. Phys. B* 7:269-287 (1974).

61. L. H. Toburen, N. Stolterfoht, P. Ziem, and D. Schneider. Electronic Screening in Heavy-Ion Collisions. *Phys. Rev. A* 24:1741-1745 (1981).

62. R. D. DuBois and S. T. Manson. Coincidence Study of Doubly Differential Cross Sections: Projectile Ionization in He^+-He Collisions. *Phys. Rev. Lett.* 57:1130-1132 (1986).

63. D. Schneider, M. Prost, N. Stolterfoht, G. Nolte, and R. D. DuBois. Projectile Ionization in Fast Heavy Ion-Atom Collisions. *Phys. Rev. A* 28:649-655 (1983).

64. S. T. Manson and L. H. Toburen. Energy and Angular Distributions of Electrons from Fast He^++He Collisions. *Phys. Rev Lett.* 46:529-531 (1981).

65. O. Heil, R. Maier, M. Huzel, K.-O. Groeneveld, and R. D. DuBois. A Systematic Investigation of Ionization Occurring in Few Electron Systems. *Phys. Rev. Lett.* (1990) (in press).

66. C. O. Reinhold, D. R. Schultz, R. E. Olson, L. H. Toburen, and R. D. DuBois. Electron Emission from Both Target and Projectile in C^+ + He Collisions. *J. Phys. B* 23:L297-L302 (1990).

67. L. H. Toburen, R. D. DuBois, C. O. Reinhold, D. R. Schultz, and R. E. Olson. Experimental and Theoretical Study of the Electron Spectra in 66.7 to 350 keV/amu C^+ + He Collisions. *Phys. Rev. A* 42:5338-5347 (1990).

68. P. H. Woerlee, Yu. S. Gordeev, H. de Waard, and F. W. Saris. The Production of Continuous Electron Spectra in Collisions of Heavy Ions and Atoms. B: Direct Coupling with the Continuum. *J. Phys. B* 14:527-539 (1981).

69. S. Schmidt, J. Euler, C. Kelbch, S. Kelbch, R. Koch, G. Kraft, R. E. Olson, U. Ramm, J. Ullrich, and H. Schmidt-Bocking. *Double Differential Stopping Powers of 1.4 Mev/amu U^{33+} in Ne and Ar Derived from Electron Production and Multiple Ionization Cross Sections.* Technical Report Number GSI-89-19, Gesellschaft für Schwerionenforschung: Darmstadt, Germany (1989).

70. R. D. DuBois. Multiple Ionization of He^+-Rare-Gas Collisions. *Phys. Rev. A* 39:4440-4450 (1989).

71. R. D. DuBois. Ionization and Charge Transfer in He^{2+}-Rare-Gas Collisions. II *Phys. Rev. A* 36:2585-2593 (1987).

72. R. D. DuBois. Recent Studies of Simultaneous Ionization and Charge Transfer in Helium Ion-Atom Collisions. *Nucl. Instrum. Meth.* B24/25:209213 (1987).

73. R. D. DuBois. Ionization and Charge Transfer in He^{2+}-Rare Gas Collisions. *Phys. Rev. A* 33:1595-1601 (1986).

74. R. D. DuBois. Coincidence Studies of Secondary Electron Emission in Light Ion-Atom Collisions. *Proceedings of the 2nd Workshop on High-Energy Ion-Atom Collisions* (Aug. 27-28, 1987, Dobrecen), p. 37. Publishing House of the Hungarian Academy of Science, Budapest (1985).

75. R. D. DuBois. Electron Production in Collisions Between Light Ions and Rare Gases: The Importance of the Charge-Transfer and Direct-Ionization Channels. *Phys. Rev. Lett.* 52:2348-2351 (1984).

76. R. D. DuBois and A. Kover. Single and Double Ionization of Helium by Hydrogen-Atom Impact. *Phys. Rev. A* 40:3605-3612 (1989).

77. R. D. DuBois and L. H. Toburen. Single and Double Ionization of Helium by Neutral-Particle to Fully Stripped Ion Impact. *Phys. Rev. A* 38:3960-3969 (1988).

78. R. D. DuBois, L. H. Toburen, and M. E. Rudd. Multiple Ionization of Rare Gases by H^+ and He^+ Impact. *Phys. Rev. A* 29:70-76 (1984).

79. T. W. Shyn and W. E. Sharp. Doubly Differential Cross Sections of Secondary Electrons Ejected from Gases by Electron Impact: 50-300 eV on Helium. *Phys. Rev. A* 19:557-567 (1979).

80. M. E. Rudd and R. D. DuBois. Absolute Doubly Differential Cross Sections for Ejection of Secondary Electrons from Gases by Electron Impact. I. 100 and 200 eV Electrons on Helium. *Phys. Rev. A* 16:26-32 (1977).

81. C. B. Opal, E. C. Beaty, and W. K. Peterson. *Tables of Energy and Angular Distributions of Electrons Ejected from Simple Gases by Electron Impact.* Joint Institute for Laboratory Astrophysics (JILA) Report No. 108, University of Colorado, Boulder, Colorado (1971).

82. C. B. Opal, E. C. Beaty, and W. K. Peterson. Tables of Secondary Electron Production Cross Sections. Atomic Data 4:209-253 (1972).

83. N. Oda, F. Nishimura, and S. Tahira. Energy and Angular Distributions of Secondary Electrons Resulting from Ionizing Collisions of Electrons with Helium and Krypton. *J. Phys. Soc. Japan* 33:462-467 (1972).

84. R. R. Goruganthu and R. A. Bonham. Secondary-Electron-Production Cross Sections for Electron Impact Ionization of Helium. *Phys. Rev. A* 34:103-126 (1986).

85. R. D. DuBois and M. E. Rudd. Absolute Doubly Differential Cross Sections for Ejection of Secondary Electrons from Gases by Electron Impact. II. 100-500 eV Electrons on Neon, Argon, Molecular Hydrogen and Molecular Nitrogen. *Phys. Rev. A* 17:843-848 (1978).

86. R. E. Mathis and D. A. Vroom. The Energy Distributions of Secondary Electrons from Ar, N_2, H_2O and H_2O with Clusters Present. *J. Chem. Phys.* 64:1146-1149 (1976).

87. T. W. Shyn, W. E. Sharp, and Y. K. Kim. Doubly Differential Cross Sections of Secondary Electrons Ejected from Gases by Electron Impact: 25-250 eV on H_2. *Phys. Rev. A* 24:79-88 (1981).

88. R. R. Goruganthu, W. G. Wilson, and R. A. Bonham. Secondary Electron Production Cross Sections for Electron Impact Ionization of Molecular Nitrogen. *Phys. Rev. A* 35:540-558 (1987).

89. T. W. Shyn. Doubly Differential Cross Sections of Secondary Electrons Ejected from Gases by Electron Impact: 50-400 eV on N_2. *Phys. Rev. A* 27:2388-2395 (1983).

90. Ce Ma and R. A. Bonham. Secondary Electron Production Cross Sections for 800 eV Electron-Impact Ionization of Carbon Monoxide. *Phys. Rev. A* 38:2160-2162 (1988).

91. T. W. Shyn and W. E. Sharp. Doubly Differential Cross Sections of Secondary Electrons Ejected from Gases by Electron Impact: 50-400 eV on CO_2. *Phys. Rev. A* 20:2332-2339 (1979).

92. C. E. Kuyatt and T. Jorgensen, Jr. Energy and Angular Dependence of the Differential Cross Section for Production of Electrons by 50-100 keV Protons in Hydrogen Gas. *Phys. Rev.* 130:1444-1455 (1963).

93. M. E. Rudd and T. Jorgensen, Jr. Energy and Angular Distributions of Electrons Ejected from Hydrogen and Helium Gas by Protons. *Phys. Rev.* 131:666-675 (1963).

94. M. E. Rudd, C. A. Sautter, and C. L. Bailey. Energy and Angular Distributions of Electrons Ejected from Hydrogen and Helium by 100 to 300 keV Protons. *Phys. Rev.* 151:20-27 (1966).

95. L. H. Toburen and W. E. Wilson. Distributions in Energy and Angle by Electrons Ejected from Molecular Hydrogen by 0.3-1.5 MeV Protons. *Phys. Rev. A* 5:247-256 (1972).

96. L. H. Toburen. Angular Distributions of Electrons Ejected by Fast Protons. In *VII International Conference on the Physics of Electronic and Atomic Collisions,* Abstracts of Papers, pp. 1120-1122. North Holland, Amsterdam (1971).

97. D. K. Gibson and I. D. Reid. Double Differential Cross Sections for Electron Ejection from Helium by Fast Protons. *J. Phys. B* 19:3265-3276 (1986).

98. N. Stolterfoht. Angular and Energy Distributions of Electrons Produced by 200-500 keV Protons in Gases II. Results for Nitrogen. *Z. Physik* 248:92-99 (1971a).

99. M. E. Rudd, L. H. Toburen, and N. Stolterfoht. Differential Cross Sections for Ejection of Electrons from Helium by Protons. *Atomic Data and Nuclear Data Tables* 18:413-432 (1976).

100. M. E. Rudd, G. L. Webster, C. A. Blocker, and C. A. Madison. Ejection of Electrons from Helium by Protons from 5-100 keV. In *IX International Conference on the Physics of Electronic and Atomic Collisions,* Abstracts of Papers, pp. 745-746. University of Washington Press, Seattle (1975).

101. M. E. Rudd and D. H. Madison. Comparison of Experimental and Theoretical Electron Ejection Cross Sections in Helium by Proton Impact from 5 to 100 keV. *Phys. Rev. A* 14:128-136 (1976).

102. G. B. Crooks and M. E. Rudd. Experimental Evidence for the Mechanism of Charge Transfer to Continuum States. *Phys. Rev. Lett.* 25:1599-1601 (1970).

103. L. H. Toburen and S. T. Manson. On the Unreliability of the Hydrogenic Approximation for Inelastic Collision of Fast Charged Particles with Atoms: Ionization of Neon by Protons. In *VIII International Conference on the Physics of Electronic and Atomic Collisions,* Abstracts of Papers, pp. 695-696. Institute of Physics, Belgrad (1973).

104. T. L. Criswell, L. H. Toburen, and M. E. Rudd. Energy and Angular Distributions of Electrons Ejected from Argon by 5 keV to 1.5 MeV Protons. *Phys. Rev. A* 16:508-517 (1977).

105. H. Gabler. *Ph.D. Thesis.* Free University, Berlin (1974).

106. T. L. Criswell, W. E. Wilson, and L. H. Toburen. Energy and Angular Distribution of Electrons Ejected from Argon by 5-2000 keV H_2^+ Impact. In *IX International Conference on the Physics of Electronic and Atomic Collisions,* Abstracts of Papers, pp. 749-750. University of Washington Press: Seattle (1975).

107. S. T. Manson and L. H. Toburen. Energy and Angular Distribution of Electrons Ejected from Ar by 1-MeV Proton Impact Ionization: Theory and Experiment. In *IX International Conference on the Physics of Electronic and Atomic Collisions,* Abstracts of Papers, pp. 751-752. University of Washington Press, Seattle (1975).

108. M. Sataka, K. Okuno, J. Urakawa, and N. Oda. Doubly Differential Cross Sections of Electron Ejection from Argon by 5-20 keV H^+, H_2^+, and He^+. In *XI International Conference on the Physics of Electronic and Atomic Collisions,* Abstracts of Papers, pp. 620-621. The Society of Atomic Collision Research, Japan (1979).

109. M. E. Rudd, L. H. Toburen, and N. Stolterfoht. Differential Cross Sections for Ejection of Electrons from Argon by Protons. *Atomic Data and Nuclear Data Tables* 23:405-442 (1979).

110. S. T. Manson and L. H. Toburen. Energy and Angular Distribution of Electrons Ejected from Kr by 1 MeV Proton Impact Ionization: Theory and Experiment. In *X International Conference on the Physics of Electronic and Atomic Collisions,* Abstracts of Papers, pp. 990-991. Commissariat A L'Energie Atomique, Paris (1977).

111. L. H. Toburen and S. T. Manson. Differential Cross Sections for Ionization of Krypton by Fast Protons: Theory and Experiment. In *XI International Conference on the Physics of Electronic and Atomic Collisions,* Abstracts of Papers, pp. 628-629. The Society for Atomic Collision Research, Japan (1979).

112. L. H. Toburen. Distributions in Energy and Angle of Electrons Ejected from Xenon by 0.3 to 2.0 MeV Protons. *Phys. Rev. A* 9:2505-2517 (1974).

113. N. Stolterfoht. Energy and Angular Distribution of Electrons Ejected from Atoms Ionized by Protons. In *VII International Conference on the Physics of Electronic and Atomic Collisions,* Abstracts of Papers, pp. 1123-1125. North Holland, Amsterdam (1971).

114. L. H. Toburen and W. E. Wilson. Energy and Angular Distributions of Electrons Ejected from Water Vapor by 0.3 - 1.5 MeV Protons. *J. Chem. Phys.* 66:5202-5213 (1977a).

115. M. A. Bolorizadeh and M. E. Rudd. Angular and Energy Dependence of Cross Sections for Ejection of Electrons from Water Vapor. II. 15-150 keV Proton Impact. *Phys. Rev. A* 33:888-896 (1986b).

116. L. H. Toburen, W. E. Wilson, and L. E. Porter. Energy and Angular Distributions of Electrons Ejected in Ionization of SF_6 and TeF_6 by Fast Protons. *J. Chem. Phys.* 4212-4221 (1977).

117. N. Stolterfoht, D. Schneider, D. Burch, H. Wieman, and J. S. Risley. Mechanisms for Electron Production in 30-MeV $O^{n+} + O_2$ Collisions. *Phys. Rev. Lett.* 33:59-62 (1974).

118. N. Oda and F. Nishimura. Energy and Angular Distributions of Electrons Ejected from He and H_2 Bombarded by Equal Velocity H_2^+ and He^+ Ions. In *XI International Conference on the Physics of Electronics and Atomic Collisions,* Abstracts of Papers, pp. 622-623. The Society for Atomic Collision Research, Japan (1979).

119. J. Fryar, M. E. Rudd, and J. S. Risley. Doubly Differential Cross Sections for Electron Production by Impact of H^o, H_2^o, H_2^o, $^3He^o$, and $^4He^o$ on Helium. In *XI International Conference on the Physics of Electronic and Atomic Collisions,* Abstracts of Papers, pp. 984-985. Commissariat A L'Energie Atomique, Paris (1977).

120. M. E. Rudd, J. S. Risley, and J. Fryar. Double Differential Cross Sections for Electron Production by Neutral Hydrogen on Helium, Ibid., pp. 986-987 (1977).

121. M. E. Rudd, J. S. Risley, J. Fryar, and R. G. Rolfes. Angular and Energy Distribution of Electrons from 15-to 150-keV H^o + He Collisions. *Phys. Rev. A* 21:506-514 (1980).

122. L. H. Toburen and W. E. Wilson. Ionization of Noble Gases by Equal Velocity He^+, He^{++}, and H^+ Ions. In *X International Conference on the Physics of Electronic and Atomic Collisions,* Abstracts of Papers, pp. 1006-1007, Commissariat A L'Energie Atomique, Paris (1977).

123. L. H. Toburen and W. E. Wilson. Differential Cross Sections for Ionization of Argon by .3-2 MeV He^{2+} and He^+ Ions. *Phys. Rev. A* 19:2214-2224 (1979).

124. M. A. Bolorizadeh and M. E. Rudd. Angular and Energy Dependence of Cross Sections for Ejection of Electrons from Water Vapor. III. 20-150 keV Neutral Hydrogen Impact. *Phys. Rev. A* 33:893-896 (1986).

125. J. Urakawa, N. Tokoro, and N. Oda. Differential Cross Sections for Ejection of Electrons and Dissociative Ionization Cross Sections for 5-20 keV H_2^+ and H_3^+ Impacts on Helium. *J. Phys. B* 14:L431-L435 (1981).

126. M. Sataka, J. Urakawa, and N. Oda. Measurements of Double Differential Cross Sections for Electrons Ejected from 5-20 keV H^+, H_2^+, and He^+ Impacts on Argon. *J. Phys. B* 12:L729-L734 (1979).

127. L. H. Toburen. Secondary Electron Emission in Collisions of 1.2 MeV C^+ Ions with He, Ne, Ar, and CH_4. In *XI International Conference on the Physics of Electronic and Atomic Collisions,* Abstracts of Papers, pp. 630-631. The Society for Atomic Collisions Research, Japan (1979).

128. L. H. Toburen. Differential Cross Sections for Electron Emission in Heavy Ion-Atom Collisions. In *Proceedings of the Fifth Conference on the Use of Small Accelerators,* pp. 1056-1061. IEEE Transactions on Nuclear Science NS-26 (1979).

129. N. Stolterfoht and D. Schneider. Double Differential Cross Sections for Electron Emission from Argon by 50- 500-keV N^+ and O^+ Impact. In *Proceedings of the Fifth Conference on the Use of Small Accelerators,* pp. 1130-1135. IEEE Transactions on Nuclear Science NS-26 (1979).

130. Yu. S. Gordeev, P. H. Woerlee, H. de Waard, and F. W. Saris. Continuous Electron Spectra Produced in Kr^{n+} - Kr Collisions. In *XI International Conference on the Physics of Electronic and Atomic Collisions,* Abstracts of Papers, pp. 746-747. The Society for Atomic Collision Research, Japan (1979).

131. Yu. S. Gordeev, P. H. Woerlee, H. de Waard, and F. W. Saris. The Production of Continuous Electron Spectra in Collisions of Heavy Ions and Atoms. A. Molecular Autoionization. *J. Phys. B* 14:513-526 (1981).

132. D. Schneider, D. DeWitt, A. S. Schlachter, R. E. Olson, W. G. Graham, J. R. Mowat, R. D. DuBois, D. H. Loyd, V. Nontemayor, and G. Schiwietz. Strong Continuum-Continuum Couplings in the Direct Ionization of Ar and He Atoms by 6-MeV/amu U^{38+} and Th^{38+} Projectiles. *Phys. Rev. A* 40:2971-2975 (1989).

133. N. Stolterfoht, D. Schneider, J. Tanis, H. Altevogt, A. Salin, P. D. Fainstein, R. Rivarola, J. P. Grandin, J. N. Scheurer, S. Andriamonje, D. Bertault, and J. F. Chemin. Evidence for Two-Center Effects in the Electron Emission from 25 MeV/amu Mo^{40+} + He Collisions: Theory and Experiment. *Europhysics Lett.* 4:899-904 (1987).

134. C. Kelbch, R. E. Olson, S. Schmidt, H. Schmidt-Böcking, and S. Hagmann. Unexpected Angular Distribution of the δ-Electron Emission in 1.4 MeV/amu U^{33+} - Rare Gas Collisions. *J. Phys. B* 22:2171-2178 (1989).

Discussion

Katz: We are talking here about the things that have been measured. We are talking about protons and helium ions incident on pure gases, particles whose initial energy is below 5 MeV per nucleon or thereabouts, and we need infinitely more information about the source function of the track structure. Whether we bombard with fast neutrons or cosmic ray projectiles on space vehicles, and so on. We have very, very little information, and we need a lot of theoretical help in even guessing what might be things we have to use further on down the line.

Christophorou: Do you expect these things to change with the energy of the particle much higher than this?

Katz: Oh, I think it changes a lot. It depends on how much clothing the particle carries; that is very important and we don't know how to handle it. Here we are only dealing with atomic physics in gases. We need to know, certainly even for neutron work, particles up to carbon, nitrogen, secondary particles.

Christophorou: Using the energy as a control, just take a particle and increase the energy just as high as it will go? Would we expect things to change?

Katz: Bare particles are not the problem. Clothed ones are a problem.

Toburen: I think that we can calculate cross sections for bare particles fairly accurately, probably even for uranium ions at GeV energies.

Varma: Let me bring the discussion back on track. I think Larry pointed this out. As long as people don't identify where these cross sections will be relevant to biology, we will be producing a lot of good data, but data not directly applicable to biology. But I think there are things that you can do to narrow down just what it is that we need the data on. I think we need to be very careful, and I agree with Larry that we don't have an unlimited amount of time to calculate all this data and produce all the cross-section data that we would like. You were talking about the double-differential cross sections, which vary by a factor of two or three between low angle and high angle, can this affect the biology?

Toburen: I think we are only now beginning to get to the point where that question might be answered; perhaps Marco may have a comment to make on that. For example, calculations by David Brenner lead to differences in the size of the relevant biological volume between high LET and low LET using track structure codes that are presently available. It may be that these differences are observed because the details of the physics in the codes are inadequate.

Zaider: This is totally dependent on the model you are using.

Toburen: The physics model or the physics itself?

Zaider: I think what Matt is talking about is some sensitivity study. What is the reflection of this about the certainty you are showing into a specific model. About five years ago there was a meeting at Argonne where we did such calculations. For some cross sections the result was not very sensitive; for others, it was sensitive. This is one point, and maybe I can make another point, relative to what Mitio said. From an experimental point of view, I know it makes a tremendous difference whether we ask for an absolute number or a relative number. In fact, for the Monte Carlo transport for most cross sections you need only relative numbers. But you need the total cross sections and there you have to have a good number. Your shapes are very important, when we talk about double-differential cross sections all we need are the shapes.

Toburen: I think the absolute numbers are probably good to within 20 percent; the shapes are probably much better.

Katz: But these are only the source functions for the Monte Carlo calculation. After that you need all kinds of electron interaction cross sections.

Zaider: If you have a total cross section which gives you the mean free path, then everything else could be relative.

Paretzke: No, be careful. Be careful, because as Larry pointed out there are different areas where different experimental methods are better. If he does only relative measurements with time-of-flight and relative measurements with the electrostatic analyzer, he never can put them together to make them relative from very low energy to very high energy, since most of the electrons are ejected with low energies. We can't scale the total energy to the total ionization cross section, so you are correct where the calculations are involved. But for the experimental data for our calculation, there we have to start from absolute values. I think 20 percent accuracy is fine for absolute double-differential cross sections.

Varma: That is just for calibration of the techniques though.

Paretzke: If you had one technique available from zero, with the same error from zero energy to the highest energy, it would be the same. Unfortunately that is not the way one gets experimental data for this range of energies.

Toburen: We have an experiment being planned with one of Herwig's students starting in January that will look at electrons ejected from DNA by proton impact. This experiment is a good example of the difficulty we have in obtaining absolute cross sections. We will use time-of-flight techniques in order to study emission of low-energy electrons; this will focus on the region in which we expect to see differences between gas and condensed phase. The time-of-flight technique provides relative cross sections and we don't have total ionization cross sections, or absolute double-differential cross sections to use for calibration. So we are going to have to come up with some other way of making these cross sections absolute. At the moment it is not clear what technique we will use or how accurate it will be. This is an example of how difficult it is to get absolute numbers in some cases.

Varma: Larry, the diagram that you showed, in which the track structure was super-imposed on the DNA, is not real since the DNA is not that simple cylinder that you have shown. How do we take that into account as we try to see what we can do in the future? What can your double-differential cross section tell you, in terms of the different orientations with respect to the DNA, about the kind of changes that may be introduced?

Toburen: You need additional information. You need such things as the oscillator strengths which represent the DNA-water complex, you need information on how these vary in order to see how the low-energy portion of the electron spectrum changes. I am convinced, although perhaps no one else is, that anytime you are looking at energy losses greater than 100 eV, it doesn't matter what the target is, it's an electron, but when you are dealing with energy losses that are smaller than that you have to worry about whether the target is a DNA, DNA-water mixture or whether there is also protein there. If the local media are protein, water, and DNA, then it is difficult to assess the energy deposition in each and certainly many questions arise regarding the molecular changes that may result. One needs to know where the interactions take place and how collective or how localized these interactions are. I think that many of these questions will require new experimental studies. In particular, I think much can be gained from Synchrotron radiation studies of oscillator strengths and molecular alterations measured in similar types of solutions.

Kasha: These pictures look beautiful, but they are aphysical in the sense that you don't have liquid water in the cell.

Weinstein: DNA doesn't look really like that; I think that's why he is saying that it is aphysical.

Varma: If you take a three-dimensional model of, say DNA, and superimpose it on the track structure, is a double-differential cross section going to make any difference? Those Are the types of questions that we need to ask.

Toburen: Looking at this picture of a track superimposed on a DNA model, you have no idea what actually has happened, nor do you know whether the actual structures have any relevance. But what I think is relevant is that it gives you some indication of the dimensions involved, say, how many excitations are within 10 angstroms or less and whether that relates to the size of the biomolecules such that it may have influence on the chemistry that follows.

Katz: The issue here is this. Given a beautiful picture like that, can you produce a cross section for a particular end point of interest? The answer is no. There are lots of reasons for measuring reaction cross sections, but that picture won't help much. There are other pictures that might.

Varma: If you have a three-dimensional picture of the DNA, and put a three-dimensional track on it, which you can calculate, then you can talk more specifically in terms of what kind of end points are produced in terms of a point mutation or something to understand it.

Katz: There is still something fundamental that you don't know. Are the changes you are looking for produced by direct or indirect effect?

Zaider: This picture actually serves what I was planning to say. I think the very notion that you can have a track which you can carry in your suitcase and impose on something else is so totally meaningless that I was surprised to see that nobody implied it. There is no such thing as a track. A track is the result of a particle interacting with a target; you cannot take a track and put it wherever you want. But what I think Herwig was getting to is that we must be very specific as to when this interaction takes place. Nothing can be learned from looking at this picture because you have a water track. Whether it is liquid or not is not important, imposed on some other drawing of DNA from a book. That is not something from which we can learn.

Paretzke: The transport term is very similar as regards DNA and water, because it is low Z material and the electron transport and the charge particle transport in this area just depend on the Z/A, the charge to atomic weight. So this is not the main problem. The main problem is the second step, the manifestation of changes that the dots infer.

Toburen: For low Z material, it doesn't really matter whether you are looking at benzene or some other hydrocarbon or nitrogen containing material insofar as the interactions leading to fast electrons are concerned. For these interactions the important parameter is the density of target electrons, and that doesn't change notably when you go from DNA to water to ice. But the production of low-energy electrons—those that are deposited locally and rapidly become subexcitation electrons—and what effect these have on the DNA, are going to depend very strongly on the electronic properties of the local media. And there we don't have an answer at this point as to the effects of molecular structure and chemical environment.

Kopelman: Let's be the devil's advocate and assume that the only thing that's important is really what ionizing radiation does to the water, H_2O molecules are all over the place. Everything after that is a secondary kind of what we call chemistry. Assuming things are that simple, does that help us? I'm not sure.

Curtis: I think it is a relative problem, relative to the ionization. How important are the subexcitation electrons? It's relative; I would suggest that the ionization for biological materials are going to be more than the slow electrons.

Toburen: I would like to suggest that there have been a number of areas where the use of track structure codes and the super position of tracks on model DNA structures have been useful. This type of analysis gives us some indication of the magnitudes of energy deposition and the volumes that are important. These analyses couldn't be done without the use of track and DNA models, even as primitive as they are. There has been good correlation observed between model calculations and measurement positions of strand breaks in the analysis of Iodine-125 incorporated into DNA. This uses only electron transport, but it makes use of the same crude DNA models and gas phase physics. Even with these crude models, we gain a good deal of insight we couldn't have gotten without them. So, in that respect they are useful. How useful such model calculations are depends on your innovation, how you use them. The accuracy that can be obtained depends on how you can modify the models to adapt to the environment that they are used in.

Christophorou: They are useful in one other way, in that they serve us in determining what the next step will be, and I think that in that regard it is a necessary step to be taken.

Toburen: I think we are only now beginning to be capable of answering these questions. With the data now available, or beginning to be available, you can actually make sensitivity tests by knowing how to change the initial parameters.

Radiation Interactions and Energy Transport in the Condensed Phase

R. H. Ritchie,[a][b] R. N. Hamm,[a] J. E. Turner,[a][b] H. A. Wright,[c] and W. E. Bolch[d]

Abstract

We review the state of knowledge about inelastic interactions of swift charged particles with condensed matter. Emphasis is placed on the properties of the dielectric response function and its representation for biologically interesting materials. Progress toward the goal of modeling electron transport and track structure in these materials is described.

Introduction

In response to the call for the precursor article on this topic at the Airlie symposium, a survey and assessment of some important physical elements of primary radiation processes in condensed matter was given.[1] In the present paper we deal quantitatively with many of the subjects covered there. We take advantage of knowledge garnered in the intervening years. Progress has been great in this area, with important stimulus from DOE-OHER. Experimental information about the response function of many biologically significant media such as liquid water, DNA, amino acids, etc., has been accumulated and refined. The dynamical response functions of relevant condensed media, suitably constrained by general requirements, have been modeled. Phenomena in the physical and prechemical stages of radiation interactions and energy transport are rich, complex and difficult to study with current experimental techniques. Much remains to be done toward a comprehensive understanding of these phenomena. Here we discuss the response function of condensed media, cross sections for transport processes, and theoretical and computational progress toward the ultimate goal of understanding the interaction of ionizing radiation in condensed media.

Progress in theoretical understanding of radiation transport in condensed matter has been based on concepts and methodology familiar from radiation physics. Current computationally heavy approaches to theoretical atomic, molecular and solid-state structure have had relatively little application in this field. The dynamical response of electrons in the condensed phase remains a difficult problem for *a priori* numerical evaluation. As is often the case in science, experiments guided by general theoretical

(a) Health and Safety Research Division, Oak Ridge National Laboratory, Oak Ridge, Tennessee
(b) Department of Physics, University of Tennessee, Knoxville, Tennessee
(c) Consultec Scientific, Inc., Knoxville, Tennessee
(d) Department of Nuclear Engineering, Texas A&M University, College Station, Texas

Physical and Chemical Mechanisms in Molecular Radiation Biology
Edited by W.A. Glass and M.N. Varma, Plenum Press, New York, 1991

considerations have offered the best route to understanding these intricate phenomena. However, theoretical evaluation of elastic and inelastic multiple scattering in simple crystalline solids is now feasible with modern high-speed computers. For example, surface structural determination of crystalline solids by low energy electron diffraction is routinely accomplished[2] using multiple scattering techniques, pseudopotentials to represent the effect of ion cores, and optical potentials, or self-energies, to represent inelastic interactions. Similar approaches to photoemission from solids are also current.[3] The application of such theoretical methods to the study of radiation interactions with biologically interesting crystalline solids would be of considerable interest for the future as computers become even faster.

At present, theoretical understanding of radiation interactions in the early physical stage (0-10^{-15}) and in the prechemical stage (10^{-15} to 10^{-12}) after irradiation comes mainly from general principles and experimental data (often incomplete and indirect), augmented somewhat by results from particular simulation codes. Several investigators have developed Monte Carlo computer codes to calculate in detail the transport of electrons and their secondaries in water.[4-9] Of these codes only our ORNL Electron Transport Code (OREC) employs cross sections that have been inferred from the optical response function of liquid water.

We are the first to have synthesized differential cross sections [or differential inverse mean free paths (DIMFPs)] from experiments on liquid water. Our approach to this synthesis has been taken over to the field of electron energy loss spectroscopy, is now state of the art in that area, and is used to infer transport parameters for electrons in many different solids from optical data on these materials (see "The Long Wavelength Dielectric Function," below). Comparisons of the results of this theoretical approach with experimental data in this area are generally quite good; thus, we believe that our synthesis for liquid water should yield reasonable results.

Charged-particle transport in condensed matter may stimulate collective motion by large numbers of electrons and ions. For this reason we espouse the use of DIMFPs rather than differential cross sections to characterize interactions in such media; the cross section concept usually refers to a reaction area per scatterer and does not seem appropriate to a many-particle system supporting collective eigenmodes. Such excitations may be coherent and unlocalized over distances larger than regions sensitive to radiation damage. Details of the localization of these unlocalized excitations through single-particle excitations may be important in determining the injury experienced by irradiated, biologically important matter. This is discussed below.

Good progress has been made since the Airlie symposium toward integrating the exciting advances of molecular biology into the field of radiation biology. But much remains to be done in understanding the complex physical, prechemical and chemical stages of radiation interactions in even the simplest, biologically relevant, condensed materials. The task of relating cross sections for isolated atoms and molecules to those that are obtained when those same atoms or molecules form condensed matter is a formidable one that can be done at present only in the crudest approximation. Hence we rely here on experimental data relating to the condensed media of interest.

For water vapor, cross sections for a number of discrete and continuous electronic transitions have been measured by many workers for incident electrons, photons and ions over a wide range of energies. For liquid water, the only experimental data are (i) reflection measurements for photons from the infrared up to 26 eV, (ii) stopping power data for fast ions, and (iii) yields of hydrated electrons for high-energy radiation. From the reflection measurements, $\epsilon(\omega)$, the long-wavelength dielectric function may be inferred, and from $\epsilon(\omega)$ the energy loss function for collisions of fast particles may be derived in the first Born approximation. In contrast to the vapor phase, no direct data exist for liquid water relative to dissociative electron attachment collisions, fragmentation and de-excitation patterns of excited molecules, and cross sections for forbidden transitions. Thus the validity of approximations in track structure calculations for liquid water cannot be tested directly against experimental data. However, reported yields of hydrated electrons at early times, together with other radiochemical information, allow for consistency checks. Available data on stopping-power modification due to phase differences are still too vague and contradictory to be of much use in obtaining DIMFPs for electrons and ions in the liquid.

We have constructed a Monte Carlo code, OREC, that embodies our syntheses of the DIMFPs for energy loss by charged particles to liquid water. We have employed experimental data to infer $\epsilon(\omega)$ up to $\hbar\omega = 26$ eV and have interpolated for larger values of ω, using the fact that for large enough ω the response of the liquid and the vapor at the same density must be identical.

As we discuss below, recent work shows that our approach to obtaining DIMFPs, incorporated into a Monte Carlo code, supplemented by models of the formation of chemically active species and the subsequent diffusion-controlled chemical reactions that take place along the path of a charged particle, gives good agreement with measurements from electron pulse radiolysis.

The paper by Werner Brandt and one of us at the Airlie symposium[1] gave an overview of the early physical stages in energy deposition by ionizing radiation in condensed matter. The scales of time and length associated with the products of these various primary energies were sketched. The terms "infratrack" and "ultratrack" (sometimes designated as the track core and the penumbra)[10] were associated with the spatial structures generated by coherent excitations and by delta-rays, respectively, in the passage of a swift ion through condensed matter. The propagation and excitation of such collective modes were discussed qualitatively. Some speculations were made about wake and nonlinear effects in such systems.

Here we discuss progress in understanding collective modes in condensed matter, especially liquids, in representing their dynamic response functions, and in characterizing the localization of initially unlocalized excitations in condensed matter. Nonlinear interactions, intrinsic and extrinsic excitations by electrons ejected from bound states in liquids, and possible future directions in the theory of condensed matter response are also discussed.

Charged-Particle Energy Losses in Matter

The dielectric function codifies in an important way the dynamical properties of matter. It has been determined for a large number of materials over a fairly wide range of infrared, optical and ultraviolet frequencies.[11,12] This complex function, $\epsilon(\omega) = \epsilon_1(\omega) + i\epsilon_2(\omega)$, characterizes the response of a medium to very long wavelength disturbance. Its variation with frequency, ω, is determined by the spectrum and strength of electronic and nuclear transitions.

Fermi[13] realized early that the response of matter to the passage of a charged particle could for many purposes be couched in terms of its response to a spectrum of equivalent electromagnetic radiation. He later analyzed[14] the Cherenkov radiation phenomenon by using $\epsilon(\omega)$ for a general medium in the context of Maxwell's equations. Subsequently the dielectric function was generalized to become a matrix operator depending on space or, equivalently, to be a function of wave vector as well as of frequency. This development and its application to charged-particle interactions was made by Soviet workers.[15] Following this, the random phase approximation (RPA) dielectric function of an isotropic, nonrelativistic electron gas was derived by

Lindhard,[16] building on the work of Klein, and independently by Hubbard.[17] This enabled a unified description of single-particle and collective aspects of an important model system that has wide utility in radiation physics and solid-state physics. We return to this point below.

Dielectric Theory of Energy Loss

Consider a classical point charge traveling at constant, nonrelativistic velocity, $\vec{v}$, through a uniform medium represented by the dielectric function $\epsilon(\omega,q)$, where q is the wavenumber and ω the frequency variable. We denote $\epsilon(\omega,o)$ by $\epsilon(\omega)$. To determine the effect of this motion, we solve Poisson's equation for $\phi(\vec{r},t)$, the resulting scalar electric potential. This may be written

$$\epsilon(\omega,q)\nabla^2\phi = -4\pi\ Ze\ \delta^3\ (\vec{r} - \vec{v}t) \tag{1}$$

where $Ze\ \delta^3(\vec{r} - \vec{v}t)$ is taken to be the electric charge density constituted by the moving charge. Expressing all quantities as Fourier integrals, we find

$$\phi_{\omega,\vec{q}} = 8\pi^2\ Ze\ \delta\ (\omega - \vec{q} \cdot \vec{v})/k^2\epsilon(\omega,q) \tag{2}$$

If $\epsilon(\omega,q) = \epsilon_o$, a constant, then

$$\phi(\vec{r},t) = \frac{1}{\epsilon_o}\ \frac{Ze}{|\vec{r} - \vec{v}t|} \tag{3}$$

by reverting to space-time representation. This, of course, is just the Coulomb potential of the charge screened by the dielectric constant of the medium. This screening is a striking manifestation of the different ways that a medium responds to a charged particle and to a photon. The photon causes a transverse disturbance that does not accumulate charge. It is thus unscreened, but not unmodified, in its travel.

When the full variation of ϵ is included, screening is still present, but may be strongly modified from that predicted by Eq. (3). To find the energy loss by the charge to the medium, one needs only to evaluate the retarding force on the charge due to the polarization it induces; one finds

$$-\frac{dE}{dx} = \frac{2}{\pi}\ \frac{Z^2e^2}{v^2}\ \int_{o}^{E/\hbar} \omega d\omega \int_{q_-}^{q_+} \frac{dq}{q}\ Im\left(\frac{-1}{\epsilon(\omega,q)}\right) \tag{4}$$

where in the present classical picture, $q_- = \omega/v$ and $q_+ = \infty$. This gives the energy loss per unit length of travel (or the stopping power of, or LET to, the medium) and, in principle, includes loss to all optically active excitations that the medium can support. Although not obvious from the derivation indicated above, Eq. (4) is applicable in the context of first Born approximation quantal theory if one uses

$$q_{\pm} = \sqrt{\frac{2m}{\hbar^2}} \left[\sqrt{E} \pm \sqrt{E - \hbar\omega} \right]$$

where m is the mass of the charged particle. However, it does not account for the possibility of exchange scattering of the incident particle with the medium if the particle is an electron. This will be discussed below. At relativistic energies, one may take

$$\hbar q_{\pm} = [2mE(1 + E/2mc^2)]^{1/2}$$

$$\pm [2mE(1 + E/2mc^2) - 2m\hbar\omega(1 + E/mc^2) + (\hbar\omega/c)^2]^{1/2} \tag{5}$$

although $\epsilon(\omega, q)$ also must be corrected at relativistic speeds (see Eqs. 24-25). The differential inverse mean free path (DIMFP) for energy loss $\hbar\omega$ and momentum transfer, $\hbar\vec{q}$, is found from the integrand as

$$\frac{d^2\Lambda^{-1}}{d\omega dq} = \frac{2Z^2e^2}{\pi\hbar v^2} \frac{1}{q} Im \left[\frac{-1}{\epsilon(\omega, q)} \right] \tag{6}$$

while the DIMFP differential in ω is found from Eq. (6) by integrating between the limits q_- and q_+. The quantity $Im(-1/\epsilon)$ is often referred to as the energy loss function.

For comparison, the attenuation coefficient (absorption probability per unit length) of a photon with energy $\hbar\omega$ traveling in the medium is given by

$$\mu(\omega) = \frac{2\omega}{c} Im \left\{ \sqrt{\epsilon(\omega)} \right\}$$

showing no screening effect in that the maxima $\mu(\omega)$ are at frequencies roughly equal to the maxima of $\epsilon_2(\omega)$.

The Long Wavelength Dielectric Function. Classical electromagnetic theory represents each electron in a given medium as a separate classical, harmonically bound charge. This procedure yields the well-known Drude-Sellmeier Formula[18]

$$\epsilon(\omega) = 1 + \sum_n \frac{f_n \omega_p^2}{\omega_n^2 - \omega^2 - i\gamma_n\omega} \tag{7}$$

where f_n is the fractional number of electrons having resonant frequency ω_n and damping constant γ_n, while

$$\omega_p = (4\pi n_e e^2/m)^{1/2}$$

is the "plasma frequency" if n_e is the average density of electrons in the medium. As is well understood, this formula is valid quantally if f_n is interpreted as the oscillator strength for a transition from the ground state to the nth excited state of the system, and if the sum is augmented by an integral over the continuously varying frequency ω' corresponding to transitions to the ionization continuum of the system, where $\omega_n \to \omega'$ and $f_n \to df/dw'$.

It has been realized for many years that there is an upward shift in the resonant frequencies of the atoms in a gas as the density of the gas is increased.[19] Williams et al.[20] have shown clearly how Coulomb interactions among electrons give rise to an upward shift in the maxima of the energy loss function. If the isolated molecule has a resonance at the frequency ω_n, the shifted frequency is given by

$$\omega_{res} = \sqrt{\omega_n^2 + f_n \omega_p^2} \qquad (8)$$

In their paper at the Airlie symposium, Brandt and Ritchie[1] found that this equation describes well the observed prominent peak in the loss function of liquid water, assuming that all eight valence electrons participate in the oscillation and that ω_n in Eq. (8) is the ionization energy of the H_2O molecule divided by $\hbar$. Williams et al.[20] point out that $\epsilon_2(\omega)$ shows resonances corresponding to single-particle transitions, while resonances in $Im[-1/\epsilon(\omega)]$ may correspond to collective modifications of these transitions. When transitions to the continuum are considered, they demonstrate that a condition for a collective resonance to occur is

$$\omega_p^2 \left| \frac{df}{d(\omega^2)} \right| \gg 1 \qquad (9)$$

which agrees with a criterion put forward by Fano.[21] However, they argue that this criterion is too stringent in that collective effects may be present in liquids and solids in various degrees even when it is not satisfied. Of course, the models used in both references are schematic and do not account for chemical effects as the isolated molecules are brought together to form the condensed phase, nor do they account for the changes in the eigenstates of the molecules due to the proximity of neighbors. Ehrenreich and Philipp,[22] on the other hand, enunciate criteria for collective effects that are based on measured or computed properties of $\epsilon(\omega)$. They point out that a maximum of $Im[-1/\epsilon(\omega)]$ corresponds to a collective resonance if $\epsilon_1(\omega)$ and $\epsilon_2(\omega)$ in the neighborhood of the peak are (i) small compared with unity; (ii) vary approximately linearly, with $d\epsilon_1/d\omega > 0$ and $d\epsilon_2/d\omega < 0$; and (iii) $(\omega/\epsilon_1)(d\epsilon_1/d\omega)$ and $(\omega/\epsilon_2)(d\epsilon_2/d\omega)$ are both $\gg 1$. Obviously it is not necessary that $\epsilon(\omega) = 0$ for a collective plasmon-type resonance. This is apparently not recognized by all workers.[23]

Examples of experimental data on $\epsilon(\omega)$ for a number of different materials are shown in Figs. 1-5. Figure 1 shows the prototypical example of collective (plasmon) effects in a real solid. The conduction band of aluminum metal is known to support plasma oscillations at a quantal energy of 15.4 eV. This is shown by the prominent peak in $Im[-1/\epsilon(\omega)]$ in Fig. 1 and is described in good approximation by the RPA dielectric function of the electron gas at a density equal to that of the conduction band in aluminum. Figure 2 shows ϵ_1, ϵ_2, and $Im(-1/\epsilon)$ for the insulator diamond.[24] The solid lines show data inferred from energy-loss measurements, while the dashed lines are from optical data. The evidence for a collective resonance is very strong. The Ehrenreich-Philipp (E-P) criteria are well-satisfied. It is worthwhile to note that there is a large energy gap (8 eV) in diamond and no electrons in the conduction band, so that real transitions are expected to play only a minor role in this collective mode, *viz.*, in giving rise to damping of the resonant state.

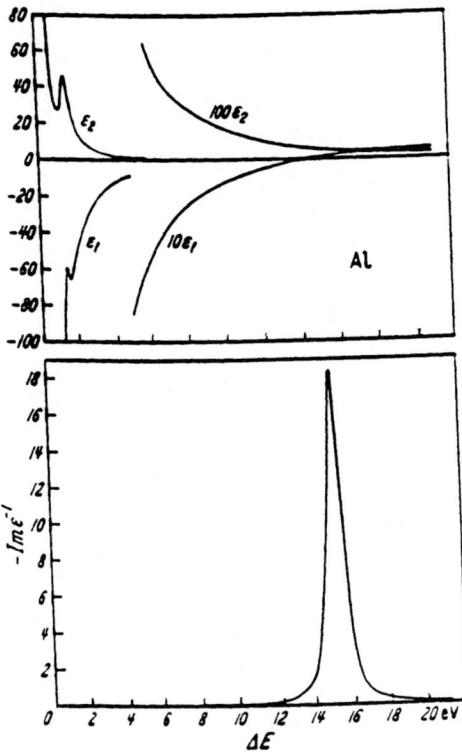

Figure 1. The spectral dependence of the real and imaginary parts of the dielectric function and the energy-loss function for aluminum metal calculated from optical data [from H. Raether, Solid State Excitations by Electrons, *Springer Tracts in Modern Physics* 38:85-157, Springer-Verlag, New York (1965)].

In an excellent recent review, Williams et al.[12] have presented data on $\epsilon(\omega)$ and $Im[-1/\epsilon(\omega)]$ for many different biologically interesting substances, comparing the properties of materials in the condensed phase with those of the corresponding vapors. They conclude that absorption spectra in the condensed phase show the same molecular excitations that are seen in that of the vapors. For the liquid the structures may be shifted slightly in energy compared with that for the vapor and always show a broadening due to damping, as expected. Further, they conclude that collective electron oscillations occur in molecular liquids. They have also reviewed the properties of biological substances of direct interest. Figure 3 shows the variation in ϵ_1, ϵ_2 and $Im(-1/\epsilon)$ for dry films of DNA (sodium salt of calf thymus DNA). One sees the characteristic upward energy shift from the largest maxima in the $\epsilon_2(\omega)$ curve to the broad maximum in $Im(-1/\epsilon)$, indicative of collective effects. The E-P criteria are approximately satisfied here. The optical properties of solid films of the nucleic acid base adenine have been studied over a wide range of energies by a number of workers.[25] These results on $\epsilon(\omega)$ and $Im(-1/\epsilon)$, together with values inferred from energy loss measurements, are plotted in Fig. 4. Discussion of the single-electron transitions contributing to the fine structure at energies < 10 eV are given in Ref. 25. Evidence for collective electronic motion in the peak in $Im(-1/\epsilon)$ at 22 eV is good; the E-P criteria are fairly well satisfied.

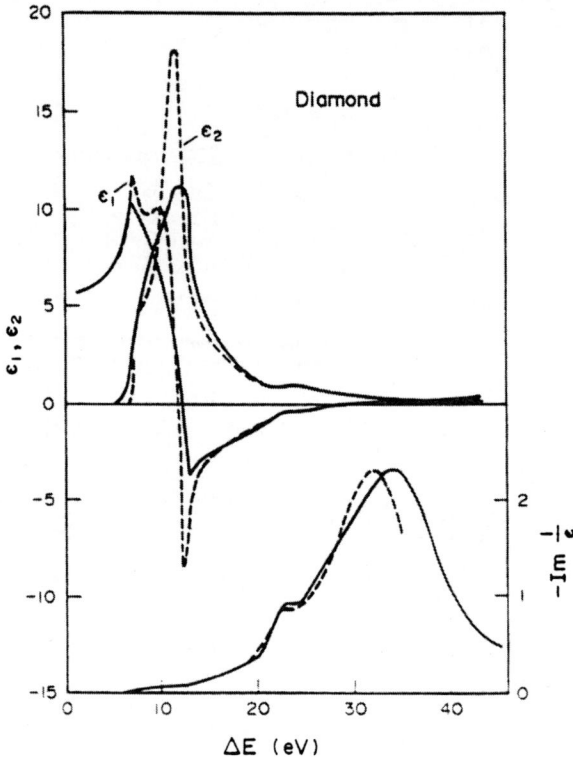

Figure 2. The dielectric functions and the energy-loss function for diamond. The solid lines show data inferred from energy-loss measurements while the dashed lines are optical data [from H. Raether (same ref. as above)].

In Fig. 5 are displayed the dielectric functions ϵ_1, ϵ_2 and *Im* $(-1/\epsilon)$ for liquid water as inferred from the measurements of Birkhoff and coworkers.[26] The energy range covered was from a few eV up to 26 eV. These are the only extant experimental data relevant to energy losses by charged particles in this important substance.

Some time ago we began a detailed study of the transport of ionizing radiation in water employing the ORNL data to construct inelastic DIMFPs. These data on $\epsilon(\omega)$ were interpolated into the q,ω plane using very general sum rule constraints and requirements on the behavior of $\epsilon(\omega,q)$ for large q and ω.[7,8] Our approach has been adopted by theorists in Electron Energy Loss Spectroscopy (EELS) to represent electron transport parameters in many different solids from detailed measurements of $\epsilon(\omega)$ in those materials.[27-31] There is generally quite good agreement between the results of this approach and experimental measurements of these parameters in a number of different media.

The Dielectric Function of Condensed Matter

A formal expression[32] may be found for the linear response function, at arbitrary frequency and wave number, of an isotropic, homogeneous, many-electron system. This may be done by equating the rate of excitation of the system under a given perturbation to that found by assuming that the system response is proportional to the inverse dielectric function $1/\epsilon(\omega,q)$. One finds

$$1/\epsilon(\omega,q) = 1 + \frac{4\pi e^2}{\hbar q^2 \Omega} \sum_n |<n|\rho_{\vec{q}}|o>|^2 \left[\frac{1}{\omega_{on}+\omega+i\eta} + \frac{1}{\omega_{on}-\omega-i\eta} \right] \quad (10)$$

where $(\hbar\omega_n, |n>)$ is the (eigenenergy, eigenvector) of the nth excited state of the perturbed system,

$$\rho_{\vec{q}} = \sum_{i=1}^{N} e^{i\vec{q}\cdot\vec{r}_i}$$

is the density operator, $\vec{r}_i$ is the position vector of the ith electron, Ω is the normalization volume of the system and η is a positive infinitesimal. Also $\omega_{on} = \omega_o - \omega_n$, and $n = o$ corresponds to the ground state of the system. Sum rules have been used to constrain approximate representations of the response function of matter in

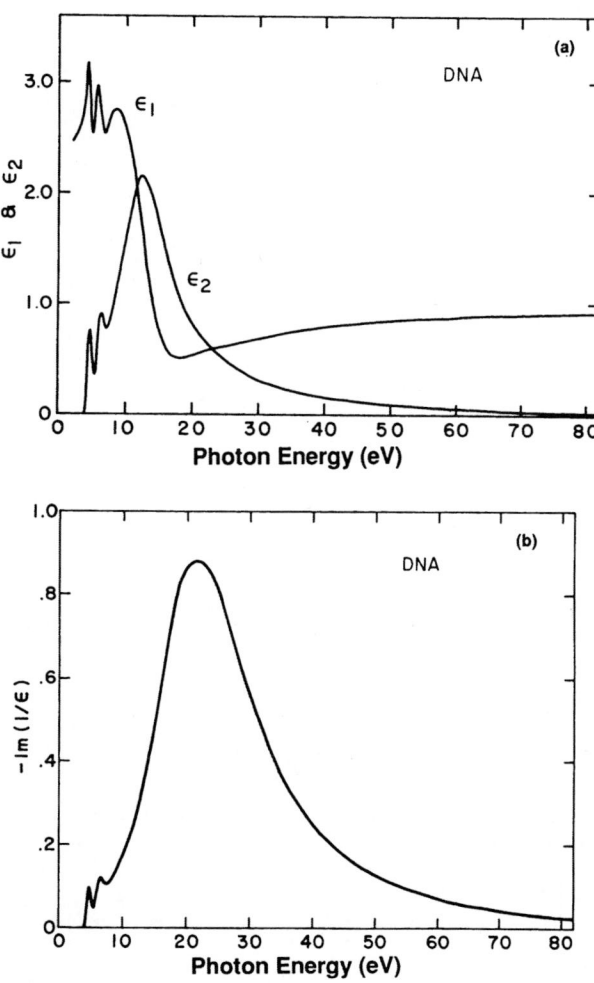

Figure 3. a) The real and imaginary parts of the dielectric function of a dry film of DNA as functions of photon energy. b) The energy loss function of a dry film of DNA versus photon energy.[12]

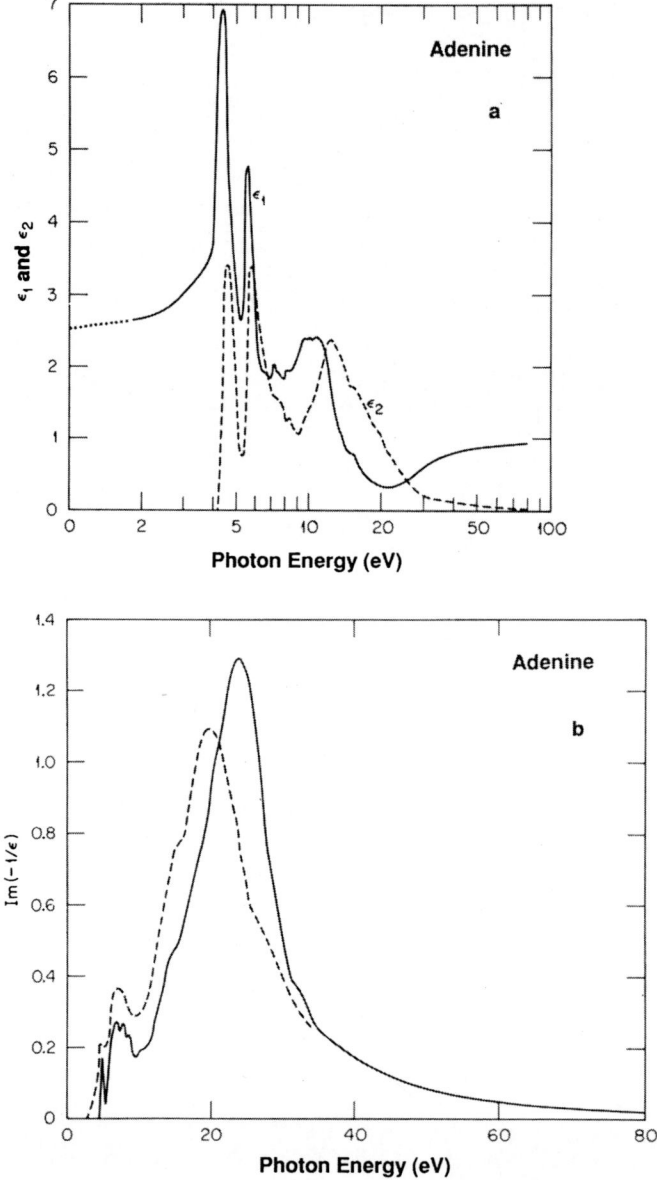

Figure 4. The spectral dependence of the real and imaginary parts of the dielectric function and the energy-loss function for adenine.[25] The solid line in the energy loss function curve is taken from optical data obtained by the authors of Ref. 25, while the dashed line shows EELS data of Isaacson [Characteristic Energy Loss of Electrons in Biological Material, *Proc. V. Int. Conf. Vac. UV Rad. Phys*, Hamburg, July 22-26, 1974, eds. E. E. Koch, R. Haensel, and C. Kunz, Pergamon Press: 826-829 (1974)].

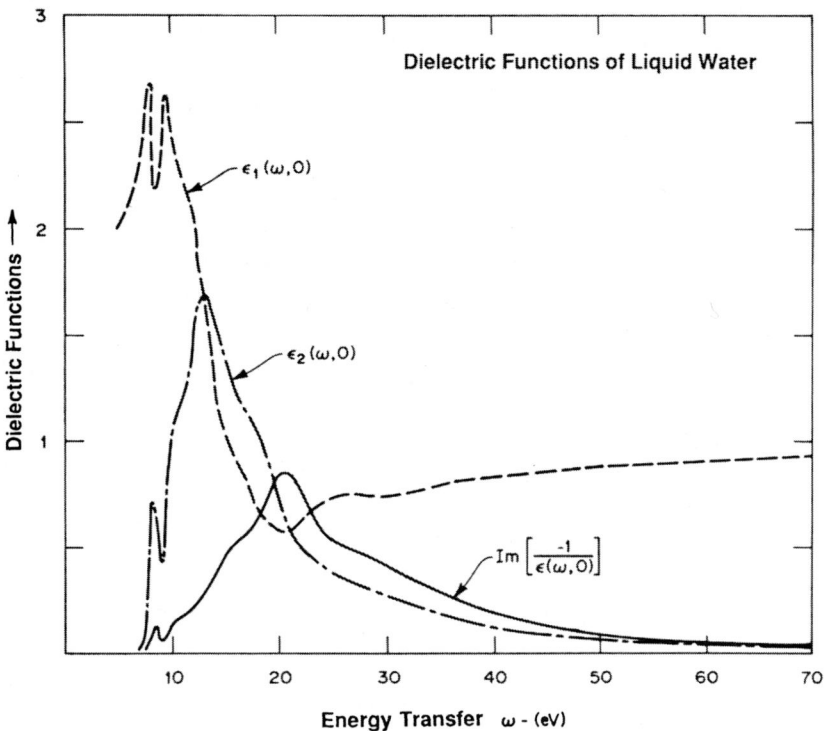

Figure 5. Dielectric functions and energy-loss function for liquid water.[26]

many applications. For example, Bethe,[33] in his pioneering theory of charged-particle energy loss, was the first to prove and use a sum rule generalized from a dipole rule employed in early optical response theory. Placzek[34] later generalized the Bethe approach, while Fano and Turner[35] made specific application of these ideas to atomic systems. Sum rule results may be found for general condensed-matter systems using similar methods as those employed for atomic systems.[36,8] Perhaps the most useful of these are the pair

$$\frac{m}{2\pi^2 nZe^2} \int_0^\infty \omega \, Im[-1/\epsilon(\omega,q)]d\omega = 1 \tag{11}$$

$$\frac{m}{2\pi^2 nZe^2} \int_0^\infty \omega \, Im[\epsilon(\omega,q)]d\omega = 1 \tag{12}$$

where m and e are the electron mass and charge, n is the density of atoms in the system, and Z is the number of electrons per atom. Higher-order sum rules defined as[8]

$$\sum_\ell(q) = \frac{m}{2\pi^2 nZe^2} \int_0^\infty \omega^{\ell+1} \, Im[-1/\epsilon(\omega,q)]d\omega = 1 \tag{13}$$

may be of value. In particular,

$$\sum_o (q) = 1$$

as in Eq. (11) and

$$\sum_1 (q) = \frac{\hbar q^2}{2m} + 4T/3\hbar + 0_2 \tag{14}$$

where T is the mean kinetic energy per electron in the ground state of the system and 0_2 represents terms originating in two-electron correlations, which go rapidly to zero for large q. The expression for Σ_2 is quite long, and Σ_3 diverges.

If one wishes to go beyond the simple isotropic, homogeneous model of a solid, it is a special task to represent the dielectric response function of each material. Considerable work has gone into approximating such functions for a few ideal crystals.[37] A general expression for the inverse dielectric function of a crystal that is characterized by the set of reciprocal lattice vectors $\{\vec{G}\}$ may be written.[38]

$$Im\left[\epsilon^{-1}(\omega,\vec{q},\vec{G})\right] = \delta_{\vec{G}}^o + \frac{4\pi e^2}{\hbar q^2 \Omega} \sum_n <o|\rho_{\vec{q}+\vec{G}}|o>\left[\frac{1}{\omega_{on}+\omega+i\eta} + \frac{1}{\omega_{on}-\omega-i\eta}\right] \tag{15}$$

where δ is the Kronecker delta symbol. The inverse dielectric function $\epsilon^{-1}(\omega,\vec{q},\vec{G})$ is not, in general, equal to its reciprocal, unlike that for an isotropic, homogeneous system. The sum rule corresponding to that given in Eq. (11) is [38,39]

$$\sum_o (\vec{q},\vec{G}) = \frac{\vec{q}\cdot(\vec{G}+\vec{q})}{q^2} <o|\rho_{\vec{G}}|o> \tag{16}$$

neglecting two-electron correlations. Higher-order sum rules are quite lengthy and will not be considered here. No theory exists for the dielectric response function of liquids or amorphous solids.

As discussed in Ref. 37, the dielectric functions of metals and semiconductors have been modeled in various approximations. Extensive use has been made of the dielectric function of the electron gas, which is the prototypical isotropic, homogeneous model of a metal.[16,17] As important as this function has been to solid-state physics in the technology of band structure and pseudopotential theory, it is still more useful in leading to qualitative ideas about general aspects of dielectric response. For example, the electron gas exhibits a collective resonance, the well-known plasmon, that exists for momenta small compared with $\hbar k_F$, the maximum momentum of electrons in the system, but is Landau damped for $q > k_F$. For momenta comparable with, or larger than, $\hbar k_F$ the plasma is then able to decay by single-particle excitation from the Fermi sea, and is thus strongly damped. Figure 6 shows a plot of $Im[1/\epsilon(\omega,q)]$ for an electron gas with parameters indicated in the caption. The Mermin form[40] of the RPA dielectric function, which allows for damping of excitations, has been used in this calculation. The prominent peak at small q is seen to spread out, to become smaller, and to move to larger values of energy transfer, $\hbar\omega$, with increasing q. The peak at small q corresponds to the existence of plasmons in the medium, while at large q the so-called Bethe ridge is centered about the line $\omega = \hbar q^2/2m$, corresponding to the fact that for large energy transfers the system responds as if it were composed of free electrons initially at rest. Figure 7 shows a logarithmic plot of the same quantity in

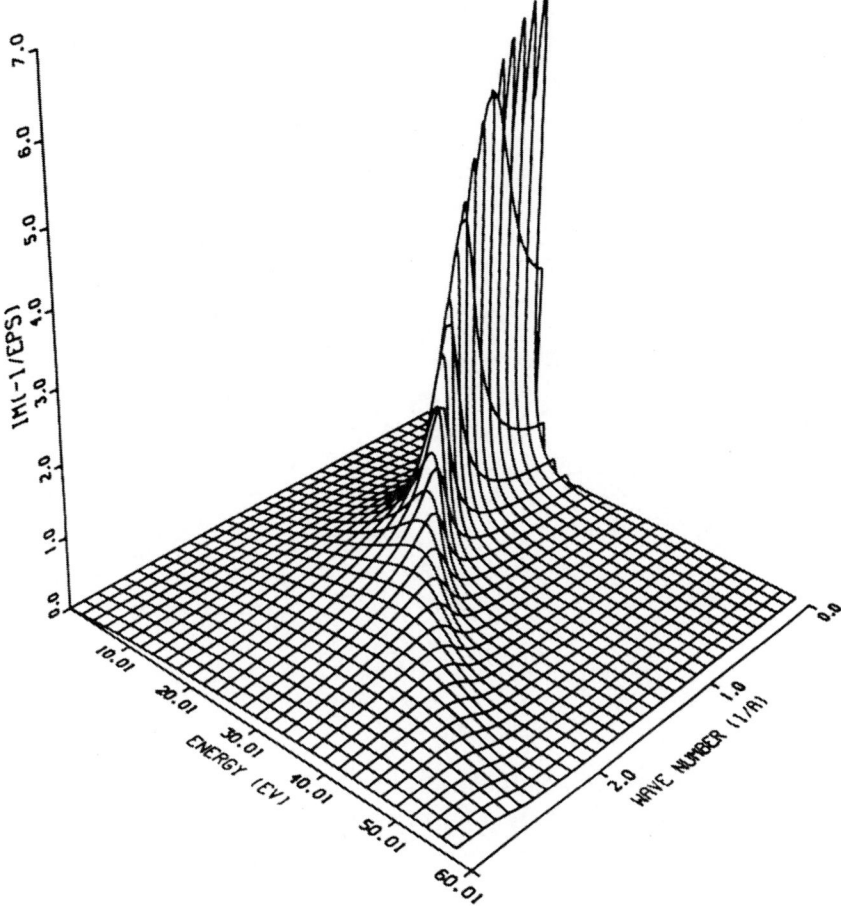

Figure 6. A contour plot of the energy-loss function of a damped electron gas as it depends on the wavenumber k in Å^{-1}, and energy transfer E in eV. The damping constant was taken to be 3 eV and the plasma energy 15.4 eV.

order to display clearly the parabolic region of single-particle excitations as q increases. This region is coincident with the Bethe ridge for large q.

The Bethe sum rule $\Sigma_o = 1$, as it must, for the function $Im[-1/\epsilon(\omega,q)]$ displayed in Figs. 6 and 7. The sum rule $\Sigma_2 = \omega_p$, the plasma frequency of the electron gas, as $q \to o$. In the limit $q \to \infty$.

$$\sum_2 (q) \to \frac{\hbar q^2}{2m} + \frac{4}{5\hbar} E_F = \frac{\hbar q^2}{2m} + \frac{4}{3\hbar} <T> \qquad (17)$$

where E_F is the Fermi energy, i.e., the maximum energy found in the system. In the neighborhood of the Bethe ridge, when q is large, the full width in energy of the Bethe ridge at half-height is $2(2E_q E_F)^{1/2}$, where $E_q = \hbar^2 q^2/2m$ is the energy at the center of the ridge. This corresponds to a Doppler broadening of the loss function corresponding to the ejection, with large momenta, of electrons from the whole Fermi sea. Such broadening is expected to be a general feature of $Im(-1/\epsilon)$ for any condensed material;

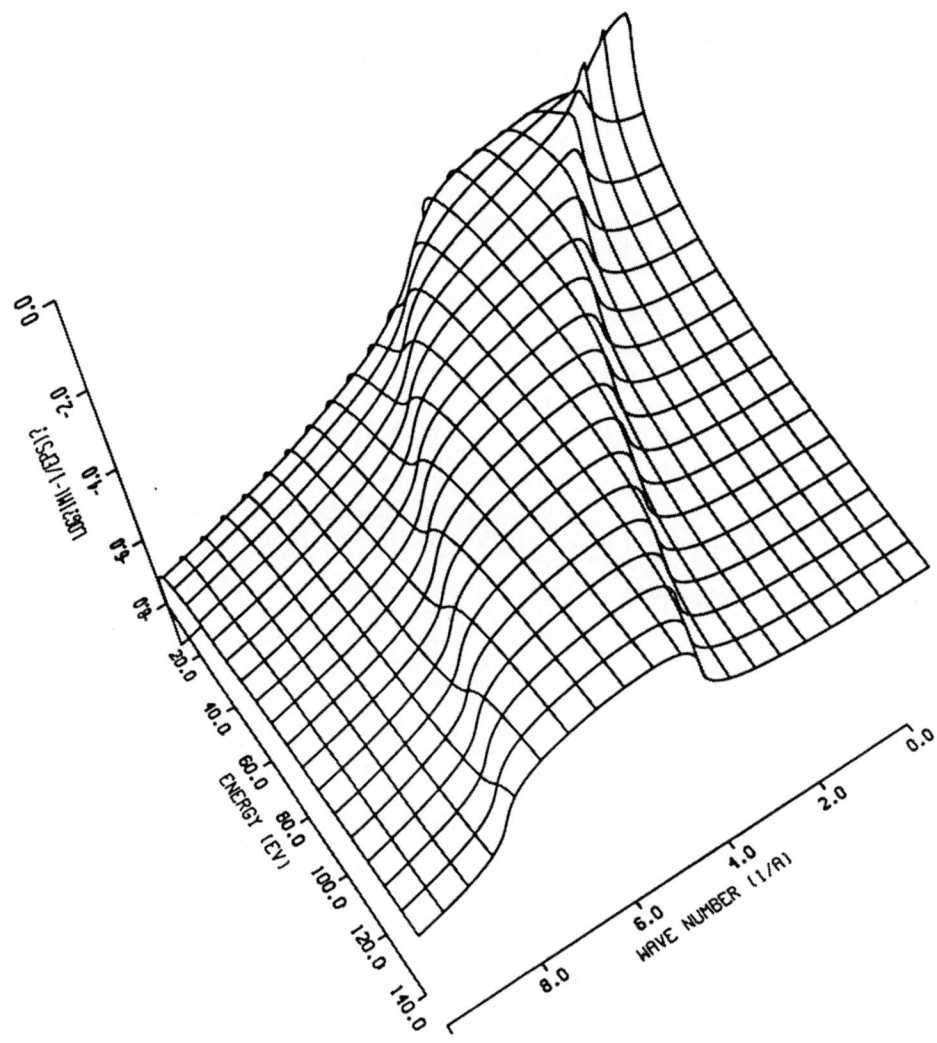

Figure 7. The same data as in Fig. 6 but with $Im[-1/\epsilon(\omega,k)]$ plotted on a logarithmic scale to emphasize the region of single particle-hole excitations.

i.e., the width of the Bethe ridge at large q for excitation of electrons from a given energy band should be proportional to the square root of the width in energy of that band in the ground state.[41] It can be incorporated into schemes for extrapolating measurements of $\epsilon(o,\omega)$ for a given medium into the q-ω plane.[8]

Electron collective resonance is displayed nicely in a simplified model of an insulator that was devised earlier by one of us (RHR).[42] The ground states of those electrons bound in an amorphous medium were represented in a tight-binding model, while excited states were represented as plane waves orthogonalized to the ground state eigenfunctions. This Orthogonalized Plane Wave (OPW) approximation has had extensive use in solid state physics.[43] The key approximation that enables one to evaluate conveniently $\epsilon(\omega,q)$ for an amorphous system is to assume that the atoms of the medium are distributed randomly in space. In this case one may express $Im\ \epsilon(\omega,q)$

analytically, and *Re* $\epsilon(\omega,q)$ may be found from the Kramers-Kronig relation. We require[42] also that *Im* $\epsilon(\omega,q)$ shall satisfy the Bethe sum rule.

Figure 8 depicts the energy loss function $Im[-1/\epsilon(\omega,q)]$ using this OPW dielectric function for a model insulator. The parameters used for this calculation are given in the figure caption. One sees now that there is a strong, narrow peak of an energy of ~1 atomic units (a.u.) or ~27.2 eV for small q, while the binding energy of the valence electrons is taken to be 0.331 a.u. or 9 eV. The approximation of Eq. (8) gives 25.3 eV for the energy of the collective resonance using the parameters assumed to describe the density of bound, valence electrons in this model. This is quite a reasonable agreement. As q increases, as in the RPA model for the electron gas, the resonance broadens and moves upward in energy. For the larger q values, the width of the resonance increases approximately as does that for the electron gas. Thus we have a clear example, from theory, of plasmonic response for model amorphous solid.

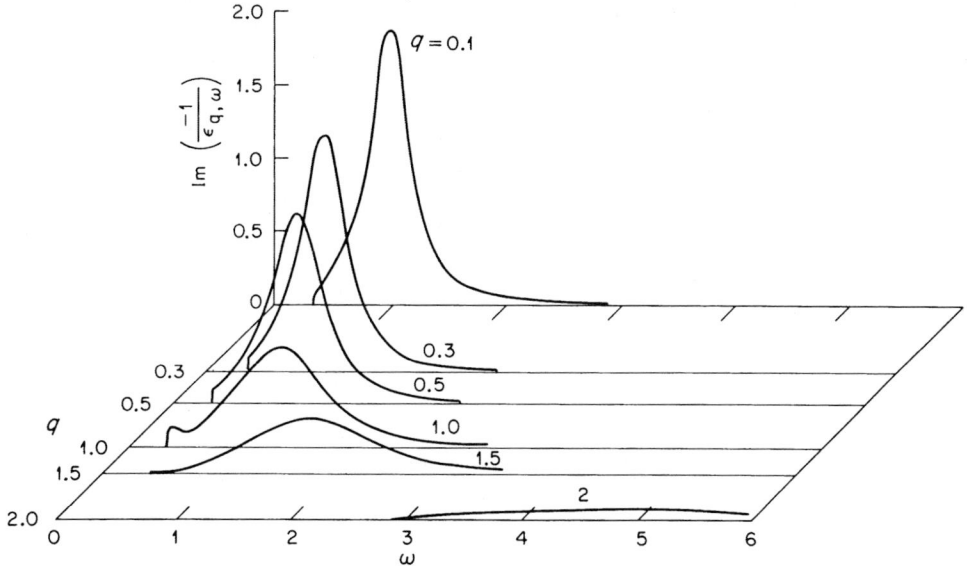

Figure 8. The energy-loss function for a tight-binding solid with a band gap of 9 eV, an atomic density of 4.05 g/cm³, and an orbital radius of 0.78 a.u.[42]

The Dielectric Function of Liquid Water

Some time ago we were faced with the task of representing the interaction of swift electrons of arbitrary energy with liquid water.[7,8] As indicated above, direct experimental data on the energy loss of charged particles in this substance are inconclusive and contradictory. In view of the extensive use that has been made of the dielectric function concept for metals and semiconductors in condensed-matter physics, it seemed quite reasonable to us to try to synthesize such a response function for liquid water, and for other biologically relevant materials. This work continues as additional experimental data become available.

Since we know of no rigorous theory that describes $\epsilon(\omega,q)$ for a liquid, we make the very plausible assumption that for water $\epsilon(\omega,q)$ depends only on the magnitude of the wave vector q at all values of q and ω. Further, we base our representation for $\epsilon(\omega,q)$ on experimental data relating to $\epsilon(\omega,o)$ for the liquid.[26]

Attempts are under way[44] to construct numerically the dielectric response function of water ice, with a view toward using these data to compute DIMFPs for swift electrons in this medium. We note that comparable numerical explorations of the response functions of the simplest solids[45,46] have not been used for such purposes because of the large amounts of computation needed, the sketchy character and the uncertain accuracy of the data produced as well as the difficulty of working with such representations. Solid-state theoretical studies of radiation interactions have proceeded mainly through the use of analytic and semi-analytic approximations to such response functions.[28-31,37,47-52]

Thus we have proceeded to fit experimental data on the complex, long-wavelength function $\epsilon(\omega) = \epsilon(\omega,o)$ of liquid water with convenient analytical functions. It is known that, for $\hbar\omega$ much larger than the important excitation energies, the response function of condensed matter must be identical with that of the atomic constituents, taken at the density of the matter. We plotted the experimental data on $\epsilon_2(\omega)$ of Ref. 24 and interpolated by drawing a curve beyond the highest experimental data points for liquid water so that it joins smoothly with a curve representing the oscillator strength of the water molecule, i.e., with

$$\pi\omega_p^2 \, \frac{df(\omega)}{d\omega} / 2\omega$$

where $df(\omega)/d\omega$ is the oscillator strength per electron of an isolated molecule in the continuum. This interpolation is constrained by the sum rule $\Sigma_0 = 1$. This curve of $\epsilon_2(\omega)$ is then fitted numerically by a sum of "derivative Drude functions" of the form

$$\epsilon_2(\omega) = \omega_p^2 \sum_n f_D(\omega,\gamma_n,\omega_n) \tag{18}$$

in which ω_n, γ_n, and f_n are fitting parameters. In the rest of this, we use atomic units (a.u.) in which $e = \hbar = m = 1$, the unit of energy is 27.2 eV, and the unit of length is 0.529 Å. The derivative Drudge function,

$$f_D(\omega,\gamma_n,\omega_n) \equiv \frac{2\gamma_n^3\omega^3}{\left[(\omega_n^2-\omega^2)^2+\gamma_n^2\omega^2\right]^2} \tag{19}$$

seems particularly well-suited for use here. One can show that

$$<\omega>_D \equiv \int_o^\infty \omega f_D(\omega,\gamma_n,\omega_n)d\omega = \pi^2 \tag{20}$$

for any γ_n and ω_n. Further $<\omega^2>_D$ and $<\omega^3>_D$ are finite, while $<\omega^4>_D$ is infinite, paralleling the behavior of Σ_o, Σ_1, Σ_2, and Σ_3.

It follows that if $\epsilon_2(\omega)$ is given by Eq. (18), then

$$\epsilon_1(\omega) = 1 + \omega_p^2 \sum_n \frac{(\omega_n^2-\omega^2)[(\omega_n^2-\omega_n^2)^2 + 3\gamma_n^2\omega^2]}{[(\omega_n-\omega^2)^2 + \gamma_n^2\omega^2]^2} \tag{21}$$

Also,

$$\int_{0}^{\infty} \omega\, \epsilon_2(\omega)\, d\omega = \frac{\pi \omega_p^2}{2} \sum_n f_n \tag{22}$$

which shows that $\sum_n f_n$ must sum to unity. It is also true that

$$\int_{0}^{\infty} \omega\, Im[-1/\epsilon(\omega,o)]\, d\omega = \pi \omega_p^2/2 \tag{23}$$

by direct numerical calculation. The data fitted using Eq. (18) are shown in Fig. 5.

To extrapolate $\epsilon(\omega,q)$ into the (q,ω) plane, we can take

$$\omega_n = \gamma_{no}[1 + g(q)] \tag{24}$$

where $\{\omega_{no}\}$ are the $\{\omega_n\}$ values fitted to the $\epsilon(\omega,o)$ data; thus, the loss function develops into a Bethe ridge as q becomes large. Another degree of freedom may be introduced by choosing

$$\gamma_n = \gamma_{no}[1 + g(q)] \tag{25}$$

where γ_{no} are the fitted damping parameters and $g(q)$ is a fitting function that is fixed from the sum rule $\Sigma_1(q)$ for large q, and can be taken to be different for the various shells. In practice a simple linear form can be used.

In the relativistic regime, the forms

$$\omega_n = \omega_{no} + c^2[(1 + q^2/c^2)^{1/2} - 1] \tag{24a}$$

$$\gamma_n = \gamma_{no}\, 1 + g\left(c^2[(1 + q^2/c^2)^{1/2} - 1]\right) \tag{25a}$$

are appropriate, where $c = 137$ is the speed of light in atomic units.

In the OREC code,[8] $\epsilon_2(\omega)$ is partitioned into fractions corresponding to the various excitations and ionization channels that are known to exist for the H_2O molecule. Information is used about the way the several threshold energies were known to shift from the molecule to the condensed state.[53-56] We take the lowest energy required to liberate free electrons in water to be 8 eV. As described below, experimental data on the time dependence of the hydrated electron are consistent with the scheme presently used in OREC in which all ionizing collisions are taken to lead to the appearance of the H_2O^+ ion and a secondary electron, while excitations are partitioned as shown in Table I.[56] The dependence of the excitation functions on ω and q are taken from Ref. 56.

Figure 9 shows a representation of the energy loss function for liquid water plotted as a function of ω and q, both expressed in atomic units. The dispersion of the response function to higher values of ω as q increases is clearly seen for those

Table I

Interaction Type	Presumed Transition	Threshold Energy (eV)	Partitioning Products at 10^{-12}s
1	$\tilde{A}^1B_1$	8.4	25% de-excite by photo emission 75% give H and OH
2	$\tilde{B}^1A_1$	10.1	H_2 and H_2O_2
3	Rydberg A+B	11.26	de-excite by photon emission
4	Rydberg C+D	11.93	
5	Diffuse Band	14.1	H_3O^+, $\dot{O}H$, e_{aq}
6	Dissociative Excitation	21.4	H, OH

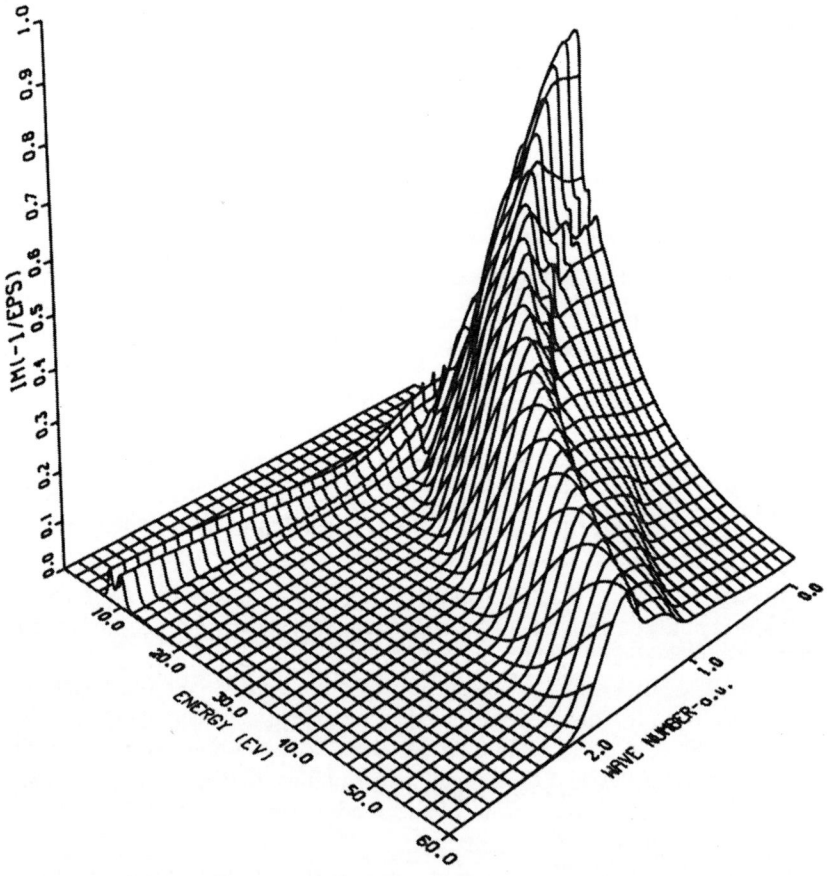

Figure 9. The energy loss function of liquid water calculated as described in the text and plotted as a function of energy loss $\hbar\omega$ in eV and as a function of wave number in a.u.

processes in which valence electrons and electrons from the K-shell are liberated into the continuum. The response function for excitations is seen to vary as q increases from zero, but does not disperse as do the ionizing processes. The response function is forced to satisfy the sum rule $\Sigma_0 = 1$ for all q. Figure 10 shows the same data but with the energy loss function data plotted on a logarithmic scale to emphasize contributions from excitations and from the K-shell.

The partial DIMFP for the ith type of interaction, where i may be any one of the excitations, or ionizations, is taken to be

$$\frac{d\Lambda_i^{-1}}{d\omega} = \frac{2}{\pi v^2} \int_{q_-}^{q_+} \frac{[\epsilon_2(\omega,q)]_i}{\epsilon_1^2(\omega,q) + \epsilon_2^2(\omega,q)} \frac{dq}{q} \qquad (26)$$

where $[\epsilon_2(\omega,q)]_i$ is the partial $\epsilon_2(\omega,q)$ function computed as described above. The relativistically corrected limits on the integral of Eq. (26) are, in atomic units,

$$q_\pm = [2E(1 + E/2c^2)]^{1/2} \pm [2E(1+E/2c^2) - 2\omega(1 + E/c^2) + \omega^2/c^2]^{1/2} \qquad (27)$$

This expression neglects the effects of exchange.

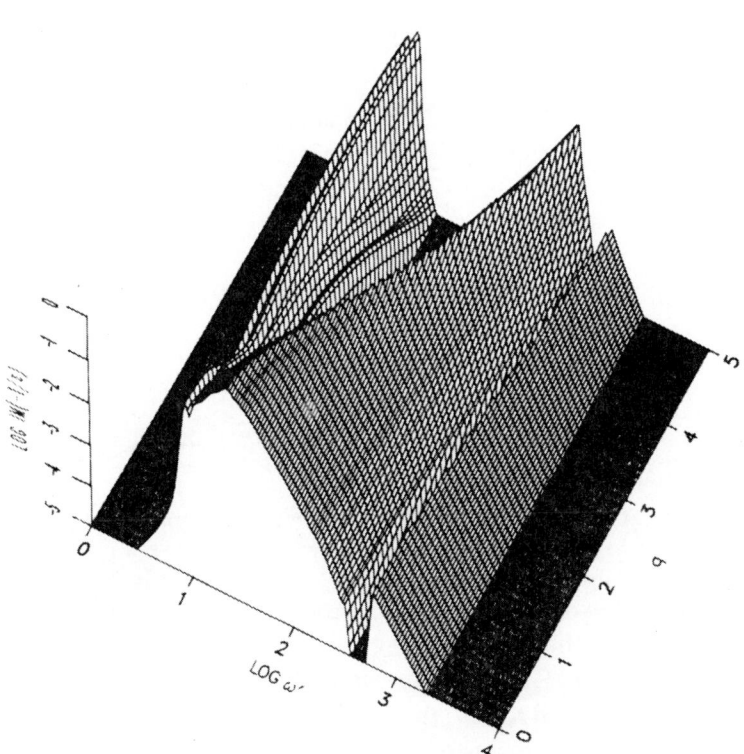

Figure 10. The same data as in Fig. 9 but with the loss function plotted on a logarithmic scale to emphasize the region of single-particle excitations.[82]

Exchange Correction to the DIMFP. For swift electrons with energies comparable with the binding energy of valence electrons in water, exchange corrections may become significant. These arise from the fundamental requirement that the total wave function of an assembly of identical particles must be antisymmetric under interchange of both spin and space coordinates. This leads to the operation of the Pauli exclusion principle in the scattering process and affects the cross section for scattering to final states.

The simplest case that illustrates this is the scattering of a non-relativistic electron with energy E on a free, stationary electron. The differential cross section for this process is[57,58]

$$d\sigma = \frac{2\pi e^4}{E} \left[\frac{1}{(E-E')^2} + \frac{1}{E^2} - \frac{1}{E'(E-E')} \right] dE' \tag{28}$$

where E' is the energy of the degraded primary and $E-E'$ is the energy of the secondary. This expression is subject to the usual condition that the less energetic of the electrons is taken to be the secondary.

The more general, relativistic version of this expression is given in Ref. 57. Exchange-corrected cross sections for electrons interacting with condensed matter are quite complicated. Ritchie and Ashley[59] have derived the exchange-corrected DIMFP for a slow electron in a free-electron gas.

We have used a semi-empirical scheme for correcting for exchange effects the DIMFP found from Eq. (6) for exchange.[60] This scheme is based on the Møller relativistic form of Eq. (28) and is used consistently in our calculations of the DIMFP. We write

$$\frac{d\Lambda^{-1}}{d\omega} = \mu(E,\omega)$$

set $\mu_B(E,\omega) \equiv \mu(E,E + \omega_B - \omega)$ and take the exchange-corrected DIMFP to be

$$\frac{d\Lambda^{-1\alpha}}{d\omega} = \mu(E,\omega) + \mu_B(E,\omega)$$

$$\tag{29}$$

$$- \left[1 - \left(\frac{\omega_B}{E} \right)^{1/2} \right] [\mu(E,\omega)\mu_B(E,\omega)]^{1/2} \frac{(1+2E/c^2)}{(1+E/c^2)^2} - \frac{\pi n}{c^4 E^3 (1+E/2c^2)}$$

where ω_B is the threshold energy for ionization and n is the density of electrons. A simple change in notation yields an expression for the nth partial DIMFP, and multiplication by ω and integration with respect to ω yields the corresponding contribution to the stopping power. Note that relativistic corrections to the exchange-corrected values are generally unimportant.

We note that in an early version of our OREC code,[8] no correction for exchange was included. Shortly after the publication of Ref. 8, the relativistic exchange correction of Eq. (29) was included in OREC and has been used consistently since then.[61] Calculations with OREC in which exchange was included (Fig. 7 of Ref. 62) are in good agreement with the results of Ashley[26] and with other estimates of the stopping power[62].

Reference 62 contains several graphs depicting DIMFPs found using the scheme described above.

Interaction Phenomena Idiosyncratic to Condensed Matter

Extant particle simulation codes are designed with many simplifying assumptions. Some of these assumptions are made out of the dearth of information about interactions in condensed matter, and some are due to the difficulty of implementing quantal processes in an essentially classical scheme.

It is commonly assumed that successive loss events are independent, even though it is known that low-energy electrons may have mean free paths of only a few angstroms in condensed matter. Such an electron is accompanied by a polarization cloud due to its interaction with ions and electrons of the medium. Its effective mass may differ appreciably from that of the bare electron due to this "dressing," and it may be so strongly damped that the view of it as a quasiparticle may lose validity. Quinn[63] has evaluated the lifetime of a low-energy electron in an electron gas. He finds that for a density comparable with that of the conduction electrons in aluminum metal, the damping rate is at most 16% of its energy in units of $\hbar$. The quasi-particle concept should thus be reasonable for this material. Similarly, one expects this concept to have some validity in condensed matter of biological significance. Energy transfer to optical phonons from low-energy electrons has been considered by Thornber and Feynman.[64] Here the mean free path for phonon emission is only a few angstroms. They evaluate the drift velocity of electrons in the presence of a uniform, constant electric field by path integral methods in order to account for quantal interference effects.

Ashley and Ritchie[65] were the first to derive quantally an expression for the distribution of multiple losses to plasmons by a swift electron in an electron gas. They summed a selective set of diagrams for the multiple plasmon excitation probability in the long-time limit. Their derivation was limited to weak coupling between the plasmon field and the electron and is thus valid primarily for fast electrons. Their work was generalized by McMullen, Bergersen and Jena[66] and by Sung and Ritchie,[67] who studied the interference between successive loss acts by a charged particle in condensed matter by modeling the medium response in terms of Bosonic excitations. A quantal generalization of the Landau distribution was derived.[67] This describes the distribution of multiple energy losses, as a function of path length, experienced by a swift charged particle interacting with condensed matter. They find a transient region at the origin of the path in condensed matter where the energy loss rate is less than the asymptotic rate given by the Bethe formula, and where the distribution of losses is also different, due to quantal interference between the basic loss acts. The length of this transient region ℓ_t is kv/ω_r, where ω_r is the mean excitation energy of the medium and k is a constant of the order of unity. Note that v/ω_r is just the Bohr adiabatic parameter.[68] For the most energetic electrons ℓ_t is of the order of a few hundred angstroms, and the actual diminution of the stopping power in this region is not expected to be large. A detailed study of the effect of this transient on the spatial distribution of excitations will be given elsewhere.[69] Note that in a real system, the appearance of an electron is either associated with its entry through a boundary, or as the product of ejection from a bound state or from the valence electron sea. Thus the picture of an isolated electron suddenly appearing in condensed matter is not realistic; one must consider the effect of boundaries on energy losses if the electron enters the system from outside, or the effect of the parent ion in case the electron is ejected from a bound state.

Quantal Interference in the Creation of a Secondary Electron. When an electron is ejected from a localized orbital in condensed matter, it may lose enough energy to recombine with its parent ion or with another ion in the system if the density of ionization is high enough. In addition to excitation of the medium by the ejected electron, the sudden appearance of the parent ion may excite the system. Quantal interference between these processes may occur. No considerations of such coherence effects have been made for biologically relevant media, although some theory has been done along these lines for radiation interactions in metals and semiconductors.[70]

We may estimate the effect of such an event using a semiclassical model. Inglesfield[70] has shown that it is a good approximation to neglect the wave properties of the ejected electron in the theory of plasmon excitation in photoemission from metals, even when the photoelectron energy is as low as 50 eV. We will employ this approximation here for purposes of orientation. Assume that the response of the medium may be represented by the function $\epsilon(\omega,q)$ and that the electron and its residual ion are equivalent to the electric charge density

$$\rho(\vec{r},t) = e\, \delta^3(\vec{r} - \vec{v}t) \tag{30}$$

for times $t > 0$. For $t < 0$ we take $\rho = 0$. Here $\delta^3(\vec{r})$ is a three-dimensional delta function. Standard theory yields an expression for the energy transfer, W, to the dielectric from the ion-electron pair; one finds that

$$W = \Delta W + \ell \cdot \frac{dW}{dx} \tag{31}$$

where dW/dx is the stopping power of the medium for the electron and ℓ is its path length. The correction term ΔW is given by the lowest order of perturbation theory as

$$\Delta W = \frac{ie^2}{4\pi^3} \int \frac{d^3q}{q^2} \int \frac{\omega[(1-\epsilon(\omega,q)]}{|\epsilon(\omega,q)|^2} \left[\frac{1}{\omega^2+\eta^2} - \frac{1}{(\omega+i\eta)(\omega-q\cdot v-i\eta)} \right]$$
$$- \left[\frac{1}{(\omega-i\eta)(\omega-q\cdot v+i\eta)} \right] d\omega \tag{32}$$

The first term represents energy absorbed due to the sudden appearance of the ion, while the effect of interference between the receding electron and the ion is contained in the last two terms. The quantity η is a positive infinitesimal. Detailed consideration of ΔW for a water medium will be given elsewhere,[69] in which the time development operator of the system in infinite order of perturbation theory will be employed. A rough estimate indicates that ΔW may amount to several eV of energy, constituting additional excitation of the medium in the vicinity of the parent ion.

Wave Properties of Transporting Electrons. At present we lack detailed understanding of the complex interactions that take place when ionization occurs in real biological systems. Besides the question of coherence between energy transfer from the ejected electron and the residual ion, no real quantitative picture is available for incorporating the effects of the wave nature of the electrons into a Monte Carlo simulation. There are essentially no experimental data on elastic scattering probabilities in the materials of interest here. Quantal effects are expected to be less important for inelastic scattering events. Ritchie[71] and Ritchie and Howie[72] have shown theoretically that the classical representation of an electron in terms of its impact parameter relative to a region in which electronic excitations may occur should be quite accurate when computing energy transfers to that region that are small compared with the electron's kinetic energy. At energies ~ a few eV errors steeming from neglect of such quantal effects are difficult to assess. However, since the range of electrons at these energies is expected to be small, such inaccuracies should have little effect on the transport process as modeled by current theoretical approaches.

The Localization of Initially Unlocalized Excitations. It has long been recognized[73] that swift charged particles may create excitations that are coherent over distances ~1000 angstroms. Further, these unlocalized excitations may localize into small regions critical to the damage or inactivation of biological materials. It has been noted[73,75] that such excitations might migrate appreciable distances before localizing. Brandt and Ritchie[1] estimated the extension of these coherent excitations radially from the track, their lifetimes, and the extent of their propagation in space for systems of biological interest.

Since then we have studied this phenomenon in detail for free-electron systems, although it is far from being well understood in biological media. Experiment furnishes little guidance about the intricacies of this process in complex molecular liquids. Nevertheless, we feel justified in adopting a simplified approach to estimate its effects in irradiated water. As implemented in OREC, the localization mechanism is modeled using a collective picture to describe excitations created by charged particles. The basis of this model can be described as follows.

Assume that a swift charged particle has charge Ze and mass M and proceeds through the condensed matter medium with speed v. The interaction Hamiltonian between the particle and collective modes in the system can be taken as

$$H_{ep} = Z \sum_{\vec{q}} \alpha_q \frac{e^{i\vec{q}\vec{r}}}{\sqrt{\Omega}} (b_{\vec{q}} + b^+_{-\vec{q}}) \tag{33}$$

where $b_{\vec{q}}$ is a destruction operator for a collective excitation with wave vector $\vec{q}$, and the coupling constant is

$$\alpha_q^2 = \frac{2\pi\hbar e^2 \omega_p^2}{\Omega \omega_q q^2}. \tag{34}$$

Ω is the normalization volume, ω_p is the plasma frequency of the electronic system, and the eigenfrequency of a plasmon with wave number q is

$$\omega_q = \sqrt{\omega_p^2 + \beta^2 q^2 + \frac{\hbar^2}{4m^2} q^4} \tag{35}$$

where β is the speed at which disturbances propagate in the valence electron gas. The q^2 term has its origin in the dispersion of electrons in the medium, and the term containing q^4 represents quantal recoil of the electrons.

To proceed with the theory, assume that the state vector of the initial state may be written:

$$|i\rangle = \frac{e^{i\vec{p}_o \cdot \vec{r}}}{\sqrt{\Omega}}|O\rangle \qquad (36)$$

and that the final state vector may be expressed as

$$|f\rangle = \frac{e^{i\vec{p}_f \cdot \vec{r}}}{\sqrt{\Omega}}b_{\vec{q}}^+|O\rangle \qquad (37)$$

where $|O\rangle$ is the vacuum state of the collective mode field.

First-order Golden Rule perturbation theory gives for the inverse mean free path, Λ^{-1}, for plasmon creation,

$$\Lambda^{-1} = \frac{2\pi}{v}\sum_{\vec{p}_f}\sum_{\vec{q}}|\langle f|H_{ep}|i\rangle|^2\delta\left(\epsilon_{\vec{p}_0} - \epsilon_{\vec{p}_f} - \omega_q\right) \qquad (38)$$

where $\epsilon_{\vec{p}} = \vec{p}^2/2M$ and we use atomic units henceforth. Evaluating the sum over $\vec{p}_f$, one finds

$$\Lambda^{-1} = \frac{2\pi Z^2}{v}\sum_{\vec{q}}\alpha_q^2\delta\left(\vec{v}\cdot\vec{q} - \frac{q^2}{2M} - \omega_q\right) \qquad (39)$$

where $v = p_o/M$ in atomic units. For simplicity and to obtain an analytical form for the radial distribution, take $\omega_q \approx \omega_p$. Then converting the sums to integrals as $\Omega \to \infty$, neglecting the second term in the energy-conserving delta function and resolving the vector q as

$$\vec{q} = \vec{Q} + q_z\left(\frac{\vec{v}}{v}\right) \qquad (40)$$

where $\vec{Q}$ is perpendicular to $\vec{v}$, we find

$$\Lambda^{-1} = \frac{Z^2\omega_p}{2\pi v^2}\int\frac{d^2Q}{Q^2 + \omega_p^2/v^2} \qquad (41)$$

This equation may be written in the equivalent form

$$\Lambda^{-1} = \frac{Z^2 \omega_p}{2\pi v^2} \int \frac{d^2 Q}{\sqrt{Q^2 + \omega_p^2/v^2}} \int \frac{d^2 Q'}{\sqrt{Q'^2 + \omega_p^2/v^2}} \delta^2(\vec{Q} - \vec{Q}') \qquad (42)$$

or, using the identity $\delta^2(\vec{Q} - \vec{Q}') = \int d^2 b e^{i\vec{b} \cdot (\vec{Q} - \vec{Q}')}/(2\pi)^2$, one has

$$\Lambda^{-1} = \frac{Z^2 \omega_p}{8\pi^3 v^2} \int d^2 b \int \frac{d^2 Q e^{i\vec{Q} \cdot \vec{b}}}{\sqrt{Q^2 + \omega_p^2/v^2}} \int \frac{d^2 Q' e^{-i\vec{Q}' \cdot \vec{b}}}{\sqrt{Q'^2 + \omega_p^2/v^2}} \qquad (43)$$

then

$$\Lambda^{-1} = \frac{Z^2 \omega_p}{8\pi^3 v^2} \int d^2 b \left| \int \frac{d^2 Q e^{i\vec{Q} \cdot \vec{b}}}{\sqrt{Q^2 + \omega_p^2/v^2}} \right|^2 \qquad (44)$$

The integral over Q may be evaluated in terms of elementary functions. We assert that one may regard the integrand of the b integration as a differential inverse mean free path (DIMFP) for interaction with the medium in the impact parameter variable b per unit volume in impact parameter space; then one writes

$$\frac{d^2 \Lambda^{-1}}{d^2 b} = \frac{Z^2 \omega_p}{2\pi v^2} \left[\frac{e^{-2\omega_p b/v}}{b^2} \right] \qquad (45)$$

A more accurate result that does not diverge as $b \to 0$ is found by retaining the full q dependence of ω_q. One can show that $d^2 \Lambda^{-1}/d^2 b$ diverges only logarithmically as $b \to 0$ and that Λ^{-1} is thus finite, as it must be. One finds

$$\frac{d^2 \Lambda^{-1}}{d^2 b} = \frac{\pi Z^2 \omega_p^3}{16\beta^2 v^2} I_0^2 \left(\frac{b\omega_-}{\beta} \right) K_0^2 \left(\frac{b\omega_+}{\beta} \right) \qquad (46)$$

where

$$\omega_\pm = \frac{\omega_p}{\sqrt{2}} \sqrt{1 \pm + \frac{\sqrt{2}\beta}{v}}$$

and I_0 and K_0 are modified Bessel functions of order 0. This expression is valid when $v >> \beta \approx 1$ *a.u.* and reduces to that given in Eq. 45 when $b >> v/\omega_p$.

For purposes of orientation, we note that the screening length v/ω_p is ~14 nm for a 100 keV electron in water, ~1.4 nm for a 1 keV electron and ~0.14 nm for a 10 eV electron.

This approach may be extended to yield a DIMFP for the creation of two excitations in the medium by the swift charged particle. Figure 11 shows Feynman diagrams representing the lowest-order interactions corresponding to this process.

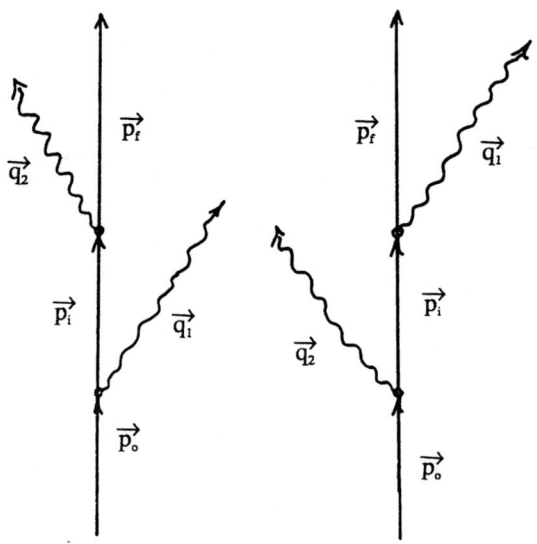

Figure 11. Feynman diagrams representing the successive creation of two collective modes of electronic oscillation by a swift charged particle.

Standard second-order perturbation theory gives the following expression for the inverse mean free path, Λ^{-1}, for the creation of two successive excitations:

$$\Lambda^{-1} = \frac{2\pi Z^2}{v\Omega^2} \sum_{\vec{q}_1} \sum_{\vec{q}_2} \sum_{\vec{p}_f} |\sum_{p_i} \frac{\alpha_{q_1}\alpha_{q_2}\delta^3(\vec{p}_i - \vec{p}_f - q_2)}{\epsilon_{\vec{p}_o} - \epsilon_{\vec{p}_i} - \omega_{q_1} + i\gamma} \tag{47}$$

$$+ \sum_{\vec{p}_i} \frac{\alpha_{q_1}\alpha_{q_2}\delta^3(\vec{p}_i - \vec{p}_f - \vec{q}_1)\delta^3(\vec{p}_o - \vec{p}_i - \vec{q}_2)}{\epsilon_{\vec{p}_o} - \epsilon_{\vec{p}_i} - \omega_{q_2} + i\gamma}|^2 \delta\left(\epsilon_{\vec{p}_o} - \epsilon_{\vec{p}_f} - \omega_{q_1} - \omega_{q_2}\right)$$

Here $\vec{p}_o$ is the wave vector of the particle in its initial state and γ is the damping rate of the collective state. Although the damping rate in general depends on the momentum of the state, we will assume for simplicity that it is constant.

Simplifying Eq. (47) with the aid of the Kronecker deltas and letting $\Omega \to \infty$, we find

$$\Lambda^{-1} = \frac{Z^2}{8\pi^3 v} \int d^3 q_1 \int d^3 q_2 \delta(\vec{v}_o \cdot \vec{q}_1 + \vec{v}_i \cdot \vec{q}_2 - \omega_{q_1} - \omega_{q_2}) |\Gamma_{\vec{q}_1, \vec{q}_2}|^2 \tag{48}$$

and

$$\Gamma_{\vec{q}_1, \vec{q}_2} = \frac{1}{\vec{v} \cdot \vec{q}_1 - \omega_p + i\gamma} + \frac{1}{\vec{v} \cdot \vec{q}_2 - \omega_p + i\gamma} \tag{49}$$

In obtaining this expression, dispersion of the collective mode ($\omega_q \approx \omega_p$) and recoil of the charged particle have been neglected. Following a procedure similar to that used above and employing approximations appropriate to large values of the coordinates, one finds

$$\frac{d^5 \Lambda^{-1}}{d^2 b_1 d^2 b_2 dz} = \frac{4\pi^2 \omega_p^2 Z^2}{v^4} \frac{e^{-2\omega_p b_1/v}}{b_1^2} \frac{e^{-2\omega_p b_2/v}}{b_2^2} e^{-2\gamma |z|/v} \tag{50}$$

This describes the probability of two interactions by the particle per unit path length in the medium per unit volume in impact parameter space for interaction 1, per unit volume in impact parameter space for interaction 2, and per unit separation, z, between the interactions. This result is neat and compact and seems quite in accord with intuition.

In work with OREC, the above results have been used to model the localization of initially unlocalized excitations created by electrons or ions in water. For simplicity in representing the localization process in a Monte Carlo calculation, it was assumed that all energy losses in amounts less than ~50 eV by swift electrons or ions were initially unlocalized and subsequently localize at radial separation b from the point at which the energy loss is found to occur in the Monte Carlo calculation. The distribution from which this radial distance is chosen is taken to be

$$P(b) db = C \frac{\exp(-\omega b/v\gamma)}{b_o^2 + b^2} b \, db \tag{51}$$

Here $\hbar\omega$ is the energy loss, γ is a constant of the order of unity, b_o is a constant ~0.1 nm, and C is a normalization constant chosen such that

$$\int_0^{b_{max}} P(b) db = 1$$

In these calculations b_{max} is chosen to be 10 nm. The constant γ is intended to represent, schematically, propagation of coherent excitations before localization as well as our lack of knowledge about the localization process in a complex molecular liquid such as water. It was taken equal to 5 to emphasize the radial extension of the localization process, and may be altered when relevant data become available.

Recent experiments indicate that coherent excitations created by swift charged particles in metallic systems have a rather narrow radial extension about the path of the particle. Scanning Transmission Electron Microscopy (STEM) experiments on plasmon losses in the vicinity of a metallic boundary show spatial resolution ~0.4 nm when the quantity v/ω appropriate to the experiments is quite large, i.e., ~4 nm.[77] Similar high resolution is achieved in band-gap spectroscopy of defects in semiconductors.[78] Recent work in STEM has shown that it is possible to obtain secondary electron signals with ~1 nm spatial resolution and 1 eV energy resolution again when the parameter v/ω corresponding to the experiments is ~4 nm.[79,80] We have given theoretical justification for the high resolution observed for free-electron-like systems.[81,76]

However, in view of the uncertainties in our knowledge of fundamental interactions in complex molecular liquids, and in view of the good agreement found[82] between calculations based in part on the OREC transport code and measurements from electron pulse radiolysis, we feel that justified in retaining the presently implemented algorithm for representing localization of initially unlocalized excitations in water until additional data become available.

The OREC Monte Carlo Scheme

As presently implemented in our OREC code, electrons are transported in a particle simulation scheme using the DIMFP values described above. Inelastic transitions are subdivided into six specific excitation events, as indicated in Table I. Ionization events are assumed to give rise to an H_2O^+ ion and a secondary, ejected electron, although in earlier versions of OREC ionization was taken to give rise to five different ionic products in fragmentation patterns not dissimilar to those observed in experiments with isolated water molecules.[8] The H_2O^+ ion is caused to migrate distances from its point of origin that are chosen from a Gaussian distribution with a mean value of 1.25 nm in a randomly chosen direction. It then combines with a water molecule to yield an OH radical at the site and an H_3O^+ ion separated by a mean distance 0.29 nm from the site. The code calculates the position and type of transition as it transports all secondary electrons until their energies fall below the assumed threshold of 7.4 eV for electronic excitation. DIMFP values for elastic scattering are also included, of course. The localization scheme codified in Eq. (51) is used.

Subexcitation electrons formed in the physical stage thermalize and become hydrated by 10^{-12} s.[82,83] Originally a hydration distance was randomly selected for each electron from a Gaussian distribution with a mean displacement of 3 nm. The electron was then replaced by a hydrated electron, e_{aq}^-, at the new position displaced in a random direction from its original position. In place of this original method, we have recently introduced an age-diffusion calculation for treating the hydration process. The inelastic cross sections used were deduced from measured dielectric functions, while the elastic cross sections were scaled from electron scattering data taken on the H_2O molecule. Fermi age theory was used to determine the r.m.s. distance of travel from subexcitation energy to thermal energy.[82,84-87]

Our work with OREC has been generalized recently[82,84-87] to model the events that extend to the completion of intratrack radical reactions at $\sim10^{-6}$ s. The reactions, rate constants, and reaction radii used in this part of the modeling process are given in Refs. 82 and 84-87, together with numerical results that have been found to date.

In developing the present version of OREC, and the extension to describe evolution of ionization products to $\sim10^{-6}$ s, we have made extensive use of experimental data for water in all three phases. We have also made the model theoretically rigorous wherever possible, such as in satisfying quantal sum rules. An advantage of using the dielectric response function as the point of departure is that it includes the collective effects of the liquid *a priori*. An important strength of the approach also lies in the fact that Monte Carlo techniques are used, thus permitting great flexibility in the assumptions made. New theoretical and experimental data can be readily incorporated into the sampling procedures, enabling improvements to be made in any part of the calculation. However, because not all of the needed information exists, our model contains many uncertainties in its attempt to represent quantitatively the complex sequence of events that occur in irradiated water.

A number of comparisons have been carried out between results found using OREC and those obtained using Monte Carlo codes based on data for the H_2O molecule. In Ref. 62 results from the code MOCA-8, developed by H. G. Paretzke, are considered in detail together with output from OREC; reasons for observed differences are understood in terms of the differences in the cross sections used.

In Fig. 12 are shown normalized first collision spectra for 5 keV electrons in the vapor and in the liquid.[62] The energy loss spectrum is harder for the liquid than for the vapor due to the observed upward shift in oscillator strength in water compared with the vapor. This fact, coupled with the assumed smaller binding energies of the valence electrons in the liquid, gives rise to the greater amount of ionization (lower W value) and smaller Fano factor for total ionization in the liquid. This difference in the number of ionizations produced is shown in detail in Fig. 13 as a function of the energy of incident electrons.

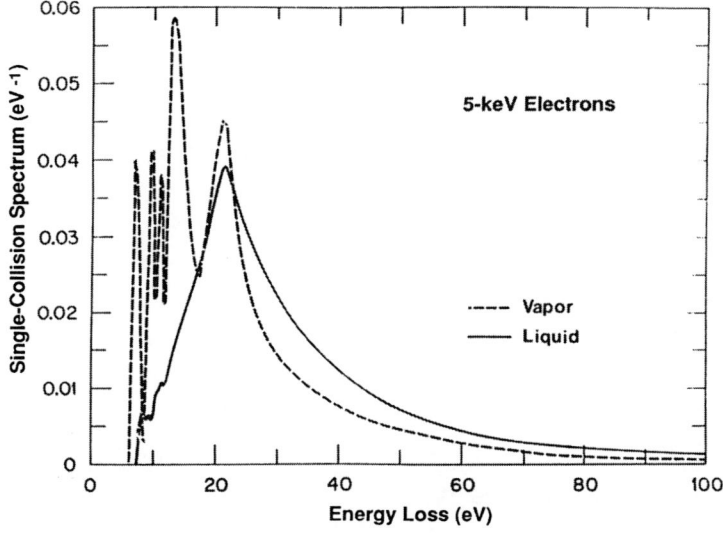

Figure 12. Normalized single-collision spectra for 5 keV electrons in water vapor and liquid water.(62)

The most important recent and direct comparisons of our calculations with experiment are provided by data from pulse radiolysis.[88a] Figure 14 shows a comparison of our calculated time-dependent yields of OH and e_{aq}^- with measurements.[88] The agreement is quite good. It results in part from assumptions made concerning two processes within the prechemical stage of track development: (i) the partitioning of high-lying bound, excited states into autoionizations, dissociations, or relaxations (see Table I); and (ii) the mean migration distance of H_2O^+ prior to its reaction with a neighboring water molecule (currently 1.5 nm). We continue to regard some details of the model as provisional and seek additional experimental data to strengthen its

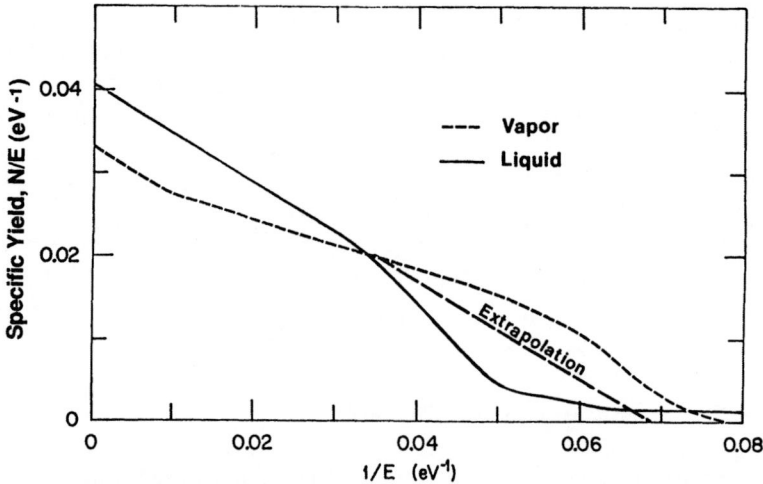

Figure 13. Specific yields of ionizations produced by electrons of different energies as a functions of the reciprocal of the initial energies of the electrons. Included are ionizations generated by all secondary, tertiary, etc., electrons. Data are shown for liquid water and for water vapor.[62b]

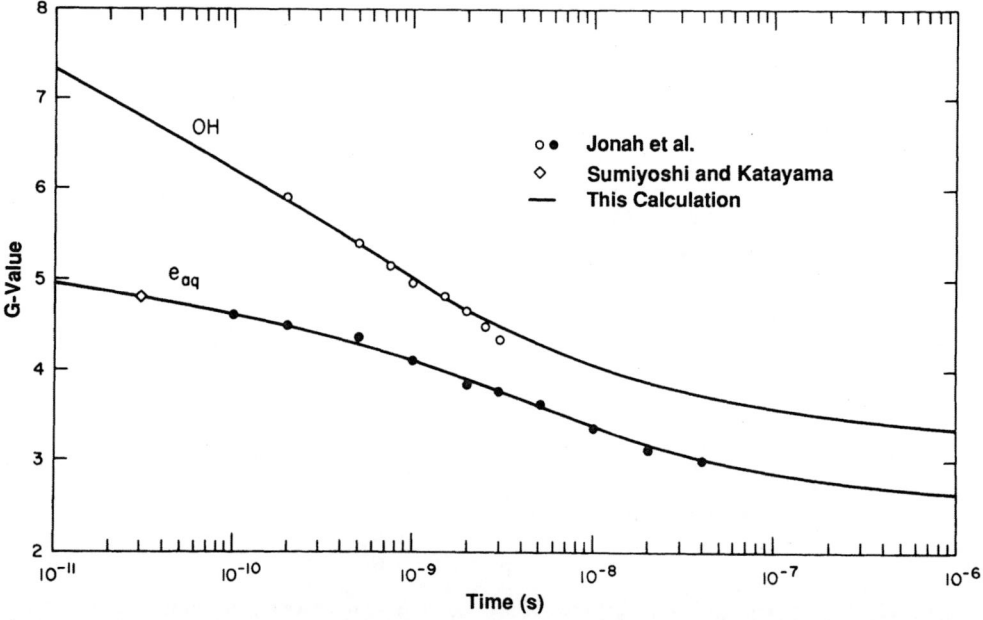

Figure 14. Comparisons of calculated and measured time-dependent yields of the OH radical and of the hydrated electron.[82] Measurements by C.D. Jonah, M.S. Matheson, J.R. Miller, and E.J. Hart [Yield and Decay of the Hydrated Electron from 100 ps to 3 ns, *J. Phys. Chem.* 80:1267-1270 (1976)] and by C.D. Jonah and J.R. Miller [Yield and Decay of the OH Radical from 200 ps to 3 ns, *J. Phys. Chem.* 81:1974-1976 (1977)] are shown as solid and empty circles, while the single point was obtained by T. Sumiyoshi and M. Katayama [The Yield of Hydrated Electrons at 30 Picoseconds, *Chem. Lett.* 1982:1887-1890 (1982)].

basis and credibility. We feel that it is very important to pursue this avenue of research in order to better our understanding of microscopic processes in irradiated biological media.

Summary and Conclusions

We have described in some detail important processes occurring in irradiated media of biological significance, and the rationale underlying our modeling of the transport of swift electrons in liquid water as implemented in the computer code OREC. Although many uncertainties exist about the basic processes that dominate the transport phenomenon in the complex, nonlinear, interacting system of charged and uncharged species liberated in water by ionizing radiation, our results to date seem quite encouraging.

Additional fundamental information of importance to improved modeling includes the following:

- Better representations of geminate recombination processes in water, including the effect of the dynamical response of the system on the transport of electrons ejected from localized sites in the medium.

- More accurate probabilities for the creation of secondary electrons in water. The interesting work of Long et al.[89] using the local plasma approximation (LPA) to generate such cross sections seems worthy of generalization to include the decay of plasmons in addition to the direct generation of electron-hole pairs. The work of Xu et al.[90] in representing the ground state density of water to compute the stopping of charged particles in the context of the LPA may be useful in this connection.

- Although quite difficult, measurements of the fragmentation and de-excitation patterns of products of ionization and excitation from small water clusters would be very valuable for modeling important transport processes in bulk water.

- Measurements of interaction cross sections, differential in both energy and momentum transfer, of swift electrons with water clusters would be very useful in verifying the details of the DIMFPs used in our work. Again, this research would require much effort to accomplish properly.

- Improvements in our knowledge of cross sections for subexcitation electrons in water. A review of current information in this area has been given recently by Voltz.[91]

- A better understanding of nonlinear interactions in biological media. In Ref. 1 it was speculated that nonlinear effects may be important in irradiated aqueous media in regions of high ionization density. Although no additional information has become available about such effects, the self-consistent, nonlinear theory of Coulombic effects in irradiated insulators recently has been discussed quantitatively.[92]

Acknowledgments

The research was sponsored by the Office of Health and Environmental Research, U.S. Department of Energy, under Contract DE-AC05-84 R21400 with Martin Marietta Energy Systems, Inc. Important contributions to this work have been made by J. L. Magee, A. Chatterjee, C. E. Klots, and H. G. Paretzke.

References

1. W. Brandt and R. H. Ritchie. Initial Interaction Mechanisms in the Physical Stage. In *Physical Mechanisms in Radiation Biology*, Proceedings of a conference held at Airlie, Virginia, 11-14 October 1972. U.S. Atomic Energy Commission, Technical Information Center, Oak Ridge, Tennessee (1974).

2. J. B. Pendry. *Low Energy Electron Diffraction: The Theory and its Application to Determination of Surface Structure*, Academic Press, New York (1974).

3. J. B. Pendry. Theory of Photoemission, *Surf. Sci.* 57:679-705 (1976).
   See also J. B. Pendry and E. Castaño, Electronic Properties of Disordered Materials: A Symmetric Group Approach, *J. Phys.* C21:4333 (1988).

4. M. J. Berger. Some New Transport Calculations of the Deposition of Energy in Biological Materials by Low-Energy Electrons. *Fourth Symposium on Microdosimetry* (J. Booz, et al. Eds.), CEC, EUR 5122 d-e-f:695-715 (1974).

5. H. G. Paretzke. Comparison of Track Structure Calculations with Experimental Results. *Fourth Symposium on Microdosimetry* (J. Booz, et al. Eds.), CEC, EUR 5122 d-e-f:695-715, pp. 141-168 (1974).

6. M. Terrisol, J.-P. Patau, and T. Eudalo. Application a la Microdosimetrie et a la Radiobiologie de la Simulation du Transport des Electrons de Basse Energie dans L'eau a L'etat Liquide. In *Sixth Symposium on Microdosimetry*, Eds. J. Booz and H. G. Ebert, CEC, EUR 6064:169-176 (1978).

7. R. N. Hamm, H. A. Wright, R. H. Ritchie, J. E. Turner, and T. P. Turner. Monte Carlo Calculation of Transport of Electrons Through Liquid Water. *Fifth Symposium on Microdosimetry*, (Eds. J. Booz, H. G. Ebert, and B.G.R. Smith, CEC, EUR 5254 d-e-f:1037-1053, (1976).

8. R. H. Ritchie, R. N. Hamm, J. E. Turner, and H. A. Wright. The Interaction of Swift Electrons with Liquid Water. *Sixth Symposium on Microdosimetry*, Eds. J. Booz and H. G. Ebert, CEC, EUR 6064:345-354 (1978).

9. M. Zaider, D. J. Brenner, and W. E. Wilson. The Applications of Track Calculations to Radiobiology. *Radiat. Res.* 95:231-247 (1983).

10. W. R. Holley, A. Chatterjee, and J. L. Magee. Production of DNA Strand Breaks by Direct Effects of Heavy Charged Particles. *Radiat. Res.*, 121:161-168 (1990).

11. E. D. Palik. *Handbook of Optical Constants of Solids*. Academic Press, New York (1985).

12. M. W. Williams, E. T. Arakawa, and T. Inagaki. Optical and Dielectric Properties of Materials Relevant to Biological Research. In *Handbook of Synchrotron Radiation*, Vol. 4, Eds. E. Rubenstein, S. Evashi, and M. Koch. Elsevier, Amsterdam (1990).

13. E. Fermi. Über die Theorie des Stosses Zwischen Atomen und Elektrisch Geladen en Teilchen. *Z. für Physik*, 29:315-327 (1924).

14. E. Fermi. The Ionization Loss of Energy in Gases and Condensed Materials. *Phys. Rev.* 57:485-493 (1940).

15. M. E. Gertsenshtein. *Zhur. Eksptl. Teoret. Fiz.* 22:303-315 (1952).
    M. E. Gertsenshtein. *Zhur. Eksptl. Teoret. Fiz.* 23:678-688 (1952).
    A. T. Akhiezer and A. G. Sitenko. *Zhur. Eksptl. Teoret. Fiz.*, 23:161-185 (1952).
    J. Neufeld and R. H. Ritchie. Passage of Charged Particles Through Plasma. *Phys. Rev.* 98:1632-1642 (1955).

16. J. Lindhard. On the Properties of Gas of Charged Particles. *Mat.-Fys. Medd., Kgl. Danske. Videnskab. Sels.* 28, No. 8 (1954).

17. J. Hubbard. The Description of Collective Motions in Terms of Many-Body Perturbation Theory: II The Correlation Energy of a Free-Electron Gas. *Proc. Roy. Soc. (Lond) A* 243:336-352 (1957).

18. L. Rosenfeld. *Theory of Electrons*, Chapter V. Dover, New York (1965) for the classical form. F. Seitz. *The Modern Theory of Solids*, Chapter 17, McGraw-Hill, New York (1940).

19. A. Bohr. Atomic Interaction in Penetration Phenomena. *Mat.-Fys. Medd., Kgl. Danske. Vidensk. Sels.* 24, No. 19 (1948).
    A. W. Blackstock, R. H. Ritchie, and R. D. Birkhoff. Mean Free Path for Discrete Electron Energy Losses in Metallic Foils. *Phys. Rev.* 100:1078 (1955).

20. M. W. Williams, R. N. Hamm, E. T. Arakawa, E. T. Painter, and R. D. Birkhoff. Collective Electron Effects in Molecular Liquids. *Int. J. Radiat. Phys. Chem.* 7:95-108 (1975).

21. U. Fano. Normal Modes of a Lattice of Oscillators with Many Resonances and Dipolar Coupling. *Phys. Rev.* 118:451-455 (1960).

22. H. Ehrenreich and H. R. Philipp. Optical Properties of Semiconductors in the Ultra-Violet. In *Proc. Int. Conf. Phys. Semiconductors*, Ed., A. C. Strickland, pp. 367-374, Bartholomew Press, Dorking, UK pp. 367-374 (1962).

23. M. Zaider, J. L. Fry, and D. E. Orr. Towards an ab initio Evaluation of the Wave Vector and Frequency Dependent Dielectric Function of Crystalline Water. *Radiat. Prot. Dosim.* 31:23-28 (1990).

24. J. Daniels, C. V. Festenburg, H. Raether, and K. Zeppenfield. Optical Constants of Solids by Electron Spectroscopy. *Springer Tracts Mod. Phys.* 54:77-135 (1970).

25. See E. T. Arakawa, L. C. Emerson, S. I. Juan, J. C. Ashley, and M. W. Williams. The Optical Properties of Adenine from 1.8 to 80 eV. *Photochemistry and Photobiology* 44:349 (1986) and the references given therein.

26. J. M. Heller, Jr., R. N. Hamm, R. D. Birkhoff, and L. R. Painter. Collective Oscillation in Liquid Water. *J. Chem. Phys.* 60:3483-3486 (1974).

27. R. H. Ritchie and A. Howie. Electron Excitation and the Optical Potential in Electron Microscopy. *Phil. Mag.* 36:463-481 (1977).

28. J. C. Ashley. Stopping Power of Liquid Water for Low-Energy Electrons. *Radiat. Res.* 89:25-31 (1982).
   J. C. Ashley. Interaction of Low-Energy Electrons with Condensed Matter: Stopping Powers and Inelastic Mean Free Paths from Optical Data. *J. Electron Spectr. Rel. Phenom.* 46:199-214 (1988).

29. D. R. Penn. Electron Mean-free-path Calculations Using a Model Dielectric Function, *Phys. Rev. B.* 35:482-486 (1987).

30. R. M. Nieminen. Stopping Power for Low-Energy Electrons, *Scanning Microsc.* 2:1917-1926 (1988).
   O. H. Crawford and C. W. Nestor, Jr. Position Dependent Stopping Power for Channeled Ions. *Phys. Rev.* A28:1260-1266 (1983).

31. S. Tougaard. Background Removal in X-ray Photoelectron Spectroscopy: Relative Importance of Intrinsic and Extrinsic Processes. *Phys. Rev.* B34:6779-6783 (1986).

32. P. Nozieres and D. Pines. A Dielectric Formulation of the Many-Body Problem: Application to the Free Electron Gas. *Nuovo Cimento* 9:470 (1958).

33. H. A. Bethe. Zur Theorie des Durchgangs Schneller Korpuskularstrahlen durch Materie, *Ann. Physik* 5:325-400 (1930).

34. G. Placzek. The Scattering of Neutrons by Systems of Heavy Nuclei. *Phys. Rev.* 86:377-388 (1952).
   G. Placzek. The Scattering of Neutrons by Systems of Heavy Nuclei. *Phys. Rev.* 94:1801 (1954) (E).

35. U. Fano and J. E. Turner. Contribution to the Theory of Shell Corrections. In *NAS-NCR Nuclear Science Series, Report 39, Committee on Nuclear Science, Publication 1133*, pp. 49-67 (1964).
   M. Inokuti, Inelastic Collisions of Fast Charged Particles with Atoms and Molecules - The Bethe Theory Revisited, *Rev. Mod. Phys.* 43:297-347 (1971); 50:23-35 (1978).

36. D. Pines. Plasma Oscillations of Electron Gases. *Physica* 26:S103-123 (1960).

37. M. A. Kumakhov and F. F. Komarov. *Energy Loss and Ion Ranges in Solids*, pp. 150-159. Gordon and Breach, New York (1981) and the references therein.

38. O. H. Crawford and C. W. Nestor, Jr. Position-Dependent Stopping Power for Channeled Ions. *Phys. Rev. A* 28:1260-1266 (1983).

39. D. L. Johnson. Local Field Effects and the Dielectric Response Function of Insulators: A Model. *Phys. Rev. B* 9:4475-4484 (1974).
   W. M. Saslow and G. F. Reiter. Plasmons and Characteristic Energy Losses in Periodic Solids. *Phys. Rev. B* 7:2955-3003 (1978).

40. N. D. Mermin. Lindhard Dielectric Function in the Relaxation-Time Approximation, *Phys, Rev. B* 1:2362-2363 (1970).

41. A. Gras-Marti and R. H. Ritchie. To be published.

42. R. H. Ritchie, C. J. Tung, V. E. Anderson, and J. C. Ashley. Electron Slowing-Down Spectra in Solids, *Rad. Res.* 64:181-204 (1975).
   C. J. Tung, R. H. Ritchie, J. C. Ashley, and V. E. Anderson. *Inelastic Interactions of Swift Electrons in Solids*, ORNL-RM-5188, Oak Ridge National Laboratory, Oak Ridge, Tennessee (1976).

43. J. L. Fry. Dielectric Function of a Model Insulator. *Phys. Rev.* 179:892-905 (1969), and references therein.

44. M. Zaider, J. L. Fry, and D. E. Orr. A Semiempirical Tight-Binding Calculation of the Dielectric Response Function of Water. *Nucl. Tracks Rad. Meas.* 16:159-168 (1989).

45. J. P. Walter and M. L. Cohen. Frequency and Wave Vector-Dependent Dielectric Function for Ge, GaAs, and ZnSe. *Phys. Rev. B* 6:3800-3804 (1972).

46. M. S. Haque and K. L. Kliever. Plasmon Properties in BCC Potassium and Sodium, *Phys. Rev. B* 7:2416-2430 (1973).

47. W. Brandt and J. Reinheimer. Theory of Semiconductor Response to Charged Particles. *Phys. Rev. B* 2:3104-3112 (1970).

48. E. Tosatti and G. P. Parravicini. Model Semiconductor Dielectric Function. *J. Chem. Phys. Solids* 32:623-626 (1971).

49. Z. H. Levine and S. G. Louie. New Model Dielectric Function and Exchange-Correlation Potential for Semiconductors and Insulators, *Phys. Rev. B* 25:6310-6316 (1982).

50. A. P. Pathak and M. Yussouf. Charged Particle Energy Loss to Electron Gas. *Phys. Stat. Solid* 649:431-441 (1972).

51. L. C. Emerson, R. D. Birkhoff, V. E. Anderson, and R. H. Ritchie. Electron Slowing-Down in Irradiated Silicon. *Phys. Rev. B* 7:1798-1811 (1973).

52. C. J. Tung, J. C. Ashley, R. D. Birkhoff, R. H. Ritchie, L. C. Emerson, and V. E. Anderson. Electron Slowing-Down Flux Spectrum in Al_2O_3. *Phys. Rev. B* 16:3049-3055 (1977).

53. B. Baron, D. Hoover, and F. Williams. Vacuum Ultraviolet Photoelectric Emission from Amorphous Ice. *J. Chem. Phys.* 68:1997-1999 (1978).

54. C. R. Brundle and M. W. Roberts. Surface Sensitivity of HeI Photoelectron Spectroscopy (UPS) for H_2O Absorbed on Gold. *Surf. Sc.* 38:234-236 (1973).

55. K. Siegbahn. *ESCA Applied to Free Molecules*. North-Holland Publishing Co., Amsterdam (1969).

56. A.E.S. Green and J. H. Miller. *Atomic and Molecular Effects in the Physical Stage in Physical Mechanisms in Radiation Biology*. CONF-721001:68-115 (1974), Eds. R. D. Cooper and R. W. Wood, Technical Information Center, Office of Information Services, U.S. Atomic Energy Commission.

57. C. Møller. Über den Stoss zweier Teilchen unter Berücksichtigung der Retardation der Kräfte. *Z. Physik* 70:786-795 (1931).

58. W. R. Ferrell, R. H. Ritchie, and T. L. Ferrell. Identical Particle States and Operators and the Case of Two-Electron Scattering. *Am. J. Phys.* 52:915-920 (1984).

59. R. H. Ritchie and J. C. Ashley. The Interaction of Hot Electrons with a Free Electron Gas. *J. Phys. Chem. Solids* 26:1689-1694 (1965).

60. C. J. Tung, R. H. Ritchie, J. C. Ashley, and V. E. Anderson. *Inelastic Interactions of Swift Electrons in Solids*. ORNL-TM-5188, Oak Ridge National Laboratory, Oak Ridge, Tennessee (1976).

61. Some time after our work was done, Ashley (Ref.28) carried out a calculation of the stopping power of water for electrons. He compared his results with those displayed in Ref. 60 and found differences at low energies that are $\leq 25\%$. He states that our early calculation of the stopping power (ref.8) of water used an exchange correction chosen to make the results agree with values recommended in ICRU Report 16 and attributes this to a private communication with one of us (RHR). This statement is based on a misunderstanding. No exchange correction was used in the results presented in Ref. 8. Further, exchange corrections according to Eq. (29) were incorporated into the OREC code soon afterwards.

62. (a) H. G. Paretzke, J. E. Turner, R. N. Hamm, H. A. Wright, and R. H. Ritchie. Calculated Yields and Fluctuation for Electron Degradation in Liquid Water and Water Vapor, *J. Chem. Phys.* 84:3182-3188 (1986).
(b) J. E. Turner, H. G. Paretzke, R. N. Hamm, H. A. Wright, and R. H. Ritchie. Comparison of Electron Transport Calculations for Water in the Liquid and Vapor Phases. *Proc, 8th Symposium on Microdosimetry*, Julich, EUR8395 en, 175-185 (1982).

63. J. J. Quinn. Range of Excited Electrons in Metals. *Phys. Rev.* 126:1453-1457 (1962).

64. K. K. Thornber and R. P. Feynman. Velocity Acquired by an Electron in a Finite Electric Field in a Polar Crystal. *Phys. Rev. B* 1:4099 (1970).

65. J. C. Ashley and R. H. Ritchie. Quantum Treatment of Real Transitions. *Phys. Rev.* 174:1572-1577 (1968).

66. T. McMullen, B. Bergersen, and P. Jena. Momentum of a Fast Particle in a Dissipative Medium with an Excitation Threshold. *J. Phys.* C9:975-989 (1976).

67. C. C. Sung and R. H. Ritchie. The Energy-Loss Spectra of Fast Charged Particles Passing Through a Thin Metallic Film. *J. Phys.* C14:2409-2421 (1981).

68. N. Bohr. The Penetration of Atomic Particles Through Matter. Eq. 3.1.8, *Mat.-Fys. Medd. Kgl. Dansk. Vidensk. Selsk.*, 18:No.8, 1-144 (1948).

69. R. H. Ritchie. Geminate Recombination in Dispersive Media. To be published.

70. J. E. Inglesfield. Plasmon Satellite in Core-Level Photoemission, *J. Phys.* C16:403-416 (1983).

71. R. H. Ritchie. Quantal Aspects of the Spatial Resolution of Energy-Loss Measurements in Electron Microscopy I. Broad-Beam Geometry. *Phil. Mag.* A44:931-942 (1981).

72. R. H. Ritchie and A. Howie. Inelastic Scattering Probabilities in Scanning Transmission Electron Microscopy. *Phil. Mag.* A58:753-767 (1988).

73. U. Fano. *The Formulation of Track Structure Theory in Charged Particle Tracks in Solids and Liquids*, Eds. G. E. Adams, D. K. Bewley, and J. W. Boag, pp. 1-8, The Institute of Physics, London (1970).

Ritchie et al.

74. M. Burton. Fundamental Processes in Radiation Chemistry: Effects of the State of Aggregation, Discuss. *Faraday Soc.* 36:7-18 (1963).

75. J. L. Magee. Elementary Processes in the Action of Ionizing Radiation. In *Comparative Effects of Radiation*, J. Wiley, 130-147 (1960).

76. R. H. Hamm, R. N. Hamm, J. E. Turner, H. A. Wright, J. C. Ashley, and G. J. Basbas. Physical Aspects of Charged Particle Track Structure. *Nucl. Tracks. Rad. Meas.* 16:141-155 (1989).

77. M. Scheinfein, A. Muray, and M. Isaacson. Electron Energy Loss Spectroscopy Across a Metal-Insulator Interface at Sub-Nanometer Spatial Resolution. *Ultramicroscopy* 16:233-240 (1985).

78. P. E. Batson, K. L. Kavanaugh, J. M. Woodall, and J. W. Mayer. Electron-Energy Loss Scattering Near a Single Misfit Location in a Dielectric Medium of Randomly Distributed Metal Particles. *Phys. Rev. Lett.* 57:2729-2732 (1986).

79. A. L. Bleloch, A. Howie, R. H. Milne, and M. C. Walls. Elastic and Inelastic Scattering Effects in Reflection Electron Microscopy. *Ultramicroscopy* 29:175-182 (1989).

80. D. Imeson, R. H. Milne, S. D. Berger, and D. McMullan. Secondary Electron Detection in the Scanning Transmission Electron Microscope. *Ultramicroscopy* 17:243-250 (1985).

81. R. H. Ritchie, A. Howie, P. M. Echenique, G. J. Basbas, T. L. Ferrell, and J. C. Ashley. Plasmons in Scanning Transmission Electron Microscopy Electron Spectra. *Scanning Microsc. Supp.* 4 (1990).

82. J. E. Turner, R. N. Hamm, H. A. Wright, R. H. Ritchie, J. L. Magee, A. Chatterjee, and W. E. Bolsh. Studies to Link the Basic Radiation Physics and Chemistry of Liquid Water, *Rad. Phys. Chem.* 32:503-510 (1988).

83. Note that J. L. Magee [Electron Energy Loss Processes at Subelectronic Excitation Energies in Liquids, *Can. J. Chem.* 55, 1847 (1977)] has considered carefully the division of energy losses between those due to 'indirect' interactions with the electric field induced in the medium by the electron, and 'direct' interactions with the short range quantum forces occurring in direct collisions. Recently U. Fano [Short- and Long- Range Interactions of Slow Electrons in Condensed Matter: Effects on Reflection and Transmission, *Phys. Rev.* A36, 1919 (1987)] has discussed subexcitation electron energy losses in a similar way. Mann and Green [The Fate of the Dry Electron - Preliminary Investigation, *Int. J. Quant. Chem.* 39:47-57 (1991)] have modeled the loss of subexcitation electrons in water as well as the hydration process.

84. W. E. Bolch, J. E. Turner, H. Yoshida, K. B. Jacobson, R. N. Hamm, H. A. Wright, R. H. Ritchie, and C. E. Klots. Monte Carlo Simulation of Indirect Damage to Biomolecules Irradiated in Aqueous Solution - The Radiolysis of Glycylglycine, ORNL/TM-10851, Oak Ridge National Laboratory, Oak Ridge, Tennessee (1988).

85. H. Yoshida, W. E. Bolch, K. B. Jacobson, and J.E. Turner, Measurement of Free Ammonia Produced by X-Irradiation of Glycylglycine in Aqueous Solutions, *Radiat. Res.* 121:257-261 (1990).

86. H. A. Wright, J. L. Magee, R. N. Hamm, A. Chatterjee, J. E. Turner, and C. E. Klots, Calculations of Physical and Chemical Reactions Produced in Irradiated Water Containing DNA, *Radiat. Prot. Dos.* 13:135-136 (1989).

87. J. E. Turner, W. E. Bolch, H. Yoshida, K. B. Jacobson, H. A. Wright, R. H. Hamm, R. H. Ritchie, and C. E. Klots, Radiation Damage to a Biomolecule: New Physical Model Successfully Traces Molecular Events, *Int. Jour. Radiat. Appl. Instrum.*, in press.

88. C. D. Jonah, M. S. Matheson, J. R. Miller, and E. J. Hart. *J. Phys. Chem.* 80:1267-1270 (1976). C. D. Jonah and J. R. Miller. *J. Phys. Chem.* 81:1974-1976(1977). T. Sumiyoshi and M. Katayama. *Chem. Phys. Lett.* 1881-1990 (1982).

89. K. A. Long, H. G. Paretzke, F. Muller-Plathe, and G.H.F. Diercksen. Calculations of Double Differential Cross Sections for the Interaction of Electrons with a Water Molecule, Clusters of Water Molecules and Liquid Water. *J. Chem. Phys.* 91:1569-1578 (1989).

90. Y. J. Xu, G. S. Khandelwal, and J. W. Wilson. Proton Stopping Cross Sections of Liquid Water. *Phys. Rev.* A32:629-632 (1985).

91. R. Voltz. Thermalization of Subexcitation Electrons in Dense Molecular Matter. In *Excess Electrons in Dielectric Media*. To be published.

92. R. H. Ritchie, A. Gras-Marti, and J. C. Ashley. The Theory of Track Formation in Insulators Due to Densely Ionizing Particles. *Proc. 12th Werner Brandt Int. Conf. on Penetration of Charged Particles in Matter*, Sept. 4-7, 1989, CONF-890921, Oak Ridge National Laboratory, Oak Ridge, Tennessee.

Discussion

Geacintov: You were discussing collective oscillations in DNA and you showed the energy loss spectrum. I was wondering if there was water included in those ideas about the mechanism?

Ritchie: This is based on dry DNA.

Geacintov: It would be interesting to see what it would look like in the wet state. And also, it might depend on the conformation of the DNA, which can assume different forms depending on the environment. I was wondering if anyone is thinking along those lines.

Ritchie: I doubt if the loss function of DNA in the wet state is very different from that of DNA in the dry state. In fact, Mike Isaacson tried to "fingerprint" some nucleic acid bases using electron energy loss spectra with the obvious aim of reading the sequences on DNA molecules [see M. Isaacson, *J. Chem. Phys.* 56:1803 and 1813 (1972)]. He found only small differences in the energy loss functions of the bases measured. The use of such a technique in genome work is made still more difficult by the fact that a swift electron interacts with matter through the long-range Coulomb force; this leads to uncertainty in the spatial precision of the measurements.

Weinstein: Is there a difference between water and ice?

Ritchie: Yes.

Geacintov: Didn't you show some data with adenine that was different, or was it in the wet state or dry state? I have forgotten. You showed two curves with adenine; one was a dotted one and the other was a solid line.

Ritchie: The one was from electron energy loss measurements, and the other was calculated using measured optical dielectric functions.

Geacintov: The absorption spectrum is very often influenced by the environment?

Paretzke: It influences the oscillator strength.

Geacintov: Are you saying that the environment would not have any influence on the energy absorption.

Paretzke: Certainly it has, but we are talking here in the frame work of radiation biology, and we are talking about differences which are seen through the effects of ionizing radiation on deactivation of cells. The difference between ^{60}Co and X-rays. We are trying to explain these big differences. We are not talking about 10 percent differences in the peak locations of those oscillator strengths that are usual in hydrogen. So in this framework where we try to understand basic mechanism, it might be of interest for the basic physics accounting for a 10 percent peaking. But, it doesn't matter in the context of this workshop, that is what I am saying.

Varma: Are you saying that these changes in the environment, whether it is dry or whether it is aqueous, do not make any difference in terms of oscillator strength?

Paretzke: The decay of the excited states caused by these perturbations might be completely different, but you don't see them.

Kopelman: Well covered. Do we still call those "collective states" or are those highly localized states, because usually it is a localized state that is independent of environment. In a collective state, in my experience, it doesn't depend very much.

Ritchie: But the electron density is about the same for all of these materials so that the upward shift from a transition with a primarily single-particle character to that with a substantial collective component would be about the same.

Kopelman: How delocalized is this collective state?

Ritchie: This delocalization may be characterized in terms of the probability of finding the collective mode as a function of distance from the path of the swift charged particle that excites the mode. The length that characterizes this spatial extension of this probability distribution is v/ω_I, where v is the speed of the particle and ω_I is the eigenfrequency of the collective motion.

Zaider: What do you mean by localized?

Ritchie: Localization of an initially unlocalized excitation is understood to consist of the decay of, e.g., a collective excitation in a condensed medium into an (or perhaps more than one) excited electron. This electron may be regarded as localizable in a region of space with dimensions much smaller than that which characterized the spatial extension of the collective state. As indicated in my talk and in the previous remarks, the length characterizing the extension of a collective mode, v/ω_f, is usually much larger than the corresponding length, the Debroglie wavelength, of the excited electron.

Ritchie: The confidence is not extremely high, but we have considerable experience in doing similar extrapolations to obtain the complete response functions of many different solids. We find good agreement with experimental data on stopping power, inelastic mean free path, and, in a few cases where data are available, with information about the distribution of energy losses for electrons penetrating thin layers of solids. In any case, we are using the only reasonable scheme that has been proposed to date for constructing the entire energy loss function; and since the dependence of our loss function on ω and q is constrained by very general sum rules, we feel that it should give quite reasonable results overall in the present application.

Paretzke: Energy transport in condensed phase relates best not only to the energy and the momentum aspect of the localization and the decay aspect of energy deposition. Would you dare to make any prediction regarding research needs, about information we could get in the next five years or so on the localization of the loss, and also on the excited states, or the decay of them in a condensed phase? These are the two things which describe much more spatial extent of the plasmon and the decay, not too much energy transfer for the excited state. It doesn't matter very much in regard to our uses of these states to analyze radiation biology effects. It is much more important to know where they localize, how large plasmons are, where they decay afterwards and how they decay. Do you see any changes of this in the next five years? Which route would you take to solve this puzzle?

Ritchie: There are movements in that direction in solid-state physics. For example, people such as Archie Howie at the Cavendish Lab are now measuring distributions of energy losses by high-energy, narrowly focused ($\sim.2$ μm) electrons in scanning transmission electron microscopy to surface excitations and to localized regions of condensed matter. Such measurements are made as a function of position in compound targets and give evidence for the existence of unlocalized excitations and their decay. Such measurements are more difficult, but not impossible, for solid H_2O targets or for biological samples; they should be, and probably will be, carried out in the future, perhaps within the next five years. Still more difficult is the experimental study of the localization products of initially unlocalized excitations. I know of only a few such attempts by workers in solid-state physics, and none for biologically interesting condensed media. The importance of such experiments for establishing the physical basis of radiation damage can hardly be overemphasized.

Paretzke: Yes, we did it for the electrons and protons to look at the loss function.

Zaider: I would like to come back to the plasmon in water. I am sure it is very important for the spatial distribution of energy deposition. There are some workers, T. Kloos, for example, who would argue that there are no plasmons.

Ritchie: I have read the paper to which you refer, but did not realize that the author expressed that opinion there. In any case, I believe that he is in a very small residual minority. The majority accept that plasmons can exist in many condensed media.

Charged-Particle Transport in the Condensed Phase

Marco Zaider[a]

Abstract

Traditionally, studies of the biological effects of ionizing radiation have rested on the triumvirate: (gas-phase) radiation physics, biophysical modeling, and radiation biology. Two technical developments, the advent of supercomputing as a routine tool in quantum solid-state material science and molecular dynamics on the one hand, and molecular biology on the other hand, have created—perhaps for the first time—the possibility of directly linking a more realistic description of the radiation field to observable events at biomolecular level. It also becomes increasingly clear that the identification of specific molecular targets imposes a challenge to the radiation physics community to be equally specific in treating the energy-deposition stage of radiation action. In this paper: a) I review—and exemplify with results from our own work—the current status in Monte Carlo simulation of gas-phase material (particle transport and stochastic chemistry); b) examine the link between these essentially geometric representations of the track and the concept of "spatial distribution of energy deposition," a staple in radiation modeling; c) advocate an effort towards developing conceptually and calculationally, the field of solid-state microdosimetry; and d) describe methods based on semi-empirical Hamiltonians or quasi-particle techniques for obtaining the frequency-dependent and wave-vector-dependent dielectric response function for biomolecular crystalline systems, which are the main ingredients for describing charged-particle transport.

> *"Composed out of scattered fragments and snatches of movements."*
> *(Epigraph set by Beethoven on one of the copies of his 14th quartet)*

Introduction

Understanding the effects of low doses of radiation on living organisms requires information on charged-particle transport in biomaterials (water, biopolymers). This assertion is based on the following: The physico-chemical events held as relevant for describing radiation quality (loosely defined as those traits of the radiation field which give it a certain relative biological effectiveness) occur in the first microsecond following energy deposition in the biological target. During this time we distinguish a physical stage (about 10^{-15} sec) and a physico-chemical stage (up to about 10^{-6} sec) where diffusion and fast interactions of radical species produced by radiation yield relevant products. The element which determines the relative effectiveness of radia-

(a) Center for Radiological Research, College of Physicians & Surgeons of Columbia University, New York

Physical and Chemical Mechanisms in Molecular Radiation Biology
Edited by W.A. Glass and M.N. Varma, Plenum Press, New York, 1991

tion is the spatial distribution of energy deposition events; this is basically the only ingredient which is different among radiations. It is during the first microsecond that radical species have a non-homogeneous spatial distribution and thus bear the "signature" of the track. Charged-particle transport yields the probability to observe certain spatial configurations of energy transfer.

A complementary—and closely related—approach to these problems is microdosimetry where, rather than transporting tracks in quasi-infinite media, one defines a volume containing the radiation-sensitive biological material and then asks for the probability of a certain energy deposition therein. The conditions under which these two methods (microdosimetry and particle transport) are providing equivalent information need to be examined carefully: a particle track (simulated event by event, as explained below) is generally not physically observable (in a quantum mechanical sense) while a microdosimetrical spectrum can be directly measured.

The notion of "spatial pattern of energy deposition" does not generally have a clear meaning. By and large, charged-particle transport is performed with Monte Carlo techniques. An essential assumption in a Monte Carlo calculation[1] is that there are well-localized (and, in most practical applications, randomly and uniformly distributed) targets against which the incident particle scatters *one target at a time*. The spatial pattern is then given by the position of scatterers where energy was deposited. This scheme describes correctly interactions in a gaseous medium where each target is represented by a molecule or an atom. In a crystalline medium (e.g., a biopolymer), on the other hand, there are strong correlations between the positions of the scattering centers (electrons, ions), as well as coherent scattering off large groups of atoms. Strictly speaking, then, one should regard the entire biopolymer as target: localizing the site of energy absorption inside the polymer becomes an exercise similar to, say, attempting to localize an ionization event within a molecule. As a corollary I note that, at least *conceptually*, microdosimetry does not depend on the possibility of defining a track as a collection of point-like energy transfers.

To date, track structure and microdosimetric information on energy deposition come mainly from gas-phase experiments (Rossi proportional counters) or gas-phase Monte Carlo transport. In the latter, the cross sections used as input are sometimes scaled to represent the strength of the interactions in a liquid (of identical composition) as well as such condensed-phase phenomena as plasmon excitation, although it is important to remember that *conceptually* the calculation remains anchored in the basic assumption characterizing a gaseous system (see above). Whether such hybrid descriptions serve well as models for charged particle interactions with otherwise highly correlated targets (e.g., DNA, structured water) depends on i) the degree of accuracy to which models of radiation action are sensitive, and ii) the degree of detail available in the description of the biological target. The extensive progress made in recent years in molecular biology, the identification of specific oncogenes, and the prospect of sequencing the DNA (the Genome project) certainly suggest that physical events should be described at a similar level of thoroughness.

In the first part of this paper, I review briefly our work on electron transport in water vapor and the stochastic (Monte Carlo) treatment of subsequent radical interactions. This is the status quo. I discuss next the implications and limitations of this basically geometric view of track structure to modeling in radiobiology as I advocate the necessity of more realistic, biopolymer-specific calculations. The main part of this paper is concerned with practical approaches to charged-particle transport in condensed media. I suggest calling this area of study "condensed-state microdosimetry."

Charged-Particle Transport in Gaseous Material

Elements of Monte Carlo Simulation

There are currently several codes capable of simulating event by event the passage of a charged particle through an extended gas target.[2-4] The objective of such a calculation, performed with Monte Carlo techniques, is to obtain samples of particle trajectories (a trajectory is a collection of straight-line segments connecting consecutive elastic or non-elastic interaction events).

The following question is addressed in a typical Monte Carlo calculation: Given the position and velocity of a particle at a certain time, what is the position of the next scattering center and the physical process responsible for the interaction? The computing procedure which solves this problem (described below) is repeated sequentially for the primary particle as well as the secondary, higher-generation particles (usually electrons) resulting from nuclear reactions or impact ionization. Particles are followed until they slow down to a preset threshold energy, or simply until a certain penetration depth has been attained.

Let σ_i be the cross section for the process "i" (e.g., excitation) and ρ the density of targets (scattering centers per unit volume). The probability of a type i scattering event along the infinitesimally short segment, dx, is

$$dp_i = \rho \cdot \sigma_i \, dx \tag{1}$$

and from the definition of a cross section, the probability of a free flight, x, followed by this type of scattering at dx about x is

$$f(x)\,dx = \exp\left[-\int_0^x \rho\,\sigma_T(t)\,dt\right]\rho\,\sigma_i\,dx \tag{2}$$

where σ_T is the total interaction cross section:

$$\sigma_T = \sum_i \sigma_i \tag{3}$$

Numerous techniques are available for generating random numbers, x, according to a given distribution, $f(x)$.[5] For the purpose of illustration, I describe here one such method. Let

$$F(x) = \int_{-\infty}^x f(x')\,dx' \tag{4}$$

be the cumulative distribution in x. The random variable

$$y = F(x) \tag{5}$$

is uniformly distributed between 0 and 1; indeed, $F(x)$ being a monotonically increasing function from 0 to 1,

$$Prob\{y \leq y_o\} = Prob\{F(x) \leq F(x_o)\} = Prob\{x \leq x_o\} = F(x) = y \qquad (6)$$

A standard feature of today's computers is a routine to generate pseudo-random numbers, ξ, uniformly distributed between 0 and 1. One obtains random numbers x by: i) generating a number ξ, and ii) from $F(x)=\xi$ calculate $x=F^{-1}(\xi)$. Obviously, to apply this procedure $f(x)$ must be integrable [to obtain $F(x)$]; also, an analytical representation, $F^{-1}(x)$, must exist.

Charged-Particle Transport

The necessary input to a Monte Carlo calculation consists of cross sections covering all particles of interest. This includes energy and angular distributions for the secondary particles. As a general rule our approach has been to use—as much as possible—experimental data represented in suitable computational form with the aid of parametric functions. Where experimental data were not available, we have used theoretical calculations.

The code DELTA[4] transports electrons in water vapor, simulating the elastic and inelastic (ionization and excitation) interactions undergone by the primary electron and all higher-generation secondaries that are produced. A detailed description of this code can be found in Ref. 4. Figures 1 and 2 exemplify the treatment of cross-sectional input data. Total elastic cross sections, $\sigma_E(T)$, for electrons of kinetic energy T between 0.3 and 500 eV are well-characterized experimentally.[6-9] Using the expression cited by Porter and Jump,[14] we have fitted Eq. (7) to the data of Fig. 1:

$$\sigma_E(T) = T_1 \left\{ \frac{T^x}{\eta(\eta+1)\left[V^{2+x}+T^{2+x}\right]} + \sum_{n=1}^{2} \frac{F_n G_n^2}{(T-E_n)^2 + G_n^2} \right\} \qquad (7)$$

Here T_1, V, x, F_n, and G_n are adjustable parameters. The screening parameter η is inversely proportional to T. The resulting fit is shown as a solid line in Fig. 1. The double differential data of Fig. 2 have been similarly represented: for energies above 0.2 keV the Rutherford formula (modified for screening) is used; at lower energies we use again an empirical analytical expression.[5] Electron ionization and excitation are treated in an analogous manner.

The computer code PROTON transports protons and heavy ions. The three main processes included in the code are ionization, excitation and charge exchange. Figure 3 shows a typical example of data on electron production by 0.3 MeV protons incident on water vapor[16] as well as their parameterization. In the case of heavy-ion transport, since only sparse data on double-differential cross sections exist, the cross sections are simply scaled for equal-velocity protons by Z^2 to obtain the cross sections for an ion of charge Z. This procedure is justified for fully ionized ions (typically when $\beta = v/c \geq 0.025 \cdot Z^{2/3}$). For lower energies an effective charge, Z^*, is used:

$$Z^* = Z\left[1 - \exp(-125 \cdot \beta \cdot Z^{-2/3})\right] \qquad (8)$$

The results of a charged-particle simulation consist of a record of positions, energy deposited, and type of interaction associated with each nonelastic event (see Fig. 4).

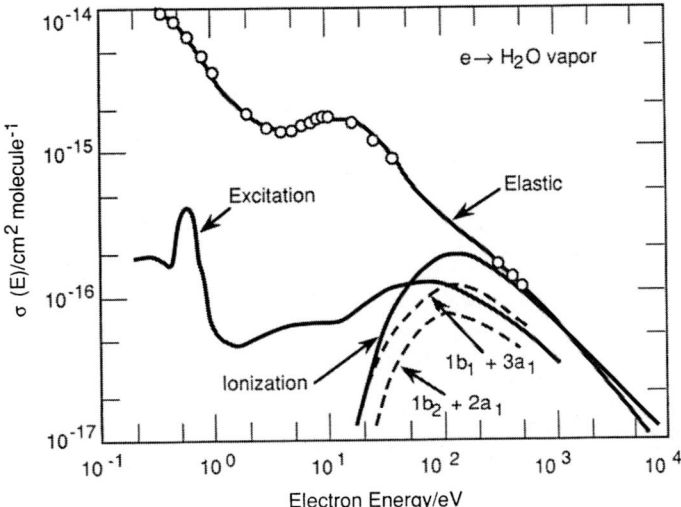

Figure 1. Total cross sections for electron interaction with water vapor. The experimental data for elastic scattering are from Refs. 6 to 9. The data shown for ionization are from Refs. 10 and 11. The solid curves are analytical representations of the cross sections as used in our Monte Carlo codes.[4]

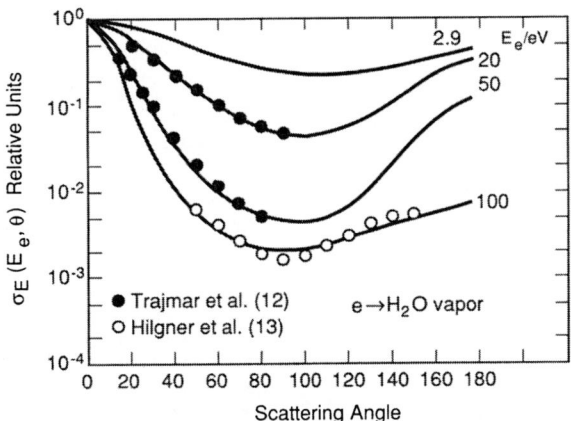

Figure 2. Differential cross sections for elastic scattering of electrons on water vapor as used in the code DELTA.[4] The data are from Refs. 9, 12, and 13.

Monte Carlo Stochastic Chemistry

The picture shown in Fig. 4 is an "imprint" left by a 1-keV electron in water vapor at about 1 femtosecond after its passage. The fate of this track during the next microsecond can be simulated using stochastic chemistry.[17]

In the time interval 10^{-15}-10^{-11} seconds, the following mechanisms involving ionization, Eqs. (9-11), and excitation, Eq. (12), are believed to occur in liquid water:[18]

$$e^- + H_2O \rightarrow H_2O + 2e^- \tag{9}$$

$$\rightarrow OH^+ + H + 2e^- \tag{10}$$

$$\rightarrow H^+ + OH + 2e^- \tag{11}$$

$$e^- + H_2O \rightarrow H_2O^* + e^- \rightarrow OH + H + e^- \tag{12}$$

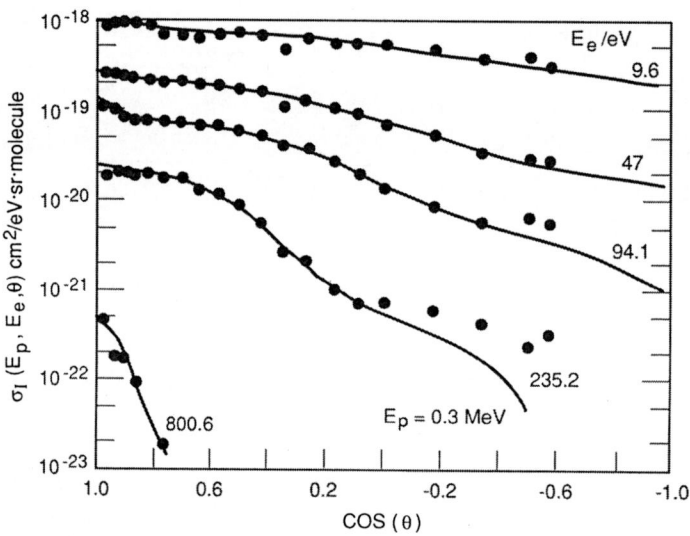

Figure 3. Differential electron production cross section for 0.3 MeV protons incident on water vapor. The data points are from Ref. 16.

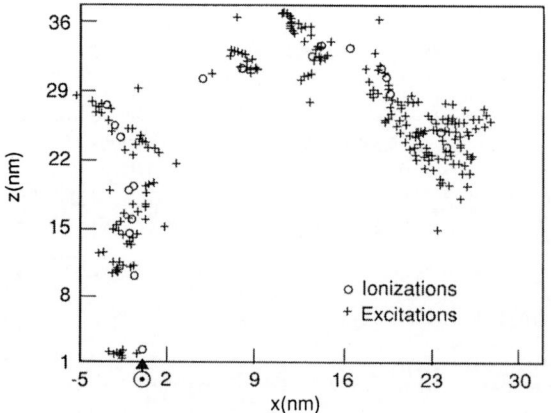

Figure 4. Projection on the x-z plane of the trajectory of a 1 keV electron. The electron starts at the origin in the positive z direction.

Furthermore, H_2O^+ and OH^+ undergo fast reactions with water:

$$H_2O^+ + H_2O \rightarrow H_3O^+ + OH \qquad (13)$$

$$OH^+ + H_2O \rightarrow H_3O^+ + O \qquad (14)$$

Electrons (secondary or primary) slow down to thermal energies and become hydrated at about 10^{-13} seconds. We are thus faced, at about 10^{-11} seconds, with tracks where non-elastic events are converted to radical or molecular species (e_{aq}, OH, H^+, H, and O). These radicals diffuse and interact among themselves or with biopolymers if those are present. Damage to biopolymers produced by radiolysis products is termed "indirect action."

The following scheme [originally proposed by Clifford et al.[19,20]] is used to describe the fast chemical processes: Assume that the type and position of all species present at 10^{-11} seconds are known. Within a good approximation the species diffuse and interact pairwise in a manner described by Smoluchowski's equation:[21]

$$F(r,t) = \frac{a}{r} \cdot erfc\left[\frac{r-a}{(4D't)^{0.5}}\right] \qquad (15)$$

This expression gives the probability that two radicals, initially a distance r apart, interact by time t. In Eq. (15) $D' = |D_1 - D_2|$ is the coefficient of relative diffusion of the two species and "a" is the encounter radius (when $r < a$ the two radicals interact). This latter quantity can be obtained from Debye's equation:[22]

$$k_{12} = 4\pi a D_{12} N_A \cdot \left[\frac{Q}{e^Q - 1}\right] \qquad (16)$$

where

$$Q = Z_1 Z_2 e^2 / a\epsilon k_B T \qquad (17)$$

and k_{12} = rate constant for the reaction between species 1 and 2 of charge Z_1 and Z_2, respectively, D_{12} = total diffusion constant ($D_1 + D_2$), ϵ = effective dielectric constant of the medium, k_B = Boltzmann's constant, T = absolute temperature, and N_A = Avogadro's number. In Table I, I list possible reactions between species, the reaction constants (k_{12}), and the encounter radii.[23]

The need for a stochastic treatment of these chemical interactions (as opposed to the usual deterministic-kinetic approach) is a result of the small number of species involved. The second crucial element is the non-homogeneous initial spatial distribution of species. As it turns out, when using stochastic treatments one cannot use as input an initial distribution of species averaged over many tracks; rather, interactions within each track have to be followed separately. Only at the end, if desired, averaged results could be calculated. The code described here is the result of these constraints.

Table I. Reaction Constants

Reaction	$k \cdot 10^{10}$ (dm^3/mole·s)	a (nm)
$OH + OH \rightarrow H_2O_2$	0.5	0.23
$OH + e_{aq} \rightarrow H_2 + 2OH^-$	3.0	0.61
$e_{aq} + e_{aq} \rightarrow H_2 + 2OH^-$	0.5	0.4
$H + OH \rightarrow H_2O$	3.2	0.42
$H + OH^- \rightarrow H_2O$	14.3	1.9
$H + H_2O_2 \rightarrow H_2O + OH$	0.016	0.002
$e_{aq} + H^+ \rightarrow H$	2.3	0.48
$H + H \rightarrow H_2$	1.3	0.21
$e_{aq} + H \rightarrow H_2 + OH^-$	3.0	0.32
$e_{aq} + H_2O_2 \rightarrow OH^- + OH$	1.2	0.27

The simulation of radical diffusion and interaction proceeds along the following steps:

1. Calculate the distribution of distances between pairs of interacting radicals.
2. Eliminate species which are already within the encounter radius and (where applicable) introduce new radicals (cf. Table I).
3. Repeat step 1 if necessary.
4. Generate encounter times for all pairs of species according to Eq. (15).
5. Consecutively, have pairs of radicals interact, repeating after each interaction (steps 1 and 2) if new reactive species are produced.

One obtains in this fashion the number of species present as a function of time post exposure.

Figure 5 shows an example of a calculated distribution of distances for the pair OH-H. This was obtained with the code DELTA for 20-MeV electrons. Also shown in the figure is the distribution which is obtained if species are assumed to have an initial Gaussian spatial distribution around the origin (this diffusion-controlled distribution is used almost exclusively by radiation chemists). The differences between these two distributions lead to very different patterns of radical decay. In Fig. 6 we compare the stochastic chemistry result for the solvated electron with measurements by Jonah et al.[24] and Sumioshi and Katayama.[25] The number of species present is normalized per 100 eV of energy deposited by the incident projectile (the G value). The two sets of numbers have very similar shapes, indicating that the decay mechanism of $G(e_{aq})$ with time (i.e., the stochastic treatment plus the initial pair distance distribution) must be correct. The failure of the calculated G values to reproduce the absolute magnitude of the measurements is not unexpected: The experimental data of Fig. 6 were obtained with liquid water while the Monte Carlo electron transport was performed in water vapor; it is known that in the liquid a larger proportion of interactions lead to ionization.[26]

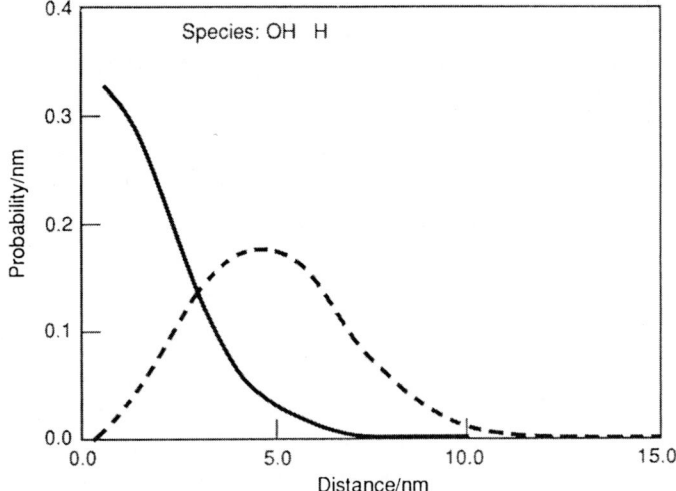

Figure 5. Distribution of distances for the radical pair *H,OH* calculated with simulated tracks (solid) or with Gaussian initial distributions (see text).

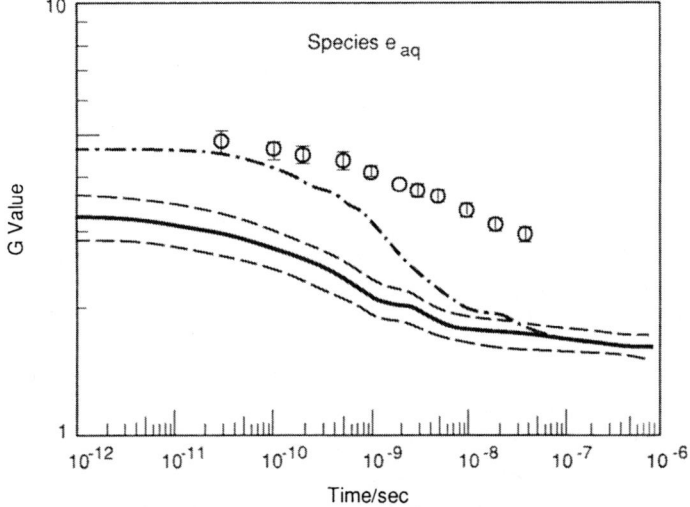

Figure 6. Measured and calculated yields (*G* values) of the species e_{aq} in the pulse radiolysis of water exposed, at low dose rates, to 22-MeV electrons. The experimental data are from Refs. 24 and 25. The solid line is the result of the stochastic chemistry calculation. The dashed curves correspond to one standard deviation in the calculated yields. Also shown are results obtained with the stochastic code but with Gaussian initial distributions; this shows the important role played by the correct description of the initial spatial distribution of species.

Radiation Field Descriptors - A Diversion

Models of radiation action are mathematical formalisms which link observable quantities describing the radiation field (I call them "field descriptors") and the biological response. At the level of primary events (ionizations, excitations, radicals) the spatial structure of a charged particle track is not amenable to measurement and, in any case, its entire informational content will be impractical to handle. The information generated in a Monte Carlo track simulation is customarily reduced to simpler quantities. Historically, the first such quantity was perhaps absorbed dose. Other examples include the linear energy transfer (LET), the radial distribution of absorbed dose, and the specific energy, z. This latter quantity is a stochastic variable defined as the energy deposited in a given volume per unit mass; it is the stochastic counterpart of absorbed dose. Measurements and calculations of the probability distribution functions of specific energy are the objective of microdosimetry.[27]

The link between Monte-Carlo-generated charged-particle tracks and, for instance, microdosimetric field descriptors is based—as was already indicated—on two fundamental approximations:

A1. The track is viewed as a geometric entity, that is as a collection of energy-transfer points.

A2. Scattering centers (targets) are randomly and uniformly distributed, or—to put it differently—one is dealing with unstructured biological material.

The origin of these approximations can be traced back to i) the fact that the precise nature of the targets is not known, and consequently that regularly-shaped volumes filled with gas are being used to model the cellular material, ii) the unavailability of tissue-equivalent solid state detectors, and iii) mathematical convenience. As a result, microdosimetry is reduced to studying the random overlap of two bodies, the track and the cell, with many results from the field of integral geometry immediately (but not obviously!) translatable into microdosimetric language. An example that epitomizes this state of affairs is the often used expression:[28]

$$z_D = \int \frac{t(r) \cdot s(r)}{4\pi r^2 m} dr \qquad (18)$$

relating the dose-averaged specific energy, z_D, to two independent functions describing, respectively, the track and the exposed volume. The point-pair distance distribution of transfer points is $t(r)dr$, and $s(r)dr$ is proportional to the point-pair distance distribution of targets inside the sensitive volume (note that the volume does not have to be convex, but inside each convex sub-volume a uniform density of targets is still required). The $t(r)$ can be obtained from Monte Carlo simulated tracks,[29] and $s(r)$ can be calculated on pure geometrical grounds if the shape of the microdosimetric volume is known. The possibility of factoring z_D into $t(r)$ and $s(r)$ is a direct consequence of the two approximations, A1 and A2, above.

In the field of biophysical modeling, the generalized Theory of Dual Radiation Action[30,31]—but not its site-model version!—is also based on these two assumptions.

Many of the current gas-phase-specific assumptions become clearly untenable (and therefore unrealistic) in a physical calculation concerned with explicit cellular targets: Biopolymers are molecular crystals with highly ordered structure and possessing the translational symmetry characteristic of condensed matter. Electrons in molecular crystals are de-localized. An electron penetrating such a crystal moves in its conduction band, rather than as a free particle. Other phenomena appear (screening, collective excitations, excitons, solitons), exclusively associated with a large system of molecules. The phenomenon of energy absorption itself must now be considered time dependent as energy can "migrate" along the biomolecule.

To the extent that the message we are getting from existing models of radiation action is correct, namely, that relevant sensitive volumes in the cellular nucleus have dimensions of the order of 10-20 nm rather than several micrometers, condensed-state microdosimetry appears quite unavoidable.

Electron Transport in Crystalline Media

The main difficulty in performing condensed-phase transport calculation is the paucity of cross-sectional data. Such calculations must then necessarily rely on theoretical evaluations (ab initio or phenomenological) of the needed input.

General Considerations

Information on electron transport in condensed materials can be obtained through the agency of frequency- and wave-vector-dependent dielectric response function of the system, $\epsilon(\vec{q},\omega)$. Thus, the rate of energy loss of an electron of charge e and sufficiently high velocity v, dE/dt, is given by[32]

$$-\frac{dE}{dt} = \frac{2e^2}{\pi v} \int \frac{dk}{k} \int \omega \, Im[-1/\epsilon(k,\omega)] \, d\omega \tag{19}$$

where $\epsilon(k,\omega)$ is the dielectric response function of the medium and k and ω are, respectively, the wave vector and frequency. The quantity $Im[-\epsilon^{-1}]$ is termed the energy-loss function of the system.

An exact expression for $\epsilon(\vec{q},\omega)$ for homogeneous, isotropic systems is[32]

$$\frac{1}{\epsilon(\vec{q},\omega)} = 1 - \frac{4\pi e^2}{q^2} \sum_n |<n|\rho_q|0>|^2 \left[\frac{1}{\omega+\omega_n+i\delta} + \frac{1}{\omega_n-\omega-i\delta} \right] \tag{20}$$

Here $\Psi_n = |n>$ is the exact eigenfunction of the crystal's Hamiltonian with eigenvalue E_n, ω_n are excitation energies ($=E_n-E_0$), ρ_q is the Fourier transform of the density operator, $\rho(\vec{r})$, and δ is an infinitesimal positive number (the limit $\delta \to 0$ is taken at the end). The calculation of $\epsilon(\vec{q},\omega)$ requires, therefore, a description of the band structure of the system.

Although the Hamiltonian is exactly known, a solution of the Schrödinger equation to obtain the N body wave function, $\Psi(\vec{r}_1,\vec{r}_2,....,\vec{r}_N)$, is generally prohibitive in terms of computer resources. The usual simplification used is the so called "single-particle approximation," according to which each electron is assumed to be an independent particle moving in the field created by all other electrons and (fixed) ions. Thus in the Hartree approximation:

$$\psi(\vec{r}_1,\vec{r}_2,...\vec{r}_N) = \psi_1(\vec{r}_1) \psi_2(\vec{r}_2)...\psi_N(\vec{r}_N) \tag{21}$$

and one solves for each electron, i, the equation:

$$[H_0(\vec{r}) + V_{H,i}(\vec{r})]\psi_i(\vec{r}) = E_i\psi_i(\vec{r}) \tag{22}$$

where H_0 contains the kinetic energy and the interaction between the electron and the ions, and V_H—the Hartree potential—is

$$V_{H,i}(\vec{r}) = \sum_j \frac{e^2|\psi_j(\vec{r}_j)|^2}{|\vec{r} - \vec{r}_j|} d\vec{r}_j \tag{23}$$

$V_{H,i}(\vec{r})$ is the potential generated by all other electrons, j $(j \neq i)$, at the position $\vec{r}$ occupied by the electron i. If the Pauli exclusion principle is taken into account, then the total wave function has to be antisymmetric relative to any interchange of two electrons. This so called Hartree-Fock (HF) approximation adds an exchange potential

$$V_{ex}(\vec{r},\vec{r}') = -\sum_j \frac{\Psi_i(\vec{r})\Psi_i*(\vec{r}')}{|\vec{r}-\vec{r}'|} \qquad (24)$$

The exchange potential is non-local (depends on both $\vec{r}$ and $\vec{r}'$) and thus significantly complicates the equation.

The HF system of equations can be solved self-consistently by consecutive iterations, each step containing Hartree and exchange potentials calculated with single electron wave functions from the previous iteration. The main defect of the HF approach is that correlations between the position of the electrons are ignored: Within the single-particle formalism, electrons can get arbitrarily close to each other with no corresponding change in energy (due to Coulomb repulsion). Omitting the correlation energy (an effect which can be as large as the exchange energy) clearly results in excitation energies which are too large, typically by 50%.[33-37]

Another popular method, the local-density approximation (LDA)[33] tends to underestimate (again by some 50% to 100%) the energy gaps. It has been suggested to rigidly displace the conduction bands until the desired (i.e., experimental) gaps are obtained. This procedure—known as applying the scissors operator[35]—is in fact not very accurate.

For our calculations we have considered two possible approaches, both exemplified below:

B1. For materials where spectroscopic data are available, one can use semi-empirical methods, that is, Hamiltonians with fitting parameters. This procedure is fast and relatively simple because self-consistency is not invoked.

B2. In a relatively recent development, a number of band calculations have been performed[33-37] by taking advantage of the fact that the HF method corresponds to the first term (zero-th approximation) in a many body perturbation series, the next term of which is the so called "GW approximation" (see below). The excitation properties of the crystals thus studied have been found to be in remarkable agreement with the experimental data, therefore raising the perspective of true *ab initio* self-consistent calculations.

I shall illustrate method B1 with results from a study[38] concerned with the dielectric properties of crystalline water (cubic ice). This particular form of water was chosen as a model for "structured water," water bound to biomolecules (there is evidence that water in biological systems contains residues with tetrahedral structure—i.e., ice—over time scales of at least 10^{-10} sec).

Method B2 is relatively new, and it appears that many of the computational aspects encountered in its actual implementation need to be systematically tested for accuracy and range of validity. As an example, the replacement in the GW approximation of the HF exchange potential with the screened Coulomb potential requires the function $\epsilon(\vec{q},\omega)$. Obtaining the dielectric function is, under the best of circumstances, a rather demanding computational enterprise and, as a general rule, the tendency has been not to calculate $\epsilon(\vec{q},\omega)$ directly but rather to use approximate, *ad hoc* models (e.g., the plasmon-pole model).[36] The consequences of this simplification remain unclear.

We demonstrate our approach to implementing the GW approximation with results from a pilot study[39,40] on a "simple" polymer, trans-polyacetylene, which (partly because of the current vogue it enjoys in material science research) is well-characterized experimentally. The combination of two elements makes our calculation somewhat different: The use of HF wavefunctions as starting point in the perturbation series (all other calculations have used the LDA), together with a full calculation of $\epsilon(\vec{q},\omega)$.

In a periodic (translationally invariant) system describable in terms of one-electron wave functions, $|n\vec{k}> = \Psi_n(\vec{k},\vec{r})$, of energy $E_n(\vec{k})$, wave vector $\vec{k}$ and band index n, explicit expressions for $\epsilon(\vec{k},\omega)$ can be obtained. Thus, within the random phase approximation (RPA) one has[41]

$$\epsilon_{GG'}(\vec{q},\omega) = \delta_{GG'} + \frac{4\pi e^2}{|\vec{q}+\vec{G}| \cdot |\vec{q}+\vec{G}'| \Omega} \sum_{\vec{k},n,m} \frac{F_{(q+G)nm}(\vec{k}) F^*_{(q+G')nm}(\vec{k})}{E_m(\vec{k}+\vec{q}) - E_n(\vec{k}) - \omega - i\delta^+} \cdot$$

$$\cdot \left[f_n(\vec{k}) - f_m(\vec{k}+\vec{q}) \right] \tag{25}$$

This expression for the dielectric matrix includes local field corrections; G and G' are reciprocal lattice vectors, $f_n(\vec{k})$ is the occupation number for the state $|n\vec{k}>$ (we shall use the zero temperature limit, $f_n=2$ for occupied states, $=0$ otherwise), δ^+ is an infinitesimal positive number, Ω is the crystal volume, and

$$F_{qnm}(\vec{k}) = <n\vec{k}|e^{-i\vec{q}\vec{r}}|m\vec{k}+\vec{q}> \tag{26}$$

If local field corrections are neglected, only the $G=G'=0$ term is retained; otherwise, to calculate the energy loss function one needs to invert the matrix $\epsilon_{GG'}$ first and then select the $G=G'=0$ term. Note the difference between the RPA expression, Eq. (25), and the exact expression, Eq. (20).

A Semi-Empirical Tight-Binding Calculation of $\epsilon(\vec{q},\omega)$ for Cubic Ice[38]

Let $\Psi_n(\vec{k},\vec{r})$ be the one-electron wave function satisfying the stationary Schrödinger equation:

$$H_0 \Psi_n(\vec{k},\vec{r}) = E_n(\vec{k}) \Psi_n(\vec{k},\vec{r}) \tag{27}$$

We express Ψ as a linear combination of Bloch functions, Φ:

$$\Psi_n(\vec{k},\vec{r}) = \sum_i C_{ni}(\vec{k}) \, \Phi_i(\vec{k},\vec{r}) \tag{28}$$

where

$$\Phi_i(\vec{k},\vec{r}) = N^{-1/2} \sum_j \exp(i\vec{k}\vec{R}_j) \, \chi_i(\vec{r}-\vec{R}_j) \tag{29}$$

Here $\chi_i(\vec{r}-\vec{R}_j)$ is the molecular orbital of a water molecule in a unit cell located at $\vec{R}_j$. There are N unit cells in the crystal. Eqs. (28,29) ensure that Ψ_n satisfy the boundary conditions (Bloch's theorem). The coefficients $C_{ni}(\vec{k})$ of Eq. (28) are found by solving the secular equation

$$\sum_j \left[H_{ij}(\vec{k}) - E_n(\vec{k})S_{ij}(\vec{k})\right] C_{nj}(\vec{k}) = 0 \tag{30}$$

for each value of $\vec{k}$. In Eq. (30):

$$H_{ij}(\vec{k}) = \int \Phi_i^*(\vec{k},\vec{r})\, H_0\, \Phi_j(\vec{k},\vec{r})\, d\vec{r} \tag{31}$$

$$S_{ij}(\vec{k}) = \int \Phi_i^*(\vec{k},\vec{r})\, \Phi_j(\vec{k},\vec{r})\, d\vec{r} \tag{32}$$

The energy bands, $E_n(\vec{k})$, are obtained from the determinental condition

$$\det |H_{ij}(\vec{k}) - E_n(\vec{k})S_{ij}(\vec{k})| = 0 \tag{33}$$

We are using a Hamiltonian which approximates the crystal potential with a sum of molecular potentials. This procedure is justified by the tight-binding nature of the system. Thus,

$$H_{ij}(\vec{k}) = (E_i + E_j)S_{ij}(\vec{k}) - K_{ij}(\vec{k}) - \delta_{\alpha\beta}\left[E_i\delta_{ij} - <\chi_i|\frac{1}{2}\nabla_r^2|\chi_j>\right] \tag{34}$$

where E_i and E_j are molecular energies and $\alpha = \beta$ only if χ_i and χ_j belong to the same molecule. K_{ij} are kinetic energy integrals similar to those of Eqs. (31,32). In setting up the Hamiltonian, Eq. (34), we treat E_6 and E_7 (corresponding to virtual orbitals) as adjustable parameters; they are obtained by comparing the calculated zero-momentum energy loss function with available data.

We use in our calculation a cubic-ice model with two water molecules in each unit cell. Each water molecule is represented by a set of seven molecular orbitals expressed in a contracted Gaussian basis.[42] The calculation of $\epsilon(\vec{q}, \omega)$ of Eq. (25) is performed with the analytic tetrahedron method.[43] Briefly, the integration (summation) over $\vec{k}$ in Eq. (25) is of the following type (a simplified but obvious notation is used):

$$I(E) = \int d\vec{k}\, \frac{A(\vec{k})}{E(\vec{k}) - E - i\delta}$$

or, since $A(\vec{k})$ is analytic,

$$Re(I) = P \int d\vec{k}\, A(\vec{k})/[E(\vec{k}) - E] \tag{35}$$

$$Im(I) = \pi \int d\vec{k}\, A(\vec{k})\, \delta[E - E(\vec{k})] \tag{36}$$

In the ATM the Brillouin zone (BZ) is divided in tetrahedra, and a linear approximation used for $A(\vec{k})$ and $E(\vec{k})$:

$$A(\vec{k}) = A(\vec{k}_O) + \vec{b}\cdot(\vec{k} - \vec{k}_O)$$

$$E(\vec{k}) = E(\vec{k}_O) + \vec{c}\cdot(\vec{k} - \vec{k}_O)$$

(37)

The values of $A(\vec{k})$ and $E(\vec{k})$ inside each tetrahedron are entirely determined by their values at the four corners of the tetrahedron. Within this approximation Eqs. (36,37) can be integrated analytically. In a typical calculation, we divide the BZ into 24240 tetrahedra (see Fig. 7).

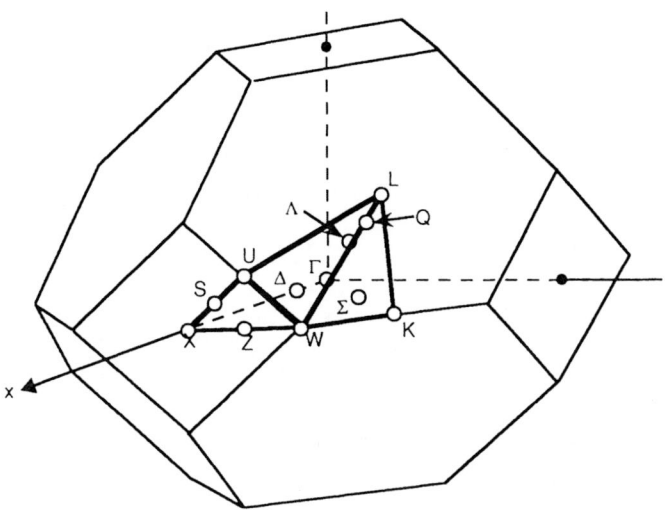

Figure 7. The Brillouin zone (BZ) for cubic ice.

Figure 8 shows a comparison between $\epsilon(\vec{q}=0,\omega)$ and some experimental data (Refs. 44,45; the data are not for cubic ice, but the difference between various forms of water is small). It is indicative for such a "fitting" procedure that detail outside the scope of the calculation could not be reproduced (for instance the 8-eV peak, which is believed to represent an exciton). It must be also realized that in no sense can the theoretical curve be considered a "best fit." The energy bands thus obtained (see Fig. 9) can be used to extend the calculation to any other value of $\vec{q}$. An example is shown in Fig. 10.

Full details on this calculation can be found in Ref. 38.

The Quasi-Particle Method

Full, *ab initio* quantum mechanical descriptions of many-body systems are exceedingly (if not prohibitively) complicated. Although the problem of N interacting particles can be treated rigorously with many-body field-theoretical techniques,[46,47] the formal result is an infinite system of coupled equations in terms of the n-particle Green's functions (n=1,2,...). Moreover, for the Coulomb potential a Born-like series would be very slowly converging (and therefore impractical) because of the long-ranging nature of the interaction.

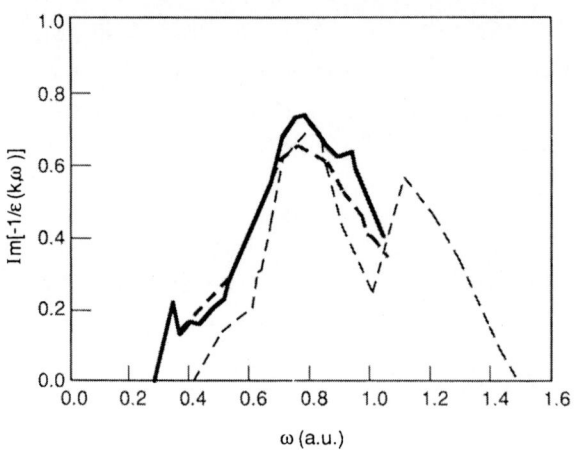

Figure 8. The energy-loss function for cubic ice at $\vec{q}=0$: our calculation (light dash) and data for amorphous ice[44] and hexagonal ice.[45]

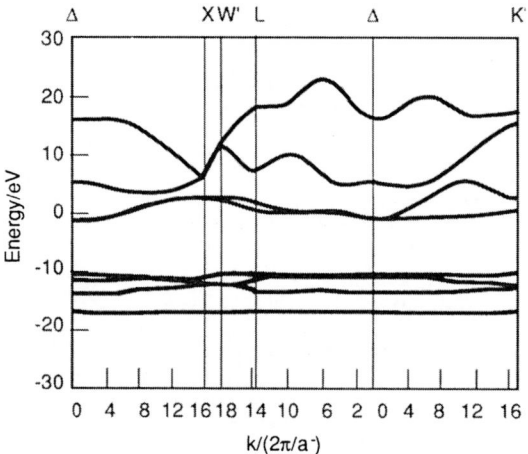

Figure 9. Valence (lower four) and conduction (upper four) bands for cubic ice. The $\vec{k}$ points are indicated in Fig. 7.

Significant progress towards a solution of this problem came with the realization that, rather than using the bare Coulomb potential, one should instead formulate the problem in terms of the screened interaction as a perturbative potential. These ideas, formalized in a 1965 paper by Hedin,[48] have been recently applied with great success to a variety of materials,[33-37] and are collectively known as quasi-particle techniques.

A quasi-particle (as opposed to an independent particle) can be thought of as an entity made of an electron and a "cloud" of distortion (produced by the Coulomb interaction) in the density of particles surrounding it. This entity has a momentum-dependent mass and moves in the screened field generated by all other particles carrying along its "cloud." A perhaps more lucid, albeit formal, definition of the concept of quasi-particle can be given in terms of single-particle Green's functions as follows:[46,47]

Let $|N>$ be the ground state of a system of N interacting particles expressed in the occupation-number representation. Let further $\Psi(\vec{r},t)$ be a field operator acting on $|N>$ to the effect of removing from the system at time t a particle positioned at

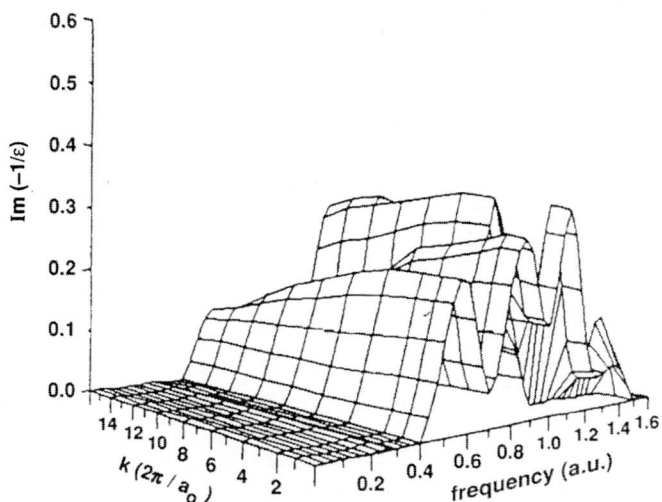

Figure 10. The energy loss function for cubic ice calculated at $k_x=8$, $k_y=8$ (in units of $2\pi/a$) as a function of frequency and wave number vector, k_z.

$\vec{r}$. The adjoint operator, $\Psi^+(\vec{r},t)$, adds a particle to the system. The single-particle Green's function is defined as follows:

$$G(\vec{r}_1,t_1;\vec{r}_2,t_2) = -i<N\,|\,T\left[\Psi(\vec{r}_1,t_1)\,\Psi^+(\vec{r}_2,t_2)\right]|\,N> \qquad (38)$$

Here T is a time-ordering operator which leaves its argument unchanged if $t_1>t_2$ but changes the sign otherwise (for fermions). $G(1,2)^{(a)}$ has the following meaning: At $(\vec{r}_2,t_2)$ a particle is added to the system. At a later time, t_1, we measure the probability that the original system $|N>$ can be recovered by removing the particle at position $\vec{r}_1$. If $G(1,2)$ remains constant over large intervals $|\vec{r}_1 - \vec{r}_2|$, $|t_1 - t_2|$, then the notion of a quasi-particle is valid. Indeed, by calculating a Fourier-transform of $G(1,2)$—"flat" in configuration space—we obtain delta-like peaks in the energy and momentum space, which is the signature of a (quasi)stable particle. The system can be described as a collection of quasi-particles.

Within the first approximation (the only one explored so far), one can set up an equation for the quasi-particle:[48]

$$\left[H_0(\vec{r}) + V(\vec{r})\right]\Psi_n(\vec{r}) + \int d\vec{r}'\,\Sigma(\vec{r},\vec{r}',E_n)\,\Psi_n(\vec{r}')\,d\vec{r}' = E_n\Psi_n(\vec{r}) \qquad (39)$$

$\Psi_n(\vec{r})$ and E_n are the quasi-particle wave function and energy, respectively. H_0 contains single-particle terms (kinetic energy, interactions with ions) and V is the Hartree potential. In this equation, $\Sigma(\vec{r},\vec{r}',E_n)$ is the self-energy which is given (again within the same approximation) by

---

(a) The notation $1=(\vec{r}_1,t_1)$, $2=(\vec{r}_2,t_2)$, etc., is used. Also $1^+=(\vec{r}_1,t_1+\delta^+)$, where δ^+ is a small positive quantity.

$$\Sigma^{(1)}(1,2) = iG^{(0)}(1,2)W^{(0)}(1^+,2) \tag{40}$$

This is the so-called GW approximation.[49] $G^{(0)}$ is the Green's function in the zero-th approximation (see below) and $W^{(0)}$ is the screened Coulomb interaction:

$$W(1,2) = \int \epsilon^{-1}(1,3)\,v(3,2)\,d(3) \tag{41}$$

$$v(1,2) = \frac{e^2}{|\vec{r}_1 - \vec{r}_2|} \cdot \delta(t_1 - t_2) \tag{42}$$

Note that:

a) Since Σ is non-local and non-Hermitian, the energies E_n will be generally complex quantities. The quasi-particles have a finite lifetime.

b) If the dielectric screening is neglected, i.e.,

$$\epsilon^{-1}(1,2) = \delta(1,2) \tag{43}$$

the HF equations obtain, with Σ replaced by the exchange term.

In the practical implementation of the GW approximation, it is preferable to Fourier-transform Eq. (40):

$$\Sigma(\vec{r},\vec{r}',E) = i \int \frac{d\omega}{2\pi} e^{i\delta^+\omega} G(\vec{r},\vec{r}',E+\omega)\,W(\vec{r},\vec{r}',\omega) \tag{44}$$

This follows from the convolution theorem. The zero-th approximation of G is

$$G^{(0)}(\vec{r},\vec{r}',E) = \sum_{n,k} \frac{\Psi_n(\vec{k},\vec{r})\,\Psi_n^*(\vec{k},\vec{r}')}{E - \left[E_n(\vec{k}) \pm i\delta_1^+\right]} \tag{45}$$

Here $\Psi_n(\vec{k},\vec{r})$ are HF wave functions of energy $E_n(\vec{k})$, and the sign $+(-)$ is for E_n representing occupied (unoccupied) states, i.e., E_n smaller (larger) than the chemical potential μ (the Fermi level).

The screened interaction term, Eq. (41), becomes

$$W(\vec{r},\vec{r}',\omega) = \sum_{q,G,G'} e^{i(\vec{q}+\vec{G})\vec{r}}\,\epsilon_{GG'}^{-1}(\vec{q},\omega)\,v(\vec{q}+\vec{G}')\,e^{-i(\vec{q}+\vec{G}')\vec{r}'} \tag{46}$$

where $v(\vec{q})$ is the Fourier transform of the Coulomb potential:

$$v(\vec{q}) = \frac{4\pi e^2}{\Omega q^2} \tag{47}$$

The expression of Eq. (25) can be used for the dielectric response function.

From Eqs. (45-48) the matrix elements of Σ in the GW approximation become

$$<n\vec{k}|\ \Sigma(\vec{r},\vec{r}',E)\ |n\vec{k}>$$

$$= i\sum_{1}\sum_{1BZ} d\vec{q}\ F_{-(q+G)nl}(\vec{k})\ F^{*}_{-(q+G')nl}(\vec{k})\ \frac{4\pi e^2}{|\vec{q}+\vec{G}|^2}\ \frac{I(E)}{(2\pi)^3} \qquad (48)$$

where

$$I(E) = \int_{-\infty}^{+\infty} \frac{d\omega}{2\pi} e^{i\delta^+\omega}\ \frac{\epsilon^{-1}_{GG'}(\vec{q},\omega)}{\omega - \left[E_1(\vec{k}-\vec{q}) \pm i\delta_1^+ - E\right]} \qquad (49)$$

In a first approximation the quasi-particle energies, $E_n^{qp}(\vec{k})$, can be obtained by solving the equation:

$$E_n^{qp}(\vec{k}) = E_n(\vec{k}) + <n\vec{k}|\ \Sigma(E_n^{qp}(\vec{k}) - V_{ex}\ |n\vec{k}> \qquad (50)$$

where we have used the fact that $E_n^{qp}(\vec{k})$ are obtainable by replacing in the HF equations the exchange potential, V_{ex}, with the self-energy operator. From Eq. (50):

$$E_n^{qp}(\vec{k}) = \frac{E_n(\vec{k}) + <n\vec{k}|\ \Sigma(0) - V_{ex}\ |n\vec{k}>}{1 - <n\vec{k}|\ d\Sigma(0)/dE\ |n\vec{k}>} \qquad (51)$$

calculated at $\Sigma(0) = \Sigma(\vec{r},\vec{r}',E=0)$. The evaluation of $I(E)$ of Eq. (49) is discussed in Ref. 40.

In Fig. 11 I show the HF calculation for the energy bands of trans-polyacetylene. This calculation (which is also the starting point in the GW scheme) was performed using the code CRYSTAL[50] recently made available through the Quantum Chemistry Program Exchange (OCPE 577) at Indiana University. This code is based on an *ab initio*, self-consistent HF linear combination of atomic orbitals (LCAO) method, with crystal orbitals (CO) expressed as combinations of Bloch functions. Four of the valence bands have a σ symmetry; that is, the eigenfunctions are even with respect to the polymer plane. The top valence state has π symmetry. The fundamental gap (at the zone boundary) is 5.9 eV--significantly larger that the measured values of 1.4-1.8 eV.[51] In terms of electron transport I compare in Fig. 12 the HF calculation of the energy loss function, $Im[-\epsilon(\vec{q},\omega)^{-1}]$, with results obtained from electron spectroscopy measurements by Fink and Leising.[52] The calculation corresponds to q=0.1 $(2\pi/a)$ (a=lattice constant=2.43 Å). It is apparent from Fig. 12 that, as a result of ignoring correlation effects in the system, a poor representation of the energy loss function obtains.

With the aid of the formalism described above, we have performed a series of calculations in the GW approximation. The quasi-particle energy bands obtained are shown in Fig. 13. The inclusion of the screened potential instead of the bare Coulomb exchange reduces the excitation energies; the fundamental gap is now 2.3 eV. A similar improvement is seen in the energy-loss function (Fig. 14), in spite of the fact that these results do not represent necessarily a convergent GW calculation (in

Zaider 155

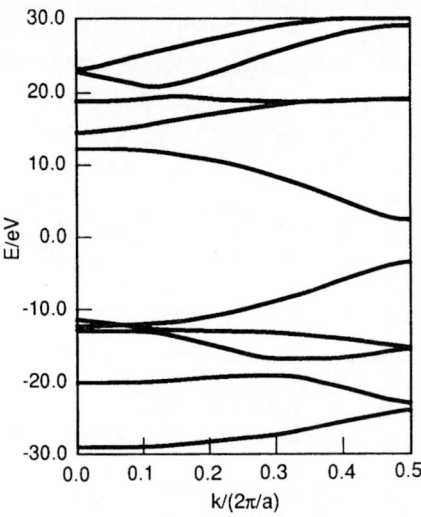

Figure 11. The valence (3-7) and conduction (8-12) energy bands for trans-polyacetylene as obtained in the HF approximation with the code CRYSTAL.[50]

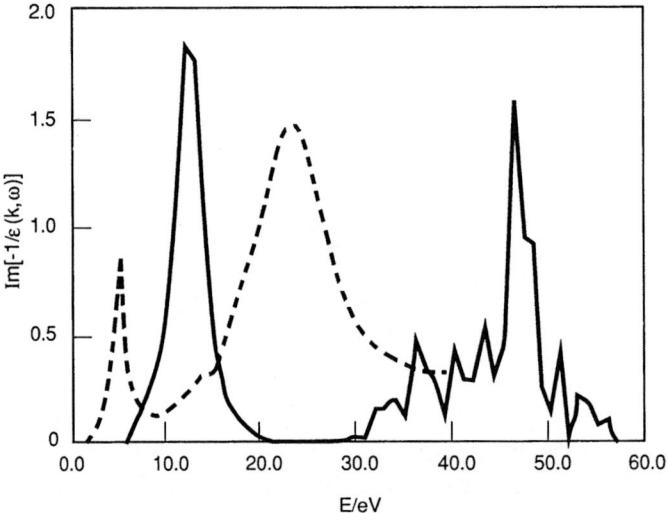

Figure 12. A comparison between calculated (HF, solid line) and measured (Ref. 52, dashed line) energy loss functions for trans-polyacetylene The wave vector ($\vec{q}$=0.05-0.1 Å^{-1}) is along the polymer's axis.

principle one should reiterate the HF calculation by replacing the exchange integrals with the self-energy operator, obtain new quasi-particle wave functions and energies, and so on).

Final Remarks

Biophysical models of radiation action are only as good as their physical input. This truism is behind many of the developments in radiation physics that occurred since the 1972 Airlie (Virginia) conference on Physical Mechanisms in Radiation Biology (the update of which is this meeting): Experimental microdosimetry and

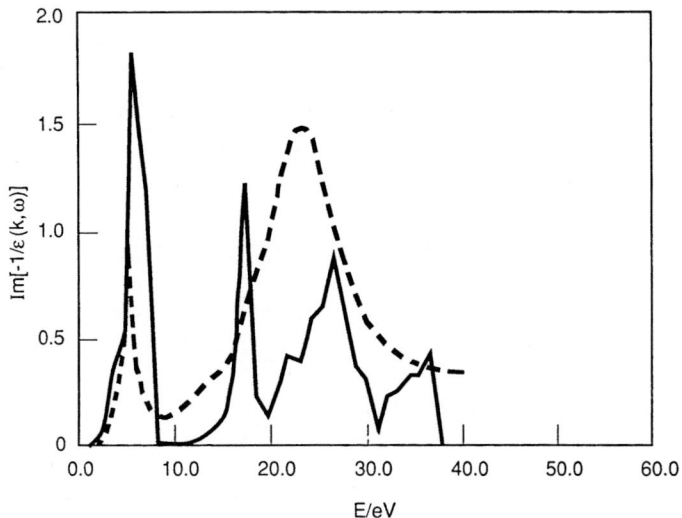

Figure 13. The quasi-particle energy bands for trans-polyacetylene.

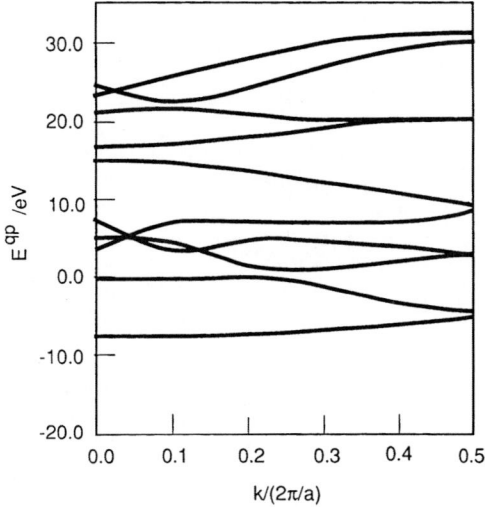

Figure 14. Same as Fig. 12, but with the calculation performed using the quasi-particle energy bands of Fig. 13.

nanodosimetry allow the measurement of energy deposition spectra in regions of cellular or sub-cellular dimensions. Monte Carlo transport codes can simulate—event by event—the passage of electrons and heavier charged particles through water vapor material, as a convenient model for the cellular material. These results are summarized and interpreted in terms of field quantities (proximity functions, radial dose distributions, specific energy spectra) and then applied to target volumes assumed to represent the sensitive sites in the cellular nucleus. We can thus predict certain "regularities" between ionizing radiation and biological response, such as the dependency of the RBE on dose or (semi-empirically) the relation between inactivation cross section and LET (or its more sophisticated counterparts).

What we can still only poorly explain is *mechanism*, that is, the relation between ionizing radiation and processes at the molecular level. The main culprit is to a large extent the *non-specificity* of our current description of the radiation field, that is, the geometrization of track structure. We should try to do better by studying the physics of well-defined biomolecular systems.

Acknowledgments

This work was performed using supercomputer time generously made available to us by the U.S. Department of Energy (in part under the "Grand Challenge" program); this is gratefully acknowledged. It is a pleasure to acknowledge collaboration with J. L. Fry, H. H. Rossi, D. J. Brenner and D. E. Orr on various aspects of the work described in this paper.

This investigation was supported in part by Grant DE-FG02 88ER60631 from the U.S. Department of Energy and by Grant CA 12536 from NCI, DHHS, to the Center for Radiological Research, Columbia University.

References

1. M. J. Berger. Monte Carlo Calculation of the Penetration and Diffusion of Fast Charged Particles. *Meth. in Comp. Phys.* 1:135-215 (1963).
2. H. G. Paretzke, G. Leuthold, G. Burger, and W. Jacobi. Approaches to Physical Track Structure Calculation. In: *4th Symposium on Microdosimetry*, CEC, Luxembourg (1974).
3. R. N. Hamm, J. E. Turner, H. A. Wright, and R. H. Ritchie. Calculated Distance Distributions of Energy Transfer Events in Irradiated Liquid Water. In: *7th Symposium on Microdosimetry, Oxford*, pp. 717-726, CEC, Harwood (1981).
4. M. Zaider, D. J. Brenner, and W. E. Wilson. The Application of Track Structure Calculations to Radiobiology. I. Monte Carlo Simulation of Proton Tracks. *Radiat. Res.* 95:231-247 (1983).
5. H. Kahn. *Applications of Monte Carlo.* The Rand Corporation AECU-3259 (1956).
6. G. Seng. *Das System (e-H_2O) Differentielle Streuexperimente bei KleinsteEnergien*, Thesis Kaiserlautern University (1975).
7. E. Bruche. Wirkungsquerschnitt und Molekelblau in der Pseudoedelgasreihe: Ne HF-H_2O-NH_3-CH_4. *Ann. Phys.* 1:93-134 (1929).
8. J. P. Bromberg. Measurement of Absolute Collision Cross Sections of Electrons Elastically Scattered by Gases. In *The Physics of Electronic and Atomic Collisions*, pp. 98-111, Univ. of Washington Press, Seattle (1975).
9. H. Nishimura. Elastic Scattering Cross Sections of H_2O by Low-Energy Electrons. In *Electronic and Atomic Collisions*, Vol. II, p. 314, Society for Atomic Collision Research, Kyoto (1979).
10. T. D. Märk and F. Egger. Cross Section for Single Ionization of H_2O and D_2O by Electron Impact from Threshold Up to 170 eV. *Int. J. Mass Spectrom. Ion Phys.* 20:89-99 (1976).
11. J. Schutten, F. J. de Heer, H. R. Moustafa, A.J.H. Boerboom, and J. Kistemaker. Gross and Partial Ionization Cross Sections for Electrons on Water Vapor in the Energy Range 0.1-20 keV. *J. Chem. Phys.* 44:3924-3928 (1966).
12. S. Trajmar, W. Williams, and A. Kuppermann. Electron Impact Excitation on H_2O. *J. Chem. Phys.* 58:2521-2531 (1973).
13. W. Hilgner, J. Kessler, and E. Steeb. Zur Spinpolarisation Langsamer Elektronen nach der Streuung an Holekulen. *Z. Physik* 221:324-332 (1969).
14. H. S. Porter and F. V. Jump. Analytic Total and Angular Elastic Electron Impact Cross Sections for Planetary Atmospheres. *Computer Science Corp. Report*, CSC/TM-78/6017 (1978).
15. D. J. Brenner and M. Zaider. A Computationally Convenient Parameterization of Experimental Angular Distributions of Low Energy Electrons Elastically Scattered off Water Vapor. *Phys. Med. Biol.* 29:23-47 (1984).
16. L. H. Toburen and W. E. Wilson. Energy and Angular Distributions of Electrons Ejected from Water Vapor by 0.3-1.5 MeV Protons. *J. Chem. Phys.* 66:5202-5213 (1977).
17. M. Zaider and D. J. Brenner. On the Stochastic Treatment of Fast Chemical Reactions. *Radiat. Res.* 100:245-256 (1984).
18. I. G. Draganic and Z. D. Draganic. *The Radiation Chemistry of Water.* Academic Press, New York (1971).

19. P. Clifford, N.J.B. Green, and M. J. Pilling. Stochastic Model Based on Pair Distribution Functions for Reactions in a Radiation-Induced Spur Containing One Type of Radical. *J. Phys. Chem.* 86:1318-1321 (1982).

20. P. Clifford, N.J.B. Green, and M. J. Pilling. Monte Carlo Simulation of Diffusion and Reaction in Radiation-Induced Spurs. Comparison with Analytic Models. *J. Phys. Chem.* 86:1322-1327 (1982).

21. S. Chandrasekhar. Stochastic Processes in Physics and Astronomy. *Rev. Mod. Phys.* 15:1-88 (1943).

22. P. Debye. Reaction Rates in Ionic Solutions. *Trans. Electrochem. Soc.* 82:265-272 (1942).

23. D. R. Short, C. N. Trumbore, and J. H. Olson. Extension of a Spur Overlap Model for the Radiolysis of Water to Include High Linear Energy Regions. *J. Phys. Chem.* 85:2328-2335 (1981).

24. C. D. Jonah, M. S. Matheson, J. R. Miller, and E. J. Hart. Yield and Decay of the Hydrated Electron from 100 ps to 3 ns. *J. Phys. Chem.* 80:1267-1270 (1976).

25. T. Sumiyoshi and M. Katayama. The Yield of Hydrated Electrons at 30 picoseconds. *Chem. Lett.* 12:1888-1890 (1982).

26. J. Bednar. Rydberg Enhancement of Total Ionization in Liquids Irradiated by Ionizing Radiation. II. Liquid Water Radiochem. *Radioanal. Lett.* 45:407-412 (1980).

27. ICRU. *Microdosimetry, Report 36.* International Commission on Radiation Units and Measurements, Washington, DC (1983).

28. A. M. Kellerer. Concepts of Geometrical Probability Relevant to Microdosimetry and Dosimetry. *Proceedings of the 7th Symposium on Microdosimetry,* eds. J. Booz, H. G. Ebert, and H. D. Hartfield. Oxford/UK, pp. 1049-1062 (1980).

29. M. Zaider and D. J. Brenner. The Application of Track Calculations to Radiobiology. III. Analysis of the Molecular Beam Experiment Results. *Radiat. Res.* 100:213-221 (1984).

30. A. M. Kellerer and H. H. Rossi. The Theory of Dual Radiation Action. *Curr. Top. Radiat. Res. 0.* 8:85-158 (1972).

31. A. M. Kellerer and H. H. Rossi. A Generalized Formulation of Dual Radiation Action. *Radiat. Res.* 75:471-488 (1978).

32. D. Pines and P. Nozieres. *The Theory of Quantum Liquids.* W. R. Benjamin, New York (1966).

33. R. W. Godby, M. Schluter, and L. J. Sham. Accurate Exchange-Correlation Potential for Silicon and Its Discontinuity on Addition of an Electron. *Phys. Rev. Lett.* 56:2415-2418 (1986).

34. R. W. Godby, M. Schluter, and L. J. Sham. Trends in Self-Energy Operators and Their Corresponding Exchange-Correlation Potentials. *Phys. Rev.* B36:6497-6500 (1987).

35. R. V. Godby, M. Schluter, and L. Sham, Self-Energy Operators and Exchange Potentials in Semiconductors. *Phys. Rev.* B37:10159-10175 (1988).

36. M. S. Hybertsen and S. G. Louie. Electron Correlation and the Band Gap in Ionic Crystals. *Phys. Rev.* B32:7005-7008 (1985).

37. M. S. Hybertsen and S. G. Louie. Electron Correlation in Semiconductors and Insulators: Band Gaps and Quasi Particle Energy. *Phys. Rev.* B34:5390-5413 (1986).

38. M. Zaider, J. L. Fry, and D. E. Orr. A Semi-Empirical Tight Binding Calculation of the Dielectric Response Function of Water. *Nucl. Tracks Radiat. Meas.* 16:159-167 (1989).

39. M. Zaider, J. L. Fry, and D. E. Orr. Calculational Aspects of the Assessment of Dielectric Response Function and Energy Loss in Biomaterials. *Int. J. Supercomp. Appl.* (in press, 1990).

40. M. Zaider, J. L. Fry, and D. E. Orr. Integral Methods in the Quasi Particle Theory of Electronic Structure. In *Proc. Int. Conf. on Integral Meth. Science and Eng.* (in press, 1990).

41. N. Wiser. Dielectric Constant with Local Field Effects Included. *Phys. Rev.* 129:62-69 (1963).

42. L. C. Snyder and H. Basch. *Molecular Wave Functions and Properties: Tabulated from SCF Calculations in a Gaussian Basis Set.* John Wiley, Chichester (1972).

43. J. L. Fry and P. C. Pattnaik. Analytic Approximation Methods of Computing Multi-Dimensional Principal Value Integrals. In *Proc. Int. Conf. Integral Meth. Sci. Eng.* F. R. Payne, C. C. Cordureanu, A. Haji-Sheikh, and T. Huang, eds. pp. 27-36, Hemisphere, Washington (1986).

44. J. Daniels. Bestimmung der Optischen Konstanten von Eis aus Energie-Verlustmessungen von Schnellen Elektronen. *Optics Comm.* 3:240-243 (1971).

45. K. Kobayashi. Optical Spectra and Electronic Structure of Ice. *J. Phys. Chem.* 87:4317-4321 (1983).

46. A. L. Fetter and J. D. Valecka. *Quantum Theory of Many-Particle Systems.* McGraw-Hill Book Co., New York (1971).

47. J. C. Inkson. *Many-Body Theory of Solids. An Introduction.* Plenum Press, New York and London (1986).

48. L. Hedin. New Method for Calculating the One-Particle Green's Function with Application to the Electron-Gas Problem. *Phys. Rev.* 139A:796-823 (1965).

49. L. Hedin and S. Lundqvist. Effects of Electron-Electron and Electron-phonon Interaction on the One-Electron States of Solids. In: *Solid State Physics* 23:1-181 (1969).

50. C. Pisani, R. Dovesi, and C. Roetti. Hartree-Fock Ab Initio Treatment of Crystalline Systems. *Lecture Notes in Chemistry* 48:1-193 (1988).

51. C. R. Fincher, C. E. Chen, A. J. Heeger, A. G. MacDiarmid, and J. B. Hastings. Electronic Structure of Polyacetylene: Optical and Infrared Studies of Undoped Semiconducting (CH)$_x$ and Heavily Metallic (CH)x. *Phys. Rev.* B20:1589-1602 (1982).

52. J. Fink and G. Leising. Momentum-Dependent Dielectric Functions of Oriented Trans-Polyacetylene. *Phys. Rev.* 34B:5320-5328 (1986).

Discussion

Paretzke: I would like to take the question farther; I think we cannot go back to dosimetric quantities because what we learned in the past twenty or thirty years is that the concept of dose has remained impotent in understanding radiation effects in biology. We should get rid of this quantity as soon as possible, so any falling back with respect to microscopic dosimetry would just prevent us from proceeding on what have achieved up until today and going any further. If you ask, most physicists would throw away a quantity like dose where you have to attach correction factors, such as quality factor, which depend on so many things which are large—two orders of magnitude—and which depend on the dose rate and many other things. We should learn from this the necessity of perfecting the basic quantity, dose. So what we should do is get rid of this quantity as soon as possible, other than for radiation protection. It is useful for practical aspects, but for radiation research itself, if you want to understand radiation effect, we just should not use this quantity any more.

Zaider: I didn't refer to dose. I talked about the stochastic distribution of energy in a given volume.

Paretzke: Not energy, not in a volume, not energies. Species and in targets which have to be modified. Just when we don't need the intermediate step, there is no quantity which characterizes the track independent of the target you are looking at. You stated this sometimes. There are other arguments, but, in principle, we just should understand what is going on in biological targets. Just forget about dose.

Kasha: You bewildered me, but perhaps I am taking the wrong convention, the wrong scale. I thought tracks were observable, from cloud chambers, bubble chambers, and nuclear emulsions.

Paretzke: It is a solid state.

Zaider: How do you compare one of these tracks with a picture in a bubble chamber? Can you compare that?

Kasha: That is where maybe the scale is different. Maybe you are talking about the atomic scale versus the macroscopic. Is that what you said?

Varma: Take a biological molecule, DNA, put a proton through. Can you see a track? He is saying that physically you cannot observe the track. Surely you can see a nuclear emulsion, and in the silver bromide you can see a track--a trail of activated silver grains.

Kasha: If you give Nicholas and me some time and money, we will put some fluorescent indicators along the whole DNA, and we will tell you where the track is. But do you need to do that?

Varma: I believe he is asking whether, if you don't see the track, you can compare it with a measurable quantity which is, in the end, a statistical distribution of energy deposition or radial dose; what do you compare it with? His point is that, if you can't see it today, what good is it to calculate these tracks in detail? But if you can create these tracks in detail--and Herwig is saying that in the last couple of years, you can include the structure of the chromosomes, etc.--then I think it becomes very critical to predict and test certain end points, taking that into account, and for that one reason, it is important to do such calculations.

Kasha: I understand that.

Zaider: But how can you test the track; how can you see it?

Varma: Consequences of the track.

Zaider: No, I don't think that you can measure it.

Paretzke: Where you can measure them in dose quantity is in gas, because you have many ionizations; it is no major problem to do that. I think that calculations are legitimate where they help me to understand the effects in biological targets. So, what we do is, we have some physical knowledge about scaling of low Z material, we know how electrons at reasonable energy range behave; then, in the condensed phase, we have transport quantities. The big problems are with the molecules in a condensed phase which absorb a certain amount of energy and decay, and nobody to my knowledge knows about this. So we made an assumption and then other biochemical assumptions and predict an outcome of a biochemical experiment or biological molecule experiment. If we are in agreement, then we don't have, for the time being, a reason to doubt that our steps are okay. At a later stage more insight might correct some of our assumptions, but that is how we have to proceed now. I don't want to calculate things in an emulsion because I am completely confident that we can do this properly, and after two years I want to focus on the more important aspects because I might run out of funding if I go the other way.

Zaider: I still say that you cannot compare one track with any one measurement. It is a meaningless thing; you can never do that.

Kasha: I have a question which I consider chemical. It may have no effect on your model, or maybe Aloke's calculations—but it may have a profound effect. We do experiments in which we try to manipulate protons. We cannot make a proton photodissociate with a molecule in the water environment unless we have water clusters greater than 10^{12}. We have a hydrocarbon environment and we have a solute; and when we excite it, the proton will never jump until we get the water clusters up to a certain size, and then we always get intermolecular proton transfer. Now, that shows that you cannot stabilize a proton if, say, with one water or two or five, you need a cluster. Does that affect your model and your calculation?

Zaider: No, this is a continuous-media model.

Weinstein: There are two methodological issues that I would like to address. Maybe because we talked about tracks, I have to understand this argument since I am very interested in the spatial localization of effects of radiation, because I only understand effects in the sense of a spatial description. There is a reason to go away from a concept or phenomenon and move to another one conceptually only if in principle it is not measurable. Are you saying that these tracks are in principle not measurable, or are they practically at this point for DNA not measurable?

Zaider: What I was trying to say is that you can not simulate a track—one track—and compare it with a measured track in anything such as an emulsion, a biological effect, etc. You cannot compare single tracks because this is a quantum wave function that you make collapse in some particular form with droplets, for example. The only way to compare this calculation with experiment is by taking a large number of tracks and obtaining something you can measure, the radial dose distribution, the microdosimetric spectrum, the LET, dose, etc. Track against track you cannot measure. This doesn't bother me so much in a condensed medium. In a condensed medium, the problem of localization makes it even more complicated. Energy loss and energy absorption are almost two different areas for research. The whole problem of assigning spatial coordinates to these phenomena becomes nearly impossible.

Weinstein: Now, if you have a quantity that has a clear definition physically, but which can be observed only as an ensemble, that is a very respectable type of quantity because it is what makes up all of thermodynamics and statistical mechanics. You don't go away from a quantity like this just because independently you cannot measure it. If you can predict the outcome of the ensemble and it is an experimental measure of the quantity, then that is a perfectly okay quantity.

Katz: That's what quantum mechanics is all about; you measure cross sections, you calculate probabilities.

Weinstein: The advantage now, as I understand it, is that it does eventually answer the question that I am interested in, namely, with spatial resolution specific at the atomic level, the occurrence of certain kinds of species which I now can understand and manipulate from a chemical point of view, and that is the answer at a statistical level, then this is the useful quantity for me.

The second question that I have is the following. If you have a formalism based on the quasi-particle, that is fine. I think it's okay to pursue it. But a Hartree-Fock calculation in a system like this, in order to understand the gap between the excited and the ground state, is not the right calculation to do. So, it is a little bit of a straw man, you see, to put this against that. I can tell you types of calculations you can do, not that I would like to do them myself because they are very time consuming, but I can tell you what they are, if you are interested, and it will give you the correct coordinate. There are hundreds, literally, of calculations of the gap in solid state that are *ab initio* with Green's Functions.

Kopelman: I want to make a comment about homogeneous versus heterogeneous structure, and there are the obvious problems, which I guess everybody here knows. What I am not so sure of, though, is what happens--not with the absorption of radiation or the interactions of radiation--but what happens after that when you have the chemical reactions like a recombination reaction? You were totally consistent; you have a homogeneous model, and you used the Schmolukowski equation, at least this is a consistent system. Now, what is not consistent is what is done all the time, that people have a heterogeneous space and still assume that the Schmolukowski equation for recombination is right, and I am going to show in my talk why this breaks down completely. I don't know how other people have done those thing, as, in your case, whether the computer is doing this just in the Monte Carlo way or whether the Schmolukowski equation is enclosed into the simulation. What I am trying to point out here is that you cannot use the Schmolukowski equation as well as those kinds of distributions for a heterogeneous structure. That is my point here, and we agree on that.

Zaider: Yes, I agree.

Kopelman: As far as the different point as Mike Kasha has suggested, about fluorescent probes, do we have a chance to see things on a smaller and smaller scale with, say, fluorescent probes, and so on? I think we could push the scale to a few hundred angstroms. The question is, would that help?

Radial Distribution of Dose

Robert Katz[a] and Matesh N. Varma[b]

Abstract

The radial distribution of dose about the path of a heavy ion, principally from delta rays, is one of the central contributions of atomic physics to the systematization of high LET radiation effects in condensed matter, whether the detection arises in chemical, physical, or biological systems. In addition to the radial distribution of dose, we require knowledge of the response of the system to X-rays or gamma-rays or to beams of energetic electrons such that the electron slowing-down spectra from these radiations can approximate the slowing-down spectra from delta rays even at different radial distances from the ion's path. A combination of these data enables us to calculate the action cross sections for heavy ion bombardments in all detectors for which this information is available. These cross sections are indispensable for the evaluation of effects caused by high LET radiations. In this paper we focus attention principally on the calculation and measurement of the radial distribution of dose and on their limitations.

Introduction

The first application of the radial dose distribution in any detector was made by Katz and Butts[1,2] in the study of the width of heavy ion tracks in electron-sensitive nuclear emulsions.

Earlier work had called attention to the significance of delta rays in track production but had assigned responsibility either to delta-ray flux[3] or to energy flux,[4] such that the track width was determined either by an appropriate flux of delta rays or of energy through the region bounding the "track width." While these criteria approximated the width of the tracks of heavy ions in the "thin-down" region in the stopping end of a particle track, neither criterion yielded a correct value of the track width over the entire range of a heavy ion. We calculated the radial distribution of dose, based on the initial energy flux calculations of Bizetti and Della Corte,[4] and revised the track-width criterion to one of a critical radial dose, which would define the "width" of the heavy ion track. This criterion was much more consistent with our track-width measurements over the entire range of heavy particle tracks than either of the two earlier criteria. We had identified the radial dose as a quantity upon which to base models of track effects.

Dry Enzymes and Viruses

In 1967, our calculations of the radial dose distribution were simplified. They were then based on the Rutherford equation for delta-ray production by heavy ions from a sea of free electrons.[5] Additionally, assumptions were made that delta rays were

---

(a) University of Nebraska, Lincoln, Nebraska
(b) U.S. Department of Energy, Washington, D.C.

Physical and Chemical Mechanisms in Molecular Radiation Biology
Edited by W.A. Glass and M.N. Varma, Plenum Press, New York, 1991

normally ejected, and that the range energy relation for electrons was linear, based on experimental data in aluminum for electrons of energies in the range of delta-ray energies from ions of energy less than about 8 MeV/amu. For heavy ions, we made use of a formula for effective charge as a function of ion velocity from Barkas,[6] which came from measurements of the range of ions from the HILAC accelerator in nuclear emulsion. These results formed the basis of the calculation of the inactivation cross section for dry enzymes and viruses, taken to be 1-hit detectors, and provided the basic structure that underlay many further developments in track theory.[7] Although at that time we had no direct experimental verification of our calculations of radial dose, we considered our calculation of the response of dry enzymes and viruses to energetic heavy ions to be an experimental confirmation of the overall procedure.

Here, we assumed that the average radial dose distribution dominated the calculation of action cross section with no need for detailed knowledge of the fluctuation in energy deposition which later became the theme of the theory of Dual Radiation Action.[8,9] We found that the significance of target size varied. With fast protons and insensitive detectors, size might be significant. There, we may be dealing with thin tracks and relatively large targets. With high LET radiations, we were dealing with thick tracks and small targets rather than the converse, so target size was of negligible importance. We found that earlier preoccupation with LET as a reference variable had been excessive, and that a much better variable was z^{*2}/β^2. And we found that there must be "hooks" in double-log plots of σ versus LET, corresponding to "thindown" of the tracks of heavy ions in nuclear emulsions as the ions approached the end of their range. Many of these observations have since been shown to apply to other physical detectors and to radiobiology. They were signaled by our innovative use of the average radial distribution of dose from delta rays as the dominating theme in a track structure model,[7] and by our use of the detector's response to gamma rays as the means of calibrating the response to heavy ions. To experimenters it said that high LET response should be measured with track-segment irradiation with heavy ions (not neutrons) and that if any model were to emerge from such measurements, they should be accompanied by measurements with gamma rays.

In later work, we have repeatedly used this sort of test of our dose calculations. Our test data have been the measured responses of detectors whose sensitive elements are of different sizes and sensitivities, and with bombardments of a range of ion charges and speeds to check on dose calculations in different regions of radial distance.

Particle Tracks in Emulsions

In an attempt to provide a firmer basis for the calculation of the radial distribution of dose, data for the range and energy dissipation of beams from electrons normally incident on different materials were systematized analytically.[10,11] That information was then applied to data on the blackness of emulsion exposed to beams of electrons[12] as a test of both the energy dissipation formulas and of the 1-hit model of detector response. It was applied to the calculation of the radial distribution of dose[10] in different materials. Of primary importance was the application to a new model of particle tracks in emulsion, where the calculations[13] could be compared with measurements of the grain-count regime.[6,14] The calculations were also compared with microdensitometric measurements of the blackness of emulsion as a function of distance from the ion's path in the track-width regime of very heavy cosmic rays, from iron to uranium.[15] The agreement of our calculations with measurement supported the validity of our calculations of the radial dose distribution, as well as our assumption that electron-sensitive emulsions were one-hit detectors. Simulations of particle tracks in nuclear emulsions making use of track theory are compared to photographs of particle tracks to good effect.[16,17]

At this time, thus, we had made analytical estimates of the radial distribution of dose in condensed matter[18] based on various approximations to the electron-energy dissipation in condensed matter and on an extension of the Rutherford formula for delta-ray production. We had tested these in application to the response of condensed-phase detectors to energetic ions. We continue to search for improved analytical expressions for the radial dose distribution in gases and in condensed matter, and to test these distributions with calculations of detector response.

Except for microdensitometric measurements of the blackness of nuclear emulsions as a function of transverse distance from the ion's path,[15,19-22] there is no possibility for direct measurement of the radial-dose distribution in solids or liquids.

Experiment

Initially stimulated by the work of Butts and Katz, a program of measurement of the radial dose distribution in gases was initiated by John Baum and co-workers at the Brookhaven National Laboratory in 1967. In the period from 1967 to 1980, a number of papers were published by this group, initially under Baum's leadership and subsequently under the leadership of Matesh Varma, with participation from S. L. Stone, A. V. Kuehner, C. L. Wingate, and J. T. Lyman.[23-31] Measurements were made of the radial dose distributions in hydrogen, nitrogen, and in tissue-equivalent gas, using the following projectiles: protons at 1 and 3 MeV/amu, helium at 0.75, 18.3, and 230 MeV/amu, oxygen at 2.4 MeV/amu, neon at 377 MeV/amu, bromine at 0.53 MeV/amu, and iodine at 0.26 and 0.49 MeV/amu. The slower projectiles were obtained at the Brookhaven National Laboratory using the Tandem Van de Graaf accelerator, while the more energetic projectiles were obtained at the Lawrence Radiation Laboratory using the 184-in. cyclotron and the BEVALAC accelerators. Some accompanying Monte Carlo calculations were made by H. G. Paretzke.

This group used essentially the same equipment for all its measurements. A collimated beam of ions was passed through a gas-filled tank. The resulting ionization was measured in a cylindrical ion chamber whose axis was parallel to, and displaced from, the beam. Together with the effect of varied pressure in the tank, one could simulate the effect of radial distance variation in a medium of constant density. A "W" value (energy per ion pair), based on ion chamber measurements with gamma rays and assumed to be constant over all radial distances, was used to convert ionization density to energy density or dose. Over a substantial range of radial distances, ions, and ion speeds, it was found that the dose varied with the square of the effective charge and inversely with the square of the product of radial distance and ion speed. These pioneering measurements provided significant support to track-structure theory. The measurements were in agreement with our calculations (see Fig. 3) and provided an important test of analytic formulations.

There are, however, some inherent difficulties with these (and other) experiments. It is impossible to approach arbitrarily close to an ion's path. It is virtually impossible to measure the dose out to the radial limit in energy deposition. Since the distribution in electron energies must be expected to vary as a function of radial distance, the W value used to convert ionization density to energy deposition should vary correspondingly. But experimentally, we do not know the electron-energy spectrum as a function of radial distance. These limitations in our knowledge at the smallest and greatest radial distances are not readily overcome. Lack of knowledge at small distances is a barrier to understanding the response of small, relatively insensitive detectors, while at large distances this lack is a barrier to the accurate evaluation of cross sections in the thin-down region.

More recent work has been done by Metting[32-34] and Toburen[35,36] with 13.0, 13.8 and 17.2 MeV/amu Ge as projectiles, with U at 5.9 MeV/amu, and with Fe at 600 MeV/amu. Kanai and Kawachi[37] used 18.3 MeV/amu α particles.

The work of Kanai and Kawachi is not different in principle from the earlier work of the Baum group, although it differed in detail, for it measured the energy deposited within a cylinder coaxial with the ion's path. The radius was scaled by varying gas pressure. The average radial dose as a function of distance from the ion's path was found by successive differences in the energy deposited in the cylinder at incremental pressures. This procedure, like that of the Brookhaven group, yielded the average radial dose.

Metting's work was conceptually different. She sought to measure the fluctuations in energy deposition in small proportional counters at different radial distances from the ion's path using equivalent micrometer-sized volumes. When these measurements were processed so as to yield the average radial dose distribution, they were found to be in excellent agreement with earlier measurements and with our analytical model.

There had been questions as to the validity of the concept of effective charge as applied to this work. As seen in the reference frame in which the projectile is at rest, electrons in the target that pass close to the projectile nucleus are partly shielded from the outermost electrons carried by the projectile, and hence should experience a different "effective charge" than that experienced by electrons passing far from the nucleus.[38] This anticipated problem did not arise in Metting's measurements. The measurements were consistent with the use of standard effective charge formulas.

Other Theory

The earliest calculations of the radial distribution of dose known to us were published by Hutchinson[39] in connection with his effort to supplement the associated volume model of Lea in relation to the measured cross sections for the inactivation of dry enzymes and viruses. Three different calculations were made, one of which preceded the later calculation made independently by Butts and Katz in connection with their theory of RBE for the inactivation of dry enzymes and viruses. To our knowledge, no quantitative application of these calculations was published by Hutchinson.

In the ensuing years, many calculations of the radial-dose distribution about the path of a heavy ion have been made by different investigators. Two different sorts of data are required. One needs a source function that describes the radial distribution of the primary excitations and ionizations in the medium from the passing ion, and the doubly differential (in both energy and angle) cross section for secondary electron (delta ray) production. Then we must know the manner of energy deposition by the delta rays themselves. The energy deposited by delta rays has been evaluated in the continuous slowing-down approximation and by Monte Carlo methods, using experimental electron collision cross sections for gases, and using mean free paths calculated from optical measurements in liquid water. The results are in reasonable agreement with each other and with such experimental data as exist.

The calculations can be tested against direct measurements of the radial dose distribution in gases. But no direct measurements can be made for the radial dose distribution in condensed matter; yet, it is condensed matter that many of the radiation effects of interest take place. Thus far, the only available tests in condensed matter are from the comparison of track theory calculations of the response of detectors to heavy ions, which were made from the calculated radial dose distributions. Even with calculations, as with measurement, there are unresolved problems in calculating the dose very close to the ion's path, (e.g., within a hundred angstroms) and out to the maximum radial penetration of delta rays.

Calculations made of the radial dose distribution in the continuous slowing-down approximation were made by Fain,[40,41] by Chatterjee,[42,43] by Hansen,[44] and by Zhang et al.[45]   In some cases, it was assumed that there was a uniform "core" of energy dissipation close to the ion's path whose radius was given by the Bohr adiabatic approximation,[5,46] an inference of questionable validity.

Other calculations have been made by Monte Carlo methods, using extensive compilations of electron interaction cross sections measured in gases,[47,48] or by using optical data to generate electron mean-free paths in water at Oak Ridge.[49,50] More recently, a direct comparison has been provided between radial dose distributions in gas-phase and liquid-phase water using Monte Carlo transport techniques (Varma and Zaider, private communications, 1990).

A point of interest is the possibility of using a description of the track in terms of its radial dose distribution as a substitute for its microdosimetric description.   A detailed study of this problem[51] indicated that the critical parameter is the ratio between the diameter of the microdosimetric volume and the maximum lateral extension of the track (accurate results obtained when this ratio is significantly larger than 1).   Radiobiological quantities, however, such as RBE appear to be described equally well in terms of either one of these two important field descriptors.

Our present best estimates of the radial dose distribution in water and other media is based on the Oak Ridge model[50] as presented in Fig. 1. The calculation, neglecting primary excitations, is based on the Rutherford formula, the approximation that delta rays are normally ejected, and a power law approximation to the range of electrons in aluminum. This approximation is taken in two segments. One fits the data well below 1 keV; the other fits much of the data reasonably well for electrons above 1 keV. In the Rutherford formula, an adjusted ionization potential is assumed for the medium (10 eV for water) so that the integrated radial dose is finite. A multiplicative correction is then applied to this formula to agree with Hamm's Monte Carlo calculations, using the Oak Ridge formulation, of radial dose in water. This yields a bump in the plotted curve at radial distances below 10 nm, presumably accommodating the primary excitations. The Waligorski formulation is an extension of the formulation of Zhang et al.,[45] which ignores the energy deposited by primary excitations and ionizations, to fit Hamm's Monte Carlo calculation and adjusted to agree with tabular values of the stopping power of protons in water at proton energies from 0.1 to 1000 meV.

A series of graphs displays the relation between the formulas of Waligorski and the experimental measurements in gases.   In Fig. 2 we compare the calculations of Zhang[45] and of Waligorski[50] (labeled "this work") to measurements with protons, deuterons, and alpha particles by Menzel and Booz[52] and Wingate and Baum.[31] In Fig. 3 we compare these calculations to the work of Varma and co-workers[53] for O, Br, and I from the Tandem Van de Graaf accelerator and for He and Ne from the Berkeley BEVALAC accelerator. Also shown in Fig. 3 are the results of Monte Carlo calculations by Paretzke for 930 MeV He ions.

In Fig. 4 we compare the Zhang and the Waligorski[45,50] calculations to the calculations by Fain,[40] made in the continuous slowing-down approximation, and to

Radial dose distribution:

$$D_1(t) = \frac{Ne^4 z^{*2}}{\alpha mc^2 \beta^2 t} \cdot \frac{(1-((t+\theta)/(T+\theta)))^{1/\alpha}}{t+\theta} \tag{1}$$

$$\theta = R(I) \qquad\qquad I = 10 \text{ eV} \tag{2}$$

$$T = R(W) \qquad\qquad W = 2mc^2\beta^2(1-\beta^2)^{-1/2} \tag{3}$$

Electron range-energy relation for aluminium:

$$R = k \cdot w^\alpha \tag{4}$$

$$k = 6 \times 10^{-6} \text{ g} \cdot \text{cm}^2 \cdot \text{keV}^{-\alpha} \tag{5}$$

$$w < 1 \text{keV} \quad \alpha = 1.079 \quad \text{for ion } \beta \leqslant 0.03 \tag{6}$$

$$w > 1 \text{keV} \quad \alpha = 1.667 \quad \text{for ion } \beta > 0.03 \tag{7}$$

Delta ray distribution:

$$dn = \frac{2\pi Ne^4 z^{*2}}{mc^2 \beta^2} \cdot \frac{dw}{(w+I)^2} \tag{8}$$

Constant for liquid water:

$$N = \frac{2\pi Ne^4}{mc^2 \beta^2} = 1.369 \times 10^{-14} \text{J} \cdot \text{cm}^{-1} = 8.5 \text{ keV} \cdot \text{mm}^{-1} \tag{9}$$

Effective charge:

$$z^* = Z(1 - \exp(-125 \cdot \beta \cdot z^{-2/3})) \tag{10}$$

Corrected radial dose distribution:

$$D_2(t) = D_1(t) \cdot (1 + K(t)) \tag{11}$$

$$K(t) = A \cdot ((t-B)/C) \cdot \exp(-(t-B)/C) \tag{12}$$

$$A = 8 \cdot \beta^{1/3} \quad \text{for } \beta \leqslant 0.03 \tag{13}$$

$$A = 19 \cdot \beta^{1/3} \quad \text{for } \beta > 0.03 \tag{14}$$

$$B = 0.1 \text{ nm} \tag{15}$$

$$C = 1.5 \text{ nm} + \beta \cdot 5 \text{ nm} \tag{16}$$

$$K(t) = 0 \quad \text{for } t < B \tag{17}$$

Figure 1. Equations for the radial dose distribution in liquid water, from Waligorski et al.[50]

Zaider's Monte Carlo calculations,[48] which were made using experimental gas-phase cross sections. In Fig. 5 we compare the Waligorski calculation to measurements by Kanai[37] for alpha particles.

In all cases, our calculated radial dose is scaled from the formulas for protons by use of the effective charge according to Eq. 10 in Fig. 1.

Note that in all cases, the data do not reflect the "bump" in the Waligorski curves (this work). Note also that the data are very sparse close to, and far from, the ion's path. Nevertheless, it is quite remarkable that expressions calculated from a wide variety of source data for condensed matter are in such good agreement with measurements and Monte Carlo calculations made for gases. It is also remarkable that the oversimplified calculation, which totally neglects the angular distribution of ejected

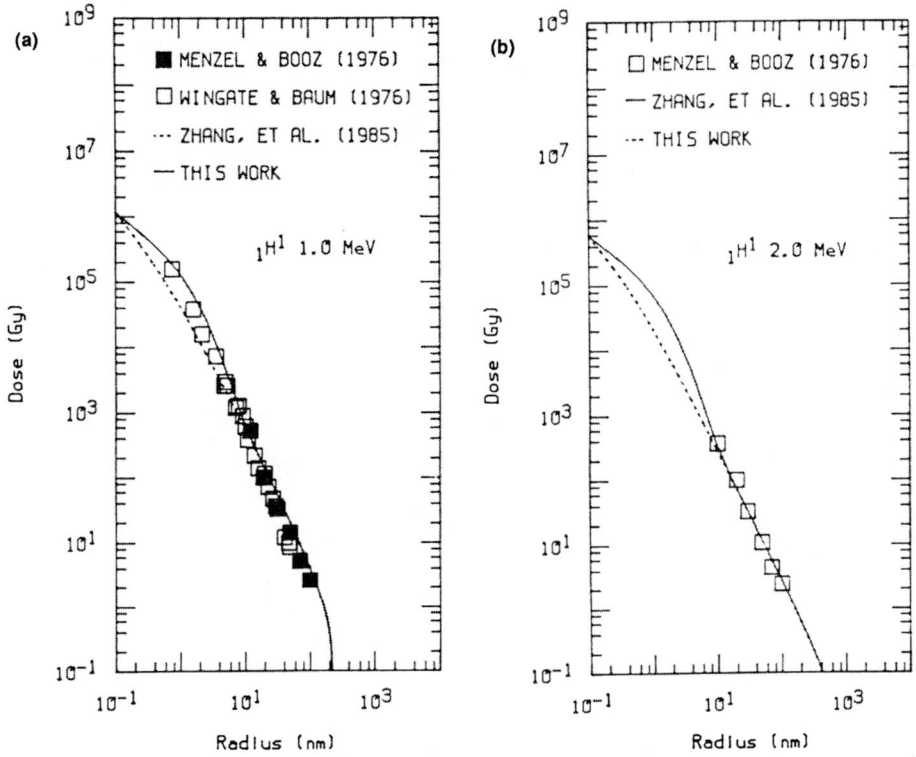

Figure 2. Data of Menzel and Booz[52] and of Wingate and Baum[31] compared to calculations from Zhang et al.[45] and of Waligorski et al.,[50] here labeled "this work".

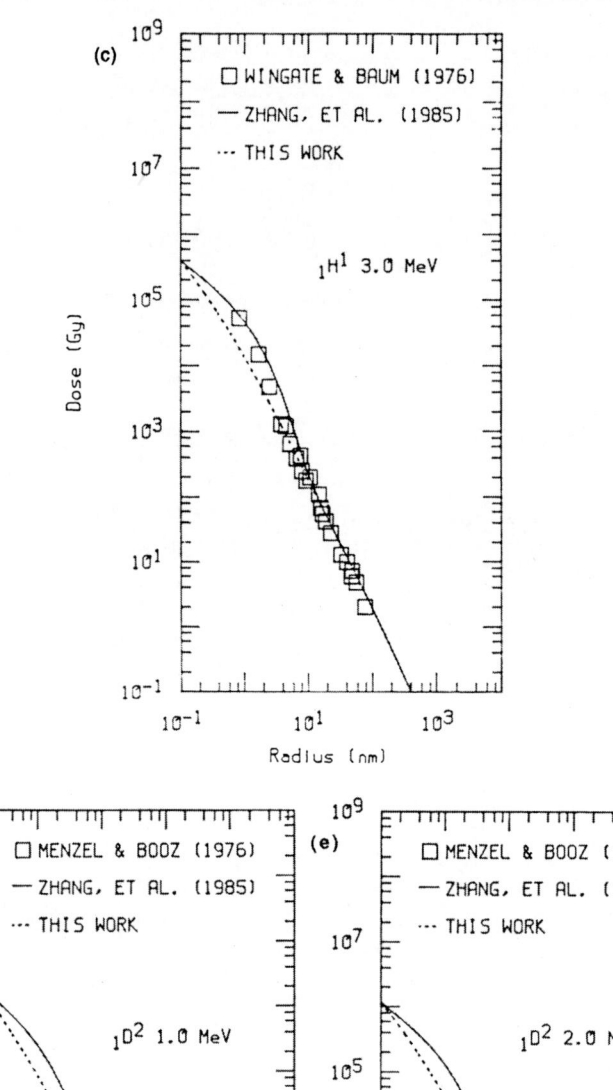

Figure 2. (Continued)

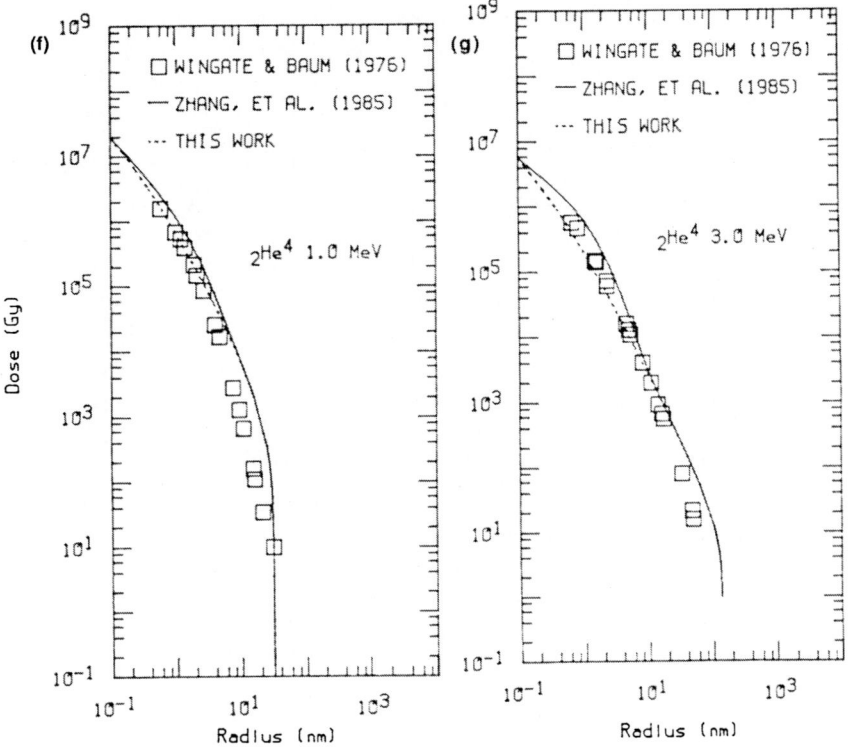

Figure 2. (Continued)

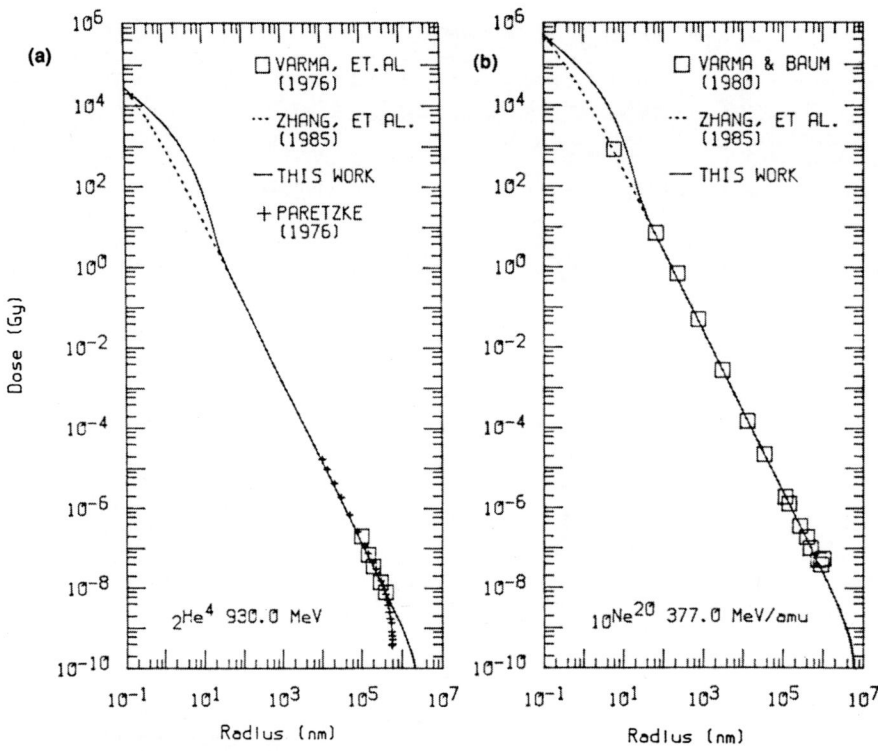

Figure 3. Data of Varma et al.[53] compared to calculations from Zhang[45] and from Waligorski.[50]

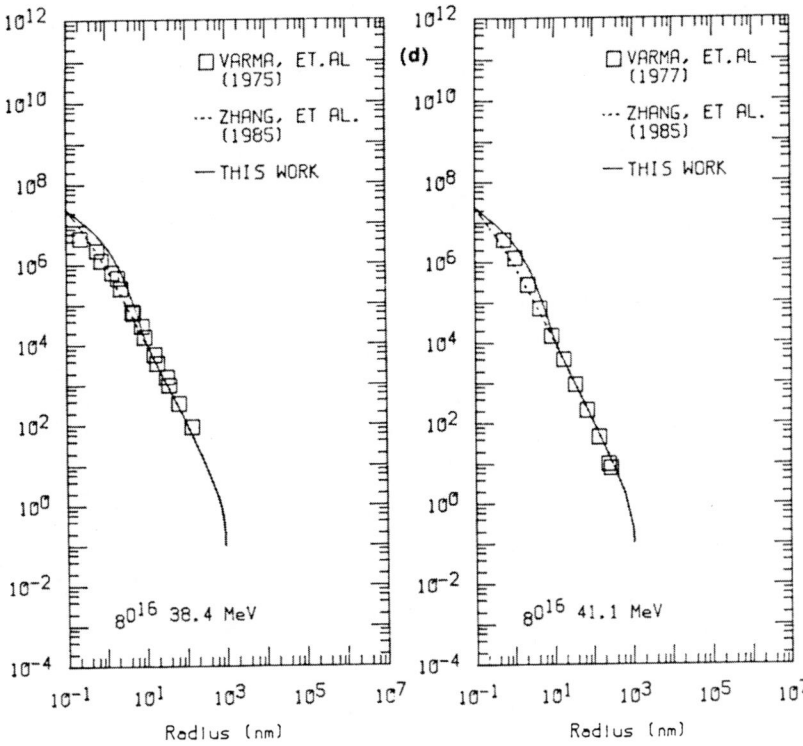

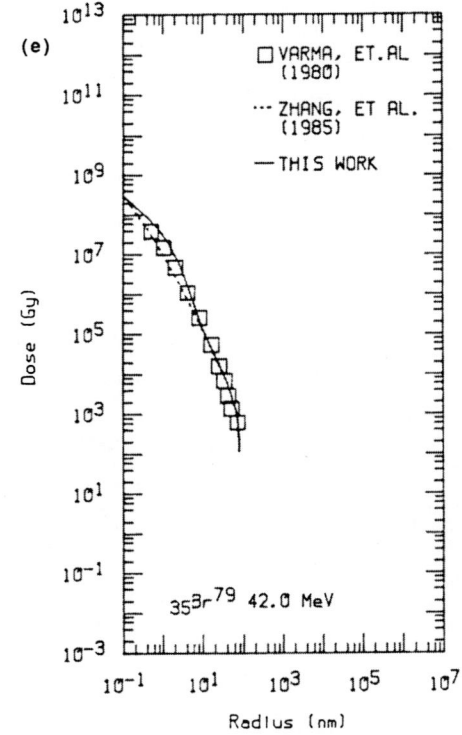

Figure 3. (Continued)

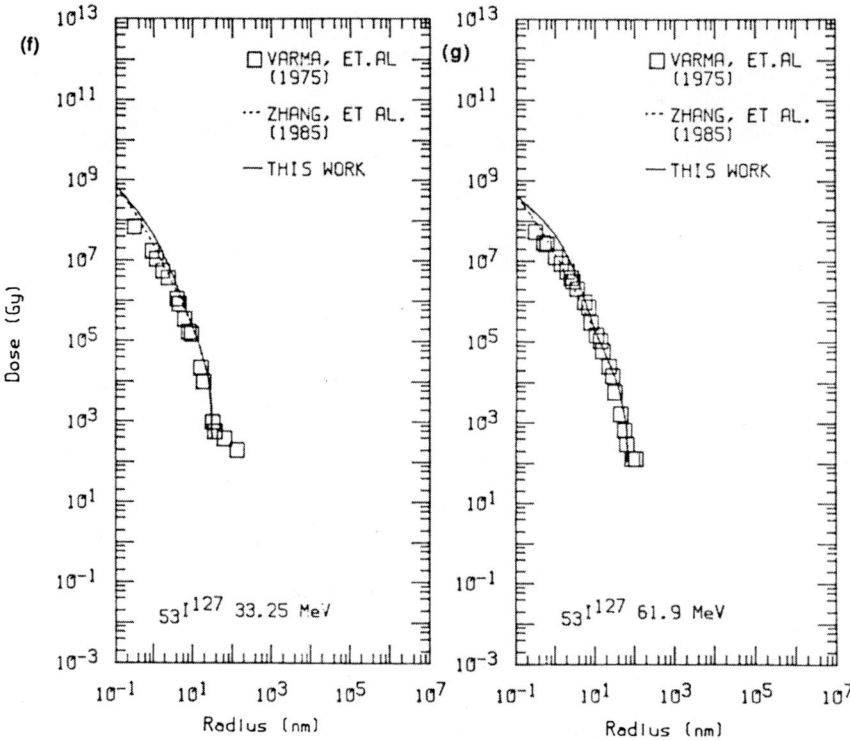

Figure 3. (Continued)

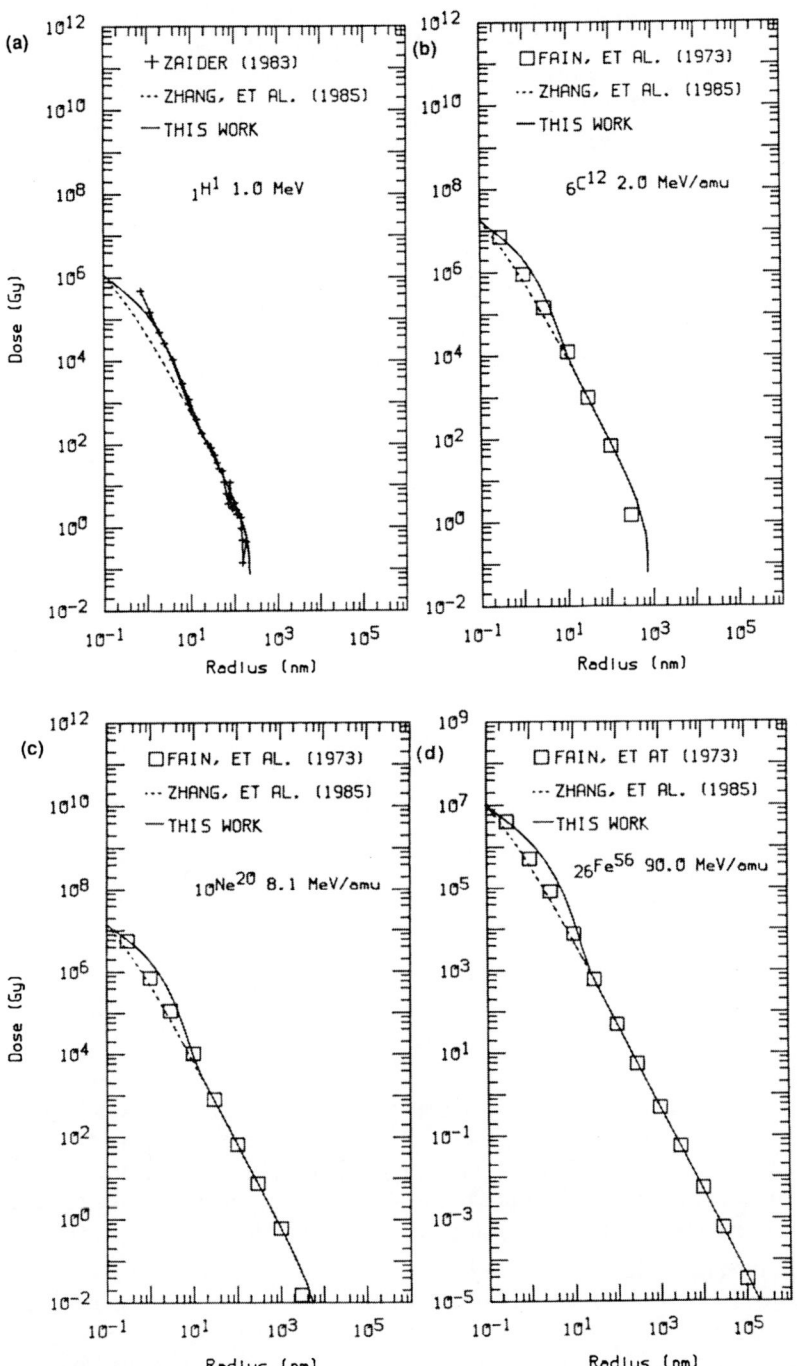

Figure 4. Calculations of Zaider et al.[48] and of Fain et al.[40,41] compared to the calculations of Zhang[45] and of Waligorski.[50]

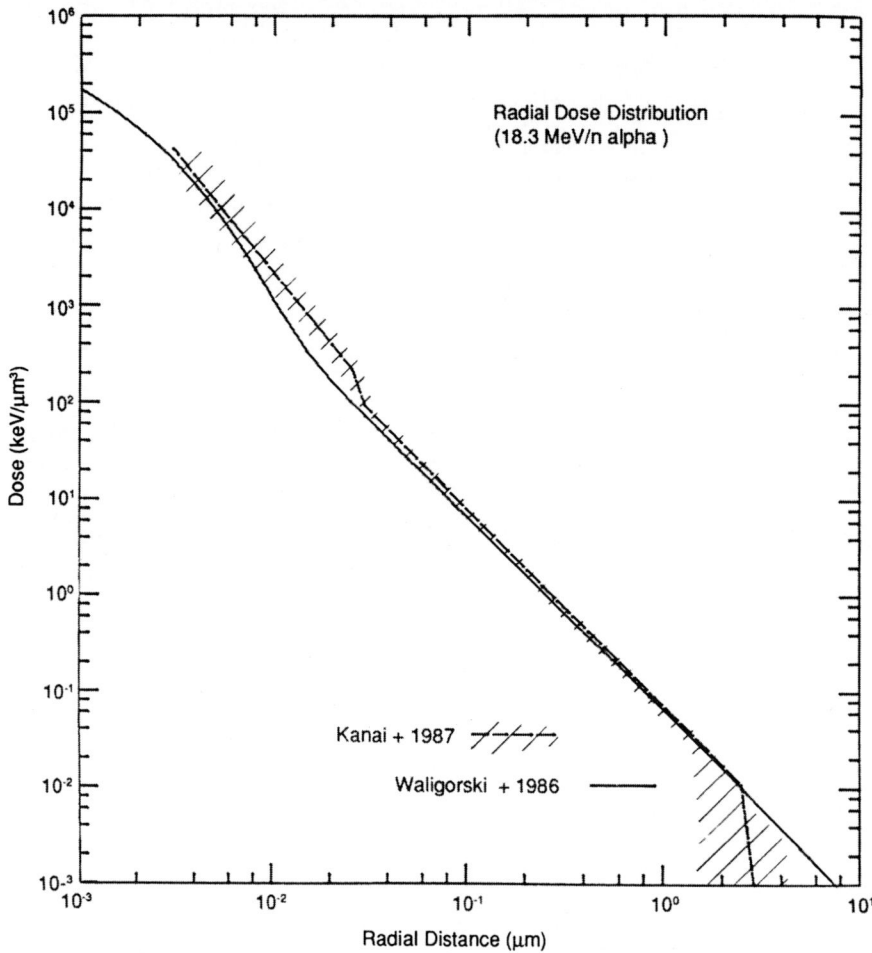

Figure 5. Measurements of Kanai and Kawachi[37] compared to the calculation of Waligorski.[50]

electrons and the difference in ionization potentials of the different shells from which the electrons are ejected, can yield such good results, not only for the radial dose distribution, but also for the subsequent calculations of cross sections.

We can anticipate that both calculations and measurements of the radial distribution of dose will continue, and that our knowledge concerning this important component of the study of the effects of high LET radiation will continually improve. To that end, we note that Rudd[54] has formulated a user-friendly model for the energy distribution of electrons from proton or electron collisions.

Continuing in the mode of our own oversimplifications, we have recently produced an analytic formulation of the radial dose distribution in several solids of interest in radiation measurement, based on an extension of our formulation for liquid water.[18] We have applied this formulation to the calculation of the response of NaI(T1) and LiF (TLD100) to energetic heavy ions.[55] In the latter case, we have shown that this model accounts for the observation of "hooks" in the response of this thermolumines-cent dosimeter to slowing heavy ions, precisely parallel to similar observations in radiobiology.[56] One value of a global parametric model of radiation effects is that parallel analyses of similar events in different detectors enable a correct attribution of these hooks to their common origin, the radial dose distribution from delta rays rather than to some imaginative mechanistic models of the behavior of particular detectors, or of biological cells.

Acknowledgment

This work was supported by the U.S. Department of Energy.

References

1. R. Katz and J. J. Butts. On the Width of Heavy Ion Tracks in Emulsion. In *Proc. of the 5th Int. Conf. on Nuclear Photography*. CERN 65-4, Vol. 2, IX-48 (1965).
2. R. Katz and J. J. Butts. Width of Ion and Monopole Tracks in Emulsion. *Phys. Rev.* 137:B198-B203 (1965).
3. J. P. Lonchamp. Contribution a l'Étude Methodologique des Emulsions Photographiques Utilisées en Physique Nucleaire. *Ann. de Phys.* 10:201-258 (1955).
4. P. G. Bizetti and M. Della Corte. On the Thinning Down of Tracks of Heavy Nuclei in Nuclear Emulsions. *Nuovo Cimento* 9:317-333 (1959).
5. J. Orear, A. H. Rosenfeld, and R. A. Schluter. Energy Loss by Charged Particles, *Chapter IIA in Nuclear Physics*, a Course Given by Enrico Fermi, pp. 27-34. University of Chicago Press, Chicago (1949).
6. W. H. Barkas. *Nuclear Research Emulsions 1.* Academic Press, New York (1963).
7. J. J. Butts and R. Katz. Theory of RBE for Heavy Ion Bombardment of Dry Enzymes and Viruses. *Radiat. Res.* 30:855-871 (1967).
8. A. M. Kellerer and H. H. Rossi. The Theory of Dual Radiation Action. *Curr. Topics Radiat. Res. Q1.* 8:85-158 (1972).
9. A. M. Kellerer and H. H. Rossi. A Generalized Formulation of Dual Radiation Action. *Radiat. Res.* 75:471-488 (1978).
10. E. J. Kobetich and R. Katz. Energy Deposition by Electron Beams and Delta Rays. *Phys. Rev.* 170:391-396 (1968).
11. E. J. Kobetich and R. Katz. Electron Energy Dissipation. *Nucl. Instr. Meth.* 71:226-230 (1968).
12. R. Katz and E. J. Kobetich. Response of Nuclear Emulsion to Electron Beams. *Nucl. Instr. Meth.* 79:320-324 (1970).
13. R. Katz and E. J. Kobetich. Particle Tracks in Emulsion. *Phys. Rev.* 186:344-351 (1969).
14. C. F. Powell, P. H. Fowler, and D. H. Perkins. *The Study of Elementary Particles by the Photographic Method.* Pergamon Press, New York (1959).
15. P. H. Fowler, R. A. Adams, V. G. Cowen, and J. M. Kidd. The Charge Spectrum of Very Heavy Cosmic Ray Nuclei. *Proc. Roy. Soc. A.* 301:39-45 (1967).
16. R. Katz and F. E. Pinkerton. Response of Nuclear Emulsions to Ionizing Radiations. *Nucl. Instr. Meth.* 130:105-119 (1975).
17. R. Katz, A. S-F. Li, Y. L. Chang, R. L. Rosman, and E. V. Benton. Tracks of Argon Ions in Ilford K-Series Nuclear Track Detectors. *The Nucleus* (Pakistan) 20:17-20 (1983).
18. R. Katz, K. S. Loh, Luo Daling, and G. R. Huang. An Analytic Representation of the Radial Distribution of Dose from Energetic Heavy Ions in Water, Si, NaI, and SiO_2. *Radiat. Effects and Defects in Solids* 114:15-20 (1990).
19. M. Jensen, L. Larsson, O. Mathiesen, and R. Rosander. Experimental and Theoretical Absorptance Profiles of Tracks of Fast Heavy Ions in Nuclear Emulsion. *Physica Scripta* 13:65-74 (1976).
20. M. Jensen and O. Mathiesen. Measured and Calculated Absorptance of Tracks of Fast Heavy Ions in Ilford G.5 Nuclear Emulsion. *Physica Scripta* 13:75-82 (1976).
21. S. Behrnetz. Application of Track Formation Theory to Calibration of Photometric Measurements on Cosmic Ray Tracks in Nuclear Emulsion. *Nucl. Instr. Meth.* 133:113-119 (1976).
22. R. Katz. Photometric Measurements of Thin Tracks in Nuclear Emulsion. *Nucl. Instr. Meth.* 169:257-259 (1979).
23. J. W. Baum. *Comparison of Distance and Energy Related Restricted Energy Transfer to Heavy Particles with 0.25 to 1100 MeV/amu.* Brookhaven National Laboratory (1967).
24. J. W. Baum, S. L. Stone, and A. V. Kuehner. Radial Distribution of Dose Along Heavy Ion Tracks, LET. In *Proc. Symp. Microdosim.*, H. G. Ebert, ed., pp. 269-281. Ispra, Italy (1968).
25. J. W. Baum, M. N. Varma, C. L. Wingate, H. G. Paretzke, and A. V. Kuehner. Nanometer Dosimetry of Heavy Ion Tracks. In *Proc. 4th Symp. Microdosim.*, J. Booz, H. G. Ebert, R. Eikel, and A. Waker, eds., 1:93-112. EUR 5122 d-e-f, Verbania Pallanza, Italy (1974).

26. M. N. Varma, J. W. Baum, and A. V. Kuehner. Energy Deposition by Heavy Ions in a "Tissue Equivalent" Gas. *Radiat. Res.* 62:1-11 (1975).

27. M. N. Varma, H. G. Paretzke, J. W. Baum, J. T. Lyman, and J. Howard. Dose as a Function of Radial Distance from a 930 MeV ^{4}He Ion Beam. In *Proc. 5th Symp. Microdosim.*, J. Booz, H. G. Ebert, and B.G.R. Smith, eds., 1:75-79. EUR 5452 d-e-f, Verbania Pallanza, Italy (1976).

28. M. N. Varma, J. W. Baum, and A. V. Kuehner. Radial Dose, LET and W for ^{16}O Ions in the N_2 and Tissue Equivalent Gases. *Radiat. Res.* 70:511-518 (1977).

29. M. N. Varma and J. W. Baum. Energy Deposition in Nanometer Regions by 377 MeV/nucleon ^{20}Ne Ions. *Radiat. Res.* 81:355-363 (1980).

30. M. N. Varma. Review of Radial Dose Measurement Technique and Data. *Nucl. Tracks. Radiat. Meas.* 16:135-139 (1989).

31. C. L. Wingate and J. W. Baum. Measured Radial Distribution of Dose and LET for Alpha and Proton Beams in Hydrogen and Tissue-Equivalent Gas. *Radiat. Res.* 65:1-17 (1976).

32. N. F. Metting. *Measurement of Energy Deposition Near High Energy Heavy Ion Tracks*. Master's Thesis, University of Washington. Pacific Northwest Laboratory, Richland, Washington (1986).

33. N. F. Metting. A Comparison of Microscopic Dose with Average Dose Near High Energy Ions. *Nucl. Instr. Meth. in Phys. Res.* B24/25:1050-1053 (1987).

34. N. F. Metting, H. H. Rossi, L. A. Braby, P. J. Kliauga, J. Howard, M. Zaider, W. Schimmerling, M. Wong, and M. Rapkin. Microdosimetry Near the Trajectory of High Energy Ions. *Radiat. Res.* 116:183-195 (1988).

35. L. H. Toburen, N. F. Metting, and L. A. Braby. Spatial Patterns of Ionization in Charged Particle Tracks. *Nucl. Instr. Meth. in Phys. Res.* B40/41:1275-1278 (1989).

36. L. H. Toburen, L. A. Braby, N. F. Metting, G. Kraft, H. Schmidt-Bocking, R. Dorner, and R. Seip. Radial Distributions of Energy Deposited Along Charged Particle Tracks. In *10th Symposium on Microdosimetry*, Rome, May 1989 (1990).

37. T. Kanai and K. Kawachi. Radial Dose Distribution for 18.3 MeV/n Alpha Beams in Tissue Equivalent Gas. *Radiat. Res.* 112:426-435 (1987).

38. L. H. Toburen, W. E. Wilson, and R. J. Popowich. Secondary Electron Emission from Ionization of Water Vapor by 0.3 to 2.0 MeV He+ and He2+ Ions. *Radiat. Res.* 82:27-44 (1980).

39. F. Hutchinson. The Interaction of Primary Cosmic Rays with Matter and Tissue. In *Symposium on Medical and Biological Aspects of the Energies of Space*, P. A. Campbell, ed., Columbia University Press, New York (1961).

40. J. Fain, M. Monnin, and M. Montret. Energy Density Deposited by a Heavy Ion Around Its Path. In *Proc. 4th Symp. Microdosim.* J. Booz, H. G. Ebert, R. Eikel, and A. Waker, eds., pp. 169-188. Verbania-Pallanza, Italy (1974).

41. J. Fain, M. Monnin, and M. Montret. Spatial Energy Distribution Around Heavy Ion Paths. *Radiat. Res.* 57:379-389 (1974).

42. A. Chatterjee, H. D. Maccabee, and C. A. Tobias. Radial Cut-Off LET and Radial Cut-Off Dose Calculations for Heavy Charged Particles in Water. *Radiat. Res.* 54:479-494 (1973).

43. A. Chatterjee and H. J. Schaefer. Microdosimetric Structure of Heavy Ion Tracks in Tissue. *Radiat. Environ. Biophys.* 13:215-227 (1976).

44. J. W. Hansen and K. J. Olsen. Experimental and Calculated Response of a Radiochromic Dye Film Dosimeter to High-LET Radiations. *Radiat. Res.* 97:1-15 (1984).

45. C. Zhang, D. E. Dunn, and R. Katz. Radial Distribution of Dose and Cross Section for the Inactivation of Dry Enzymes and Viruses. *Radiat. Prot. Dosim.* 13:215-218 (1985).

46. W. Brandt and R. H. Ritchie. Primary Processes in the Physical Stage. In *Physical Mechanisms in Radiation Biology*, R. D. Cooper and R. W. Wood, eds., pp. 20-46. U.S. Atomic Energy Commission CONF-721001 (1974).

47. H. G. Paretzke, G. Leuthold, G. Burger, and W. Jacobi. Approaches to Physical Track Structure Calculations. In *Proc. 4th Symp. Microdosim.*, J. Booz, H. Ebert, R. Eickel, and A. Waker, eds., 1:75-79. EUR 5122 d-e-f, Verbania Pallanza, Italy (1976).

48. M. Zaider, D. J. Brenner, and W. E. Wilson. The Applications of Track Calculations to Radiobiology 1. Monte Carlo Simulation of Proton Tracks. *Radiat. Res.* 95:231-242 (1983).

49. R. H. Ritchie, R. N. Hamm, J. E. Turner, H. A. Wright, J. C. Ashley, and G. J. Basbas. Physical Aspects of Charged Particle Track Structure. *Nucl. Tracks Radiat. Meas.* 16:141-155 (1989).

50. M.P.R. Waligorzki, R. N. Hamm, and R. Katz. Radial Distribution of Dose Around the Path of a Heavy Ion in Liquid Water. *Nucl. Tracks Radiat. Meas.* 11:309-319 (1986).

51. N. M. Varma and M. Zaider. The Radial Dose Distribution As a Microdosimetric Tool. *Radiat. Prot. Dosim.* 31:155-160 (1990).

52. H. G. Menzel and J. Booz. Measurement of Radial Energy Deposition Spectra for Protons and Deuterons in Tissue Equivalent Gas. In *5th Symposium on Microdosimetry*, J. Booz, H. G. Ebert, B.G.R. Smith, eds. Commission of the European Communities, Luxembourg (1976).

53. M. N. Varma, J. W. Baum, and A. V. Kuehner. Stopping Power and Radial Dose Distribution for 42 MeV Bromine Ions. *Phys. Med. Biol.* 25:651-656 (1980).

54. M. E. Rudd. User Friendly Model for the Energy Distribution of Electrons from Proton or Electron Collisions. *Nucl. Tracks Radiat. Meas.* 16:213-218 (1989).

55. R. Katz and Luo Daling. Response of NaI(T1) and TLD(100) to Energetic Heavy Ions. *9th Int. Conf. on Solid State Dosimetry*, Vienna 1989 (in press).

56. R. Katz, E. D. Dunn, and G. L. Sinclair. Thindown in Radiobiology. *Radiat. Prot. Dosimetry* 13:281-284 (1985).

57. R. Katz. Cross Section. *Appl. Radiat. Isot.* 41:563-567 (1990).

Discussion

Ward: What is the range of energies per grain that we were looking at in these discussions?

Katz: It could be something like 100 eV. It might be much more than that. Insensitive materials might take much more than that. The most sensitive, I think, would be something between 100 eV and a keV. The size is gram comparable to the radius of a cell when you take the density difference into account.

Ward: How do you account for this? If it takes 100 eV to a keV to produce a grain, how does this relate to the molecules and the radicals in water or biological media?

Katz: I don't think you make that kind of comparison. What you can see from this is how large the brush of delta rays is. You have an illustration here of what a typical track looks like and how the character of the track changes with the sensitivity of the material. When it was suggested that we might be able to actually see a track in cells, I thought that would be wonderful, because then we wouldn't make all these foolish statements about the importance of LET, and we wouldn't talk quality factor nonsense. We would see what was going on instead of inferring from the prejudices of our past.

Zaider: The associated volume concept contains the radial distribution as a particular case, and the track contains all the cases that are associated with the track, e.g. larger than the target, the same size as the target, or smaller than the target. So I don't think that you should call this a false concept; it is not a mistaken concept. It is more powerful concept than is radial distribution, and the reason for that (correct me if you think I am wrong) is that from the associated volume concept, you can produce radial distributions. You cannot go back.

Katz: The associated volume concept was the original motivation for experiments on dry enzymes and viruses. That concept produced data in the form of cross sections attributed to the sizes of the enzymes and viruses as corrected by Lea's correction, which turned out to be wrong by orders of magnitude. The reason that they were wrong by orders of magnitude is what you see in these tracks.

Zaider: Because it involved radial distribution. But not because the concept is wrong.

Katz: My view is that the concept is wrong if it gives you wrong measurements or wrong calculations.

Moolgavkar: What do you mean by the cross section for enzyme inactivation?

Katz: This is the way that it is measured. You have a beam of particle striking a layer of enzymes or viruses, and you measure the surviving fraction, which declines exponentially and is represented as $e^{-\sigma F}$, where F is the fluence and σ is a probability. A cross section is always a probability.

Moolgavkar: Okay. It is a matter of semantics here.

Katz: No, no. Physical cross sections are probabilities, nothing else.

Moolgavkar: All this exponential behavior, to what does it apply?

Katz: Oh, when the process is a one-hit process. Well, a one-hit process is one where either a simple event in the target or a single incident particle produces the effect so that you can get this exponential response with heavy particles; even when the response to gamma rays has a great big shoulder, as for mammalian cells.

Moolgavkar: Excuse me, one more question. You are making the assumption here that you already have a large number of particles or whatever so that you have a Poisson approximation for something like a binomial distribution. Do you understand the problem?

Katz: The character of the response is due to both fluctuations in energy deposition to the character of the target. Cross section is the measure of the probability of an effect being observed after an infinite number of identical repeated trials (57).

Curtis: It is the probability per unit of fluence?

Katz: Well, that is what probability means. The probability for a single particle....

Curtis: The reason that this quantity has the units of area is because it is the probability per unit fluence.

Katz: Well, that is the same as saying "per incident particle."

Varma: I want to bring the discussion back to what you talked about with radial dose distribution. What you didn't mention (and it is very important) is that what we should have in the record is the fact that all the data that you showed really are from, say, 10 to 20 angstroms away from the track and to a large extent where you have 90 to 95 percent of the energy deposited. The interesting part might be if you looked at 10 angstroms and above, where about 95 percent of the energy is deposition. Experimentally that is very hard to measure. Theoretically, when Herwig calculates or when PNL people calculate, it takes a tremendous amount of time on computers. The point that we need to make (and I don't know the answer to this) is that dose deposition at a very short range, say at 10 angstroms, is probably very important because the secondary electrons are very high in energy. As Harel pointed out, once you go outside the track, you might have an energy deposition that is very large, which might be doing the damage.

Katz: But they are from single electrons, a single electron passing through an emulsion grain or a microdosimeter.

Varma: But sometimes a single electron produces a large amount of energy deposition and sometimes a small energy deposition.

Katz: I think they produce what single electrons produce. What we are doing is normalizing these responses to single electrons at low doses of gamma rays. Otherwise, this wouldn't work at all. But the whole basis of this is that we have a means of normalization. We have a means of calibrating the effects of small dose. That is why we are not doing anything *ab initio*. It is all phenomenological. Lest you turn your nose up at phenomenology; keep in mind that if there hadn't been Kepler's phenomenology, there wouldn't be Newton's gravitation, and if there hadn't been Balmer's phenomenology, there wouldn't have been Schröedinger. Phenomenology precedes mechanism very often in physics, so don't turn your nose up at it. It will lead you to mechanistic conceptions more frequently than the other way around.

Early Chemical Events

Radiation Interactions in High-Pressure Gases

Loucas G. Christophorou[a]

Abstract

This paper concerns basic radiation interaction processes in dense fluids and interphase studies aimed at interfacing knowledge on radiation interaction processes in low-pressure gases and knowledge on such processes in liquids. Microscopic and macroscopic properties of—and processes in—matter in the intermediate density range between the low-pressure gas and the condensed phase are discussed. Results of recent studies on the effect of the density and nature of the medium on electron production in irradiated fluids and on the state, energy, transport, and attachment of slow excess electrons in dense fluids (high-pressure gases and dielectric liquids) are described. The possible significance of electron-molecule interactions in dense gases in establishing mechanisms of radiobiological action is indicated.

Introduction

The understanding of radiobiological effects and mechanisms necessitates knowledge of i) the basic phenomena and processes involved in the interaction of ionizing radiation with matter, and ii) the effects of the density, nature, and state of matter on these basic phenomena and processes. Our knowledge in this area—especially on gases at or below atmospheric pressures—advanced phenomenally in the last two decades, and has illuminated the fields of radiobiological science with a broad impact on pure science, applied science and engineering. The previous papers concentrated on the former; we shall now focus on aspects of the latter.

The effects of the nature, density, and state of the medium are generally small for low-lying valence states, but they are profound for quasi-charge-separated states (e.g., high-n Rydberg states), charge-separated states (electrons; positive and negative ions) transient negative ion states, and the physical quantities which describe their behavior and reactions. Thus, for example, the following attributes are all strong functions of the nature and density of the medium in which these elementary physical processes occur: electron energy ϵ, the quasi-free (excess) electron "ground state" energy V_o, the electron drift velocity w, the electron mobility μ, the cross section for electron scattering σ_{sc}, electron attachment σ_a, dissociative attachment σ_{da}, autodetachment σ_{ad}, ionization σ_i, and the associated physical quantities [e.g., the electron affinity EA, the vertical attachment energy VAE, the vertical detachment energy VDE, the ionization threshold energy I, and the polarization energy of the positive (P^+) and the negative (P^-) ions]. While these physical quantities have well-defined values at low-gas number densities, they assume a spectrum of values in dense media. Understanding these

(a) Health and Safety Research Division, Oak Ridge National Laboratory, Oak Ridge, Tennessee; also Department of Physics, The University of Tennessee, Knoxville, Tennessee

Physical and Chemical Mechanisms in Molecular Radiation Biology
Edited by W.A. Glass and M.N. Varma, Plenum Press, New York, 1991

183

effects is a prerequisite for the successful use of physico-chemical knowledge in establishing mechanisms of radiobiological action and in developing novel radiobiological instrumentation.

This paper concerns basic radiation interaction processes in dense fluids and interphase studies aimed at the interfacing of knowledge on radiation interaction processes in the gaseous and the liquid state of matter. It concentrates on slow electrons since these are the most abundant and most reactive species produced in matter by the action of ionizing radiation. It is specifically focused on the effect of the density and nature of the medium on electron production in irradiated fluids and on the state, energy, transport, and attachment of slow excess electrons in dense fluids—especially dielectric liquids which possess excess-electron conduction bands (energy of the quasi-free electron at the bottom of the conduction band of the fluid $V_o < 0$ eV).[1-4] Studies over the past two decades have shown that the interactions of low-energy electrons with molecules embedded in dense media depend not only on the molecules themselves[1-7] and their internal state of excitation,[8-10] but also on the electron state and energy in—and the nature and density of—the medium in which the interactions occur.[3,4,11,12]

Electron Production in Dense Fluids

Total and Free Electron Yields

The processes by which isolated atoms and molecules are ionized by radiation, and the corresponding cross sections, have been well studied and are reasonably well understood.[1,2,5-7] Measurements of electron impact ionization cross sections $\sigma_i(\epsilon)$ in gases are, for example, abundant. This, however, cannot be said of dense fluids. In these, there exist measurements of only the density, N, normalized ionization coefficient α/N as a function of the density-reduced electric field E/N (Fig. 1) at relatively low pressures ($\lesssim 5 \times 10^{19}$ molecules cm^{-3}). In this gas density range, α/N is generally independent of N. This, however, is expected not to be the case for dense gases ($N \gtrsim 10^{20}$ molecules cm^{-3}) and liquids. An indirect measurement[15] of $\alpha(E/N)$ for liquid Xe, for example, has indicated that $\alpha(E/N)$ does not scale from its gaseous value by considering the density difference between the gas and the liquid.

While the effect of N on $\sigma_i(\epsilon)$ and $\alpha/N(E/N)$ is largely unknown, many studies[2-4,16] have shown that the total, G_{te}, and especially the free, G_{fe}, electron yield are strong functions of the density and state - gaseous (G), liquid (L) or solid (S) - of matter. At least for the heavier rare gases (Ar, Kr, Xe)

$$\left(G_{te}\right)_L > \left(G_{te}\right)_G \tag{1}$$

and for all liquids

$$\left(G_{fe}\right)_L << \left(G_{fe}\right)_G \tag{2}$$

The inequality [Eq. (2)] is especially profound for densely ionizing (high-LET) particles (Table I). The primary reason for (1) is the lowering of the isolated atom's minimum ionization energy I_G when the atom is embedded in the dense medium. The primary reason for (2) is the profound effect of geminate recombination on $(G_{fe})_L$, especially

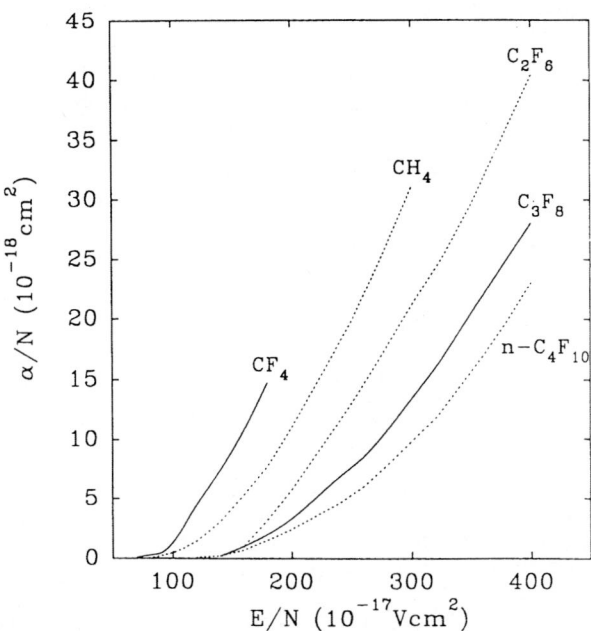

Figure 1. Density-normalized electron impact ionization coefficients α/N versus E/N for CF_4, CH_4, C_2F_6, C_3F_8, and n-C_4F_{10}.[13,14] Note that the $\alpha/N(E/N)$ curves shift to higher E/N values as the molecular size is increased. This is because α/N is related to $\sigma_i(\epsilon)$ and $f(\epsilon, E/N)$ by $\alpha/N\ (E/N) = (2/m)^{1/2} \int_0^\infty \sigma_i(\epsilon)\ \epsilon^{1/2} f(\epsilon, E/N)d\epsilon$; and as the molecule becomes more complex, $f(\epsilon, E/N)$ shifts to lower energies for a fixed E/N due to the increase in the energy loss processes.

Table I. Electron Yields in Dielectric Liquids with Excess Electron Conduction Bands ($V_o < 0$ eV)

Liquid	Electron Yield (Electrons/100 eV Energy Absorbed)				Gas: $G_{te} \approx G_{fe}$[c]
	$G_{te}(e,\gamma,x)$[a]	$G_{fe}(e,\gamma,x)$[a]		$G_{fe}^E(\alpha)$[b]	
		$E^{[d]} = 0$	E > 0	E > 0	
Ar(87K)	4.2[e]	2.3[f]	4.4[f,g]	0.45[f,g]	3.8
		2.7[h]	4.15[f,g]	0.46[f,g]	
				0.38[f,g]	
Kr(129K)	6.0[h]	4.0[h]	-	-	4.2
	4.9[i]				
Xe(165K)	6.4[e]	4.4[h]	-	-	4.6
$C(CH_3)_4$ (~296K)	4.3[j]	1.1[f]	1.18[f,g]	0.036[f,g]	-
		1.0[k]			
$Si(CH_3)_4$ (~296K)	-	0.74[f,l]	1.19[f,g]	0.029[f,g]	4.2°
			0.98[m,g]	0.075[n]	-

(continued)

Table I. (continued)

Liquid	Electron Yield (Electrons/100 eV Energy Absorbed)			Gas:
	$G_{te}(e,\gamma,x)^{(a)}$	$G_{fe}(e,\gamma,x)^{(a)}$	$G_{fe}^{E}(\alpha)^{(b)}$	$G_{te} \approx G_{fe}^{(c)}$
$Ge(CH_3)_4$				
($\sim$296K)	-	-	$0.95^{(m,g)}$	-	-
$Sn(CH_3)_4$				
($\sim$296K)	-	$0.62^{(l)}$	$1.15^{(m,g)}$	-	-
$(CH_3)_3 CCH_2$				
$C(CH_3)_3$	-	$0.73^{(f,g)}$	$1.1^{(f,g)}$	$0.025^{(f,g)}$	-
		$0.83^{(p)}$	$1.14^{(m,g)}$		

(a)   For low-ionization density radiation [electrons (e), γ-rays (γ), x-rays (x)].
(b)   For high-ionization density radiation [α-particles (α)].
(c)   Value for low-pressure gas determined from the W-data (eV per ion pair) in Ref. 1 unless otherwise noted.
(d)   Applied electric field.
(e)   [11]; (f) [17]; (g) [E = 10^4 V cm^{-1}], [18]; (h) [19]; (i) [20]; (j) [16]; (k) ([21]; (l) [22]; (m) [23]; (n) [24]; (o) [25]; (p) [26].

for high-LET particles.   Understanding electron production in dense fluids, then, requires an understanding of the density dependence of the ionization cross sections and energetics and electron-positive ion recombination.

Ionization Threshold Energy and Its Dependence on the Nature and Density of the Medium

In contrast to the almost complete absence of studies on electron-impact ionization of atoms and molecules in dense fluids, a number of investigations have focused on the photoionization processes and energetics of molecules in dense media[11,27-34].   In Table II are listed the ionization threshold energies I_L of a number of dielectric liquids, and of two organic molecules in dielectric liquids for which V_o is known.   The data on pure liquids were obtained using ultraviolet photoconductivity techniques and assumed photoionization threshold laws.   The data for the two organic molecules in dielectric liquids were obtained using a laser multiphoton ionization conductivity method and carefully monitoring the order of the multiphoton ionization process(es).   Figure 2 shows this for the case of azulene in 2,2,4,4-tetramethylpentane (TMP).   When the two-photon energy lies well above I_L [at 5.70 eV; (Fig. 2b)], the probability of electron escape is large and two-photon ionization is the predominant multiphoton ionization mechanism.   When, however, the two-photon energy approaches I_L, the geminate electron-ion pair recombines, and ionization occurs predominantly via three-photon ionization.   This latter process (Fig. 2b) occurs via two-photon absorption to a high-lying state at energies $\lesssim I_L$, which converts internally to a lower-lying long-lived state (second excited π-singlet state S_2 for azulene) from which a third photon leads to the ionization continuum high above I_L.   The two-photon ionization onset I_L was taken to be at that laser wavelength where the two- and three-photon ionization processes contribute, respectively, 10% and 90% of the total photoionization signal, i.e., when the overall order of multiphoton ionization S = 2.9.   This method is accurate, but requires proper identification of the ionization mechanisms that often become complicated depending on the characteristics of the laser pulse and those of the excited electronic

Table II. Ionization threshold energies I_L for a number of dielectric liquids and corresponding low-pressure gas values I_G along with the V_o values of these liquids. I_G values for azulene and fluoranthene and the I_L values for these two molecules in various dielectric liquids.

Liquid	I_L (eV)[a]		I_G (eV)[b]	V_o (eV)[c]
Tetramethylsilane [$(CH_3)_4Si$]	8.1[d]	8.05[e]	9.65	-0.55
Tetramethylgermanium [$(CH_3)_4Ge$]	7.6[d]		9.35	-0.64
Tetramethyltin [$(CH_3)_4Sn$]	6.9[d]		8.89	-0.75
Neopentane [$(CH_3)_4C$]	8.55[d,e]		10.23	-0.43
n-Pentane [n-C_5H_{12}]	9.15[d]	8.86[e]	10.28	+0.01
Cyclopentane [c-C_6H_{14}]	8.82[e]		10.53	-0.22
3-Methylpentane [$(C_2H_5)_2CHCH_3$]	8.85[d]		10.08	+0.01
Neohexane [$(CH_3)_3CCH_2CH_3$]	8.73[d]	8.49[e]	10.06	-0.22
n-Hexane [n-C_6H_{14}]	8.70[e]		10.22	+0.07
Cyclohexane [c-C_6H_{12}]	8.75[d]	8.43[e]	9.87	+0.01
2,2,4-Trimethylpentane [$(CH_3)_3CCH_2CH(CH_3)_2$]	8.38[e]		9.86	-0.17
2,2,4,4-Tetramethylpentane [$(CH_3)_3CCH_2C(CH_3)_3$]	8.2[f]		9.5[f]	-0.36
n-Tridecane [$C_{30}H_{62}$]	9.25[d]		10.03	+0.21
Tetramethylethylene [$(CH_3)_2CC(CH_3)_2$]	6.80[d]		8.30	-0.24
Azulene in				
Tetramethyltin	5.33[g]		7.42[g]	
Tetramethylsilane	5.45[g]			
2,2,4,4-Tetramethylpentane	5.70[g]			
n-Pentane	6.12[g]			
n-Tridecane	6.28[g]			
Fluoranthene in				
Tetramethlysilane	5.70[h]		7.57[h]	
Ar	~14.1[87K][i]		15.755[j]	-0.20[k]
Kr	11.55[121K][l]		13.996[j]	[k]
Xe	9.2[165K][l]		12.127[j]	-0.61[k]
	8.9[m]			

(a)   T ≈ 295K unless otherwise indicated.
(b)   Average of photoionization and photoelectron values given in Ref. 35.
(c)   Unless otherwise indicated, from Refs. 11, 36 and 37.
(d)   [32,38]; (e) [31]; (f) [39]; (g) [33]; (h) [40]; (i) [11]; (j) [1]; (k) [see Fig. 5]; (l) [30]; (m) [41].

states involved, especially their lifetimes, and intramolecular relaxation pathways.[33] The transition to the continuum by absorption of a photon by the excited molecule is both a function of the energy of the absorbed photon and of the particular electronic/vibrational state involved in the process.

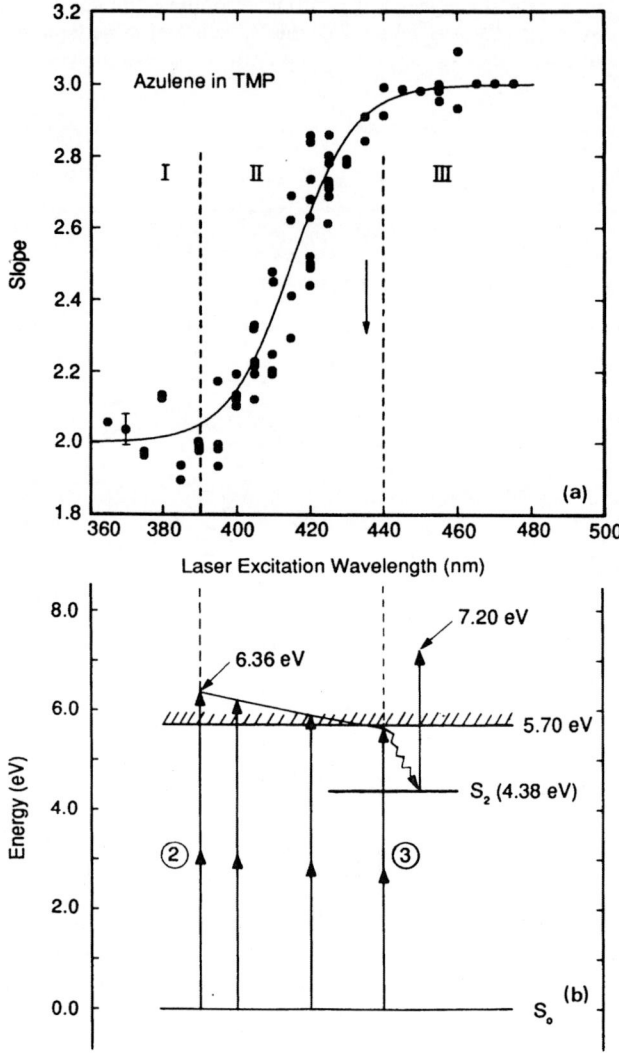

Figure 2. (a) The Slope (order of multiphoton ionization) as a function of laser excitation wavelength for the azulene molecule in liquid 2,2,4,4-tetramethylpentane (TMP). Regions I, III, and II, correspond respectively to wavelength ranges where two-photon, three-photon and both two- and three-photon ionization occurs. The arrow points to the laser wavelength where the two-photon ionization onset energy is located.[33] (b) Schematic illustration of the energetics and mechanisms involved (see the text).

The data in Table II (and other similar measurements)[11,28,42] show that I_L is related to I_G[43](a) by

(a) By analogy, the electron affinity of a molecule in the liquid, $(EA)_L$ is related to that, $(EA)_G$, in the low-pressure gas by $(EA)_L = (EA)_G + V_o - P^-$ where P^- is the difference in the solvation energies of the neutral molecule and its anion in the liquid. Since P^- is always a negative quantity and, as a rule, much larger in absolute magnitude than V_o, $(EA)_L > (EA)_G$. Photodetachment ($M_L^- + h\nu \to M_L + e_L$) studies in dense fluids are scarce and must be undertaken.

$$I_L = I_G + V_o + P^+ \tag{3}$$

The polarization energy of the positive ion in the medium P^+ is a negative quantity which is usually approximated by the Born charging energy[44] (see, however)[33,45]

$$P^+ = -\frac{e^2}{2R}\left(1 - \frac{1}{\epsilon}\right) \tag{4}$$

where ϵ is the optical dielectric constant of the medium and R is the effective radius of the positive ion cavity. The fact that the measurements follow (3) shows that in the dense fluid the polarization dynamics are very fast ($<$ps). For the systems in Table II, the average values of P^+ and ϵ are -1.273 $\pm$ 0.156 eV and ~1.94, respectively, which give a mean value for the effective radius R of the positive ion cavity of ~2.74 Å.

The energy difference $I_L - I_G$ (~-1 to -3 eV for organic molecules in nonpolar liquids) is, then, given by $P^+ + V_o$, and the effect of the medium and its density on I_L is a manifestation of that on P^+ and V_o.

The gradual lowering of I_G to its I_L value in going from the low-pressure gas to the liquid can be seen from Fig. 3.[29,30-34] Here the ionization onset energy I_F of the molecule TMPD (N,N,N´,N´-tetramethyl-p-phenylenediamine) mixed with C_2H_6 was measured[34] as a function of the density of the latter from the low-pressure gas to the liquid and in the liquid itself. The I_F decreases with increasing gas density and temperature.[a] The dependence of I_F on N has been shown[b] to be due to the N-dependence of P^+ and V_o, viz.,

$$I_F(N) = I_G + P^+(N) + V_o(N) \tag{5}$$

The density dependence of P^+ is dominated by the N-dependence of ϵ, and can be calculated[34] or be experimentally determined[3,11,29-34,37,38,42] via Eq. (3) when the other quantities are known or assumed. The $V_o(N)$ function is more complicated; it has been calculated for a few cases[34,46,47] using the Springett-Jortner-Cohen (SJC) model[46] or has been estimated via Eq. (5) when the other quantities are known or assumed.[29,30,48] In Fig. 4 are shown values of P^+ for TMPD in C_2H_6 calculated[34] as a function of the C_2H_6 density; they clearly show that P^+ is a monotonically decreasing function of the fluid density ρ. This decrease of P^+ with increasing ρ is primarily due to the increase of ϵ with ρ and to a lesser extent due to changes in R with increasing ρ.[34] Also plotted in Fig. 4 is $V_o(\rho)$ for C_2H_6 calculated[34] by using the SJC model or by relaxing some of the model's basic assumptions.[34] The $V_o(\rho)$ values determined by the modified SJC model[34] agree rather well with the experimental measurements.[48]

To understand the function $V_o(N)$, let us refer briefly to the SJC model. In this model each molecule occupies a sphere whose radius ("the Wigner-Seitz (WS) radius") r_{ws} is equal to $(\frac{4}{3}\pi N)^{-1/3}$; each molecule, also, has a hard-core radius $\tilde{a}$ such that the

---

(a) At a given medium density the observed decrease in I_F with increasing T (Fig. 3) was understood from the T-dependence of V_o.[34]

(b) In Eq. (5) the broadening of the valence levels of the isolated atom (molecule) in the dense gas or the liquid has been neglected.

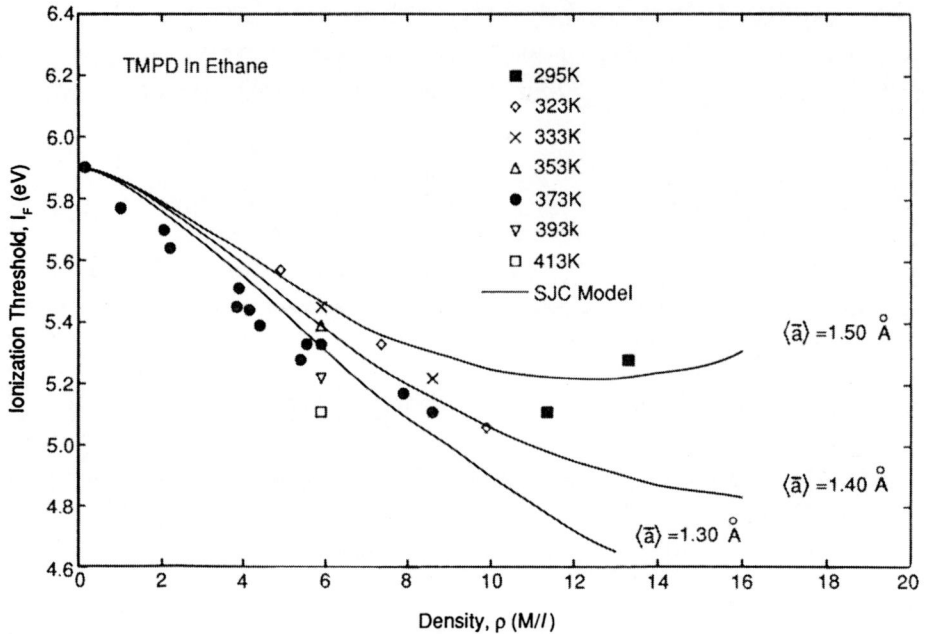

Figure 3. Measured ionization threshold energy I_F of the TMPD molecule in ethane as a function of ethane density ρ at various temperatures T. The solid lines represent the predicted values of I_F based on the SJC model[46] for hard core radii $<\bar{a}>$ = 1.30, 1.40, and 1.50 Å; the best fit to the experimental data is for $<\bar{a}>$ values between 1.45 and 1.50 Å. Densities $\rho > M\ell^{-1}$ are for liquid ethane.[34]

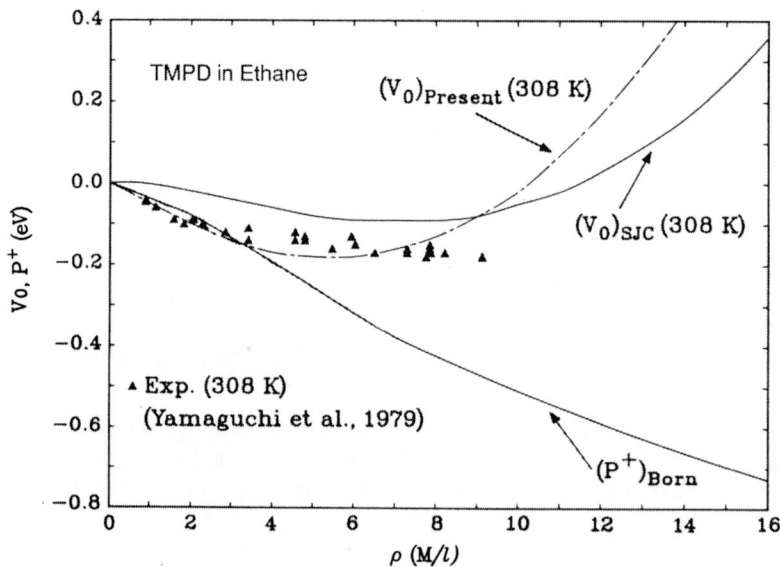

Figure 4. V_o of C_2H_6 and P^+ of TMPD$^+$ in C_2H_6 as a function of the ethane density ρ. The P^+ was determined using Eq. (4) and the V_o using either the SJC model (———) or as modified in Ref. 34 (— · —). The experimental points (▲) are from Ref. 48.

Christophorou

sum of the Hartree-Fock atomic potentials $U_{HF}(r) = \infty$ for $r < \bar{a}$ and $U_{HF}(r) = 0$ for $r > \bar{a}$.[34,46,47] The calculated values of $V_o(N)$ (Fig. 4) are rather strong functions of these quantities. For TMPD in C_2H_6 both theory and experiment[34,48] suggest a value of 1.45 to 1.50 Å for $\bar{a}$. The data in Fig. 4 show that $V_o(N)$ goes through a minimum; this is more dramatically exemplified in the case of the rare gases Ar, Kr, and Xe whose scattering length is negative[a] as can be seen for the data in Fig. 5. This is understood by noting that V_o is a sum of two terms: the polarization energy U_p and the kinetic energy K. That is,

$$V_o = U_p + K \qquad (6)$$

where

$$U_p = U_p(r) + \left\langle \Sigma U_p(r-r_i) \right\rangle$$

$$= -\left(\frac{3}{2} \frac{\alpha e^2}{r_{WS}^4}\right) x \left[\frac{8}{7} + \left(1 + \frac{8}{3}\pi a N\right)^{-1}\right] \qquad (7)$$

and

$$K = \frac{\hbar^2 k_o^2}{2m} \qquad (8)$$

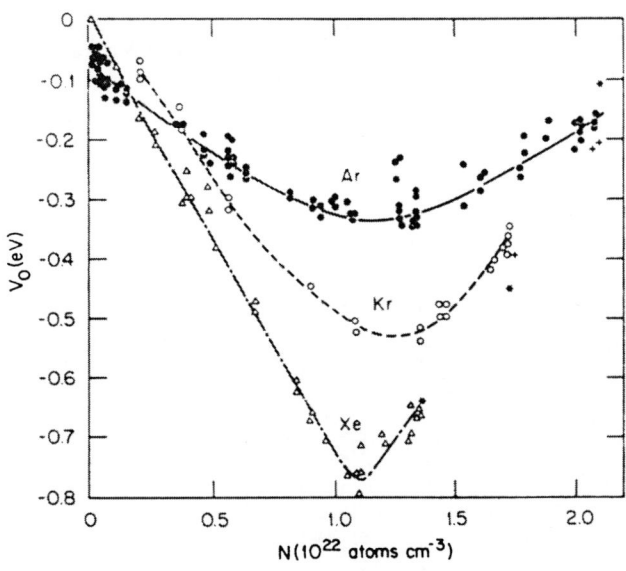

Figure 5. V_o versus N for Ar, Kr, and Xe.[12,49]

(a) For He, whose scattering length is positive, V_o increases with increasing N.[49]

In the above equations $U_p(r)$ is the polarization energy due to the single molecule whose Wigner-Seitz sphere the electron occupies, $<\Sigma U_p(r\text{-}r_i)>$ is the structure-average polarization energy due to the surrounding molecules, α is the molecular polarizability, a $(\simeq \bar{a})$ is the scattering length, and k_o is the electron wave number determined by

$$\tan k_o(r_{ws} - \bar{a}) = k_o r_{ws} \tag{9}$$

Since U_p is negative and decreases with ρ, and K is positive and increases with ρ, the V_o has a minimum whose value and position depends mainly on the hard core radius $\bar{a}$. The functions $P^+(N)$ and $V_o(N)$ in Fig. 4 were used by Faidas et al.[34] along with the value, 5.9 eV, they measured for I_G(TMPD) to estimate the $I_F(N)$ of TMPD in C_2H_6 using Eq. (5). Their results for three values of $\bar{a}$ are represented by the solid lines in Fig. 3. They obtained the best fit to their measurements at 373 K with $\bar{a} = 1.4$ Å. This understanding, gratifying as is, has not as yet been extended to polar media. Knowledge on the photoionization and photodetachment mechanisms and energetics is essential for the understanding of electron transfer reactions in biological matter.

As $I_F(N)$ is lowered from its low-pressure value I_G, excited states lying in the range $I_G - I_F(N)$ become autoionizing. The effect of the medium density on such high-lying excited electronic states has been investigated, especially for high-n Rydberg states. For example, high n-states of CH_3I and C_6H_6 in H_2 or Ar were found[50] to shift linearly with the number density of atomic Ar (0.6 to 7.5 x 10^{20} atoms cm^{-3}) and molecular H_2 (0.9 to 10.5 x 10^{20} molecules cm^{-3}). The density-induced energy shift (ΔN) of atomic or molecular high-n Rydberg states was reported[50] to decrease linearly with the density of the medium to $N \simeq 1$ x 10^{21} molecules cm^{-3} and to agree with the modified Fermi model.[51,52] This model expresses $\Delta(N)$ as

$$\Delta(N) = \pm (2\pi\hbar^2/m)\, aN - 9.87(\alpha e^2/2)^{2/3}(h\nu)^{1/3}N \tag{10}$$

where a is the scattering length and ν is the relative thermal velocity of the colliding partners. In Eq. (10) the first term is due to scattering and the second due to polarization.

Recombination

Geminate recombination. While in low-pressure gases $G_{fe} \simeq G_{te}$, in liquids $G_{fe} < G_{te}$. The two quantities are related to each other and to the applied electric field E by

$$G_{fe}^o = p_{esc}^o\, G_{te} \tag{11}$$

$$G_{fe}^E = p_{esc}^E\, G_{te} \tag{12}$$

In Eqs. (11) and (12), p_{esc}^o and p_{esc}^E are, respectively, the escape probabilities (i.e., the probabilities that an electron at an initial separation distance r would escape recombination with its sibling cation) in the absence ($E = 0$) and in the presence ($E > 0$) of an applied electric field. For isolated ionizations (low-LET particles) p_{esc}^o and p_{esc}^E are normally expressed as[3,53]

$$p_{esc}^{o} = \exp\left(-r_c/r\right) \tag{13}$$

$$p_{esc}^{E} = \exp\left(-r_c/r\right)\left[1 + \frac{er_c}{2kT}E\right] \tag{14}$$

The quantity r_c—the "Onsager length"—is the distance at which the Coulomb energy of the electron-cation pair equals kT; vis.,

$$r_c = \frac{e^2}{\epsilon kT} \tag{15}$$

where ϵ is the dielectric constant of the medium. Actually, since there is a distribution of electron-cation thermalization distances r the fraction of the electrons that escape (for $E = 0$) is

$$\int d^3r\, g(r) \exp\left(-r_c/r\right) \tag{16}$$

where $g(r)$ is the probability density of the electron thermalization distances.[a]

In low-pressure gases, r is very large and $p_{esc} \rightarrow 1$. In liquids, however, r is short [100 to 200 Å for dielectric liquids[3] and ~11 Å for water[54]] and thus $p_{esc} \ll 1$. Of the parameters which govern the process of free electron production in liquids ϵ appears to be of primary significance; G_{fe} increases with increasing ϵ.[3,55]

In Eq. (12), G_{fe}^{E} is the free electron yield when an electric field E is applied across the volume in which the electrons are generated; G_{fe}^{E} exceeds G_{fe}^{O} by an amount which depends on E and the liquid.[3,53] This is especially the case for densely ionizing particles (e.g., α-particles, Table I) due to the low value of p_{esc} in such instances. The p_{esc} increases with increasing drift velocity w; the w—as well as the electron thermalization length—increases with decreasing electron scattering cross section of the liquid. For pure liquids ($V_o < 0$ eV), then, a low I_L and a large w and E are desirable for a large G_{fe}^{E}.

In Fig. 6 the G_{fe}^{O} and the density-normalized thermalization distance $b\rho$ are plotted as a function of the medium density ρ (from 0.01 to 0.6 g cm^{-3} along the vapor/liquid coexistence curve) for n-pentane (n-Pt) and neopentane (neo-Pt).[16] The G_{fe}^{O} decreases continuously with increasing ρ in Pt, but it goes through a maximum in neo-Pt (and other "spherical"-molecule liquids) with conduction bands ($V_o < 0$ eV). The $b\rho$ also goes through a maximum in neo-Pt (and other "spherical"- molecule liquids) but remains relatively constant for n-Pt. The maximum in G_{fe}^{O} and $b\rho$ for "spherical"-molecule liquids correlates with that of the density-normalized thermal electron mobility μN.

Finally, attention is drawn to recent femtosecond studies of the kinetics and dynamics of geminate recombination of electron-cation pairs formed in liquids (water, alkanes)[54] by photoionization. In the case of water it was reported[54] that within ~60 ps of electron solvation, ~50% to 60% of the solvated electrons undergo geminate recombination.

(a) Geminate recombination can, of course, involve the whole spectrum of incompletely relaxed to completely relaxed states.

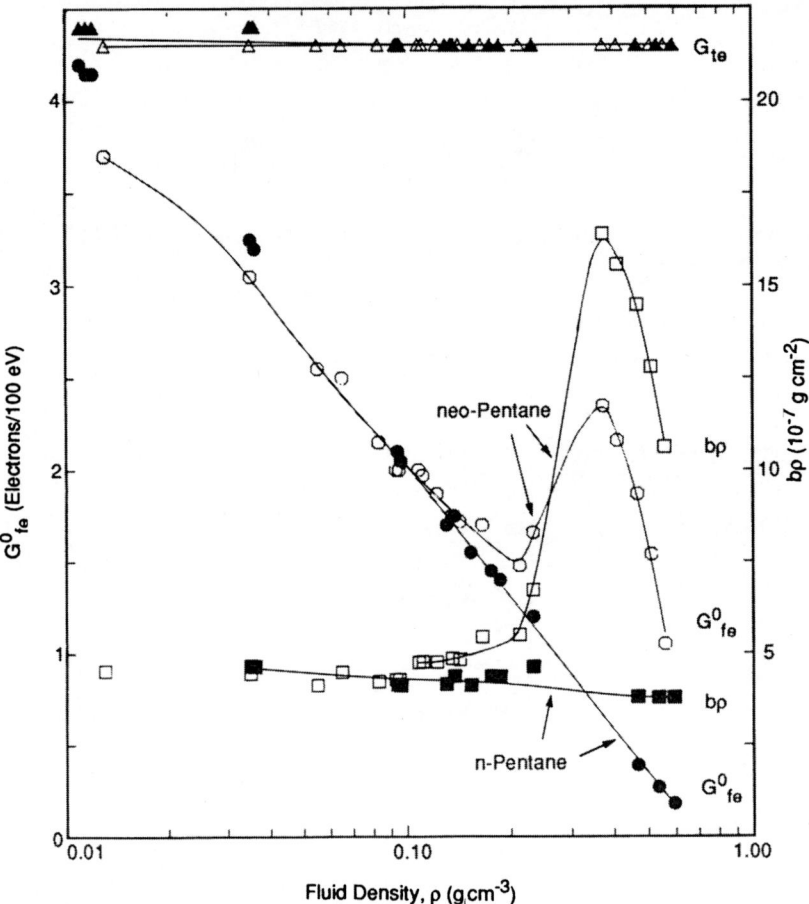

Figure 6. G_{fe}^0 and $b\rho$ versus fluid density ρ for n-pentane and neo-pentane. Also shown are the $G_{te}(\rho)$ for n-pentane (▲) and neopentane (△). The critical densities and temperatures are, respectively, 0.232 g cm^{-3} and 434 K for neo-pentane and 0.237 g cm^{-3} and 470 K for n-pentane[16] (see the text).

Volume or Bulk. In low-pressure gases, volume electron-cation recombination is normally an overall three-body process with a recombination rate constant k_r that can be expressed as[57]

$$k_r = k_2 + k_3 N \tag{17}$$

where k_2 and k_3 are the two- and three-body coefficients and N is the medium number density. At high gas pressures and in nonpolar liquids (at least in those with electron mobilities μ in the range 0.09 to 300 cm^2 V^{-1} s^{-1}),[58] recombination occurs on a time scale many orders of magnitude longer than geminate recombination, and k_r assumes diffusion-controlled values

$$k_D = 4\pi \, e\mu/\epsilon \tag{18}$$

that is, k_r is limited by the rate at which the electrons diffuse toward the ions.

In Fig. 7 is shown[58] the variation of the k_r for CH_4 in the density range 2×10^{20} to 1.9×10^{22} molecules cm^{-3} including the critical region and the liquid-solid phase change. The N-dependence of k_r follows closely that of $\mu(N)$ (Fig. 7b). The abrupt increase in k_r and μ on the phase transition from liquid to solid is common to other liquids with $V_o < 0$ eV and may be due to the decrease of the isothermal compressibility on phase change.[58]

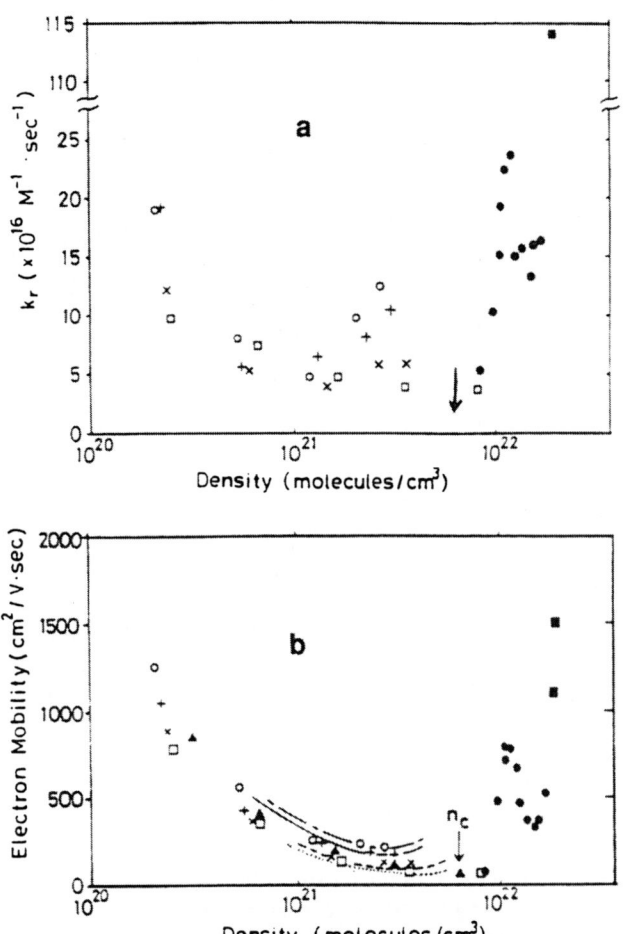

Figure 7. Density dependence of k_r (Fig. 7a) and μ (Fig. 7b) in methane.[58] Solid, ■; liquid, ●; gas, o (295 K), + (254 K), x (222 K), □ (193 K). n_c ≡ critical density (6.11 x 10^{21} molecules cm^{-3}). ▲ (194 K);[59] —·— (295 K), ——— (273 K), - - - (206 K), ... (196 K).[60]

The Electron State and Energies (Under Steady-State Conditions) in Low-Pressure Gases, Dense Gases, and Dielectric Liquids

In low-pressure gases the electron mean free path ℓ ($\approx[N\sigma_{sc}]^{-1}$; σ_{sc} = total electron scattering cross section) is much longer than the electron de Broglie wavelength $\lambda(= 2\pi\lambda)$, and the electrons in such media are free, interacting with single atoms and single molecules. Their electron transport properties have been, in many instances, successfully treated using the Boltzmann transport equation. The electron kinetic energies have been found to be well in excess of thermal by an amount that depends on the gas and the value of E/N.[1,2] In Fig. 8 the mean electron energy $<\epsilon>$ is plotted as a function of E/N for a number of low-pressure gases to illustrate the gas

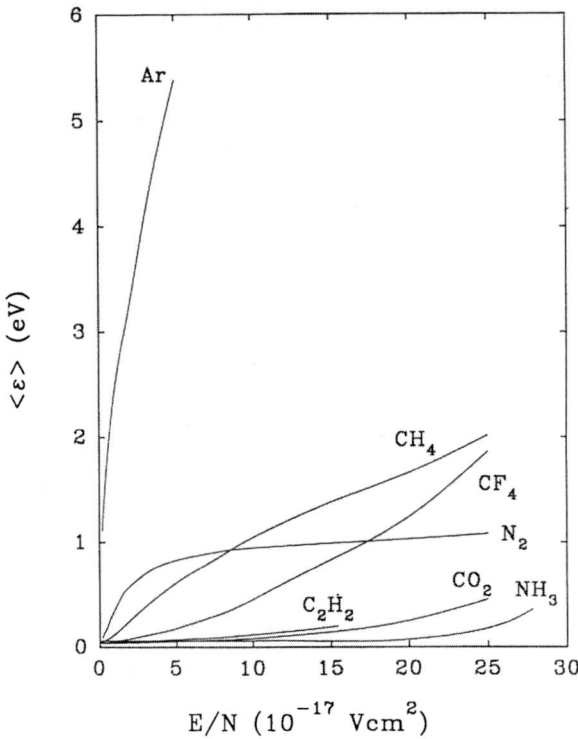

Figure 8. $<\epsilon>$ versus E/N for Ar, CH_4, CF_4, N_2, C_2H_2, CO_2, and NH_3. The data for Ar and N_2 are from Ref. (61); those for CH_4, CF_4, and CO_2 are computed values[62] using published cross section data for elastic and inelastic scattering and ionization;[2,63] the data for C_2H_2 and NH_3 are characteristic energies.[1,2]

properties which determine their ability to slow-down subexcitation electrons: negative ion states for N_2 and CO_2 (the low-lying[1,2] negative ion states of CO_2 are responsible for its exceptional slowing-down properties, while the 2.3 eV negative ion state of N_2 explains the flat portion of the $<\epsilon>$ versus E/N function for this gas), dipole scattering for NH_3, multiple bonds for C_2H_2. The lower-lying vibrational thresholds of CF_4 compared to CH_4 explain the superior thermalizing ability of CF_4 at low energies. The steady-state electron energy distribution functions $f(\epsilon, E/N)$ are non-Maxwellian (except at very low E/N) and—depending on the gas, E/N and T—they peak at energies ranging from 1.5 kT to $\simeq$ 10 eV ([1,2]; Fig. 9).

In dense gases and liquids, $\ell < \lambda$, the electrons interact with more than one species simultaneously, and they are not free but quasi-free (e_{qf}) and/or localized (e_ℓ) depending on the medium, N and T. Excess electrons are generally localized—upon thermalization—in dense media whose $V_o > 0$ eV and they are quasi-free in those with $V_o < 0$ eV. The V_o itself can be a function of N (Fig. 4 and 5). Quasi-free electrons have much higher mobilities than localized electrons.[3,11,12,64] The energies of the excess electrons are generally thermal in dense media. However, in dense media with $V_o < 0$ eV, the excess electrons can—especially at high E/N—attain energies well above thermal. This can be seen from the data shown in Figs. 10 to 12 for liquid Ar and liquid Xe. Evidence for steady-state excess-electron energy distributions energetically lying above thermal has been obtained also for "spherical"-molecule liquids with $V_o < 0$ eV.[72,73] While the accuracy and the quality of such knowledge is still very

                                           Christophorou

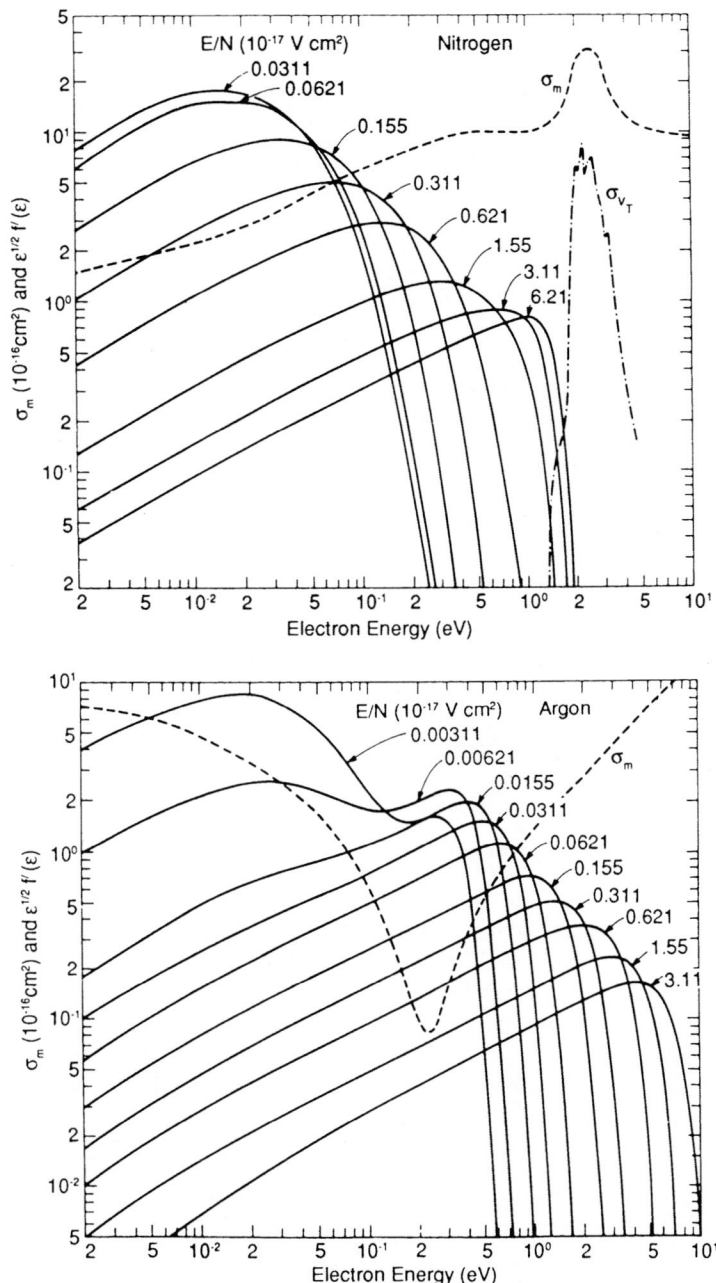

Figure 9. Normalized electron energy distribution functions $f(\epsilon,E/N) \equiv f'(\epsilon)\epsilon^{1/2}$ for several E/N values in N_2 and Ar obtained using a two-term Boltzmann solution and the cross sections shown in the figure.[61]

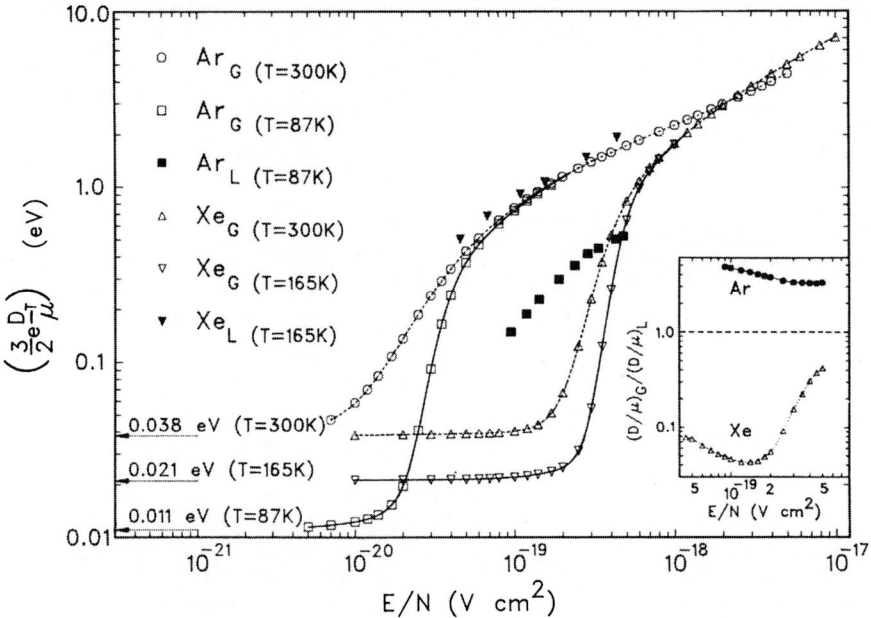

Figure 10. Calculated[65] $(3eD_T/2\mu)_G$ versus E/N for gaseous Ar at $T = 300$ K (o) and $T = 87$ K (□) and gaseous Xe at $T = 300$ K (Δ) and 165 K (∇). The experimental $(3eD_T/2\mu)_L$ versus E/N for liquid Ar(□:)[66] and for liquid Xe (▼:)[67] respectively at 87 and 165 K. Inset: Ratio $(D_T/\mu)_G/(D_T/\mu)_L$ versus E/N for Ar (●) and Xe (Δ) [from Ref. 65].

limited,[a] it clearly shows the existence of energetic excess electrons in dense fluids with conduction bands ($V_o < 0$ eV). The state of the electron and its energy crucially depend on the medium and profoundly affect the magnitude of the interaction cross sections of excess electrons in dense media.

Electron Drift and Scattering in Low-Pressure Gases, Dense Gases, and Dielectric Liquids

The electron transport coefficients—electron drift velocity w and transverse (D_T) and longitudinal (D_L) diffusion coefficient—are functions of the gaseous medium, E/N, and T.[1,2,74,75] For sufficiently low N where $\ell \gg \lambda$, w and D_T (D_L) are independent of N. Under these conditions well-developed expressions relate w and D_T (D_L) to f(ϵ,E/N) and the cross section $\sigma_{sc}(\epsilon)$ of the various collision processes.[1,2,74] Thus, the low-density low-field value, $(\mu N)_o$, of the density-normalized electron mobility μN is expressed as

$$(\mu N)_o = \frac{4}{3} \frac{e}{(2\pi m k T)^{1/2}} \frac{1}{<\sigma_{sc}>} \tag{19}$$

where $<\sigma_{sc}>$ is an average of the scattering cross section at thermal energies. Accurate measurements of $w(E/N)$ and $D_T/\mu(E/N)$,[1,2,74,75] along with the development of the Boltzmann transport and Monte Carlo computational code analyses in the past two

---

(a) The E/N dependence of the D_T/μ measurements for liquid Ar and liquid Xe in Fig. 10 lends support to the $<\epsilon>$ versus E/N data of Ref. 65 (Figs. 11 and 12).

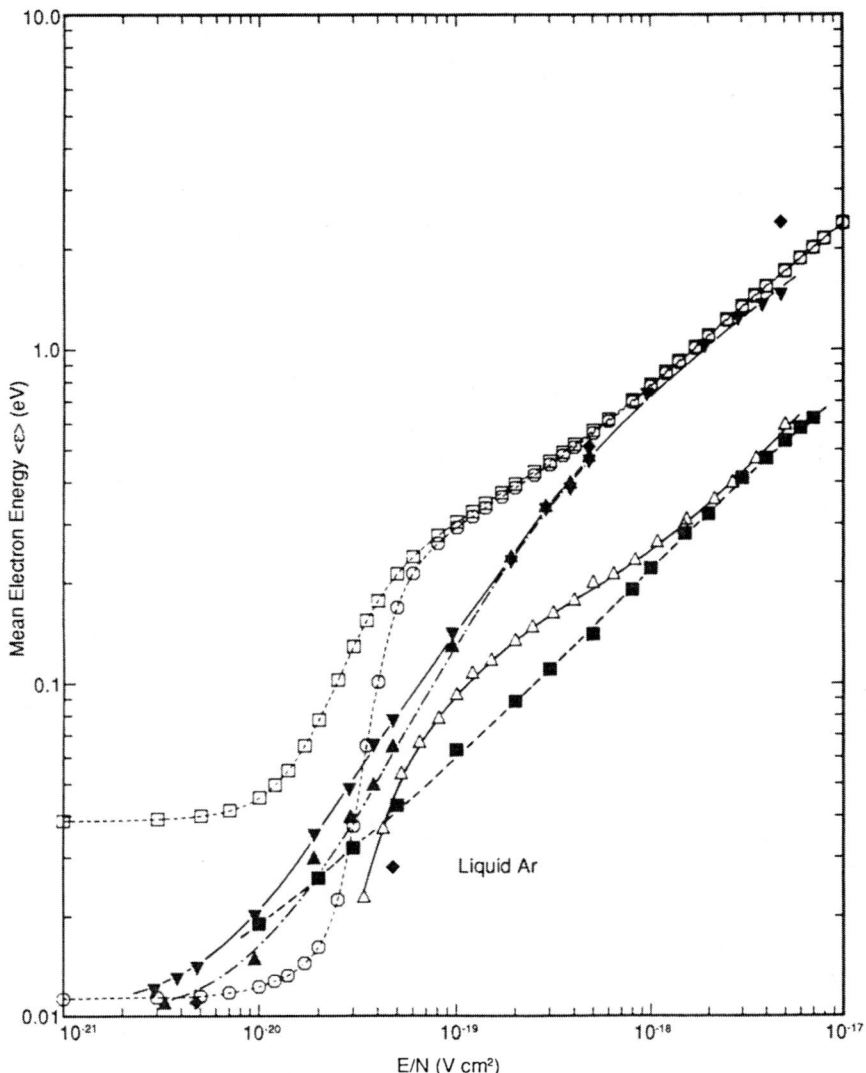

Figure 11. Calculated $<\epsilon>_L$ versus E/N for liquid Ar: ($\blacktriangledown$);[68] $\square$;[69] ($\blacktriangle$);[70] $\blacklozenge$,[71] in comparison with the calculated values of (65) for liquid Ar at $T = 87$ K ($\triangle$) and gaseous Ar at $T = 87$ K (o) and 300 K ($\square$) [from Ref. 65].

decades, have provided a substantial body of information on—and an improved understanding of—electron motion in gases and the cross sections for the various energy-loss processes.[2,63,74-79] In Fig. 13 an example of the collision cross sections obtained from such analyses[83] is shown. As the density of the medium is increased, however, w decreases or increases depending on the medium and the type and cross sections of the electron-interaction processes involved.[a] Thus, in the density range over which $N\sigma_{sc}\lambda \lesssim 0.5$, w (or μ or μN) decreases slowly with increasing N for nonpolar gases with small σ_{sc} (e.g., H_2, N_2, C_2H_6). An understanding of these dependences of w on N was

---

(a) The author is not aware of any experimental observations on the dependence of D_T on N, although such a dependence has been hypothesized.[84]

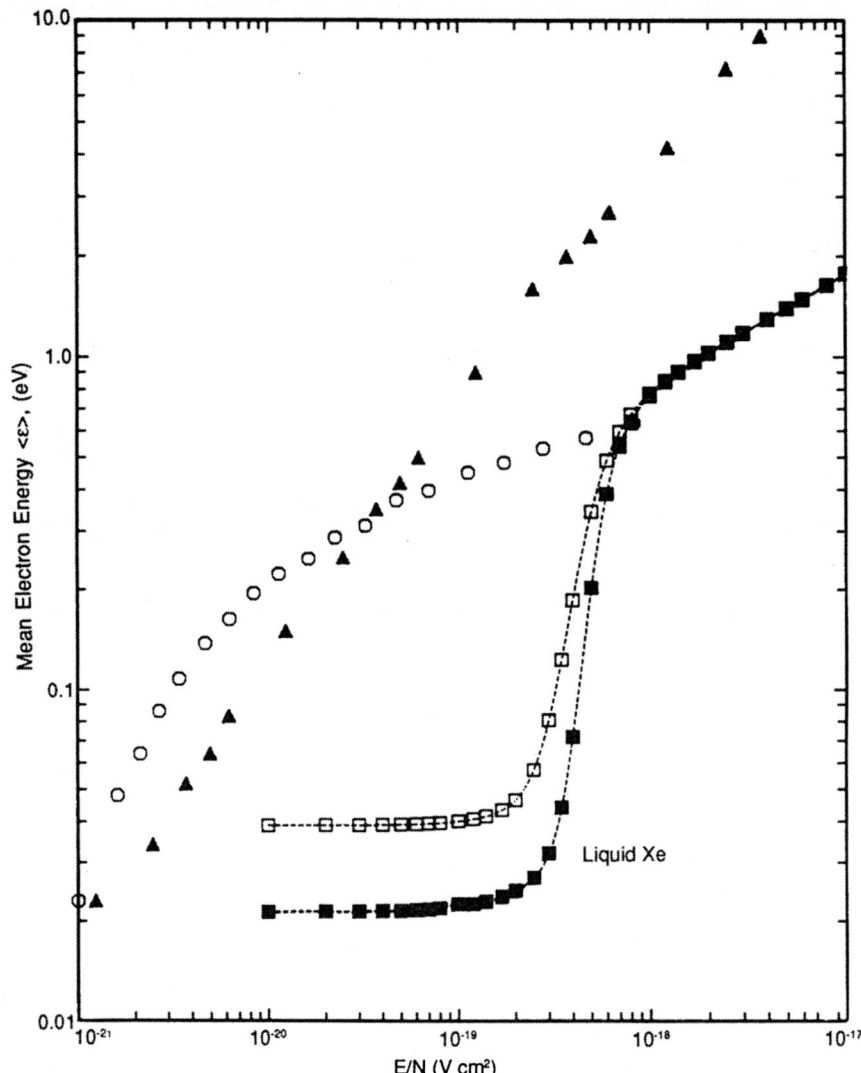

Figure 12. Calculated $<\epsilon>_L$ versus E/N for liquid Xe($T \simeq 165$ K): ▲;[70] o.[65] For comparison $<\epsilon>_G$ versus E/N is shown for gaseous Xe: ■ (165 K); □ (300 K) [from Ref. 65].

attempted[2,71,85-88] by considering the effects of multiple scattering on w and by introducing phenomenological density corrections to the low-N expressions for drift velocity w [or mobility μ, Eq. (19)]. The dependence of w on N seems to be energy dependent. For example, it was found[89] that the maximum change in w with pressure occurs at thermal energies (~0.038 eV) for H_2 and C_3H_8 but at ~0.06 eV for CH_4 and at ~0.07 eV for C_2H_6.

The electron drift velocity (or the electron mobility) has been shown[2,12] to decrease substantially with increasing medium density in systems (e.g., He, CO_2, 1-C_3F_6) where transient or permanent anions are increasingly formed as the medium density is increased, and for polar media where, in addition, the electron scattering cross section is large (due to the long-range electron-electric dipole interaction). Thus

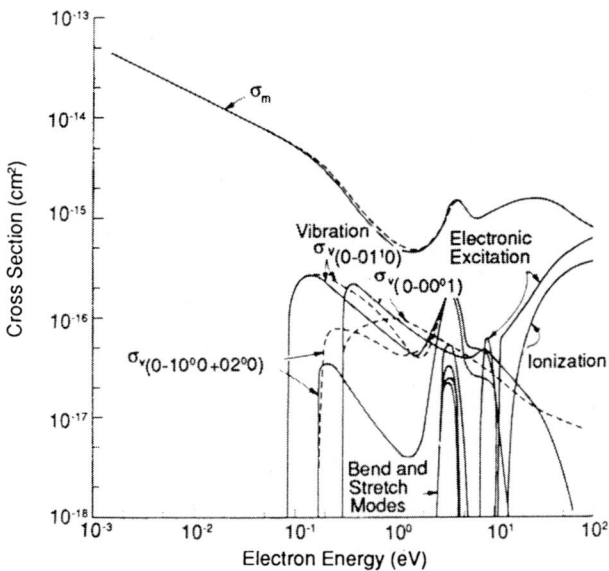

Figure 13. Cross sections for momentum transfer and inelastic electron scattering in CO_2 calculated[80,81] from an analysis of electron transport data. The ionization cross sections are from Ref. 82 [from Ref. 83].

the w in CO_2 decreases by ~3 orders of magnitude when N is increased to ~10^{21} molecules cm^{-3};[90] large decreases in w at even lower number densities ($\lesssim 5$ x 10^{18} molecules cm^{-3}) were reported for 1-C_3F_6.[91] In polar media $N \sigma_{sc} \lambda > 1$ at relatively low N, and the delay in electron drift begins at relatively low values of N (~2 x 10^{19} molecules cm^{-3}). This can be seen from Figs. 14a and 14b where the dependence of $w(E/N)$ on $N(T = 300$ K$)$ and $(\mu N)/(\mu/N)_o$ on N (at various T) are given for the polar molecule NH_3. For NH_3 the $\langle\sigma_{sc}\rangle$ at thermal (T=300 K) energies is ~1.2 x 10^{-13} $cm^{2(92)}$ and $\ell \simeq \lambda$ at ~5.5 x 10^{19} molecules cm^{-3}. As N increases beyond the range of values in Fig. 14, permanent or transient electron trapping occurs and causes a rather sharp decrease in w or μ as can be seen from the data in Fig. 15. These latter processes are, obviously, a function of T; their effect decreases with increasing T.

On the other hand, $w(\mu$ or $\mu N)$ was found to increase with increasing N for the heavier rare gases and the "spherical"-molecule hydrocarbon dielectric fluids for which $V_o < 0$ eV. An example of this type of behavior is shown in Fig. 16 for Xe. The increase in w up to ~3 x 10^{21} atoms cm^{-3} is associated with a reduced contribution to electron scattering from the polarization component of the interaction potential by overlapping of the fields of adjacent atoms. The evolution of the N-dependence of the low-field (thermal) electron mobility for Xe in the density range from the low-pressure regime to the liquid is shown in Fig. 17. Below ~3 x 10^{20} atoms cm^{-3} (Fig. 17) μN is independent of N since in this range $N\sigma_{sc}$ is low enough for single-atom scattering to prevail. Beyond this value, μN decreases, passes through a minimum, increases to a maximum and falls again. The decrease in the density range from ~3 to ~40 x 10^{20} atoms cm^{-3} was attributed[11] to enhanced electron scattering due to multibody interactions and the large subsequent increase due to interference effects. Certainly the structure of the medium and its effect on electron transport needs to be considered at these high densities. Multiple scattering theories[85-88] attempted to account for these changes for gases with negative scattering lengths by considering the attenuation of the electron scattering by density effects due to the screening of the long-range polarization interaction potential. In general, however, the theoretical treatments of

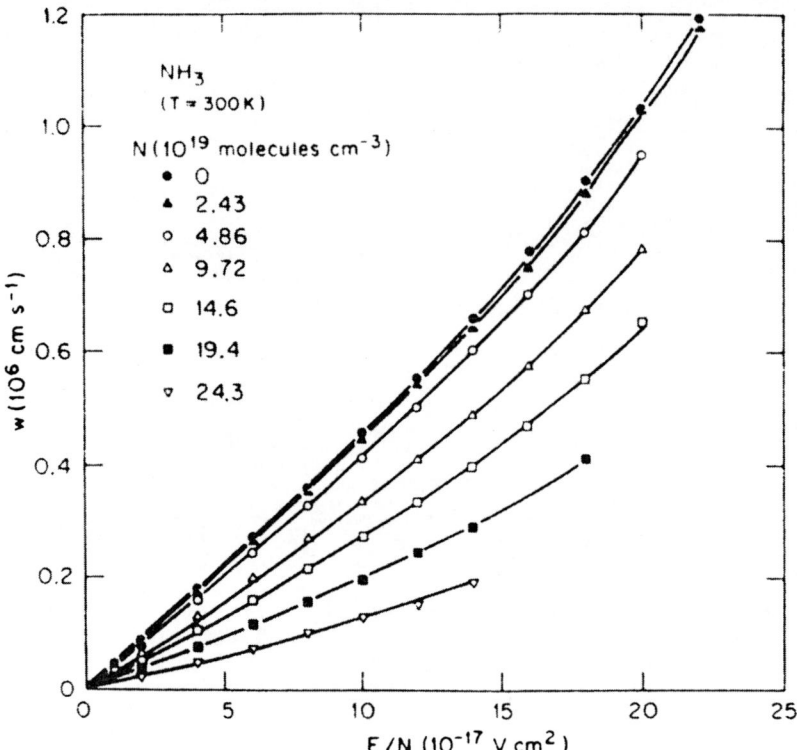

Figure 14a. w *versus* E/N *for* NH$_3$ at various values of N at $T = 300$ K.

$w(N)$ fail at high N, *especially* in the transition region. Part of the difficulty lies in the fact that the theory often retains the atomic scattering picture and introduces corrections to the low-density scattering as the N is increased. Another serious problem is the inability to properly describe the electron scattering potential and its screening by the medium, and to account for the effects on w of temporary electron trapping.

In the liquid, $N\sigma_{sc}\lambda >> 1$ and the electron is strongly influenced by the structure of the medium. The thermal electron mobility $(\mu_L)_{th}$ in the liquid has been expressed by

$$(\mu_L)_{th} = \frac{2}{3}\frac{1}{N}\left(\frac{2}{\pi mkT}\right)^{1/2}\frac{e}{4\pi a_{ef}^2 S(0)} \tag{20}$$

where $a_{ef}(N)$ is the effective scattering length at a density N, $S(0) = NkT\chi$ is the structure factor at vanishing momentum transfer $(K \to 0)$, and χ is the isothermal compressibility. [$S(0) \simeq 0.03 - 0.05$ near the triple point of simple fluids; $S(K) \to 1$ for $\epsilon \geq 4$ eV.] It is emphasized that the success of any theoretical treatment of electron motion in liquids depends on the assumed form of the scattering potential and the proper correction for its screening by the medium.

That the electron scattering processes at low energies are strongly affected by the liquid structure can be seen from Figs. 18 and 19 where $w(E/N)$ for gaseous (low-pressure) and liquid Ar and Xe (Fig. 18) and gaseous (low-pressure) and liquid

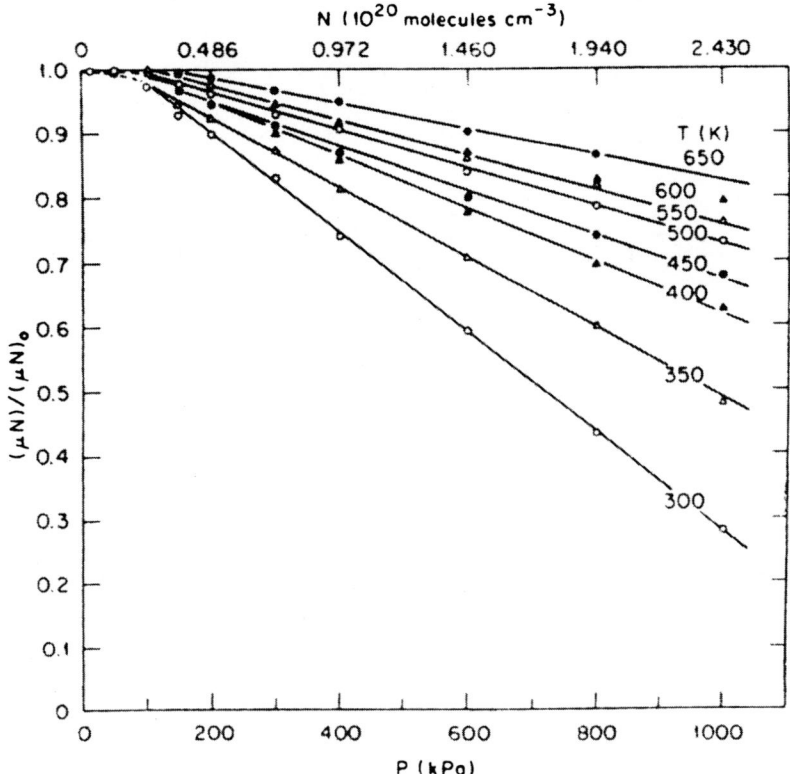

Figure 14b. $\mu N/(\mu N)_0$ versus P (or N) for NH_3 at a number of T. For a given value of E/N, the w and μN data were plotted as a function of N and extrapolated to $N \to 0$. These values are designated in (a) by the solid circles ($N = 0$) and in (b) by $(\mu N)_0$ [from Ref. 92].

tetramethylsilane (TMS), neo-pentane (TMC) and 2,2,4,4,-tetramethylpentane (TMP) (Fig. 19) are shown. The polyatomic molecules of TMS, TMC, and TMP are "spherical," and their liquids—as those of Ar and Xe—have excess electron conduction bands (Table II). It is clearly seen (Figs. 18 and 19) that at a given value of E/N, the w is very much larger in the liquid than in the corresponding gas, indicating that $\sigma_{sc}(\epsilon)$ is much smaller at low energies ϵ in the liquid. Indeed, the $w(E/N)$ measurements for liquid Ar and Xe in Fig. 18 were used[99]—in a manner analogous to that for gases[83]—in a Boltzmann transport equation analysis to determine a set of scattering cross sections and an electron energy distribution function $f(\epsilon, E/N)$ consistent with the $w(E/N)$ measurements. The cross sections for elastic energy loss $\sigma_0(\epsilon)$, elastic momentum transfer $\sigma_1(\epsilon)$, and inelastic electron scattering for liquid Ar and Xe obtained this way[99] are shown in Fig. 20. Clearly $\sigma_0(\epsilon)$ and $\sigma_1(\epsilon)$ are lower than $\sigma_m(\epsilon)$ below ~0.2 eV, and they exhibit a shallower Ramsauer-Townsend minimum which is shifted to lower ϵ compared to that of $\sigma_m(\epsilon)$.[89,100] At $\epsilon \gtrsim 4$ eV, σ_0 and $\sigma_1 \to \sigma_m$. A quantitative calculation of σ_{sc} in any liquid is still lacking.

Finally, it is interesting to observe that for the molecular liquids in Fig. 19, w increases linearly with E up to a "critical" electric field E_c beyond which the E/N dependence of w becomes sublinear, indicating that at $E > E_c$ the excess electrons have mean kinetic energies $<\epsilon>_L > 1.5$ kT. Such media have potential applications in radiation detectors,[17,101-104] pulsed power switches,[105] and other technologies.[4]

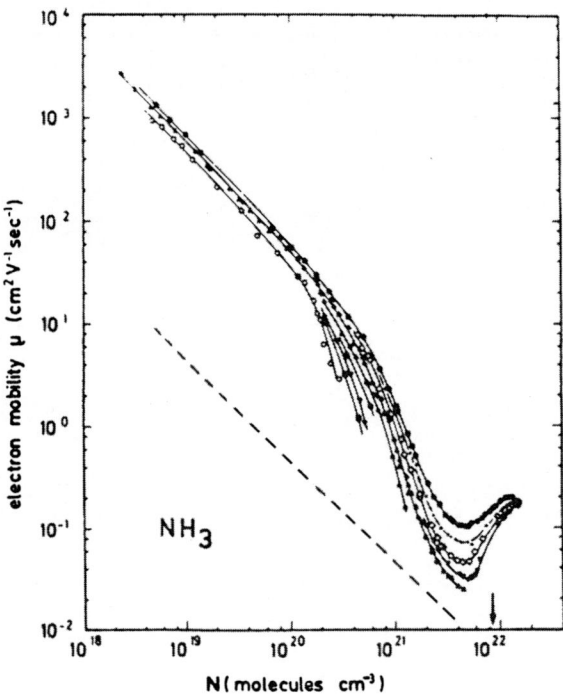

Figure 15. μ versus N in subcritical and supercritical NH_3 vapor at various T:300 (o), 320 (■), 340 (▼), 360 (◉), 380 (▲), 400 (Δ), 410 (∇), 420 (◇), 440 (+), and 460 (●). The arrow indicates the critical density of NH_3, and the dashed line represents the averaged mobility of unidentified impurity ions ($T < 400$ K).[93]

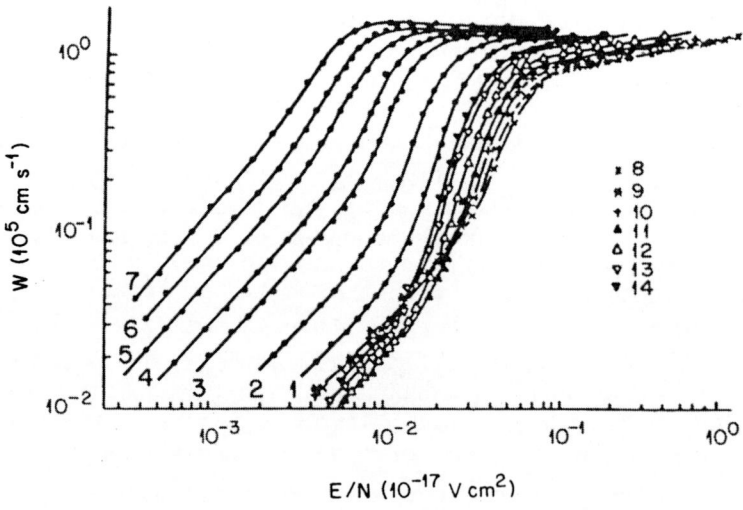

Figure 16. w versus E/N for Xe ($T = 298$ K) at various values of N. Curves 1 through 14 correspond to N (in units of 10^{21} atoms cm^{-3}) of the following: 4.24, 4.97, 5.38, 6.3, 6.97, 7.34, 7.75, 0.1, 0.438, 0.91, 1.92, 2.74, 3.54, and 3.92, respectively.[94]

Christophorou

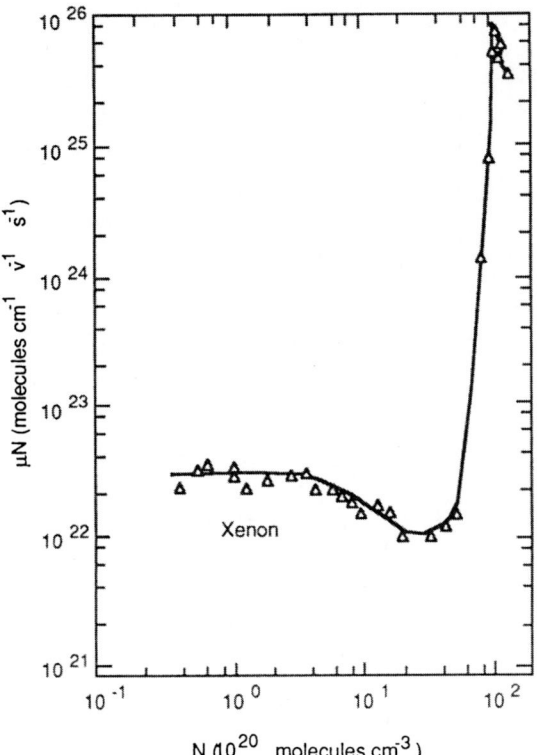

Figure 17. μN versus N for Xe.[95]

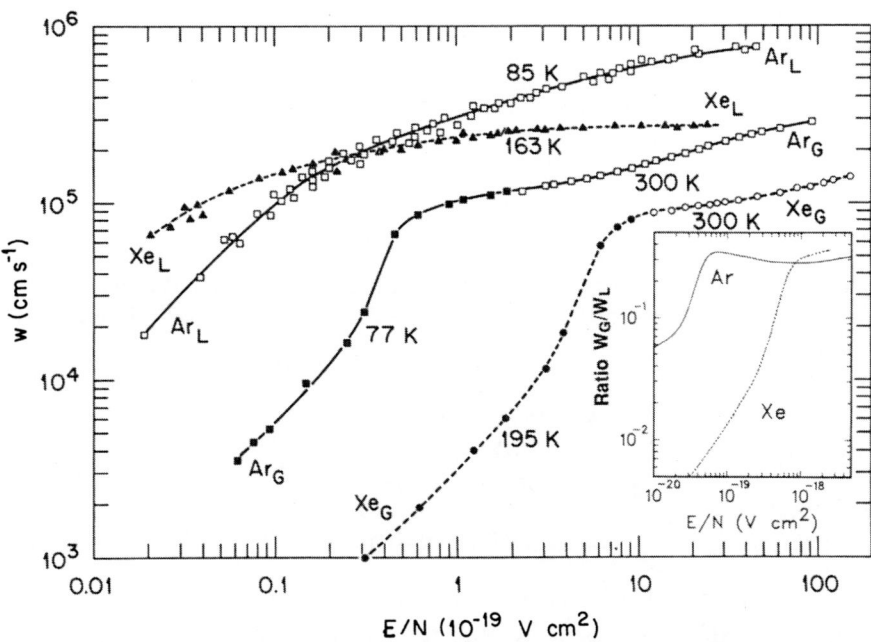

Figure 18. w versus E/N for gaseous Ar (■, □),[1,96] gaseous Xe (●, o),[1,97] liquid Ar (□),[98] and liquid Xe (▲).[98] Inset: Ratio w_G/w_L versus E/N for Ar and Xe.[65]

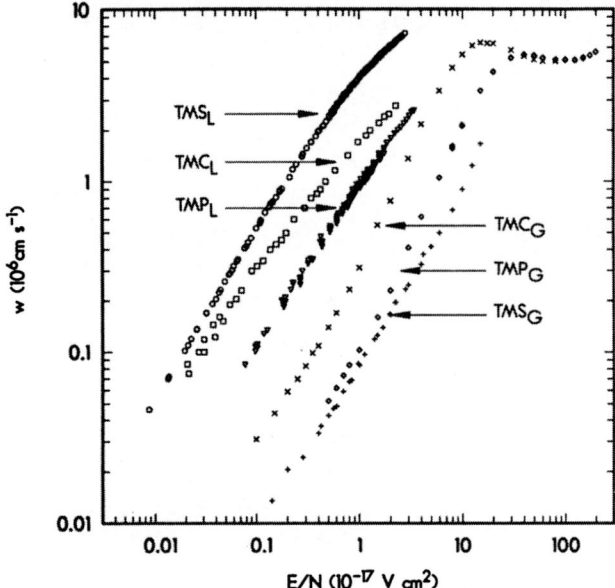

Figure 19. w versus E/N ($T \simeq 295$ K) for gaseous and liquid TMS, TMC, and TMP.[73]

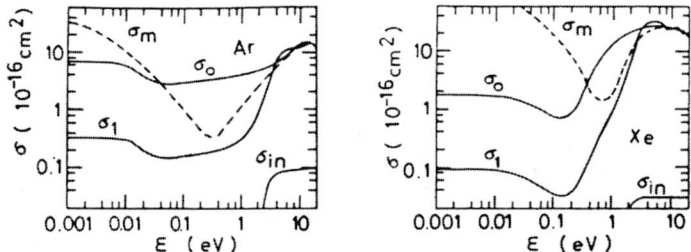

Figure 20. Cross sections $\sigma_0(\epsilon)$, $\sigma_1(\epsilon)$. and $\sigma_{in}(\epsilon)$ for liquefied Ar and Xe (see the text); σ_m is the low-density gaseous momentum transfer cross section.[99]

In Table III are listed values of E_c, μ_{th}, maximum, w_{max}, electron drift velocity measured and the corresponding, E_{max}, applied, for dielectric liquids with excess electron conduction bands.

Effect of the Medium on Negative Ion States (Transient Anions)

Studies of negative ion states (NISs) of isolated molecules are abundant.[1,2] None exists on liquids, to the author's knowledge. Notable changes are expected, however, in the resonance energy, cross section, and lifetime of the NISs of atoms and molecules embedded in dense fluids as the fluid density is increased.[2,11,12,65,69] Figure 21 shows schematically the increase in the vertical attachment energy (VAE) and the electron affinity (EA) in going from a low-pressure gas to the liquid. Figure 22 shows how the position of the NO_2^{-*} resonance responsible for the reaction

$$e + N_2O \rightarrow N_2O^{-*} \rightarrow O^- + N_2 \qquad (21)$$

Table III. E_c, μ_{th}, w_{max}, and E_{max} for a Number of Dielectric Liquids with $V_o < 0$ eV

Liquid[a]	E_c (10^3V cm^{-1})	μ_{th} (cm^{2}s^{-1}V^{-1})	w_{max} (10^6cms^{-1})	E_{max} (10^3V cm^{-1})
Ar(87K)	0.3[b]	400[b]	0.65[b,c]	--
Xe(165K)	0.05[b]	2000[b]	0.26[b,c]	--
$C(CH_3)_4$[d]	3.5[e]	71.5[e]	3.3[e,f]	116[e]
$Si(CH_3)_4$[d]	7[e]	119.3[e]	7.2[e,f]	125[e]
$Ge(CH_3)_4$[d]	15[e]	114.7[e]	7.4[e]	109[e]
$Sn(CH_3)_4$[d]	30[e]	85.7[e]	6.0[e]	75[e]
$(CH_3)_3CCH_2C(CH_3)_3$[d]	15[e]	31.8[e]	2.6[e,f]	115[e]

(a) See Table II for values of I_L, I_G, and V_o for these liquids.
(b) Ref. [101].
(c) See Fig. 18.
(d) ~296K.
(e) Refs. [102, 73].
(f) See Fig. 19

shifts from ~2.3 eV when the reaction occurs in a low-pressure gas of Ar to ~0.3 eV when the reaction occurs in liquid Ar.[69] Similar downward shifts in the energies of the NISs of isolated molecules have been observed for solid films of molecules such as H_2, N_2, O_2, C_6H_6.[106] In general, these downward shifts in the resonance energies of transient anions can be accounted for by considering the polarization of the dense medium (fluid or solid) by the temporarily localized electron. The gradual downward shift in the energy position of a NIS (and the associated changes in the electron attachment cross section) with increasing N of a dense gas, have first been observed for the case of O_2^- in a buffer gas of N_2 whose density was increased from ~1 x 10^{19} to ~1 x 10^{21} molecules cm^{-3}.[107-109] Later work on the EA of cluster negative ions as a function of their size[110] is consistent with these early findings.

Under isolated-molecule (single-collision) conditions, the autodetachment lifetimes τ_a of transient anions vary from ~10^{-16} to > 10^{-4} s.[1,2,111] Similarly, in low-pressure gases (multiple-collision conditions) the attachment cross sections σ_a for negative-ion formation vary from molecule to molecule (and the position of the negative ion states) by over 11 orders of magnitude.[1,2,111] The transition from a low-pressure gas to the liquid can result in changes of both τ_a and σ_a. In general such changes would depend on whether the electron affinity of the molecule in the gas $(EA)_G$ is negative (< 0 eV) or positive (> 0 eV) and on whether a negative $(EA)_G$ becomes positive in the condensed phase.

In connection with the changes in τ_a, studies of the lowest [$(EA) < 0$ eV] NISs of N_2, CO, and H_2 in solid films have shown[106] that the τ_a of the NIS in the solid is decreased from its value, $(\tau_a)_G$, in the low-pressure gas; in the condensed phase, the centrifugal barrier is greatly distorted due to symmetry changes, which are effected by the medium. One might, thus, expect $(\tau_a)_G$ to be longer than the lifetime $(\tau_a)_L$ of the NIS in the liquid. When $(EA)_G > 0$ eV (or when the EA of a molecule is negative in the gas but positive in the liquid), $(\tau_a)_G < (\tau_a)_L$ due to the faster energy relaxation in the liquid.

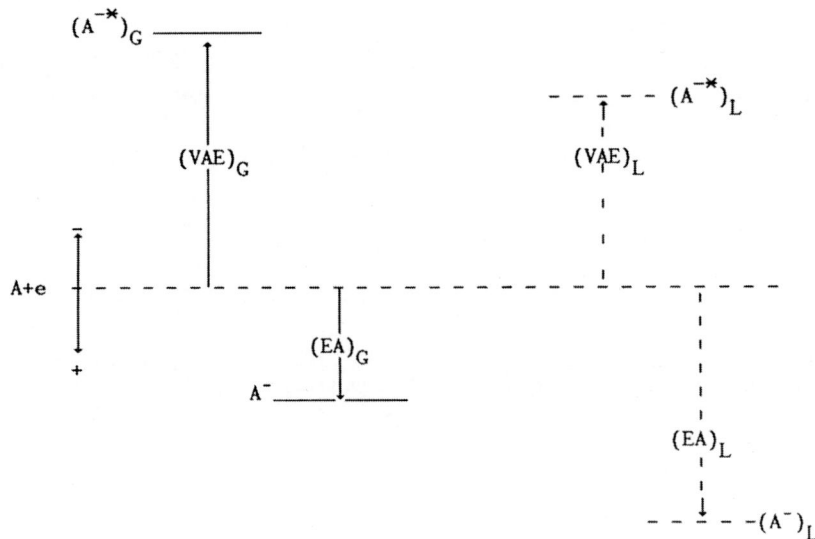

GAS   **LIQUID**

- <u>NIS: ENERGIES</u>

Figure 21. Schematic illustration of the relative value of *EA* and *VAE* in a gas and a liquid.

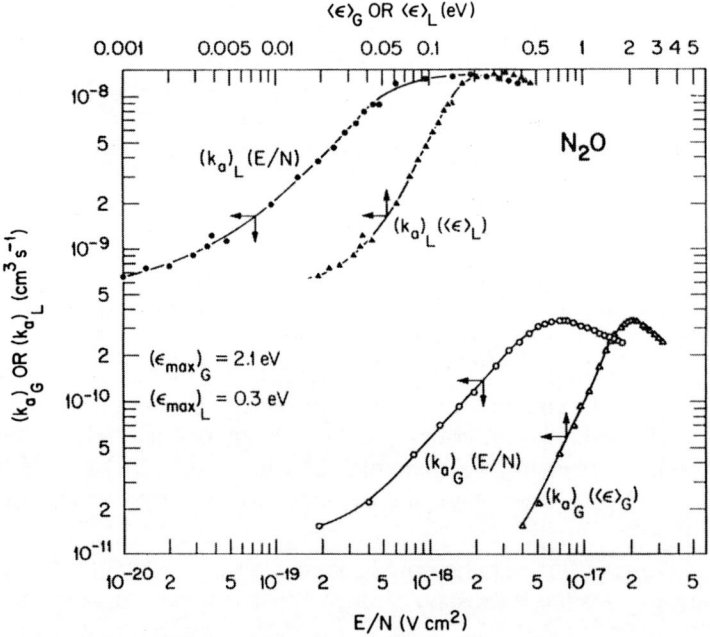

Figure 22. Electron attachment rate constant for N_2O in gaseous, $(k_a)_G$, and liquid, $(k_a)_L$, argon plotted versus E/N and $<\epsilon>_G$ or $<\epsilon>_L$.[69] The attachment is due to the reaction (21). Note the shift of the resonance to lower energies—and the increase in the rate constant—in the liquid.

               Christophorou

In connection with the changes in σ_a, the electron state in the liquid (or the dense gas) crucially determines both the magnitude of $(\sigma_a)_L$ and the relation of $(\sigma_a)_L$ to the corresponding value $(\sigma_a)_G$ in the gas. As will be shown in the next section, $(\sigma_a)_L$ (e_{qf}) $>> (\sigma_a)_L$ (e_ℓ). It is only for liquids for which the electron is in the quasi-free state that a comparison with gaseous data is meaningful. When, however, $(EA)_G > 0$ eV the electron attachment cross section in the liquid $(\sigma_a)_L$ is maximum when the captured electron is quasi-free (e_{qf}) and $(\sigma_a)_L$ is close to its diffusion-controlled value when the captured electron is localized (e_ℓ).

In connection with the decomposition of the NIS via dissociative attachment, it is clear that since $(\tau_a)_L < (\tau_a)_G$ for molecules such as O_2 ($O_2 + e \rightarrow O_2^{-*} \rightarrow O^- + O$; resonance peak at 6.7 eV in the gas) the cross section for dissociation attachment σ_{da} would be smaller in the liquid than in the low-pressure gas due to the decrease in the survival probability.[1,2] It should, however, be noted that since the position of the resonance is lower in the liquid than in the gas, and since the magnitude of the dissociative attachment cross section is larger as the energy position of the resonance is lowered,[1,2] the cross section for a dissociative attachment process may actually be much larger in the liquid than in the gas. This is certainly the case for reaction (21) (see Fig. 22). Fast energy relaxation in dense fluids may reduce $(\sigma_{da})_L$ when the dissociative attachment process is exoergic.

Electron Attachment in Dense Gases and Liquids

Studies of electron-molecule attaching collisions in dense gases are the domain of electron swarm methods.[1,2,5,112,113] The last two decades have seen the "maturity" of swarm methods and their unique contributions to the understanding of electron interactions in fluids. A most distinct advancement in this area has been the development of experimental methods to study electron attachment to molecules embedded in dense buffer gases for which the electron energy distribution function $f(\epsilon,E/N)$ can be calculated over a wide range of E/N values[1,2] (Fig. 9). These methods allowed measurement of the absolute (total) electron attachment rate constant k_a as a function of E/N or as a function of the mean electron energy $<\epsilon>$ since $<\epsilon>$ (E/N) can be computed once $f(\epsilon,E/N)$ is known. Furthermore, the measured $k_a(<\epsilon>)$ have been used to determine the absolute total electron attachment cross section $\sigma_a(\epsilon)$ from

$$k_a(<\epsilon>) = (2/m)^{1/2} \int_0^\infty \sigma_a(\epsilon)\epsilon^{1/2} f(\epsilon,E/N) d\epsilon \qquad (22)$$

Figure 23 shows the $k_a(<\epsilon>)$ and $\sigma_a(\epsilon)$ obtained[61] by these methods for the perfluoroalkanes $n\text{-}C_N H_{2N+2}$ ($N = 1$ to 6). The attachment cross sections increase with decreasing resonance energy, ϵ_{res}, approaching $\pi \lambda^2$ as $\epsilon_{res} \rightarrow kT$.[1,8] The wealth of knowledge that has been obtained from such studies has had a profound impact on the basic understanding of slow electron-molecule interactions (especially indirect electron scattering and molecular fragmentation by electron capture; it demonstrated, for example, the extreme fragility of many organic structures toward slow electrons, including those with thermal energies), and on the development of energy and pollutant-monitoring technologies (e.g., laser,[114] radiation and chemical detection devices,[114-116] gaseous dielectrics,[114,117] and pulsed power switches.[114,117,118] The development of radiosensitizing agents and radiopharmaceuticals benefited greatly from measurements of the electron attachment capabilities and electron affinities of biochemicals (see Ch. 5 of Vol. 2 of Ref. 2).

The nature and number density of a gaseous medium in which electron attachment reactions occur can have a profound effect on such reactions. The effect is a function of the medium and its density, the mode (dissociative or nondissociative) of electron attachment, and the anionic state involved in the electron attachment process—especially its lifetime and resonance energy. For dissociative electron attachment processes, the effect of the medium is insignificant at low N because fragment anions do not normally require collisional stabilization. For nondissociative electron attachment, however, the effect of the medium—even at low N—can be profound.

Quite generally, at low gas number densities ($\lesssim 5 \times 10^{19}$ molecules cm^{-3}) electron attachment to a molecule AX in a gas M can be represented by the reaction scheme.

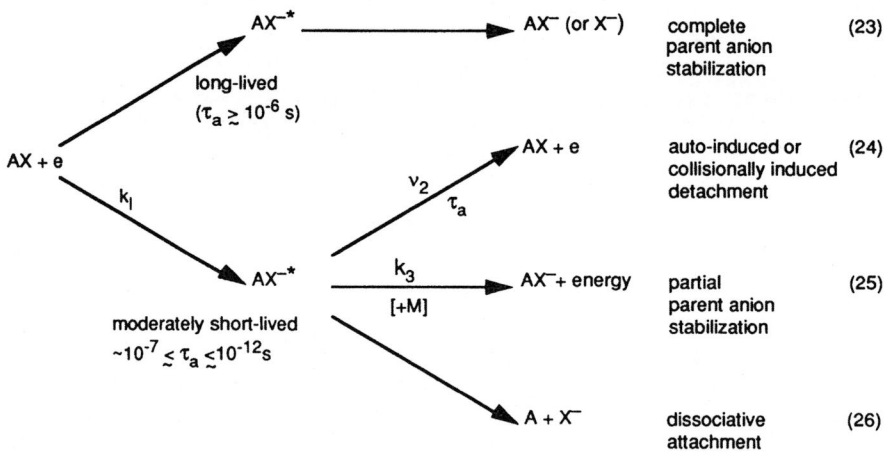

When only long-lived ($\tau_a \gtrsim 10^{-6}$ s) anions [process (23); e.g., SF$_6^{-*}$ at near-zero energy] and/or dissociative attachment fragment anions [process (26); e.g., fragment anions of C_3F_8 and n-C_4F_{10} (Fig. 23)] are formed by electron capture, k_a is independent of N. In the absence of processes (23) and (26), competition between auto-induced or collisionally-induced detachment [process (24)] and stabilization of AX^{-*} by collision with M [process (25)] can result in a pressure-dependent k_a. For the above reaction scheme

$$k_a = k_1 \left[\frac{k_3 N}{\nu_2 + k_3 N} \right] \qquad (27)$$

At low N or short τ_a such that $\nu_2 = \tau_a^{-1} >> k_3 N$, $k_a \propto N$ if dissociative attachment and other processes are absent. Alternatively, if the τ_a is long or N is large (i.e., $\tau_a^{-1} << k_3 N$), then $k_a = k_1$ and is independent of N or the nature of M. Considerable variations in the ability of the molecules M to collisionally stabilize AX^{-*} have been reported[107,109,119-121] and show that k_3 increases with increasing complexity of the stabilizing third body M. At relatively low N ($\lesssim 5 \times 10^{19}$ molecules cm^{-3}) a number of molecules [e.g., O_2, SO_2, N_2O[1,8,107,109,119-122]] have been found to attach electrons by a three-body process which is well-represented by (24) and (25). At higher N, however, the $k_a(N)$ of a number of molecules (e.g., O_2, 1-C_3F_6)[107,109,120-123] does

Christophorou

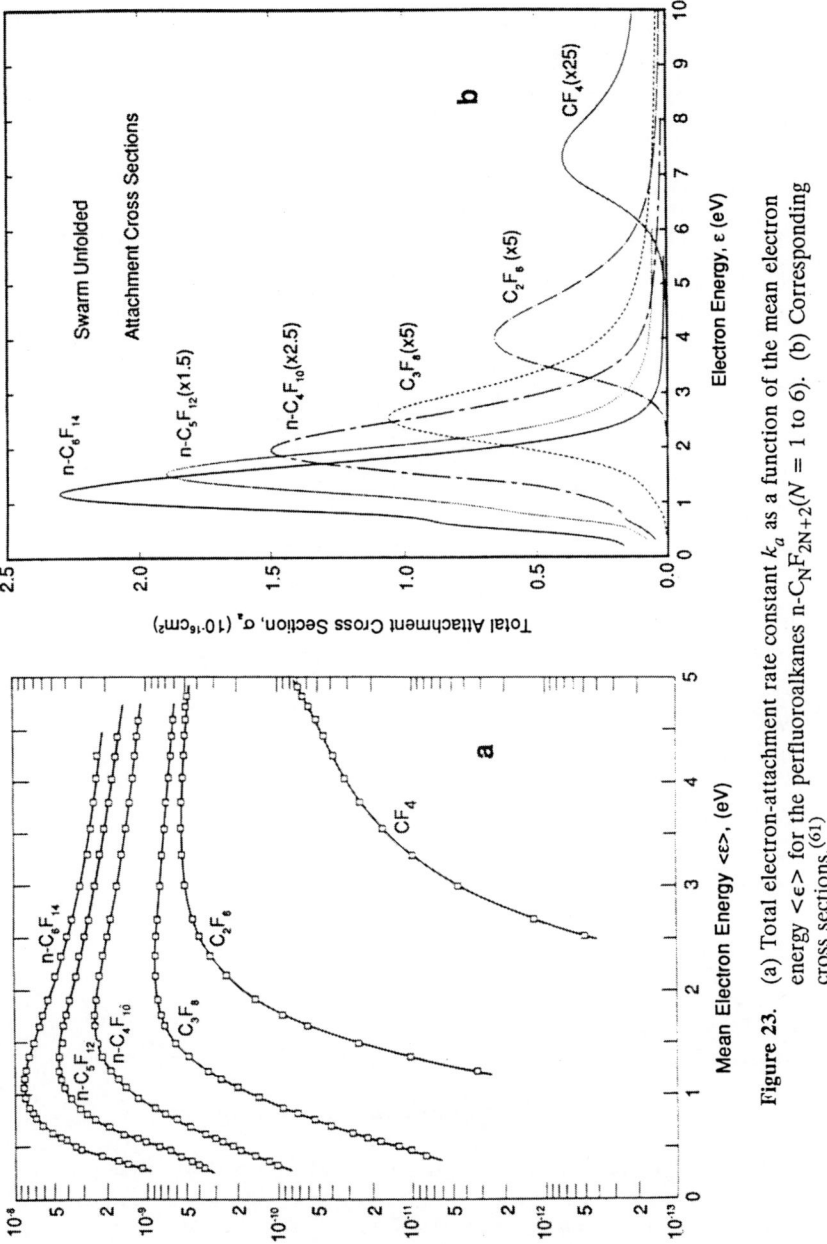

Figure 23. (a) Total electron-attachment rate constant k_a as a function of the mean electron energy $\langle \epsilon \rangle$ for the perfluoroalkanes n-$C_N F_{2N+2}$ (N = 1 to 6). (b) Corresponding cross sections.[61]

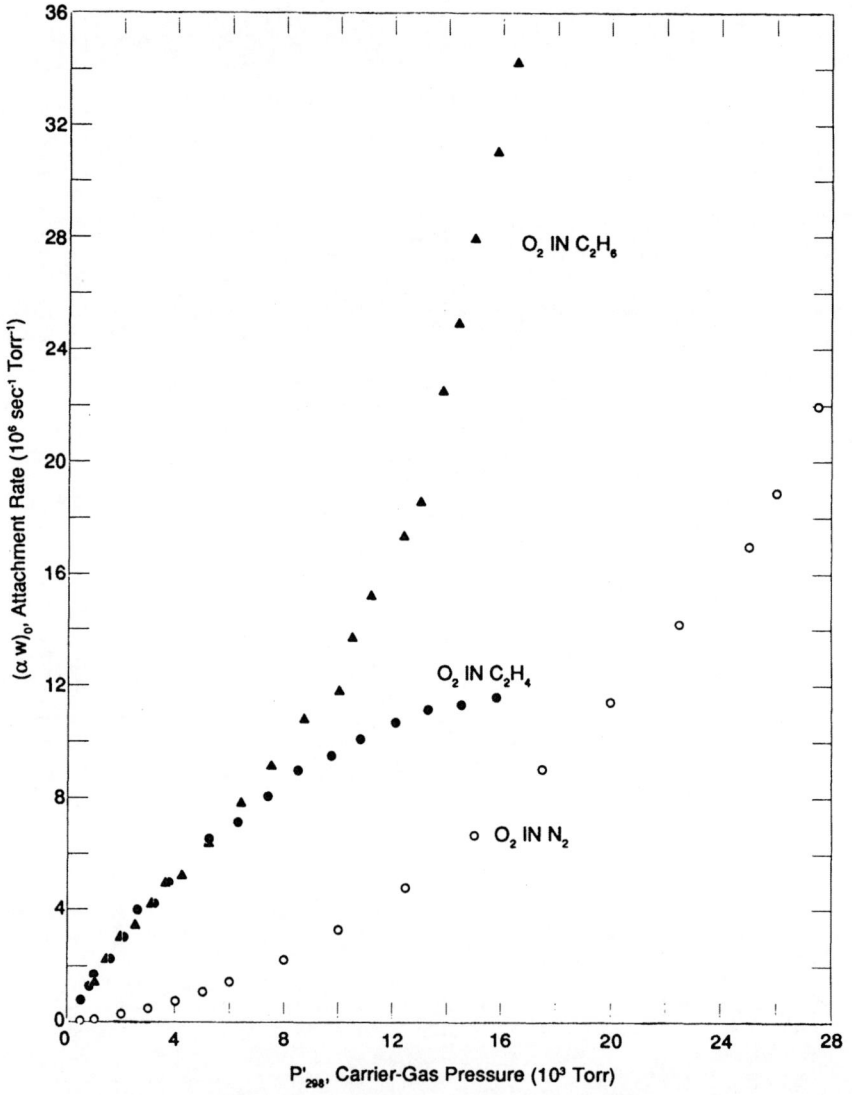

Figure 24. Electron attachment rate constant for O_2 in N_2 (o), C_2H_4 (●) and C_2H_6 (▲) as a function of the pressure (corrected for compressibility) of these buffer gases. These rate constants correspond to a value of $<\epsilon> \simeq 0.05$ eV.[107]

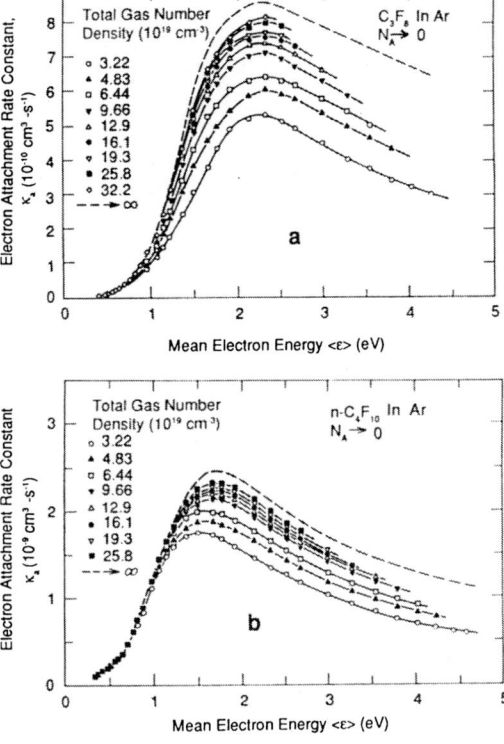

Figure 25. Electron attachment rate constant k_a as a function of mean electron energy $<\epsilon>$ and total gas pressure for C_3F_8 (a) and n-C_4F_{10} (b) in Ar buffer gas.[61]

not follow that predicted by a three-body process. This is exemplified by the $k_a(N)$ data in Fig. 24 for the formation of O_2^- in the buffer gases N_2 and C_2H_6, which show the involvement of more than one buffer gas molecules in the electron attachment process.[107,109] Other studies[121,122] contended that electron attachment to Van der Waals dimers is a major electron capture process; for example, electron attachment to O_2 forming O_2^- in a dense buffer gaseous medium M was suggested to be principally due to electron capture by Van der Waals molecules of the form $(O_2 \cdot M)$.

When, over a given energy range, reactions (25) to (26) occur concomitantly, the $k_a(<\epsilon>)$ has both a density independent [due to processes (26) and/or (23)] and a density-dependent [due to process (25)] component; the latter is a function of M and τ_a. This can be seen from the data in Fig. 25 where the $k_a(<\epsilon>)$ are shown for C_3F_8 and n-C_4F_{10} for several buffer gas pressures. The increase in k_a with N is due to process (25) and the N-independent component due to process (26). For these molecules, as the size of the transient anion increases its τ_a increases;[61] hence, the effect of N on $k_a(<\epsilon>)$ is less for n-C_4F_{10} compared to C_3F_8.

To illustrate the effect of the state of matter on electron attachment and the relevance of electron attachment to molecules in low-pressure gases to those in liquids, we focus on the simple case where the electron-attaching molecule (SF_6) in low-pressure gases captures thermal and epithermal energy electrons forming predominantly long-lived ($\tau_a > 10^{-5}$ s) parent anions (SF_6^{-*}) and the liquid medium (liquefied Ar) has a conduction band ($V_o < 0$ eV; e_{qf}). In Fig. 26 the rate constant $(k_a)_G$ for electron attachment to SF_6 in low-pressure ($\lesssim 3$ atm) gaseous Ar ($T \simeq 300$ K) as a function of E/N or $<\epsilon>_G$ (the mean electron energy in the gas) is compared with that,

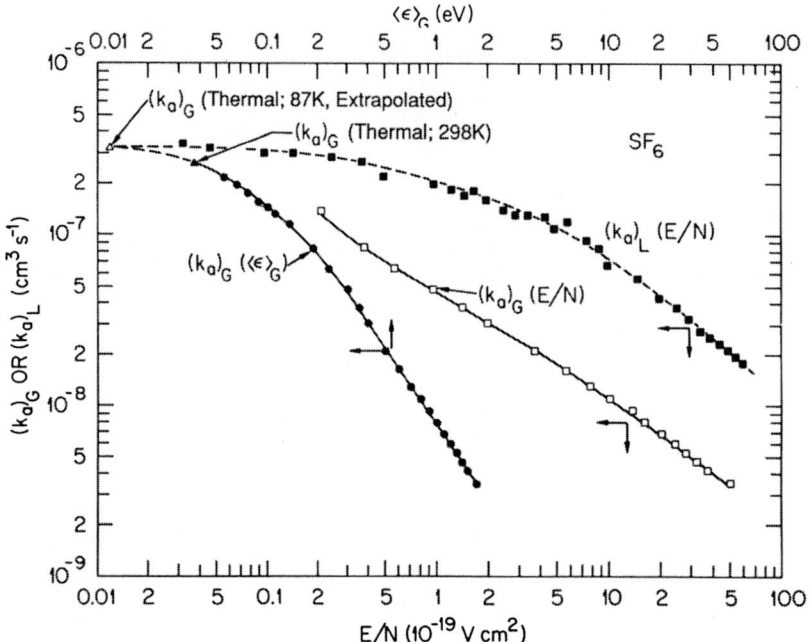

Figure 26. $(k_a)_G$ versus $<\epsilon>_G$ or E/N for SF_6 in gaseous Ar and $(k_a)_L$ versus E/N for SF_6 in liquid Ar; ▲, thermal value of $(k_a)_G$ at 298 K; Δ, thermal value of $(k_a)_G$ extrapolated to 87 K [from Ref. 69].

$(k_a)_L$ (E/N) in liquefied Ar$(T \sim 87$ K). The thermal value of $(k_a)_G$ at 298 K and at 87 K (extrapolated) agree well with the liquid $(k_a)_L$ data for the lowest E/N, showing that $(k_a)_G \simeq (k_a)_L$ when the electron energy distribution is thermal. Indeed, in liquefied rare gases the attachment of slow electrons to SF_6 forming SF_6^- is similar to that in gases both in magnitude and energy dependence.[65] That this is actually the case for other molecules can be seen from Fig. 27, where the k_a for SF_6, O_2 and N_2O measured[124] as a function of the electric field strength E is plotted as a function of $<\epsilon>_L$.[65]

In general[11,12,65,69,109] for high mobility dielectric liquids ($\mu >> 1$ cm^2 V^{-1} s^{-1}; $V_o < 0$ eV) the electron is quasi-free, and its attachment to a molecule AX embedded in the liquid can be viewed—as in gases—as a vertical transition between the initial $(e + AX)_L$ and the final $(AX^*)_L$ state; the attachment process depends on the properties of AX and the medium (especially V_o) and a comparison of $(k_a)_L$ with $(k_a)_G$ is possible. However, in liquids in which the electron is initially in a localized state, the rate-determining step is the diffusive motion of the electron, and $(k_a)_L$ depends only weakly on the medium and varies little with AX. In such cases $(k_a)_L$ can attain diffusion-controlled values as long as a negative-ion state of AX exists at thermal energies; this condition seems to be satisfied for most liquids when $(\epsilon_{max})_G \lesssim 1$ eV.[109] In these cases $(k_a)_L$ can be expressed as

$$(k_a)_L \simeq 4\pi R D_e \qquad (28)$$

where R is the encounter radius, and D_e is the electron diffusion coefficient. Since, moreover, $D_e = (kT/e)\mu$, $(k_a)_L$ is predicted to increase linearly with μ, a behavior observed experimentally in some instances.[124,125] Finally, when the electron drifts part of the time as quasi-free and part of the time as localized, $(k_a)_L$ can be expressed as

$$(k_a)_L = (k_a)_\ell \, p + (k_a)_f (1 - p) \tag{29}$$

where $(k_a)_\ell$ and $(k_a)_f$ are, respectively, the attachment rate constants involving e_ℓ and e_{qf} and p is the probability of finding the electron in the localized state.[11]

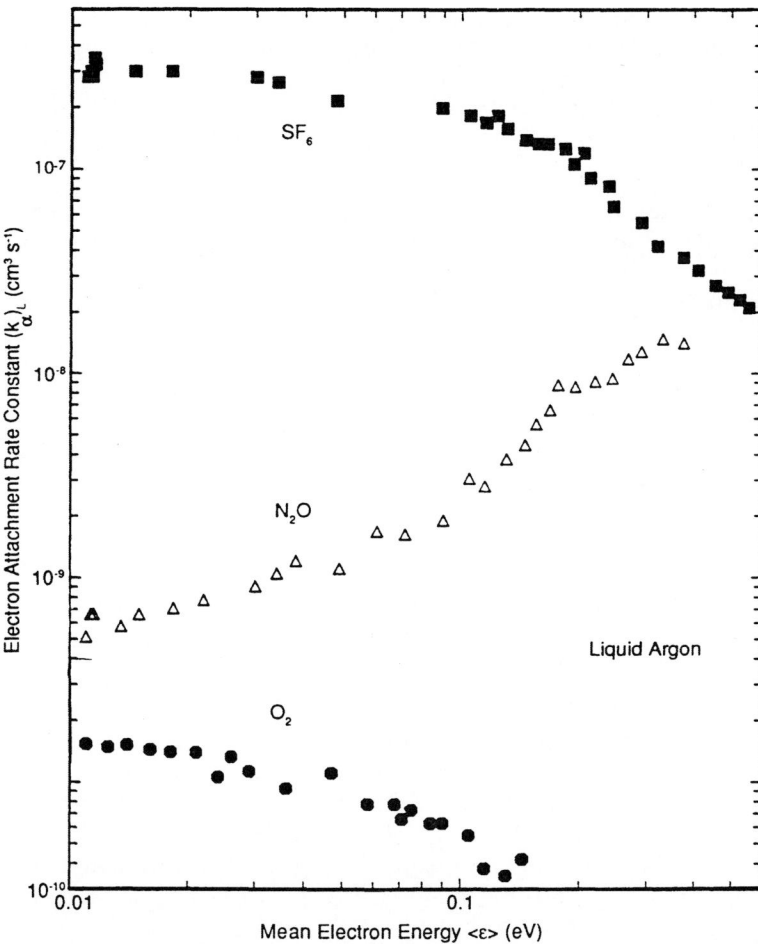

Figure 27. Rate constant, $(k_a)_L$ for electron attachment to SF_6, N_2O and O_2 measured in liquid Ar[124] plotted versus $<\epsilon>_L$.[65]

Electron-Excited Molecule Interactions

The interactions of slow electrons with molecules depend not only on the electron energy, but also on the internal-energy content of the molecules themselves. While the study of electron-ground state molecule interactions traces back many decades, the study of electron-excited molecule interactions is just beginning. Indeed, little is known about the scattering of electrons from excited molecules in spite of their implicit significance in radiation and life sciences—especially the initial stages of radiation action on matter—and in many applied areas such as lasers and plasmas.

Electron transport in "hot" (vibrationally/rotationally excited) gases has been shown to be influenced by the internal vibrational/rotational states,[1,2,74,126] and a number of recent studies[8,9,112,113,127-133] have unraveled delicate and often large effects of the internal energy of molecules (e.g., freons, halocarbons, perfluorocarbons) on the decomposition of their transient anions by dissociative electron attachment and auto-detachment. Dissociative electron attachment to molecules[8,9,112,113,127-131] has been shown to increase with increasing internal energy of molecules (Fig. 28a). On the other hand, nondissociative electron attachment to many molecules which form long-lived parent negative ions at ambient temperature (e.g., the perfluorocarbons C_6F_6, n-C_4F_{10}, c-C_4F_8) has been shown[8,9,129,131-133] to decrease when the gas temperature is increased above ambient (Fig. 28b). While the preponderance of the observations concerning the effect of internal energy of molecules on their dissociative electron attachment cross sections involved electron attachment to thermally excited vibrational/rotational states of the ground electronic states of molecules, similar observations have been reported for vibrationally excited molecules produced by laser irradiation.[134]

Electron scattering data on electronically excited atoms and molecules is very limited indeed. The little information that exists on electronic excitation from metastable states suggests that the cross sections are substantially higher than those for the ground states. Thus, the scattering cross section for the reaction

$$O_2\left(a^1\Delta_g\right) + e\,(4.5\;eV)\xrightarrow{\;2.3\,x\,10^{-17}cm^2\;} O_2^*\left(b^1\Sigma_g^+\right) + e\,'$$

has been reported[135] to be ~10 times larger than that for the ground state; viz.,

$$O_2\left(X^3\Sigma_g^-\right) + e\,(4.5\;eV)\xrightarrow{\;2.1\,x\,10^{-18}cm^2\;} O_2^*\left(b^1\Sigma_g^+\right) + e\,'$$

In keeping with this trend, the low-lying (excitation energy ~ 0.98 eV) electronically excited state O_2^* $(a^1\Delta_g)$ produced in a microwave discharge was shown to have 3 to 4 times larger cross section for dissociative electron attachment compared to the $O_2(X^3\Sigma_g^-)$ ground-electronic state.[136]

Recently a number of novel techniques have been developed[10,137,138] for the study of electron attachment to electronically excited molecules using lasers. The first observation of optically enhanced electron attachment to electronically excited molecules was reported[137] in 1987. Thiophenol (C_6H_5SH) molecules were indirectly excited (via laser light absorption to high-lying optically allowed states which undergo rapid internal conversion and efficient intersystem crossing) to their long-lived (lifetime ~8 ms) first-excited triplet states; at near-zero electron energies ~5 to 6 orders of magnitude larger electron attachment coefficients (due to dissociative attachment via these triplet states) were measured compared to the ground electronic states of these molecules (Fig. 29).[10,137] Quite similar to the case of thiophenol, 5 to 7 orders of magnitude enhancement in electron attachment has been reported for the first-excited triplet states of p-benzoquinone and its methylated derivatives.[139]

Very recently, Pinnaduwage et al.[138] used a newly developed technique for the study of electron attachment to short-lived ($<10^{-8}$ s) species. In this technique the same laser pulse that produces (via multiphoton absorption) the electronically excited species also produces in their vicinity concomitantly (via multiphoton ionization of the same gas or of an additive gas) the attaching electrons. They claim to have observed electron attachment to superexcited states of molecules occurring with enormous ($\gtrsim 10^{-11}$ cm^2) cross sections which (for the triethylamine molecules investigated) are

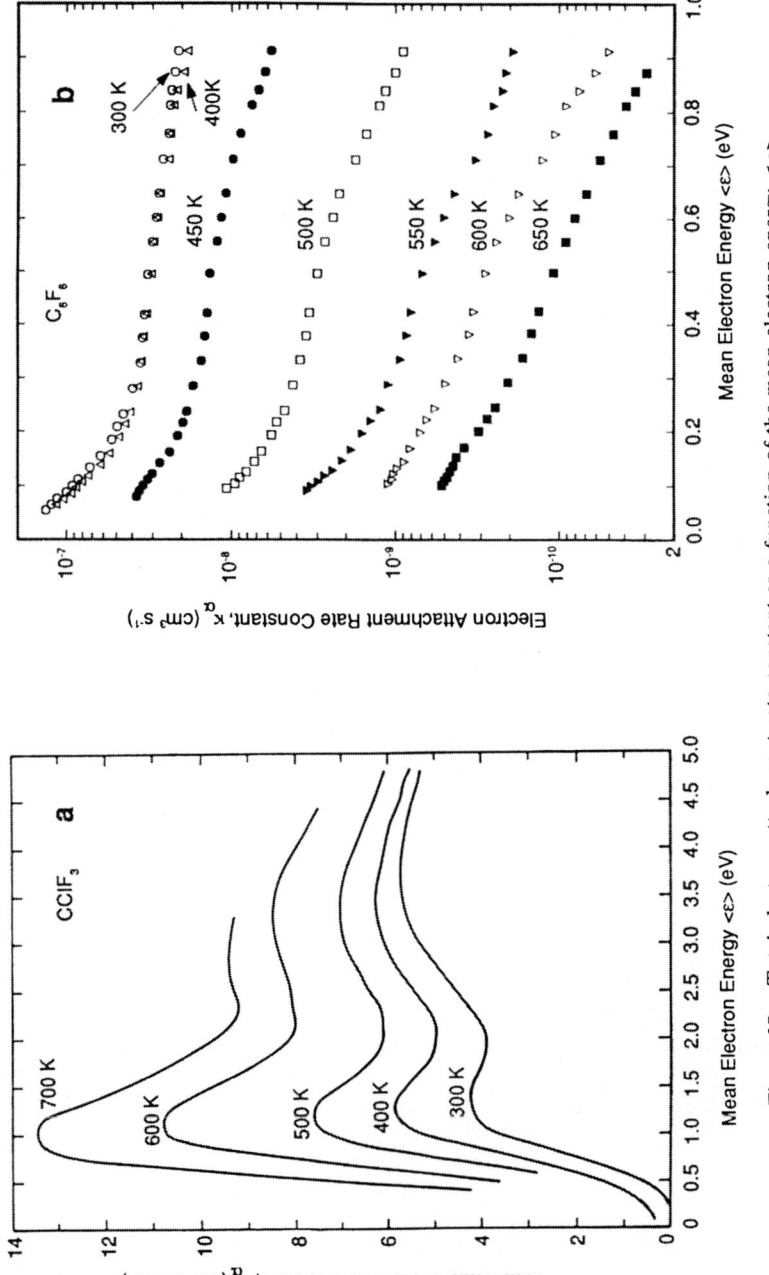

Figure 28. Total electron attachment rate constant as a function of the mean electron energy $\langle \epsilon \rangle$ and T, $k_a(\langle \epsilon \rangle, T)$ for (a) freon CClF$_3$, which attaches slow electrons dissociatively;[128] and (b) C$_6$F$_6$, which attaches slow electrons nondissociatively.[132]

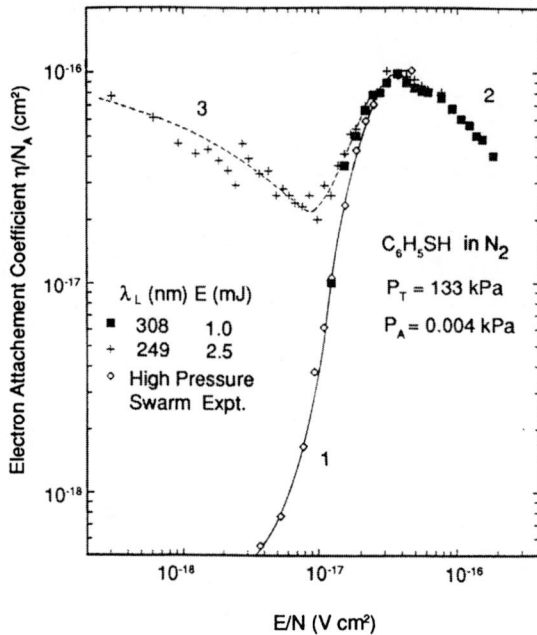

Figure 29. Electron attachment coefficient η/N_A versus E/N for thiophenol (C_6H_5SH) in N_2 buffer gas. Curve 1 was obtained without laser irradiation and depicts electron attachment to the ground state. Curves 2 and 3 were obtained with XeCl and KrF laser lines, respectively. The photon energy of the XeCl line is not sufficient to excite electronically the molecule monophotonically and therefore only the ground state attachment is observed; however, electronic excitation and enhanced electron attachment occurs at the KrF line. [Note that η/N_A^* is ~ 100 times larger than η/N_A since the excited molecule number density N_A^* is about one percent of N_A.] The photoenhanced electron attachment (Curve 3) was attributed to electronically excited first triplet states populated indirectly via laser irradiation.[10,137]

~10^7 times larger than those for the ground state molecule. These incredibly larger cross sections may involve high-lying Rydberg states.[138]

It is, therefore, clear that slow electrons colliding with electronically excited molecules have electron attachment cross sections many (as high as 10^7) times larger than those for the ground state (unexcited) molecules. The optical control of the electron-molecule collision cross sections opens up new frontiers and new possibilities of optically controlling the impedance characteristics of (gaseous) matter at times in the µs to *ns* range. The excited species are very reactive toward slow electrons. Slow electrons, by transferring their energy to molecules, make the molecules and themselves more reactive.

Concluding Remarks

Our understanding of the basic processes of radiation interaction with dense gaseous matter has advanced considerably over the last two decades. Progress has also been made in our efforts to link knowledge of radiation interactions in low-pressure gases with knowledge of such processes in dense fluids. The basic knowledge acquired has illuminated broad areas of pure and applied science, led to new radiobiological and environmental-monitoring instrumentation, and aided the development of many energy technologies.

In spite of the impressive progress, our knowledge still remains incomplete in a number of important areas (see a partial list in Table IV). Foremost among these are the interfacing of the gaseous and condensed phases of matter and the interactions of radiation (especially slow electrons and photons) with energy-rich (excited) atoms and molecules. The understanding of radiobiological effects and mechanisms from basic knowledge remains a challenge.

Table IV. Radiation Interaction with Dense Fluids (Quasi-Liquids and Liquids): Future Research Areas

- Electron interactions with electronically excited species
- Electron diffusion and energies in dense fluids with conduction bands
- Electron-impact ionization and excitation cross sections
- Electron dynamics and fast (sub-picosecond) medium responses
- Theoretical understanding of electron scattering and dynamics
- Photon absorption by electronically excited species
- (Multi) photon ionization and energetics especially in polar media
- Transient photoionization mechanisms
- Ionic and neutral decomposition mechanisms
- Relation of microscopic to macroscopic properties

Acknowledgment

The long-range programs of the Office of Health and Environmental Research at the U.S. Department of Energy contributed fundamentally to these developments.

References

1. L. G. Christophorou. *Atomic and Molecular Radiation Physics*. Wiley-Interscience, New York (1971).
2. L. G. Christophorou, ed. *Electron-Molecule Interactions and Their Applications,* Vol. 1 and 2. Academic Press, New York (1984).
3. G. R. Freeman, ed. *Kinetics of Nonhomogeneous Processes*. Wiley-Interscience, New York (1987).
4. E. E. Kunhardt, L. G. Christophorou, and L. H. Luessen, eds. The Liquid State and Its Electrical Properties. *NATO ASI Series B: Physics,* Vol. 193, Plenum Press, New York (1987).
5. H.S.W. Massey. *Electronic and Ionic Impact Phenomena*, Vol.I-IV. Oxford Press (1969).
6. M. Inokuti. In Applied Atomic Collision Physics. H.S.W. Massey, E. W. McDaniel and B. Bederson, eds., Vol. 4, Ch. 3 (1983).
   I. Shimamura and K. Takayanagi, eds., *Electron-Molecule Collisions*, Plenum Press, New York (1984).
7. E. W. McDaniel. *Atomic Collisions*. John Wiley & Sons, New York (1989).
8. L. G. Christophorou, D. L. McCorkle, and A. A. Christodoulides. Ref. 2, Vol. 1, Chapt. 6.
9. L. G. Christophorou. Electron Attachment and Detachment Processes in Electronegative Gases. *Plasma Physics* 27:237-281 (1987).
10. L. A. Pinnaduwage, L. G. Christophorou, and S. R. Hunter. Optically Enhanced Electron Attachment to Thiophenol. *J. Chem. Phys.* 90:6275-6289 (1989).
11. L. G. Christophorou and K. Siomos. Ref. 2, Vol. 2, Chapt. 4.
12. L. G. Christophorou. Ref. 4, pp. 283-316.
13. S. R. Hunter, J. G. Carter, and L. G. Christophorou. Electron Attachment and Ionization Processes in CF_4, C_2F_6, C_3F_8, and n-C_4F_{10}. *J. Chem. Phys.* 86:693-703 (1987).
14. S. R. Hunter, J. G. Carter, and L. G. Christophorou. Electron Transport Measurements in Methane Using an Improved Pulsed Townsend Technique. *J. Appl. Phys.* 60:24-35 (1986).

15. S. E. Derenzo, T. S. Mast, H. Zaklad, and R. A. Muller. Electron Avalanche in Liquid Xenon. *Phys. Rev. A* 9:2582-2591 (1974).

16. I. György and G. R. Freeman. Ionization and Electron Thermalization Distances in Isomeric Pentanes: Effects of Molecular Shape and Density. *J. Chem. Phys.* 86:681-687 (1987).

17. R. A. Holroyd and D. F. Anderson. The Physics and Chemistry of Room-Temperature Liquid-Filled Ionization Chambers. *Nucl. Instr. Meth. Phys. Res.* A236:294-299 (1985).

18. T. G. Ryan and G. R. Freeman. Electron Mobilities and Ranges in Methyl-Substituted Pentanes Through the Liquid and Critical Regions. *J. Chem. Phys.* 86:5144-5150 (1978).

19. S.S.-S. Huang and G. R. Freeman. Effect of Density on the Total Ionization Yields in X-Irradiated Argon, Krypton, and Xenon. *Can. J. Chem.* 55:1838-1845 (1977).

20. T. Takahashi, S. Konno, and T. Doke. The Average Energies, W, Required to Form an Ion Pair in Liquefied Rare Gases. *J. Phys.* C7:230-240 (1974).

21. P. G. Fuochi and G. R. Freeman. Molecular Structure Effects on the Free-Ion Yields and Reaction Kinetics in the Radiolysis of the Methyl-Substituted Propanes and Liquid Argon: Electron and Ion Mobilities. *J. Chem. Phys.* 56:2333-2341 (1972).

22. W. F. Schmidt and A. O. Allen. Free-Ion Yields in Sundry Irradiated Liquids. *J. Chem. Phys.* 52:2345-2351 (1970).

23. S. Geer, R. A. Holroyd and F. Ptohos. Field Dependent Free Ion Yields of Room Temperature Tetramethyl Liquids and Their Mixtures. *Nucl. Instr. Meth. Phys. Res.* A287:447-451 (1990).

24. R. C. Munoz, J. B. Cumming, and R. A. Holroyd. Ionization of Tetramethylsilane by Alpha Particles. *Chem. Phys. Lett.* 115:477-480 (1985).

25. I. Lopes, H. Hilmert, and W. F. Schmidt. Ionization of Some Molecular Gases by ^{60}Co-γ-Radiation: W-Values. *Radiat. Phys. Chem.* 29:93-95 (1987).

26. J.-P. Dodelet and G. R. Freeman. Mobilities and Ranges of Electrons in Liquids: Effect of Molecular Structure in C_5-C_{12} Alkanes. *Can. J. Chem.* 50:2667-2679 (1972).

27. B. S. Yakovlev and L. V. Lukin. *In Photodissociation and Photoionization*. K. P. Lawrey, ed., p. 99. John Wiley & Sons, New York (1985).

28. R. A. Holroyd and R. L. Russell. Solvent and Temperature Effects in the Photoionization of Tetramethyl-p-phenylenediamine. *J. Phys. Chem.* 78:2128-2135 (1974).

29. R. Reininger, V. Saile, P. Laporte, and I. T. Steinberger. Photoconduction in Rare Gas Fluids Doped with Small Organic Molecules. *Chem. Phys.* 89:473-479 (1984).

30. R. Reininger, V. Saile, G. L. Findley, P. Laporte, and I. T. Steinberger. *In Photophysics and Photochemistry Above 6 eV*, F. Lahmani, ed. Elsevier Science Publishers, Amsterdam, 253.
    U. Asaf and I. T. Steinberger, Photoconductivity and Electron Transport Parameters in Liquid and Solid Xenon. *Phys Rev. B* 10:4464-4468 (1974).

31. J. Casanovas, R. Grob, D. Delacroix, J. P. Guelfucci, and D. Blanc. Photoconductivity Studies in Some Nonpolar Liquids. *J. Chem. Phys.* 75:4661-4668 (1981).

32. E.-H. Böttcher and W. F. Schmidt. Photoconductivity of Nonpolar Liquids Induced by Vacuum-Ultraviolet Light. *J. Chem. Phys.* 80:1353-1359 (1984).

33. H. Faidas and L. G. Christophorou. Determination of the Ionization Threshold of Azulene in Hydrocarbon Liquids by Multiphoton Ionization. *J. Chem. Phys.* 88:8010-8011 (1988).
    Laser Multiphoton Ionization of Aromatic Molecules in Nonpolar Liquids. *Radiat. Phys. Chem.* 32:433-438 (1988).

34. H. Faidas, L. G. Christophorou, P. G. Datskos, and D. L. McCorkle. The Ionization Threshold of N,N,N´,N´-Tetramethyl-p-phenylenediamine. *J. Chem. Phys.* 90:6619-6626 (1989).

35. R. D. Levin and S. G. Lias. Ionization Potential and Appearance Potential Measurements 1971-1981. NSRDS-NBS-71, U.S. Department of Commerce, NBS, Washington, D.C. (1982).

36. A. O. Allen. Drift Mobilities and Conduction Band Energies of Excess Electrons in Dielectric Liquids. NSRDS-NBS 58, U.S. Department of Commerce, Washington, D.C. (1976).

37. R. A. Holroyd, S. Tames, and A. Kennedy. Effect of Temperature on Conduction Band Energies of Electrons in Nonpolar Liquids. *J. Phys. Chem.* 79:2857-2861 (1975).

38. E.-H. Böttcher. Experimentelle Untersuchung der photoelektrischen Leitung reiner und mit aromatischen Molekülen dotierter organischer Flüssingkeiten. GmbH HMI-B406, Berlin (1984).

39. K. Buschick and W. F. Schmidt. Vacuum Ultraviolet Photoconductivity of 2,2,4,4-Tetramethyl-pentane and Bis (Trimethylsilyl) Ethane. *IEEE Trans. Electr. Insul.* 24:353-356 (1989).

40. H. Faidas and L. G. Christophorou. Multiphoton Ionization of Fluoranthene in Tetramethyl-silane. *J. Chem. Phys.* 86:2505-2509 (1987).

41. I. Roberts and E. G. Wilson. The Intrinsic Photoconductivity of Liquid Xenon. *J. Phys. C* 6:2169-2183 (1973).

42. R. A. Holroyd, J. M. Preses, and N. Zevos. Single-Photon Induced Conductivity of Solutes in Nonpolar Solvents. *J. Chem. Phys.* 79:483-487 (1983).

43. B. Raz and J. Jortner. Energy of the Quasi-Free Electron State in Liquid and Solid Rare Gases. *Chem. Phys. Lett.* 4:155-158 (1969).

44. M. Born. Volumen und Hydratationswärme der Ionen. *Z. Phys.* 1:45-48 (1920).

45. I. Messing and J. Jortner. Adiabatic Polarization Energy in a Simple Dense Fluid. *Chem. Phys.* 24:183-189 (1977).

46. B. E. Springett, J. Jorner, and M. H. Cohen. Stability Criterion for the Localization of an Excess Electron in a Nonpolar Fluid. *J. Chem. Phys.* 48:2720-2731 (1968).

47. S. Noda, L. Kevan, and K. Fueki. Conduction State Energy of Excess Electrons in Condensed Media: Liquid Methane, Ethane, and Argon and Glassy Matrices. *J. Phys. Chem.* 79:2866-2874 (1975).

48. Y. Yamaguchi, T. Nakajima, and M. Nishikawa. Conduction Band Energy in Dense Ethane Fluid. *J. Chem. Phys.* 71:550-551 (1979).

49. U. Asaf and I. T. Steinberger. The Energies of Excess Electrons in Helium. *Chem. Phys. Lett.* 128:91-94 (1986).
    R. Reininger, U. Asaf, I. T. Steinberger, and S. Basak, Relationship Between the Energy V_0 of the Quasi-Free Electron and Its Mobility in Fluid Argon, Krypton, and Xenon. *Phys. Rev. B* 28:4426-4432 (1983).
    U. Asaf, R. Reininger, and I. T. Steinberger. The Energy V_0 of the Quasi-Free Electron in Gaseous, Liquid, and Solid Methane. *Chem. Phys. Lett.* 100:363-366 (1983).
    J. T. Steinberger, in Ref.[4], p. 235.

50. U. Asaf, W. S. Felps, K. Pupnik, S. P. McGlynn, and G. Ascarelli. Density Effects of High-n Molecular Rydberg States: CH_3I and C_6H_6 in H_2 and Ar. *J. Chem. Phys.* 91:5170-5174 (1989).

51. E. Fermi. Sopra Lo Spostamento per Pressione Delle Righe Elevate Delle Serie Spettrali. *Nuovo Cimento* 11:157-166 (1934).

52. V. A. Alekseev and I. I. Sobel'man. A Spectroscopic Method for the Investigation of Elastic Scattering of Slow Electrons. *Sov. Phys. JETP* 22:882-888 (1966).

53. L. Onsager. Initial Recombination of Ions. *Phys. Rev.* 54:554-557 (1938).

54. H. Lu, F. H. Long, R. M. Bowman, and K. B. Eisenthal. Femtosecond Studies of Electron-Cation Geminate Recombination in Water. *J. Phys. Chem.* 93:27-28 (1989).

55. C. Ferradini and J.-P. Jay-Gerin. Radiolysis of Liquids with High Static Dielectric Constant: An Estimate of the Total Ionization Yield, Electron Thermalization Distance, and Contribution of Heterogeneous Reactions. *J. Chem. Phys.* 89:6719-6722 (1988).
    J.-P. Jay-Gerin and C. Ferradini. On the Variation of the Free-Ion Yield with the Static Dielectric Constant in the Radiolysis of Liquids. *Radiat. Phys. Chem.* 33:251-253 (1989).

56. R. M. Bowman, H. Lu, and K. B. Eisenthal. Femtosecond Study of Geminate Electron-Hole Recombination in Neat Alkanes. *J. Chem. Phys.* 89:606-608 (1988).

57. J. M. Warman, E. S. Sennhauser, and D. A. Armstrong. Three-Body Electron-Ion Recombination in Molecular Gases. *J. Chem. Phys.* 70:995-999 (1979).
    E. S. Sennhauser, D. A. Armstrong, and J. M. Warman. The Temperature Dependence of Three-Body Electron Ion Recombination in Gaseous H_2O, NH_3, and CO_2. *Radiat. Phys. Chem.* 15:479-483 (1980).

58. Y. Nakamura, K. Shinsaka, and Y. Hatano. Electron Mobilities and Electron-Ion Recombination Rate Constants in Solid, Liquid, and Gaseous Methane. *J. Chem. Phys.* 78:5820-5824 (1983).

59. N. Gee and G. R. Freeman. Density and Temperature Effects on Electron Mobility in Fluid Methane. *Phys. Rev.* A20:1152-1161 (1979).

60. N. E. Cipollini, R. A. Holroyd and M. Nishikawa. Zero-Field Mobility of Excess Electrons in Dense Methane. *J. Chem. Phys.* 67:4636-4639 (1977).

61. S. R. Hunter and L. G. Christophorou. Electron Attachment to the Perfluoroalkanes $n\text{-}C_NF_{2N+2}(N = 1 \text{ to } 6)$ Using High-Pressure Swarm Techniques. *J. Chem. Phys.* 80:6150-6164 (1984).

62. L. G. Christophorou, P. G. Datskos, and J. G. Carter. (To be published.)

63. M. Hayashi. In Swarm Studies and Inelastic Electron-Molecule Collisions. L. C. Pitchford, B. V. McKoy, A. Chutjian, and S. Trajmar, eds. pp. 167-187. Springer-Verlag, New York (1987).

64. W. F. Schmidt. Electron Conduction Processes in Dielectric Liquids. *IEEE Trans. Electr. Insul.* EI-19:389-418 (1984).

65. L. G. Christophorou, S. R. Hunter, and J. G. Carter. Electron Attachment to SF_6 in Gaseous Ar and Xe; Comparison to Results in Liquid Ar and Xe and Energy of Excess Electrons. *Radiat. Phys. Chem.* 34:819-827 (1989).

66. E. Shibamura, T. Takahashi, S. Kubota, and T. Doke. Ratio of Diffusion Coefficient to Mobility for Electrons in Liquid Argon. *Phys. Rev. A* 20:2547-2554 (1979).

67. S. Kubota, T. Takahashi, and J. Ruangen. Hot Electron Relaxation in Solid and Liquid Argon, Krypton and Xenon. *J. Phys. Soc. Japan* 51:3274-3277 (1982).

68. S. Nakamura, Y. Sakai, and H. Tagashira. Effective Momentum Transfer Cross Section for Excess Electrons in Liquid Argon. *Chem. Phys. Lett.* 130:551-554 (1986).

69. L. G. Christophorou. Mean Energy of Excess Electrons in Liquid Ar as a Function of E/N; Electron Attachment to N_2O in Gaseous and Liquid Ar. *Chem. Phys. Lett.* 121:408-411 (1985).

70. E. M. Gushchin, A. A. Kruglov, and I. M. Obodovskii. Electron Dynamics in Condensed Argon and Xenon. *Sov. Phys. JETP* 55:650-655 (1982).

71. J. Lekner. Motion of Electrons in Liquid Argon. *Phys. Rev.* 158:130-137 (1967).

72. G. Bakale and G. Beck. Field-Dependent Electron Attachment in Liquid Tetramethylsilane. *J. Chem. Phys.* 84:5344-5350 (1986).

73. H. Faidas, L. G. Christophorou, D. L. McCorkle, and J. G. Carter. Electron Drift Velocities and Electron Mobilities in Fast Room Temperature Dielectric Liquids and Their Corresponding Vapors. *Nucl. Instr. Meth. Phys. Res. A* 294:575-582 (1990).

74. L.G.H. Huxley and R. W. Crompton. *The Diffusion and Drift of Electrons in Gases.* Wiley-Interscience, New York (1974).

75. W. L. Morgan. *A Bibliography of Electron Swarm Data 1978-1989.* JILA Data Center. Report No. 33, NIST, Boulder, Colorado, July (1990).

76. L. C. Pitchford and A. V. Phelps. Comparative Calculations of Electron Swarm Properties in N_2 at Moderate E/N Values. *Phys. Rev. A* 25:540-554 (1982).

77. G. L. Braglia, L. Romano, and M. Diligenti. Comment on "Comparative Calculations of Electron Swarm Properties in N_2 at Moderate E/N Values." *Phys. Rev. A* 26:3689-3694 (1982).

78. M. Yousfi, P. Ségur, and T. Vassiliadis. Solution of the Boltzmann Equation with Ionization and Attachment: Application to SF_6. *J. Phys. D* 18:359-375 (1985).

79. S. Yachi, Y. Kitamura, K. Kitamori, and H. Tagashira. A Multi-Term Boltzmann Equation Analysis of Electron Swarms in Gases. *J. Phys. D* 21:914-921 (1988).

80. J. J. Lowke, A. V. Phelps, and B. W. Irwin. Predicted Electron Transport Coefficients and Operating Characteristics of CO_2—N_2—He Laser Mixtures. *J. Appl. Phys.* 44:4664-4671 (1973).

81. B. R. Bulos and A. V. Phelps. Excitation of the 4.3 μm Bands of CO_2 by Low-Energy Electrons. *Phys. Rev. A* 14:615-629 (1976).

82. D. Rapp and P. Englander-Golden. Total Cross Sections for Ionization and Attachment in Gases by Electron Impact: I Positive Ionization. *J. Chem. Phys.* 43:1464-1479 (1965).

83. S. R. Hunter and L. G. Christophorou. Ref. 2, Vol. 2, p. 202.

84. T. F. O'Malley. Electron Diffusion and the Einstein Relation in High-Density Gases. *Phys. Lett.* 95A:32-34 (1983).

85. V. M. Atrazhev and I. T. Yakubov. The Electron Drift Velocity in Dense Gases. *J. Phys. D* 10:2155-2163 (1977).
Electron Mobility in Liquids and Dense Gases. *High Temp.* 18:966-985 (1980).

86. G. L. Braglia and V. Dallacasa. Theory of the Density Dependence of Electron Drift Velocity in Gases. *Phys. Rev. A* 18:711-717 (1978); Theory of Electron Mobility in Dense Gases, *Phys. Rev. A* 26:902-914 (1982).

87. M. H. Cohen and J. Lekner. Theory of Hot Electrons in Gases, Liquids, and Solids. *Phys. Rev.* 158:305-309 (1967).
J. Lekner. Mobility Maxima in the Rare-Gas Liquids. *Phys. Lett.* A27:341-348 (1968).
S. Basak and M. H. Cohen. Deformation-Potential Theory for the Mobility of Excess Electrons in Liquid Argon. *Phys. Rev. B* 20:3404-3414 (1979).
H. T. Davis, L. D. Schmidt, and R. M. Minday. Kinetic Theory of Excess Electrons in Polyatomic Gases, Liquids, and Solids. *Phys. Rev. A* 3:1027-1037 (1971).

88. T. F. O'Malley. Multiple Scattering Effect on Electron Mobilities in Dense Gases. *J. Phys. B* 13:1491-1504 (1980).

89. L. G. Christophorou. Mobilities of Slow Electrons in Low- and High-Pressure Gases and Liquids. *Intern. J. Radiat. Phys. Chem.* 7:205-221 (1975).

90. H. Lehning. Resonance Capture of Very Slow Electrons in CO_2. *Phys. Lett.* 28A:103-104 (1968).

91. Th. Aschwanden. *In Gaseous Dielectrics III.* L. G. Christophorou, ed., p. 32, Pergamon Press, New York (1982).

92. L. G. Christophorou, J. G. Carter, and D. V. Maxey. Electron Motion in High-Pressure Polar Gases: NH_3. *J. Chem. Phys.* 76:2653-2661 (1982).

93. P. Krebs and M. Heintze. Migration of Excess Electrons in High Density Supercritical Ammonia. *J. Chem. Phys.* 76:5484-5492 (1982).
P. Krebs. Localization of Excess Electrons in Dense Polar Vapors. *J. Phys. Chem.* 88:3702-3709 (1984).

94. V. V. Dmitrenko, A. S. Romanyuk, S. I. Suchkov and Z. M. Uteshev. Electron Mobility in Dense Xenon Gas. *Sov. Phys. Tech. Phys.* 28:1440-1444 (1983).

95. G. R. Freeman. In *Electron and Ion Swarms,* L. G. Christophorou, ed., p. 93, Pergamon Press, New York (1981).

96. A. G. Robertson. Drift Velocities of Low-Energy Electrons in Argon at 293 and 90 K. *Aust. J. Phys.* 30:39-49 (1977).

97. S. R. Hunter, J. G. Carter, and L. G. Christophorou. Low-Energy Electron Drift and Scattering in Krypton and Xenon. *Phys. Rev.* A38:5539-5551 (1988).

98. L. S. Miller, S. Howe, and W. E. Spear. Charge Transport in Solid and Liquid Ar, Kr, and Xe. *Phys. Rev.* 166:871-878 (1968).

99. Y. Sakai, S. Nakamura, and H. Tagashira. Drift Velocity of Hot Electrons in Liquid Ar, Kr, and Xe. *IEEE Trans. Electr. Insul.* EI-20:133-137 (1985).

100. L. G. Christophorou and D. L. McCorkle. Experimental Evidence for the Existence of a Ramsauer-Townsend Minimum in Liquid CH_4 and Liquid Ar (Kr and Xe). *Chem. Phys. Lett.* 42:533-539 (1976).

101. W. F. Schmidt. In Ref. 4, p. 273.

102. H. Faidas, L. G. Christophorou, and D. L. McCorkle. Electron Transport in Fast Dielectric Liquids at High Applied Electric Fields. *Proceedings 10th Intern. Conf. on Conduction and Breakdown in Dielectric Liquids,* Grenoble, France, September 10-14 (1990).
Drift Velocities of Excess Electrons in 2,2,4,4-Tetramethylpentane and Tetramethylsilane: A Fast Drift Technique. *Chem. Phys. Lett.* 163:495-498 (1989).

103. C. Brassard. Liquid Ionization Detectors. *Nucl. Instr. Meth.* 162:29-47 (1979).
J. Engler and H. Keim. A Liquid Ionization Chamber Using Tetramethylsilane. *Nucl. Instr. Meth. Phys. Res.* 223:47-51 (1984).

104. M. G. Albrow, et al. Performance of a Uranium/Tetramethylpentane Electromagnetic Calorimeter. *Nucl. Instr. Meth.* A265:303-318 (1988).

105. L. G. Christophorou and H. Faidas. Dielectric Liquids for Possible Use in Pulsed Power Switches. *Appl. Phys. Lett.* 55:948-950 (1989).

106. J. E. Demuth, D. Schmeisser, and Ph. Avouris. Resonance Scattering of Electrons from N_2, CO, O_2, and H_2 Adsorbed on a Silver Surface. *Phys. Rev. Lett.* 47:1166-1169 (1981).
L. Sanche and M. Michaud. Resonance-Enhanced Vibrational Excitation in Electron Scattering from O_2 Multilayer Films. *Phys. Rev. Lett.* 47:1008-1011 (1981).
Vibrational Excitation Via Shape Resonances in Electron Scattering from N_2 Multilayer Films. *Chem. Phys. Lett.* 84:497-500 (1981).
L. Sanche. Investigation of Ultra-Fast Events in Radiation Chemistry with Low-Energy Electrons. *Radiat. Phys. Chem.* 34:15-33 (1989).
Low-Energy Electron Scattering from Molecules on Surfaces. *J. Phys. B* 23:1597-1624 (1990).

107. R. E. Goans and L. G. Christophorou. Attachment of Slow (< 1 eV) Electrons to O_2 in Very High Pressures of Nitrogen, Ethylene, and Ethane. *J. Chem. Phys.* 60:1036-1045 (1974).

108. D. L. McCorkle, L. G. Christophorou, and V. E. Anderson. Low-Energy (< 1 eV) Electron Attachment to Molecules at Very High Gas Densities: O_2. *J. Phys. B* 5:1211-1220 (1972).

109. L. G. Christophorou. Intermediate Phase Studies for Understanding Radiation Interaction in Condensed Media: The Electron Attachment Process. *J. Phys. Chem.* 76:3730-3734 (1972).
Electron Attachment to Molecules in Dense Gases ("Quasi-Liquids"). *Chem. Rev.* 76:409-423 (1976).

110. T. D. Märk and A. W. Castleman, Jr. *In Advances in Atomic and Molecular Physics,* D. R. Bates and B. Bederson, eds., 20:65-172. Academic Press, Orlando, Florida (1985).
A. W. Castleman, Jr., and R. G. Keesee. Gas-Phase Clusters: Spanning the States of Matter. *Science* 241:36-42 (1988).
R. G. Keesee and A. W. Castleman, Jr. *In Atomic and Molecular Clusters,* E. R. Bernstein, ed., pp. 507-550. Elsevier Scientific Publishing Company, Amsterdam (1990).
R. N. Compton and J. N. Bardsley. In Electron-Molecule Collisions, I. Shimamura and K. Takayanagi, eds., pp. 275-349. Plenum Press, New York (1984).

111. L. G. Christophorou. The Lifetimes of Metastable Negative Ions. *Adv. Electron. Electron Phys.* 46:55-129 (1978).

112. H.S.W. Massey. *Negative Ions.* Cambridge University Press, Cambridge (1976).

113. B. M. Smirnov. *Negative Ions.* McGraw-Hill, New York (1982).

114. L. G. Christophorou and S. R. Hunter. Ref. 2, Vol. 2, Chapt. 5.

115. L. G. Christophorou, D. L. McCorkle, D. V. Maxey, and J. G. Carter. Fast Gas Mixtures for Gas-Filled Particle Detectors. *Nucl. Instr. Meth.* 163:141-149 (1979).
M. K. Kopp, K. H. Valentine, L. G. Christophorou, and J. G. Carter. New Gas Mixture Improves Performance of ^{3}H Neutron Counters. *Nucl. Instr. Meth.* 201:395-401 (1982).
L. G. Christophorou, H. Faidas and D. L. McCorkle. In *Nonequilibrium Effects in Ion and Electron Transport*, J. W. Gallagher, D. F. Hudson, E. E. Kunhardt, and R. J. Van Brunt, eds., pp. 313-328, Plenum Press, New York (1990).
116. A. Zlatkis and C. F. Poole, eds. Electron Capture: Theory and Practice in Chromatography. *Journal of Chromatography Library,* Vol. 20. Elsevier Scientific Publishing Company, Amsterdam (1981).
L. G. Christophorou, D. L. McCorkle, and I. Sauers. Tagging Materials for Detection of Explosives. *Analytical Chimica Acta* 135:179-192 (1982).
117. L. G. Christophorou and D. W. Bouldin, eds. *Gaseous Dielectrics V.* Pergamon Press, New York (1987).
L. G. Christophorou and M. O. Pace, eds., Gaseous Dielectrics IV, Pergamon Press, New York (1984).
L. G. Christophorou and L. Pinnaduwage. Basic Physics of Gaseous Dielectrics. *IEEE Trans. Electr. Insul.* 25:55-74 (1990).
118. A. Guenther, M. Kristiansen, and T. Martin, eds. *Opening Switches.* Plenum Press, New York (1987).
L. G. Christophorou. Electron Collisions in Gas Switches. In *Nonequilibrium Processes in Partially Ionized Gases*, M. Capitelli and J. N. Bardsley, eds. Plenum Press, New York (1990).
S. R. Hunter, J. G. Carter, and L. G. Christophorou. Electron Transport Studies of Gas Mixtures for Use in e-Beam Controlled Diffuse Discharge Switches. *J. Appl. Phys.* 58:3001-3015 (1985).
119. G. E. Caledonia. A Survey of the Gas-Phase Negative Ion Kinetics of Inorganic Molecules. Electron Attachment Reactions. *Chem. Rev.* 75:333-351 (1975).
120. L. G. Christophorou. Interactions of O_2 with Slow Electrons. *Radiat. Phys. Chem.* 12:19-34 (1978).
121. Y. Hatano and H. Shimamori. In *Electron and Ion Swarms*, L. G. Christophorou, ed., pp. 103-116. Pergamon Press, New York (1981).
122. H. Shimamori and Y. Hatano. Mechanism of Thermal Electron Attachment in O_2-N_2 Mixtures. *Chem. Phys.* 12:439-445 (1976).
H. Shimamori and R. W. Fessenden. Thermal Electron Attachment to Oxygen and Van der Waals Molecules Containing Oxygen. *J. Chem. Phys.* 74:453-466 (1981).
123. S. R. Hunter, L. G. Christophorou, D. L. McCorkle, I. Sauers, H. W. Ellis, and D. R. James. Anomalous Electron Attachment Properties of Perfluoropropylene (1-C_3F_6) and Their Effect on the Breakdown Strength of This Gas. *J. Phys. D* 16:573-580 (1982).
124. G. Bakale, U. Sowada, and W. F. Schmidt. Effect of an Electric Field on Electron Attachment to SF_6, N_2O, and O_2 in Liquid Argon and Xenon. *J. Phys. Chem.* 80:2556-2559 (1976).
125. G. Bakale and W. F. Schmidt. Effect of an Electric Field on Electron Attachment to SF_6 in Liquid Ethane and Propane. *Z. Naturforsch.* 36a:802-806 (1981).
126. S. R. Hunter and L. G. Christophorou. *Basic Studies of Gases for Fast Switches.* Oak Ridge National Laboratory Report ORNL/TM-10844, August (1988).
127. T. F. O'Malley. Calculation of Dissociative Attachment in Hot O_2. *Phys. Rev.* 155:59-63 (1967).
J. N. Bardsley and J. M. Wadehra. Dissociative Attachment and Vibrational Excitation in Low-Energy Collisions of Electrons with H_2 and D_2. *Phys. Rev.* 20:1398-1405 (1979).
Dissociation Attachment in HCl, DCl, and F_2. *J. Chem. Phys.* 78:7227-7234 (1983).
128. S. M. Spyrou and L. G. Christophorou. Effect of Temperature on the Dissociative Electron Attachment to $CClF_3$ and C_2F_6. *J. Chem. Phys.* 82:2620-2629 (1985).
129. E. Alge, N. G. Adams, and D. Smith. Rate Coefficients for the Attachment Reactions of Electrons with c-C_7F_{14}, CH_3Br, CF_3Br, CH_2Br_2, and CH_3I Determined Between 200 and 600 K Using the FALP Technique. *J. Phys. B* 17:3827-3833 (1984).
130. S. M. Spyrou and L. G. Christophorou. Effect of Temperature of the Dissociative and Nondissociative Electron Attachment to C_3F_8. *J. Chem. Phys.* 83:2829-2835 (1985).
P. G. Datskos and L. G. Christophorou. Variation with Temperature of the Electron Attachment to SO_2F_2. *J. Chem. Phys.* 90:2626-2630 (1989).
P. G. Datskos, L. G. Christophorou, and J. G. Carter. Temperature-Enhanced Electron Attachment to CH_3Cl. *Chem. Phys. Lett.* 168:324-329 (1990).
P. J. Chantry and C. L. Chen. Ionization and Temperature Dependent Attachment Cross Section Measurements in C_3F_8 and C_2H_3Cl. *J. Chem. Phys.* 90:2585-2592 (1989).

131. P. G. Datskos and L. G. Christophorou. Variation of Electron Attachment to n-C_4F_{10} with Temperature. *J. Chem. Phys.* 86:1982-1990 (1987).
132. S. M. Spyrou and L. G. Christophorou. Effect of Temperature on Nondissociative Electron Attachment to Perfluorobenzene. *J. Chem. Phys.* 82:1048-1049 (1985).
133. A. A. Christodoulides, L. G. Christophorou, and D. L. McCorkle. Effect of Temperature on Low-Energy (< 1 eV) Electron Attachment to Perfluorocyclobutane (c-C_4F_8). *Chem. Phys. Lett.* 139:350-356 (1987).
134. C. L. Chen and P. J. Chantry. Photon-Enhanced Dissociative Electron Attachment in SF_6 and Its Isotopic Selectivity. *J. Chem. Phys.* 71:3897-3907 (1979).
    I. M. Beterov and N. V. Fateyev. Laser Optogalvanic Effects Caused by Formation of Negative Ions. *J. de Phys. (Paris)*, Colloq., C7:447 (1983).
    M. W. McGeoch and R. E. Schlier. Dissociative Attachment in Optically Pumped Lithium Molecules. *Phys. Rev. A* 33:1708-1717 (1986).
135. R. I. Hall and S. Trajmar. Scattering of 4.5 eV Electrons by Ground ($X^3\Sigma\bar{g}$) State and Metastable ($a^1\Delta_g$) Oxygen Molecules. *J. Phys. B* 8:L293-L296 (1975).
136. P. D. Burrow. Dissociative Attachment to $O_2(a^1\Delta_g)$ State. *J. Chem. Phys.* 59:4922-4931 (1973).
    D. S. Belic and R. I. Hall. Dissociative Electron Attachment to Metastable Oxygen ($a^1\Delta_g$). *J. Phys. B* 14:365-373 (1981).
137. L. G. Christophorou, S. R. Hunter, L. A. Pinnaduwage, J. G. Carter, A. A. Christodoulides, and S. M. Spyrou. Optically-Enhanced Electron Attachment. *Phys. Rev. Lett.* 58:1316-1319 (1987).
138. L. A. Pinnaduwage, L. G. Christophorou, and A. P. Bitouni. Enhanced Electron Attachment to Superexcited States of Saturated Tertiary Amines. *J. Chem. Phys.* 95:274-287 (1991).
139. R. S. Mock and E. P. Grimsrud. Optically-Enhanced Electron Capture by p-Benzoquinone and Its Methylated Derivatives. *J. Phys. Chem.* 94:3550-3553 (1990).

Discussion

Weinstein: Is the transfer of the resonant energy of the interaction of the electron with N_2O in the liquid to lower energies due to the solvation of the transition state?

Christophorou: Polarization energy around the transient anion.

Weinstein: Correct me if I am wrong, but that is likely to be small compared to the solvation energy of the electron which you are now losing, so in balance, the energy doesn't seem to want to go to lower energies.

Christophorou: I haven't lost an electron. The electron attaches to the molecule. The electron that attaches to the N_2O molecule is in liquid argon and is quasifree; the V_0 of liquid argon is -0.2 eV, and the negative ion state is lowered by an amount equal to the polarization energy of the transient N_2O^{-*}.

von Sonntag: Dr Weinstein is speaking about pure N_2O, not the aqueous solution of N_2O.

Christophorou: In the experiments I have described, N_2O is in argon.

Weinstein: Oh, I see.

von Sonntag: You were mentioning that electrons with energies slightly above thermal energy may cause DNA damage. I don't see exactly how, but in this context I would like to ask you a question. The ESR-experts tell us that in DNA irradiated at low temperatures only the pyrimidine electron adducts can be observed, and that no purine electron adducts can be detected. At least two possibilities could account for this. Either there is a rapid electron transfer from the purine electron adducts to the pyrimidines, or the pyrimidines have a higher cross section for scavenging thermal, but not yet solvated, electrons. Could you comment on this?

Christophorou: No, because I don't know much about these chemicals in those environments.

Kopelman: I think the answer to this would be to measure them separately in purines and separately in pyrimidines.

Ward: This would not answer the question; when the compounds are irradiated separately there's nowhere else for the electrons to go.

Kopelman: Yes, but then you know that it's transfer or not transfer.

von Sonntag: We know that the hydrated electron reacts equally well with purines and pyrimidines. The question is whether there is now a substantial preference for the reaction with the pyrimidines in the case of the thermalized but dry electron. Can this question be answered?

Kopelman: I was going to ask if there are any measurements for critical liquids or gases. Are there any measurements at the critical point where the gases and the liquid properties should come together?

Christophorou: Yes, there are. There isn't any discontinuity, if that is what you are asking. You can go smoothly from one into the other without any particular discontinuities, at least as far as electron mobilities, ionization threshold energies, and V_0 values are concerned.

Kasha: I have been hearing about electron transport in gases versus liquids. You gave us differences in rates, but do you have a mechanism? How is the electron carried in a gas? Is it the mean free path and the velocity of the gas molecule carrying the electron?

Christophorou: Indeed, the changes in the rates and the electron transport properties I mentioned relate to the basic mechanisms involved when electrons interact with individual molecules or with the bulk medium. You have to keep in mind that these experiments are performed under an applied electric field. When the field is small, the electrons are in thermal equilibrium with the gas molecules. When the field is high, however, it is possible for the electrons to have energies in excess of thermal and thus to lead to reactions which require energy. This, of course, depends not just on the applied electric field, but also on the nature of the medium. For low-pressure gases, electrons can have steady-state electron energy distributions well in excess of thermal rather easily, but for liquids this is only possible for media with $V_0 < 0$ eV. Is this what you are asking?

Kasha: Yes, but that is my very question. That seems to treat the electrons as if there is no carrier gas, as if there is no gas interacting with the electron.

Christophorou: No, there is; the electron undergoes a large number of collisions with the gas as it drifts.

Kasha: How do we know that the electron doesn't attach itself and migrate under the field with the attached gas?

Christophorou: That process indeed happens and we know of it—and study it—by its effect on the electron transport coefficients which we measure as a function of the electric field, medium density, temperature, etc. For example, if electrons are captured and released quickly (i.e., many times during their transit time) then this process increases their drift time. If the electrons are captured to form long-lived or stable negative ions, then they drift as anions and their mobility is very small (ionic). Such processes occur in many systems (e.g., polar media, CO_2)--especially at high densities due to various mechanisms which include electron-dipole capture, resonance capture, electron attachment to transient species (including clusters) or potential fluctuations in the bulk medium, and so on. As the density of the medium increases, multiple scattering processes affect the electron transport behavior.

Weinstein: Is it correct that the major differences you see between liquid and gas--and especially dilute gases--is for substances with molecules that are more polar?

Christophorou: Not necessarily so, although for those that are polar the changes are expected to be large.

Weinstein: So, therefore, with water for example, you expect a very large difference.

Christophorou: Yes, also water is notorious for other reasons. For example, water molecules readily cluster around the charged species which are present in the system. Electrons in polar media will be solvated quickly, and they will then be localized; hence, their mobility will be low (ionic) and their energies thermal.

Weinstein: So that means that the application of those relationships that we have seen from the gas phase to liquid to water aqueous solution provides surprises.

Christophorou: You have to approach them with extreme caution. Usually the effect of the medium on a basic physical process or phenomenon is large whenever charged species are involved (electrons, positive or negative ions, quasi-charge separated states). The analogy of basic processes in gases to those in liquids, which I have indicated, mostly applies to those liquids with conduction bands ($V_o < 0$ eV; e.g., heavier rare-gas liquids, spherical hydrocarbon liquids) where the excess electrons are quasifree. If we go into systems such as water, ammonia, or other polar dense fluids where electrons are localized, then the effects of the medium and its polarization are very profound (e.g., electron mobilities are ionic, reaction rates of electrons with scavengers are much lower than in media where electrons are quasifree). The comparisons between the behavior of electrons in these liquids and their low-pressure vapors are not so evident.

Weinstein: In addition, these are homogeneous. Now if one adds the type of strong local fields that one would have, say, when you have charged molecules that come back to DNA and counter ions, then the picture will become even more clear.

Christophorou: That is why I believe that one of the areas we need to look at is the study of these processes in carefully selected polar media.

Varma: You pointed out the advancement in our knowledge acquired from the gas phase and with high pressure-gases. How can we relate that information to what is needed to relate to the polar liquid?

Christophorou: As I have already pointed out, in a polar liquid the electron is localized and, therefore, its mobility is ionic; the electron is slow. Its reactions with substances dissolved in the polar liquid will likely be diffusion controlled (or slower) and would, thus, be expected to increase with the electron mobility in that system. But let me go a step further. If I follow some arguments about the structure of water around DNA correctly, namely that there is a conduction band, then the electron need not be localized in that band. If the electron is not localized it can be quite mobile, and if it is mobile its reactions with molecules (chemicals present) could perhaps be understood by the kind of knowledge in dense gases/non-polar liquids I presented. So we have to know to what kind of a system we are actually referring or are applying this knowledge, before we can proceed. Is there such structured water around DNA?

Varma: We know, in DNA for example, that the water molecules are attached to DNA are going to be the backbone of the structure; the question is, even in that structured water composition, what differences you would expect....

von Sonntag: We also know that it doesn't solvate; this water is not capable of solvation.

Chatterjee: But that is not important if the electrons are solvated in the backbone.

Ward: Can I just take this question and play the devil of a devil's advocate? Clemens has shown quite clearly that slow electrons do no damage to DNA or DNA constituents, so why do we need to know what the electron is doing in the backbone?

Varma: That is a fair question to ask, and I think it should be discussed.

Weinstein: When you say slow electrons you mean thermal electrons?

Ward: Subionization, subionization electrons.

von Sonntag: What do you refer to?

Ward: You said that you looked for breakage; this is a bond where you get no breakage at all.

von Sonntag: You can break the phosphate bond, the tri-methyl-phosphate will be irritated as such, but it would not break the phosphate bond. Let's say that if it is more negatively charged, the effect will be quite negligible. You don't have those disassociated electron attachments which you might expect that this doesn't happen in solution.

Varma: The question of electron mobility that you also showed us in spherical liquids and non-spherical liquids, I believe it was neo-pentane and pentane. Now, if you take that structure, how much difference is it going to make in the polar case and polar liquid case?

Christophorou: In polar liquids it would be very different.

Varma: Is there an experiment that you can do to tell us that?

Christophorou: We know that an electron in a polar medium would be localized, would be ionic in nature. It won't be a quasifree electron; it won't be highly mobile; it won't have energies in excess of thermal; and its reactivity will be limited/determined by its mobility and perhaps be a function of the electric dipole moment of the polar medium.

Varma: That would be smaller then. What would you expect in case of these critical....

Christophorou: The interactions, say, of electrons with molecules dissolved in water would be slower because you are dealing with slower species that you have in nonpolar media with conduction bands where the electrons are more mobile.

Inokuti: I think what he says is about the electrons at the very end, at the low-energy end of the spectrum. But under irradiation conditions, the electrodes have to start at much higher energy and get through. This remark applies to the von Sonntag's observation. You see, what is apparently observed is something about the interaction of the lowest energy electron. It doesn't say anything about electrons above a few eV.

Varma: But, Mitio, when you start out with ionizing radiation with high- energy electrons, that eventually will go down to the energies that Loucas is talking about. The mechanisms we understand up to a point, of the high-energy electrons before they become subexcitation and thermal electrons. You see big differences in the case of neo-pentane and pentane as Loucas points out. Does that magnitude of difference exist in polar liquids where you can't have spherical symmetry?

Inokuti: But all this statement is about the extreme end.

Christophorou: The difference between neopentane and pentane relates, I believe, to the fact that the molecules of the former are spherical and those of the latter are not; the former has a conduction band ($V_0 \approx -0.43$ eV) and the latter has not ($V_0 \approx +0.12$ eV). Thus, the electron is more mobile in neopentane than in pentane. For polar media such differences are unlikely since in those media there are no conduction bands and since the electron-molecule interaction is dominated by the long-range electron-electric dipole interaction.

But let me come to Mitio's comment. Mitio is correct; my discussion refers to electrons which have reached the final stage; they are either in thermal equilibrium

Christophorou

with the medium or—when an applied electric field is imposed—they may (for certain dense media) have a steady-state electron energy distribution in excess of thermal. However, we need to keep in mind that it takes very little time for high-energy electrons to be thermalized in the complex systems we are speaking of. So, it is really a matter of how much time a 2 or a 3 or a 6 eV electron has to interact as such, and certainly the thermalized electron has all the time in the world to do so.

Inokuti: Yes, timing is one thing but availability is another. Which you remember, say, the dissociative attachment of an electron in the gas phase and they occur at certain fixed energy.

Christophorou: But I would say that in a medium such as that the most important dissociative attachment processes are those which lie close to zero.

Zaider: If we take seriously this model for the structure of water surrounding the DNA molecule, the existence of the conduction band--it must exist, I assume. Then you wouldn't have any such electrons that are just... they get to one of those bands and that would be the energy of the electron.

Christophorou: Yes.

Zaider: These are typical gas phase energies that we are talking about. They will not be relevant to electron energies in a conduction band, would they? I want to ask a different question, if I may. You talk about the scattering of electrons off excited molecules. Isn't this the same as multiplying two very small numbers?

Christophorou: In the initial stages of interaction of radiation with matter, the density of excited species may be high, and the generated electrons and excited species are in close proximity to each other.

Zaider: If it is a second order process....

Christophorou: I am talking about a process which has a cross section over one million times larger than for the corresponding ground-state species. That is a pretty big difference.

Ward: I don't think you said anything about this. This is just getting into the literature that one mechanism of sensitization is the electron under global circumstances; it is ionizing the DNA and then it can go back, neutralizing the parental cation, and sensitization occurs by scavenging of that electron. It is a possibility. What concentration of scavengers will be needed to cause this?

Christophorou: The concentration of the scavenger depends, of course, on the magnitude of its electron attachment rate constant which for efficient scavengers and quasifree thermal electrons can be $> 10^{-7}$ cm^3 s^{-1}. The type of sensitization you referred to has been discussed by G. E. Adams and others. The mechanism proposed relies on the existence of quasifree electrons in the system and the ability of strongly electron-attaching chemicals present to capture these electrons and irreversibly change the mobile (reactive) electrons into immobile (less-reactive) anions. In general, electron capture can (i) hinder a reaction which relies on electron availability or, (ii) enhance a reaction that relies on the lack of electrons. Such an electron capture process (leading to parent or to fragment stable anions), by depleting the available electrons in the system, may hinder critical steps in biological action mechanisms. The slow electron can be a really disruptive creature for most organic structures.

von Sonntag: No, that is not enough energy for doing this.

Christophorou: But you yourself this morning expressed doubt about statements as to how the energy inside a molecule is dissipated; is it dissipated as radiation or does it provide the energy required for a particular reaction, in this case dissociation?

von Sonntag: But this dissociation requires at least one strength, and one strength is about four electron volts.

Christophorou: Except when there are charged particles involved, in which case you are capitalizing on the solvation energy of the medium and—in cases of negative ion formation—also on the electron affinity of the electron attaching species; this will decrease the energy required.

von Sonntag: You are talking about electrons, electrons causing this, all right. So you at least require four electron volts of energy in order to break the single strand....

Weinstein: It doesn't matter, the activation energy will still be very high, so the probability of this occurring will be one, even if thermodynamically; eventually it will be stabilized, the activation energy, because the breaking of the bond will be recognized. The probability is low, even though thermodynamically it would be more feasible. The probability for it depends on the activation energy.

Varma: I think you are saying that if the one electron is at 40 eV energy, and if the molecule dissociation energy is equal to .2 eV, you get 20.

Weinstein: Loucas is saying that it is not possible that two 20 eV electrons could do the same change as the 40 eV electron. If you reduced the probability.

Varma: Is there any experiment that will answer this question?

von Sonntag: No, in DNA we know that DNA strand breakage is not induced by electrons, for example. The problem is that in the systems that we are studying, the time scale of electrons from this subexcitation electrons to thermal electrons, the time is so short that we only have a chance of observing such a reaction very next neighborhood to DNA. If we study such reactions of sorbate electrons interacting with DNA in aqueous solution, all the electrons are already solvated before they get a chance of having enough energy to interact with DNA. So in this case, we just do not know whether there is a dissociative electron attachment to the phosphate backbone and can cause a disfunction. This we do not know.

Christophorou: What is the current thinking? Is there or is there not a conduction band around DNA?

Ward: This work was done by John Warman and colleagues [*Biopolymers* 22:807-810 (1983)], who found conductivity in the water shell around the DNA. They found similar effects using collagen or momomeric adenosine monophosphate.

Energy Transfer, Charge Transfer, and Proton Transfer in Molecular Composite Systems

Michael Kasha[a]

Abstract

Models for the fundamental mechanisms of excitation energy transfer, including cases involving singlet oxygen states, twisting-intramolecular-charge-transfer (TICT) states, and intramolecular proton transfer, are described in terms of elementary concepts and energy diagrams. Three limiting cases of energy transfer are distinguished, Davydov free excitons (Simpson and Peterson strong coupling) and localized excitons (weak coupling), and the Förster mechanism of vibrational-relaxation energy transfer. The prominent rôle of the singlet molecular oxygen states is described, together with the rôle of simultaneous transitions for molecular oxygen pairs. The origin of sudden polarization via the TICT-state potential is discussed and the generality of this phenomenon emphasized. The intramolecular proton-transfer phenomenon is outlined, and its role in molecular excitation transient phenomena is described. The complex interaction of all of these excitation mechanisms in determining photochemical and radiation chemical pathways is suggested.

Introduction

The fundamental mechanisms underlying energy transfer, charge transfer (or electron transfer), and proton transfer are relatively well understood, and are described by highly developed and sophisticated theoretical treatments. Currently many applications of these theories are appearing, coupled to novel examples of general chemical and biological interest. In this chapter a skeletal outline is given of the basic qualitative ideas of these theories as an index to the principal literature and a sampling of current examples. Mathematical details are left to the references, and a review of the contributions of this laboratory to these topics is integrated with the general literature.

Energy Transfer

In this presentation *excitation energy transfer* is described, i.e., transfer of energy of electronic excitation from one molecular unit to another. Such energy transfers are easily accessible to experimental study, especially in this age of pulsed laser excitation technology.

(a) Institute of Molecular Biophysics and Department of Chemistry, Florida State University, Tallahassee, Florida

Physical and Chemical Mechanisms in Molecular Radiation Biology
Edited by W.A. Glass and M.N. Varma, Plenum Press, New York, 1991

231

The Molecular Exciton Model

Tight-Binding Approximation. The *molecular exciton* is defined as a "particle" of excitation which (in the time-dependent picture) travels through a structured molecular array, without electron migration. In the stationary-state picture (time independent) the *molecular exciton states* are defined as coherently excited delocalized states of an entire molecular array.[1]

The molecular exciton model offers a mechanism for describing the rapid delocalization of excitation energy among identical molecular units. The general need for such a mechanism was first recognized by J. I. Frenkel,[2] who sought a model to explain the rapid delocalization of energy when an energetic particle impacted upon a solid surface without immediately melting a hole at the point of impact. J. Franck and E. Teller presented a general discussion of types of energy migration in molecular systems, with some specific applications, and discussed qualitatively the two limiting cases involving electronic and vibrational coupling.[3]

A. S. Davydov[4] developed the concept of delocalization of excitation into a quantum mechanical form, in which calculable molecular parameters were introduced. Although Davydov was concerned mainly with interactions in molecular crystals, he extended his methodology to the para-polyphenyl molecular case as a composite molecule, i.e., phenyl rings connected (but not conjugated) in the 1,4-position. He assumed a tight-binding approximation for both molecular crystals and the linked phenyls, i.e., no electron migration from one molecular unit to another (thus, no conjugation between phenyl rings). In contrast, in the weak-binding Wannier atomic exciton, as observed in atomic and simple ionic crystal lattices, the electron and hole may separate during their migration through a lattice. The electron is not tightly bound in that case.[5]

The Davydov Molecular Exciton Model has now been used to treat many molecular systems[6-10] assumed to conform to the tight-bonding approximations of electron localization: H-bonded dimers and higher-order H-bonded aggregates; van der Waals dimers and higher aggregates, up to high-molecular-weight van der Waals polymers (dyes); co-valently linked molecular composite molecules such as diphenylmethane and triphenylmethane (with conjugation-blocking methylene groups); van der Waals molecular lamellae of various geometries; helical polypeptides and helical polynucleotides; and other systems in which strongly-light-absorbing molecular sub-units are electronically separated. In the tight-binding approximation, the molecular entities preserve their individuality (or identity), in Davydov's language.[4]

Dipole-Dipole Coupling Approximation. In the quantum-mechanical treatment of a molecule, the Hamiltonian operator representing the energy of the system contains kinetic energy terms for each particle of the molecule, and potential energy terms for each pair-interaction of the particles of the system. The potential energy terms in the energy operator are coulombic repulsion or attraction terms, and are by definition 1/r dependent where r is the distance between each pair of particles.

When two molecules are brought near each other to form a molecular composite system, the extra terms needed in the Hamiltonian operator are the extra coulombic terms for inter-molecular particle repulsion and attraction. Formulating the composite molecule system interactions then requires the solution of what could be very complicated intermolecular integrals for the composite super-molecule if the coulombic potential were used as an operator. In the molecular exciton model, where hundreds or thousands of molecules can form a composite structured aggregate, impossible mathematical complexity would arise. Fortunately, a mathematical simplification is available which reduces the composite molecule problem to a product of single molecule solutions.

The classical coulombic interaction between two electrons can be expanded in an infinite series as a point-multipole expansion, in which the 1/r coulombic dependence is preserved for each term. The quantum mechanical calculation of energy of interaction then uses the classical potential energy as an operator.

Then, for a strongly *electric-dipole-allowed* electronic transition the exact coulombic potential operator is approximated by the dipole-dipole term as the leading term in the point-multipole expansion. In cartesian form each dipole-dipole term has the form $q_i q_j / r_{ij}^3$, i.e., of dimension 1/r as required, in the operator for the potential energy. The q's are assimilated in the transition moments for each molecule.

Thus, it is considered that two electronically excited molecules near each other interact by the long range $(1/r^3)$ dipole-dipole interaction, and that the transition moment electric dipoles of one molecule interact with the transition dipoles of the nearby molecule. In this manner the excited states of the separate molecular units are coupled.

Parameters of the Molecular Exciton Model: Dimers. For each molecular like-pair, the *interaction energy* for a pair of molecules in their like excited states proves to be proportional to the product of the electronic transition moments (absolute value of transition moment squared), inversely proportional to the cube of the molecular center distance, and directly dependent on the dipole-dipole orientation geometry.[4,6-10] The unusual feature in molecular exciton theory is the finding that the exciton energy splitting (bandwidth) is proportional to the *intensity* of a molecular absorption band.

The geometrical factor is one of the most valuable dividends of the molecular exciton method. Consider the three geometric cases of molecular pairing shown in Fig. 1.

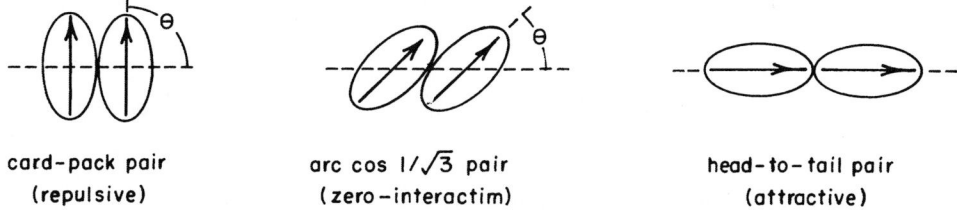

card-pack pair   arc cos 1/√3 pair   head-to-tail pair
(repulsive)   (zero-interactim)   (attractive)

Figure 1. Molecular Pair (In-Phase) Dipole-Dipole Interactions

In a molecular exciton problem, there are two aspects which may be described by classical vector diagram analogs. The *energy* of dipole-dipole interaction is pictured as a classical dipole-dipole vector interaction (attraction or repulsion). The *selection rule for light absorption* (electric dipole radiation transition moment) to an exciton state is given by the vector sum of the individual molecule transition moments.

The card-pack pair (sandwich pair) obviously represents a repulsive interaction, and the head-to-tail pair an attractive dipole-dipole interaction. These have finite transition moment sums, and both are allowed for electric dipole transitions (arrows to allowed excited states in Fig. 2). Note that the other quantum-mechanically required dipole-dipole interaction pairs (Fig. 3) result in electric-dipole forbidden (net vector sum equals zero) transitions to the molecular exciton states (dashed curve in Fig. 2).

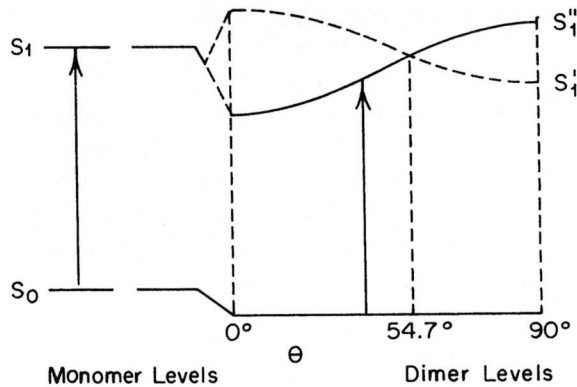

Figure 2. Energy Level Diagram for Co-Planar Inclined Transition Dipoles

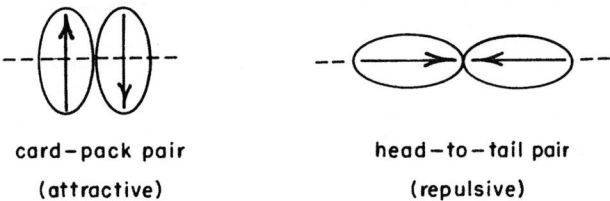

Figure 3. Molecular Pair (Out-of-Phase) Dipole-Dipole Interactions

If the transition dipoles of Fig. 1 are inclined so that the transition moments make an angle $\theta = \mathrm{arc\ cos\ } 1\sqrt{3} = 54.7°$, the geometry is reached for which the dipole-dipole interaction is zero. In Fig. 2 this corresponds to a crossing of the allowed and forbidden molecular exciton states. So here we have a remarkable physical result: two very strong electric dipoles adjacent to each other, and yet having zero net repulsion or attraction!

The idea of zero-interaction arrays of composite molecule systems is important, and as described later, leads to a model for antenna chlorophyll in the photosynthetic unit.

Molecular Excitons in Extended Molecular Arrays. In the last section, the simple description of the molecular exciton model for a molecular pair or dimer was given in a *stationary-state model,* i.e., a coherently excited pair. It is possible to describe the same system as an excitation hopping back and forth very rapidly in a *time-dependent model.* Such a model would suggest that the excitation resides half of the time on one molecule or the other. It is fascinating to find that the vibrational frequency change upon excitation for a covalently-bound "dimer," e.g., diphenylmethane[11] does indeed show *half* the frequency change per phenyl as is observed in an isolated benzene ring upon excitation.

Suppose that a very large structured array of dye molecules bound by van der Waals forces was excited. In a stationary state description, an N-fold exciton band of considerable width (~ 2000 cm^{-1}) could ensue, and only one or a few allowed exciton states could be observed (Davydov rule). Suppose we had a simple stacking (Fig. 4) corresponding to two molecules per unit cell. This would result[6,7,9] in an exciton band

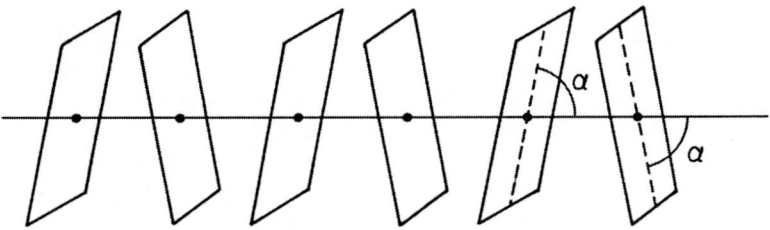

Figure 4. Linear Chain Polymer of Alternate Translational Structure

in which the *bottom* and *top* of the exciton band could be reached by light absorption, but all exciton states in between would be forbidden and would not appear (Fig. 5) (two molecules per unit cell).

What would we see as a spectrum of such an array? Consider the time-dependent model, and that the jumping time is 10^{-15} sec, a femtosecond (vide infra). This is realistic for a dye molecule aggregate in which the oscillator strength is about 1 or greater. If a molecule is excited, excitation would jump to a neighbor before the ground-state geometry could respond, because vibrational relaxation is in the 10^{-12} or picosecond range, a thousand times slower. So the initially excited molecule had to pass on its excitation before it could readjust its geometry. The molecular exciton travels down the chain of thousands of molecules (and reflects back), and no molecule "felt" the excitation vibrationally through a skeletal distortion. Such a case is the *free-exciton case* of Davydov.

In dye spectroscopy, if one salts out a pseudoisocyanine dye with sodium chloride from a strong solution of dye in water, van der Waals polymers of molecular weight up to 10 million are formed (determined by rotatory-diffusion coefficient). The normal single molecule absorption band appears suppressed, and is replaced by an astonishingly narrow "J" band[12-14] which appears on the long-wavelength side of the normal absorption band. The J-band of cyanine-dye polymers is a pure molecular electronic band without vibrational band-width. The Franck and Teller[3] description of the band narrowing is the one given here. Further discussion of the possible electronic structure of this narrow band is given later, in the section on helical polymers.

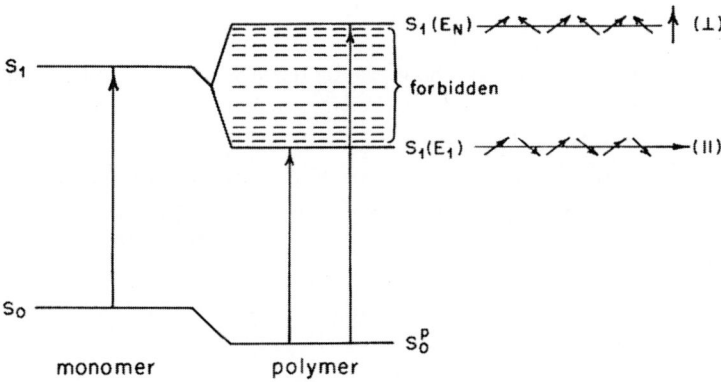

Figure 5. Exciton Band Structure for Alternate Linear Polymer

Energy Transfer Mechanisms - The Three Different Coupling Cases. A molecule which is electronically excited, coupled to its neighbors by strong dipole-dipole interaction, could transfer its excitation before vibrational relaxation could occur, as discussed in the previous section. However, at the other extreme if the intermolecular interaction is weak, complete vibrational relaxation could take place before the electronic excitation energy was transferred. Franck and Teller[3] gave a qualitative description of the consequence of the two rate inequalities. Davydov[4] used these considerations in his classification of *free excitons* and *localized excitons* for the two limiting cases.

The quantum mechanical criteria for the two coupling cases were deduced by Simpson and Peterson,[15] who labelled them as *strong-coupling* and *weak-coupling* cases, corresponding to the two Davydov molecular exciton limits.

The most detailed quantum mechanical treatment of these two energy limiting cases was given by Förster,[16,17] who compared them with his own model of *vibrational relaxation resonance transfer.*[18] Förster diagrammed the dependence of rates of transfer for the three different models against the energy of interaction.[16] In Fig. 6 this diagram is reproduced with specific labelling. The three coupling cases differ dramatically in molecular parameters:

- *Strong-coupling exciton* rate of transfer depends directly on the energy of dipole-dipole interaction between molecules (inverse-cube distance dependence).

- *Weak-coupling exciton* rate of transfer depends directly on the energy of dipole-dipole interaction between molecules (inverse-cube distance dependence) and on the square of the Franck-Condon integral over the vibronic levels of the excited and non-excited molecule summed over respective populations.

- *Vibrational-relaxation resonance transfer* (very weak coupling)[16-18] has a rate dependent on the square of the energy of dipole-dipole interaction (inverse sixth power dependence on distance) between molecules, and the fourth power of the Franck-Condon integral sums. The quantum-mechanical formulation of this case[16-17] resembles the interaction of continua, which can be used also as a name for Förster transfer.[18]

The third case Förster considered is that which he originated,[19,20] known variously as "Förster transfer," "resonance-transfer," "resonance dipole-dipole transfer," etc. All of these energy transfer cases involve "resonance transfer" and may involve "dipole-dipole transfer" (if not higher multipoles). The phenomenological formulation by Förster,[19,20] in which the rate of transfer is dependent on overlap of the fluorescence vibronic envelope of the donor molecule with the absorption vibronic envelope of the acceptor molecule, suggests the nomenclature for vibrational relaxation resonance transfer.

In Fig. 6 it is thus evident that the two exciton theory plots show parallel dependence of rate of transfer of energy of dipole-dipole interaction (with a Franck-Condon retardation for the vibrational localization case); whereas the Förster transfer or vibrational-relaxation resonance transfer has a different slope, and is much slowed down by the Franck-Condon fourth-power factor.

We might conclude that the strong-coupling exciton case could dominate energy transfer phenomena, because the rates can be very high (up to 10^{15} sec^{-1}) and the coupling is very long range (inverse third). However, exact quantum mechanical resonance is needed. This may prevail in a structured molecular aggregate, but a colliding pair of molecules may lose exact resonance owing to environmental perturbations. In such a case, the interaction could collapse to a much weaker case.

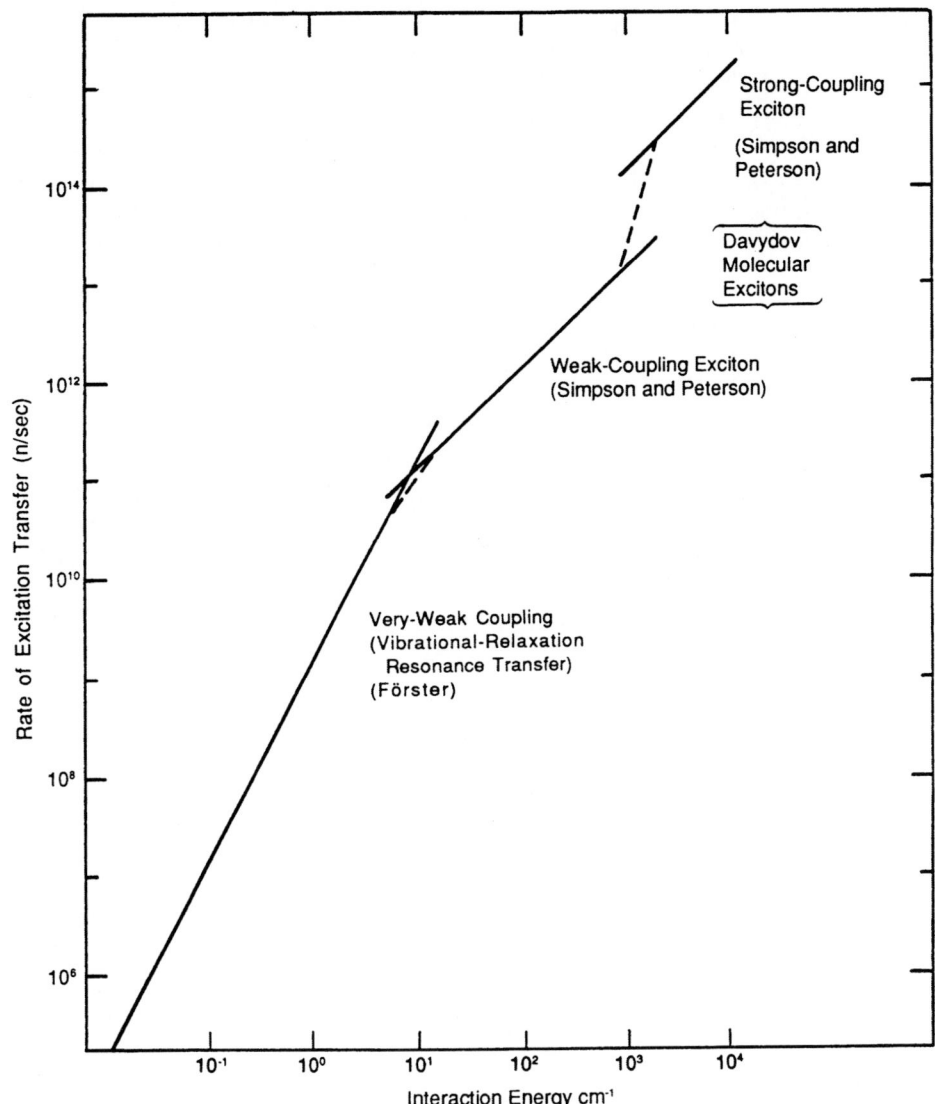

Figure 6. Förster Rates of Energy Transfer Diagram

The Förster Transfer case achieves resonance by vibrational relaxation to the fluorescence emission, and the resonance requirement is much more liberal. The inverse sixth-power dependence is still considered to be a long-range interaction, even though a much shorter range interaction than molecular exciton cases.

Applications of Molecular Exciton to Helical Polymers and Lamellar Systems

Helical polymers. If N molecules are structured in a helical array so that transition moments have projection components parallel to the polymer axis and perpendicular to the helical axis, we could picture that for the ∥ component the molecular exciton state with dipoles in an in-phase, attractive array would yield a lowest allowed exciton state, and all other N components would yield exciton states forbidden for electric dipole transitions (Fig. 7). Thus, a long-wavelength shifted component should be observed for the polymer compared with the absorption band for the individual molecular chromophores components.

If there are, e.g., four molecules per helix turn, we could consider the nature of the molecular exciton states generated by projecting the four transition dipoles in a plane $\perp$ to the polymer axis. The full interaction would require pair-wise dipole-dipole sums over the full helix, in which the inverse-cube dependence from the reference plane would diminish the effect of distant pairings.

Figure 7 illustrates the vector diagrams for the $\parallel$ and $\perp$ dipole vector components for the classes of molecular exciton states for four transition dipoles per helix turn. For the $\perp$ projection, two of the exciton states have a net transition moment of zero, whereas the single-noded molecular exciton functions[9] have equal energy (are energy degenerate).

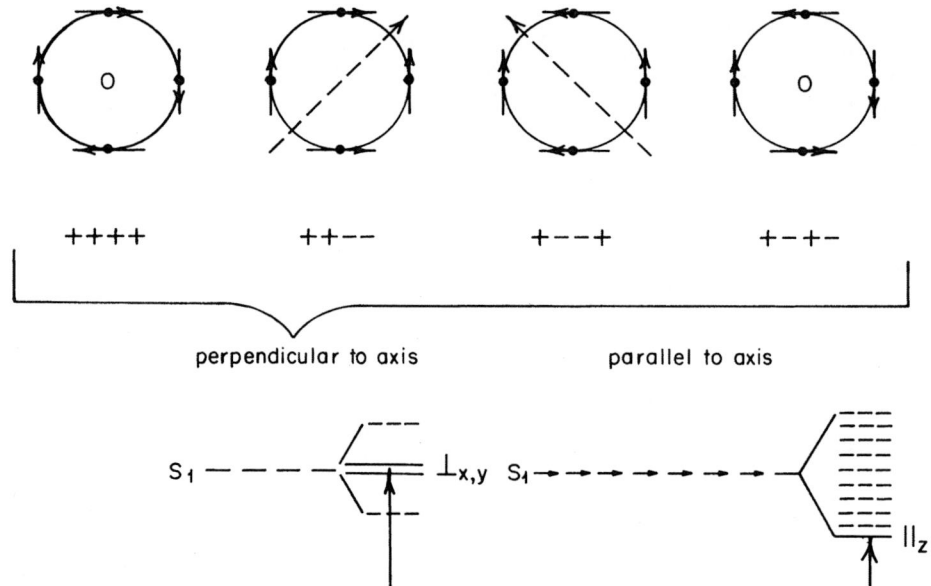

Figure 7. Exciton Band Structures for Helical Polymers

Cyanine dye polymers. J-band polymers with van der Waals binding conform to the "tight-binding" approximation and the free-exciton (strong coupling) model. These aspects arise from the exceedingly high oscillator strength for the lowest $S_0 \rightarrow S_1$ absorption, f-values exceeding unity being common.

In the J-band spectroscopy[12-14] a single pure-electronic transition is observed at room temperature for cyanine dye polymers. The very high molecular weight of the observed van der Waals polymers implies unusual hydrodynamic stability, which suggests a helical configuration. It is then expected that the molecular exciton states for a helical polymer should prevail. We have found[21] that at 77 K the cyanine dye polymer J-bands, as narrow as they seemed to be in 298 K solution, narrowed even more remarkably and split into two mutually perpendicularly polarized components. Those sharp bands are certainly pure electronic components and conform to the molecular exciton selection rules.

Helical polypeptides. The theory for this case was developed by Moffitt,[22] with particular attention to optical rotatory dispersion. Later it was shown that the absorption spectrum in the phase transition from random coil to alpha-helix exhibited a splitting into two distinct peaks.[23] The polarization of the absorptions with respect to the polymer axis conformed to the helical exciton coupling selection rules.[22] Further confirmation of these results was obtained in the circular dichroism studies for alpha-helix polypeptides.[23]

Helical polynucleotides. DNA from various biological sources has been studied extensively by various optical methods, with a search for theoretical correlation. This is a heavily studied field driven by the evolution of molecular biology. Three essential features of the random-to-helix transition will be outlined.

The fibre X-ray diffraction studies of polynucleotides have established the structure of B-DNA as a helical configuration with H-bonded base-pairs internally arranged in a staircase pattern, with base-pair planes $\perp$ to the polymer axis.

The experimental observations and their interpretation are as follows:

1. The absorption spectrum of both random-coil and helix-coil polynucleotides and DNA are closely matched by the composite absorption spectra of the individual bases. Explanation: excitation coupling must be in the weak-coupling molecular exciton limit.

2. A strong hypochromism is observed[25] in the $(\pi \rightarrow \pi^*)$ absorbance in the phase transition random-coil to helix. What is the nature of this suppression of electric-dipole absorption? Contrary to early expectation, the molecular exciton model predicts no change in net absorbance even if molecular units are strongly coupled. Rhodes[26] demonstrated that a second-order dispersion effect is the origin of the light absorbance suppression. This can be seen in a simple way if one examines the field in which a transition moment of a molecule must be excited, with neighbor molecules opposing (Fig. 8) the transition dipole generation, by virtue of excited higher excited state dispersion effects.[26,27]

3. A hyperchromism is observed on the long-wavelength side of the polynucleotide absorption spectrum, in the phase-transition random-coil to helix coil.[28] The presence of $n \rightarrow \pi^*$ electronic transitions in this region of absorption for N-heterocyclic bases,[27,29] which have lone-pair n-orbital excitations $\perp$ to the base plane. This results in an in-line array of higher excited-state transition moments with respect to the reference molecule, thus enhancing (Fig. 8) the reference molecule absorbance in the helical array.[27]

Lamellar Molecular Arrays: Antenna Chlorophyll and Reaction Center Pairs. In the photosynthetic unit of plants, it is understood that the chlorophyll molecule serves two distinct functions:

1. Quasi-systematic arrays of chlorophyll must serve as an antenna for collecting photons.

2. The excitation energy must migrate to a trapping center where the redox steps of photosynthesis are initiated.

The physical puzzle in these requirements lies in the fact that the large aggregate of chlorophylls must interact relatively weakly, and still transfer energy efficiently, and the trapping center (consisting of a pair of chlorophyll molecules) must interact strongly so as to lower the excitation energy sufficiently to serve as an energy trap. Molecular exciton theory provides models both for antenna chlorophyll and for reaction center pairs.

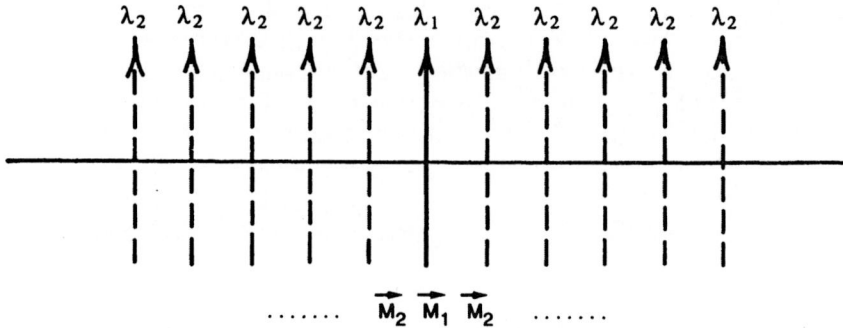

Hypochromic Array: $\vec{M}_1$ Retarded by $\vec{M}_2$ Fields

Hyperchromic Array: $\vec{M}_1$ Augmented by $\vec{M}_2$ Fields

Figure 8. Hypochromic and Hyperchromic Transition Dipole Arrays Illustrating Second-Order Dispersion Effects [Rhodes[26,27]]

Lamellar molecular excitons have been treated quantum-mechanically by Hochstrasser and Kasha.[30] Six different molecular packing geometries were studied, yielding energy expressions showing exciton state dependence on the number of molecules per unit cell, transition moment, inter-center molecule spacing, and especially on the geometrical factor. Dipole-dipole interaction sums were made over an infinite lattice in a monolayer molecular array as a prototype for molecular interactions in lamellae.

The extreme variations in energy of interaction and exciton band structure with geometry are illustrated in the models chosen. For example, the pin-cushion array with all transition moments normal to the molecular laminar plane yields 8.4 times the dipole-dipole interaction of a simple dimer of the same molecule. However, if the transition dipoles (in each unit cell of the lamella) are each inclined to arc $\cos 1/\sqrt{3}$ to the respective Cartesian axes, the zero-interaction lamella is achieved. The allowed transition for the aggregate is at the same energy as that for a single molecule, except for van der Waals differences in the states of the lamella. Nevertheless, there is a finite exciton bandwidth of forbidden levels.[30]

Antenna chlorophyll must correspond to a near approach to the zero-interaction lamellar exciton case. Bulk packing with a similar result is just as possible. The spectral observation on leaf chlorophyll showing almost identical spectra as solution chlorophyll corresponds to a weak interaction case in exciton theory. Nevertheless, as seen from Fig. 6, rates of transfer from 10^{11} to 10^{13} sec^{-1} are still possible, as required.

Reaction-center chlorophyll as a dimer exciton model was first discussed by Brody et al.,[31] using the McRae and Kasha formulas.[6] A current research with modern structural elaboration of reaction-center-chlorophyll has been published by Maruyama and Osuka.[32] Their calculations use the oblique geometry for chlorophyll dimers. Figure 9 illustrates a model for coupling of antenna chlorophyll exciton states to reaction-center-chlorophyll dimers.

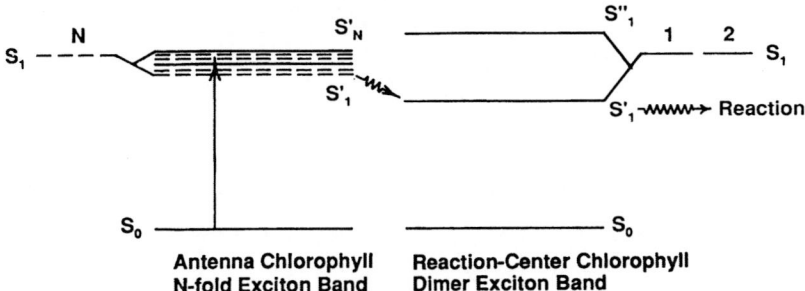

Figure 9. Antenna-Chlorophyll to Reaction-Center-Chlorophyll Excitation Transfer

Exciton-State Sensitization of Triplet States. The original impetus for the McRae and Kasha development of the molecular exciton model for small aggregates was to explain the disappearance of fluorescence in an aggregated molecule such as a dimer, and the simultaneous appearance of an enhanced phosphorescence.[6] As Fig. 10 illustrates, a dye molecule which has very efficient $S_1 \to S_0$ fluorescence and comparatively little $T_1 \to S_0$ phosphorescence can have its spectroscopic transitions entirely altered by dimerization: (i) the absorption peak for a card-pack van der Waals dimer should strongly blue shift, and (ii) the metastable lowest exciton state (metastable, or forbidden, even though a singlet state) now very efficiently undergoes intersystem crossing to a lower triplet state. The triplet state of the dimer is usually little displaced from the monomer triplet, as dipole-dipole coupling is absent.

Thus, the molecular exciton model supplies a direct mechanism for dye-dimer sensitization of triplets. Photochemical studies of sensitized oxidation have relied on the efficiency of this triplet excitation mechanism. It is noteworthy that no special spin-orbital effect is present in the dimer; the enhancement of triplet excitation arises simply from advantageous kinetic competition.

Singlet Oxygen Energy Transfer

Singlet Oxygen as a Chemical Species

Singlet molecular oxygens as excited spectroscopic species have been recognized by spectroscopists since G. N. Lewis'[33] and R. S. Mulliken's[34] early speculations. The exact spectroscopic characteristics of $^1\Delta_g$ and $^1\Sigma_g^+$ were established by L. Herzberg and G. Herzberg.[35] In this period these exotic oxygens were considered to exist as metastable species in the stratosphere, and indeed appeared only as infrared absorption lines for observations at low sun or occasionally in emission in *aurora borealis*.

The chemical era of singlet molecular oxygen research was initiated by Khan and Kasha,[36] who showed that the red-chemiluminescence accompanying the reaction of hypochlorite with hydrogen peroxide in alkaline solution corresponded to light emission from excited oxygen molecules. Subsequent research revealed a richness of spectroscopic states, both in absorption and emission for the unusual singlet oxygen excitation.[37]

Fig. 11 illustrates the rich manifold of states which can be observed. The infra-red excitations to $^1\Delta_g$ and $^1\Sigma_g^+$ states involve extremely forbidden transitions; these states have mean lifetimes of 2600 and 7.0 seconds, respectively. Such highly metastable states make these oxygens available for chemical reactions or for novel radiation mechanisms.

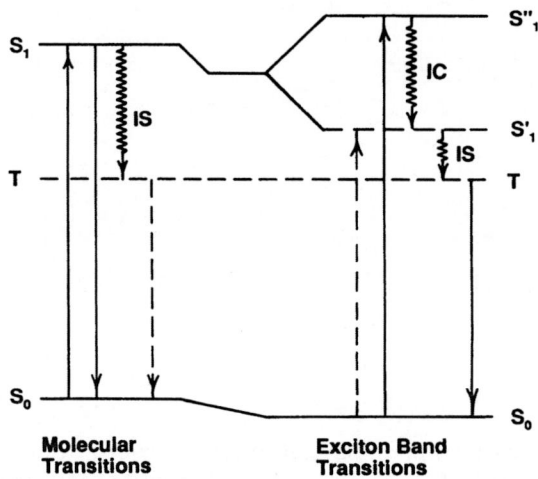

Figure 10. Molecular Exciton Sensitization of the Lowest Triplet State

Simultaneous Transitions

Molecular pairs under appropriate conditions can undergo electronic transitions with a relaxation of the prohibition against radiation in the single molecules. In molecular exciton pairs, excitation can be considered to reside on one molecule or the other, e.g., A*A or AA*, and a quantum mechanical resonance splitting can occur. In the case that both molecules are excited, A*,A*, no resonance phenomenon can exist; instead, a simultaneous transition can occur in which two excited molecules emit one photon, A*,A* → A,A + hν. The photon energy closely approximates twice the energy of the single molecule transition A* → A + hν. In order for this to occur, the simultaneous transition must have a much more rapid relaxation path than that for the single molecule transition.

There are many different cases of simultaneous transitions, but in the singlet oxygen case, e.g., $(^1\Delta_g)(^1\Delta_g) \rightarrow (^3\Sigma_g^+)(^3\Sigma_g^+)$ or S + S → {T + T} = S + T + Q, the highly forbidden $^1\Delta_g \rightarrow {}^3\Sigma_g^-$ or T → S transition, with an intrinsic mean lifetime of 2600 seconds (43.3 minutes), develops an S → S pathway for the colliding pair. This requires an electron exchange collision, with very small interaction energy, of the order of 20 cm^{-1}. So the spectroscopy for the simultaneous transitions shown in Fig. 11 involves almost simple arithmetic additivity of the single molecule energies.

The chemistry of singlet oxygen proves to be very rich, and many symposia and research volumes[38] have appeared on the ramifications of this previously neglected subject.

Sensitization of Luminescence

Sensitization of luminescence and of chemical reactions involves many different singlet oxygen mechanisms. Khan and Kasha[39] proposed that singlet oxygen may be the initiator of many bioluminescences and chemiluminescences. Figure 12 illustrates one of five mechanisms[40] which have been proposed for various energy schemes offered by different molecular systems. The red-fluorescing dye violanthrone is excited to vivid fluorescence in the hypochlorite-peroxide reaction mixture. Obviously, here, the single molecule excitation would have insufficient energy for the red photon emission.

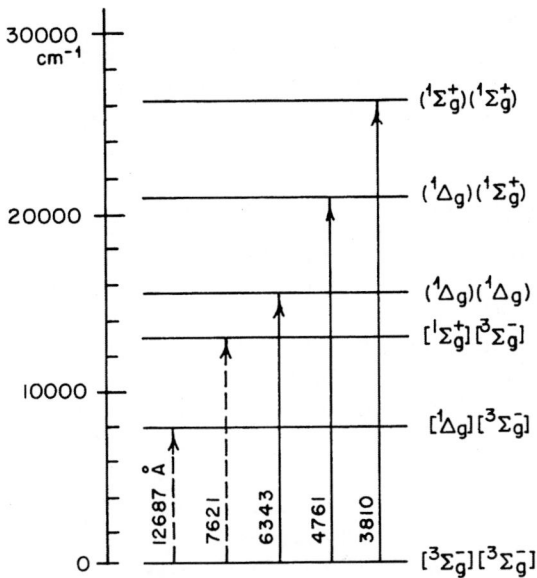

Figure 11. Simultaneous Transitions in Molecular Oxygen-Pairs

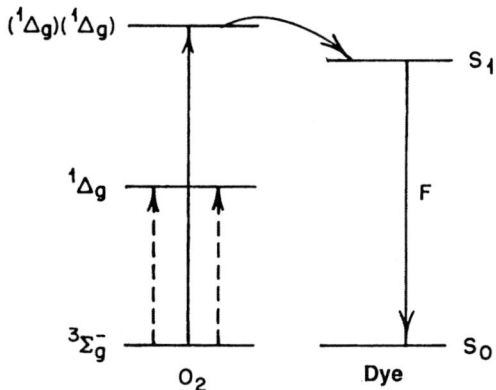

Figure 12. Photosensitization of Dye Molecule Fluorescence by Molecular Oxygen-Pair States

Cancer Tumor Singlet Oxygen Sensitization

T. Dougherty[41] has developed a method for treatment of consolidated single tumors by light-induced sensitization of singlet oxygen. Hematoporphyrin was chosen as a sensitizing dye (a physiologically non-exotic pigment), and after tumor injection, red light from a laser was introduced to the tumor center via a light pipe. Dibenzo-furan (scavenger for singlet oxygen) was shown to nullify the necrotic effect of the treatment. Direct spectroscopy[42] showed the effectiveness of dye-singlet-oxygen excitation transfer. The method has now been approved by the FDA and is being introduced in many hospitals. This case represents a remarkable example of technology transfer, from the realms of pure spectroscopy, quantum chemistry, and photo-chemistry, to a powerful medical advance.

Intramolecular Charge Transfer

Intermolecular charge transfer[43] interactions and intramolecular charge transfer[44-46] interactions are highly developed subjects, and are important clues to reaction pathways. Here we shall limit our attention to the recent research on the *Twisting-Intramolecular-Charge-Transfer* or TICT-state phenomenon.

TICT-State Configuration Requirements

The initiation of this field is connected with the discovery of dual fluorescence in para-dimethylaminobenzonitrile and its elucidation.[47] The configurational aspects of the TICT mechanism are illustrated in Fig. 13. If two components (e.g., a phenyl ring and p-dimethyl group) of a torsional molecular system are co-planar, π-electron states frequently are the lowest-lying excited states of the system. At the 90° configuration, a charge-transfer state may correspond to the lowest excited-state minimum. A full electron transfer is considered to exist in the charge-transfer perpendicular configuration. However, the 90° angle between the two molecular components would, by orthogonality of their component orbitals, isolate the component electron systems. Thus, there must be some intermediate angle at which the action of electron transfer takes place. The earlier researches of Murrell et al.[48] suggest a spectroscopic mechanism for the TICT-state electron transfer. Their calculation indicates that the torsional potential for twisting of an amino group in an aromatic system could make a higher-lying charge-transfer state cross the vibrational potential for the π-electron excitation.

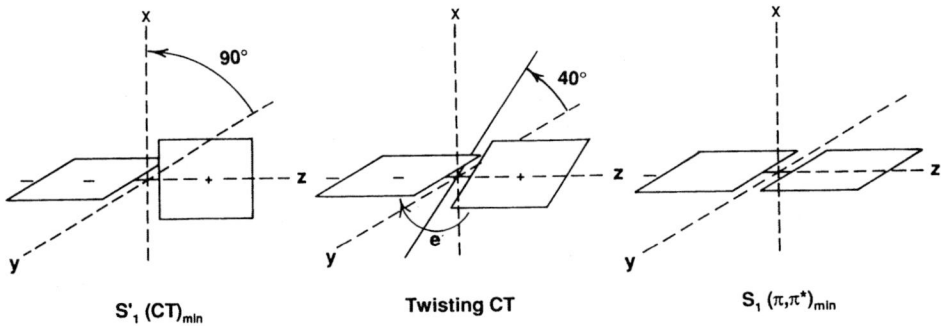

Figure 13. Mechanism of Twisting Intramolecular Charge Transfer

The Avoided-Crossing Mechanism for TICT Excitation

The *avoided crossing* between the charge-transfer upper-state potential and the lowest excited state potential (Fig. 14) offers a quantum-mechanical mechanism for the TICT-state phenomenon. The potential curves of Fig. 14 satisfy the known facts of TICT-state formation:

- A thermal barrier exists from the S_1 state minimum to TICT-state formation.[47]

- Dual fluorescence may be observed[47] in competition for the primary absorption.
- A sudden polarization[49] as a result of the avoided crossing produces a highly dipolar excited state.

- The TICT-state fluorescence proves to be very sensitive to dielectric constant solvent effects.[50]

- No absorption to the TICT minimum is observed because of the extremely small Boltzmann factor in the ground-state vibrational activation required.

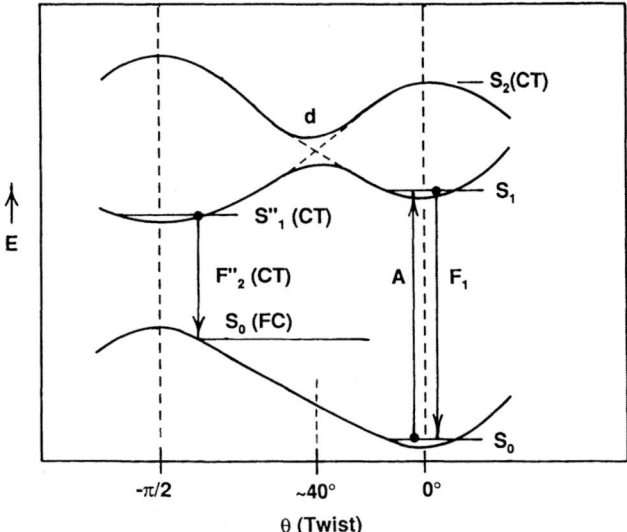

Figure 14. Twisted Intramolecular Charge Transfer

Generality of TICT-State Occurrence

The important lesson from TICT-state spectroscopy is its wide occurrence,[44-46] often in previously unsuspected molecular cases. This phenomenon offers an important excitation pathway, in competition with normal excitation mechanisms. The TICT-state excitation must be carefully monitored for a full understanding of complex molecular systems whenever internal torsion is a possibility, either between large molecular components, e.g., the two phenyls of biphenyl, or between side torsional groups such as amino, dimethylamino, ethoxy, vinyl, etc., twisting with respect to an aromatic ring system. The TICT phenomenon has been disregarded in quantum chemical calculations on excited states in general.

Intramolecular Proton Transfer

Proton transfer, excitation transfer and electron transfer constitute the classes of fundamental processes governing critical steps in chemical and biological reactions. Ground-state and excited-state proton transfer are observable, but the advent of ultrafast pulsed laser spectroscopy has focused intense new interest on excited state proton transfer. This is exemplified by the recently published special issue of *Chemical Physics* on dynamics and spectroscopy of proton-transfer systems.[51]

Mechanisms of Intramolecular Proton Transfer

A large number of examples can now be found in the literature on proton-transfer spectroscopy. The use of spectroscopic methods allows detailed study of the dynamics of formation of each tautomeric species and its reversion to the more stable form. In older studies of tautomerism, only equilibrium information on tautomeric species was available.

Figure 15 illustrates a typical example of intramolecular proton transfer, 3-hydroxyflavone (R = phenyl) and 3-hydroxychromone (R = methyl). This molecule is a much-studied species and has become a prototype for detailed mechanistic study of *excited-state intramolecular proton-transfer* (ESIPT in some vocabularies).

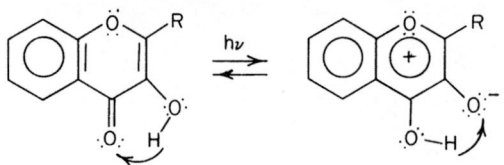

Figure 15. Intramolecular Proton-Transfer in Flavonols

The molecule 3-hydroxyflavone exhibits a high sensitivity to solvent perturbations[52] which introduced considerable complexity in mechanistic studies. This high sensitivity arises from the 5-membered H-bonded ring in the stable ground state 4-pyrone structure. In the tautomer reached by proton-transfer, a pyrylium aromatic oxygen ring is considered to be present, the anthocyanine-like molecular overall skeleton accounting for the shift of the fluorescence of the normal molecule from the ultraviolet range to the green range of the tautomer species.[53] The complexity of the system, especially on the role of the triplet manifold in proton transfer, makes this molecule a centerpiece of continued research.[54]

In order to systematize the study of intramolecular proton transfer, a system of classification has evolved, dividing the phenomenon into specific molecular cases:[55,56]

Case (I)   symmetrical intramolecular proton transfer (tautomers identical),

Case (II)   intrinsic intramolecular proton transfer (tautomers electronically distinguished),

Case (III)   concerted biprotonic transfer (cooperative transfer of two protons in dimers or H-bonded solvents),

Case (IV)   strong-catalysis proton transfer (requires mediation of strong acid-base catalysis),

Case (V)   proton-relay transfer (solvent H-bonded bridge chains).

The study of proton-transfer dynamics over the range of cases outlined above is just beginning, even though various cases have been recognized as fundamental in important processes [e.g., Case (V) in proton-pump mechanisms in biological transport phenomena].

The Proton-Transfer Potential

The schematic potential considered to apply generally to intramolecular proton transfer is illustrated in Fig. 16. After normal molecule excitation $S_0 \rightarrow S_1$, a picosecond-range proton transfer occurs with population of the excited tautomer state S_1'. This acts as a bottleneck state because its fluorescence mean lifetime for the transition $S_1' \rightarrow S_0'$ in the normal nanosecond range yields a population inversion under the laser pumping $S_0 \rightarrow S_1$. Thus, transient transitions such as $S_1' \rightarrow S_2'$ can be studied by picosecond laser spectroscopy.[56]

The population inversion of state S_1' (relative to S_0') permits the generation of the extraordinarily efficient *proton-transfer laser*.[57] The existence of the four-level laser system[58] makes the population inversion more complete, especially because of the continual rapid depletion of the S_1' state[59] by the $S_0' \rightarrow S_0$ reverse proton transfer.

Much exploration of proton-transfer potentials is currently in progress as femtosecond and other techniques are brought to bear on the detailed dynamics of these complex systems.

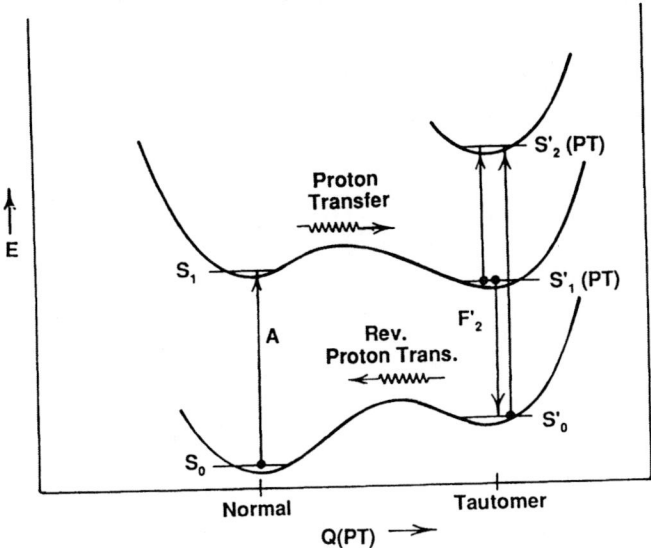

Figure 16. Proton-Transfer Potential

Interferences in Proton-Transfer Spectroscopy

Current research has revealed two major interferences with the observation of proton-transfer fluorescence $S_1' \to S_0'$. Most of the discussions to date have concentrated on the characteristics of the singlet-state potentials. As in the case of normal molecule excitation, *intersystem crossing to triplet potentials* offers important competition to $S_1 \to S_0$ fluorescence, the subsequent $T_1 \to S_0$ emission becoming dominant for many cases.

The observation of the participation of triplet potentials in proton-transfer spectroscopy was first made by Mordzinski and Grellman.[60] They considered a case in which the T_1 and T_1' minima were quasi-equivalent in energy. Of the three energetic possibilities for the absolute energy values of the T_1 and T_1' minima, Gormin et al.[61] explored the possibility of normal phosphorescence ($T_1 \to S_0$) enhancement via the proton-transfer potentials, i.e., $S_0 \to S_1, S_1 \to S_1', S_1' \to T_1', T_1' \to T_1, T_1 \to S_0$ (the $\to$ representing radiationless steps,[61]) with the energy relation $T_1 < T_1'$.

It is important to have information on the participation of the triplet potentials in proton transfer; yet, hardly any cases have been deciphered. Certainly any understanding of proton-transfer phenomena will require a knowledge of competing radiationless pathways.

An example of the complication which can arise is in the slow reverse proton transfer (microsecond and longer) claimed by Itoh et al.[62] Although there is unresolved controversy on this, a detailed mechanism has been proposed by Chou et al.[54] to account for a side channel, explaining the microsecond reverse proton step (in 3-hydroxyflavone) decay as illusory. Numerous other puzzles involving triplet potentials are hiding in the literature, ready for exploration.

A second interference which has recently arisen[56] is the competition to intramolecular proton transfer given by the TICT phenomenon. The complex situation arises of triple fluorescence in molecules for which both ESIPT and TICT are both possible (Fig. 17). Two such cases have been described, substituted aminosalicylates and benzanilides. Other cases are popping up in current research.

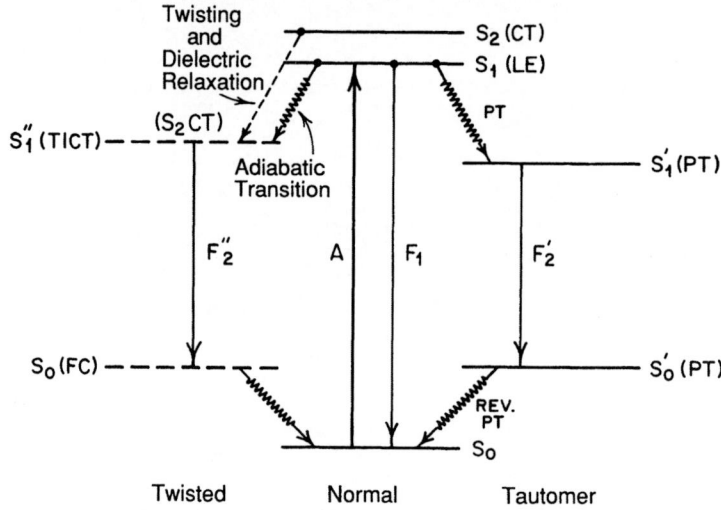

Figure 17. Triple Fluorescence Competition: LE, PT, and CT

All of these examples indicate that we are beginning to recognize the complexity of molecular excitation in complicated organic molecule systems. This is especially the case for composite molecules, in which the components exhibit individual excitation behavior in addition to the composite molecule cooperative behavior.

Photochemistry and radiation chemistry of complex organic molecule systems can now be explored by transient spectroscopy, and forced or channeled behavior into new product pathways can be considered. As one example, in an alpha-helix peptide, after excitation energy migrates via molecular excitons to a trapping center, proton transfer in these H-bonded amino-to-carbonyl systems could give rise to a transient imidol structure,[63] —C(OH)=N—, with novel chemistry.

Acknowledgments

Work was done under Contract No. DE-FG05-87ER60517 between the Office of Health and Environmental Research, U. S. Department of Energy, and the Florida State University.

References

1. M. Kasha. Relation Between Exciton Bands and Conduction Bands in Molecular Lamellar Systems. *Rev. Modern Phys.*, 31:162 (1959).
2. J. I. Frenkel. On the Absorption of Light and the Trapping of Electrons and Positive Holes in Crystalline Dielectrics. *Phys. Rev.* 37:17, 1276 (1931);
   J. I. Frenkel. *Physik. Z. Sowjetunion*, 9:158 (1936).
3. J. Franck and E. Teller. Migration and Photochemical Action of Excitation Energy in Crystals. *J. Chem. Phys.* 6:861 (1938).
4. A. S. Davydov. *Theory of Molecular Excitons* (transl. by M. Kasha and M. Oppenheimer, Jr.), McGraw-Hill Book Co., Inc., New York (1962), 174 pp.
5. R. S. Knox. *Theory of Excitons*, Academic Press, New York (1963), 207 pp.
6. E. G. McRae and M. Kasha. (Note) The Enhancement of Phosphorescence Ability Upon Aggregation of Dye Molecules. *J. Chem. Phys.* 28:721 (1958).
7. E. G. McRae and M. Kasha. The Molecular Exciton Model. In *Physical Processes in Radiation Biology* (Augenstein, Rosenberg and Mason, eds.), p. 23-42. Academic Press, New York (1964).
8. M. Kasha, H. R. Rawls and M. A. El-Bayoumi. The Exciton Model in Molecular Spectroscopy. *Pure and Appl. Chem.* 11:371 (1965).

9. M. Kasha. Molecular Excitons in Small Aggregates. In *Spectroscopy of the Excited State* (B. diBartolo, ed.), pp. 337-363. Plenum Press, New York (1976).

10. R. M. Hochstrasser and M. Kasha. Application of the Exciton Model to Molecular Lamellar Systems. *Photochem. Photobiol.* 3:317 (1964).

11. D. S. McClure. Energy Transfer in Molecular Crystals and in Double Molecules. *Can. J. Chem.* 36:59 (1958).

12. E. E. Jelley. Spectral Absorption and Fluorescence of Dyes in the Molecular State. *Nature* 138:1009 (1936);
E. E. Jelley. Absorption Bands in the Spectrum of ψ-Isocyanine Dyes. *Nature* 139:378 (1937);
E. E. Jelley. Molecular, Nematic and Crystal States of 1,1'-Diethyl-ψ-Cyanine Chloride. *Nature* 139:631 (1937).

13. G. Scheibe et al. Über die Veränderlichkeit des Absorptionsspektrums einiger Sensibilisierungsfarbstoffe und deren Ursache, *Z. angew. Chem.* 49:563 (1936);
G. Scheibe et al. Über die Veränderlichkeit des Absorptionsspektren in Lösung und die van der Waalsschen Kräfte als ihre Ursache, *Z. angew. Chem.* 50:51 (1937).
G. Scheibe et al. Über die Veränderlichkeit des Absorptionsspektren in Lösung und die Nebenvalenzen als ihre Ursache, *Z. angew. Chem.* 50:212 (1937).

14. S. E. Sheppard. The Effects of Environment and Aggregation on the Absorption Spectra of Dyes, *Rev. Mod. Phys.* 14:303 (1942).

15. W. T. Simpson and D. L. Peterson. Coupling Strength for Resonance Force Transfer of Electronic Energy in Van der Waals Solids, *J. Chem. Phys.* 26:588 (1957).

16. Th. Förster. Excitation Transfer. In *Comparative Effects of Radiation* (M. Burton, J. S. Kirby-Smith and J. L. Magee, eds.), p. 300-341. John Wiley and Sons, Inc., New York, (1960).

17. Th. Förster. Delocalized Excitation and Excitation Transfer. In *Modern Quantum Chemistry* (O. Sinanoğlu, ed.), Vol. III, p. 93-137. Academic Press, New York (1965).

18. M. Kasha. Energy Transfer Mechanisms and the Molecular Exciton Model for Molecular Aggregates. *Radiation Res.* 20:53, 55 (1963).

19. Th. Förster. Zwischenmolekulare Energiewanderung, *Naturwiss.* 33:166-175 (1946).

20. Th. Förster. Zwischenmolekulare Energiewanderung und Fluorescenz, *Ann. Physik* 2:55-73 (1948).

21. P. DeTar and M. Kasha. Unpublished Work, Florida State University, Tallahassee, Florida.

22. W. Moffitt. The Optical Rotatory Dispersion of Simple Polypeptides. II, *Proc. Nat. Acad. Sci.* (US), 42:736 (1956).
W. Moffitt. Optical Rotatory Dispersion of Helical Polymers, *J. Chem. Phys.* 25:467 (1956).

23. W. B. Gratzer, G. M. Holzwarth and P. Doty. Polarization of the Ultraviolet Absorption Bands in Alpha-Helical Polypeptides. *Proc. Nat. Acad. Sci. (USA)* 47:1785 (1961).

24. G. Holzwarth, W. B. Gratzer and P. Doty. The Optical Activity of Polypeptides in the Far Ultraviolet. *J. Am. Chem. Soc.* 84:3194 (1962).

25. P. Doty, J. Marmur, J. Eigner and C. Schildkraut. Strand Separation and Specific Recombination in Deoxyribonucleic Acids: Physical Chemical Studies. *Proc. Nat. Acad. Sci. (USA)* 46:461 (1960).

26. W. Rhodes. Hyperchromism and Other Spectral Properties of Helical Polynucleotides, *J. Am. Chem. Soc.* 83:3609 (1961).

27. M. Kasha, M. A. El-Bayoumi and W. Rhodes. Excited States of Nitrogen Base-Pairs and Polynucleotides. *J. Chim. Phys.* 58:816 (1961).

28. A. Rich and M. Kasha. The n → π* Transition in Nucleic Acids and Polynucleotides. *J. Am. Chem. Soc.* 82:6197 (1960).

29. M. Kasha. The Nature and Significance of n → π* Transitions. In *Light and Life* (W. D. McElroy and B. Glass, eds.), pp. 31-64. Johns Hopkins University Press, Baltimore, Maryland (1961).

30. R. M. Hochstrasser and M. Kasha. Application of the Exciton Model to Molecular Lamellar Systems. *Photochem. Photobiol.* 3:317 (1964).

31. S. S. Brody and M. Brody. Spectral Characteristics of Aggregated Chlorophyll and Its Possible Role in Photosynthesis. *Nature* 189:547 (1961).
S. S. Brody and M. Brody. Fluorescence Properties of Aggregated Chlorophyll In Vivo and In Vitro. *Trans. Faraday Soc.* 58:416 (1962).

32. K. Maruyama and A. Osuka. A Chemical Approach Toward Photosynthetic Reaction Center, *Pure and Appl. Chem.* 62:1511-1520 (1990).

33. G. N. Lewis. The Magnetochemical Theory, *J. Am. Chem. Soc.* 38:762 (1916); *Chem. Rev.* 1:233 (1924).

34. R. S. Mulliken. Interpretation of the Atmospheric Oxygen Bands; Electronic Levels of the Oxygen Molecule, *Nature* 122:505 (1928).

35. L. Herzberg and G. Herzberg. Fine Structure of the Infrared Atmospheric Oxygen Bands, *Astrophys. J.* 105:353 (1947).
    H. D. Babcock and L. Herzberg. Fine Structure of the Red System of Atmospheric Oxygen Bands, *Astrophys. J.* 108:167 (1948).
36. A. U. Khan and M. Kasha. The Red Chemiluminescence of Molecular Oxygen in Aqueous Solution. *J. Chem. Phys.* 39:2105 (1963).
    A. U. Khan and M. Kasha. Rotational Structure in the Chemiluminescence Spectrum of Molecular Oxygen in Aqueous Systems. *Nature* 204:241 (1964).
37. A. U. Khan and M. Kasha. Chemiluminescence Arising from Simultaneous Transitions in Pairs of Singlet Oxygen Molecules. *J. Am. Chem. Soc.* 92:3293 (1970).
38. E.g., A. Paul Schaap (ed.), *Singlet Molecular Oxygen*, 399 pp. Dowden, Hutchinson and Ross, Inc., Stroudsberg, Pennsylvania (1976).
    H. H. Wasserman and R. W. Murray (eds.), *Singlet Oxygen*, 684 pp. Academic Press, New York (1979).
    A. A. Frimer (ed.), *Singlet O_2*, Vols. I-IV. CRC Press, Inc., Boca Raton, Florida (1985).
39. A. U. Khan and M. Kasha. (Note) Physical Theory of Chemiluminescence in Systems Evolving Molecular Oxygen, *J. Am. Chem. Soc.* 88:1574 (1966).
40. M. Kasha and D. Brabham. Singlet Oxygen Electronic Structure and Photosensitization. In *Singlet Oxygen* (H. Wasserman and R. W. Murray, eds.), p. 1-33. Academic Press, New York (1979).
41. T. Dougherty. Activated Dyes as Antitumor Agents, *J. Nat. Cancer Inst.* 52:1333 (1974).
    T. Dougherty, C. J. Gomer and K. R. Weishaupt. Energetics and Efficiency of Photoinactivation of Murine Tumor Cells Containing Hemotoporphyrin, *Cancer Res.* 36:2330 (1976).
    T. Dougherty, J. E. Kaufman, A. Goldfarb, K. R. Weishaupt, D. Boyle and A. Mittleman. Photoradiation Therapy for the Treatment of Malignant Tumors, *Cancer Res.* 38:2628 (1978).
42. A. U. Khan and M. Kasha. Direct Spectroscopic Observation of Singlet Oxygen Emission at 1268 nm Excited by Sensitizing Dyes of Biological Interest in Liquid Solution. *Proc. Nat. Acad. Sci. (USA)*, 76:6047 (1979).
43. R. S. Mulliken and W. B. Person. *Molecular Complexes*, Wiley Interscience, New York (1969), 498 pp.
44. W. Rettig. Charge Separation in Excited States of Decoupled Systems - TICT Compounds and Implications Regarding the Development of New Laser Dyes and the Primary Processes of Vision and Photosynthesis, *Angew. Chem. Int. Ed.* 25:971-988 (1986).
45. E. Lippert, W. Rettig, V. Bonačić-Koutecký, F. Heisel and J. A. Miehé. Photophysics of Internal Twisting. In *Advances in Chemical Physics* (I. Prigogine and S. A. Rice, eds.), p. 1-173. Wiley-Interscience, New York (1987).
46. V. Bonačić-Koutecký, J. Koutecký and J. Michl. Neutral and Charged Biradicals, Zwitterions, Tunnels in S_1, and Proton Translocation: Their Role in Photochemistry, Photophysics, and Vision, *Angew. Chem. Int. Ed.* 26:170-189 (1987).
47. Z. R. Grabowski, K. Rotkiewicz, A. Siemiarczuk, D. J. Cowley and W. Baumann. Twisted Intramolecular Charge Transfer States (TICT). A New Class of Excited States with a Full Charge Separation, *Nouv. J. Chim.* 3:443-454 (1979).
48. J. N. Murrell. The Electronic Spectrum of Aromatic Molecules. VI: The Mesomeric Effect, *Proc. Phys. Soc. (London)* A68:969-975 (1955).
    Cf. M. Godfrey and J. N. Murrell. Substituent Effects on the Electronic Spectra of Aromatic Hydrocarbons. III. An Analysis of the Spectra of Amino- and Nitrobenzenes in Terms of the Localized-Orbital Model, *Proc. Roy. Soc.* A278:57, 71 (1969).
49. L. Salem and C. Rowland. The Electronic Properties of Diradicals, *Angew. Chem. Int. Ed.* 11:92-111 (1972).
50. J. Heldt and M. Kasha. Dialectric Medium Interactions in Proton-Transfer and Charge-Transfer Molecular Electronic Excitation. *J. Molec. Liquids* 41:305 (1989).
51. P. F. Barbara and H. P. Trommsdorff (eds.). Spectroscopy and Dynamics of Elementary Proton Transfer in Polyatomic Systems *Chem. Phys. Special Issue* 136:153-160 (1989).
52. D. McMorrow and M. Kasha. Intramolecular Excited-State Proton Transfer of 3-Hydroxyflavone. H-Bonding Solvent Perturbations. *J. Phys. Chem.* 88:2235 (1984).
53. P. J. Sengupta and M. Kasha. Excited State Proton-Transfer Spectroscopy of 3-Hydroxyflavone and Quercetin. *Chem. Phys. Lett.* 68:382 (1979).

54. W. E. Brewer, S. L. Studer and P.-T. Chou. Temperature-Dependent Study of the Ground-State Reverse Proton Transfer of 3-Hydroxyflavone, *Chem. Phys. Lett.* 158:345-350 (1989).
W. E. Brewer, S. L. Studer, M. Standiford and P.-T. Chou. Dynamics of the Triplet State and the Reverse Proton Transfer of 3-Hydroxyflavone, *J. Phys. Chem.* 93:6088 (1989).
M. L. Martinez, S. L. Studer and P.-T. Chou. Direct Evidence of the Triplet State Resulting in the Slow Reverse Proton-Transfer Reaction of 3-Hydroxyflavone, *J. Am. Chem. Soc.* 112:2427-2429 (1990).
55. M. Kasha. Proton-Transfer Spectroscopy. Perturbation of the Tautomerization Potential. *J. Chem. Soc. Faraday Transactions, II* 82:2379 (1986).
56. J. Heldt, D. Gormin and M. Kasha. A Comparative Picosecond Spectroscopic Study of the Competitive Triple Fluorescence of Aminosalicylates and Benzanilides. *Chem. Phys.* 136:321 (1989).
57. P.-T. Chou, D. McMorrow, T. J. Aartsma and M. Kasha. The Proton-Transfer Laser. Gain Spectrum and Amplification of Spontaneous Emission of 3-Hydroxyflavone. *J. Phys. Chem.* 88:4596 (1984).
D. Parthenopoulos and M. Kasha. Coherent Pulse and Environmental Characteristics of the Intermolecular Proton-Transfer Lasers Based on 3-Hydroxyflavone and Fisetin. *Chem. Phys. Lett.* 146:77 (1988).
58. A. U. Khan and M. Kasha. Mechanism of Four-Level Laser Action in Solution Excimer and Excited State Proton-Transfer Cases. *Proc. Nat. Acad. Sci. (US)* 80:1767 (1983).
59. P.-T. Chou and T. J. Aartsma. The Proton-Transfer Laser. Dual Wavelength Lasing Action in Binary Dye Mixtures Involving 3-Hydroxyflavone, *J. Phys. Chem.* 90:721-723 (1986).
T. Dzugan, J. Schmidt and T. J. Aartsma, *Chem. Phys. Lett.* 127:336 (1986).
60. A. Mordzinski and H. K. Grellman. Excited-State Proton-Transfer Reactions in 2-(2-Hydroxyphenyl)benzoxazole. Role of Triplet States, *J. Phys. Chem.* 90:5503-5506 (1986).
61. D. Gormin, J. Heldt and M. Kasha. Molecular Phosphorescence Enhancement via Tunneling Through Proton-Transfer Potentials. *J. Phys. Chem.* 94:1185 (1990).
62. M. Itoh, Y. Tanimoto and K. Tokumura. Transient Absorption Study of the Intramolecular Excited-State and Ground-State Proton Transfer in 3-Hydroxyflavone and 3-Hydroxychromone, *J. Am. Chem. Soc.* 105:3339-3340 (1983).
M. Itoh and Y. Fujiwara. Two-Step Laser Excitation Fluorescence Study of the Ground- and Excited-State Proton Transfer in 3-Hydroxyflavone and 3-Hydroxychromone, *J. Phys. Chem.* 87:4558-4560 (1983)
M. Itoh and Y. Fujiwara. The Ground State Tautomer in the Excited State Relaxation Process of 3-Hydroxyflavone at Room Temperature, *Chem. Phys. Lett.* 130:365-367 (1986).
63. G.-Q. Tang, J. MacInnis and M. Kasha. Proton-Transfer Spectroscopy of Benzanilide. The Amide-Imidole Tautomerism. *J. Am. Chem. Soc.* 109:2531 (1987).

Discussion

Kopelman: Mike, in the beginning you started with J-bands, and the peculiar thing was that not only are they sharp but they are also intense and, I think that to this day people are trying out new theories concerning the emissions and absorptions of these compounds. Then you went to the opposite extreme where there was a reduction of intensity—I forget the name you used for this phenomenon.

Kasha: Hypochromism (depression) or hyperchromism (enhancement), geometry-dependent intensity effects.

Kopelman: When do we get one and when do we get the other? Can you go back into this?

Kasha: Let's see if I can connect those. Now if you have weak coupling vibronic excitons, the only thing that is observable, if there is any coupling, is hypochromism, because unless you had high resolution you could not discern that fine structure splitting. In fact, spectroscopically in weak coupling there is no shift at all; you just have the possibility of hypochromism or hyperchromism. Now, these bands, the J-bands, are they hyperchromic? They would have to be hyperchromic from the nature of the shift because they are in line transitional dipoles, so they would be *hyperchromic*. No one has precisely tested that because you have to have the exact molecular weight to know how many dye monomers are involved to calculate what the oscillator strength should be for the n-fold polymer, so there is some missing information, but I would be confident that they are hyperchromic. That's all I can tell you right now. In other words, there's really strong coupling because of a big shift, 2,000 to 3,000 wavenumber splittings. That's pretty large, and they correspond to the oscillator strength and the rough knowledge of the system, but the exact addition of oscillator strengths is not known yet.

Kopelman: The term "super radiance" came up with the J-bands.

Kasha: Super radiance, to me, implies Einstein-induced emission, which means you have to populate an upper state more than a lower state. Now I don't know anyone who has made a laser from J-band emission, but our systems are super radiant when we get emission of the induced type. Most people know about lasers, but they don't know that you don't need a resonant cavity to get laser-like action. In fact, in a standard way, we induce super radiance with a non-laser cavity in order to measure the exponentiality of the intensity. We measure the super radiance. We measure the intensity at half-cell length and full-cell length, and the ratio of those two should be 2 if there is normal fluorescence. When the ratio is more than 2, we know we have super radiance. Then, when we have measured the alpha-coefficient, the gain coefficient, we can put the sample into a cavity and determine laser pulse and other characteristics.

Kopelman: If you have too large a system, like a crystal, you can't really get super radiance from all of the molecules. If it is too small there are not enough of them. Is there an optimum number of monomers?

Kasha: The big worry about too large an aggregate is that something will act as a trap. For example, in the photographic industry, you have to put in a "spoiler" to break up the J-band effect in the sensitizer dye. If any component acts as a trap, then suddenly one gets emission from it. I think that is the usual thing in a crystal. The surface effects, perfections, surface defects, isotope defects, and trapping ensue by a series of agents.

Christophorou: Sometimes you do not see lasing action, not because it cannot occur but because the excited state absorbs another photon and goes into the continuum.

Kasha: Well, that's the trouble, because we frequently get beautiful proton transfer fluorescence and we would like to develop lasers. Well, we are disappointed much of the time. It took us two years to find out what was the parasitic pathway leading to reduction of the population inversion in one case. If any upper transient coincides even partially with the lasing frequency, you are unable to observe lasing action.

Christophorou: But my impression from your diagrams was that you did not seem to have considered the dissociation of the molecule from those excited states. Why is that?

Kasha: It depends on the molecule. The 3-hydroxyflavone in ordinary solvent at room temperature, after five minutes of radiation in the presence of oxygen, the lasing is gone. It is now known that there is a powerful oxidation pathway. But all you have to do is bubble in argon or nitrogen, then you can observe lasing for six hours without photo-destruction. So the proton-transfer cycle is internal, is very tight, is very efficient; as long as we prevent a competitive chemical reaction, it lases powerfully, but the minute you have a photo-reaction it collapses. We have studied some more complicated systems, and they always go to some electron transfer reaction. So you frequently have competition which is chemical for the proton transfer tautomer lasing.

Christophorou: Can you comment on the dissociation, the molecular dissociation?

Kasha: That is a well-studied subject in certain kinds of systems where they have identified fragments of every kind: hydrogen atoms, all sorts of radicals, in fact even symmetical dissociation. One of the topics that is now being studied by photochemists is how to control ionic versus radical dissociation, and how the solvent influences this. That is a red-hot research field. The object is how to modify the nature of this dissociation, and that is one of the biggest topics in laser photochemistry.

Hall: You left us with the impression that photodynamic therapy is the greatest invention since sliced bread, and that it is only held up by the entrenchment of the FDA, but as I understand it there are two problems that I would like you to comment on. First of all, what is the penetration of red light into tissue? If you want to get to a deep-seated tumor, you've got to penetrate to 10, 15, or 20 centimeters. Secondly, how does the dye cleverly know the difference between a tumor cell and a normal cell? If indeed a substance existed that would distinguish between normal and cancer cells, the whole cancer problem would be solved!

Kasha: I spent a couple of days in Thomas Dougherty's lab in Buffalo, and I learned that at first they used a projection lantern as the light source. They severely burned the whole area, normal skin and tumor, so they had to learn how to more carefully control the light. They got the idea of using a light pipe for internal radiation. Now, the big question that came up is what the dye-sensitized singlet molecular oxygen does. They did extremely careful experiments for scavenger action by simultaneously injecting dibenzfuran; they killed all the effect of radiation and then the tumor grew without cessation. Then the second question is, why is singlet oxygen so toxic? It is so reactive under ordinary circumstances you would think that it would probably kill all cells. And it evidently does have that effect. There are two different notions about this. One is that it kills all cells in the region of sensitization. Dougherty had another idea in that the oxygen supply to the tumor was the locus of sensitization. What he thought he was doing was sealing off the capillary flow and thereby causing the necrosis of the tumor, not by direct cytotoxicity, but in fact, by stopping the oxygen supply. So those are questions that remain, especially the after-effects of the dye injection. There is now extensive study of other systems, and I understand that from experience in ear, eye, and throat hospitals, they have been successful in treating the voicebox and esophagal cancer with this because it doesn't damage the organ once the tumor is killed.

Chatterjee: What about the penetration of light?

Kasha: I don't know how to answer that, Aloke. When I studied Dougherty's results at the time I was there, I got the impression that this guy really understands light flux, all kinds of critical questions, did terrific experiments over a ten-year period on animal tumors until he felt secure. But, also, I didn't believe much until I went up there, and I saw the medical pictures. The one that really impressed me was of an old man, who was around 87. He had a big lump on his cheek and they were treating it. They showed me pictures of him, and about a month later he had scar tissue left. What a fantastic thing if it works that way!

von Sonntag: It is surprising how far this light can enter into the body. It is only very weakly absorbed.

Kasha: It is scattered.

von Sonntag: It is scattered, but since it is only weakly absorbed, it can penetrate surprisingly far.

Hall: Twenty centimeters?

von Sonntag: Well, maybe not that far, but certainly up to between 5 and 10 cm. Maybe longer-wavelength light penetrates even deeper. There is another point which is interesting. Some of these dyes are picked up by the tumor more readily than by healthy tissue. Often, a tumor is also relatively sensitive, which increases the promise of the technique.

Kopelman: About this point that light can penetrate deeply, I was going to say the same. Scattering is somehow not as serious as one would think. George Weiss of NIH has a theory that fits very well with the experiment about this penetration of light.

Kasha: I want to make one comment. The literature before the 1960s on oxygen effects was that oxygen gave reactions directly, as a peroxide, yielding electron transfer. With the development of knowledge of singlet molecular oxygen, especially by dye-sensitization, physical instead of chemical first steps became known. Once you've got singlet oxygen, it has the specific reactivities, and then it in itself can cause cancer or steps initiating cancer inducers.

Inokuti: I want to ask a different question, on discussing the energy transfer and migration in DNA. That is the scientist's idea many times. I never quite understand.

Kasha: Well, here is how I think about it. Surely it is not energy transfer in DNA by strong coupling in the free-exciton range. Surely it is within the weak-coupling range in DNA. The hypochromism is so powerful that it certainly indicates that there is exciton coupling but of the weak-coupling type.

Inokuti: But how do you see it?

Kasha: Oh, I picture it as using the model that I know from Förster. It is a hopping process, not a delocalization of the free exciton.

Inokuti: No, I mean, how do you get the evidence?

Kasha: Oh, hypochromism in the random-to-helix transformation experiments is the evidence, and the hypochromism theory (W. Rhodes). Then you can do trapping experiments. You can put on a dye which will fluoresce while exciting DNA, a dye fluorescence. You can sensitize transfer from the excited bases, distal from where you put the excitation.

Inokuti: In a real cell there are all kinds of other things there; don't they work also as interferences?

Kasha: It depends on where the energy states are. You are saying that there are traps that steal energy. Yes, for example the cytochromes of enzymes in the cells will act as traps as they have low-lying states.

Inokuti: Oh yes.

Varma: All these energy tranfers from the excited state and the twisted molecules that you talked about in terms of energy transfer processes never basically changed to something that reflected itself into a biological end point. You have to have a dissociation. Does this kind of transformation lead to a dissociation?

Kasha: I foresee from the description I've given that there could be reactive tautomers produced as a result of proton transfer, because they are available. There is a finite probability of getting excited tautomers.

Varma: Would the excited tautomer affect, say the base pair chain in DNA? Because you have to dissociate the molecule, is there sufficient energy available to do this?

Kasha: Yes, I believe you get abnormal base pairs as tautomers, they have been discussed in literature, and they would have abnormal geometry of hydrogen bonding, and reactivity would be different.

Chatterjee: How big is the energy transfer in DNA?

Kasha: How many bases? I don't know the answer to that. Hypochromism is the only clue we have to the magnitude of the interaction. We could look at Rhodes' sums and find out how many molecules it would take to give that effect. It is a large number, like hundreds.

Kopelman: This number of bases should be sequence-dependent, right? This should be so because the bases with low-energy levels will act as traps while others will act as anti-traps. So details of the sequence should define the range of energy transfer.

Kasha: Absolutely yes. The exciton level requires exact resonance, and if you lose it you are out of the model. Now, we know that there are four kinds of DNA bases, and we know that their electronic structures are different with the pyrimidines and the purines, having quite different electronic states. Strangely enough, they come in about the same spectral region, so I don't believe anyone has investigated what would happen with, first of all, pure purine-pyrimidine base pairs of single-sequence type, and what would happen when you favor the GC ratio. We have those available from nature. Julius Marmur and Jacques Fresco have studied the melting temperature of different DNAs from different biological sources so we have those available and one could begin to study, what is the effect there on energy transfer as seen by any process, either hypochromism, or especially by an energy transfer trapping experiment.

Geacintov: I would just like to point out that the end products of ultraviolet irradiation of DNA are usually pyrimidine dimers whether energy is transferred or not. That is the bottom line.

Kasha: So they essentially act as the trap.

Ward: On this point, we measured thymine yields in irradiated DNA. We found that the dimers formed in runs of thymine. I don't know if that means that the dimer can be used to reflect the excitation, ... a mirror.

von Sonntag: You mentioned that you see low-lying states in DNA. Might they possibly act as a sink?

Kasha: Oh, very much so, because they are a low oscillator strength. They act as sensitizing sinks because they are the most likely to capture the energy and transfer it to other components.

von Sonntag: But when the other component has no such low-lying states, it wouldn't be able to transfer to it.

Kasha: But in all these photochemical processes, if you get to a state which holds the excitation because it is quasi-stable, then anything can happen here. The excited state could be capable of a chemical reaction, e.g., electron transfer, dissociation, or a radical reaction, or something along that line. In other words, it could yield a chemical step, not an energy transfer step, if the reactive state is coupled to an approaching receptor molecule, e.g., one able to capture an electron. (Moreover, electron transfer in the condensed state is a function of host-guest interaction, so that the magnitude of the energy required can be much diminished over that required for electron ejection or capture in a gaseous state.)

Early Chemical Events and Initial DNA Damage

Aloke Chatterjee and William R. Holley[a]

Abstract

Early chemical events (between 10^{-15} and 10^{-6} seconds) as they relate to the evolution of damage in radiation biology have been described in terms of a theoretical model. DNA is the target of concern in this model, and both indirect and direct effects have been explicitly accounted for in evaluating yields of strand breaks. In the indirect-effect considerations, a quantitative estimation of the time decay of water radical species—beginning with their production at 10^{-14} seconds and leading to the interactions of hydroxyl radicals with DNA—has been a major focus. A method based on stopping-power theory and the Bragg rule has been described to account for direct effects. However, no attempt is made to follow all the chemical events that take place between the creation of initial (10^{-6} seconds) damage and the observable strand break yields. The theoretical calculations refer to a simple aqueous system containing DNA molecules and scavenger (Tris). The theoretical results of strand break yields by different heavy charged particles are in good agreement with experimental cellular data under conditions of minimal enzymatic repair.

Introduction

About eighteen years ago, in Airlie, Virginia, a group of scientists assembled to review what was known at that time about the physical and chemical mechanisms associated with damage to a cell exposed to ionizing radiation. Two years later, a detailed account of the entire proceedings was published in a book,[1] which has been very influential in formulating the future direction of research. In the overall evolution of radiobiological damage, the importance of "early chemical events" has been recognized for a long time; and two chapters in this book, one by Platzman and the other by Kuppermann, provided the basis for critical research in this area. Due to the concerted efforts of physicists, chemists, and biologists, we have made significant advances along these lines in the last eighteen years, and a partial review of the progress made is being reported in this chapter. In the Airlie conference, the discussion of early chemical events was focused entirely on damage to water molecules, whereas in the present chapter we have also included a discussion of DNA damage. This addition has been made possible by the recent success in understanding the mechanisms associated with the initial radiation damage to a DNA molecule, especially the formation of DNA strand breaks. Most researchers believe that double-strand breaks are precursors to observable mutation and/or cell transformation effects which, as a result of exposure to ionizing radiation, can eventually lead to detrimental health effects such as carcinogenesis.

(a) Division of Cell and Molecular Biology, Lawrence Berkeley Laboratory, University of California at Berkeley, Berkeley, California

Physical and Chemical Mechanisms in Molecular Radiation Biology
Edited by W.A. Glass and M.N. Varma, Plenum Press, New York, 1991

257

The early chemical events are very fast processes, spanning a wide range of time scale, from 10^{-15} to 10^{-6} seconds. However, it should be noted that the overall chemical stage in an irradiated cell can persist as long as a second or more. In this chapter, discussion is restricted to early chemical events and hence covers the time period between 10^{-15} and 10^{-6} seconds.

The initiation of radiation damage to a living cell begins with the physical stage of energy deposition in the target medium. The primary processes of ionization and excitation are extremely fast and generally take place in about 10^{-15} seconds. As a result of the energy deposition, new chemical species such as hydroxyl radicals (radiolysis of water) are produced within the cellular complex. These species attain the surrounding temperature in about 10^{-12} seconds. This period is called the "prethermalization stage." At this point in the time-scale, diffusion of radicals begins and various chemical reactions follow, some of which result in biological damage. Since DNA inside a cell is the most critical target, an evaluation of damage to this important molecule is of primary concern. Water-radical-induced damage is generally over in about 10^{-6} seconds, and the resulting chemical change in a DNA molecule may produce cell death, mutation, or transformation. Damage to a DNA polymer can also result as a consequence of direct energy deposition on the various sites (sugars and bases) of the molecule. This process also takes place in about 10^{-15} seconds. However, the damage to DNA (e.g., strand breaks, base release) from these fast processes can evolve over several orders of magnitude in time. Not much is currently known about the kinetics of these processes.

Once the damage has been chemically stabilized, the cell responds by trying to repair the lesions through enzymatic processes and the success of this effort depends upon the nature of the chemical change. Thus, evaluation of the early (10^{-6} seconds) damage to a DNA molecule is of crucial importance in radiation biology.

Early in the development of radiation chemistry, it was realized that understanding the radiolysis of water is vital to the study of mechanisms associated with DNA damage. Since then, extensive effort has been expended by many investigators to provide a quantitative analysis of the steady-state and time-dependent yields of chemical species following absorption of energy by water molecules. In the second section, an account of the present status in this area of research is presented along with a physical description of the pattern of energy deposition. Most of these studies are theoretical in nature, and it has become increasingly clear that improvements in experimental techniques are not likely to reduce the time resolution to less than a few picoseconds. This limit, referred to as the picosecond barrier to early real-time observations, may be imposed either by the time required to make the deposit or the transit time of the analyzing light. For the purpose of this consideration, a minimal reaction cell can be taken as a cube, 3 mm on a side. If the energy is deposited by a δ pulse of electrons traveling with the velocity c of light in a vacuum, the beam requires 10 ps to travel through the cube. The analyzing light requires a slightly longer time to traverse the sample (because the refractive index of the liquid is > 1.0) and inevitably records effects arising from energy deposits with a time spread of 10 ps or more. However, even the limited availability of experimental data above 10 ps has proven to be extremely useful in developing theoretical models for quantitating the evolution of early chemical events. In the third section, a theoretical model is presented which describes the evolution of DNA strand breaks by water-radical species. The model considers a very simple DNA aqueous solution system also containing a scavenger of water radicals. This seemingly simple DNA system gives results which are quantitatively similar to these observed in a cellular complex. As mentioned earlier, direct energy deposition on a DNA molecule can also result in the formation of strand breaks. This process, which is initiated through excitation and ionization in about 10^{-15} seconds, is described in the fourth section. In the fifth section, results of theoretical calculations based on our present understanding of early chemical stages in a simple DNA aqueous solution system are compared with experimental data derived from observations of mammalian cell systems. In the final section, we present a brief discussion of the various limits of the current theoretical model along with some suggestions on future studies.

Energy Deposition and Water Radiolysis

There are several theoretical approaches which deal with the calculation of yields of chemical species produced as a result of radiolysis of water. For example, Zaider et al.[2] have developed a Monte Carlo scheme using gas-phase cross sections for electronic energy loss. Then they extrapolate to unit density for applicability in the liquid phase. Their strategy is based on the fact that much more extensive experimental data are available for the vapor phase than for the liquid phase. It is not clear whether this procedure adequately describes the liquid-phase cross sections. However, their time-dependent decay curves for $\cdot OH$ as well as e_{aq}^- have similar shapes to those observed experimentally by Jonah et al.[3,4] In a different approach, Brenner and Zaider[5] use Monte Carlo methods to obtain the spatial distributions of water radicals at $\sim 10^{-12}$ seconds, which become the input to another code, based on Smoluchowski's equation, that computes events to later times. In contrast to these approaches, Chatterjee and Magee[6] initiate the computation with an assumed Gaussian distribution following the passage of a charged particle track. In their method, they utilize track entities called "spurs," "blobs," and "short tracks" for electron irradiation and "core" and "penumbra" for heavy-charged particles. This approach has now been extended to evaluate the yields of DNA strand breaks. However, as will be seen later, in the Chatterjee and Magee model, the stage between 10^{-15} and 10^{-12} seconds is treated very approximately.

Perhaps the most complete model available at present is the one developed by Turner et al.[7] In this model, attempts are made to account for every ionization or excitation event in a statistical manner that follows through to the essential completion of intratrack radical reactions at 10^{-7} seconds. In liquid water, they represent the initial physical processes by writing

$$H_2O \rightarrow H_2O^+, H_2O^*, e^-$$

They have developed a Monte Carlo computer code, OREC, to perform the detailed transport and energy loss calculations of the primary charged particle and all of its secondary electrons. For a primary or secondary particle of given energy, the code selects the following: a flight distance to the next event, the type of event that occurs, the energy lost, the scattering angle, and the energy and angle of the secondary electron when an ionization occurs. The inelastic cross sections are derived from a complex dielectric response function developed specifically for liquid water.[8] Turner et al. explicitly consider the physico-chemical and chemical stages. From their methods of computation, they obtain the same ferric yield for tritium β-rays as has been measured experimentally. In addition, their theoretical decay curves for $\cdot OH$ and e_{aq}^- are in good agreement with the experimental data of Jonah et al.,[3,4] measured between 3×10^{-11} and 4×10^{-8} seconds.

In spite of these exciting developments, the method has several limitations. First, it has not been extended to high-energy (~ 600 MeV/n) heavy charged particles, nor has it yet been used to calculate the yields of DNA strand breaks. We now proceed to describe the model of the present authors, since that is the only one which has been applied to evaluate DNA strand break yields based on chemical and biochemical mechanisms.

Track Entities of Energetic Electrons: Spurs, Blobs, Short Tracks, and Branch Tracks

Stopping-power theory is based on an implicit assumption that is called the CSDA, or the continuous-slowing-down approximation. From this theory one gets quantitative information on the average rate of energy loss (i.e., the linear energy transfer, or LET). The usefulness of this information is limited, and for an understanding of the radiation chemical processes it can actually be misleading if one assumes that all track segments receive an energy corresponding to the average energy deposit. This would be the situation if a continuous-slowing-down mechanism were actually applied. Glancing collisions create excitations to energy levels that have oscillator strength for ordinary optical excitation. In materials of low atomic number, the oscillator strength distributions are essentially contained in the region 6 to 100 eV. These low-energy losses create track entities called "spurs." If the particle is fast and lightly ionizing, the spurs are widely spaced and develop independently, that is, without interference from the neighboring spurs. If the particle is slow enough to be heavily ionizing, the spurs overlap and form a continuous column.

Half of the energy deposit of a heavy particle is in the spurs. If the particle is lightly ionizing, such as a proton with energy of about 100 MeV, the spurs are well separated and isolated spurs form a good approximation of the track. An average spur may contain two or three radical pairs, and so we see that even the approximation of isolated radical pairs for a track may be reasonable.

The part of the spectrum arising from the knock-on electrons consists of electrons with energies from about 100 eV to a maximum value of $2 mv^2(1 - \beta^2)^{-1}$. These recoil electrons are created in processes that depend on the Rutherford scattering cross-section, which varies as ϵ^{-2}. Thus, low energies are favored. The low-energy-loss events (near 100 eV) are like spurs whereas the very high-energy-loss events (e.g., in the multi keV range) produce branch tracks. Thus, the behaviors of individual loss events in the knock-on spectrum vary widely.

The problem of describing the radiation chemistry of these events is the same as that of describing the radiation chemistry of the primary track itself. The 100-eV upper limit for spurs is somewhat arbitrary. Track entities formed with energies a little above 100 eV are like spurs; however, for some higher value of the energy loss, a recoil electron energetic enough to create a series of spurs is formed and a short track results. Mozumder and Magee[9] considered these matters for electrons in water and decided that for an energy loss of 500 eV, the track entity should be classified as a short track. The track entities in the range of 100 to 500 eV are called "blobs." This noncommittal name is used because the shape of the entities varies from essentially spherical to rodlike as the energy increases from 100 to 500 eV.

The "short track" category of track entity also has an upper limit in energy. Mozumder and Magee argued that at an energy of 5 keV, a knocked-out electron forms several spurs before ending up in the continuous distribution called a short track. Thus, 5 keV was chosen as the upper limit for the short track.

The treatment of the radiation chemistry of electron tracks depends on the system absorbing the energy, its reaction sequences, and so on. However, to a large extent the track entities are the same for all materials of low atomic number. Mozumder and Magee suggested that all electron tracks can be considered as composed of three track entities:

Spur	Energy deposit of ~ 6 to 100 eV
Blob	Energy deposit of ~ 100 to 500 eV
Short track	Energy deposit of ~ 500 to 5000 eV

Heavy Particle Tracks: Core and Penumbra

Ionizing radiations in the form of energetic heavy charged particles are now available in the Lawrence Berkeley Laboratory's compound accelerator called the BEVALAC.[10] A heavy ion linear accelerator (the SuperHILAC) and a synchrotron (the BEVATRON) are joined by a transfer line. Particles from protons through uranium ions are accelerated to 8.5 MeV/n in the SuperHILAC and then injected into the BEVATRON ring to produce intense beams of heavy charged particles with energies up to 1 GeV/n. Because of their varying rate of energy deposition (-dE/dx), these particles constitute an excellent class of probes for unfolding the mechanisms involved in the induction of damage such as DNA strand breaks by ionizing radiation.

The best known physical quantity associated with the energy loss phenomenon is the stopping power, -dE/dx, frequently referred to as linear energy transfer (LET). According to the Bethe theory,[11] this can be written non-relativistically as

$$LET = \frac{2\pi z^2 e^4}{mv^2} \, NZ \ln \frac{2mv^2}{I} \tag{1}$$

where z is the charge number of the incident particle, e is the electronic charge, m is the mass of an electron, v is the particle velocity, NZ is the number of electrons per unit volume of the target medium, and I is the mean excitation potential of the medium. For water, $I = 65$ eV.[11] A plot of LET in water versus energy for several representative particles from hydrogen to fermium is shown in Fig. 1.

Many biological end points have been studied as a function of the LET values. However, experiments[12] have shown that a given value of LET does not uniquely determine the measured values. Experimental data in radiation biology and in radiation chemistry have clearly demonstrated that different ions having the same LET yield quantitatively different values for biological end points in a given system. The reason for such differences has been attributed to the differences in the microscopic pattern of energy distribution around the trajectory of a charged particle track. For high-LET charged particles, one way to describe the microscopic pattern of energy distribution is through the concept of "core" and "penumbra." The r_c of the core is defined as the radius within which all of the energy loss due to glancing collisions is contained. In addition energy is also deposited in the core by those low-energy knock-on electrons which cannot penetrate this radius successfully. The radius of the core, r_c, is given by

$$r_c = \hbar v / 2\epsilon_1 \tag{2}$$

where v is the velocity of a charged particle, ϵ_1 ($\approx$ 6 eV for water) is the lowest electronic transition energy, and $\hbar = h/2\pi$ is the modified Planck's constant. The core radius is independent of the charge on the projectile. The energy density within the core, not its size, varies with the charge. For 600 MeV/n, the radius of the core is 82 Å; this goes down to 15 Å for a 10 MeV/n particle. Somewhat more than 50% of

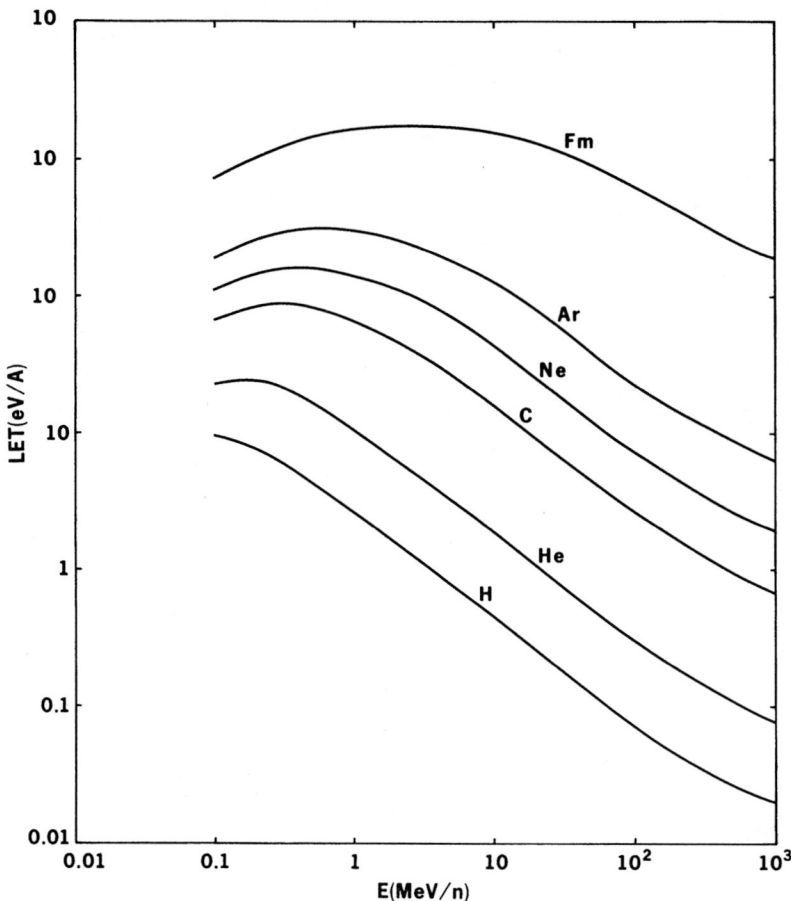

Figure 1. The variation of LET in water as function of the specific energy for six selected particles has been plotted. Except Fm, all the particles can be accelerated to high energies at the Berkeley BEVALAC.

the total LET is contained within r_c, and the rest of the energy is carried out to very large distances by energetic secondary electrons. The average energy density deposited by these electrons is at least an order of magnitude less than the energy density inside the core region. The space traversed by energetic δ-ray electrons outside the core is called the "penumbra." The average radius, r_p, of a penumbra is given by the empirical relation

$$r_p = 396 \, (v)^{2.7} \qquad (3)$$

where v is in units of 10^9cm·s^{-1} and the radius is in angstroms. It should be noted that r_p is not the range of the maximum energy knock-on electron because r_p refers to the radial extension of the track, and, in a classical description, maximum energy electrons are always ejected in the forward direction. The empirical relation given by Eq. (3) has been obtained by fitting values from detailed calculations based on scattering theory and diffusion models for electron propagation.[13,14] A cross-sectional view of a core and penumbra is shown in Fig. 2. Particles of different charge having the same LET values have different values of velocity, resulting in different

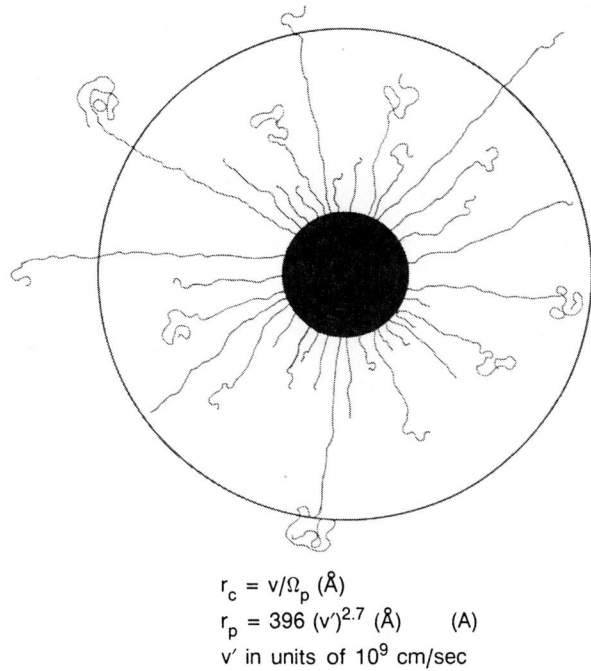

$$r_c = v/\Omega_p \ (\text{Å})$$

$$r_p = 396 \ (v')^{2.7} \ (\text{Å}) \qquad (A)$$

$$v' \text{ in units of } 10^9 \text{ cm/sec}$$

Figure 2. A cross sectional view of a long segment of a heavy particle track is shown. The central dense region is called the "core," and it is surrounded by an envelope containing the tracks of energetic δ-rays or secondary electrons. The region which contains the δ-rays is called the "penumbra."

radii of core and penumbra, which then result in different energy densities as a function of radial distance from the trajectory of the incident particle. The average energy density within the core ($0 < r < r_c$) can be expressed by

$$\rho_{core} = \frac{(LET/2)\left[1 + \ln\left(r_p/r_c\right)\right]}{\pi r_c^2 \left[1 + 2\ln\left(r_p/r_c\right)\right]} \tag{4}$$

and the corresponding expression for the penumbra region ($r_c < r < r_p$) is given by

$$\rho_{pen}(r) = \frac{LET/2}{\pi r^2 \left[1 + 2\ln\left(r_p/r_c\right)\right]} \tag{5}$$

It is not experimentally feasible at the present time to measure ρ_{core}; however, $\rho_{pen}(r)$ has been measured by Varma and Baum[15] with good agreement with the theory. From these expressions, one can see that the energy density within the core is constant, and within the penumbra it varies as the reciprocal of the square of the radial distance from the trajectory of the penetrating charged particle. If one sums the integrals of these energy density expressions within their radial regions, the total value of the LET is obtained.

The sharp separation between core and penumbra given by Eqs. (4) and (5) are only for calculational purposes. In reality, such a fine demarcation between the two regions does not exist. However, this procedure has been used successfully in studies of other aqueous systems.[6,16]

The concept of core and penumbra has been extensively used in the analysis of the data for water radiation chemistry with different heavy charged particles.[17] In this paper, a procedure is presented to model DNA strand break production as an extension of our present understanding of the early chemical events in radiation chemistry. The aim is to describe qualitative and quantitative aspects of the yields of DNA strand breaks as functions of the velocity and charge of an incident particle so that the biophysical and biochemical basis of the initial radiation damage by high-LET charged particles can be understood.

The Chemical Stage: The Formation of Strand Breaks by Water Radicals

When mammalian cells are subjected to radiation exposure, energy is deposited on various molecules (e.g., proteins, enzymes, water), including the DNA. These molecules are present inside the cellular nucleus. Energy absorbed by water molecules constitutes a significant fraction of the energy deposited. It has been demonstrated experimentally that the products of water radiolysis can indirectly cause biochemical changes in a DNA molecule. These processes are called the "indirect effects" of radiation damage.[18,19] Energy can also be deposited directly on the DNA molecule, creating ionized and excited states of the various molecules (such as sugars, bases, or phosphates). These physical processes, which can also lead to DNA damage, are generally known as the "direct effects" of radiation.[18] In order to understand them, it is necessary to model the various types of damage to a DNA molecule from the fundamental laws of atomic and molecular physics and then to correlate them with the biochemical and biomolecular changes.

The various targets inside the nucleus of a cell represent a very complex system. It is not clear at present how the energy deposition on molecules such as proteins and enzymes leads to DNA damage. It is generally assumed that since the relative concentrations of these molecules are small compared with that of water molecules, their role in causing indirect damage to a DNA molecule, particularly with respect to the formation of strand breaks, is insignificant. However, the various radical products formed as a result of energy deposition on water molecules can be prevented from interacting with the DNA molecule by the proteins, enzymes, etc., and this phenomenon is called the "scavenging effect." This effect cannot be neglected because the concentration of the water radicals produced is also very small. If we keep these facts in mind and recognize the complexities involved in a cellular environment, we can simplify the problem by considering an aqueous solution of DNA containing, for example, Tris buffer. The purpose of Tris is to maintain the proper structural conformation of the DNA molecule and also to act as a scavenger of water radicals. This system has proven to be convenient for theoretical studies as well as experimental measurements, and the results seem at present general enough for applications in cellular radiobiology.

The specific systems we have chosen to study are aqueous solutions of Simian Virus (SV40) DNA with either 10 mM Tris buffer (in which indirect effects dominate) or 500 mM Tris buffer (which simulates cellular conditions). The concentration of SV40 is typically 10 μg/mL, although higher concentrations have also been used. In a cellular complex, the average migration distance of a water radical is about 30 Å.[18] Thus, if such a radical is produced significantly beyond this distance from a DNA molecule, the chance that it will reach the DNA is extremely small because the proteins or enzymes will scavenge it. With 500 mM of Tris, the average migration distance of such radicals is also about 30 Å.

In order to assess the damaging effects of water radicals, it is important to consider the various stages in the evolution of damage. Because the water molecules absorb energy, the initial processes are excitations and ionizations, which occur in about 10^{-15} seconds as dictated by the uncertainty principle. The first ion molecule reaction takes place in about 10^{-14} seconds. The "primary" chemical species produced are $\cdot$H (hydrogen), $\cdot$OH (hydroxyl), H_3O+ (hydronium ion) and e_{aq}^- (solvated electron). These species quickly reach the temperature of the surrounding medium ($\sim$ kT, where k is Boltzman's constant). This process is called "thermalization" and takes place in about 10^{-12} seconds. At this time they undergo diffusive motion in three-dimensional space, and there are three ways that their fates are determined: (i) chemical reactions between themselves (e.g., $\cdot$OH $+ \cdot$OH $= H_2O_2$, $\cdot$OH $+ \cdot$H $= H_2O$ and so on); (ii) reactions with scavengers (e.g., Tris), and (iii) reactions with sugars or bases of DNA molecules. Unless a radical is produced within about 30 Å from a DNA molecule, the first two reactions dominate. However, even beyond 30 Å, where the probability of a reaction with DNA is low, it is important to consider the fates of all the radicals, irrespective of their positions.

On an average, about 17 eV is needed to create a hydroxyl radical in water. The initial (10^{-12} seconds) yields[20] of the respective species are expressed in terms of G-values, which can be defined as the number of a given radical produced per 100 eV of deposited energy: $G_{OH} = 5.88$, $G_H = 0.88$, $G_{e_{aq}^-} = 5.0$, and $G_{H_3O^+} = 5.00$. How these initial values depend upon LET is not clear at the present time. For operational purposes they have been considered to be independent of LET. The initial distribution of these species is assumed to be Gaussian, and the various standard deviations have been reported.[6]

How close to, or how far from each other the radicals are produced depends upon the track structure. In the region of the core, they are generally formed very close to each other, depending upon the energy density for a given particle charge and velocity. In the penumbra, they are generally much more spread out. Hence, compared to the core, more radicals survive reaction (i) in the penumbra. It is possible to simulate the random diffusion process by a Monte Carlo technique.[21,22] The computer simulation of the actual diffusive motion by a finite number of steps is carried out by allowing the reactants to "jump" in three dimensions over a distance ℓ during a time interval Δt. The time interval and distance are interrelated by the diffusion constant, D, such that

$$\ell^2 = 6 D \Delta t \qquad (6)$$

The step size, ℓ, has been taken to be ≤ 1 Å to ensure that the reactants cannot diffuse far enough in one step to "jump through" each other without having a chance to react. A pair of species are allowed to react with each other when they are within a critical distance, characteristic of the species. This idea is based on a model proposed by Smoluchowski.[23] For $\cdot$OH-$\cdot$OH reactions, this distance is 2.83 Å; for e_{aq}^- and $\cdot$H reactions, it is 2.56 Å, and so on.[20] These reaction radii, R_{AB}, have been calculated from the available experimental data on the rate constants, k, according to the formula

$$k = 4\pi (D_A + D_B) R_{AB} \qquad (7)$$

where D_A and D_B are diffusion constants and k is the rate constant for a chemical reaction (defined as the rate at which the concentration of a reactant decreases for unit concentrations of other chemical species). For hydroxyl radicals, $D = 2 \times 10^{-5}$ cm$^2 \cdot$sec^{-1}; for hydrogen radicals it is 8×10^{-5} cm$^2 \cdot$sec^{-1}. The rate constant k for water radical reactions is of the order of 10^9 l mole^{-1} sec^{-1}.[22]

Based on these procedures, decay curves (loss of water radicals from the system) in pure water have been obtained by developing a suitable algorithm for energetic electrons and heavy charged particles. The accuracy of these theoretical curves cannot be determined at the present time due to the lack of experimental data for heavy charged particles. For these measurements, nanosecond or shorter duration pulses are needed, and the BEVALAC does not have this capability. However, at the Argonne National Laboratory, Jonah and Miller[3] have obtained experimental data for a 20 MeV pulsed electron beam; hence, a limited comparison is available for electron irradiation, as will be seen below in "Results."

The presence of Tris affects the pure-water decay curves and has been accounted for by reducing the survival probability after each jump time, Δt, corresponding to the jump distance, ℓ, [see Eq. (6)] by a factor given by $\exp(-\Delta t/3 \times 10^{-8})$, where 3×10^{-8} seconds is the characteristic absorption time (reduction of radical by e^{-1}) for 10 mM Tris and 6×10^{-10} seconds for 500 mM Tris.

In order to calculate the effects of water radicals on the production of strand breaks, the following assumptions have been made. Only hydroxyl radicals produce strand breaks, and they do so by extracting hydrogen from the sugar moiety (see Fig. 3). Attacks on the bases lead to point mutations, not strand breaks.[24] Solvated electrons react only with the bases (adenine, guanine, thymine, and cytosine). In the presence of dissolved oxygen in water (always present unless care is taken to remove it), hydrogen radicals react predominantly with molecular oxygen. A few react with the bases. These considerations are based on experimental data.[24]

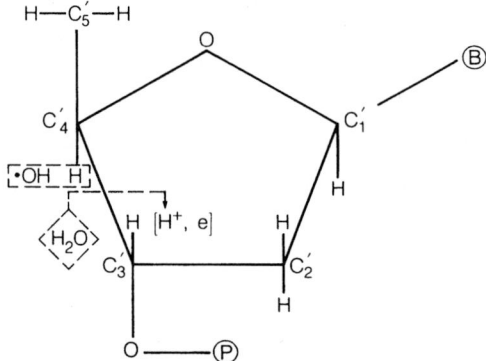

Figure 3. A schematic view of a sugar ring attached to a base is shown. According to the presently understood mechanisms, there are three ways to produce strand breaks: (i) abstraction of an ·H by an ·OH (indirect effect), (ii) ionization of a hydrogen atom followed by deprotonation (direct effect), and (iii) cleavage of C-C bond (direct effect). From the point of view of sugar damage, the first two processes are similar, namely, the loss of an ·H from one of the five carbon positions.

If the various sugar and base molecules are randomly distributed according to their concentrations, and if their structural organization is not taken into account, the rate constants for reactions with water radicals do not agree with observed experimental data. For example, when isolated sugar molecules are present in a solution, the rate constant for .OH reaction is about five to ten times greater than when they are present in the structural configuration of DNA.[25] Hence, to calculate the production of DNA strand breaks, the coordinates of each sugar molecule and each base molecule have been positioned in three-dimensional space using X-ray diffraction data of Arnott and Hukins.[26] Once these coordinates are fixed in space, the diffusive motion of each surviving hydroxyl radical has been followed (for each Δt) and, based on Smoluchowski's theory, a radical has been allowed to react with a sugar or a base molecule, depending upon its critical distance of approach. These distances are obtained from Eq. (6) with $D_{DNA} = 0$; i.e., diffusion of the large DNA polymer has been neglected. The respective reaction radii for hydroxyl radicals and the various sites are as follows: $\cdot OH$ + sugar, 1.0 Å; $\cdot OH$ + adenine, 3.6 Å; $\cdot OH$ + thymine, 3.6 Å; $\cdot OH$ + guanine, 5.2 Å; and $\cdot OH$ + cytosine, 3.6 Å.[20]

For each OH radical whose diffusive motion is to be followed, initial position coordinates (distributed uniformly in space around the DNA) are determined using random numbers and the assumed Gaussian distribution. The radicals are followed in space and time by simulating the diffusive motion as described earlier. Corresponding to an interval, Δt, of jump time, the probability P of survival for an OH radical under consideration is determined from the previously obtained decay curves containing Tris molecules. Next, a random number between zero and one is selected, and, if it happens to be larger than P, the $\cdot OH$ is removed from further consideration. If the number is less than or equal to P, the next jump for $\cdot OH$ is taken. In this manner a surviving OH radical is followed until it is found to be within the reaction radius of a sugar or a base. We keep a record of how many OH radicals are lost from the initial number on their way to a DNA molecule, how many have reacted with the sugar moiety, and how many have reacted with the bases. The total number of trials (for a hydroxyl radical) is typically about 200,000; this seems to reproduce results within ± 2%. The use of the decay curves for the Tris solutions has saved an enormous amount of computer time. A pictorial representation of the $\cdot OH$ attack on a DNA molecule is shown in Fig. 4.

Based on these procedures, the D_{37} values have been calculated for single-strand breaks (damage to one of the strands only). The D_{37} value is defined as the dose required to reduce the fraction of undamaged DNA molecules to 37% of the initial unirradiated number. Using appropriate decay curves, separate calculations have been made for the core and the penumbra. From the D_{37} values, we have computed the cross sections for single-strand breaks for different LET particles according to the formula

$$\sigma_{SSB} (\text{Å}^2) = K \, LET \left[f_{core}/D_{37}^{core} + \left(1 - f_{core} \right)/D_{37}^{pen.} \right] \qquad (8)$$

where f_{core} is the fraction of energy deposited in the core, $K = 16.02$ when the LET is in eV/Å and the D_{37} values are in kilorads.[16]

Double-strand breaks have been estimated by assuming that 2.0% of the single-strand breaks (experimental result) also result in breaks on the opposite strand. This procedure is somewhat unsatisfactory. However, the method has been adopted in the absence of a mechanism for the production of double-strand breaks due to indirect effects. Two hydroxyl radicals attacking opposite strands in these calculations do not produce enough breaks to agree with the experimental data. This mechanism yields smaller values by about two orders of magnitude.

Energy Deposition on a DNA Molecule: The Formation of Strand Breaks by Excitation and Ionization

The main features of the procedure for calculating the yields of strand breaks due to direct effects have been reported previously[27] and are summarized below. Simple elements of track structure and stopping-power theory are combined with the detailed geometry of DNA molecules to calculate the average energy deposition by fast charged

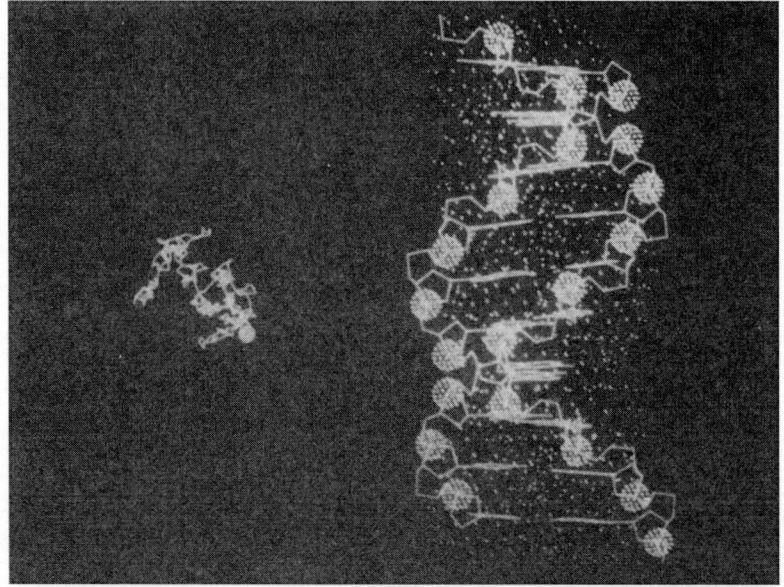

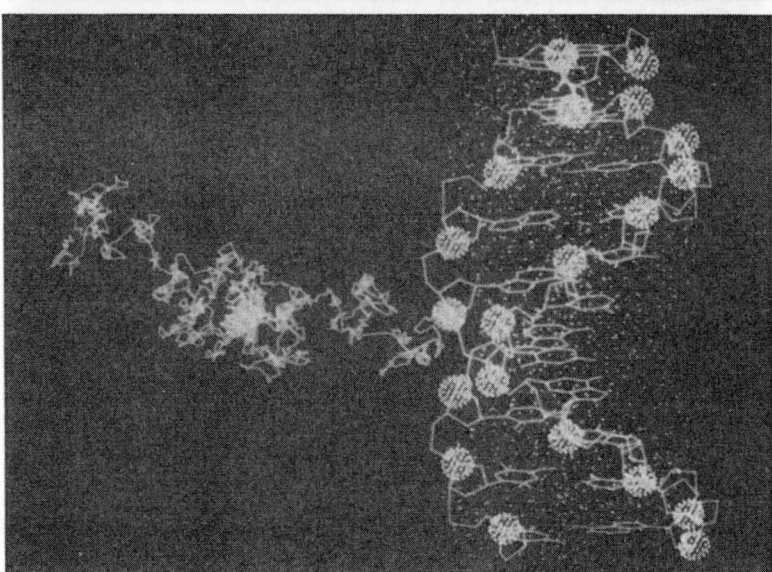

Figure 4. Monte Carlo simulations of the random motion of an OH radical formed in water in the vicinity of a DNA molecule. In the upper figure, the radical is destroyed by reaction with one of its siblings or the tris buffer. In the lower figure, the radical reacts with a deoxyribose unit of the DNA, which leads to strand break formation.

particles on DNA. Since the actual physical energy deposition processes are stochastic in nature, we use Poisson statistics to obtain a detailed description of the distribution of the resulting states of molecular excitation and ionization. The average energy of excitation (or ionization) on the DNA has been calculated from existing oscillator strength data[28] for biological molecules to be approximately $<E> = 30$ eV. Excitation of an atom on the DNA sugar-phosphate backbone is assumed to be followed by

Chatterjee and Holley

one of two different types of biochemical change, deprotonation or direct dissociation, both of which then lead to DNA strand breaks. Base damage has been assumed not to lead to the formation of strand breaks.

In these calculations, also, we have used a three-dimensional geometric model of DNA. The spatial orientation of a given track can be represented by its impact parameter or distance from the DNA central axis, the angle between the track and DNA axis, and the core radius, r_c. The coordinates of a particle track and the DNA model can be used to determine which sugar-phosphate molecular groups lie within the core as shown in Fig. 5. By restricting ourselves to glancing collisions only and using the Bragg Rule[29] in combination with the stopping-power formula (as discussed in detail in Ref. 27), it has been possible to calculate the average energy, E_g, lost to a collection of atoms (sugar-phosphate) inside the track core. Since energy density in the core is a sum of glancing collisions and low-energy knock-on collisions, we have used Eq. (4) to calculate the average total energy deposited on a particular site inside the core to be

$$E_d(r \leq r_c) = E_g\{1 + 1/[1 + 2\ln(r_p/r_c)]\} \tag{9}$$

Similarly, for the penumbra region (to account for the δ-rays) we use Eq. (5) to obtain

$$E_d(r_c \leq r \leq r_p) = E_g r_c^2 / \{r^2[1 + 2\ln(r_p/r_c)]\} \tag{10}$$

The average number of excitations (and/or ionizations) on a sugar-phosphate molecule lying at a radial distance, r, from the trajectory is

$$n = E_d(r) / <E> \tag{11}$$

Use of Poisson statistics results in the respective probabilities of no excitation, one excitation, two excitations, etc., of

$$P_i = \frac{n^i e^{-n}}{i!} \tag{12}$$

where i = 0,1,2,.....

The probability of one or more excitations is given by

$$P_{\geq 1} = 1 - P_0$$
$$= 1 - e^{-n} \tag{13}$$

For each incident particle track we use Eqs. (9), (10), and (11) to calculate the probability of ionization, $P_{\geq 1}$, at each reaction site within the track penumbra radius r_p. We generate an "event" by choosing a random number, r#, for each of these sites. If r# < $P_{\geq 1}$ for a particular site, we consider an ionizing reaction to have occurred

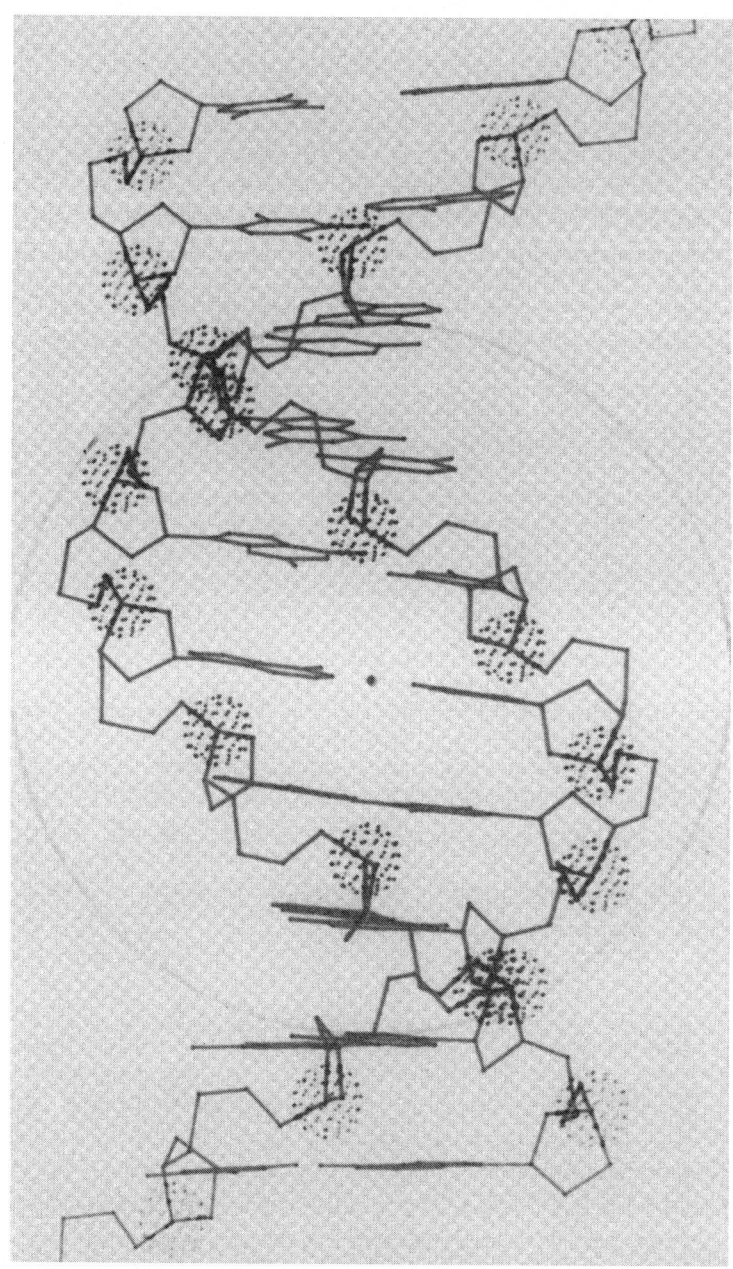

Figure 5. On a one-to-one scale, the core of a 10 MeV/n Ne particle (r_c = 15 Å) has been superimposed at normal incidence both to the plane of the figure and the axis of a structured DNA molecule with a strand separation distance of 20 Å. It is seen very clearly that within the core there are several sites on the DNA molecule that can be affected by energy deposition.

270 *Chatterjee and Holley*

with the ultimate production of a DNA strand break. Breaks occurring close to each other on opposite strands (i.e., within 10 base pairs) lead to double-strand breaks. An event is then classified into one of the following categories:

- No breaks (NB)
- One or more single-strand breaks (SSB)
- One or more double-strand breaks (DSB).

If we consider N incident particle tracks uniformly distributed over an area, A, yielding NSSB single-strand breaks and NDSB double-strand breaks on DNA of molecule weight M, the corresponding normalized cross sections are given by

$$\sigma_{SSB} = N_{SSB} / (M\ N/A) \qquad (14)$$

$$\sigma_{DSB} = N_{DSB} / (M\ N/A) \qquad (15)$$

It is frequently more convenient to discuss strand break production in terms of yields of breaks per unit radiation dose rather than cross sections. The yields, Y_x, correspond to the normalized cross sections obtained from Eqs. (8), (14), and (15), including indirect and direct effects. They are given by the following expressions:

$$Y_{SSB} = \sigma_{SSB}\rho / LET \qquad (16)$$

$$Y_{DSB} = \sigma_{DSB}\rho / LET \qquad (17)$$

where ρ is the mass density of the medium and LET is the total linear energy transfer of the incident particle. In terms of conventional units of μm^2 for σ, keV/μm for LET, and rad for dose, the expression for the strand break yield per rad per dalton becomes

$$Y = 0.062\ \sigma / LET \qquad (18)$$

In the expression for Y, it should be kept in mind that σ depends upon track structure. For simplicity in our calculations, we have taken σ to represent the sum of σ_{Ind} and σ_{Dir}. We recognize that at high LET, this assumption of simple additivity may overestimate the total cross sections because it does not take into account possible spatial correlation between direct-effect and indirect-effect damage. Based on preliminary studies of correlated damage sites, we estimate that this effect introduces less than 10% error.

Results

In order to examine the validity of theoretical calculations, comparison with experimental data is essential. However, the kinetics of early chemical events have been measured in only a few experiments. These measurements have been carried out in pure water by Jonah et al.[3,4] for 20 MeV electrons. Decay curves are now available between about 100 picosecond and 4×10^{-8} seconds for e_{aq}^- and between 200 picosecond and 3×10^{-9} seconds for $\cdot$OH. In Fig. 6, experimental values ($\bullet$) along with theoretical calculations (solid line) for $\cdot$OH decay are presented for intercomparison. This limited comparison indicates that the Monte Carlo technique

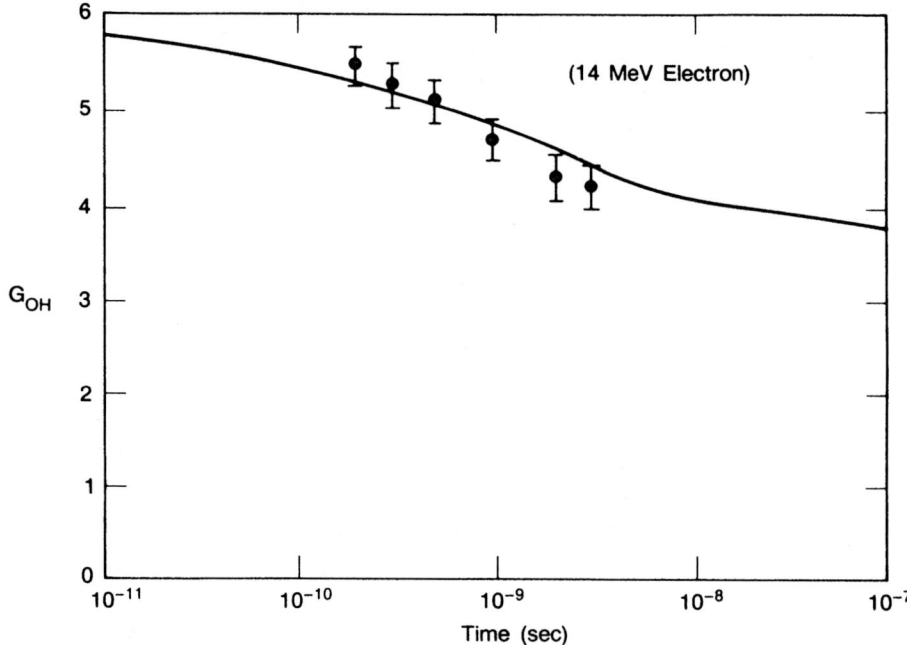

Figure 6. Decay of hydroxyl radicals in water with time. The solid line is a result of the present model, and the solid circles are the experimental points.

simulates the early chemical events in a manner which is compatible with the experimental data. Similar agreements have also been reported by Turner et al.[30]

The corresponding ·OH decay curves for energetic heavy charged particles are presented in Fig. 7, based on theoretical calculations only. Since the presence of an ·OH scavenger is an important consideration, the results are also presented, including effects due to the presence of Tris-HCl (500 mM) in the aqueous system. When the energy of a neon particle is 40 MeV/n or even 670 MeV/n, the decay rates are much faster than the corresponding curve for 14 MeV electron. For 670 MeV/n, the Tris has little scavenging effect before 10^{-9} seconds; i.e., the water radical species do not feel the presence of scavengers until they are sufficiently dispersed in about 10^{-9} seconds. When the energy of the neon particle is 40 MeV/n, the energy density is very high in the core, as well as in the penumbra. This results in the formation of high radical density, which favors the formation of molecular products of water. Hence, Tris molecules do not effectively compete with radical-radical interactions until about 10^{-8} seconds. As mentioned earlier, these results seem plausible, although the absence of any experimental data prohibits verifying the accuracy of simulating the kinetic processes. However, these processes have been used to calculate yields of single and double-strand breaks, which are comparatively easier to measure.

Since the production of electrons is an essential feature of any quality of ionizing radiation, it is of fundamental importance to obtain yields of strand breaks over a wide range of electron energies, 100 eV to 1 MeV. Details of the calculations are published elsewhere.[31] The results are based on indirect as well as direct mechanisms and are presented in Figs. 8 and 9. The concentration of Tris-HCl for these calculations is 500 mM to mimic the cellular conditions from the point of view of radical migration distance.

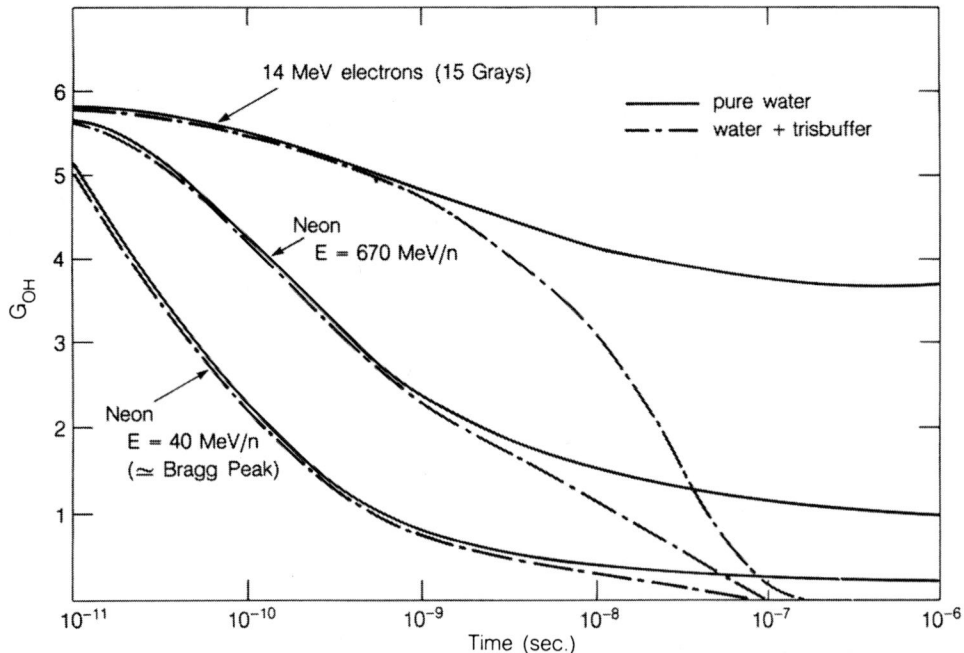

Figure 7. The decay of hydroxyl radicals with time in pure water and water containing Tris when irradiated with 14 MeV electrons and 40 and 670 MeV/n neon ions. The full lines represent the decay in pure water, and the broken lines are for decay in Tris water.

In Fig. 8, the yields of single-strand breaks and double-strand breaks from indirect effects have been plotted as a function of electron energy expressed in keV. The respective yields are normalized to breaks/rad/dalton. Over the span of electron energies between 100 eV and 10^3 keV, both of these breaks go through a wide minimum. At lower electron energies (less than 600 eV), as the energy decreases, fewer and fewer water radicals are produced; hence, the relative importance of sibling reactions (reactions between water radicals) becomes less. This results in higher yields of single- and double-strand breaks. As electron energy increases above about 600 eV, spur production increases as well, but, because of the reduction in the stopping power values, the inter-separation distance between spurs becomes progressively larger. Thus, the water radical species have less and less chance to interact from adjacent track entities, and the yields of strand breaks show a steady increase with electron energy.

The results of calculations based on the direct mechanism for the production of single- and double-strand breaks by electrons in the energy range from 100 eV to 1 MeV have been plotted in Fig. 9. Some aspects of yield behavior can be understood in the following manner. First, let us consider the double-strand break production. At the lowest energies (~ 100 eV), the tracks are short, the total energy available is small, and the probability of hitting both strands and producing independent breaks leading to a double-strand break is also small. The yield increases as more energy is available for deposition on the DNA until the track length becomes longer than the typical distance separating strand breaks on opposite strands, which would lead to a double-strand break. This separation has been taken to be about ten base pairs or 30 to 40 Å. As the energy increases, the LET goes down and the energy available for deposition within this "critical" length also decreases. Consequently, the yield of

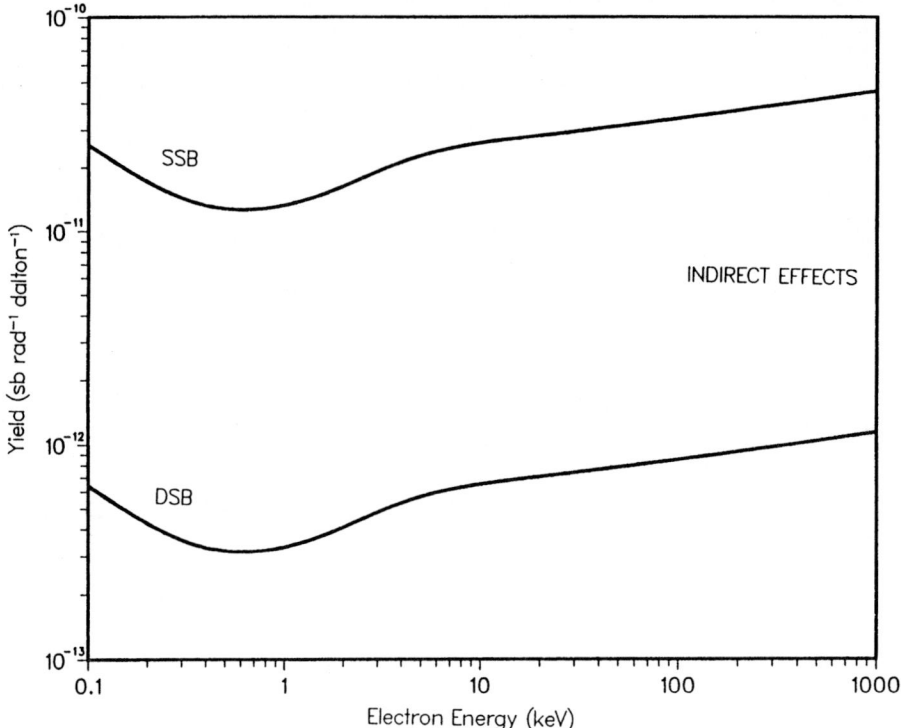

Figure 8. Yields of strand breaks (contributed by the indirect mechanism only) have been plotted against electron energies. The yields of double-strand breaks are about 2 to 3% of the single-strand breaks.

double-strand breaks decreases. The single-strand break curve is, to a good approximation, a simple reflection of the double-strand break curve. In a very simple picture, all energy deposited in the DNA backbone leads to strand breaks (i.e., ~ one strand break for every 30 eV deposited). Single-strand break production should be very roughly independent of electron energy except for the fact that sometimes two (or more) strand breaks become a double-strand break and do not count as single-strand breaks.

Irradiation with different heavy charged particles can be of very special interest with respect to early chemical events. By varying the energy and atomic number of a charged particle, one can also vary the track structure. This helps to manipulate the importance of the water radical reactions with targets such as DNA. For example, if the energy density is low, the radicals can escape earlier reactions and are available to interact with DNA sites. On the other hand, if the energy density is high, the radicals do not contribute much to DNA damage and the significance of early chemical events is reduced.

Results of our calculated strand break cross sections for representative particles, He and Ne, are shown in Fig. 10. Single-strand break cross sections vary approximately linearly with LET, and the corresponding dependence for double-strand scissions is LET^2. For single-strand breaks, it appears that both the indirect and the direct mechanisms have roughly equal cross sections, which introduces a possible new phenomenon. Namely, one strand can be broken by the indirect mechanism and the other by the direct mechanism. This effect has not been included in the present

Chatterjee and Holley

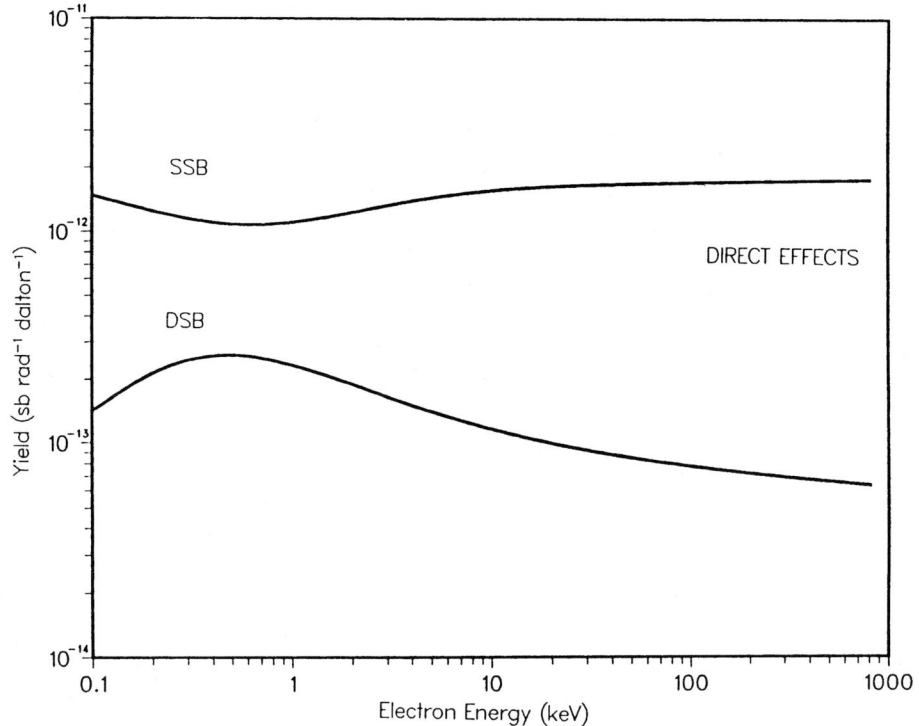

Figure 9. Calculated strand break yields due to the direct deposition of energy on a DNA molecule are plotted versus incident electron energy. The double-strand break yields are about 10% (less at higher electron energies) of the single-strand break yields.

model. Below 30 keV/μm, radical mechanisms seem to create double-strand breaks with a greater efficiency than the deprotonation events. Above this LET value, the reverse is true.

One aim of this work is to compare our calculations with strand break data when cells are irradiated with various qualities of radiation. When the LET is high, the contribution to strand breaks from the indirect mechanism is typically less than 20%.[32] In Fig. 11, such a comparison has been made by using data from reported measurements of several investigators. In these measurements, the experimental conditions were manipulated so that no (or minimal) enzymatic repair of strand breaks was allowed. It can be seen from the comparison that the theoretical and experimental results are in good qualitative agreement, providing confidence in the calculational procedure described in this paper.

Discussion and Future Research

The various stages in the evolution of damage in radiobiology can be understood in rational terms if expressed in the language of chemistry. It provides the most important connection between physical events, such as energy deposition, and biological effects, such as mutation, transformation, and cell death. The events related to the chemical stage can start as early as 10^{-14} seconds and last as late as a few seconds. In this paper, we have attempted to account for events which are over in about 10^{-6} or 10^{-7} seconds. It is believed that the initial damage to a DNA molecule inside a cell nucleus is generally completed by this time and that it is this initial damage which, if not repaired, can lead to permanent change or what is sometimes known as "stable

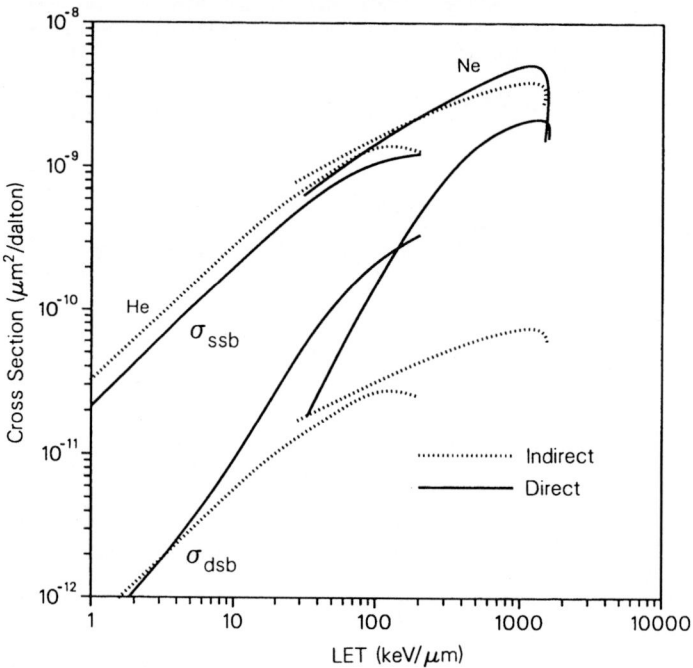

Figure 10. Calculated cross sections for strand break formation by direct and indirect effects for He and Ne as functions of LET. The total cross section is the sum of the two contributions. At the lowest LETs, the indirect effect cross sections are larger and at the highest LETs, the direct effect dominates.

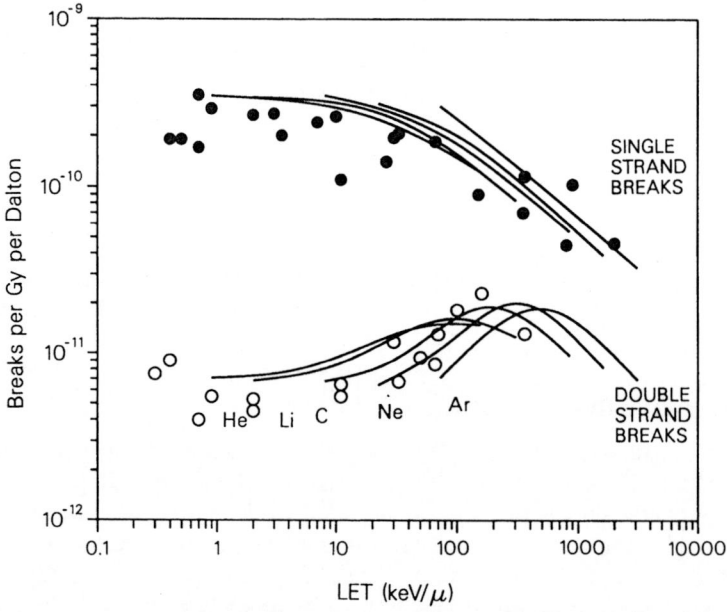

Figure 11. The smooth curves are calculated yields (including both direct and indirect effects) of single and double-strand breaks plotted versus LET. For comparison, a selection of experimental measurements from the literature of radiation-induced initial (enzymatic-repair-inhibited) strand break yields for a variety of mammalian cell types is also plotted.

Chatterjee and Holley

biomedical change." Several orders of magnitude in time are involved between the formation of initial damage and the eventual observable permanent changes. No attempt has been made to include them here.

Most of our present understanding of early chemical events is based on a theoretical approach. The "picosecond barrier" will not allow us to observe events earlier than 10^{-12} seconds; hence, between 10^{-15} and 10^{-12} seconds, we have to rely on theoretical approaches. It is for this reason that the detailed computational developments at Oak Ridge National Laboratory by Turner and his colleagues are extremely important. It is hoped that in the future their expertise will enable them to generate computer codes that are suitable for energetic (~ 600 MeV/n) heavy charged particles with atomic numbers up to 26 or more. Their existing codes for low-energy protons and alpha particles and energetic electrons can account for events which last until about 10^{-6} seconds. It is expected that similar features will also be available with their codes for heavy charged particles. Ultimately, we would like to see how these early events relate to the formation of initial DNA damage.

Most of the present paper deals with events which begin at 10^{-15} seconds. However, several assumptions have been made with respect to developments which take place earlier than 10^{-12} seconds (prethermal stage). The kinetics that evolve after this time have been described in terms of diffusion processes. In the past, several theories have accounted for chemical reactions that result from diffusion of primary chemical species. Most of these theories are based on the following considerations.

The nature of the initial spatial inhomogeneity of the chemical species is that their concentrations are high in the vicinity of the sites of substantial energy deposition (spurs or tracks). These species then proceed to react with each other or to diffuse away from each other and subsequently react with radical scavengers. This strong spatial inhomogeneity and the consequent competition between the reaction and recombination processes constitute the real essence of the diffusion model. In addition, it is assumed that the macroscopic laws of diffusion and reaction are valid when expressed in terms of ensemble averages of large numbers of spurs or tracks.

These considerations have been generally expressed in terms of partial diffusion equations.[1] Those diffusion models have been quite successful in calculating steady state yields. However, they have not been able to predict accurately the time history of decay of chemical species as measured by Jonah et al.[3,4] The reason for this discrepancy is not well understood.

In view of this failure of the diffusion equations to adequately describe the course of time decay, several investigators in the last 10 years or so proceeded with the Monte Carlo approach as described in the Introduction. It seems that this approach results in better agreement with the measured decay curves for water radical species. Furthermore, an extension of this procedure to account for DNA strand breaks has also yielded encouraging results.

Despite the agreement between the theoretical results and the experimental data, much work is needed before we can have confidence in the present approach. It is quite clear that many more experiments are needed with different qualities of radiation to measure the decay of the radical species associated with the radiolysis of water. At present no data are available for heavy charged particles. This lack is primarily due to the unavailability of pulsed beams.

The model of DNA strand breaks presented in this paper seems to have some validity with respect to the production of initial damage. In spite of this success, however, we believe that there are several drawbacks in the model in its present form. For example, energy migration along the DNA chain is a well-known phenomenon; the site of damage may be quite different from the location where energy was deposited. Available experimental data indicate that theoretical considerations must be made to account for this phenomenon. Without that, it will be difficult to have full confidence in the model.

Similarly, no attempt has been made in the proposed model to understand the effects of structured water associated with a DNA molecule. It is quite possible that structured water behaves differently than bulk water. It is, therefore, important to understand how a hydroxyl radical propagates through the bound water before it gets to a particular DNA site (sugar or base).

Another weakness in the present model is the lack of a suitable mechanism for the production of double-strand breaks by hydroxyl radicals. As discussed in the text, reactions by two or more hydroxyl radicals from the same incident track do not yield an appreciable amount of double-strand breaks. It is quite possible that a single ·OH may be responsible for the production of these types of alterations in a DNA molecule. If this is so, we need to understand the mechanisms involved before they can be incorporated in the model. In the present model, no account has been taken of the creation of double-strand breaks by a cooperative phenomenon in which one strand is broken by indirect effects and the other by direct effects. We do not expect this mechanism to introduce a large contribution to the overall yield of double-strand breaks. Nevertheless, inclusion of this phenomenon is expected to improve the accuracy of the results presented in this report.

From these and other considerations, it is quite clear that a mechanistic model such as the one presented here needs further improvement. It is hoped that other investigators will interest themselves in this problem and provide added insight. At present the authors are considering deficiencies of the model presented here, with particular reference to the time evolution of DNA damage following the creation of the initial lesions.

Acknowledgment

This research was supported by the Office of Health and Environmental Research, U.S. Department of Energy, under contract DE-AC03-76SF00098.

References

1. R. D. Cooper and R. W. Wood, eds. *Physical Mechanisms in Radiation Biology.* Technical Information Center, Office of Information Services, United States Atomic Energy Commission (1974).
2. M. Zaider, D. J. Brenner, and W. E. Wilson. The Application of Track Calculations to Radiobiology. I. Monte Carlo Simulation of Proton Tracks. *Radiat. Res.* 95: 231-247 (1983).
3. C. D. Jonah and J. R. Miller. Yield and Decay of the OH Radical from 200 ps to 3 ns. *J. Phys. Chem.* 81:1974-1976 (1977).
4. C. D. Jonah, M. S. Matheson, J. R. Miller, and E. J. Hart. Yield and Decay of the Hydrated Electron from 100 ps to 3 ns. *J. Phys. Chem.* 80:1267-1270 (1976).
5. D. J. Brenner and M. Zaider. Stochastic and Deterministic Treatments of the Time Decay of Species Created by Heavy-Charged Particle Interactions. *Radiat. Prot. Dosimetry* 13: 127 (1985).
6. A. Chatterjee and J. L. Magee. Radiation Chemistry of Heavy-Particle Tracks. 2. Fricke Dosimeter System. *J. Phys. Chem.* 84:3537-3543 (1980).
7. J. E. Turner, J. L. Magee, R. N. Hamm, A. Chatterjee, H. A. Wright, and R. H. Ritchie. Early Events in Irradiated Water. In *Seventh Symposium on Microdosimetry*, Oxford, U.K., J. Booz, H. G. Ebert, and H. D. Hartfield, eds., p. 507. Commission of the European Communities, Hardwood, London (1981).
8. R. H. Ritchie, R. N. Hamm, J. E. Turner, and H. A. Wright. The Interaction of Swift Electrons with Liquid Water. In *Sixth Symposium on Microdosimetry*, Brussels, Belgium, J. Booz and H. G. Ebert, eds., pp. 345-354. Commission of the European Communities, Hardwood, London (1978).

9. A. Mozumder and J. L. Magee. The Early Events of Radiation Chemistry. *Int. J. Radiat. Phys. Chem.* 7:83 (1975).

10. H. A. Grunder, W. D. Hartsough, and E. J. Lofgren. Acceleration of Heavy Ions at the Bevatron. *Science* 174:1128-1129 (1971).

11. A. Chatterjee. Interaction of Ionizing Radiation with Matter. In A Textbook of Modern Radiation Chemistry, Farhataziz and M.A.J. Rodgers, eds., pp. 1-28. *Verlag Chemie Internationa* (1986).

12. C. A. Tobias, E. A. Blakely, P. Y. Chang, L. Lommel, and R. Roots. Response of Sensitive Human Ataxia and Resistant T1 Cell Lines to Accelerated Heavy Ions. *Br. J. Cancer 49*, Suppl. VI:175-185 (1984).

13. A. Chatterjee, H. D. Maccabee, and C. A. Tobias. Radial Cutoff LET and Radial Cutoff Dose Calculations for Heavy Charged Particles in Water. *Radiat. Res.* 54:479-494 (1973).

14. A. Chatterjee and H. J. Schaefer. Microdosimetric Structure of Heavy Ion Tracks in Tissue. *Radiat. Environ. Biophys.* 13:215-227 (1976).

15. M. N. Varma and J. W. Baum. Energy Deposition in Nanometer Regions by 377 MeV/Nucleon 20 Ne Ions. *Radiat. Res.* 81:355-363 (1980).

16. J. L. Magee and A. Chatterjee. Radiation Chemistry of Heavy Particle Tracks. 1. General Considerations. *J. Phys. Chem.* 84: 3529-3536 (1980).

17. J. L. Magee and A. Chatterjee. The Track Reactions of Radiation Chemistry. In Kinetics of Nonhomogeneous Processes, Gordon R. Freeman, ed., pp. 171-214. John Wiley and Sons, Inc. (1986).

18. F. Hutchinson. Chemical Changes Induced in DNA by Ionizing Radiation. Progress in Nucleic Acid Research and Molecular Biology 32:115-154 (1985).

19. J. F. Ward and M. M. Urist. γ-Irradiation of Aqueous Solutions of Polynucleotides. *Int. J. Radiat. Biol.* 12:209 (1967).

20. A. Chatterjee, P. Koehl, and J. L. Magee. Theoretical Consideration of the Chemical Pathways for Radiation-Induced Strand Breaks. *Adv. Space Res.* 6(11):97-105 (1986).

21. J. Turner, J. L. Magee, H. A. Wright, A. Chatterjee, R. N. Hamm, and R. H. Ritchie. Physical and Chemical Development of Electron Tracks in Liquid Water. *Radiat. Res.* 96:437-449 (1983).

22. H. A. Wright, R. N. Hamm, J. E. Turner, J. L. Magee, and A. Chatterjee. Physical and Chemical Structure of Charged Particle Tracks in Liquid Water. In *Proc. Third Workshop on Heavy Charged Particles in Biology and Medicine,* GSI, Darmstadt, Germany, B1, (1987).

23. M. V. Smoluchowski. Drei Vorträge über Diffusion, Brownsche Molekular-bewegung und Koagulation von Kolloidteilchen. *Physik. Zeitschr.* 17:557 (1916).

24. F. Hutchinson and J. Arena. Destruction of the Activity of Deoxyribonucleic Acid in Irradiated Cells. *Radiat. Res.* 13:137 (1960).

25. H. B. Michaels and J. W. Hunt. Reactions of the Hydroxyl Radical with Polynucleotides. *Radiat. Res.* 56:57-70 (1973).

26. S. Arnott and D.W.L. Hukins. Optimized Parameters for A-DNA and B-DNA. *Biochem. Biophys. Res. Comm.* 47:1504-1509 (1972).

27. W. R. Holley, A. Chatterjee, and J. L. Magee. Production of DNA Strand Breaks by Direct Effects of Heavy Charged Particles. *Radiat. Res.* 121:161-168 (1990).

28. T. Inagaki, R. N. Hamm, E. T. Arakawa, and L. R. Painter. Optical and Dielectric Properties of DNA in the Extreme Ultraviolet. *J. Chem. Phys.* 61:4246-4250 (1974).

29. A. Mozumder. Charged Particle Tracks and their Structure. In *Advances in Radiation Chemistry*, M. Burton and J. L. Magee, eds., 1:1-102. Wiley-Interscience, New York (1969).

30. J. E. Turner, R. N. Hamm, H. A. Wright, R. H. Ritchie, J. L. Magee, A. Chatterjee, and Wesley E. Bolch. Studies to Link the Basic Radiation Physics and Chemistry of Liquid Water. *Radiat. Phys. Chem.* 32(3):503-510 (1988).

31. A. Chatterjee and W. Holley. Energetic Electron Tracks and DNA Strand Breaks. *Nucl. Tracks and Radiat. Meas.* 16(2/3):127-133 (1989).

32. R. Roots, A. Chatterjee, P. Chang, L. Lommel and E. A. Blakely. Characterization of Hydroxyl Radical Induced Damage after Sparsely and Densely Ionizing Radiation. *Int. J. Radiat. Biol.* 47:157-166 (1985).

Discussion

Geacintov: You said that the normal diffusion theory didn't work very well for the radical recombination in solution and that you had to use Monte Carlo methods to obtain reasonable fits; yet, when it came to the DNA, you used the Smoluchowski bimolecular encounter diffusion theory.

Chatterjee: No, we used Monte Carlo for the DNA. The ·OHs are moved through jump distances and jump times. What we do in the calculation is drop tracks at random, and wherever the tracks fall we position the ·H, hydrated electrons, H_3O^+, etc., based on a certain Gaussian distribution. Once we position them, we ask whether or not they will be within the reaction radius as determined from experimentally measured rate constants and Smoluchowski equation. If they are within the radius, we make them react. We do Monte Carlo for radical reactions with various DNA sites.

Ward: You have reaction radii located on molecules which are founded on rate constants.

Chatterjee: Yes.

Ward: How do you account for the fact that possibly the ·OH cannot physically get to that site? It is sterically hindered from getting to the site. [*J. Mol. Biol.* 132:411 (1979)]

Chatterjee: Yes, it is true that in some instances ·OH cannot physically get to a site. If one uses experimentally determined rate constants of ·OH reaction with monomers (sugars or bases) and calculates the respective reaction radii; then, just by positioning the monomers in a structured DNA molecule, one finds that the rate constants go down by a factor of 5 or 7 [similar to the result obtained by Hunt *et al.* (*Radiat. Res.* 1972)]. This decrease in rate constant is due to inaccessibility.

Ward: John Hunt's interpretation is that the diffusion constant of the DNA molecule is decreased and that is why you got a lower constant. What I'm asking is, you've got bonds on thymine, for instance, which the ·OH cannot physically get to. There is no space in the structure for the ·OH to get to those points, and you draw a reaction radius around them.

Chatterjee: Our interpretation is not the same as that of Hunt *et al.* In our case the DNA molecule is stationary, and we think that it is reasonable because in 10^{-9} seconds a DNA molecule cannot diffuse appreciably. With respect to ·OH reaching a specific bond, I have to say that our calculation does not have that kind of resolution. We cannot say where on a molecule ·OH has reacted. However, if there is a site which is obscure, it is treated, we believe, through our consideration of DNA structure because we can reproduce observable rate constant with a DNA molecule.

Kopelman: I wanted to ask about the difference between the diffusion theory and the Monte Carlo simulations. First, you had the tracks where Kooperman's diffusion reactions didn't fit. I was going to ask about this, being a newcomer here. The Monte Carlo, you say, fits. The way I see it, the reason the diffusion theory didn't fit was that it was a three-dimensional Smoluchowski-type diffusion theory. As long as you are in a track, you really have more like a one-dimensional behavior giving anomalous diffusion, which should be less efficient than expected as long as the interactions are inside the track. Is it intra-track or inter-track?

Chatterjee: This is intra-track.

Kopelman: This is fine because inside a track you have certain directions which are preferred to others. You don't have a homogeneous distribution. The track itself defines a line or something like that, so, as a matter of fact, I would love to try it some time to fit those data with some other kind of diffusion equation where it is adapted to a lower dimension. On the last point where this simple diffusion equation does fit, this appears to be again consistent for the following reason, that once you have hindrance, once it is difficult to get in, this is really going into the so-called reaction-limited regime—rather, diffusion-limited regime—and it's there that those simple equations always work, i.e., where the classical equations work. The numbers may go down, I mean, the cross section may come down, but the form of the equation works, so maybe this is why there is reasonable agreement.

Chatterjee: Your comment seems very reasonable.

Zaider to Kopelman: I would like to answer your question. In the 1972 Airlie meeting, there is a beautiful paper by Kooperman that I enjoyed very much and which educated me in radiation chemistry, in which he quite frankly explains that the kinetic reaction diffusion approach is grossly overparameterized. This means you can fit anything by changing initial distributions, reaction rates, etc.

Chatterjee: Except that he had to change to spur size too big to agree with the experimental data.

Zaider: Exactly. In our stochastic chemistry calculation, the whole problem was suddenly reduced to determining one single parameter. Everything else was obtained *ab initio*, and this was, in my view, a fantastic step forward. This is Point One. The second point is our demonstration that you cannot take an average spur, on which you perform the fast chemistry, and expect to get correct results. You have to take each spur individually, do the chemistry first, and then average. That is another important conclusion that we drew. So I don't think that you should worry too much about the Kooperman approach because it could probably be used to describe almost any system, given enough parameters.

Chatterjee: I agree with you and want to add that DNA structure also plays an important role.

Weinstein: This relates to structure. At the level at which I am looking at the reactivity of the sugars, I find that each sugar and each atomic site in them is different in the probability of reaction with a radical. Obviously each site in the base is also different in the reactivity, and you measured that. In fact, once you have one lesion at some point, all the other sites in that neighborhood of the DNA will have a different reactivity. So the kind of average description you use is probably excellent in order to give you an order of magnitude, and after that I will need the details to compare with exact chemical reactions because the results that a chemist would give you from a clear identification of the products will probably require a little bit of a refinement of the model.

Chatterjee: How many base pairs are affected in the vicinity of a strand break in terms of their shifts in coordinates?

Weinstein: About seven base pairs, right?

Chatterjee: Thirty angstroms is about eight base pairs.

Weinstein: You will see in my talk some information regarding structural changes in that range. The accessibility of sites will definitely have changed due to these structural changes, and we can discuss the implications for the Monte Carlo models when I show you the actual structures.

Ward: What is the probability that the strand break opens before another ·OH gets there? That strand break isn't going to form for one millisecond or something. The probability of a second hydroxyl getting to that site is very small.

Weinstein: I'm not even talking about strand break; I'm talking about change in the electronic structure at the site of interaction, which now produces an enormous change in reactivity, in susceptibility to attack by the species produced by radiation. I will show how much of an effect on conformation one finds with just the radical attack on a base.

Varma: I think this is an important point. When you do the Monte Carlo calculations with 500 tracks or more, you consider where each track goes, and you don't know which sites really are responsible for producing the double- strand breaks? Is that true?

Weinstein: That is correct. What I'm talking about is sensitive sites of the DNA in the cell.

Varma: When you do the detailed calculation of conformation changed in a single electronic site, you determine the probability of particular damage. You cannot say anything about that because you average it over eight or nine base changes.

Weinstein: But relating the detailed structural results to such findings could obtain a more realistic parameterization of the model.

Rossi: I just was going to continue my remarks of yesterday. The concept of a homogeneous track is adequate if one is dealing with first-order reactions. But, as I think you pointed out, LET is really the wrong concept to use with more complex interactions. They depend on LET, but you shouldn't relate them to LET in any quantitative way.

Chatterjee: We don't. We mention LET only for experimental measurement purposes, but for calculational purposes we do not use LET.

Paretzke: Aloke, in your Monte Carlo chemistry program, how far did you position the H and OH radicals apart after break up of the water molecule? What is the scientific basis for your choice?

Chatterjee: Your question refers to the time period between 10^{-15} and 10^{-12} seconds. Our model does not address this region of the physico-chemical process in a rigorous way primarily because so little is known. We position the basic four chemical species ($\cdot H$, $\cdot OH$, e_{aq}^-, and H_3O^+) according to Gaussian distribution with respective σ values for each of the species. This has been our practice for the last fifteen years, and we have kept the same σ values whether it is a Fricke system or DNA aqueous solution.

Paretzke: At the point of origin in time and space after the ionization of a water molecule, all atoms of this water molecule are still close together. How do the $\cdot H$ and $\cdot OH$ fragments physically move apart?

Chatterjee: The source for H_3O^+ and $\cdot OH$ is the ion molecule reaction between H_2O^+ and H_2O. For about one vibration time period ($\sim 10^{-14}$ seconds), the positive charge diffuses to an average distance of about 15 angstroms and then quickly reacts with neutral H_2O. H_3O^+ replaces the H_2O^+ and $\cdot OH$ is displaced by the size of a water molecule (~ 2.9 angstroms). $\cdot H$ is generally produced as a dissociative process of H_2O^*. $\cdot H$ and $\cdot OH$ reform water most of the time. However, there is a finite probability that they escape this recombination process, and what coordinates they acquire are anybody's guess at this time.

Paretzke: They are created from one and the same H_2O molecule.

Chatterjee: No, they are not formed from the same H_2O molecule.

Paretzke: Since they are born from one molecule, they must initially be close together!

Chatterjee: $\cdot H$ and $\cdot OH$ can be close together. The hydrated electron can form at a distance depending upon its thermalization distance. The positive charge on a given H_2O^+ can also migrate.

Paretzke: How far?

Chatterjee: Oh, I would say an average of 15 angstroms or so.

Paretzke: That is a long distance; how do you justify this scientifically?

Chatterjee: Yes, this is a long distance and comes from indirect evidences of charge transfer process to explain initial yields.

Paretzke: So, you assume that your proton is hopping 14 angstroms in 10^{-12} seconds.

Chatterjee: About 15 angstroms.

Paretzke: That's a lot.

Chatterjee: It is a fitting parameter, I do not know *a priori* the number, because a fitting parameter must give the right ferric yield, and for the DNA calculations I did not change the parameter. That's all I can say.

von Sonntag: I think your data are very beautiful to model the observations. It represents the big progress that has been made in this area. I wonder, you mentioned that 20 percent of the OH radicals attack the sugar moiety. Is it an experimental fact? Which work do you relate to?

Chatterjee: It was in one of your papers actually. It was your poly uracil work.

von Sonntag: In poly uracil you only get 7 percent.

Chatterjee: But that's single-stranded.

von Sonntag: Yes, it is single-stranded, but and with double-stranded DNA we don't have good data. We always rely on the paper by Wilson, Ebert, and Scholes, but this is sort of a good guess, not much more.

Ward: I calculated this ratio from base release measurements [*Radiat. Res.* 75:278 (1978)]. The numbers are base damage to sugar damage in double-stranded DNA is between 2.8 and 4 for ·OH attack.

Kasha: Just a comment about proton impairing diffusion, we would regard proton motion as an inductive relay through a water chain, so that the same proton does not appear at the end, and a rapidity of transfers unrelated to other normal diffusion processes.

Chatterjee: Yes, we don't have diffusion before picosecond.

Kasha: In a picosecond range.

Chatterjee: You are thinking of photochemistry.

Kasha: An excited-state proton transfer occurs so rapidly in water chains that they are inducted through a hydrogen body network.

Chatterjee: This happens in hydrogen transfer as well in 10^{-14} seconds or so—one vibration transfer.

Varma: You made a point about the sophistication in terms of the DNA structure that you are putting into the code. In terms of a mechanistic point of view, how does this affect the initial chemical species production and resulting molecular damage you see? What can you tell us about the influence of the structure on initial chemical species or initial damage?

Chatterjee: In our model the inclusion of DNA structure does not alter the production of initial chemical species. At least for those water molecules which are very close to different DNA sites, this may be a simplification but a necessary one.

Varma: Would you change this parameter of 17 eV?

Chatterjee: I would not change.

Varma: But that's not the number to use, then, in case of DNA because you don't have pure water there.

Chatterjee: Yes, we do not have pure water. However, if our current understanding of the mechanism of water radical formation is correct (my answers to Paretzke's questions), then for most of the bulk water it is very difficult to see how DNA is going to influence this because it involves a few neighboring water molecules.

Weinstein: Would this number—17 or 18 eV—change in ionic solutions? Would this change as a function of ionic strength?

Chatterjee: The pH may change this number, but I will guess that the impact may be small.

Weinstein: My question concerns the specific energy value of 17 eV. Will it be affected by the fact that the water is organized around ions that surround the DNA, by the fact that you have both water, some of which is bulk, but a lot of which is organized by the ionic strength? I'm not even talking about the water bound to the DNA. I'm really talking about the fact that DNA is floating, if you want, in a very highly concentrated solution containing a variety of ions and a variety of other proteins; it is a very structured solution. And the question is, what effects we can expect from this?

Chatterjee: All I can say at this point is that our model does have several approximations.

Weinstein: That is the point; is that a good approximation?

Kasha: I'd like to comment on Harel's question. It seems to me illegitimate to compare the molar concentration DNA with the molar concentration of water. You should give atomic concentration ratios because you are dealing on the atomic level and you have a poly electrolyte in DNA; you are really going to affect water for a long distance with coulombic orientation forces. I think that is a very legitimate question, and if you compare the DNA atomic concentration with water atomic concentration you will get a different answer.

Chatterjee: It seems to me that because of the presence of counter ions, the DNA may not influence the water radiation chemistry. However, I do have one concern in our approach, and that has to do with electron yields. Since bases can act as nice "sinks" for electrons, the initial yield of electrons may be different. Whether this will have a large impact, I do not know.

Kopelman: I have a question on this same point. Assume that you had ice instead of liquid water. How much difference would it make? You say here you believe in structure and so on...ice is about as structured as water will get.

Chatterjee: I don't know about ice completely, but I think that a long time ago, Ed Hart tried to measure radiation chemistry by lowering the temperature as close as he could get to zero. He did not find any difference in the yields, so temperature did not have much effect.

Kasha: I don't think that ice is a good model. It is known from the X-ray diffraction studies of salt solutions that you get a very different water structure around a charged ion. And I think that is a better model.

Ward: I'd like to ask Michael Kasha: If you have structured water, what is the possibility of hole transfer by a similar kind of mechanism to the proton transfer that you were talking about?

Kasha: I haven't thought about that; I just don't know.

Inokuti: My question is also related to the same question. I wondered what you would predict for the following situation. Suppose you have D_2O and suppose you have the same DNA and radiation, what do you predict? How much change would there be?

Chatterjee and Holley

Chatterjee: Except for the changes in diffusion constants of the species, I do not see any need for making conceptual changes.

Inokuti: But on the other hand, everything is related to conformation.

Chatterjee: I do not understand your comment.

Inokuti: But certainly I think that the migration speeds would be very different.

Chatterjee: So, I may have to change the sigmas for the Gaussian distributions.

Inokuti: Mine is a testable question.

Chatterjee: Yes, testable, that is a good point.

Varma: I think the more general question that relates to all this is how closely the biochemistry in a cell relates to a model system.

Chatterjee: We have compared the results of our model system with measured values in yeast, bacteria, and mammalian cells. In these measurements the enzymatic repair processes were inhibited; hence, a comparison with our estimates was a fair one. The agreement has been quite reasonable when proper normalization with respect to DNA length is taken into account.

The Chemistry of Free-Radical-Mediated DNA Damage

Clemens von Sonntag[a]

Abstract

In the living cell, ionizing radiation can cause DNA damage by the direct effect (ionization of DNA) and the indirect effect (reaction of radicals formed in the neighborhood of DNA with DNA, e.g., OH, e_{aq}^-, H, protein- and glutathione-derived radicals). Properties of the base radical cations have been studied in model systems using SO_4^- radical to oxidize the nucleobases in aqueous solution. The pK_a values of some nucleobase radical cations are reported, so are the ensuing reactions of the thymidine radical cation with water. The products of reactions are compared with those formed by OH radical attack. The reaction of e_{aq}^- with the nucleobases yields radical anions. Protonation at heteroatom sites and at carbon are discussed, and some recent results regarding the electron transfer to adjacent nucleobases as well as to 5-bromouracil are reported. A brief account is given on the reaction of carbon-centered radicals with the nucleobases. These reactions may mimic the reactions of protein-derived radicals with DNA. Glutathione is present in cells at rather high concentrations and is expected to act as an H- or electron-donor in repairing radiation-induced DNA damage (chemical repair). As thiyl radicals are known to also undergo the reverse reaction, i.e., H-abstraction from suitable solutes, some experiments are reported which probe this type of reaction with dilute DNA solutions.

In some polynucleotides radical transfer from the base radical to the sugar moiety occurs with the consequence of strand breakage and base release. Some currently held mechanistic concepts are discussed.

Attention is drawn to some important open questions which should be addressed in the near future.

Introduction

After the discovery of the X-rays by Röntgen in 1895, it took only a few weeks to realize the damaging effects on the tissue of the hands of the persons operating the X-ray lamps. Based on this observation, the first attempt was made at the time to cure cancer by an X-ray treatment.[1] Also in these early days (1909), Schwarz recognized the sensitizing effect of oxygen.[2,3]

As we know now, on a molecular level these effects are caused mainly by the damage done by the ionizing radiation to the DNA of the living cell.[4] From the chemist's point of view, it is useful to distinguish two ways in which this damage can be set: the direct effect, where the energy is absorbed by the DNA molecule itself (Reaction 1); and the indirect effect, where the energy is absorbed in the close vicinity of the DNA. This leads to the formation of reactive intermediates (in the surrounding water ·OH, H· and e_{aq}^- are formed (Reaction 2), which can attack the DNA (Reactions 3 through 7).

(a) Max-Planck-Institut für Strahlenchemie, Mülheim a.d. Ruhr, Germany

Physical and Chemical Mechanisms in Molecular Radiation Biology
Edited by W.A. Glass and M.N. Varma, Plenum Press, New York, 1991

287

$$DNA \xrightarrow[\text{radiation}]{\text{ionizing}} DNA \cdot^+ + e^- \tag{1}$$

$$H_2O \xrightarrow[\text{radiation}]{\text{ionizing}} \cdot OH, e_{aq}^-, H\cdot, H^+, H_2, H_2O_2 \tag{2}$$

$$\cdot OH + DNA \longrightarrow DNA(+OH)\cdot \tag{3}$$

$$\cdot OH + DNA \longrightarrow DNA(-H)\cdot + H_2O \tag{4}$$

$$e_{aq}^- + DNA \longrightarrow (DNA)\cdot^- \tag{5}$$

$$DNA\cdot^- + H_2O \longrightarrow DNA(+H)\cdot + OH^- \tag{6}$$

$$H\cdot + DNA \longrightarrow DNA(+H)\cdot \tag{7}$$

$$DNA\cdot^+ + H_2O \longrightarrow DNA(+OH)\cdot + H^+ \tag{8}$$

$$DNA\cdot^+ \longrightarrow DNA(-H)\cdot + H^+ \tag{9}$$

The water radicals react with the DNA molecule in various ways. The OH radicals and the H atoms can add to the nucleobases to produce a large number of different OH- and H-adduct radicals (Reactions 3 and 7), collectively denoted as $DNA(+OH)\cdot$ and $DNA(+H)\cdot$.

Further, the OH radical (but not the H atom) abstracts an H atom from the sugar moiety with a notable efficiency (Reaction 4). In these H-abstraction reactions, five different sugar radicals are formed, collectively denoted as $DNA(-H)\cdot$. There is good reason, from our present knowledge of carbohydrate free-radical chemistry,[5,6] to expect the OH radicals not to be very selective, and if all positions were equally accessible, H-abstraction should be more or less random. In the case of the single-stranded polynucleotide poly(U), a proportion of only 7% of the OH radicals attack the sugar moiety.[7] For single-stranded DNA this value may be similar. In double-stranded DNA, however, the bases are hidden inside the double helix and the sugar moieties are relatively more exposed to OH radical attack. Hence, OH radical attack at the sugar moiety may be more important in double-stranded than in single-stranded DNA. The effect of a reduced OH radical accessibility of a given part of the DNA is very well shown by the data obtained from OH radical-induced strand breakage of a model of the Holliday junction.[8]

The solvated electron reacts readily with DNA (Reaction 5, $k_5 \approx 1.4 \times 10^8$ dm^3 mol^{-1} s^{-1}, based on nucleotide units). It will be shown below that the radical anions of some of the free monomeric nucleobases react very rapidly with water (A$\cdot^-$ and C$\cdot^-$) while T$\cdot^-$ is rather stable (on the microsecond timescale). Hence, also, part of the DNA radical anions must undergo a very rapid reaction with water (Reaction 6) while others are more persistent. Some of the protonated electron adduct radicals can undergo rearrangement reactions whereby the redox properties of the radicals change drastically. All these different radicals are formally H-adducts and therefore are also denoted DNA(+H)·.

It must be considered in addition to this whether an electron transfer might take place from the base which has originally trapped the solvated electron (a reaction which should be about equally fast for all nucleobases) to a nucleobase with a higher electron affinity [for the state of the art in solid DNA and at very low temperatures, see Refs. (9) and (10)]. In aqueous medium at the appropriate pH, the electron adduct radicals become protonated.

The DNA radical cations (DNA·$^+$) formed in Reaction 1 may undergo a variety of reactions, e.g., reaction with water (Reaction 8) whereby OH-adduct radicals are formed. These various DNA(+OH)· radicals are not necessarily the same as those formed upon OH radical attack (Reaction 3), and if they are, the proportion of the different isomers might be substantially different (cf. Scheme 1). Moreover, there are deprotonation reactions (Reaction 9) which can lead to sugar-derived radicals as discussed for Reaction 4, but deprotonation at the NH and CH_3 groups (in thymine) is also possible, yielding base-centered radicals. Here again, model systems have been most helpful in elucidating some fundamental aspects; e.g., Reactions 10 through 18 in Scheme 1 are derived from such studies.

Much of our present understanding regarding the extent to which the direct effect and the indirect effect could contribute to DNA damage comes from computer calculations and modeling studies. This is largely the work of the groups of Aloke Chatterjee,[11-15] John Ward,[16-18] and more recently also of the group of Harel Weinstein and Roman Osman,[19-22] and is addressed by them elsewhere in these proceedings.

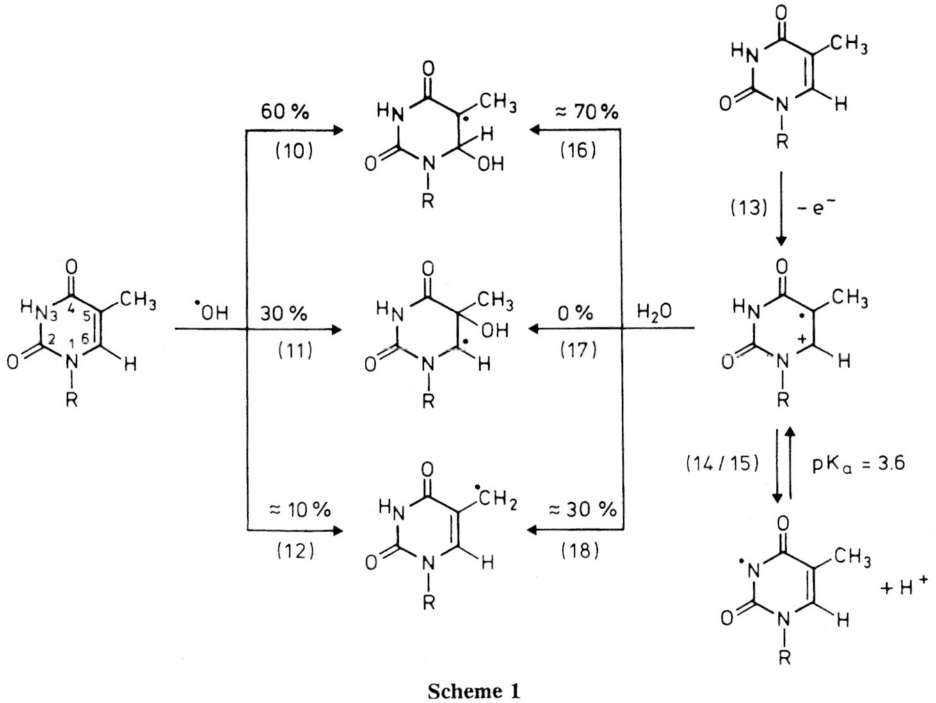

Scheme 1

It is obvious that much of the mechanistic detail cannot at present be studied with DNA—or only with extreme difficulty. Thus, much of our present knowledge is based on investigations of model systems such as the nucleobases, nucleosides or methylated nucleobases. The use of the latter has an advantage over the natural nucleobases in that the relevant nitrogen is substituted similar to the situation in DNA. This is often very useful because the nitrogen-bound hydrogen (in the natural nucleobases) can have a drastic influence on the ensuing free-radical chemistry, which may be much different from that of the nucleosides, and hence in the DNA. On the other hand, they do not contain the analytically inconvenient sugar moiety. Of course, the mechanism of DNA strand breakage cannot be studied by looking only at the bases. Suitable low-molecular-weight model systems have been used, such as alkyl phosphates, sugar phosphates, nucleotides, and polynucleotides, in particular homopolynucleotides. The latter have an advantage over DNA in that they carry a single type of base. Unfortunately, polynucleotides at reasonable cost are only available in the ribo-series (analogs of RNA). On the basis of such model systems, several aspects of the mechanism of radiation-induced DNA damage are fairly well understood. These are briefly discussed in the following chapters. It is clear that the detailed knowledge of the chemical properties (H-abstractive power, redox behavior, pK) of the reactive intermediates within and near the DNA molecule is of high importance if one seeks methodically to forestall, alleviate, or enhance DNA damage (as in cancer therapy).

To help focus future research, it may be equally useful to point up questions that to date have remained open. A brief remark regarding this point will be made below.

Reactions of the Solvated Electron

With the free nucleobases, the solvated electron reacts at rates which are close to diffusion controlled; but with the negatively charged macromolecule, DNA the rate constant is two orders of magnitude lower.[4] In low-molecular-weight model systems it shows practically no reactivity towards the sugar moiety or the phosphate group.[4] In single-stranded biologically active DNA, solvated electrons do not induce strand breakage, which is a lethal damage in this system[23] (cf. also Ref. 24). However, they induce base damage, which in single-stranded biologically active DNA causes inactivation with an efficiency of about 8%.[25]

Thymine and Uracil

Our knowledge of the reactions of the nucleobase radical anions formed in the reactions of the solvated electron with these substrates is fairly well advanced in the case of thymine and uracil. These pyrimidine radical anions react with protons ($k \approx 5 \times 10^{10}$ dm^3 mol^{-1} s^{-1}) at O^4, yielding the corresponding radicals, often termed "protonated radical anions." These O-protonated radical anions have pK_a values around 7.3 (equilibrium 19/20).[26]

With oxygen they react rapidly ($k_{22,23} \geq 2 \times 10^9$ dm^3 mol^{-1} s^{-1}) by restoring the nucleobase (Reactions 22 and 23). The other product is $HO_2 \cdot / O_2 \cdot^-$.[27]

In competition with the protonation at oxygen, there is a much slower protonation at C(6) (Reaction 21).[28-30] This reaction is practically irreversible (i.e., the pK_a value of the C(6)-H-adduct radical must be very high). This radical has considerably different properties. It shows no reducing properties [(e.g., with tetranitromethane (TNM) it does not yield the nitroform anion (NF^-)], and also in its reactions with oxygen (Reaction 24) the pyrimidine is no longer restored.[27]

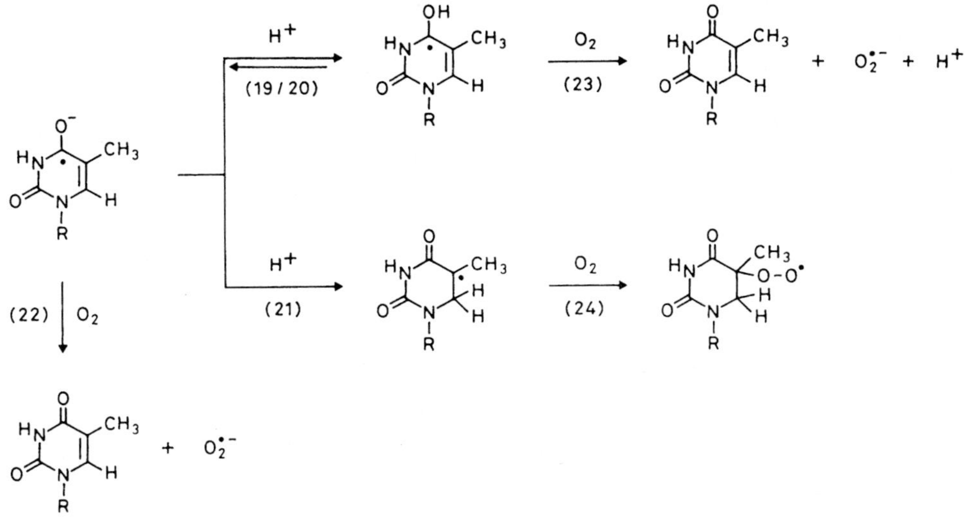

H⁺

(19 / 20)

O₂

(23)

+ O₂•⁻ + H⁺

(22) | O₂

H⁺

(21)

O₂

(24)

+ O₂•⁻

Scheme 2

Cytosine

Comparatively little is known about the cytosine radical anion. This intermediate is extremely short-lived and reacts very rapidly with water ($t_{1/2} \leq 100$ ns).[31] It is not yet fully clear how to formulate the reaction product(s). There are conflicting results: on the one hand, these protonated radical anions react quantitatively with p-nitroacetophenone (PNAP) by forming PNAP•⁻ with an electron transfer rate constant of 5×10^9 dm³ mol⁻¹ s⁻¹.[31] On the other hand, in their reaction with oxygen it seems that the yield of the superoxide radical anion is considerably less than quantitative.[31] There are now some additional unpublished experiments from our laboratory which appear to confirm this conclusion (gamma radiolysis of cytidine 10^{-3} mol dm⁻³, N₂O/O₂-(4:1)-saturated; O₂•⁻ formation has been probed[4] with TNM 5×10^{-5} mol dm⁻³ G(NF⁻) = 4.0×10^{-7} mol J⁻¹; cytidine 3×10^{-3} mol dm⁻³, Ar/O₂-(4:1)-saturated, TNM 3×10^{-5} mol dm⁻³, G(NF⁻) = 4.4×10^{-7} mol J⁻¹. These data appear to suggest that the protonated cytosine electron adducts yield O₂•⁻ with an efficiency of only $\leq 80\%$). Usually protonation at oxygen and nitrogen is much faster than at carbon.[32] Hence, one would expect that the primary products are the O- and N-protonated radical anions. However, such species are expected to re-form cytosine and produce superoxide radicals in their reaction with oxygen (like the corresponding uracil and thymine intermediates).

Adenine and Guanine

The adenine radical anions (A•⁻) show a behavior very similar to the cytosine radical anions. They are also very rapidly protonated by water,[33,34] preferably at nitrogen (ANH•). The more recent data[34] indicate a value of $k \geq 1.4 \times 10^8$ s⁻¹. Again, this N-protonated adenine radical anion (ANH•) can transfer an electron to PNAP.[33] In basic solutions this protonated electron adduct isomerizes into another radical (ACH•) with different properties; it no longer transfers an electron to PNAP, but forms an adduct. These results point to the possibility that a similar process as described for the thymine and uracil systems (cf. Reactions 19-21) might take place. First, a rapid protonation occurs at a heteroatom (i.e., $k \geq 1.4 \times 10^8$ s⁻¹). The resulting N-protonated radical anion (ANH•) has a pK$_a$ value of 12.1.[35] The radical anion, A•⁻, in equilibrium is then more slowly protonated by water at carbon (formation of

ACH·; $k = 3.6 \times 10^6$ s^{-1}).[35] Two sites must be envisaged, C-2 and C-8, but the ratio of the yields of these two (and possibly other) potential products is as yet unknown.[35]

Preliminary data have been presented,[36] but to our knowledge no detailed study of the fate of the guanine radical anion is available to date. Preliminary results from our laboratory (X.-M. Pan, unpublished) indicate, however, that the guanine radical anion is also very rapidly ($t_{1/2} < 1$ μs) protonated by water.

Electron Transfer from Nucleobase Radical Anions to Other Nucleobases

There is a long-standing question as to whether or not, in DNA, nucleobase radical anions can transfer an electron to a neighboring base. Such a transfer would eventually lead to one single kind of radical anion or possibly its protonated form (see above) on condition that the electron affinity of one of the nucleobases in DNA is much higher than the electron affinities of the rest. If the electron affinities of the nucleobases are very close to one another, equilibrium will occur. Theory predicts that the electron affinity of the bases follows the sequence T > C > A > G.[37]

This question has recently been addressed experimentally by the pulse radiolysis technique for adenine/pyrimidine pairs, using dinucleotides and mixtures of mononucleotides.[38] It has been found that there is efficient electron transfer from the adenine radical anion (A·⁻) to the pyrimidine (either uracil, thymine or cytosine). Neither the N-protonated radical anion (ANH·) nor the C-protonated radical anion (ACH·) is capable of transferring an electron. Thus the protonation by water prevents the electron transfer. In the dinucleotides, however, the two bases are so close to each other that electron transfer is quantitative.

Another way of testing the occurrence of electron transfer between different bases is the transfer from a nucleobase radical anion (and/or its protonated form) to another nucleobase which can undergo a rapid chemical reaction, e.g., 5-bromouracil (5-BrU). The 5-bromouracil radical anion (5-BrU·⁻) thus formed rapidly loses a bromide ion with a very high efficiency.[4] The resulting uracilyl radical is a very good hydrogen abstractor, and when an H-donor such as t-butanol is offered, this radical is converted into uracil. Thus the uracil yield of a given mixture can be used as a measure for the electron transfer reaction (C. Nese, Z. Yuan, and C. von Sonntag, to be published). A somewhat different approach has been taken by Adams and Willson[39] by measuring pulse-radiolytically the competition between PNAP and 5-BrU for a given pyrimidine radical anion. They observed that the electron adducts of thymine and uracil, as well as the (protonated) adenine radical anion (ANH·), readily transfer an electron to 5-BrU (k ranging from 1.1×10^9 to 3.5×10^8 dm^3 mol^{-1} s^{-1}), but that the reaction of the cytosine radical anion was too slow to be measured ($k < 5 \times 10^7$ dm^3 mol^{-1} s^{-1}).

In Figs. 1 and 2, unpublished data from our laboratory are presented (Nese, Yuan, and von Sonntag). They show the G value of uracil formation as a function of the ratio of nucleobase versus 5-BrU concentration (for experimental details see legend). The dotted lines are the expected G(uracil) values if no electron transfer were to take place (calculations based on known rate constants for the reaction of the solvated electron with thymine and 5-BrU, respectively). It is obvious that in the thymine/5-BrU system (data similar to the adenine/5-BrU system, cf. Fig. 1), an electron transfer must occur. This is to be expected because the thymine radical anion is a long-lived species. More surprising is the observation that there is also electron transfer between the adenine electron adduct and 5-bromouracil (Fig. 1). From the rate constants given above, it is clear that this transfer cannot be due to the (short-lived) radical anion (A·⁻) itself but must be attributed to a reaction of its protonated form (·ANH). It has been mentioned above that there is no transfer from ·ANH to thymine or cytosine on

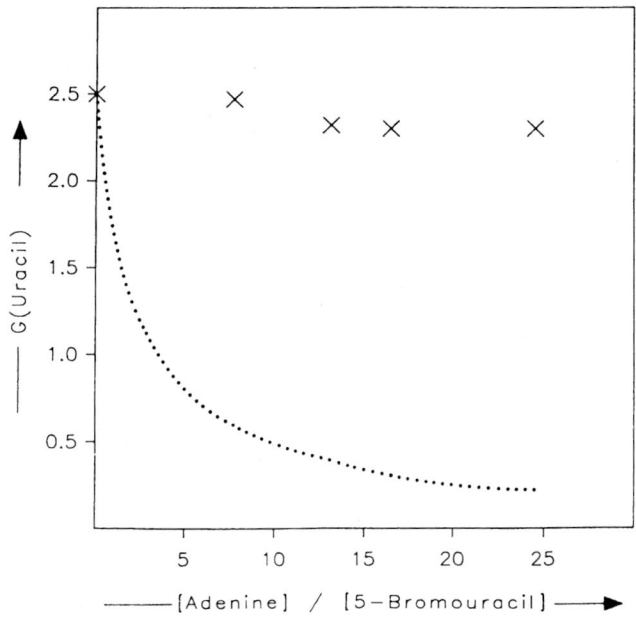

Figure 1. Gamma Radiolysis of Ar-saturated aqueous solutions of adenine 5-bromouracil mixtures (total concentration 10^{-3} mol dm^{-3}) containing t-butanol to scavenge OH radicals. G(uracil) as a function of the [adenine]/[5-bromouracil] ratio.

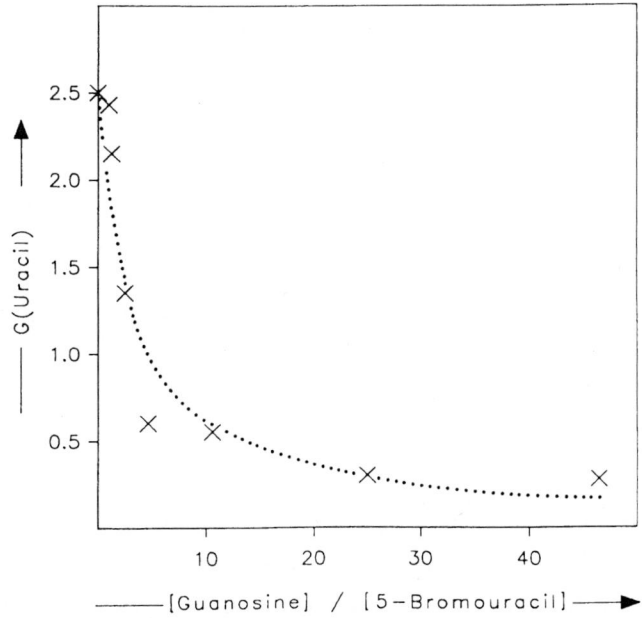

Figure 2. Gamma Radiolysis of Ar-saturated aqueous solutions of guanosine 5-bromouracil mixtures (total concentration 10^{-3} mol dm^{-3}) containing t-butanol to scavenge OH radicals. G(uracil) as a function of the [guanosine]/[5-bromouracil] ratio.

the pulse radiolysis time scale. If it is not a question of time scale (pulse radiolysis versus steady state gamma radiolysis), it follows that the 5-bromouracil must be a better electron acceptor than the other pyrimidines.

The data on guanosine/5-BrU mixture (Fig. 2) are quite different. It is evident that no transfer occurs. Based on the theoretical calculations of the sequence of the electron affinity of the nucleobases given above, one would expect that the guanine radical anion should be a much better electron donor than e.g., adenine and thymine. Since the guanine electron adduct does not transfer an electron to 5-BrU, it must have been protonated (at carbon?) at times ≤ 1 μs (assuming a transfer rate of 10^9 dm^3 mol^{-1} s^{-1} and taking the first indicative point in Fig. 2, i.e., 5 x 10^{-4} mol dm^{-3}). In agreement with earlier data,[39] no efficient electron transfer occurs from the (protonated) cytosine radical anion to 5-bromouracil, but the situation is not as clear-cut as in the case of guanine. Work is in progress to probe such transfer reactions also with dinucleotides.

Reactions of the OH Radical and the H Atom

It has been mentioned above that OH radicals and H atoms react very rapidly with the nucleobases. The rate constants for these reactions are in the order of 5 x 10^9 dm^3 mol^{-1} s^{-1} for the OH radical and in the range of 1 to 5 x 10^8 dm^3 mol^{-1} s^{-1} for the H atom.[4] Both radicals add preferentially to the double bonds of the nucleobases and only to a minor extent abstract carbon-bound H atoms, e.g., from the methyl group in thymine or the sugar moiety (in nucleosides and nucleotides). This selectivity is much more pronounced in the case of the H atom. Thus in its reaction with thymine, $\leq 3.5\%$ of the H atoms react with the methyl group,[40] while about 10% of the OH radicals undergo this reaction.[41] Both radicals are electrophilic and hence add preferentially to the position of highest electron density. For the OH radical, steric effects may also play a role.

With DNA the observed rate constant for the reaction of OH radicals is much lower. The reasons for this are now fairly well understood and are mentioned in the section "Prevention of DNA Damage."

Pyrimidines

This preference of addition to the position of highest electron density is well expressed in the ratios of addition to the 5-position versus the 6-position (cf. Reactions 11 vs 10) which are 7.3, 4.6, and 2.0 for 6-methyluracil, uracil, and thymine (5-methyluracil), respectively (cf. Ref. 41).

In the uracil derivatives, the C(5)-adduct and C(6)-adduct radicals can be distinguished by their redox properties.[41] While the 5-adduct radicals have reducing properties (for example, they react rapidly with tetranitromethane, giving rise to the nitroform anion), the C(6)-adduct radicals display more oxidizing properties (they oxidize N,N,N´,N´-tetramethylphenylenediamine to its radical cation). These oxidizing properties are more pronounced in the case of the OH-adducts than in the corresponding H-adducts (cf. Refs. 41 and 42). What applies to the uracil derivatives (including thymine) also holds for cytosine and its derivatives.[43]

For some of these systems (e.g., 1,3-dimethyluracil),[44] detailed product studies are available which support the conclusions drawn by the pulse radiolysis experiments. This material has been compiled in Ref. 4.

In the presence of oxygen, the primary radicals formed either by ·OH (H·) addition, or H-abstraction are converted into the corresponding peroxyl radicals. In the case of the N(1)-unsubstituted nucleobases, the C(6)-peroxyl radicals undergo an

OH⁻ induced release of $O_2^{\cdot-}$ yielding an isopyrimidine as a short-lived intermediate. The kinetics of this $O_2^{\cdot-}$ release and the subsequent chemistry of the isopyrimidines have been elucidated.[45-47]

In DNA, the bases are bound to the sugar moiety; i.e., in DNA the pyrimidine nucleobases have no hydrogen atom at N(1), which is required for the formation of the isopyrimidine intermediate. Hence other mechanisms must operate. Under such conditions (i.e., when the N(1)-position is substituted or the OH⁻ induced reaction is inhibited in acid solution) the pyrimidine peroxyl radicals must decay by second-order kinetics. These processes have been elucidated with uracil as a substrate.[47] It would exceed the scope of this paper to go into the details of these peroxyl radical reactions. They are reviewed in Ref. 4.

Purines

Our knowledge of the reactions of the OH radical and the H atom with the purines is still rather limited.[4,48] There are several reasons for this lack. So far, despite the efforts made, nobody has yet achieved a reasonable material balance in product studies (see Tables I and II). In the absence of oxygen, purine consumption as a rule fell short of what was expected on the basis of the yields of OH radicals and H atoms reacting with the purines. This indicates that there must be some highly reactive (non-radical) intermediate(s) capable of reacting with adenine radicals and hence restoring some of the starting material. This is in line with the observation of D. J. Deeble (cited in Ref. 4) that in the reaction of adenine and hypoxanthine OH-adduct radicals with ferricyanide, the G values of iron reduction are $\gg 6 \times 10^{-7}$ mol J⁻¹, the observed values being 19×10^{-7} and 12.5×10^{-7} mol J⁻¹, respectively.

We have recently done some high-dose-rate experiments with N^6,N^6,N^9-trimethyladenine using the pulse radiolysis facility to achieve high radical densities. Under such conditions radical-radical reactions should be favored over radical-product reactions. We did indeed observe new products that are not present at the low dose rates of gamma radiolysis; their identification is pending (C. Nese and C. von Sonntag, unpublished observations). However, these preliminary experiments indicate that even under conditions of high dose rates, we are unable to account for all OH radicals and H atoms in the products and are still left with the problem of low substrate consumption.

A second difficulty lies in the fact that some of the purine radicals do not react with oxygen.[50,51] Thus, an often-convenient procedure to trap intermediates with oxygen fails with these systems.

Table I. Gamma Radiolysis of N_2O-Saturated Aqueous Solutions of Adenine $(2 \times 10^{-3}$ mol dm⁻³)[49]

Product	G value
4,6-Diamino-5-formamido pyrimidine (FAPy-A)	0.2
8-Hydroxyadenine	0.35
6-Amino-8-hydroxy-7,8-dihydropurine	0.1
Adenine consumption	1.0

Table II. Gamma Radiolysis of 2´-Deoxyguanosine (5 x 10^{-3} mol dm^{-3}) in Deaerated and N_2O-Saturated Solutions. Products and their G values.[48]

Product	N_2	N_2O
N^6-(2´Deoxy-α-D-erythro-pentopyranosyl)-2,6-diamino-5-formamidopyrimid-4-one	0.26	0.25
N^6-(2´-Deoxy-β-D-erythro-pentopyranosyl)-2,6-diamino-5-formamidopyrimid-4-one	0.08	0.09
5´,8-Cyclo-2´,5´-dideoxyguanosine	0.05	0.06
8-Hydroxy-2´-deoxyguanosine	---	0.24
Guanine	0.19	0.38
9-(2´-Deoxy-α-D-erythro-pentopyranosyl)guanine	0.02	0.03
9-(2´-Deoxy-β-D-erythro-pentopyranosyl)guanine	0.01	0.02
9-(2´-Deoxy-α-D-erythro-pentofuranosyl)guanine	0.02	0.02
9-(2´-Deoxy-α-L-threo-pentofuranosyl)guanine	0.02	0.02
9-(2´-Deoxy-β-D-erythro-pento-1,5-dialdo-1,4-furanosyl)guanine	0.07	0.08
2´-Deoxyribonolactone	*	*
2´-Deoxyguanosine consumption	0.81	1.50

* Observed but not quantitatively determined.

A third difficulty is experienced in pulse radiolysis studies. The redox properties of the purine radicals are not as clearly distinct as those of the pyrimidine radicals (redox ambivalence).[35] There, this very convenient technique for probing various radicals is not so easily applicable.

A fourth problem, not encountered with pyrimidines, concerns rapid rearrangement reactions.

Although these problems persist, recent pulse radiolysis studies[35,52-54] have yielded very valuable information. Scheme 3 shows the sites where the OH radicals can add, using adenine as an example.

Of special interest are two reactions: (i) the elimination of water, which leads to the N-centered radical (e.g., Reaction 29)[52] in the unsubstituted compound, but to a radical cation in the case N^6,N^6-dimethyladenosine,[53] which cannot easily deprotonate, and (ii) a rapid ring-opening undergone by the C(8)-OH-adduct radical (Reaction 30).

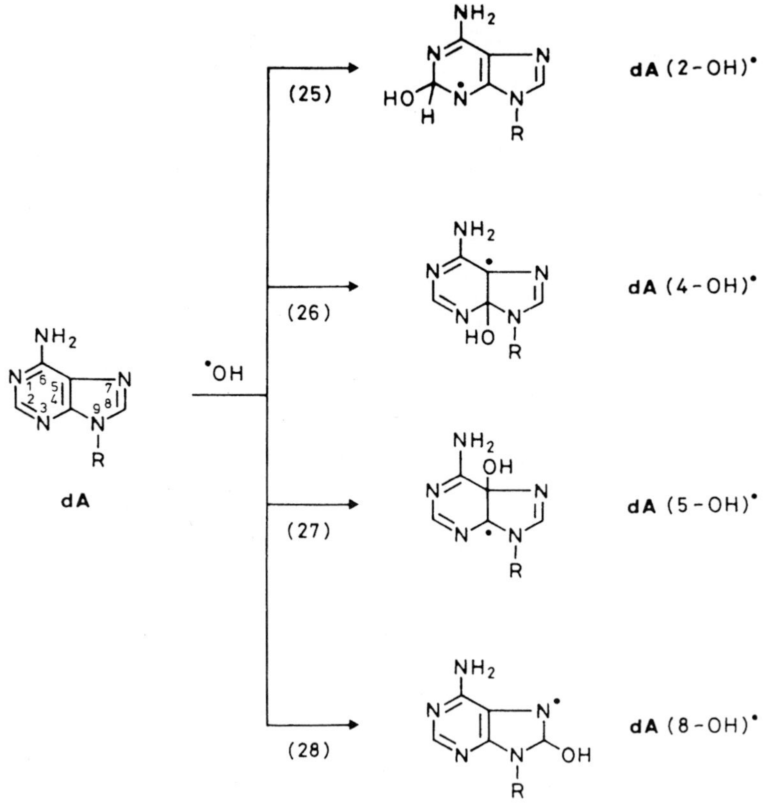

Scheme 3

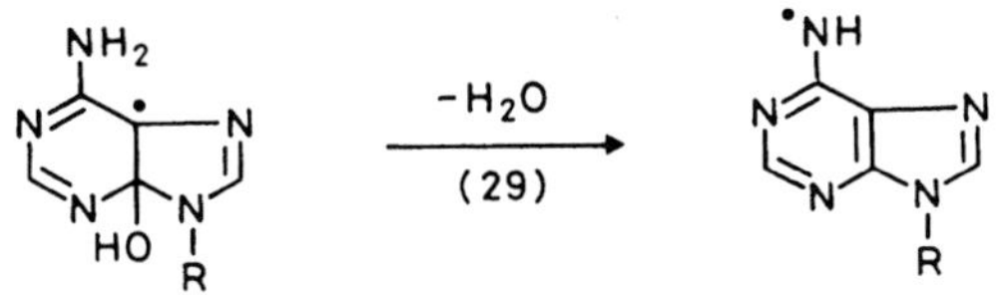

Scheme 4

For adenine the rate constant for this ring-opening reaction has been given as 1.3×10^5 s^{-1}.[52] Thus, it is so fast that bimolecular reactions of radicals cannot compete with it. However, as can be seen from Table I, two products are found that have the C(8)-OH-adduct radical as their precursor: 8-hydroxyadenosine [an oxidation product of the C(8)-OH-adduct radical, Reaction 31] and 4,6-diamino-5-formamidopyrimidine (FAPy-A), which is formed when the ring-opened radical is reduced (Reaction 33 and 34). We therefore conclude that the ring-opening reaction must be reversible; otherwise, the formation of 8-hydroxyadenine (in water, its tautomer 8-oxo-7,8-dihydroadenine prevails) cannot be explained, unless oxidation of the ring-opened radical yields a product which closes again thereby forming 8-hydroxyadenine (Reactions 36 and 37).

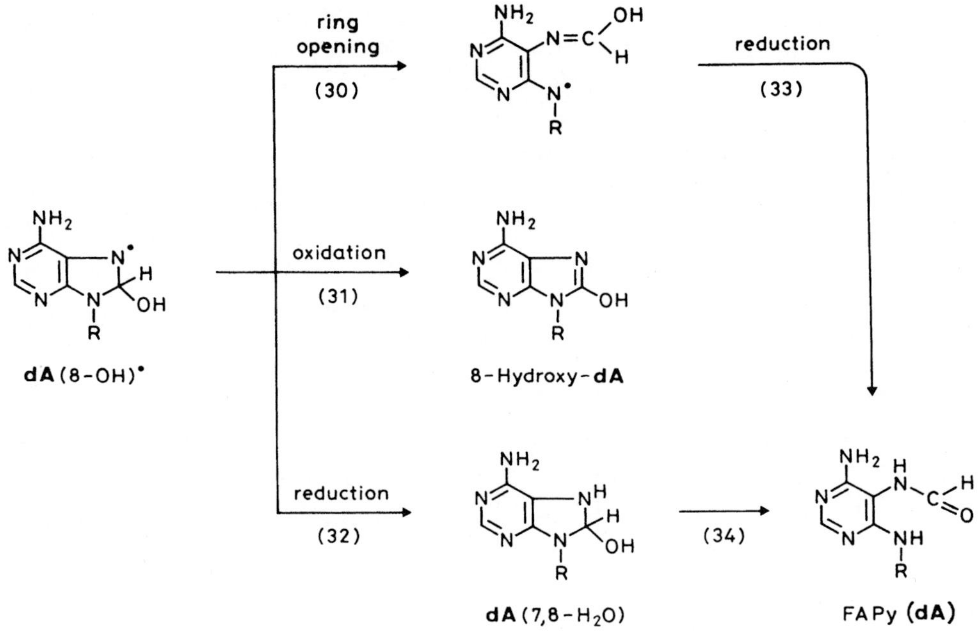

Scheme 5

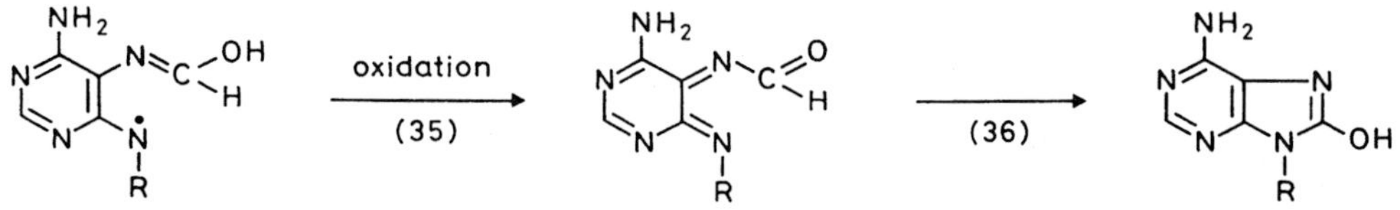

Scheme 6

Recently it has been reported that 2′-deoxyguanosine in the presence of oxygen yields 2,2-diamino-4-(β-2-deoxy-D-erythro-pentofuranosyl)amino-oxazol-5-one (Reaction 37).[55] The mechanism of its formation is not yet elucidated.

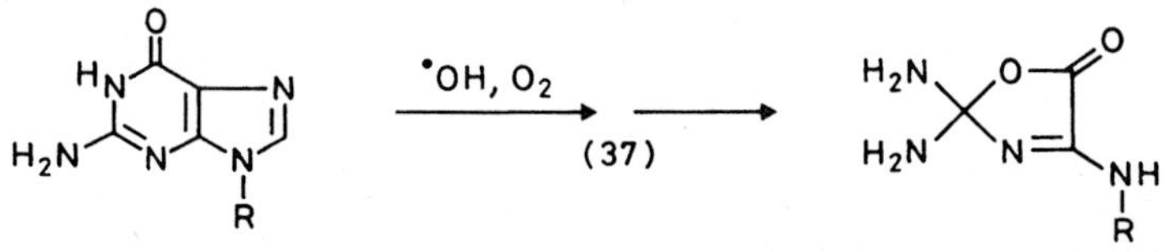

Scheme 7

Reactions of Carbon-Centered Radicals with the Nucleobases

In the living cell, proteins (and possibly other cell constituents) are in close proximity to the DNA. Hence, DNA-protein crosslinks are well-documented lesions. Such crosslinks might occur by the trivial mechanism, i.e., the combination of a DNA radical with a protein radical. Much more interesting is the question whether protein radicals can add to the DNA bases (and DNA radicals to aromatic amino acids). There is, however, an additional problem. Base radicals, such as formed by OH- and H-addition to pyrimidines, are also carbon-centered radicals as are the sugar-centered radicals. In DNA such radicals could, in principle, add to an unaltered base, and hence could lead to an intramolecular DNA-DNA crosslink. Very little is known about such reactions.[4] This is why we have to rely on some model systems which are still very distant from the real situation in the cell.

Pyrimidines and Alkyl Radicals

There is only one detailed study which deals with the reaction of alkyl radicals with pyrimidines, i.e., the reactions of hydroxyalkyl radicals with 1,3-dimethyluracil and 1,3-dimethylthymine.[56] It became obvious that, in contrast to the water radicals $\cdot$OH and $\cdot$H, the $\cdot$CH$_2$OH radical preferentially adds at the C(6)-position (Reaction 38), i.e., at the position of lowest electron density. This is readily understood, since α-hydroxyalkyl radicals are nucleophilic in contrast to the electrophilic OH radicals and H atoms. The reactivity of the $\cdot$CH$_2$OH radicals towards the pyrimidines is four to five orders of magnitude lower than the reactivity of $\cdot$OH and $\cdot$H. With thymine they also react by H-abstraction from the methyl group (Reaction 39). A very interesting observation of possibly more general importance has been made: in the disproportionation reactions with other radicals present, the C(6)-CH$_2$OH,C(5)-yl radicals derived from thymine can disproportionate, yielding an intermediate with an exo-methylene group (Reaction 40). This intermediate does not accumulate to measurable steady-state concentrations because it is a much better radical scavenger than thymine. This raises the question as to whether such intermediates can take part in the cross-linking process (at sites of high radical density, i.e., in spurs or blobs).

In dilute aqueous solutions of 1,3-dimethyluracil (10^{-3} mol dm^{-3}), there is no indication that the OH-adduct radicals add to the unaltered base with any efficiency at all ($k < 10^4$ dm^3 mol^{-1} s^{-1} (cf. data reported in Ref. 44). However, this does not rule out such a reaction in DNA where another base is in close proximity.

Purines and Alkyl Radicals

Hydroxyalkyl radicals also add to purines. There is considerable material available on this topic (for references see Ref. 4) but no attempt has yet been made to estimate rate constants and evaluate details of the mechanism. It is noted that the main product observed in these reactions is the C(8)-alkylated compound, which is reminiscent of the 5′,8-cyclonucleotides formed in DNA.[57-59] The C(8)-alkylating reaction is not restricted to radicals derived from alcohols, but also observed with radicals derived from amino acids,[60] which suggests that this is a potential route to crosslinking.

Reactions of the Nucleobase Radical Cations

Nucleobase radical cations are produced by the direct effect. Their fate is not only of interest because of the subsequent alterations of the base moiety, but they might also cause sugar damage, i.e., base release and strand breakage.

Several approaches can be taken to produce radical cations and study their reactions in dilute aqueous solutions. Biphotonic photoionization of the nucleobases yields a base radical cation and a solvated electron (for additional references, see Refs. 61 and 62). Photoexcited menadione is a good oxidant but a poor hydrogen abstractor; and in its reaction with the nucleobases, radical cations are also formed (together with the menadione radical anion). One can also make use of the pulse radiolysis technique, which allows to produce strong oxidants such as Br$_2$·$^-$, Tl^{2+} and SO$_4$·$^-$.[63] In our hands, the SO$_4$·$^-$ system has yielded very interesting results regarding the spectra, pK$_a$ values and reactions of the pyrimidine radical cations (see Ref. 63 for additional references).

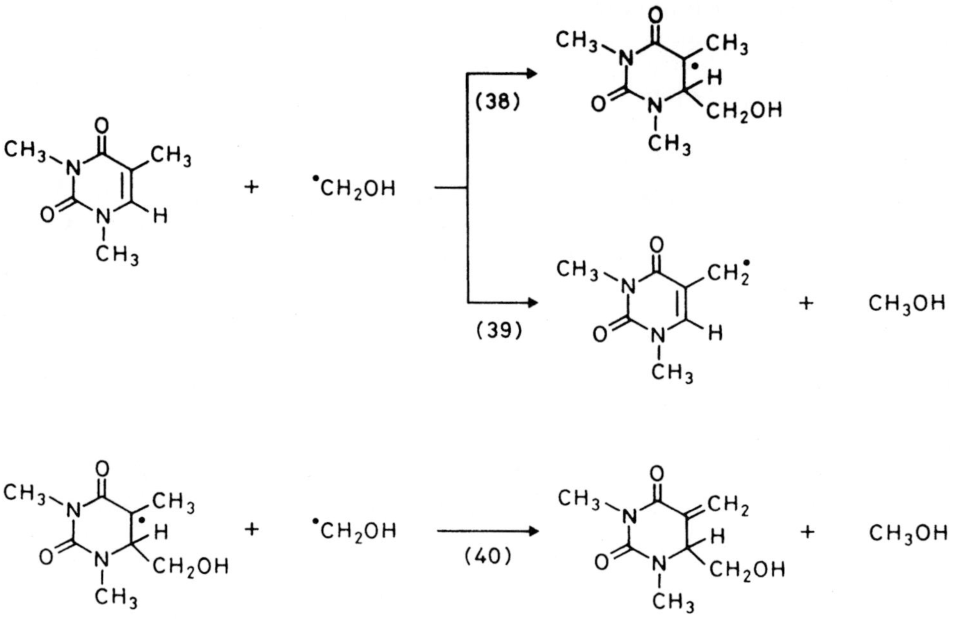

Scheme 8

Pyrimidine Radical Cations

The $SO_4^{.-}$ radicals are generated by reacting the solvated electrons and the H atoms with peroxodisulfate (Reaction 41), while the OH radicals are scavenged by t-butanol (the resulting t-butanol-derived radicals do not absorb in the wavelength region of interest).

$$S_2O_8^{2-} + e_{aq}^- \ (H\cdot) \longrightarrow SO_4^{.-} + SO_4^{2-} \ (HSO_4^-) \qquad (41)$$

The sulfate radicals react with the nucleobases with rate constants that are practically diffusion-controlled.[63] In the first step, a very short-lived adduct is most likely formed, never observed in these systems by pulse radiolysis but stable in some olefins. These adducts eliminate the sulfate dianion, and in favorable cases such as 1,3,5,6-tetramethyluracil, 1,3-dimethylthymine and 3-methylthymidine, the first observable intermediate is the corresponding radical cation.[63,64] In the case of the 1,3,5,6-tetramethyluracil, the radical cation has a lifetime of about 20 μs. In the other thymine derivatives, there is less alkyl substitution at the C=C double bond to stabilize the radical cation by hyperconjugation. This must shorten the lifetime of the radical cations relative to 1,3,5,6-tetramethyluracil; accordingly, their lifetimes are only about 2 μs. In 1,3-dimethyluracil, which carries no methyl group at the C(5)-C(6) double bond, we were unable to detect a radical cation intermediate on the μs time scale.[65] From the above reasoning, we would indeed expect that a uracil radical cation must be considerably shorter lived than 1 μs, about the detection limit in these experiments. In 1-methylthymine and in thymidine too, radical cations of similar lifetimes (2.8 and 0.9 μs, respectively) can be detected. Here, however, these deprotonate at N(3) (Reaction 14). The pK_a values of the radical cations of these compounds are at 3.8 and 3.6, respectively.

It has been mentioned above that the radical cations of the N(1)-substituted thymine derivatives react rapidly with water. Irrespective of how this reaction proceeds (Reactions 14, 16 or 18), a proton is formed besides a neutral radical. Since protons have a much higher equivalent conductance than the radical cations, the result is an increase in conductivity.[63]

The question now arises as to the identity of the resulting species. Here we recall that radicals such as the C(5)-OH-adducts have reducing properties and react rapidly with tetranitromethane, yielding the nitroform anion. With 1,3-dimethyluracil as the substrate, the yield of the nitroform anion is quantitative[65,66] Thus, only the reducing C(5)-OH,C(6)-yl radical is formed. As a consequence, in the presence of the strongly oxidizing $S_2O_8^{2-}$, a chain reaction sets in at the low dose rates of ^{60}Co-gamma-radiolysis, which shows some very interesting peculiarities.[65] This is also true for 1,3,6-trimethyluracil.[66,67]

However, when the methyl substituent is in the 5-position, as in the thymine derivatives, the chemistry is completely different. There is no reducing radical formed (hence, also, no chain reaction induced with $S_2O_8^{2-}$). What is even more interesting is the fact that considerable deprotonation occurs at the C(5)-methyl group (Reaction 18). The resulting allylic radical (for its UV spectrum see Ref. 68) can be trapped by oxygen and (some of) its products detected.[67] From their yield it is concluded that deprotonation at C(5)-methyl must indeed be important, perhaps 30%. About 70% of the 1,3-dimethylthymine-derived radical cations react with water, adding to the C(6)-position (Reaction 17). In basic solution the formation of the allylic radical is suppressed; i.e., OH$^-$ does not speed up the deprotonation reaction at the methyl group. Instead, the preferred reaction of the OH$^-$ is to combine with the radical cation.

Upon the freezing of aqueous thymidine solutions, the thymidine molecules aggregate as the water crystallizes. When the frozen solutions are irradiated, much of the chemistry observed is due to the direct effect, since only very few OH radicals and H atoms become available for reaction with the solute. Under such conditions, products were observed which involve not only the allylic radical but also the N(3)-centered radical mentioned above.[69]

A look at Scheme 1 shows that OH radicals and radical cations can lead to the same intermediates. However, the reducing C(5)-OH,C(6)-yl radical generated in 30% yield by OH radicals is not formed in the reaction of the radical cation with water. On the other hand, the allylic radical is only a minor product of OH radical attack, but at about 30% it is an important product in the decay of the radical cation. In addition there is the puzzling N(3)-centered radical. In neutral solutions it is an important intermediate (note that the pK_a value of the thymidine radical cation is 3.6). Coexisting in equilibrium with the radical cation, it eventually decays via this intermediate when no substrates are present with which to react. Otherwise, its reactivity is unknown.

Photoexcited menadione is a convenient oxidant for generating nucleobase radical cations. This approach has been used to produce radical cations of 2´-deoxycytidine and to measure the resulting products.[70] Kinetic measurements giving a detailed picture are still lacking. There is, however, good experimental evidence that besides reacting with water, the 2´-deoxycytidine radical cation deprotonates at C(1´). This observation is very interesting, because in DNA such a process can lead to base release and a 2-deoxyribonolactone unit, i.e., an alkali-labile site.

Purine Radical Cations

The guanine radical cation has been produced by reacting guanosine with $Br_2^{\cdot-}$, $SO_4^{\cdot-}$ and with Tl^{2+}.[54,71] The guanine radical cation is stable only in acidic media; its $pK_{a,1}$ value being 3.9 (deprotonation at N(1); equilibrium 48/49). In basic solution the amino group also deprotonates ($pK_{a,2} = 10.8$; equilibrium 50/51).[54] Details are shown in Scheme 9.

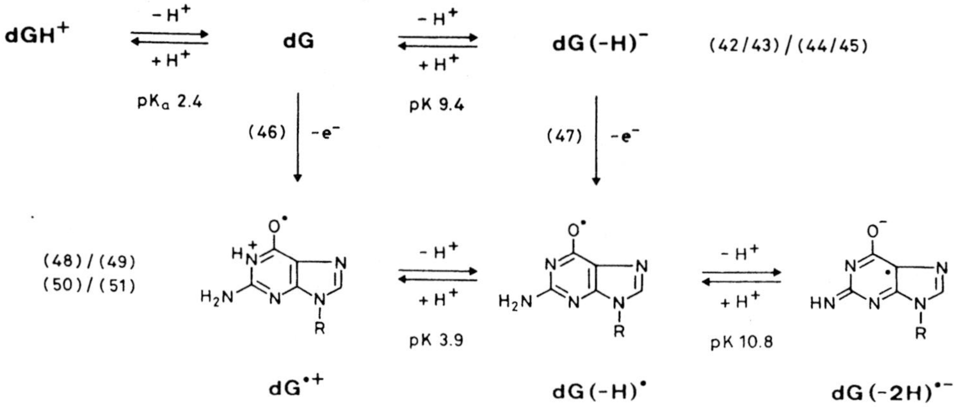

Scheme 9

A similar situation prevails in the case of the radical cation of 2′-deoxyadenosine where the radical cation deprotonates at N(1) (Reaction 53).[35] The pK_a value is <1; i.e., it is more than 12 units more acidic than its parent compound (Table III).

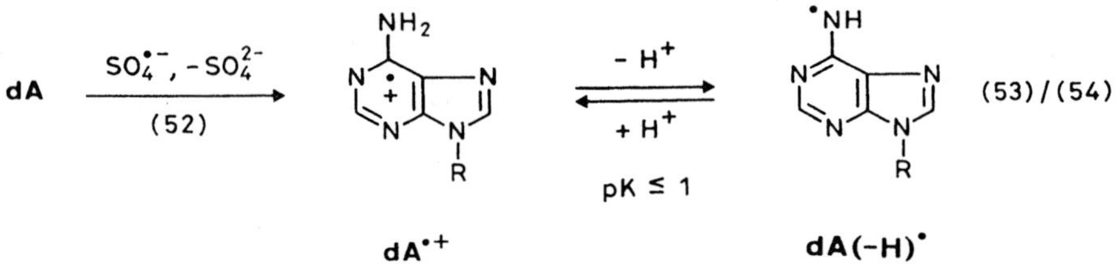

Scheme 10

It is worth mentioning that the C(4) and C(5) OH-adduct radicals of the purines yield the same intermediates as the purine radical cations (cf. Reaction 29).

Radical Transfer Reactions in Polynucleotides and DNA

There is now unequivocal evidence that in poly(U), base radicals are capable of attacking the sugar moiety and causing strand breakage[24,72-75] as well as base release.[7,76,77] These processes are observed both in the absence and presence of oxygen. In the latter case, peroxyl radicals must play a role—as clearly evidenced by the high G values of oxygen uptake[52] and hydroperoxide formation.[77] More recently, ESR studies have been carried out in the absence of oxygen, and a radical has been observed which gives evidence for an attack at the C(2′)-position.[78,79] In the ribo-series, this type of radical carries an OH group in α-position and a phosphate group in β-position. Such radicals are known to eliminate the phosphate group very rapidly,[80,81] most likely at times ≤2 μs.[4] In contrast to the rate constants of

Table III. pK$_a$ Values of Nucleobases and Their Radical Cations (according to Refs. 35 and 63)

Nucleobase	pK$_a$(parent)	pK$_a$(radical cation)	Δ(pK$_a$)
dA	>13.75	$\leq$1.0	>12.75
dG	9.4	3.9	5.5
dC	$\approx$14.0	<4.0	>10.0
dT	9.9	3.6	5.3

phosphate eliminations from α-alkoxy-β-phosphatoalkyl radicals[82] (e.g., the C(4´) radical in DNA), the rate constants of phosphate eliminations from such OH-substituted β-phosphate radicals have not been found to depend strongly on pH. However, in poly(U) the observed rate constant of strand breakage strongly depends on pH (it increases with decreasing pH).[83] Thus, an H$^+$-catalysed reaction precedes strand breakage. In addition, strand breakage proceeds with the same rate constant as the decay of the base radicals. Hence, the precursor radicals, the C(5)-OH,C(6)-yl (70%) and/or the C(6)-OH,C(5)-yl (23%)[7] must undergo the H$^+$-catalyzed reaction which speeds up the transfer reaction. It has been suggested[79] that it could be an H$^+$-catalyzed rearrangement from the C(5)-OH,C(6)-yl to the C(6)-OH,C(5)-yl (Reactions 55 and 58), which then undergoes the H-abstraction reactions. There is some experimental evidence, based on the low yield of the uracil hydrate (6-hydroxy-5, 6-dihydrouracil),[7] which argues against this hypothesis. Alternatively, the H$^+$-induced formation of the radical cation (Reactions 55 and/or 57) as well as the N(3)-centered radical (Reaction 59) may be envisaged as more reactive intermediates.[79] If it were the N(3)-centered radical which would undergo the H-abstraction reaction, only a low G(chromophore loss) would be expected while the observed value is 5.1 x 10^{-7} mol J^{-1},[76] very close to the G value of OH and H attack at the base moiety of poly(U) in N$_2$O-saturated solutions.

There are two further interesting aspects. The H atom does not react in any significant amount with the sugar moiety;[40] nevertheless, it also induces strand breakage. This must again be by a radical transfer from a base radical. In contrast to the reaction of the OH-adducts, H-atom-induced strand breakage is not H$^+$-catalyzed.[84]

Solvated electrons induce no strand breaks.[24] A trivial reason would be a scavenging of solvated electrons by unavoidable impurities (recall the low rate constant of e$^-_{aq}$ with poly(U)). But if it reacts, one must assume that the uracil electron adduct should be converted into the C(6)-H,C(5)-yl radical (see above). This radical appears to be non-reactive. It must therefore be concluded that it is the C(5)-H,C(6)-yl radical which induces the H-transfer reaction.

More recently, results obtained with poly(dT) and other thymidine derivatives[85] were interpreted as being indicative of an H-transfer reaction similar to that observed in poly(U). In contrast to poly(U), the observed yields of the relevant processes, base release and 5-hydroxy-5,6-dihydrothymine formation, are very low. This is also reflected in low G values of strand breakage in poly(dT)[86] (see Table IV). Thus, we believe that such an H-transfer is not yet fully apparent in the poly(dT) system.

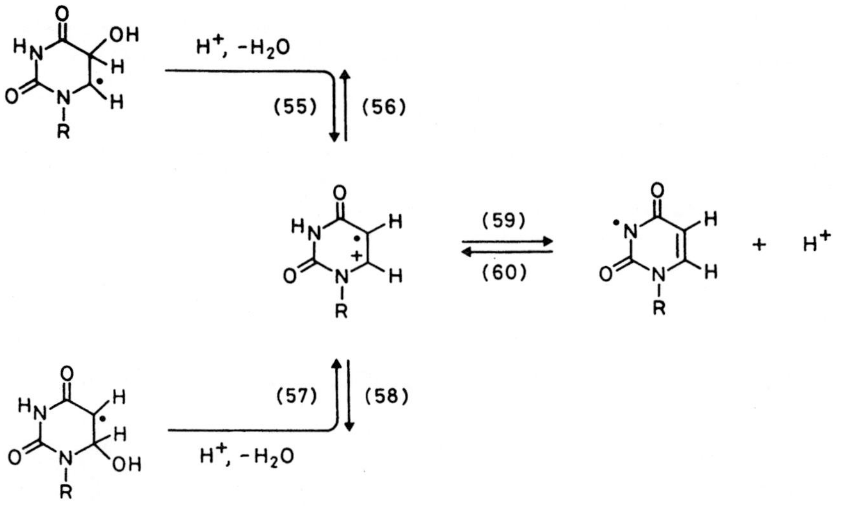

Scheme 11

Table IV. Strand break formation in the gamma radiolysis of polynucleotides and DNA in N_2O- and N_2O/O_2-(4:1)-saturated aqueous solutions. G values in units of 10^{-7} mol J^{-1}.[86]

Polynucleotide and DNA	G Values	
	N_2O	N_2O/O_2
Poly(U)	2.4*	2.6*
Poly(C)	2.2*	1.25
Poly(A)	0.44	1.04
Poly(G)	< 0.2	< 0.2
Poly(dU)	0.42	---
Poly(dT)	0.57	---
Poly(dC)	1.4	---
Poly(dA)	2.3*	---
Poly(dG)	0.09	---
ssDNA	0.84	0.62

* G (strand breakage) is significantly in excess of 20% of OH radical attack (G = 1.1 x 10^{-7} mol J^{-1}).

G(strand breakage) for a number of polynucleotides has been measured both in the ribo and in the 2-deoxyribo series[86] (cf. also Ref. 87). The data are compiled in Table II. If one accepts that OH radical attack at the sugar moiety does not exceed 20% [in poly(U) a value of 7% has been estimated],[7] G(strand breakage) by this route cannot exceed a value of 1.1×10^{-7} mol J^{-1}. It can then be seen from the data in Table IV that, in the absence of oxygen, poly(U), poly(C) and poly(dA) are in fact the only obvious cases where strand breakage must be induced by base radicals. Most interestingly, poly(dU) does not give this reaction, nor does poly(A). In poly(G) and poly(dG), G(strand breakage) is very low indeed (which also holds for phosphate end groups).[87] This raises the question whether, in these systems, sugar attack by OH is actually very minor. It has been suggested that in DNA the guanine radical cation might be a precursor of DNA strand breakage.[88] It has been shown above that guanine-OH-adduct radicals and the guanine radical cation are related to one another. In view of these poly(G) and poly(dG) results, the hypothesis that the guanine radical cation plays a decisive role in strand break formation[88] may have to be revised.

Two important points must be mentioned here. Photoionization of poly(U) causes strand breakage which follows the same kinetics (pH dependence of rate constants) as does the OH radical-induced reaction.[89] The kinetics of strand breakage in single-stranded DNA and in poly(dA) show very striking similarities.[86,90] The first point is readily understood because of the rapid reaction of the uracil radical cation with water, which yields the C(5)-OH,C(6)-yl radical (see above). The other point is very puzzling: Does this result mean that in single-stranded DNA strand breakage is mainly induced by the adenine moiety? Work is in progress to elucidate this question.

Radical Reactions of the DNA Sugar Phosphate Backbone

In the living cell, hydroxyl radicals from the indirect effect and base radicals in radical transfer reactions will create sugar-derived radicals which can lead to strand breakage and base release. Concerning the direct effect, i.e. energy absorbed by the sugar and the phosphate moieties, very little is known. A likely process is the ionization of the phosphate group (Reaction 61).

$$(RO)_2 P(O)O^- \xrightarrow[radiation]{ionizing} (RO_2)_2 P(O)O \cdot + e_{aq}^- \qquad (61)$$

Similar to SO_4^- radicals phosphate radicals have oxidizing properties (for references see Ref. 4). Thus, they could oxidize the nucleobase to their radical cations. In addition, such radicals have also a certain H-abstractive power. In DNA such phosphate radicals would have the C(4′)-H in a convenient distance (a six-membered ring can be written as transition state), and H-abstraction from this position is known to produce a strand break (see below).

Detailed model studies have been carried out to elucidate the mechanism of DNA strand breakage (for references see Ref. 4). From these models, it is concluded that the C(4′) radical should give rise to strand breakage by eliminating the phosphate group, thereby forming a sugar radical cation (Reactions 62 and 63). When this mechanism is applied rigorously to DNA, four products are expected in the absence of oxygen and after reduction of the resulting dephosphorylated radicals. In fact, all the four products (1, 2, 6, 7 in Figure 3) have been found in gamma-irradiated DNA.[4] The various steps are depicted in Scheme 12.

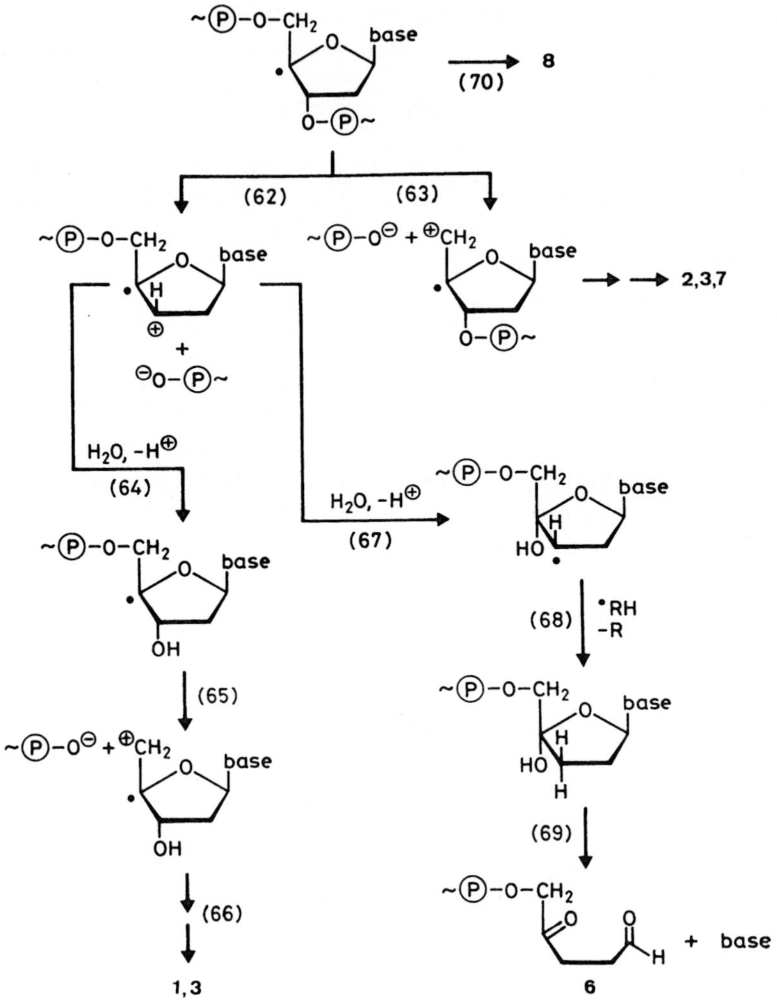

Scheme 12

In a recent paper, theoretical calculations have been put forth which deal with the question whether a radical cation is an intermediate or whether the reaction proceeds by a concerted mechanism involving one water molecule.[20] This study came to the conclusion that the latter might operate. It is not quite clear whether this mechanism can account for all four observed products.

In the ribo-series, similar reactions as depicted for the C(4')-position can be written for the C(2')-position; i.e., rapid strand breakage will occur (see above). Low-molecular-weight models provide no indication that rapid strand breakage in DNA can occur from the C(2')-position. This is important to realize when modeling H-transfer reactions from the base radicals to the sugar moiety.

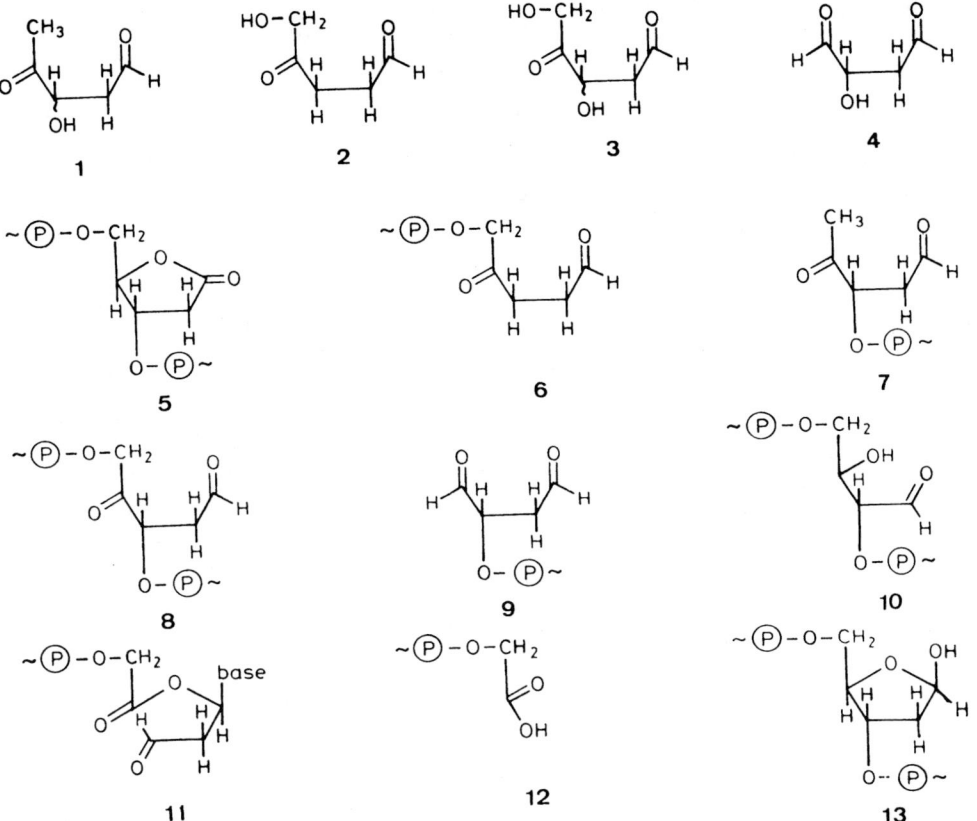

Figure 3. DNA sugar lesions detected in gamma-irradiated aqueous solutions of DNA in the absence and presence of oxygen.[4]

Base and Sugar Damage in DNA

A large number of radiation products, first observed with model systems, have also been found to be formed in DNA (for references see Ref. 4). Most recently, hydrolysis of DNA, trimethylsilyl derivatization of the nucleobase products, and subsequent gas chromatography combined with mass spectrometry (GC-MS) using the single-ion monitoring (SIM) technique gave the most detailed analysis so far available.[91,92] They are compiled in Table V.

It can be seen from this table that a good number of products have indeed been detected and quantitatively determined. However, examination of the total yields show that under N_2O, G(detected products) is only 0.8×10^{-7} mol J^{-1}, while in the presence of oxygen (N_2O/O_2) this yield is 1.4×10^{-7} mol J^{-1}. Compared to G(·OH) = 5.6×10^{-7} mol J^{-1} and taking into account sugar attack of about 10%, only 16% (N_2O) to 28% (N_2O/O_2) of the base damage is found in these products. These values would not change drastically (18% and 32%) when 20% sugar attack is taken into consideration.

In Fig. 3 the various altered sugars observed so far in irradiated DNA are compiled.[4] Although attempts have been made to determine these products quantitatively,[93-98] the data are far from being reliable, and at present it is not yet possible to see whether we have a full material balance.

Table V. Gamma Radiolysis of double-stranded DNA in aqueous solution. Products and their G values (in units of 10^{-7} mol J^{-1}). Saturating gases are N_2O and N_2O/O_2.[92]

Product	G Value	
	N_2O	N_2O/O_2
5,6-Dihydrothymine	0.056	---
5-Hydroxy-5,6-dihydrothymine	0.077	<0.001
5-Hydroxy-5,6-dihydrocytosine	0.026	<0.001
Cytosine glycol	0.107	0.256
Thymine glycol	0.102	0.434
FAPy-adenine*	0.087	0.059
8-Hydroxyadenine	0.053	0.158
FAPy-guanine**	0.124	0.036
8-Hydroxyguanine	0.162	0.467
Total	0.794	1.410

*   FAPy-adenine = 4,6-Diamino-5-formamido pyrimidine
** FAPy-guanine = 2,6-Diamino-4-hydroxy-5-formamido pyrimidine

Prevention of DNA Damage

In vivo, DNA and low-molecular-weight material will compete for the water radicals ($\cdot OH$, $H\cdot$ and e_{aq}^-) formed in the neighborhood of DNA. This situation has been simulated by a large number of *in vitro* studies (cf. Ref. 4). The kinetics of this situation are by no means simple, and a number of models have been developed (see Refs. 4 and 98). The most refined is the cylinder model.[98]

In bacterial cells, very high concentrations of scavengers such as alcohols or dimethylsulfoxide (up to 20% in weight) are required to reduce DNA damage by about 40% (see Ref. 4 and references therein). The remaining DNA damage is usually attributed to the direct effect (for a cautionary remark see Ref. 4).

Repair of DNA Damage

Besides the scavenging of the water radicals (mainly $\cdot OH$ and $H\cdot$) in the close vicinity of DNA, which converts the reactive water-derived radicals into less-reactive organic radicals discussed in a preceding paper, an additional effect is exerted by thiols.[99-101] Here it is recalled that the cell's glutathione (GSH) pool can reach levels as high as 10^{-2} mol dm^{-3}.

The S-H bond in thiols is only about 91 kcal mol^{-1}, weaker than many C-H bonds. For this reason thiols can undergo the so-called repair reaction (Reaction 70).

$$DNA(-H)\cdot + GSH \longrightarrow DNA + GS\cdot \qquad (70)$$

For the radicals derived from 2,5-dimethyltetrahydrofuran, it has been shown that not only the forward reaction has to be considered ($k_{71} = 6 \times 10^8$ dm^3 mol^{-1} s^{-1}), but that the reverse reaction occurs as well, in this case with a rate constant of $k_{72} \approx 5 \times 10^3$ dm^3 mol^{-1} s^{-1}.[102] Thus, the thiyl radicals formerly considered to be inert are not so by any means. There are now additional data available that point in the same direction.[103,104] When the radical R$_3$C· in the equilibrium (Reaction 71/72) is a reducing radical such as the ·C(CH$_3$)$_2$OH radical derived from 2-propanol, low concentrations of an oxidant such as p-nitroacetophenone can quantitatively oxidize the 2-propanol radical despite the fact that the thiyl radical predominates in the equilibrium.[103]

$$R_3C\cdot + RSH \rightleftharpoons R_3CH + RS\cdot \qquad (71/72)$$

If this observation is transferred to the cellular system, one might speculate that the radicals formed in the vicinity of as well as at the DNA are "repaired" by GSH. But then, the GS· radicals may abstract an H-atom from the sugar moiety of DNA, e.g., from the C(4´) position (Reaction 73; cf. the similarity of this radical with the radical derived from 2,5-dimethyltetrahydrofuran). This reaction is of course reversed either by proper repair (Reaction 74) or by a pseudo-repair yielding the threo conformation instead of the erythro (Reaction 75).[102] Such a reaction is already a DNA

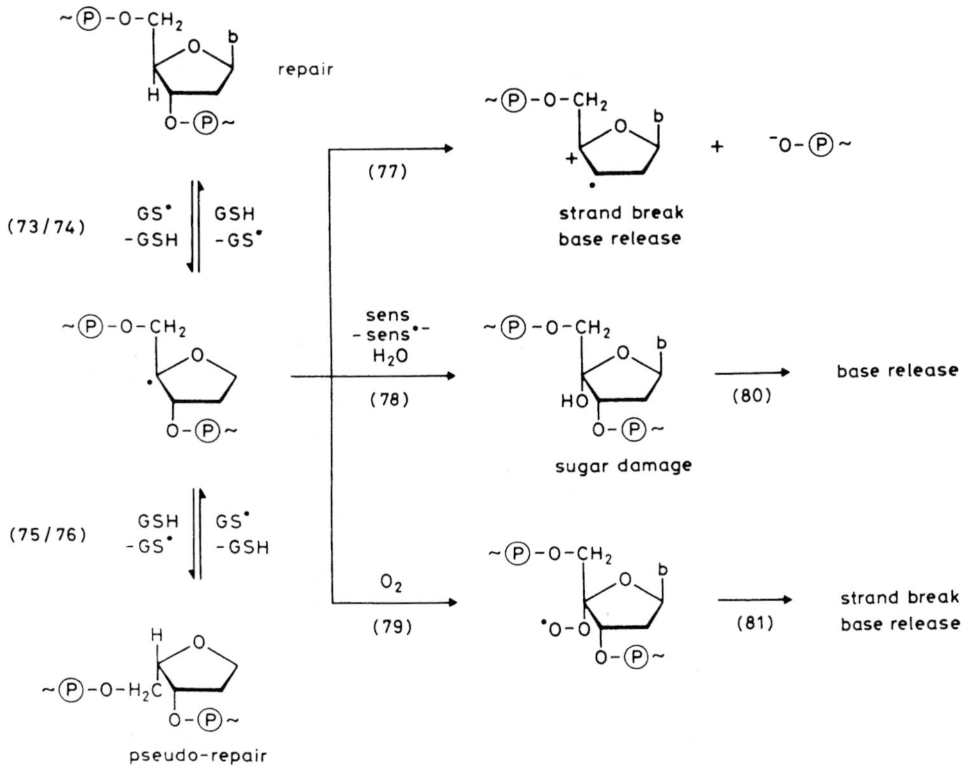

Scheme 13

damage and has been observed to occur at least on the nucleoside level.[105] However, in competition a strand break may occur from the 4´-radical, or, in the presence of a radiosensitizer, it may be oxidized to the corresponding keto compound, a well-known type of DNA sugar damage (product 8 in Figure 3). However, in the cell with glutathione as the thiol and DNA, the situation becomes very complex; both are negatively charged, and their interactions strongly depend, among other factors, on the salt concentration present. It is well documented that with DNA and polynucleotides, the rate constants of the repair reactions strongly depend on the charge of the thiol, the negatively charged glutathione being especially slow.[106]

In our laboratory we have undertaken some experiments in order to investigate the possibility of thiyl radicals reacting with the sugar moiety and hence causing strand breakage (M. Wala and E. Bothe, unpublished results). At pH 7, neutral thiyl radicals (which should be more reactive towards DNA than the negatively charged thiyl radicals derived from glutathione, cf. Ref. 106) were generated by reacting $(HOCH_2CH_2S)_2$ (2×10^{-4} mol dm^{-3}) with the solvated electron (Reactions 82 and 83).

$$HOCH_2CH_2SSCH_2CH_2OH + e_{aq}^- \longrightarrow (HOCH_2CH_2SSCH_2CH_2OH)^{\cdot -} \quad (82)$$

$$(HOCH_2CH_2SSCH_2CH_2CH_2OH)^{\cdot -} + H_2O \longrightarrow HOCH_2CH_2S\cdot$$

$$+ HOCH_2CH_2SH + OH^- \quad (83)$$

The DNA was single stranded (heat-denatured, sodium perchlorate 10^{-2} mol dm^{-3}) at a concentration of 4×10^{-4} mol dm^{-3} in nucleotides. At a dose rate of 4×10^{-2} Gy s^{-1}, strand breaks in excess of those calculated for the attack of OH radicals were not observed. This indicates that the bimolecular decay of the thiyl plus the disulfide-derived radicals (formed by OH-attack) must have been faster than the (slow) attack at the DNA sugar moiety. In this experiment the thiol concentration (recall the repair reaction, Reaction 74) was nil at the beginning of the experiment, and during irradiation only very little thiol (compared with the thiol content of cells) could have been built up. Although the DNA concentration in this experiment was very low, these results seem to suggest that thiyl radical attack will not contribute significantly to DNA strand breakage in cells.

The repair reaction is usually written as an H-donation. But there is increasing evidence that for many oxidizing radicals this reaction is surprisingly slow, if it occurs at all. However, such oxidizing radicals are readily reduced by thiolate ions. A simple example is the (reducing) 1,2-dihydroxyethyl radical, which is readily reduced by thiols ($k = 2.6 \times 10^7$ dm^3 mol^{-1} s^{-1}). When this radical eliminates water, an (oxidizing) formylmethyl radical is formed. This radical no longer reacts with thiols (at least at the pulse radiolysis time scale, i.e., $k << 10^7$ dm^3 mol s^{-1}), but it is readily reduced by the thiolate ion ($k = 1.2 \times 10^8$ dm^3 mol^{-1} s^{-1}).[107]

This observation has also a bearing on the repair of DNA radicals. It has been reported for some (oxidizing) purine-OH-adduct radicals that with measurable rate constants they react only with the thiolate ions (but not with the thiols).[108-110] It has also been reported that the (reducing) C(5)-OH,C(6)-yl radical derived from thymine does react with thiols by H-abstraction, but no such product was found for the (oxidizing) C(6)-OH,C(5)-yl radical, at least not in the case of uracil[111] (see however data given in Ref. 112). It should, however, be possible to reduce the latter radical with the thiolate ion. The resulting intermediate carbanion could either lose an OH$^-$ ion and the pyrimidine would be restored (proper repair), or it could be protonated by water at C(5) yielding the hydrate. This system warrants further, more detailed investigation.

Problems to Be Addressed

Research often causes us to be a bit shortsighted because of its (necessarily) step-by-step approach. When we pause to ask ourselves what we do not yet understand, we see clearly only the very next problem but cannot well define the more important problem which lies beyond. Of course, one can ask weighty questions such as "How are double-strand breaks formed by ionizing radiation?" "What is the mechanism of the oxygen effect in cell-killing?" "How do the repair enzymes operate on the DNA damage?" "Which are the lesions that lead to mutations?" — also very practical questions such as "How can radiation therapy be improved?"

Even obviously important questions could not have been asked some time ago with any hope of finding the answer, because the necessary knowledge was not then available. For example, only when we had learned that an unrepaired double-strand break is lethal for a cell could we address the question of how such strand breaks are formed and how the repair enzymes cope with the rest (i.e., the majority) of the double-strand breaks. Most of the questions posed in this paragraph will be of the more myopic kind, but before these are solved scant progress can be made regarding the fundamental aspects of radiation biology and the radiotherapy of cancer.

We have seen that the reactions of the thymine radical anions are fairly well understood. This does not hold for the other three bases. In adenine the site(s) of protonation at carbon is (are) not yet elucidated. In cytosine there is this 20% uncertainty regarding the reaction of the protonated radical anion with oxygen. In the case of guanine derivatives practically nothing is known about the reactions subsequent to electron addition.

An open and very important area relates to the question of intermolecular electron transfer between nucleobase electron adducts and adjacent nucleobases, which mainly centers on whether there is a preferred sink. In the solid state, at low temperatures and with oriented DNA fibers, this is a problem tackled by ESR for many years and is still controversial.[9,10]

The products formed upon OH- and H-attack at the purines are not yet fully elucidated. At present, one must fairly say that in a higher sense our knowledge about these reactions is close to nil, although a number of products are known. I expect that in the near future, research will concentrate on the free-radical chemistry of purines, especially since in DNA transition-metal-ion-mediated (e.g., copper), damage is thought to center on guanine. It is noted that the DNA scaffolding protein contains copper.[4]

As concerns DNA-protein crosslinking, it is far from established how fast the radicals derived from proteins react with DNA. Also the question remains as to whether purines or pyrimidines are more reactive. Livneh et al.[113] studied adenine C(8)-alkylation in mixed dA/dPy dinucleotides, but reported no pyrimidine alkylation. It is still unclear whether this is due to a much higher reactivity of the purine as compared with the pyrimidine or whether the pyrimidine products escaped detection.

In the reaction of OH radicals with thymine, 30% of the radicals formed give rise to C(6)-OH,C(5)-yl radicals (Reaction 10). This raises the question whether in subsequent reactions the corresponding exo-methylene compound may be formed by disproportionation (cf. Reaction 40) and whether such a species contributes to subsequent crosslinking reactions.

Similar to the question of electron transfer reactions of the nucleobase electron adducts to adjacent nucleobases in DNA, one may ask whether there is a similar reaction of the nucleobase radical cations. Because of the high rate of the deprotonation reactions, especially in the purines, there is also the question whether the heteroatom-centered radicals can also undergo an electron transfer reaction and which reactions this would entail.

A large field only partially understood is that of radical transfer reactions in the polynucleotides and in DNA. There are also remarkable oxygen effects which should be looked at.

Because of spatial constraints, we have not, in this paper, gone into a presentation of the knowledge regarding peroxyl free-radical chemistry (for a recent reviews see Refs. 4 and 115), but it is obvious that oxygen plays a very important role in radiobiology and cancer therapy. It thus follows that peroxyl radical reactions will be another focus of interest in the near future. The competition of thiols (repair) and oxygen (oxygen fixation) for radicals in vivo is a salient topic which might help to understand details of the oxygen effect. First attempts to link radiation chemistry and radiation biology at this point have already been made (Ref. 100 and references cited therein).

It seems also worthwhile to investigate the fate of phosphate-centered radicals, i.e., to follow up the question whether they undergo facile oxidation of the nucleobases or, rather, abstract an H atom from the sugar moiety. The use of peroxodiphosphates would allow study of this question.

More quantitative data (G values of base and sugar products) are required to understand the details of DNA radiolysis. As it stands, these values do not add up to the expected total.

Already, the information presently available allows computer modeling. Studies on polymer dynamics might help to improve our understanding of the interesting radical transfer reactions. Our own experience indicates that present concepts to define the properties and the reactions of nucleobase radical cations[64,67] are not yet adequate. This is another challenge for the theoreticians.

References

1. E. H. Grubbe. Priority in the Therapeutic Use of X-rays. *Radiology* 21:156-162 (1933).
2. G. Schwarz. Über Desensibilisierung gegen Röntgen- und Radiumstrahlen. *Münchner Medizinische Wochenschrift* 56:1217-1218 (1909).
3. G. Schwarz. Zur genaueren Kenntnis der Radiosensibilität. *Wiener Klin. Wochenschr.* 11:397-398 (1910).
4. C. von Sonntag. *The Chemical Basis of Radiation Biology.* Taylor and Francis, London (1987).
5. M. N. Schuchmann and C. von Sonntag. Radiation Chemistry of Carbohydrates. Part 14. Hydroxyl Radical-Induced Oxidation of D-Glucose in Oxygenated Aqueous Solution. *J. Chem. Soc. Perkin Trans.* 2:1958-1963 (1977).
6. C. von Sonntag. Free Radical Reactions of Carbohydrates as Studied by Radiation Techniques. *Adv. Carbohydr. Chem. Biochem.* 37:7-77 (1980).
7. D. J. Deeble, D. Schulz, and C. von Sonntag. Reactions of OH Radicals with Poly(U) in Deoxygenated Solutions: Sites of OH Radical Attack and the Kinetics of Base Release. *Int. J. Radiat. Biol.* 49:915-926 (1986).
8. M.E.A. Churchill, T. D. Tullius, N. R. Kallenbach, and N. C. Seeman. A Holliday Recombination Intermadiate is Twofold Symmetric. *Proc. Natl. Acad. Sci. USA* 85:4653-4656 (1988).
9. J. Hüttermann. Free Radicals from Solid Oriented DNA-fibers: Facts and Fancy. *Free Radical Res. Comms.* 6:103-104 (1989).
10. M.C.R. Symons. ESR Spectra for Protonated Thymine and Cytidine Radical Anions: Their Relevance to Irradiated DNA. *Int. J. Radiat. Biol.* 58:93-96 (1990).
11. J. E. Turner, R. N. Hamm, A. H. Wright, R. H. Ritchie, J. L. Magee, A. Chatterjee, and W. E. Bolch. Studies to Link the Basic Radiation Physics and Chemistry of Liquid Water. *Radiat. Phys. Chem.* 32:503-510 (1988).

12. A. Chatterjee. Radical-Induced DNA Damage and Specific Implications in Evaluating Genetic Changes. *Free Radical Res. Comms.* 6:189-190 (1989).

13. A. Chatterjee and W. Holley. Energetic Electron Tracks and DNA Strand Breaks. *Nucl. Tracks Radiat. Measur.* 16:127-133 (1989).

14. W. R. Holley, A. Chatterjee, and J. L. Magee. Production of DNA Strand Breaks by Direct Effects of Heavy Charged Particles. *Radiat. Res.* 121:161-168 (1990).

15. A. Chatterjee and W. R. Holley. A General Theory of DNA Strand Break Production by Direct and Indirect Effects. *Radiat. Prot. Dosimetry* (in press).

16. J. F. Ward. Molecular Mechanisms of Radiation-Induced Damage to Nucleic Acids. *Adv. Radiat. Biol.* 5:181-239 (1975).

17. J. F. Ward. Biochemistry of DNA Lesions. *Radiat. Res.* 104:103-111 (1985).

18. J. F. Ward. Radiation Chemical Mechanisms of Cell Death. In Radiation Research. *Proc. 8th Int. Congr. Radiat. Res. Edinburgh,* E. M. Fielden, J. F. Fowler, J. H. Hendry, and D. Scott (eds.), pp. 162-168, Vol. 2. Taylor and Francis, London (1987).

19. R. Osman, W. J. Clark, A. P. Mazurek, and H. Weinstein. Theoretical Studies of Molecular Mechanisms of DNA Damage Induced by Hydroxyl Radicals. *Free Radical Res. Comms.* 6:131-132 (1989).

20. L. Pardo, R. Osman, J. Banfalder, A.P. Mazurek, and H. Weinstein. Molecular Mechanisms of Radiation Induced DNA Damage: H-Abstraction and β-Cleavage. *Free Radical Res. Commun.* 12/13 (pt. II):461-463 (1991).

21. R. Osman, L. Pardo, J. Banfelder, A. P. Mazurek, L. Shwatzman, R. Strauss, and H. Weinstein. Molecular Mechanism of Radiation Induced DNA Damage: H-Addition to Bases, Direct Ionization and Double Strand Break. *Free Radical Res. Commun.* 12/13 (pt. II):465-467 (1991).

22. H. Weinstein, R. Osman, G. A. Mercier, A. P. Mazurek, L. Pardo, and L. A. Rubenstein. Theory and Computation of Molecular Mechanisms in Biological Processes: Radiation-Induced Damage to DNA and Neurotransmitter Function. *Computer Assisted Analysis and Modeling on the IBM 3090,* H. U. Brown, ed., pp. 629-673, MIT Press, Boston (1990).

23. J. Blok and H. Loman. Bacteriophage DNA as a Model for Correlation of Radical Damage to DNA and Biological Effects. In *Mechanisms of DNA Damage and Repair,* M. G. Simic, L. Grossman, and A. D. Upton (eds.), pp. 75-87. Plenum Press, New York (1986).

24. D.G.E. Lemaire, E. Bothe, and D. Schulte-Frohlinde. Yields of Radiation-Induced Main Chain Scission of Poly U in Aqueous Solution: Strand Break Formation Via Base Radicals. *Int. J. Radiat. Biol.* 45:351-358 (1984).

25. F. J. Nabben, J. P. Karman, and H. Loman. Inactivation of Biologically Active DNA by Hydrated Electrons. *Int. J. Radiat. Biol.* 42:23-30 (1982).

26. E. Hayon. Optical-Absorption Spectra of Ketyl Radicals and Radical Anions of Some Pyrimidines. *J. Chem. Phys.* 51:4881-4892 (1969).

27. D. J. Deeble and C. von Sonntag. Radioprotection of Pyrimidines by Oxygen and Sensitization by Phosphate: a Feature of Their Electron Adducts. *Int. J. Radiat. Biol.* 51:791-796 (1987).

28. S. Das, D. J. Deeble, M. N. Schuchmann, and C. von Sonntag. Pulse Radiolytic Studies on Uracil and Uracil Derivatives. Protonation of Their Electron Adducts at Oxygen and Carbon. *Int. J. Radiat. Biol.* 46:7-9 (1984).

29. D. J. Deeble, S. Das, and C. von Sonntag. Uracil Derivatives: Sites and Kinetics of Protonation of the Radical Anions and the UV Spectra of the C(5) and C(6) H-atom Adducts. *J. Phys. Chem.* 89:5784-5788 (1985).

30. H. M. Novais and S. Steenken. ESR Studies of Electron and Hydrogen Adducts of Thymine and Uracil and Their Derivatives and of 4,6-Dihydroxypyrimidines in Aqueous Solution. Comparison with Data from Solid State. The Protonation at Carbon of the Electron Adducts. *J. Am. Chem. Soc.* 108:1-6 (1986).

31. A. Hissung and C. von Sonntag. The Reaction of Solvated Electrons with Cytosine, 5-Methylcytosine and 2´-Deoxycytidine in Aqueous Solution. The Reaction of the Electron Adduct Intermediates with Water, p-Nitroacetophenone and Oxygen. A Pulse Spectroscopic and Pulse Conductometric Study. *Int. J. Radiat. Biol.* 35:449-458 (1979).

32. M. Eigen. Protonenübertragung, Säure-Base-Katalyse und enzymatische Hydrolyse. Teil I: Elementarvorgänge. *Angew. Chem.* 75:489-508 (1963).

33. A. Hissung, C. von Sonntag, D. Veltwisch, and K.-D. Asmus. The Reaction of the 2´-Deoxyadenosine Electron Adduct in Aqueous Solution. The Effects of the Radiosensitizer p-Nitroacetophenone. A Pulse Spectroscopic and Pulse Conductometric Study. *Int. J. Radiat. Biol.* 39:63-71 (1981).

34. K. J. Visscher, M. P. de Haas, H. Loman, B. Vojnovic and J. M. Warman. Fast Protonation of Adenosine and of Its Radical Anion Formed by Hydrated Electron Attack; a Nanosecond Optical and dc-Conductivity Pulse Radiolysis Study. *Int. J. Radiat. Biol.* 52:745-753 (1987).

35. S. Steenken. Purine Bases, Nucleosides and Nucleotides: Aqueous Solution Redox Chemistry and Transformation Reactions of Their Radical Cations, e- and OH Adducts. *Chem. Rev.* 89:503-520 (1989).

36. P. N. Moorthy and E. Hayon. Free-Radical Intermediates Produced from the One-Electron Reduction of Purine, Adenine and Guanine Derivatives in Water. *J. Am. Chem. Soc.* 97:3345-3350 (1975).

37. B. Rakvin, J. N. Herak, K. Voit, and J. Hüttermann. Free Radicals from Single Crystals of Deoxyguanosine 5´-monophosphate (Na Salt) Irradiated at Low Temperatures. *Radiat. Environ. Biophys.* 26:1-12 (1987).

38. K. J. Visscher, H.J.W. Spoelder, H. Loman, A. Hummel, and M. L. Hom. Kinetics and Mechanism of Electron Transfer Between Purines and Pyrimidines, Their Dinucleotides and Polynucleotides After Reaction with Hydrated Electrons; A Pulse Radiolysis Study. *Int. J. Radiat. Biol.* 54:787-802 (1988).

39. G. E. Adams and R. L. Willson. On the Mechanism of BUdR Sensitization: A Pulse Radiolysis Study of One Electron Transfer in Nucleic-acid Derivatives. *Int. J. Radiat. Biol.* 22:589-597 (1972).

40. S. Das, D. J. Deeble, and C. von Sonntag. Site of H Atom Attack on Uracil and Its Derivatives in Aqueous Solution. *Z. Naturforsch.* 40c:292-294 (1985).

41. S. Fujita and S. Steenken. Pattern of OH Radical Addition to Uracil and Methyl- and Carboxyl-Substituted Uracils. Electron Transfer of OH Adducts with N,N,N´,N´-Tetramethyl-p-Phenylenediamine and Tetranitromethane. *J. Am. Chem. Soc.* 103:2540-2545 (1981).

42. M. N. Schuchmann, S. Steenken, J. Wroblewski, and C. von Sonntag. Site of OH Attack on Dihydrouracil and Some of Its Methyl Derivatives. *Int. J. Radiat. Biol.* 46:225-232 (1984).

43. D. K. Hazra and S. Steenken. Pattern of OH Radical Addition to Cytosine and 1-,3-,5-, and 6-Substituted Cytosines. Electron Transfer and Dehydration Reactions of the OH Adducts. *J. Am. Chem. Soc.* 105:4380-4386 (1983).

44. M. Al-Sheikhly and C. von Sonntag. γ-Radiolysis of 1,3-Dimethyluracil in N_2O-Saturated Aqueous Solutions. *Z. Naturforsch.* 38b:1622-1629 (1983).

45. M. I. Al-Sheikhly, A. Hissung, H.-P. Schuchmann, M. N. Schuchmann, C. von Sonntag, A. Garner, and G. Scholes. Radiolysis of Dihydrouracil and Dihydrothymine in Aqueous Solutions Containing Oxygen; First- and Second-Order Reactions of the Organic Peroxyl Radicals; the Role of Isopyrimidines as Intermediates. *J. Chem. Soc. Perkin Trans. II* 601-608 (1984).

46. M. N. Schuchmann, M. Al-Sheikhly, C. von Sonntag, A. Garner, and G. Scholes. The Kinetics of the Rearrangement of Some Isopyrimidines to Pyrimidines Studied by Pulse Radiolysis. *J. Chem. Soc. Perkin Trans. II* 1777-1780 (1984).

47. M. N. Schuchmann and C. von Sonntag. The Radiolysis of Uracil in Oxygenated Aqueous Solutions. A Study by Product Analysis and Pulse Radiolysis. *J. Chem. Soc. Perkin Trans. II* 1525-1531 (1983).

48. J. Cadet and M. Berger. Radiation-Induced Decomposition of the Purine Bases Within DNA and Related Model Compounds. *Int. J. Radiat. Biol.* 47:127-143 (1985).

49. J. J. van Hemmen and J. F. Bleichrodt. The Decomposition of Adenine by Ionizing Radiation. *Radiat. Res.* 46:444-456 (1971).

50. R. L. Willson. The Reaction of Oxygen with Radiation-Induced Free Radicals in DNA and Related Compounds. *Int. J. Radiat. Biol.* 17:349-358 (1970).

51. M. Isildar, M. N. Schuchmann, D. Schulte-Frohlinde and C. von Sonntag. Oxygen Uptake in the Radiolysis of Aqueous Solutions of Nucleic Acids and Their Constituents. *Int. J. Radiat. Biol.* 41:525-533 (1982).

52. A.J.S.C. Vieira and S. Steenken. Pattern of OH Radical Reaction with 6- and 9-Substituted Purines. Effect of Substituents on the Rates and Activation Parameters of Unimolecular Transformation Reactions of Two Isomeric OH Adducts. *J. Phys. Chem.* 91:4138-4144 (1987).

53. A.J.S.C. Vieira and S. Steenken. Pattern of OH Radical Reaction with N6,N6-Dimethyladenosine. Production of Three Isomeric OH Adducts and Their Dehydration and Ring Opening Reactions. *J. Am. Chem. Soc.* 109:7441-7448 (1987).

54. L. P. Candeias and S. Steenken. Structure and Acid-Base Properties of One-Electron-Oxidized Deoxyguanosine, Guanosine, and 1-Methylguanosine. *J. Am. Chem. Soc.* 111:1094-1099 (1989).

55. J. Cadet, M. Berger, C. Decarroz, J.-F. Mouret, J. E. van Lier and R. J. Wagner. Reactions Radicalaire Photo- et Radio-induites Des Bases Puriniques et Pyrimidiniques Des Acides Nucleiques. *J. Chim. Phys.* (in press).

56. H.-P. Schuchmann, R. Wagner, and C. von Sonntag. The Reactions of the Hydroxymethyl Radical with 1,3-Dimethyluracil and 1,3-Dimethylthymine. *Int. J. Radiat. Biol.* 50:1051-1068 (1986).

57. H. Steinmaus, I. Rosenthal, and D. Elad. Photochemical and γ-Ray-Induced Reactions of Purines and Purine Nucleosides with 2-propanol. *J. Am. Chem. Soc.* 91:4921-4923 (1969).

58. R. Ben-Ishai, M. Green, E. Graff, D. Elad, H. Steinmaus, and J. Salomon. Photoalkylation of Purines in DNA. *Photochem. Photobiol.* 17:155-167 (1973).

59. H. Steinmaus, I. Rosenthal, and D. Elad. Light- and γ-Ray-Induced Reactions of Purines and Purine Nucleosides with Alcohols. *J. Org. Chem.* 36:3594-3598 (1971).

60. D. Elad and I. Rosenthal. Photochemical Alkylation of Caffeine with Amino-Acids. *Chem. Commun.* _:905-906 (1969).

61. M. Wala, E. Bothe, H. Görner, and D. Schulte-Frohlinde. Quantum Yields for the Generation of Hydrated Electrons and Single-Strand Breaks in Poly(C), Poly(A) and Single-Stranded DNA in Aqueous Solution on 20 ns Laser Excitation at 248 nm. *J. Photochem. Photobiol. A: Chem.* 53:87-108 (1990).

62. D. Schulte-Frohlinde, M. G. Simic, and H. Görner. Laser-Induced Strand Break Formation in DNA and Polynucleotides. *Photochem. Photobiol.* 52:1137-1151 (1990).

63. D. J. Deeble, M. N. Schuchmann, S. Steenken, and C. von Sonntag. Direct Evidence for the Formation of Thymine Radical Cations from the Reaction of $SO_4^{\cdot-}$ with Thymine Derivatives: A Pulse Radiolysis Study with Optical and Conductance Detection. *J. Phys. Chem.* 94:8186-8192 (1990).

64. D. Schulte-Frohlinde and K. Hildenbrand. Electron Spin Resonance Studies of the Reactions of ·OH and $SO_4^{\cdot-}$ Radicals with DNA. In *Free Radicals in Synthesis and Biology*, F. Minisci, (ed.), pp. 335-359. Dordrecht: Kluwer, (1989).

65. H.-P. Schuchmann, D. J. Deeble, G. Olbrich, and C. von Sonntag. The $SO_4^{\cdot-}$-Induced Chain Reaction of 1,3-Dimethyluracil with Peroxodisulphate. *Int. J. Radiat. Biol.* 51:441-453 (1987).

66. E. Bothe, D. J. Deeble, D.G.E. Lemaire, R. Rashid, M. N. Schuchmann, H.-P. Schuchmann, D. Schulte-Frohlinde, S. Steenken, and C. von Sonntag. Pulse-Radiolytic Studies on the Reactions of $SO_4^{\cdot-}$ with Uracil Derivatives. *Radiat. Phys. Chem.* 36:149-154 (1990).

67. R. Rashid, F. Mark, H.-P. Schuchmann, and C. von Sonntag. The $SO_4^{\cdot-}$-Induced Oxidation of 1,3-Dimethylthymine and 1,3,6-Trimethyluracil by Potassium Peroxodisulphate and Oxygen in Aqueous Solution: An Interesting Contrast. *Int. J. Radiat. Biol.* 59:1081-1100 (1990).

68. C. von Sonntag. New Aspects in the Free-Radical Chemistry of Pyrimidine Nucleobases. *Free Rad. Res. Comms.* 2:217-224 (1987).

69. A. A. Shaw and J. Cadet. Radical Combination Processes Under the Direct Effects of Gamma Radiation on Thymidine. *J. Chem. Soc. Perkin Trans.* 2:2063-2070 (1990).

70. C. Decarroz, J. R. Wagner, and J. Cadet. Specific Deprotonation Reactions of the Pyrimidine Radical Cation Resulting from the Menadione Mediated Photosensitization of 2´-Deoxycytidine. *Free Radical Res. Commun.* 2:295-301 (1987).

71. M. G. Simic and S. V. Jovanovic. Free Radical Mechanisms of DNA Damage. In *Mechanisms of DNA Damage and Repair*, M. G. Simic, L. Grossman, and A. D. Upton (eds.), pp. 39-50. Plenum Press, New York (1986).

72. D. Schulte-Frohlinde and E. Bothe. Identification of a Major Pathway of Strand Break Formation in Poly U Induced by OH Radicals in Presence of Oxygen. *Z. Naturforsch.* 39c:315-319 (1984).

73. E. Bothe, G. Behrens, E. Böhm, B. Sethuram, and D. Schulte-Frohlinde. Hydroxyl Radical-Induced Strand Break Formation of Poly(U) in the Presence of Oxygen. Comparison of the Rates as Determined by Conductivity, ESR and Rapid-Mix Experiments with a Thiol. *Int. J. Radiat. Biol.* 49:57-66 (1986).

74. D.G.E. Lemaire, E. Bothe, and D. Schulte-Frohlinde. Hydroxyl Radical-Induced Strand Break Formation of Poly(U) in Anoxic Solution. Effect of Dithiothreitol and Tetranitromethane. *Int. J. Radiat. Biol.* 51:319-330 (1987).

75. E. Bothe, M. Adinarayana, and D. Schulte-Frohlinde. Rate and Yield of OH-Induced Strand Break Formation of Polynucleotides and DNA. *Free Radical Res. Commun.* 6:139 (1989).

76. D. J. Deeble and C. von Sonntag. γ-Radiolysis of Poly(U) in Aqueous Solution. The Role of Primary Sugar and Base Radicals in the Release of Undamaged Uracil. *Int. J. Radiat. Biol.* 46:247-260 (1984).

77. D. J. Deeble and C. von Sonntag. Radiolysis of Poly(U) in Oxygenated Solutions. *Int. J. Radiat. Biol.* 49:927-936 (1986).

78. K. Hildenbrand and D. Schulte-Frohlinde. E.S.R. Studies on the Mechanism of Hydroxyl Radical-Induced Strand Breakage of Polyuridylic Acid. *Int. J. Radiat. Biol.* 55:725-738 (1989).

79. K. Hildenbrand and D. Schulte-Frohlinde. ESR Studies on the Mechanism of ·OH-Induced Strand Breakage of Poly(U). *Free Radical Res. Comms.* 6:137-138 (1989).

80. A. Samuni and P. Neta. Hydroxyl Radical Reaction with Phosphate Esters and the Mechanism of Phosphate Cleavage. *J. Phys. Chem.* 77:2425-2429 (1973).

81. S. Steenken, G. Behrens, and D. Schulte-Frohlinde. Radiation Chemistry of DNA Model Compounds. Part IV. Phosphate Ester Cleavage in Radicals Derived from Glycerol Phosphates. *Int. J. Radiat. Biol.* 25:205-210 (1974).

82. G. Behrens, G. Koltzenburg, A. Ritter, and D. Schulte-Frohlinde. The Influence of Protonation or Alkylation of the Phosphate Group on the E.S.R. Spectra and on the Rate of Phosphate Elimination from 2-Methoxyethyl Phosphate 2-yl Radicals. *Int. J. Radiat. Biol.* 33:163-171 (1978).

83. E. Bothe and D. Schulte-Frohlinde. Release of K^+ and H^+ from Poly U in Aqueous Solution Upon γ and Electron Irradiation. Rate of Strand Break Formation in Poly U. *Z. Naturforsch.* 37c:1191-1204 (1982).

84. E. Bothe and H. Selbach. Rate and Rate-Determining Step of Hydrogen-Atom-Induced Strand breakage in Poly(U) in Aqueous Solution Under Anoxic Conditions. *Z. Naturforsch.* 40c:247-253 (1985).

85. L. R. Karam, M. Dizdaroglu, and M. G. Simic. Intramolecular H Atom Abstraction from the Sugar Moiety by Thymine Radicals in Oligo- and Polydeoxynucleotides. *Radiat. Res.* 116:210-216 (1988).

86. M. Adinarayana, E. Bothe, and D. Schulte-Frohlinde. Hydroxyl Radical-Induced Strand Break Formation in Single-Stranded Polynucleotides and Single-Stranded DNA in Aqueous Solution as Measured by Light Scattering and by Conductivity. *Int. J. Radiat. Biol.* 54:723-737 (1988).

87. C. P. Murthy, D. J. Deeble, and C. von Sonntag. The Formation of Phosphate End Groups in the Radiolysis of Polynucleotides in Aqueous Solution. *Z. Naturforsch.* 43c:572-576 (1988).

88. P. J. Boon, P. M. Cullis, M.C.R. Symons, and B. W. Wren. Effects of Ionizing Radiation on Deoxyribonucleic Acid and Related Systems. Part 1. The Role of Oxygen. *J. Chem. Soc. Perkin Trans. II* 1393-1399 (1984).

89. D. Schulte-Frohlinde, J. Opitz, H. Görner, and E. Bothe. Model Studies for the Direct Effect of High-Energy Irradiation on DNA. Mechanism of Strand Break Formation Induced by Photoionization of Poly U in Aqueous Solution. *Int. J. Radiat. Biol.* 48:397-408 (1985).

90. E. Bothe, G. A. Qureshi, and D. Schulte-Frohlinde. Rate of OH Radical Induced Strand Break Formation in Single Stranded DNA Under Anoxic Conditions. An Investigation in Aqueous Solutions Using Conductivity Methods. *Z. Naturforsch.* 38c:1030-1042 (1983).

91. A. F. Fuciarelli, B. J. Wegher, E. Gajewski, M. Dizdaroglu and W. F. Blakely. Quantitative Measurement of Radiation-Induced Base Products in DNA Using Gas Chromatography-Mass Spectroscopy. *Radiat. Res.* 119:219-231 (1989).

92. A. F. Fuciarelli, B. J. Wegher, W. F. Blakeley, and M. Dizdaroglu. Yields of Radiation-Induced Base Products in DNA: Effects of DNA Conformation and Gassing Conditions. *Int. J. Radiat. Biol.* (in press).

93. F. Beesk, M. Dizdaroglu, D. Schulte-Frohlinde, and C. von Sonntag. Radiation-Induced DNA Strand Breaks in Deoxygenated Aqueous Solution. The Formation of Altered Sugars as End Groups. *Int. J. Radiat. Biol.* 36:565-576 (1979).

94. M. Dizdaroglu, C. von Sonntag, and D. Schulte-Frohlinde. Strand Breaks and Sugar Release by γ-Irradiation of DNA in Aqueous Solution. *J. Am. Chem. Soc.* 97:2277-2278 (1975).

95. M. Dizdaroglu, D. Schulte-Frohlinde, and C. von Sonntag. Radiation Chemistry of DNA, II. Strand Breaks and Sugar Release by γ-Irradiation of DNA in Aqueous Solution. The Effect of Oxygen. *Z. Naturforsch.* 30c:826-828 (1975).

96. M. Isildar, M. N. Schuchmann, D. Schulte-Frohlinde and C. von Sonntag. γ-Radiolysis of DNA in Oxygenated Aqueous Solutions: Alterations at the Sugar Moiety. *Int. J. Radiat. Biol.* 40:347-354 (1981).

97. M. Dizdaroglu, D. Schulte-Frohlinde, and C. von Sonntag. Isolation of 2-deoxy-D-erythro-pentonic Acid from an Alkali-Labile Site in γ-Irradiated DNA. *Int. J. Radiat. Biol.* 32:481-483 (1977).

98. M. Dizdaroglu, D. Schulte-Frohlinde, and C. von Sonntag. γ-Radiolysis of DNA in Oxygenated Aqueous Solution. Structure of an Alkali-Labile Site. *Z. Naturforsch.* 32c:1021-1022 (1977).

99. F. Mark, U. Becker, J. N. Herak, and D. Schulte-Frohlinde. Radiolysis of DNA in Aqueous Solution in the Presence of a Scavenger: A Kinetic Model Based on a Nonhomogeneous Reaction of OH Radicals with DNA Molecules of Spherical or Cylindrical Shape. *Radiat. Environ. Biophys.* 28:81-99 (1989).

100. T. Alper. *Cellular Radiobiology.* Cambridge: Cambridge University Press, (1979).

101. D. Schulte-Frohlinde and E. Bothe. Determination of the Rate Constants of the Alper-Formula from Kinetic Measurements on DNA in Aqueous Solution and Comparison with Data from Cells. *Int. J. Radiat. Biol.* 58:603-611 (1990).

102. C. von Sonntag and H.-P. Schuchmann. Sulfur Compounds and Chemical Repair in Radiation Biology. In *Sulfur-Centered Reactive Intermediates in Chemistry and Biology,* C. Chatgilialoglu and K.-D. Asmus, eds., pp. 409-414, Plenum Press, London (1990).

103. M. S. Akhlaq, H.-P. Schuchmann, and C. von Sonntag. The Reverse of the "Repair" Reaction of Thiols: H-Abstraction at Carbon by Thiyl Radicals. *Int. J. Radiat. Biol.* 51:91-102 (1987).

104. C. Schöneich, M. Bonifacic, and K.-D. Asmus. Reversible H-atom Abstraction from Alcohols by Thiyl Radicals: Determination of Absolute Rate Constants by Pulse Radiolysis. *Free Radical Res. Commun.* 6:393-405 (1989).

105. C. Schöneich, K.-D. Asmus, U. Dillinger, and F. von Bruchhausen. Thiyl Radical Attack on Polyunsaturated Fatty Acids: a Possible Route to Lipid Peroxidation. *Biochem. Biophys. Res. Commun.* 161:113-120 (1989).

106. M. Berger and J. Cadet. Isolation and Characterization of the Radiation-Induced Degradation Products of 2´-Deoxyguanosine in Oxygen-Free aqueous solutions. *Z. Naturforsch.* 40b:1519-1531 (1985).

107. S. Zheng, G. L. Newton, G. Gonick, R. C. Fahey, and J. F. Ward. Radioprotection of DNA by Thiols: Relationship Between the Net Charge on a Thiol and Its Ability to Protect DNA. *Radiat. Res.* 114:11-7 (1988).

108. M. S. Akhlaq, S. Al-Baghdadi, and C. von Sonntag. On the Attack of Hydroxyl Radicals on Polyhydric Alcohols and Sugars and the Reduction of the So Formed Radicals by 1,4-Dithiothreitol. *Carbohydr. Res.* 164:71-83 (1987).

109. P. O'Neill. Pulse Radiolytic Study of the Interaction of Thiols and Ascorbate with OH-Adducts of dGMP and dG. Implications for DNA Repair Processes. *Radiat. Res.* 96:198-210 (1983).

110. P. O'Neill. Hydroxyl Radical Damage: Potential Repair by Sulphydryls, Ascorbate and Other Antioxidants. In *Oxidative Damage and Related Enzymes*, G. Rotilio and J.V. Bannister, Chur: Harwood, eds., pp. 337-341. Life Chem. Rep. Suppl. Ser. 2 (1984).

111. P. O'Neill and P. W. Chapman. Potential Repair of Free Radical Adducts of dGMP and dG by a Series of Reductants. A Pulse Radiolytic Study. *Int. J. Radiat. Biol.* 47:71-80 (1985).

112. S. A. Grachev, E. V. Kropachev, and G. I. Litvyakova. The Reaction of Hydrogen Atom Transfer from the SH Group of Thiols to the OH-Adducts of Uracil in Radiolysis. *Izv. Akad. Nauk SSSR, Ser. Khim.* 2746-2752 (1988).

113. J. Cadet, M. Berger, P. Demonchaux, and J. Lhomme. Modifying Effects of Cysteine and Aromatic Sulfhydryl and Disulfide Agents on the Radiation-Induced Decomposition of Thymidine. *Radiat. Phys. Chem.* 32:197-202 (1988).

114. E. Livneh, S. Tel-Or, J. Sperling, and D. Elad. Light-Induced Free-Radical Reactions of Purines and Pyrimidines in Deoxyribonucleic Acid. Effect of Structure and Base Sequence on Reactivity. *Biochemistry* 21:3698-3703 (1982).

115. C. von Sonntag and H.-P. Schuchmann. Elucidation of Peroxyl Radical Reactions in Aqueous Solution with Radiation-Chemical Techniques. *Agnew. Chem. Int. Ed. Engl.* (in press).

Discussion

Geacintov: What is known about various probabilities of reactions?

von Sonntag: We know that the accessibility of the sugar must have a considerable effect on whether the OH radical reacts with the bases or with the sugar moiety. Such experiments have been done by Neville Kallenback [M.E.A. Churchill, T. D. Tullius, N. R. Kallenbach, and N. C. Seeman, *Proc. Natl. Acad. Sci. USA* 85:4653 (1988)] when probing a model of the Holliday junction with OH radicals. In fact his data might be used by you, Aloke, to test your nice programs.

Ward: This question refers to the diffusion distance of the electron in DNA. Beach and Zimbrick have published work [*Radiat. Res.* 91:328, (1982)] in which bacterial DNA with different levels of BU substitution was irradiated and the bromide yield measured (Br⁻ is presumably released as a result of dissociative electron attachment). They concluded that the electron only moves 3-4 base pairs. Does your data agree with this?

von Sonntag: There must be something in between.

Ward: A second question. If the adenine or guanine radical anion protonates, can it still transfer an electron to an electron sink?

von Sonntag: We know that the adenine radical anion rapidly protonates at nitrogen, much faster than any transfer to 5-bromouracil, and yet there is full electron transfer. This means that the N-protonated adenine radical anion is capable of undergoing this electron transfer reaction. However, when you convert this into a C-protonated radical anion (an OH⁻-catalyzed reaction), efficient electron transfer to 5-bromouracil no longer takes place. The same holds for the thymine/5-bromouracil system. Thymine radical anion (and possibly its O-protonated form) transfers an electron to 5-bromouracil. However, C-protonation [which can be speeded up by phosphate buffer; cf. D. J. Deeble, S. Das and C. von Sonntag, *J. Phys. Chem.* 89:5784 (1985)] inhibits this electron transfer. In this context I should mention an interesting experiment. Irradiating aqueous solutions containing 10^{-2} mol dm⁻³ thymine, 5×10^{-4} guanosine and 5×10^{-5} 5-bromouracil at pH 5.7 resulted in partial electron transfer (ca. 70 percent). If the transfer from the primarily formed thymine radical anion (and/or its oxygen-protonated form) to guanine were very fast, no such transfer would have been observed because the one-electron-reduced guanosine does not transfer an electron to 5-bromouracil.

Ward: You mentioned that with poly(dA) the G(strand breaks) is 2.4 but that the G value for adenine release in N_2O is 1.0. Therefore, some of the adenine radicals which are precursors for strand breaks don't end up as damaged adenine. Are OH radical reactions with deozyribose responsible for some of the breaks?

von Sonntag: I was referring to the work of Adinarayana, Bothe, and Schulte-Frohlinde [*Int. J. Radiat. Biol.* 54:723 (1988)]. With poly(dA) they measured only strand breakage, since there was not enough material to also measure adenine release. As regards the radical transfer from the adenine to the sugar moiety, D. Langfinger and I have published a study on 2-deoxyadenosine where we *tentatively* suggest that radical transfer from the base to the sugar moiety may indeed occur [*Z. Naturforsch.*, 40c:446 (1985)]. If this conclusion is correct, a similar reaction may occur in poly(dA) which then could explain the high yield of strand breakage.

Weinstein: Since I am only going to use the implications of that work that you cite today, I wanted to give you the opportunity to say what you wanted to say about the transfer of the radical to the sugar because I am not going to go into it in detail. So, if you want to present this, I want you to have the opportunity to do so.

von Sonntag: Thank you very much. There are two points. There is now very good evidence that we have a radical transfer in poly(U) from the base to the sugar, most likely by abstraction of an H-atom from C(2′) [and possibly to some extent from C(4′)]. The C(2′) radical in poly(U) rapidly eliminates the phosphate group at C(3′) as is typical of α-hydroxy-β-phosphato alkyl radicals.

On the other hand, if a similar reaction would occur in DNA, the resulting C(2′) radical no longer carries an OH group at this carbon; hence, phosphate elimination will *not* occur. This we conclude from experiments with simple alkyl phosphates.

Another point is the interesting mechanistic question concerning the mode of hydrolysis of the C(4´) radical in DNA. We believe that a radical cation is an intermediate, while your theoretical calculations suggest that it should proceed via an SN_2 reaction with water as the nucleophile. In our experiments we observe two kinds of products: free sugars (both phosphate groups released), altered sugars bound by the C(5´) or the C(3´) phosphate group to the broken DNA strand. It is easy to account for these products by a radical cation mechanism, but an SN_2 mechanism cannot explain the formation of these two types of products equally well unless two different types of SN_2 reactions are invoked. I have the impression that more-detailed calculations are required to establish that an SN_2 mechanism also allows to account for the elimination of both phosphate groups.

Varma: Clemens, without getting bogged down into details of the chemistry, I have a simple question. I read a paper that came out from Schulte-Frohlinde's laboratory where they talk about the sticky end and the blunt end of double-strand breaks. When you are talking about sticky and blunt ends, what does it mean? Can you identify those in terms of chemical species?

von Sonntag: A double-strand break with sticky ends is formed when two single-strand breaks occur on opposite strands a few base pairs apart. The strength of the hydrogen bridges of the remaining overlapping base pairs is not sufficiently strong, and this type of lesion can open and yield a double-strand break, with some probability of reclosure (sticky ends). On the other hand, when two single-strand breaks are formed exactly on opposite sides, a double-strand break with blunt ends is formed. Such different types of DNA damage can be made enzymatically and have been shown to be treated quite differently by repair enzymes [M. Bien, H. Steffen, and D. Schulte-Frohlinde, *Mutation Res.* 194:193 (1988)].

Varma: Chemically, what is the difference?

von Sonntag: DNA double-strand breaks with sticky ends stand a better chance of becoming repaired by the repair enzymes than those with blunt ends. It is noted that in competition with repair, there is always a more-or-less complete degradation of the DNA.

Chatterjee: Can we say anything about the chemical products?

Varma: This is only a qualitative picture.

von Sonntag: The studies which I mentioned have been done with well-defined enzymes, but some results concerning the effect of ionizing radiation are also reported.

Varma: That was my second question. Could you tell the difference between these sticky ends and the blunt ends as it relates to the radiations?

von Sonntag: Not from the experiments which we have been doing so far. You recall that we get practically only single-strand breaks when we irradiate DNA in aqueous solutions and that there is only a very minor component which yields double-strand breaks [M. A. Siddiqi and E. Bothe, *Radiat. Res.* 112:449 (1987)] the nature of which are not yet elucidated.

Ward: I will definitely say something about that tomorrow night. The point that one might make here is what Clemens has shown, or practically shown, is that wherever you get a strand break, you lose a base.

Varma: In single-strand breaks?

Ward: That would be the same in double-strand breaks. So, if you've got the one break and you lost a base from one side and you lost a base from the other side and that mostly went back together, you would then have a point mutation. It is not possible to determine where the released bases come from—they could come from single-strand breaks or double-strand breaks. Their production at each break of a double-strand break must be inferred from the single-strand break mechanism.

Hall: But John, is the implication that every time you have a translocation you have a mutation? Is that what you are saying?

Ward: If this mechanism is correct, then if a translocation occurred at the site of a radiation-induced double-strand break (where the two breaks are close together), a point mutation would occur in one of the resultant chromosomes.

von Sonntag: What John was referring to is the observation that whenever you induce a strand break radiolytically, the base at this damaged sugar also gets lost. Thus, with a single-strand break one base is lost, and with a double-strand break two bases are lost, one on each strand. Repair of such a double-strand break may well lead to a mutation.

Curtis: I would just suggest that looking at the relative number of sticky breaks as opposed to blunt end breaks as a function of LET would be a very interesting experiment to do.

Varma: If you can not identify what the sticky end and the blunt end mean chemically, how would you do it? Another question I have is on this first reaction that you have written: The interpretation of the ESR spectra of the thymine radical anion and the interpretation of corresponding pulse radiolysis data. The ESR studies were done at very low temperatures?

von Sonntag: Exactly.

Varma: And, therefore, the reaction rates that you are seeing might be very much different. And how do you distinguish between those two experiments?

von Sonntag: The pulse radiolysis data (room temperature) were interpreted by Hayon [*J. Chem. Phys.* 51:4881 (1969)] as being due to the thymine radical anion and its oxygen-protonated form ($pK_a \approx 7$), while the ESR experiments (low temperature) clearly showed that the species formed must have been protonated at $C(6')$. This apparent discrepancy has been resolved by showing that protonation at oxygen is fast and reversible, and protonation at $C(6')$ slow but irreversible [D. J. Deeble, S. Das and C. von Sonntag, *J. Phys. Chem.* 89:5784 (1985)]. Thus, your expectation is fully correct: On the faster timescale of the pulse radiolysis experiment of Hayon, protonation only occurs at oxygen while in the ESR experiments, although done at much lower temperatures, the time lapse between irradiation and measurements was sufficiently long to allow the slow protonation at $C(6')$ to proceed.

Varma: Is that what Mike is working on?

Christophorou: I am glad you asked that question because Mike and I were just pondering it. Electron transfer reactions as well as electron attachment reactions are very strong functions of temperature, so that is really quite a building point in the discussion.

von Sonntag: This is a very interesting comment. We can do pulse radiolysis experiments and product studies in the temperature range between about 0 and 80 degrees. We should indeed try to investigate the temperature dependence of some of these electron transfer reactions.

Ward: I understand also the recent picture about the ESR experiments that you get both the C^- and the T^-. But also that the C^- does not end up as $TH^.$. It goes away somewhere else and they don't know where it ends up. Well, let me ask you another question. The people who do ESR DNA in all kinds of shapes and sizes in their systems do not see any sugar radicals at all. Would you care to comment on that?

von Sonntag: As you know, with ESR some radicals may not be observed because their lines are too broad. But this may not be the main problem in DNA. It continues to puzzle me that the total G value of radical formation in DNA is only 2 while we have to expect that, in the primary processes, radicals with a G value of at least 6 should be formed. Where are those missing ≥ 67 percent?

Ward: Would they have a lot of recombination?

von Sonntag: Yes, I am sure. But what our ESR experts claim is that there is no radical combination because they work at such low termperatures. Martyn Symons claims that the guanine radical cation may be the precursor of the DNA strand breaks. Now take a system which has only guanines such as poly(G). Hence, you would get a large yield of guanine radical cations, but in this system practically no strand breaks are formed. I feel that the guanine radical cation may be excluded as a source of sugar radicals and hence strand breaks.

Ward: If we can have another hour we can discuss the models of DNA.

von Sonntag: We better not.

Models of Radiation Effects

The Radiobiological Significance of Spatial and Temporal Distribution of Energy Absorbed from Ionizing Radiations

Harald H. Rossi

Abstract

The cells of higher organisms respond in a non-linear fashion to the energy absorbed from ionizing radiation. However, there appears to be no indication of a dependence that is of a higher power than the square of the absorbed energy. This relatively simple alternative permits operational definitions of two types of injuries, termed lesions and sublesions, and a basic description in terms of dual radiation action. There are, however, various complicating factors and uncertainties. Further progress requires the development of a modified microdosimetry that incorporates energy transport, a more complete treatment of saturation and especially a specific identification of what is probably damage to DNA.

Introduction

There are various hazards confronting theoretical radiobiologists. They almost always have to rely on data that were obtained by experimental radiobiologists or by radiation epidemiologists, and these data may be deficient in various ways.

To begin with, the dosimetry may be wrong. A recent example is the revision of the dosimetry for the Japanese Atomic bomb survivors.[1] In another study, in which groups of caged mice huddling together in a cold room were irradiated with reactor neutrons, absorbed dose variations exceeded a factor of three.[2] Like most people concerned with radiation dosimetry, I know of a number of other instances in which it was performed incorrectly. In the case of neutrons and other high LET radiations, there is an additional possible source of uncertainty in the specification of radiation quality. The term "fission neutrons" is vague unless it refers to the unusual case of an unmoderated source.

There are further problems with questionable reliability of the biological data which can reach the point where they are contradictory. All data are subject to a variety of acknowledged, and often also unrecognized, statistical and other errors and usually do not preclude a substantial range of underlying functional dependencies.

The limitation of the experimental information can be compounded with naive assumptions made by the theoretician. He often assumes that the cells irradiated are alike (although any cursory view through a microscope indicates otherwise), or that they do not change their characteristics with time during irradiation in dose rate experiments.

In view of the variety of confounding factors, it is quite possible that the reported and analyzed results of radiobiology or epidemiology can contradict a valid theory. Conversely, a theory should not be considered proven if it agrees with one, or even a

Physical and Chemical Mechanisms in Molecular Radiation Biology
Edited by W.A. Glass and M.N. Varma, Plenum Press, New York, 1991

325

few, experiments. It is therefore desirable that a theoretical treatment be based on very broadly accepted findings.

Many complicated and largely unknown steps intervene between the physical process of energy absorption and the manifest biological effects; it may seem altogether doubtful whether these phenomena can be correlated in a meaningful way. That this may nevertheless be possible is signalled by the fact that alterations of the spatial or temporal pattern of energy deposition cause not only drastic, but also consistent, changes in the probability of virtually all effects of ionizing radiation on the cells of higher organisms. Thus the effectiveness regularly increases with increasing energy concentration up to a maximum that can be roughly specified by an LET near 100 keV/μm with a decline at higher concentrations. An equivalent observation is that the RBE of neutrons is routinely found to be maximal near 400 keV. Fractionation of moderate absorbed doses almost always reduces the effectiveness of low LET radiation but not of high LET radiation; likewise, the influence of oxygen concentration seems to be quite universal—especially in its dependence on radiation quality. These well-established facts indicate a common target and basically similar lesions. Since the common characteristic of eukaryotic cells is the diameter and general organization of the DNA helix, which is also known to be the most critical component of cells, it can be assumed that whatever major damage is caused by various ionizing radiations depends on their production of DNA lesions.

Lesions and Sublesions

Some confusion in the literature arises from employment of the term "lesion" with a variety of meanings. In the following pages, it will refer to a deleterious change that causes the effect with incoherent action;[3] i.e., the effect probability associated with a lesion does not depend on the existence of other lesions. This criterion is in accord with the observation that although it may require nuclear traversals by more than 10 high-LET particles to "kill"[a] mammalian cells, the survival curve is exponential and there are no dose rate effects.[4,5,6] In this one-event modality, the incoherent action of lesions is obvious: They act like damaged links in a chain that cause rupture with independent probabilities. With incoherent action lesion repair is not necessarily inconsistent with lack of dose rate dependence, but it would merely decrease the apparent cross section for lesion formation.

Any repair can thus not alter the first-order kinetics of cell inactivation in incoherent action by lesions. If λ denotes the mean number of unrepaired lesions per cell, the survival probability, S, is given by

$$S = e^{-c\lambda} \qquad (1)$$

where c is the mean number of lesions produced per inactivation. It should be recognized that in incoherent action this is not a number *required*. In general there can be several lesions that are produced in several events; but if a lesion produced in one event (which may not be the last one of a series) kills the cell, the corresponding event was the killing event and such a mechanism contributes to the first-order (exponential) portion of the survival curve. This definition also does not exclude the existence of varying severity (i.e., killing probability) of lesions. Of course the simplest condition

---

(a) The simple term "killing" is employed rather than a more precise expression specifying limited cell proliferation.

would be $c = 1$; i.e., a single lesion is sufficient for the observed cellular effect. There are arguments in favor of this thesis,[7] but it cannot be considered to be established.

Sublesions are impairments that cannot singly cause the effect; they must combine to do so by forming lesions. If sublesions are produced in separate events (especially by low-LET radiation), the dose effect relation must be non-linear and repair can lessen effectiveness. On the basis of this operational definition, it can be stated that all lesions are formed by combinations of sublesions because the existence of relative biological effectiveness (RBE) indicates that effectiveness depends on local specific energy, (z); and in order for the local energy concentration to influence effectiveness, at least two molecular changes--arising from at least two energy deposits—must be involved.

Quadratic Dependence on Energy Concentration

The basic kinetics of lesion formation are most obvious in track segment experiments in which cells are irradiated with particles of substantially single-valued LET. The term "kinetics" refers to the mode rather than the degree of action, and equal kinetics require the same shape of dose effect relations as revealed in logarithmic plots. The example in Fig. 1 indicates that the differences in the configuration of DNA during the cell cycle affect the sensitivity of cells, and therefore its absolute dependence on LET as shown in numerous other experiments. However, the basic relative dependence is preserved. This logarithmic graph depicts the cross section of the cells for killing in single events (as given by the initial slope of the survival curve) versus LET. Its outstanding characteristic is a slope near 2 over a wide range of LET. Since at a given absorbed dose, the particle fluence is inversely proportional to LET, this implies proportionality of RBE to LET with a decline at very high LET.

The essentially quadratic dependence of cross section on LET was noticed many years ago,[8] and it has been confirmed in other experiments.[5,9,10] Since it is clearly non-linear, LET, which is a linear average of energy deposition, is basically an unsuitable quantity for the characterization of radiation quality. For the same reason, effect

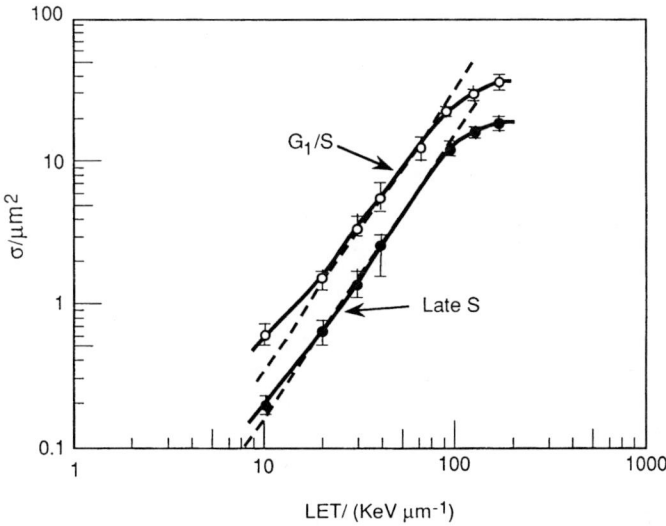

Figure 1. Cross section for killing of Chinese Hamster Cells in G_1/S and late S phase. The dashed lines represent a slope of 2 (quadratic dependence) [after Bird et al.].[5]

probabilities should also not be interpreted in terms of other simple averages relating to energy deposition. The most important example is, of course, the absorbed dose. Another example is the mean energy deposited at various radial distances from a particle trajectory, although this quantity is of considerable utility in track structure calculations.[11]

Linkage between radiation characteristics and biological effects can employ the concept of *sites* as convex regions where concentration of absorbed energy, as specified by lineal energy (y) or specific energy (z), determines the probability of lesion production. This is not a rigorous concept because it must be assumed that the initial separation of sublesions affects the probability of lesion formation. Although the site radius can only be considered to represent characteristic distance of interaction, the site concept is convenient in semi-quantitative considerations. A more advanced concept, that of the *matrix*, is considered below. The term "site" needs to be distinguished from what may be termed a complex site which is the intermediate concept of a non-convex and hypothetical region in which the yield of lesions is actually proportional to z^2.[12]

In large sites delta ray escape and straggling may be negligible; and therefore E, the energy lost in the site, may differ little from the properly weighted value, $\bar{\bar{\eta}}$, given by

$$\bar{\bar{\eta}} = 1/\bar{\eta} \int \eta^2 f(\eta)\, d\eta = \eta_D \qquad (2)$$

where $f(\eta)$ is the probability distribution of energy actually imparted within the site. One may then make the further approximation

$$E = L_\Delta \bar{x} \qquad (3)$$

where L_Δ is the LET with a "suitable" energy cutoff and $\bar{x}$ is the mean diameter of the site.

For small E straggling becomes more important, and this should account for the initial slope of somewhat less than 2. However, if the site involved is quite small, straggling can be important in all cases, and there is the additional complication of delta ray escape. The actual energy loss can be expressed in terms of the lineal energy, and in the case of a quadratic dependence the appropriate quantity is

$$y_D = \eta_D/\bar{x} \qquad (4)$$

It is of interest to evaluate $\bar{y}_D$ for various site diameters and the charged particles (protons, deuterons and helium-3 nuclei) employed in the experiments leading to the data in Fig. 1. Based on calculations in the thesis of Chmelevsky,[13] Kellerer and Chmelevsky have provided a suitable method (A. M. Kellerer and D. Chmelevsky, personal communication). While this is considered to be an approximation, it is sufficiently accurate for present purposes.

Figure 2 shows the cross sections given in Fig. 1 for late S phase cells versus $\bar{y}_D$ in sites ranging from 1 to 100 nm. (The dependence is similar for the G_1/S interface.) It would appear that a quadratic dependence obtains somewhere between 20 and 100 nm. However, this subject will be reconsidered later.

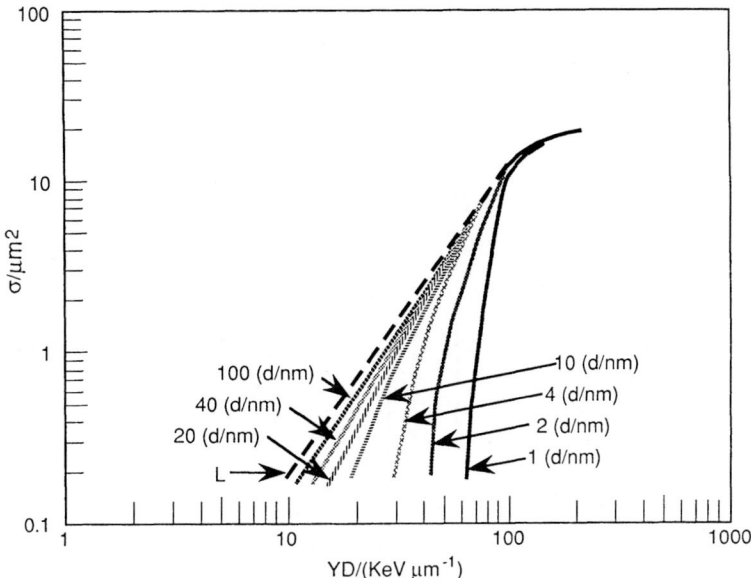

Figure 2. Data for late S in Fig. 1 plotted as function of y_D in sites ranging in diameter from 1 to 100 nm. The dashed line is the LET dependence in Fig. 1.

A simple interpretation of the relation shown in Fig. 2 is that cell killing depends on the square of the energy deposited within distances of the order of tens of nanometers. The plateau with its implied reduction of RBE at very high energy density can then be regarded to be due to saturation, i.e., excessive energy deposition.

The term "saturation" has also been employed in a different meaning. In theories centered on what is presumed to be a limited repair capacity of the cell, this other kind of saturation is important even at moderate LET. It is assumed that "lesions" are produced at a rate that is proportional to the absorbed dose. On the basis of a further assumption, the rising portion of the curve is attributed to limited repair that reduces the cross section at low lineal energy but is increasingly saturated at higher ones. This appears to require that in essentially all cells saturation is inversely proportional to η and operates within short distances.

I have suggested the term "cislesions" for the lesions arising from short-range interactions in order to distinguish them from "translesions" produced in interactions of sublesions that are initially separated by distances of the order of a micrometer. The need for a separation of at least several tenths of a micrometer was rigorously proven[14] and is required on the basis of microdosimetric grounds. The downward bend of exponential cell survival curves beyond absorbed doses of low-LET radiation that are of the order of 1 Gy indicates action by more than one event; and at such absorbed doses, the mean intratrack distance is of the order of 1 μm. Φ^*, the event frequency for Co^{60} gamma radiation is about 3 Gy^{-1} in a 1 μm diameter site. The sublesions are subject to repair as indicated by the lesser effectiveness when irradiation is protracted. In this case the interacting sublesions are produced in separate events.

Like cislesions, translesions appear to have the dual interaction characteristic; i.e., a pair of sublesions forms a lesion. The evidence for this is that in a logarithmic plot of RBE versus absorbed dose of the higher LET radiation, the slope has apparently always been found to be between -1/2 and 0. For neutrons producing protons of maximum LET, first-order kinetics predominate up to moderate absorbed doses. As shown

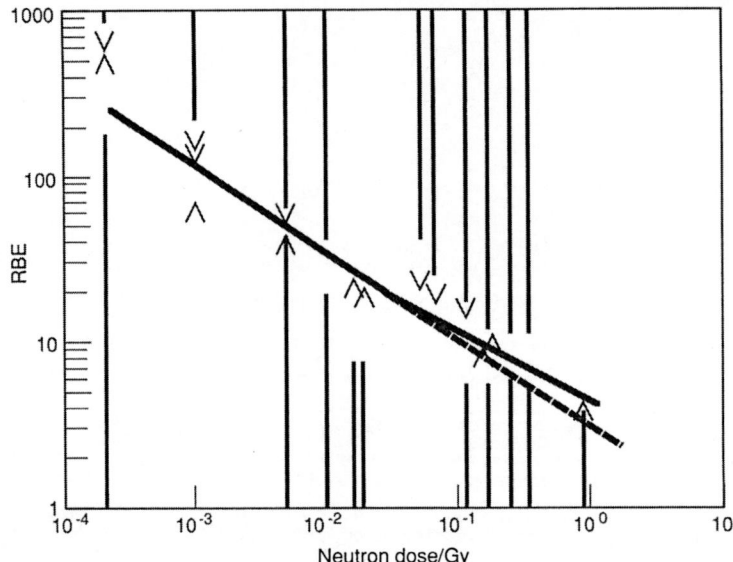

Figure 3. RBE of 430 KeV neutrons relative to 250 k Vp X-rays for a wide range of lenticular opacifications of the mouse. Solid bars indicate values excluded at the 99% level, light bars exclude at the 95% level. Arrowheads indicate non-significant differences. The dashed line has a slope of -1/2. The solid curve is in accord with dual radiation action. (After Bateman et al.).[15]

in the example in Fig. 3, a slope of -1/2 is maintained for nearly three decades of neutron dose, and agreement with dual radiation action extends for more than three decades.[15] This example shows that even in a complex biological response that must be quantified by an arbitrary scale, the basic kinetics of lesion formation are evident. The same dependence has been found for other phenomena that would seem to be indirectly related to the basic pattern of energy deposition such as life shortening[16] and various defects of plant growth.[17]

At both low and high absorbed doses, the slope of the curve is less; in the former case because single-event action predominates even at low LET, and in the latter case because multi-event action becomes appreciable at high LET. A slope as steep as -1/2 is not approached when the LET difference between the radiations is moderate.[18]

While microdosimetric considerations make it highly improbable that cislesions are produced in two events, there is no *a priori* reason why translesions cannot be produced as a result of one event. As derived by the simple site model, the average initial separation of the sublesions involved is of the order of one micrometer. As pointed out before,[19] it must require some period of time for sublesions to encounter each other. During this period they may be assumed to be subject to repair; the reason why the effective site diameter is less than the nuclear diameter may be that even with essentially uniform formation of sublesions in the nucleus, only relatively proximal sublesions can combine before they are repaired. It would then also seem possible that some sublesions formed by a single charged particle within a distance shorter than the site diameter could combine before they are repaired. Differences in RBE_M, the maximum RBE which is obtained when both of the radiations subject to comparison act in single events, may be due to variations in translesion production by the low LET radiation. It can be presumed that more pairs of sublesions can cause cell killing than some more specific damage (i.e., defective fiber formation causing lenticular opacification), and that translesion formation in single events is relatively more probable for killing. This would lower the RBE and account for the relatively low RBE_M for cell killing.

There is probably no sharp division between cislesions and translesions. A treatment that encompasses both is the generalized theory of dual radiation action (TDRA).[20]

Rather than presenting the details of the TDRA, it would seem appropriate here to merely consider its essentials. The basic premise is that the cell contains a sensitive region, termed the "matrix," in which sublesions can be produced by charged particles. The yield of lesions is determined by the overlap of two patterns described by their proximity functions.[11] The proximity function, $s(x)$ of the matrix specifies the average amount of matrix material as a function of distance from a point in the matrix; the proximity function of energy transfers, $t(x)$, specifies the average energy absorbed as a function of distance from an energy transfer point. A third function, $g(x)$, represents the probability that two sublesions initially separated by x form a lesion. The function $t(x)$ is determined by the pattern of energy deposition while $s(x)$ and $g(x)$ are determined by the characteristics of the cell. Their product is proportional to a quantity that has been denoted by $\gamma(x)$.[a] The quantity ξ is defined by

$$\xi = \int_0^\infty t(x)\, \gamma(x)\, dx \qquad (5)$$

and appears in the principal formula

$$\xi = k\,(\xi\, D + D^2) \qquad (6)$$

where ξ is the yield of lesions, k a constant and D the absorbed dose. In this expression the first or "intratrack" term refers to lesions produced in a single event. It includes all cislesions and some translesions. The second, "intertrack," term refers only to translesions.

The molecular ion experiment in which pairs of charged particles with varying separation are traversing cells[21] has been designed to determine $\gamma(x)$ and has shown that the function is highly skewed and reaches very low values beyond 100 nm.[22,23]

Based on the results of experiments utilizing very low-energy X-rays that generate quite short electron tracks,[24] an even steeper dependence was found.[25] A possible reason for the discrepancy is considered below.

The so-called L-Q dependence on the first and second power of the absorbed is a feature of a number of other "models" that provide somewhat different rationales[26-31] and it also is expressed in the formula

$$S = \exp(-\alpha D - \beta D^2) \qquad (7)$$

according to which cell survival data are often analyzed.

(a) The factor of proportionality is $1/4\pi\rho x^2$ where ρ is the density.

Energy Transport

According to the TDRA, the production of sublesions is determined by the overlap of the pattern of energy transfer points and that of the matrix. This is, however, incorrect if energy can migrate into the matrix from transfer points located outside of it. Various mechanisms have been implicated in this process; the most frequently cited mechanism involves active radicals (e.g., OH˙). Energy transport leads to a proximity function that is more diffuse. Kellerer originally pointed out that this requires a modification of $t(x)$ but does not alter the linear-quadratic dependence of lesion formation on absorbed dose (A. M. Kellerer, personal communication). This has been confirmed in a more recent analysis.[32]

Indirect action caused by energy transport must be particularly important at the small scale characterizing cislesions. It should therefore modify the estimates of $\gamma(x)$ as obtained in the two kinds of experiments from which it has been derived. As schematically indicated in Fig. 4, energy transport should have an opposite effect on the two approaches because calculations based on the inchoate energy distribution should lead to underestimates in the soft X-ray experiments and to overestimates in the molecular ion experiment: This is in line with the somewhat discordant results of the analysis of the respective results. Transport over tens of nanometers would particularly affect the soft X-ray experiment, and the site size of tens of nanometers indicated in Fig. 2 supports this interpretation.

It should, however, be appreciated that indirect action must lead to the concept of a site (as the region from which energy instrumental in causing the effect is deposited) that is larger than the target. This somewhat vague concept is more accurately replaced by the more advanced geometrical concepts of the generalized TDRA in terms of a proximity function of energy deposits that is modified by diffusion.[32] In particular it is not clear whether it is appropriate to consider the data in Fig. 2 in terms of the same "site diameter" at all values of y_D. Further considerations based on varying diffusion coefficients are necessary.

Numerical evaluation of the pattern of available energy which results from energy transport from the inchoate (microdosimetric) distribution of energy transfer points should be a major objective of theoretical radiobiology. It must involve consideration of two variables: the available energy(ies) leaving the transfer point and the diffusion length(s).

Saturation

While energy transport can affect the yield of sublesions, saturation affects the yield of lesions. Increase of the local specific energy must result in an increase of the production of sublesions and hence that of a lesion with a probability that cannot be larger than one. This saturation must lessen the yield of both cislesions and translesions. In the case of cislesions, it results in reduction of the slope of initial portion of the (exponential) survival curve. In the case of translesions, it has the same effect when these are produced in single events. In the generally more important situation where translesions are caused by multiple events, however, the result is a departure from their simple quadratic dependence, and at high absorbed doses, the exponential survival curve decreases linearly rather than quadratically.

Partial and approximate treatments of saturation have been given,[12,18,33] but there is as yet no comprehensive analysis available. Further progress is desirable because of the significance of this process. At low absorbed doses, it reduces the effectiveness of very high LET radiations; and at high absorbed doses, it reduces the

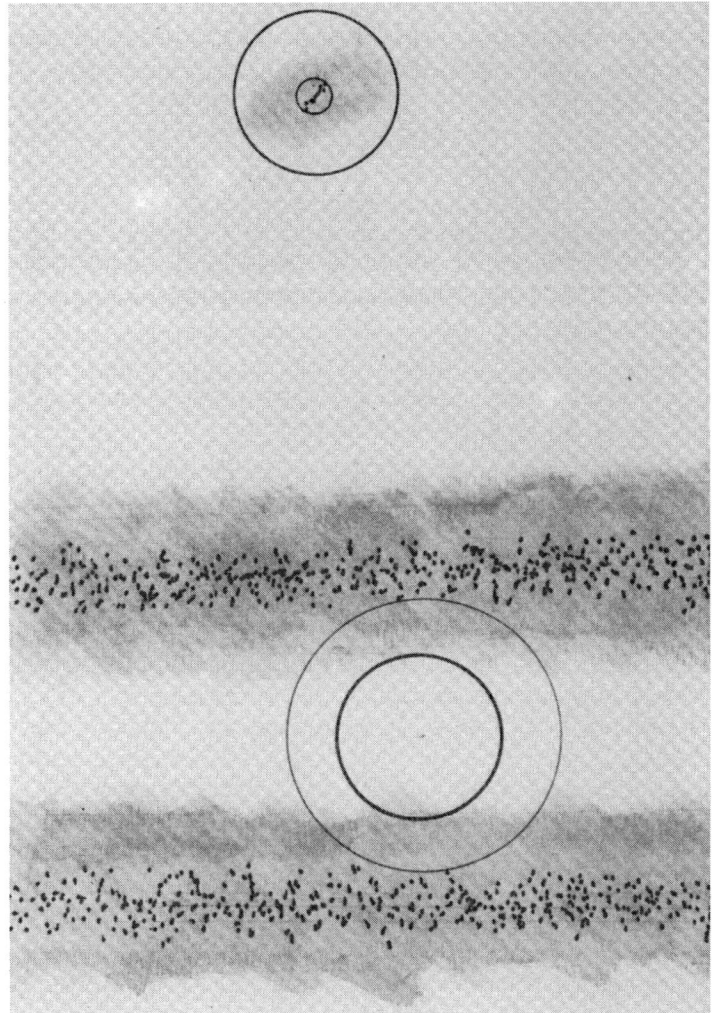

Figure 4. Energy transport extends beyond the range of low energy electrons and within the separation of ion tracks. Consequently the size of the sensitive site can be underestimated in soft X-ray experiments and overestimated in molecular ion experiments (light circles) with the actual site diameter being the same (heavy circles).

effectiveness of all ionizing radiations and invalidates the "α-β" or "L-Q" models. It is thus important in both radiation protection and in radiotherapy and limits the simple numerical analysis of interactions between ionizing radiations.

Further Uncertainties and Limitations Regarding Lesion Formation

In line with the principle of Occam's razor, the TDRA contains various simplifications that are unproven. A basic assumption is that all ionizing radiations produce sublesions with the same effectiveness; i.e., there is equal yield of sublesions per unit of absorbed dose regardless of radiation quality. There are two publications reporting that β, which is proportional to K, depends on radiation quality;[5,34] and, while there are statistical uncertainties, both indicate higher values at high LET. As indicated in Fig. 2, straggling results in minimal dependence of lineal energy on LET in very small sites, but the conclusion that the site for sublesion formation must be relatively small may not be justified because of energy transport. The possible variability of K remains a matter of uncertainty. It would however not change the linear-quadratic dependence of lesion yield on absorbed dose of a given type of radiation.

The formulation of the TDRA might be considered to imply identity of cislesions and translesions, but differences in the type of damage (rather than in its effect) would not change the mathematical expressions involved. Estimates of the radiation risks at low doses are based on the findings of conventional radiobiology and epidemiology that are mostly obtained at high doses of low LET radiation. It would be an oddity if the lesions of concern in the former (cislesions) differed in nature from those involved in the latter (translesions).

It is plainly desirable to go beyond a "black box" approach and to attempt to identify the molecules involved in lesion formation that depends on pairs of injuries. On this subject there also remains uncertainty.

The most obvious assumption is that sublesions are single-strand breaks in DNA. With this interpretation it is evident that a cislesion is a double-strand break. Since translesions can result from sublesions that are initially separated by micrometers, it might be hypothesized that the two ends of a single-strand break that belong to one part of the genome might misjoin with those of another one that is encountered before neither single strand break is repaired.[7]

Another possibility is that sublesions are double-strand breaks. This would be in accord with the linear dependence of "one-hit" chromosome aberrations (deletions) and the L-Q dependence of "two-hit" types (exchanges). One may even speculate that cislesions arise from pairs of adjacent double-strand breaks in nucleosomes while translesions involve two double-strand breaks in portions of the DNA molecules that are not supercoiled. Variations of K with LET could also account for changes in the effectiveness in double-strand break production.

At this point the second alternative looks perhaps more credible, but any conclusions must be based on definitive information regarding DNA damage on radiation quality. It is essential that the in vivo situation then not be distorted by extraction procedures. Further considerations should also be made by investigators who are more knowledgeable in these matters than I.

Cellular Effects

Although there is proportionality between lesion number and cell killing, the lesion yield varies unless populations are homogenous and constant.

Sensitivity variation with cell age must introduce positive ("hockey-stick") curvature by superposition of exponential survival curves. This tends to mask negative curvature introduced by multi-event killing. However, even with a two-fold variation of the linear component through the cell cycle, the departure from exponential dependence is difficult to discern.[35]

In protracted irradiation, cells can (especially at incubator temperature) progress through the cycle with changes of sensitivity that invalidate an analysis based on the survival parameters for an asynchronous population. This could modify dose-effect relations in transformation and cause a paradoxical dose rate effect.[36]

Another major complication in transformation experiments is the interrelation between transformation and killing. This is often characterized by

$$T = (aD + bD^2) \exp{-} (\alpha D + \beta D^2) \qquad (8)$$

where a and b refer to transformation and α and β to killing. While the equation has already four parameters, this may still not be enough because statistical independence between transformation and killing is implied as well as equal killing rates for transformed and untransformed cells. Presumably many more pairs of sublesions can kill than can transform. But must both sublesions in transformation be of a special kind, or (as might be more probable) need only one of them belong to a specific type, while the other merely fixes it in a lesion? It is evident that either possibility causes complications beyond those expressed by the equation.

Tissue and Organ Effects

Many types of radiation injury, termed *non-stochastic* involve multicellular (coherent) action, and they are generally characterized by sigmoid dose effect curves. However the most important, *stochastic*, radiation hazards, genetic damage and carcinogenesis, have often been considered to follow the L-Q scheme with the implication that they are due to deleterious changes in autonomous cells (i.e., cells that act incoherently). There is, at this time, no solid evidence to challenge this opinion with regard to genetic effects.

There are, however, strong reasons for discounting the notion of a proportional relationship between carcinogenesis and transformation of autonomous cells. While there is evidence that at least some cancers are monoclonal (i.e., arise from a single cell), this does not eliminate the possibility that their proliferation is controlled by a dose-dependent systematic reaction. In fact many dose-effect curves of carcinogenesis have various shapes, including negative initial slopes (reduction of the "natural" incidence of low doses).[3]

The only characteristic of dual radiation action that extends beyond unicellular action is the relation between RBE and effect level which, as stated above, involves the inverse square root of the absorbed dose of high LET radiation. This is equivalent to a dependence on the reciprocal of the absorbed dose of low LET radiation. A likely reason is that the processes and interactions which determine manifest effects are generally the same following production of the same number of lesions.

Conclusion

Despite various complications and uncertainties, the concept of dual radiation action seems well established. Its fundamental justification is that there is no evidence that cellular effects depend on a power larger than the square of the absorbed dose or, more precisely, the square of the specific energy in a complex site. Instances of a linear or the intermediate L-Q dependence can be simply accounted for. Thus at very low doses proportionality must obtain in single-event action. At high doses, saturation must cause proportional elimination of unaffected entities by a further increment of dose.

While it is possible to bridge the wide gap between energy absorption and biological effects in general terms, it is plainly necessary to provide more specific information at both sides. On one hand, this involves energy transport and saturation. At the other end, more specific information is needed on the nature of DNA damage and its biological consequences.

Acknowledgments

I am indebted to Dr. D. Chmelevsky and Dr. A. M. Kellerer for provision of numerical information and formulas required for Fig. 2. I also wish to thank Dr. Kellerer and Dr. M. Zaider for critical reviews of an earlier draft.

References

1. DS 86. *Radiation Effects Research Foundation,* Hiroshima (1987).
2. L. J. Goodman and M. Pearlman. *Depth Dose Studies at the ORNL DOSAR Facility.* U.S. Atomic Energy Commission, Washington, DC (1966).
3. H. H. Rossi. Limitation and Assessment in Radiation Protection. L.S. Taylor Lecture No. 8, NCRP, Bethesda, Maryland (1984).
4. E. L. Lloyd, M. A. Gemmell, C. B. Henning, D. S. Gemmell and B. J. Zabransky. Cell Survival Following Multiple-Track Alpha Particle Irradiation. *Int. J. Radiat. Biol.* 35:23-31 (1979).
5. R. P. Bird, N. Rohrig, R. C. Colvett, C. R. Geard and S. Marino. Inactivation of Synchronized Chinese Hamster V-79 Cells with Charged Particle Track Segments. *Radiat. Res.* 82:277 (1980).
6. E. A. Blakely, F. Q. Ngo, S. B. Curtis and C. A. Tobias. Heavy-Ion Radiobiology: Cellular Studies. *Adv. in Rad. Biol.* 11:295, Academic Press (1984).
7. H. H. Rossi. Microdosimetry and Radiobiology. *Radiat. Prot. Dosim.* 13(1-4):259-265 (1985).
8. E. L. Powers, J. T. Lyman and C. A. Tobias. Some Effects of Accelerated Charged Particles on Bacterial Spores. *Int. J. Radiat. Biol.* 14:313-330 (1968).
9. G. W. Barendsen. Mechanism of Action of Different Ionizing Radiations on the Proliferative Capacity of Mammalian Cell. *Theoretical and Experimental Biophysics* 1:167 (1967).
10. P. W. Todd. Reversible and Irreversible Effects of Ionizing Radiations on the Reproductive Integrity of Mammalian Cells Cultured In Vitro. Thesis, University of California, Lawrence, Rad. Lab., UCRL 11614, Berkeley, California (1964).
11. A. M. Kellerer and D. Chmelevsky. Concepts of Microdosimetry III. *Rad. Environ. Biophys.* 12:321 (1975).
12. H. H. Rossi and M. Zaider. Saturation in Dual Radiation Action, Quantitative Mathem. *Models in Radiat. Biol.,* 111-118 Springer, New York (1987).
13. D. Chmelevsky. Distributions et Moyennes des Grandeurs Microdosimetriques A L'Echelle Du Nanometre - Methodes De Calcul Et Resultats, Thesis, U de Toulouse, France (1976).
14. O. Hug and A. M. Kellerer. Stochastik der Strahlenwirkung, Springer, New York (1966).
15. J. L. Bateman, H. H. Rossi, A. M. Kellerer, C. V. Robinson and V. P. Bond. Dose-Dependence of Fast Neutron RBE for Lens Opacification in Mice. *Radiat. Res.* 51:381-390 (1972).
16. J. B. Storer, L. J. Serrano, E. B. Darden, Jr., M. C. Jerrigan and R. L. Ullrich. Life Shortening in RFM and Balb/c Mice as a Function of Radiation Quality. Dose and Dose Rate. *Radiat. Res.* 78:122-161 (1979).
17. H. H. Smith, H. H. Rossi and A. M. Kellerer. Relation Between Mutation Yield and Cell Lethality Over a Wide Range of X-ray and Fission Neutrons in Maize, Biological Effects of Neutron Radiation, IAIA (1974).
18. A. M. Kellerer and H. H. Rossi. The Theory of Dual Radiation Action. *Curr. Topics Radiat. Res. Q.* 8:85-158 (1972).
19. H. H. Rossi and A. M. Kellerer. Biological Implications of Microdosimetry: I, Temporal Aspects. In *Proc. 4th Symp. on Microdosimetry,* ed. J. Booz, pp. 315-326, EUR 5122. Verbania Pallanza, Italy, Commission of the European Communities, Brussels (1973).
20. A. M. Kellerer and H. H. Rossi. A Generalized Formulation of Dual Radiation Action. *Radiat. Res.* 75:471-488 (1978).
21. H. H. Rossi. Biophysical Studies with Spatially Correlated Ions, 1, Background and Theoretical Considerations. *Radiat. Res.* 78:185-291 (1979).
22. A. M. Kellerer, Y-M. P. Lam and H. H. Rossi. Biophysical Studies with Spatially Correlated Ions, 4, Analysis of Cell Survival Data for Diatomic Deuterium. *Radiat. Res.* 83:511-528 (1980).
23. M. Zaider and D. J. Brenner. The Application of Track Calculations to Radiobiology, III Analysis of the Molecular Beam Experiment Results. *Radiat. Res.* 100:213-221 (1984).
24. D. T. Goodhead, J. Thacker and R. Cox. Effectiveness of 0.3 keV Carbon Ultrasoft X-rays for the Inactivation and Mutation of Cultured Mammalian Cells. *Int. J. Radiat. Biol.* 36:101-114 (1979).
25. D. J. Brenner and M. Zaider. Modification of the Theory of Dual Radiation Action for Attenuated Fields, II, Application to the Analysis of Soft X-ray Result. *Radiat. Res.* 99:492-501 (1984).

26. W. C. Roesch. Models of the Radiation Sensitivity of Mammalian Cells. In *Proc. Third Symp. Neutron Dosimetry in Biology and Medicine,* pp. 1-27. Commission of the European Communities, Brussels (1977).

27. K. H. Chadwick and H. P. Leenhouts. The Molecular Model for Cell Survival Following Radiation. In *The Molecular Theory of Radiation Biology,* pp. 25-50. Springer-Verlag, New York (1981).

28. C. A. Tobias. The Repair-Misrepair Model in Radiobiology: Comparison to Other Models. *Radiat. Res. Suppl.,* 8:104, S-77 (1985).

29. S. B. Curtis. Lethal and Potentially Lethal Lesions Induced by Radiation - A Unified Repair Model. *Radiat. Res.* 106:252 (1986).

30. W. Sontag. A Cell Survival Model with Saturable Repair After Irradiation. *Radiat. Environ. Biophys.* 26:63 (1987).

31. D. Harder. Pairwise Lesion Interaction - Extension and Confirmation of Lea's Model. In *Proc. Eighth Int. Congr. Radiation Research,* p. 318, Edinburgh. Taylor and Francis, New York (1987).

32. M. Zaider and H. H. Rossi. Indirect Effects in Dual Radiation Action. *Radiat. Phys. Chem.* 32:143-148 (1988).

33. M. Zaider and H. H. Rossi. Saturation Effects for Sparsely Ionizing Particles. DOE Progress Report 10/1/79-9/30/80, Technical Information Center, U.S. Department of Energy, Oak Ridge, Tennessee (1980).

34. A. M. Kellerer, E. J. Hall, H. H. Rossi and P. Teedla. RBE as a Function of Neutron Energy II, Statistical Analysis. *Radiat. Res.* 65:172-186 (1976).

35. E. J. Hall, W. Gross, R. F. Dvorak, A. M. Kellerer and H. H. Rossi. Survival Curves and Age Response Functions for Chinese Hamster Cells Exposed to X-rays or High LET Alpha Particles. *Radiat. Res.* 52:88-98 (1972).

36. H. H. Rossi and A. M. Kellerer. The Dost Rate Dependence of Oncogenic Transformation by Neutrons May Be Due to Variation of Response During the Cell Cycle. *Int. J. Rad. Biol.* 50(2):353-361 (1986).

Phenomenological Models

L. A. Braby[a]

Abstract

The biological effects of ionizing radiation exposure are the result of a complex sequence of physical, chemical, biochemical, and physiological interactions which are modified by characteristics of the radiation, the timing of its administration, the chemical and physical environment, and the nature of the biological system. However, it is generally agreed that the health effects in animals originate from changes in individual cells, or possibly small groups of cells, and that these cellular changes are initiated by ionizations and excitations produced by the passage of charged particles through the cells. One way to begin a search for an understanding of health effects of radiation is through the development of phenomenological models of the response. Many models have been presented and tested in the slowly evolving process of characterizing cellular response. Different phenomena (LET dependence, dose rate effect, oxygen effect etc.) and different end points (cell survival, aberration formation, transformation, etc.) have been observed, and no single model has been developed to cover all of them. Instead, a range of models covering different end points and phenomena have developed in parallel. Many of these models employ similar assumptions about some underlying processes while differing about the nature of others. An attempt is made to organize many of the models into groups with similar features and to compare the consequences of those features with the actual experimental observations. It is assumed that by showing that some assumptions are inconsistent with experimental observations, the job of devising and testing mechanistic models can be simplified.

Introduction

Models, ranging from vague conceptual descriptions to detailed mathematical expressions, play a major part in the development of most sciences. Models serve many different purposes, and take on different properties in accordance with those purposes. When previously unknown phenomena are observed, scientists find it natural to look for their causes and to devise models which will help in that effort. The process of organizing observations implies some model which provides a structure for the resulting list. This model may only be a vague concept of the relationship between the observed end point and changes in the system, or its environment, which may effect that end point. It may be no more than an effort to fit mathematical relationship to the effect and one or more variables that are thought to be related to it, but even the act of selecting the related variables necessitates some thought about the underlying mechanisms.

The effect of ionizing radiation on biological systems has been found to be the result of an extremely complicated combination of physical, chemical and biological processes. Thus the primary function of many models of the effects of ionizing

---

(a) Pacific Northwest Laboratory, Richland, Washington

Physical and Chemical Mechanisms in Molecular Radiation Biology
Edited by W.A. Glass and M.N. Varma, Plenum Press, New York, 1991

radiation on cells has been to aid in the organization of the data so that some pattern can be recognized in the complex collection of causes and effects. As the data are organized, trends become visible, and a model can be used to predict effects under conditions which have not previously been tested experimentally. These predictions, which may be derived from purely conceptual models but which are more likely to be quantitative if the model is given a mathematical formulation, can be used to help in the design of experiments to test the validity of the assumptions of the model. Thus model building and testing becomes an iterative process. Refinements in the model provide new predictions, which can be used to devise new experiments, which produce new data, which provide the information needed to refine the model.

The use of models to predict effects under conditions which have not been explored experimentally has an important practical application as well. Radiation exposure limits for the protection of radiation workers and the general public are based, as much as possible, on real human health effects data. However, those data are quite limited with respect to the range of doses, the variety of effects observed, and the statistical reliability. Thus it is necessary to utilize available models to guide the extrapolation to the effects of low doses and to predict possible effects under different conditions of exposure. To some extent data obtained with experimental animals can be used to estimate effects in humans, but even animal experiments are limited to relatively high doses and dose rates by the cost and experimental difficulty involved in trying to obtain statistically reliable results at doses which are relevant to the majority of actual human exposures. Thus it is necessary to model the effects on experimental animals and then to devise an additional model that relates the animal response to health effects in humans.

The greatest value of models may be their inherent relationship to the mechanisms which are responsible for the effects being modeled. The health effects of irradiation involve not only the production of biochemical damage by different radiations under different physical and chemical environments, but also the modification of that damage by a variety of biochemical systems which may be involved in modifying damage produced by other environmental insults and the normal metabolic activity of the cell. Furthermore, both the production of specific biochemical changes and the response of cells to those changes may depend on the current and past exposure of that cell to environmental factors which alter the levels of protective compounds and the activity of the repair systems. In order to interpret or predict interactions between radiation exposure and other health risk factors such as smoking and industrial chemical exposure, we must know considerably more about the mechanisms governing these effects.

Background

Limitations on Models

All efforts to model natural systems face a dilemma. In order to describe the complex processes which occur in nature, the model must include many variables. However, to be testable and useful in any predictive way, the model must not have more variables than the experiments which are used to test it. For example, a very simple target model is sufficient to describe the inactivation of phage by low LET radiations, but this is not very helpful when considering the risk to mammalian cells with their complex capacity to repair DNA strand breaks and to attenuate the killing effect during a post-irradiation holding period. On the other hand, a detailed model of mammalian cell survival involving coefficients for the production of half a dozen different forms of DNA damage and the repair kinetics for all of the known DNA repair systems would probably not be useful because there are no experiments which can be used to test so many variables simultaneously.

Much effort goes into devising models which can be evaluated using currently available experimental techniques. Thus as new experimental Southern blot analyses become available, new biological effects such as DNA fragment deletion have been modeled. These new phenomena, which can be included in a model, make it possible to test models which incorporate a wider range of features. A detailed model of mechanisms may require more parameters than can be tested, but by carefully choosing potential experiments it is possible to devise definitive comparisons between alternative models. Models with only two parameters may describe the dose effect relationship for acute exposures to low LET radiations, but the effects of dose-rate, split-dose, and delayed-plating schedules can be related to one or more repair processes to provide an opportunity to evaluate additional parameters. By considering the effects of oxygen and differences in LET on the probability of producing damage, it may be possible to evaluate even more.

It is not possible to design an experiment, or even a set of experiments, which proves a specific model. A more fruitful approach is to contrast two or more alternative models which predict a different and mutually exclusive effect under specific conditions and determine which of the models are incorrect.[1] As the variety of models which have been shown to be incorrect increases, confidence in the alternative improves. Similarly it is more advantageous to test a variety of different models, using one or more carefully studied biological systems, than to determine the number of different systems which behave in a way consistent with a specific model. In addition, testing large numbers of different systems provides information on which characteristics affect the parameters used in the model. This also may provide information on the sources of biological variability, and on how to control it. For example, it has long been known that the position of the cell in the cell cycle can have a dramatic effect on the radiation sensitivity, so by limiting the experimental population to a specific age, i.e., specific point in the cell cycle, this source of biological variability is minimized and the effects of the parameter being tested can be seen more easily.

Attempts to limit biological variability and to obtain more precise measurements have often led to the use of experimental conditions and radiation doses which are far removed from the conditions which prevail in normal tissues at reasonable doses. The results obtained with such experiments are valuable in determining the nature and effectiveness of specific mechanisms, but they may not be appropriate for the determination of the significance of those mechanisms at realistic doses. Evaluation of the significance of specific mechanisms is one of the primary challenges in modeling. Unusual biological systems or extreme doses may also provide the most powerful test of specific aspects of models, but such conditions may introduce competing processes which are not significant at realistic doses. In order to avoid testing models which are only relevant to the extreme situations, it is important that a model can also be confirmed at a dose where multiple energy deposition events are rare.

Phenomenological Versus Mechanistic Models

It is always difficult to establish a boundary between phenomenological and mechanistic models; almost all involve some elements of both. The extremes are easy to identify. A few models are almost completely phenomenological, with a minimum of underlying assumptions about mechanisms, while some are nearly entirely mechanistic with only a few parameters which must be determined by fitting to experimental data. However, the majority of models lie between these two extremes, with a significant basis in mechanistic assumptions and a phenomenological approach to fitting those mechanisms together into a complete picture. In preparing this and the accompanying paper (by Curtis), we have attempted to divide the known models in a logical fashion and minimize overlap. However, the division is often arbitrary, and we do not claim that we can properly determine whether a specific model is mechanistic or phenomenological. In fact some features of specific models are mechanistic while other features are phenomenological. As a result, different aspects of a single model may be presented in both papers.

Biological End Points

Reproductive death of individual cells, roughly following the procedures of Puck and Marcus,[2] is the most common end point described by phenomenological models. By this end point, we measure the fraction of cells which have not lost their potential for unlimited proliferation, i.e., cells which are capable of passing on genetic heritage through their progeny. This end point has been studied for many years, and is accessible for most types of cultured cells so that a great deal of experimental data is available. It is readily applicable to new types of experiments since the fraction killed can be measured over a wide range with relatively good statistical precision. Furthermore, when finally extrapolating to human health effects, it is important to understand the effects on cell growth potential so that other effects can be appropriately evaluated in the presence of the loss of cells from the population.

However, it is clear that cell killing is not primarily responsible for the health effects of concern in radiation protection situations. Although cell killing is believed to be correlated with the deterministic effects of irradiation (protection levels are set so that those effects do not occur), the stochastic effects of radiation are believed to result from nonlethal modification of single cells. Thus the more relevant end points are malignant transformation, specific locus mutation, chromosome aberration, and other inheritable genetic alterations. However, reproductive death, interphase cell death, and mutation marker salvage interfere with the measurement of these more relevant end points. It may be possible to argue that malignant transformation is the most important end point for the study of health effects, but malignant transformation is probably the end point of a long sequence of unidentified specific changes within the cell. Since many, if not all, of these changes are probably DNA deletions, rearrangements, or point mutations, and since similar DNA changes are usually responsible for cell death, it can also be argued that for many tests of models, the end point used can be selected for experimental efficiency.

The Phenomena to be Modeled

A remarkable variety of factors affects the response of cells and more complex biological systems to ionizing radiation. Often the effect of these factors can provide additional information for the testing of the models, while in other situations these factors are of interest in their own right. The phenomena which have most often been included in models are i) the response to acute doses of low LET radiation, ii) the relative biological effectiveness as a function of radiation quality (often expressed in terms of LET), iii) the change in effect with change in dose rate, and iv) the interval between two doses when the dose is split or the time between irradiation of the initiation of progression through the cell cycle. The difference in the responses of cells in different parts of the cell cycle at the time of irradiation, the effects of changes in the chemical environment such as the oxygen concentration or radical scavenger, and the effect on cell survival as a function of the time between irradiation and assay for cell survival have also been discussed. Most models do not treat all of these phenomena explicitly; they normally cover only a portion of them. The more phenomena that are described, the more precisely a model can be related to underlying mechanisms.

Every biological system has a unique response to radiation, but the differences are frequently only a matter of degree; most responses are qualitatively similar. Those systems with unique characteristics deserve special attention. Although the exception may be the result of an unusual adaptation to a specific environment or an unusual mutation during the adaptation of a cell line to continuous culture, it is often the exception to the rule which provides the critical information. However, we will limit this discussion to the typical response and leave the investigation of exceptions to others. Figure 1 illustrates the typical relationship between survival, dose, and dose rate for eukaryotic cells exposed to low LET radiations. The reduction in slope with decreasing dose rate is assumed to result from repair during irradiation. Here repair refers specifically to observed changes in survival (or other end point); any association between this and specific changes at the molecular level, e.g., double-strand break rejoining, is still only an assumption. Repair is observed in other types of experiments as well. In Fig. 2, split-dose repair occurs during the interval between exposures in a

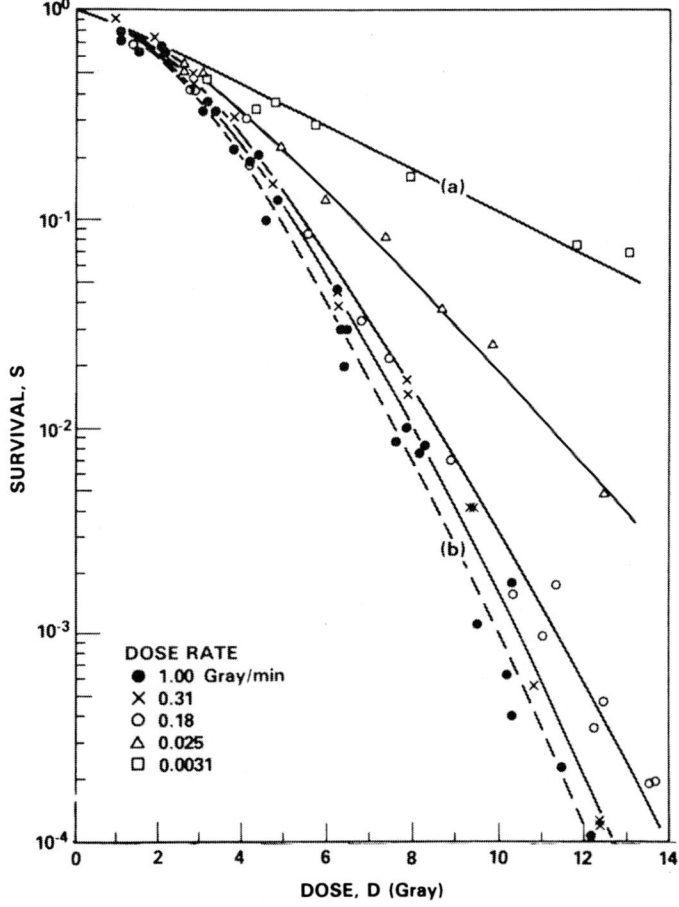

Figure 1. An example is shown of the typical dose and dose rate response of mammalian cells, in this case plateau phase CHO cells exposed to 250 KV X-rays.

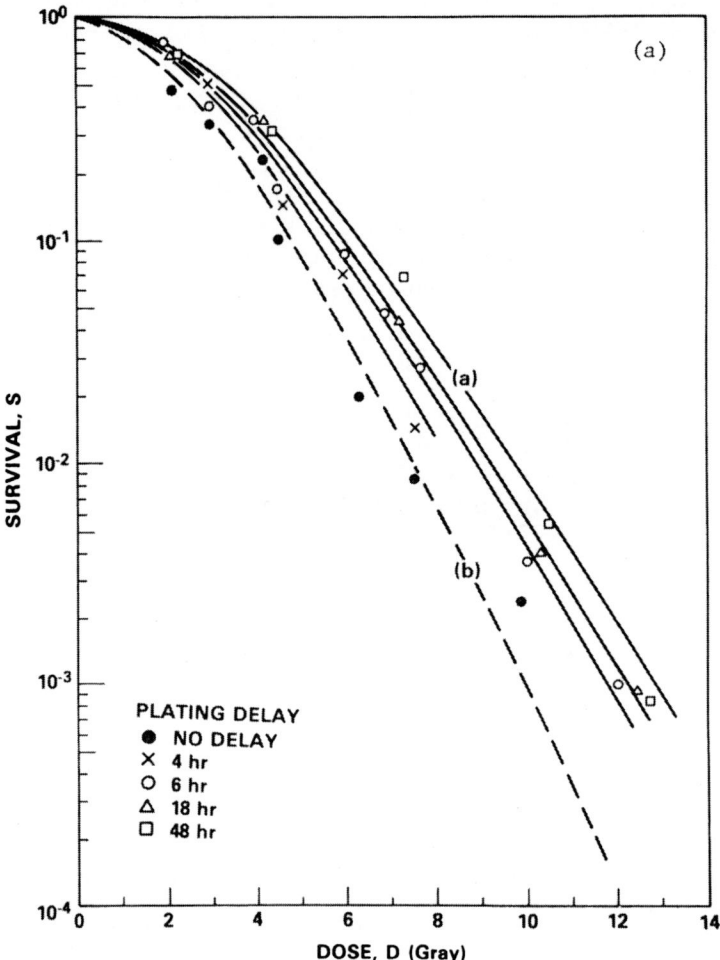

Figure 2a. Cell survival typically increases with time held in stationary phase after irradiation. The shoulder of the survival curve is recovered between doses in a split-dose experiment.

fractionated irradiation, while the delayed-plating repair occurs after the irradiation is completed. This difference becomes significant when some damage can become more effective at higher concentration; opportunity for repair during irradiation would then be more effective than repair time which occurred after the exposure. The effect of oxygen, the most common radiosensitizer, is illustrated in Fig. 3. Oxygen seems to react with initial damage in such a way as to prevent the damage from undergoing a spontaneous change to an innocuous condition.

The cell-cycle effect for reproductive survival following low LET radiation is illustrated in Fig. 4. Although this type of response is typical, many variations are possible with different cell types and experimental conditions. Both the sensitivity of DNA to damage and the ability to repair damage vary throughout the cell cycle. It should be remembered that the assay used to determine reproductive death is somewhat arbitrary. Irradiated cells do not simply continue to grow undisturbed, or to die. Delay in cell progression, slow growth mutation, sectored colonies, and delayed lethal effects all result in different measures of lethality with different assay procedures.

One of the most extensively studied phenomena in radiation biology is the effect of different radiation qualities. The typical response for cell survival is shown in Fig. 5.

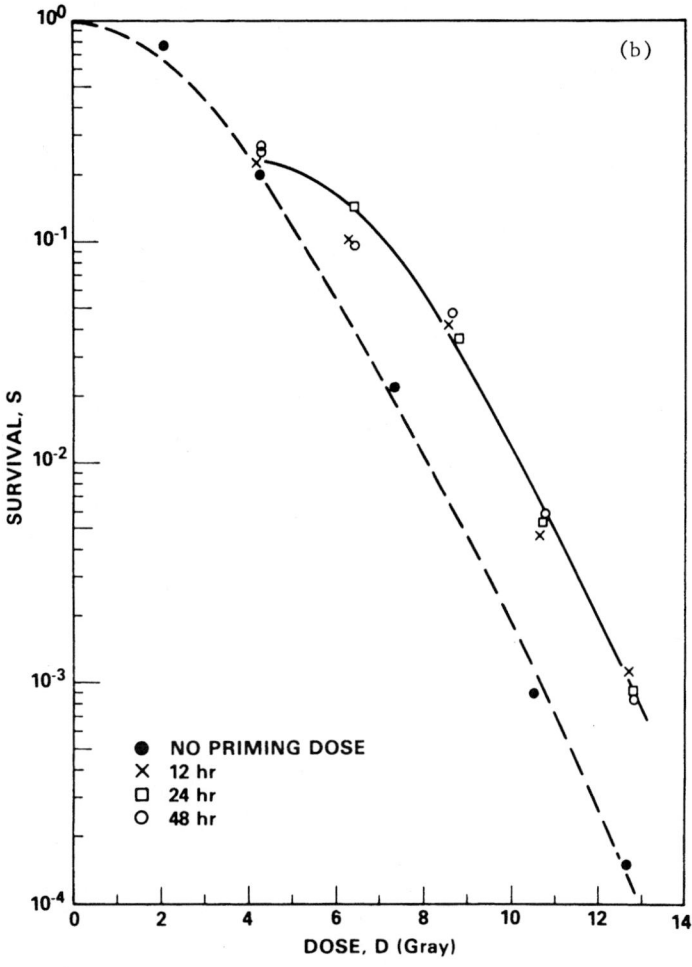

Figure 2b. (continued)

However, radiation quality also alters the effects of repair, oxygen, radical scavengers, and cell cycle age. Due to the differences in the spacing of ionizations along the tracks of the charged particles involved, different radiations can produce effects which differ qualitatively as well as quantitatively.

Models for Reproductive Survival

Radiation Quality Models

The effects of changes in radiation quality and the methods for characterizing that elusive property have dominated much of radiation biophysics modeling. Not only is the quality factor a major concern in radiation protection, there is obviously something of fundamental importance involved when the response of cells to the same amount of energy can vary by a factor of 20 or more depending on the velocity of the particles that deposit the energy. We can approach the effects of changes in the radiation alone by using experimental data for a limited set of radiations to establish the response of cells as a function of selected characteristics of those radiations, and then by using measured or calculated values of those characteristics for additional radiations to predict the response.

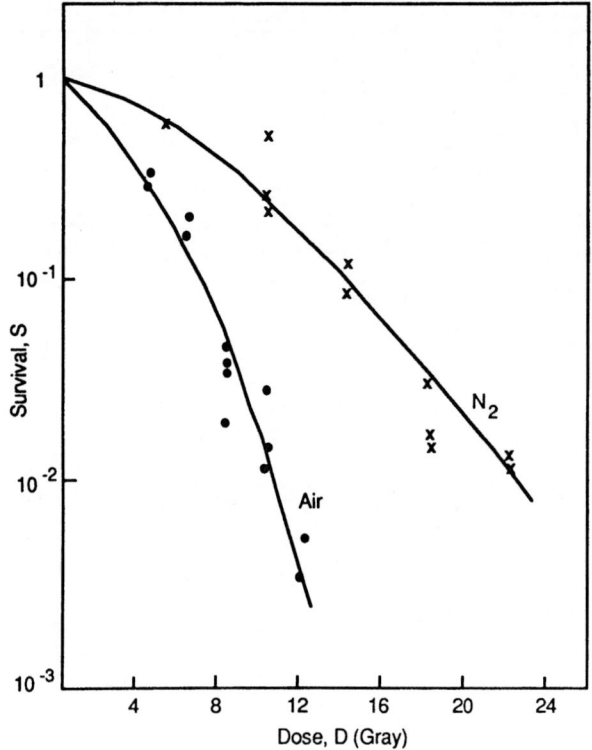

Figure 3. Cells irradiated in air and in N_2 at low oxygen concentration seem to have the same response, with a change only in the effectiveness of the radiation.

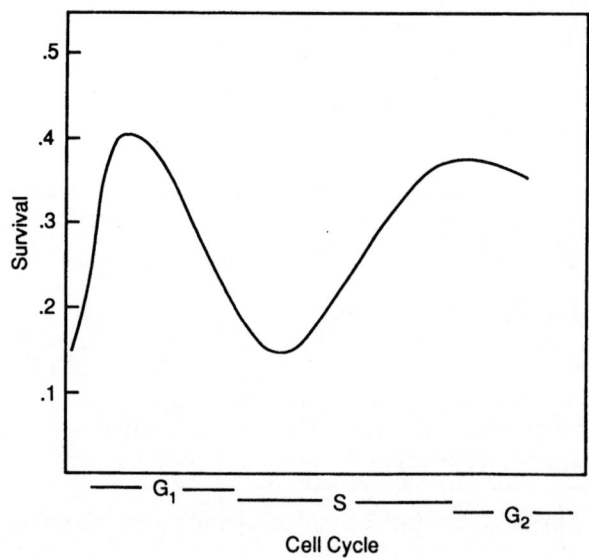

Figure 4. Cell cycle position can cause large changes in the survival following a fixed radiation exposure.[3]

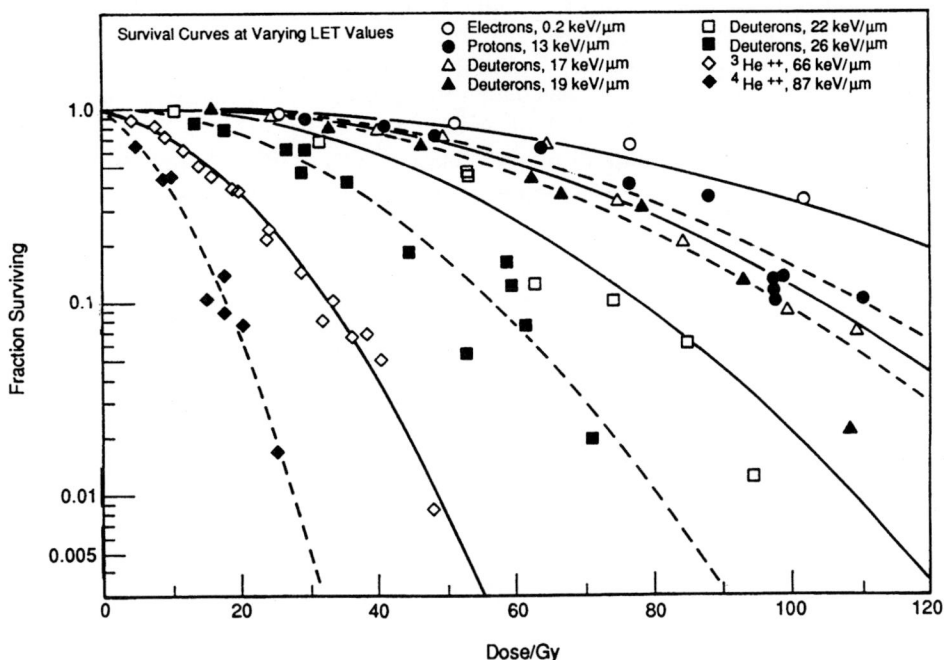

Figure 5. Changes in charged particle velocity and charge have a dramatic effect on shape of the survival relationship as well as the sensitivity.

One highly successful model of this type was proposed by Katz and co workers[4] and has been extended to cover a wide range of radiations.[5,6] In this model it is assumed that the response of any small radiation detector, including a DNA molecule or a cell, to high-energy ion irradiation can be divided into a response to the interaction of the ion with the target (called "ion kill") and the interaction of the delta rays with the target (called "gamma kill"). Four parameters, σ_0, κ, m, and D_0, are used to describe the cell. D_0 comes directly from the final slope of the survival of cells exposed to gamma rays, and the other parameters are determined by fitting the survival expression

$$S = e^{-\sigma D L}\left\{1 - \left[1 - e^{-(1-P)(D/Do)}\right]^m\right\}$$ (1)

where $\sigma = \sigma_o(1 - e^{Z^{*2}/\kappa\beta^2})^m$

$P = \sigma/\sigma_o$

D = absorbed dose

L = stopping power

Z^* = effective charge of the charged particle

β = the velocity of the particle relative to that of light to the experimental data for a few radiation types. Clearly, the same irradiation conditions, specifically single-exposure acute irradiation, must be used in all of the experiments to determine the parameters, and the model only applies to similar high dose rate exposures. Consequently, this model deals only with the effects of radiation quality and makes no attempt to deal with other phenomena such as dose rate effect or repair after irradiation. The model, using parameters which were determined soon after it was originally formulated, has been quite successful in predicting the response of cells to new radiations as experiments with new accelerators have been performed. Figure 6 shows the fit obtained for mammalian cells exposed to a variety of ions. The relationship of this model to some others has been explored by Curtis[7] who showed that it becomes linear-quadratic in form at low dose if m=2.

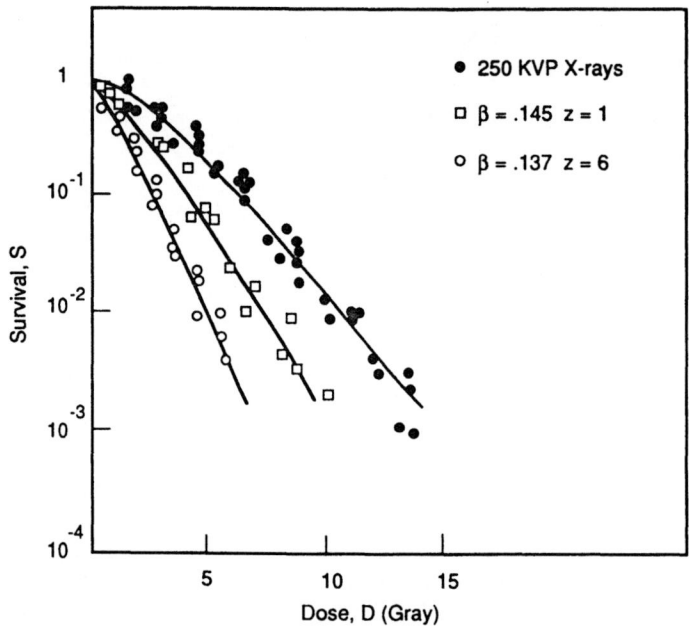

Figure 6. The Katz model provides fits to charged particle irradiation data based on the response of the cell to low LET irradiation and a fit to data for other high LET particles.[5]

Another approach to determining the effects of radiation of different qualities is based on the assumption that the response of a cell to irradiation depends on the energy deposited in a small volume such as the cell nucleus. Although the actual volume which is important in determining the response is not known, we assume that it can be simulated by a sphere of one or a few micrometers in diameter. The frequency distribution of energy deposited, which can be expressed as the microdosimetric quantity, y (the energy deposited divided by the mean chord length of the sample volume), is considered a reasonable indicator of radiation quality. Several different approaches have been taken to evaluate the relationship between lineal energy and probability of effect. Initially, Varma and Bond[8,9] limited the analysis to low-dose or low-dose-rate experimental data where the observed biological effect was the product of the interaction of a single charged particle track and could be related to the single event distribution, y. They assumed a relatively simple relationship between y and effectiveness, and were able to evaluate the coefficients of the relationship for tradescantia mutation using data obtained for radiations with a wide range of LET spectra. Later Varma and coworkers[10] showed that the hit size effectiveness function determined in this way could be used to predict the effects of high doses of low LET radiation delivered at high dose rate so that biological repair need not be considered. Morstin and coworkers[11] and Zaider and Brenner[12] took more flexible approaches to determining the effective relationship, avoiding the need to assume general form of the relationship. The resulting effectiveness functions for a wide range of cell types and end points generally resemble Fig. 7 (derived from Morstin).[11]

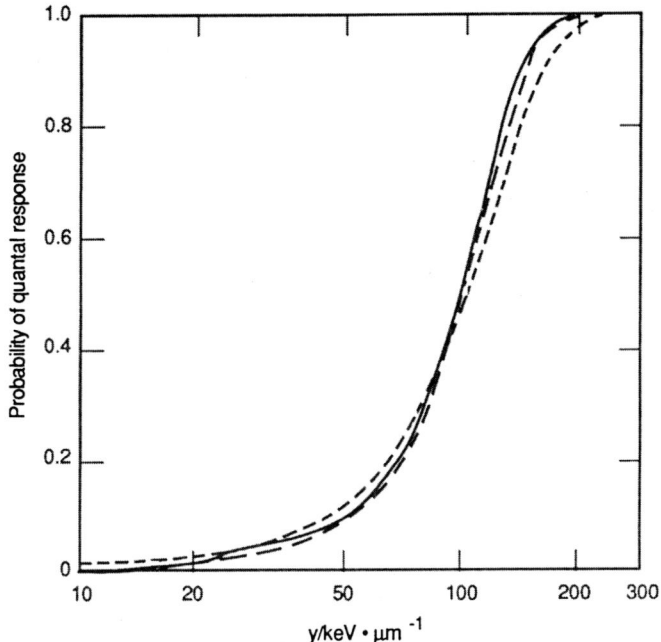

Figure 7. Hit size effectiveness functions are similar for end points such as HPRT mutations in human fibroblasts, chromatid exchanges, and abnormal metaphase in CH2B2 hamster cells.[11]

The basis for this approach is the assumption that the spatial distribution of energy deposition within the cell nucleus is not critical in determining the effect, and the elimination of the effects of repair, by limiting the application to single-event effects or high-dose-rate effects where repair does not occur. Within these limitations, this approach has been applied to a variety of cell types and end points such as chromosome aberration as well as cell survival. Its applicability to low dose effects makes it particularly relevant to the needs of radiation protection. The results obtained for different cell types and by different mathematical approaches are remarkably consistent. The differences between cell lines suggest that this approach may help in identifying those characteristics of cells which are related to their response to radiation.

Cell Response Models

The second major grouping of models involves those which attempt to describe the response of cells in terms of the probability of energy being deposited, the organization of the sensitive material in the cell, and the response of the cell to specific concentrations of damage. The initial goal of these models is to describe the shape of the acute-exposure dose-response curve, with refinements to add dose-rate and fractionation effects, and RBE. There are many different models in this group (discussed later), ranging from quite simple to a substantial involvement of biological processes, but they are all based on the same basic assumptions and are all constrained by the consequences of those assumptions. Although most of the models have been formulated in terms of reproductive survival of cells, the underlying assumptions deal with effects in general. The models may be applicable to other end points as well. However, the specific mechanisms involved may be different for each end point, and the coefficients derived from a model for cell survival may not be applicable to mutation or a specific step in malignant transformation. Models of the mechanisms leading to these specific end points may eventually indicate the relationships between end points, making it possible to predict mutation and transformation rates from survival data, or it may be necessary to fit a model directly to data for each end point.

A reasonable starting point for a model of radiation effects on single cells is to assume that if a charged particle passes through some region, usually referred to as the target, there is a finite probability of doing critical damage. If this probability is assumed to be the same for all charged-particle events, either because the radiation is controlled so that all events are identical or because the cell is not sensitive to differences in the energy deposition, and if all events act independently in the cell, then from Poisson statistics the probability that there will be one or more critical events in the cell is

$$f = 1 - e^{-\lambda D} \tag{2}$$

where λ is the mean number of critical hits per cell per unit of dose, D. Since the probability of a cell surviving irradiation is equal to the probability of its not having lethal damage,

$$S = 1 - f = e^{-\lambda D} \tag{3}$$

In some systems, such as enzymes irradiated in the dry state and some simple viruses, the only significant type of damage is a single ionization at a critical location, and the conditions required for the validity of the above equations are met. However this is not the case for most biological systems. It seems unlikely that an isolated ionization in a large molecule would be as likely to do critical damage as would a cluster of closely spaced ionizations, which would disrupt more bonds and possibly produce larger distortions in the higher level structure of the molecule. The increase in relative biological effectiveness with increasing LET of the radiation is consistent with this assumption. Furthermore, it seems unlikely that the products of separate charged particle events within a single target would act entirely independently. Cells are known to repair most radiation-induced damage, but the capacity to repair is probably not unlimited. It is also likely that the process of replication prevents further repair, so there will be competition between repair and replication of damaged sites. The process of repair requires information on the "correct" structure, but as the concentration of damage increases, the chance that this information will be lost increases as well. Damaged DNA molecules may undergo interactions, such as formation of exchanges or deletions, which result in damage that is less easily repaired. These, and probably many other types of interactions, cause an increase in effect, but such interactions may also decrease the effectiveness of damage. For example, some cell types are known to activate additional repair systems in response to damage. The effects of multiple-track processes can be included by a simple polynomial expansion of Eq. (2):

$$S = \exp - (a_1 D + a_2 D^2 + a_3 D^3 \ldots) \qquad (4)$$

Most of the models developed in recent years incorporate some combination of features such as repair of sublesions and interaction that blocks repair. The diversity of these combinations complicates comparison of models or progression from basic models to more advanced versions. However, many models share specific features, and these models can be grouped according to shared features, even though each model may also possess other, quite distinct characteristics.

One of the major differences between phenomenological models in this group has been the assumption made about the nature of the interaction process which is assumed to be responsible for the curvature of the dose response relationship. It has generally been assumed, based primarily on Elkind's work on the split-dose recovery effect,[3] that the shoulder of the survival curve, the dose-rate effect, and repair of damage are interconnected. However, there has been considerable controversy about the nature of the connection. The difference in the alternative assumptions about interaction has been expressed in terms of the effectiveness of unrepaired damage produced by a single charged particle track. Some models assume that the damage produced by a single (low LET) charged particle track is insufficient to produce an effect and that the damage elements produced by two or more tracks must interact in order to produce the effect. This sublethal damage is generally assumed to be repaired, but if the elements do not interact, it makes no difference whether they are repaired or not, they will not be detectable in terms of an effect on the cell. The alternative approach assumes that damage produced by a single low LET track is lethal to the cell unless it is repaired, and that the interaction is some type of process which limits the amount of repair or its accuracy.

If some damage is assumed to be sublethal in nature, then there are at least two different ways in which the interaction might come about. Several early models[13-17] assume what amounts to a target with two damaged states, sublethally damaged and lethally damaged. Sublethally damaged cells must receive additional damage to be converted to lethally damaged, but they are also subject to repair, which converts them back to undamaged condition. For example the expression for cell survival for the Swann-del Rosario model is

$$S = [R_1 \exp(- R_2 D) - R_2 \exp(- R_1 D)]/(R_1 - R_2) \qquad (5)$$

with

$$R_1, R_2 = (1/2)\left\{K_1 + K_2 \pm \left[(K_1 + K_2 + K)^2 - 4 K_1 K_2\right]^{0.5}\right\} \qquad (6)$$

where K_1 is the rate of cell damage, K_2 is the rate of damaged cell killing, T is the mean recovery time for damaged cells, and K is $(D'T)^{-1}$ where D' is the dose rate.

The alternative assumption is that is that bits of damage, sometimes referred to as "sublesions," are produced by radiation interactions and are repaired by a process which has a fixed rate for each bit of damage. The damage accumulates in the cell with increasing dose until an equilibrium, which depends on the dose rate, is reached. New damage has a probability of interacting with existing damage to produce a lethal effect which is proportional to the concentration of damage. Models of this general type were developed by Roesch,[18] Kellerer and Rossi,[19] and Payne and Garrett.[20] They lead to the expression for survival

$$S = \exp\left\{-CD - A(D'\tau)^2\left[\exp[-D/(D'\tau)] - 1 + D/(D'\tau)\right]\right\} \qquad (7)$$

where C and A are coefficients for single-event and two-event damage respectively. The characteristic time, τ, is the mean time required to repair a lesion.

Both versions of the sublethal damage repair model can be made to fit typical survival curves to within the precision of the experimental measurements, yet they differ in a fundamental way. Equation (7) suggests plotting dose rate data as $-\ln S/D_2 - C/D$ versus the duration of the irradiation, D/D'.[21] When this is done, the accumulation models result in a single curve independent of the dose rate, while the promotion models result in families of curves. Experimental evidence for a variety of mammalian cell lines[22] clearly shows the single curve rather than a family (Fig. 8) when stationary phase cells are exposed to a wide variety of dose rates.

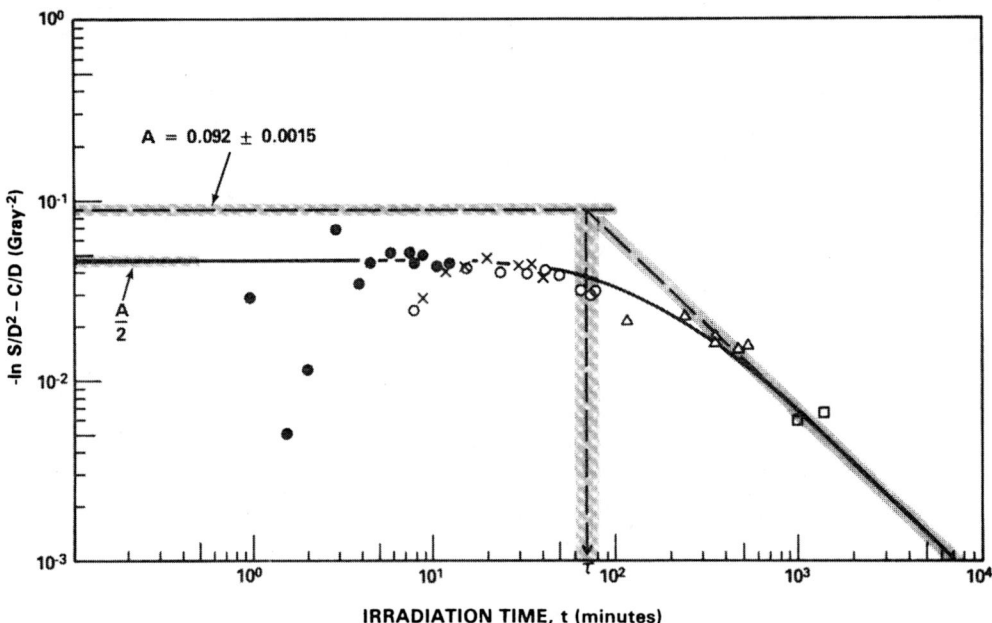

Figure 8. When plotted in a way which allows discrimination between models which assume that cells are promoted from undamaged to damaged to lethally damaged states and models which assume that damage accumulates in cells, the data for plateau phase CHO cells are consistent with the accumulation assumption. Each symbol represents a specific dose rate.

An important parameter in models which describe dose rate and fractionation experiments is the mean time required to repair a lesion. This time can be obtained from the $-\ln S/D^2$ curve, but a substantial amount of low-dose-rate irradiation is required to obtain τ with reasonable precision. It is much easier to obtain data for a wide range of repair times by using split-dose methods. The accumulation model can be solved for irradiation which consists of two fractions given at high dose rate and separated by a variable interval t_r.[23,24] This analysis shows that the repair time can be determined from the slope of a plot of $-\ln S_m/S$ versus t_r, where S_m is the maximum survival, i.e., the survival following an infinite repair time. The model assumes that any repair which might occur after the second irradiation is constant. In an attempt to meet this requirement, cells are plated immediately after the second dose. Experimental results for *Chlamydomonas reinhardi*, a eukaryotic algae which is amenable to

Braby

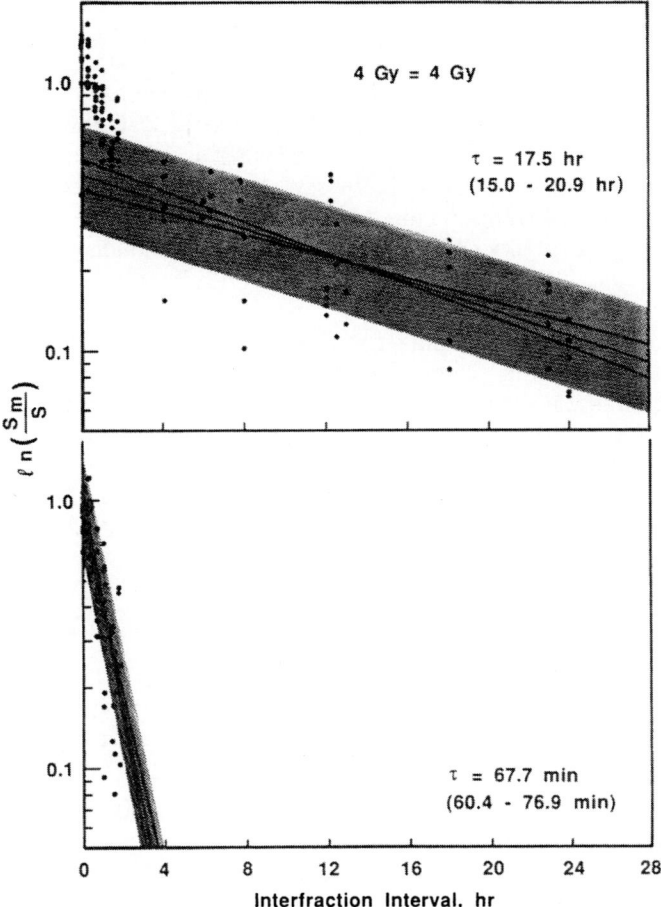

Figure 9. Split-dose data for plateau phase CHO cells shows two repair rates, indicating two types of damage, which differ in some way that alters their repair.

determination of τ by the analysis of dose-rate data, showed that split-dose and dose-rate experiments resulted in the same repair rate. When the repair rate of plateau-phase CHO cells was determined,[25] the results (Fig. 9) indicate that a model with a single repair process is too simple to describe the actual biological situation. Plateau-phase cells which could be held for three days with less than 0.1% contamination of cycling cells were used in order to obtain a good value for S_m. Dose rates as high as 10 Gy/second and intervals ranging from 10 seconds to 24 hours were used in order to detect the widest possible range of repair rates. The majority of the damage was found to be repaired with a mean time of about 68 minutes, with an additional component being repaired with a characteristic time of about 18 hours. No repair with a characteristic time between 10 seconds and 50 minutes could be detected.

The split-dose results suggest that at least two distinct types of damage are being repaired, and that both of them involve some form of interaction so that expression as reproductive death becomes more likely with increasing dose. However, many different types of interactions might be responsible for the observations, and the models discussed so far have only dealt with one, the interaction of sublethal damage to produce lethal or potentially lethal damage. The main alternative to this type of model assumes that all relevant damage is lethal unless repaired. This approach follows naturally from the observation of repair in delayed plating experiments,[26] a phenomenon which is not predicted by the sublethal damage models. However, if a fixed probability for repair exists, the shape of the survival curve is not changed. Several different approaches can be taken to include a decrease in efficiency of repair with increasing dose. One approach, used by a number of authors,[27-31] essentially assumes that radiation can inactivate repair enzymes as well as damage the structures that they repair. Thus, as the dose increases, repair becomes less efficient due to a reduction in the repair system. A closely related approach is to assume that irradiation reduces the supply of energy (ATP) within the cell, thus reducing the effectiveness of repair.[27,32] Although the depletion of ATP has not been found to be sufficient to account for the apparent reduction in repair in any system, the depletion of repair enzymes is consistent with the data for simple systems with relatively small target volumes and small amounts of repair enzyme. For more complex mammalian cells, the effect of interaction is evident at a few Gray, which would require that the repair enzymes themselves must be very large. A more realistic alternative is the assumption that the supply of repair enzyme is quite limited in ordinary cells, and that at high doses the repair system is saturated. This approach was developed by Goodhead[33,34] with the further assumption that repair molecules are not recycled, and leads to the expression

$$S = \exp{-p(aD - c_0)}/\{-(c_0/aD)\exp[kt(c_0 - aD)]\} \tag{8}$$

where a gives the number of potentially lethal lesions per unit dose, c_0 is the initial number of repair molecules, p is the fraction of the damage which is capable of producing the specific biological effect, and k is the rate constant for repair. This model provides good fits to acute survival data (Fig. 10) and delayed-plating experiments but requires an additional component, the rate of replacement of repair molecules, for application to dose-rate or split-dose experiments.

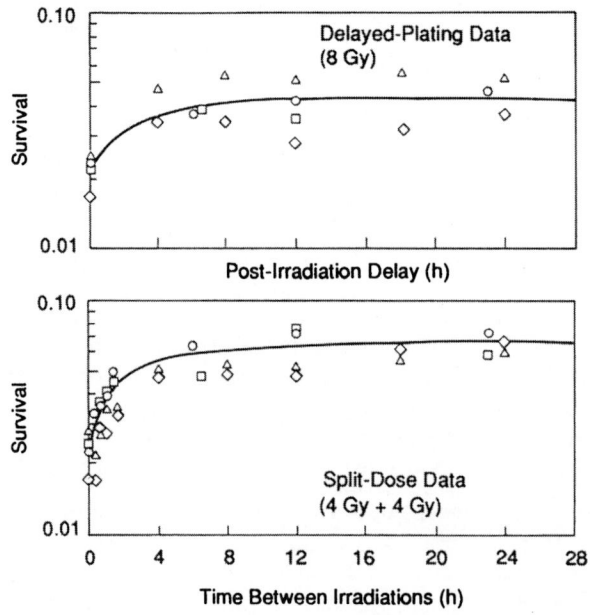

Figure 10. The repair saturation hypothesis is consistent with survival data for different radiations.[34]

One consequence of the type of interaction characterized by repair saturation models is that the total amount of repair which occurs depends on the amount of damage, the rate of replenishment of the repair material, and the total time for repair. It is not a function of the time-dependent concentration of the damage. This is significantly different from the accumulation models and some mechanistic models[35] where the total amount of possible repair is reduced by interactions which depend on the square of the damage concentration. This difference can be tested by comparing the survival of cells receiving the same total dose in a split-dose experiment or in a delayed-plating experiment. Results for this type of comparison for plateau-phase CHO cells are shown in Fig. 11. The split-dose exposure resulted in higher final survival, even though cells were plated immediately after the second dose. If the cells exposed to the split dose were held for a few hours after the second dose, their survival was still higher. Thus this experiment indicates that the process responsible for the shoulder of the survival curve is not simply the saturation of a repair process; it may be any of the interaction processes, which become more probable as damage concentration increases.

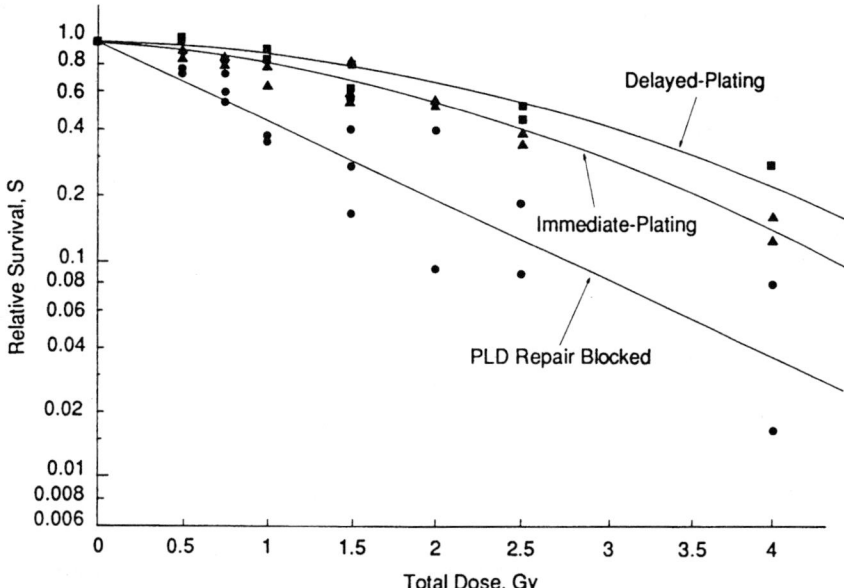

Figure 11. Split-dose and delayed-plating data show higher final survival in the split-dose case, where the damage concentration is relatively lower, the opposite of the prediction of repair saturation models.

Those models which are consistent with the split-dose versus dose-rate data for CHO cells include the accumulation models and some potentially lethal damage models. These two types differ in a fundamental way which will be important in the future development of mechanistic models, as well as in the prediction of effects at low doses and low dose rates. The essence of the difference is whether the interaction process converts relatively innocuous damage to potentially lethal damage or whether it alters the probability that potentially lethal damage will be repaired. If the damage is sublethal, dose fractionation will become less effective with decreasing dose, i.e., decreasing concentration of damage. This is because unrepaired sublethal damage cannot be detected. However, if the damage is potentially lethal, delayed plating will become more effective with decreasing dose because there will be less of the interaction that can block repair. In order to distinguish between these possibilities, the total change in survival due to repair as a function of dose is needed. However, cells are known[35] to repair damage for about three hours after replating in growth medium, so delayed plating experiments do not detect all of the damage being repaired. Drugs such as β-ara-A are known to block repair, but may also alter the configuration of the sensitive molecules and thus alter the production of damage if they are present during irradiation. Recent experiments in which the repair inhibitor was added after irradiation (Fig. 12) show an exponential survival curve for repair-inhibited cells, and a corresponding increase in the effectiveness of repair with decreasing dose. If this result is representative of the situation for other cell types and for cells in other culture conditions, it suggests that biochemical mechanisms which can be characterized as resulting from the interaction of sublethal damage to produce lethal damage play a very small part in the effect of radiation at low doses. However, mechanisms which can be characterized as altering the repairability of potentially lethal damage or the effectiveness of repair may play major roles in the response at low doses.

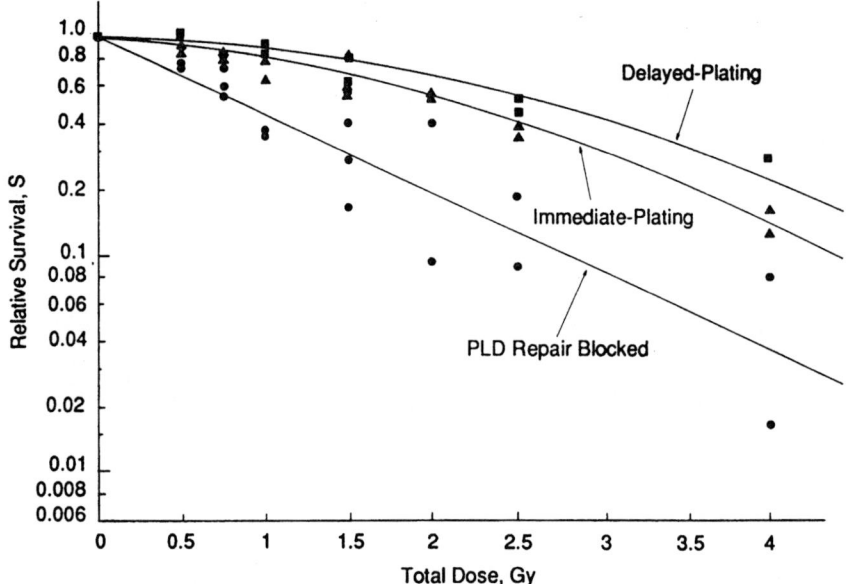

Figure 12. Plateau phase CHO cells treated with ara-A for 4 hours starting immediately after irradiation show an exponential survival curve, compared with the shouldered curve resulting from 24-hour plating delay or from immediate plating in growth medium.

Models for Other End Points

Chromosome Aberrations

The development of models for the production of chromosome aberrations as a function of radiation quality has roughly paralleled the development of models for cell survival. A model based on the interaction of sublesions was developed by Neary.[36] In this model the sublesions are assumed to be produced by a single ionization, and interaction is assumed to occur within a relatively small site within the nucleus. Many sites, possibly overlapping, exist in each nucleus. The probability of an aberration, Y, is

$$Y = NE\{1 - \exp(-mgk^2)[1 - \{1 - \exp(-mk + mgk^2)\}^2]\} \qquad (9)$$

where N is the number of sites in the nucleus; E is the probability of sublesions within the site interacting; m is the mean number of tracks through the projected area, A, of the chromosome segment within a site; g is the probability that a track which has traversed one target chromosome of a site will also traverse the other; and k is the probability that one track through a target produces a primary lesion. K is a function of p, the probability that a single energy loss event in a target chromosome produces a sublesion, and of the LET. This model can be used to predict the general shape of the curve representing the probability of producing an interchange by a single-track interaction (Y_1) or through interaction of two tracks (Y_2) as a function of LET (Fig. 13). However, due to the large number of parameters in the model, it is necessary to assign mechanistic significance to them and to try to evaluate them independently in order to test the validity of this approach.

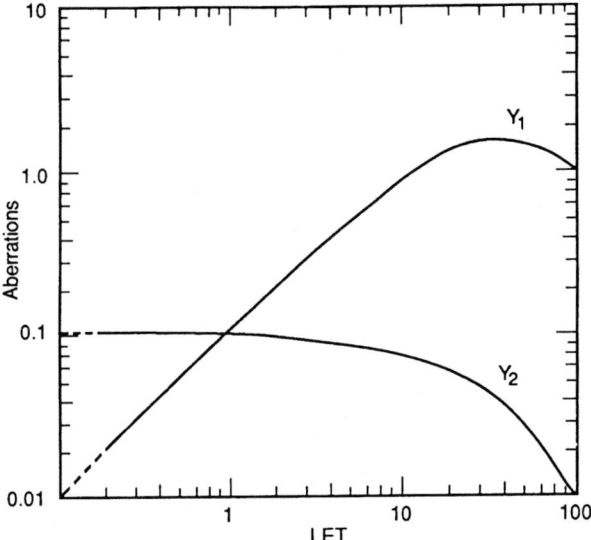

Figure 13. The LET dependence of the yield of chromosome aberrations produced by single tracks and by two track interactions can be predicted qualitatively.[36]

More recently, the success of repair depletion and related models for cell survival has suggested that the interaction of sublesions may not be required for chromosome aberration.[37] However, the experimental evidence does not seem to be consistent with a single hit mechanism at low LET.

Oxygen Effect

The dependence of the oxygen enhancement ratio (OER) (Fig. 3) on oxygen concentration has been studied extensively. Alper and Howard-Flanders[38] proposed the relationship

$$OER = [m(O_2) + k]/[(O_2) + k] \qquad (10)$$

where m and k are constants and $[O_2]$ is the oxygen concentration. As an understanding of radiation chemistry of cells developed,[39,40] it become generally accepted that oxygen reacts with radiation-induced DNA radicals to produce products which are more damaging than those which would otherwise result. Kiefer[41] showed that if a single type of initial damage is converted into either type x, if it reacts with oxygen, or type y if it reacts with a hydrogen donor such as glutathione, with rates k_x and k_y, respectively, and if these two types of damage are repaired by processes with rates k_{xo} and k_{yo} or converted to lethal damage with rates k_{xx} and k_{yy}; then, the constants in Eq. (9) become

$$k = k_y/k_x \tag{11}$$

and

$$m = \left(1 + k_{yo}/k_{yy}\right)/\left(1 + k_{xo}/k_{xx}\right) \tag{12}$$

This provides a starting point for an understanding of the mechanisms involved in the repair of damage formed in the presence and absence of oxygen, but as Chapman and Koch pointed out,[40] there are likely to be many different DNA radicals formed, and there may be different repair processes involved as a result. Alper[42] has shown that if more than one type of oxygen-dependent damage does exist, it will be very difficult to demonstrate it on the basis of OER measurements.

Future Directions

The comparison of model predictions with experimental results has led to a significant evolution in our thinking about the nature of the response of biological systems to ionizing radiation. Attempts to model the relative biological effectiveness of different radiations have dramatized the importance of the stochastics of energy deposition at the cellular and subcellular levels. Tests of dose response relationships under different conditions of irradiation have shown that modification of the effectiveness of damage by cellular processes is a major factor in determining final effect. Modeling the oxygen effect has shown how a single type of initial damage can result in different DNA alterations, and how differences in the rates of repair of these alterations can change cell survival. It is probably not possible to change any physical property of the irradiation without some corresponding change in the biological modification of the resulting damage.

Models of cell survival are only one step in understanding the health effects of ionizing radiation. However, they are an important step because reproductive death is probably the result of some of the same processes that lead to malignant transformation and other undesirable changes in growing cells. Refinements are still needed in models of the reproductive survival in order to determine the role of multiple repair processes and to understand the long-term effects of repaired cells in the population. Models now include the assumption that "repaired" DNA is the same as undamaged DNA, but there is some evidence[43] that additional changes occur many cell generations later, or that repair only converts damage to a form that cannot be detected for many generations. However, enough is now understood about cell survival that we should be able to apply similar modeling concepts to the individual steps in the complex process of transformation. Due to the limited number of variables which can be tested in realistic biological experiments, it is likely that different cell types and transformation assays will be needed to investigate each step in a multi-step process.

This will be difficult, because there is conflicting evidence about the number of steps involved in carcinogenesis, and it may be that different combinations of changes are required to produce transformation in cells of different tissues. However, it is reasonable to speculate that DNA deletions, rearrangements, and point mutations may all play roles in different steps of the process. It may also be possible for radiation to affect DNA indirectly through damage to molecules involved in critical feedback mechanisms or through damage to structural components of the cell which are involved in regulation of growth or expression of specific genes (cf. Hall, these proceedings).

Both the degree of understanding of the phenomenology of radiation effects and the availability of biochemical techniques for detecting the rare changes produced in cells by ionizing radiation make it practical to attempt to identify the specific mechanisms involved in radiation damage and repair. Several chemical repair processes, for example double-strand break rejoining, have been studied extensively (cf. Ward, these proceedings), but it is difficult to establish quantitative relationships between them and end points such as mutation. It is likely that there are several steps between damage to DNA and the expression of that damage as a deletion or rearrangement. The fact that some phenomenological models can be rejected reduces the range of mechanisms which have to be considered for these steps. The results discussed above suggest that we should be looking for mechanisms that alter the effectiveness of repair, rather than those that might convert innocuous damage into a more serious form. However, the results still indicate that this alteration of repair effectiveness must occur over distances comparable to the diameter of the nucleus of the cell. This finding probably eliminates mechanisms such as interaction of damage at different points on a single DNA molecule, but leaves open many other possibilities for mechanisms involving the information encoded in the DNA and in its structural relationship with the rest of the cell.

Acknowledgment

Work was supported by the Office of Health and Environmental Research (OHER), U.S. Department of Energy, under Contract DE-AC06-76RLO 1830.

References

1. J. R. Platt. Strong Inference. *Science.* 146:347-353 (1964).
2. T. T. Puck and P. I. Marcus. Action of X-rays on Mammalian Cells. *J. Experimental Med.* 103:653-666 (1956).
3. M. M. Elkind and G. F. Whitmore. *The Radiobiology of Cultured Mammalian Cells.* Gordon and Breach, New York (1967).
4. J. J. Butts and R. Katz. Theory of RBE for Heavy Ion Bombardment of Dry Enzymes and Viruses. *Radiat. Res.* 30:855-871 (1967).
5. R. Katz, B. Ackerson, M. Homayoonfar, and S. C. Sharma. Inactivation of Cells by Heavy Ion Bombardment. *Radiat. Res.* 47:402-425 (1971).
6. R. Katz, D. E. Dunn and G. L. Sinclair. Thindown in Radiobiology. *Radiat. Prot. Dos.* 13:281-284 (1985).
7. S. B. Curtis. The Katz Cell-Survival Model and Beams of Heavy Charged Particles. *Nucl. Tracks Radiat. Meas.* 16:97-103 (1989).
8. V. P. Bond and M. N. Varma. Low Level Radiation Response Explained in Terms of Fluence and Cell Critical Volume Dose. In *Proceedings of the Eighth Symposium on Microdosimetry*, pp. 423-437. Julich, West Germany. Harwood Academic, London (1982).
9. M. N. Varma and V. P. Bond. Empirical Evaluation of a Cell Critical Volume Dose vs. Cell Response Function for Pink Mutations in Tradescantia. In *Proceedings of the Eighth Symposium on Microdosimetry*, pp. 439-450. Julich, West Germany. Harwood Academic, London (1982).
10. M. N. Varma, V. P. Bond and G. Matthews. Hit-Size Effectiveness Theory Applied to High Doses of Low Let Radiation for Pink Mutations in Tradescantia. *Radiat. Prot. Dos.* 13:307-309 (1985).

11. K. Morstin, V. P. Bond, and J. W. Baum. Probabilistic Approach to Obtain Hit-Size Effectiveness Functions Which Relate Microdosimetry and Radiobiology. *Radiat. Res.* 120:383-402 (1989).

12. M. Zaider and D. J. Brenner. On the Microdosimetric Definition of Quality Factors. *Radiat. Res.* 103:302-316 (1985).

13. W.F.G. Swann and C. del Rosario. The Effect of Radioactive Emanations Upon Euglena. *J. Franklin Inst.* 211:303-317 (1934).

14. A. Kellerer and O. Hug. Zur Kinetik der Strahlenwirkung. *Biophysik* 1:33-50 (1963).

15. G. J. Dienes. A Kinetic Model of Biological Radiation Response. *Radiat. Res.* 28:183-202 (1966).

16. B. Rajewsky and H. Danzer. Uber einige Wirkungen von Strahlen. VI. Eine Erweiterung der Statistishen Theorie der Biologischen Strahlenwirkung. *Z. Physik.* 89:412-420 (1934).

17. D. E. Lea. A Theory of the Action of Radiations on Biological Materials Capable of Recovery. I. The Time-Intensity Factor. *Br. J. Radiol.* 11:489-497 (1938).

18. W. C. Roesch. A Model for the Action of Radiation on Simple Biological Systems. In *Proceedings of the First International Symposium on the Biological Interpretation of Dose from Accelerator-Produced Radiation*, pp. 297-305. CONF-670305 U.S. Atomic Energy Commission, Washington, D.C. (1967).

19. A. M. Kellerer and H. H. Rossi. The Theory of Dual Radiation Action. *Curr. Top. Radiat. Res. Q.* 8:85-158 (1972).

20. M. C. Payne and W. R. Garrett. Some Relations Between Cell Survival Models Having Different Inactivation Mechanisms. *Radiat. Res.* 62:388-394 (1975).

21. L. A. Braby and W. C. Roesch. Testing of Dose-Rate Models with Chlamydomonas reinhardi. *Radiat. Res.* 76:259-270 (1978).

22. N. F. Metting, L. A. Braby, W. C. Roesch, and J. M. Nelson. Dose-rate Evidence for two Kinds of Radiation Damage in Stationary Phase Mammalian Cells. *Radiat. Res.* 103:204-218 (1985).

23. L. A. Braby, J. M. Nelson, and W. C. Roesch. Comparison of Repair Rates Determined by Split-Dose and Dose-Rate Methods. *Radiat. Res.* 82:211-214 (1980).

24. J. M. Nelson, L. A. Braby, and W. C. Roesch. Rapid Repair of Ionizing Radiation Injury in Chlamydomona reinhardi. *Radiat. Res.* 83:279-289 (1980).

25. J. M. Nelson, L. A. Braby, N. F. Metting, and W. C. Roesch. Multiple Components of Split-Dose Repair in Plateau-Phase Mammalian Cells: A New Challenge for Phenomenological Modelers. *Radiat. Res.* 121:154-160 (1990).

26. R. A. Phillips and L. J. Tolmach. Repair of Potentially Lethal Damage in X-Irradiated HeLa Cells. *Radiat. Res.* 29:413-432 (1966).

27. R. H. Haynes. The Interpretation of Microbial Inactivation and Recovery Phenomena. *Radiat. Res.* S-6:1-29 (1966).

28. R. H. Haynes. The Influence of Repair Processes on Radiobiological Survival Curves. In *Cell Survival After Low Doses of Radiation*, T. Alper, ed., pp. 197-208. Institute of Physics/Wiley, London (1975).

29. W. Pohlit. The Shape of Dose-Effect Curves for Diploid Yeast Cells Irradiated With Ionizing Particles. In *Cell Survival After Low Doses of Radiation*, T. Alper, ed., pp. 190-196. Institute of Physics/Wiley, London (1975).

30. A. Kappos and W. Pohlit. A Cybernetic Model for Radiation Reactions in Living Cells. I. Sparsely-Ionizing Radiations; Stationary Cells. *Int. J. Radiat. Biol.* 22:51-65 (1972).

31. W. R. Garrett and M. G. Payne. Applications of Models for Cell Survival: The Fixation-Time Picture. *Radiat Res.* 73:201-211 ((1978).

32. W. Pohlit and I. R. Heyder. The Shape of Dose-Survival Curves for Mammalian Cells and Repair of Potentially Lethal Damage Analyzed by Hypertonic Treatment. *Radiat. Res.* 87:613-634 (1981).

33. D. T. Goodhead. Models of Radiation Inactivation and Mutagenesis. In *Radiation Biology in Cancer Research*, R. E. Meyn and H. R. Withers, eds., pp. 231-247. Raven Press, New York (1980).

34. D. T. Goodhead. Saturable Repair Models of Radiation Action in Mammalian Cells. *Radiat. Res.* 104:S-58 - S-67 (1985).

35. S. B. Curtis. Lethal and Potentially Lethal Lesions Induced by Radiation-A Unified Repair Model. *Radiat. Res.* 106:252-270 (1986).

36. G. J. Neary. Chromosome Aberrations and the Theory of RBE 1. General Considerations. *Int. J. Radiat. Biol.* 9:477-502 (1965).

37. M. N. Cornforth. Testing the Notion of the One-Hit Exchange. *Radiat. Res.* 121:21-27 (1990).

38. T. Alper and P. Howard-Flanders. Role of Oxygen in Modifying the Radiosensitivity of E. Coli. *Nature* 178:978-979 (1956).

39. M. Quintiliani. The Oxygen Effect in Radiation Inactivation of DNA and Enzymes. *Int. J. Radiat. Biol.* 50:573-594 (1986).

40. J. D. Chapman and C. J. Koch. Comment on the Paper by van der Schans et al. *Int. J. Radiat. Biol.* 50:467-470 (1986).
41. J. Kiefer. Theoretical Aspects and Implications of the Oxygen Effect. *Radiation Research*, O. F. Nygaard, H. I. Adler, and W. K. Sinclair, eds., pp. 1025-1037. Academic, New York (1975).
42. T. Alper. Adding Two Components of Radiosensitization by Oxygen. *Int. J. Radiat. Biol.* 46:569-585 (1984).
43. C. B. Seymour and C. Mothersill. Lethal Mutations, the Survival Curve Shoulder and Split-Dose Recovery. *Int. J. Radiat. Biol.* 56: 999-1010 (1989).

Discussion

Hall: I had a couple of comments that I jotted down along the way, Les. One was the repair time; we have found over recent years that the two components of three pair, the fast and the slow, seem to be more common, more obvious, in human cell lines than in the rodent cell lines. You get a lot smaller dose rate effect with human cells than rodent cells in general, and time and time again this fast repair comes up in cells in culture and in some *in vivo* situations where the human is being looked at. And the other thing that is worth noting was first reported, I think, by Gordon Steele, and I think that we would have to compare with it; if you modeled dose rate and split doses experiments and work out half times for repair, you don't get the same half times for whatever reason.

Braby: We have done similar experiments comparing the dose rate results with split dose repair times, and our results are just the opposite. We find that the repair timing is the same with the split dose and the dose rate assay. The use of dose rate data to get a repair time is very difficult. It requires a huge amount of low dose rate data which takes a long time to accumulate. The split dose experiments are much quicker to do. We have done it, particularly for *Clamydomonas* again because that's an easy system to do quite precisely, and they matched very well. We have made the comparison with the mammalian cell data and they are consistent with equal repair rates, but I wouldn't say that the rates are the same because the uncertainties are quite large. The fact that you see the two repair processes, the two repair times, in the human cell more easily that in the rodent cell is interesting. I am not sure that the reduced dose rate effect is in any way a contradiction, because we are probably talking about several different repair processes, and it may be that the human cells have a particularly long time for repair after replating in the delayed plating experiment, so that there is more repair occurring during the "fixation delay" and less of it evident in the result.

Zaider: I wanted to make a statement. There is a multitude of models, and I suppose there could be more since everybody likes to make models. Perhaps we can assign some criteria, comparable to what Mitio did for data, as to what criteria models should satisfy in order to be totally acceptable for discussion and consideration. I would ask them to be comprehensive in the sense of fitting much of the available data and to be falsifiable, I don't remember anybody ever saying that this model, my model, doesn't fit something. They all appear to fit everything.

Braby: I just said that our model does not fit the latest experiments we have done as well as the LPL model.

Zaider: Of course.

Braby: It is rare, but it does happen.

Varma: I think Marco's point is a good one. As you mentioned, especially in terms of the phenomonological models, you need to identify the purpose for which it is to be used. If you want to use a phenomonological model for interpretation of data for radiation protection, that is a different matter. If you want to use a phenomonological model that gives you some idea about the fundamental mechanisms that might be involved, like LET effects and cell damage and repair, it is a different kind of model that might be used. Therefore, we need to classify as different those models that ought to used for practical purposes, as opposed to models that are to be used for the understanding of the basic mechanisms.

Braby: I don't agree completely on that because there may be simplified models which are appropriate for some aspect of the radiation protection problem. The more comprehensive the model, the more adequately it can handle the entire radiation protection problem. One example that comes to mind is that many models deal with the LET effect without dealing with the dose rate effect. This becomes a complication when much of the data you are using to help guide radiation protection standards is accumulated at different dose rates. One of the controversies now about the appropriate values for quality factors stem from differences in the dose rate for the neutron and low LET radiations that were used in determining RBEs, so models that are more comprehensive, though they may be slower to develop, will probably offer more, even for the practical applications, in the long run.

Zaider: I just wanted to say that models should be predictive. Certain models are not meant to be predictive; they are meant to describe dose-effect data. However, mechanistic models certainly need to make a prediction. There are a number of mathematical techniques for setting up model-free constraints on biophysical models. For instance, one can show quite easily that linear-quadratic survival curves cannot remain valid at large doses. Similarly, there is a relation between the initial slope of the dose-effect curve and the size of the sensitive volume in the cell. Many of those constraints are poorly known or understood in our community.

Braby: Those are good points. One of the things that we need to do is to test the models with respect to as many different phenomena as possible in a single system in order to really compare the models. It is probably less useful to take one model and find out how many different experimental systems it is consistent with. You don't prove anything about the models by showing consistency; the only time you gain anything is by showing inconsistency, then you can eliminate something.

Varma: Marco, you talk in terms of comparison of the models in predictive capabilities. But here is another question: Les, why are we still using the absorbed dose on the x axis when we have a better quantity which is the microdosimetric distribution of energy? It seems to me that even after twenty years now in microdosimetry, we still keep on plotting the absorbed dose and LET and keep saying that they are not good parameters. It seems to me that we need to analyze this data in terms of the advancement that we have made. Why aren't we doing it?

Paretzke: That's an easy answer, because microdosimetry didn't solve the problem. It just introduced a different quantity because one doesn't know the shape of sensitive volume or even the size of the volume, so you just shift the problem from one quantity to the other. I just remind you of the problem with the carbon X-rays. You don't know which size to use or which volume shape to use; you solve one problem and you buy another problem.

Varma: I think you are making this too negative, because I think that carbon X-rays pointed out that you probably had to look into smaller volumes, right? And you have the capability now to look into this through your own calculations, can now look at nanometer volumes. You should be able to use those.

Paretzke: There is not one single volume, and microdosimetry can't cope with the problem that you might have to deal with the changes in very small volumes in the chromosome area. The effectiveness of this might depend on the energy or changes on an even smaller scale, say the namometer size, so you have the atomic or the molecular scale; at the same time the number of elements changes, and there is a spectrum of element changes in the whole cell that will affect the outcome. So you have to deal with energies in at least three different dimensions of volumes. You can't deal with this properly in any present terms, and that is why we don't know which quantity to use there. It is trivial to convert charged particle fluence to dose, and it has been done over the past forty years or so; that was one of the reasons why I plot my data with dose. I would rather like to see data plotted in terms of fluence.

Varma: Yes, that brings up another question. Why don't you plot in terms of fluence to analyze the data and maybe answer the type of question that John raised?

Braby: There's another answer. I agree with everything that you said, but there is another aspect. You could plot the fluence, but that wouldn't change anything in the case of high LET radiations—it is how the dose was calculated to start with. The other thing, though, is to really take advantage of the microdosimetric information; you have not just the mean energy deposited, but you actually have the distribution and variance of the energy deposition. That provides so much more information that the biology isn't up to the task. We can't resolve those small differences from the biological side, so it hasn't been feasible to do that kind of experiment. But the microdosimetry does contribute a great deal to the understanding of the models and actually is input to the models.

Kasha: Matt Varma has the dream that we can someday have a convergence of opinion and the concept and methods of treatment from the organism to the cellular and down to the atomic or molecular scale. Now, I am getting the vision that we are actually working in two different channels which may never converge, and it is sort of like putting molecular mechanisms into thermodynamics; microscopic physics and gross observations are in such a different realm, even though you get down to nanometer dimensions, you are not using the specificity. I am wondering if Marco had a comment on this; it looks to me like there are two parallel things which may never really converge because the whole thought process is different.

Varma: We know now that we can determine energy deposition in small sites that show inherent fluctuations, which is the field of microdosimetry, so we can identify the energy deposition in small sites. If we know that the DNA molecule is the important one biologically, then we need to use that quantity rather than average it. On the other side, I agree with you that you have to start at the atomic and molecular level, and progress through to look at initial events that might eventually be the cause of carcinogenesis. Models must make that connection from the atomic and molecular species that are produced, chemical species that are produced, and molecular biological end points such as base pair changes and point mutations. So those two lines are parallel; I think we agree on that. But why can't we use the sophistication that we have developed in physical characterization of the energy deposition?

Katz: Because it would be wrong to plot things that way. One should not plot digested information; one should report accurate experimental information. You want to digest that information and plot in terms of digested information, and I am very much opposed to that. There is altogether too much digested information plotted which is very difficult to bring back to the actual experimental observation. I want to see experimentalists' report, actual hard experimental observation, and then you digest it if you will in the interpretation. But do not digest the information in the reporting.

Braby: The quantity "Y" is not a digested quantity; it is directly observable.

Katz: Only because you take the size. Once you take the size, you have digested it.

Zaider: I would like to actually support what has been said. It is absolutely true that we do not know the shapes of the sensitive volumes. However, I see going back to fluence to be really bad for the future because this really comes, if I can say it, from the arrogance of the physicist in thinking he can totally ignore the biological object. Then at least instead of using dose, just an average quantity, we would use energy in a cellular nucleus as we know that things must happen there. I think we can use that amount of energy and the distribution, and that this is what Matt is saying. I think that this practice will help build better models.

I would also like to comment on what you said [referring to Kasha's comment]. You are absolutely right when you say that we are dealing with thermodynamics here and this is the only reason we are so ignorant yet so successful. The reason we are somewhat successful in describing such complicated events is just because they are. If they were not so complicated, we would be absolutely lost.

Paretzke: There is no way, and I want it in the records. If we stay with quantities like dose, microdose or macrodose, energy in any volumes, we will never have any progress. I had hoped—it was one of the reasons I came here—that we would be here at the funeral of the quantity dose, but apparently the time is not yet right that we understand this need. We will not have any progress, and that is one of the reasons why we know so little, we don't have good models, so I really disagree with what you are saying. That is one of the main reasons, because we still think in terms of energy in volumes; that is the wrong concept, and it has been for the past forty years. As long as we stay with it—and you just made a plea for it—we will not have any advance in understanding in radiation research. We really deserve what is going to happen with that one.

The second point is we should just get rid of the *trying to analyze* and understand relative biological effectiveness. I started to evolve the theory that this is the radiobiologists' biggest error because if we don't understand dose effect curves or exposure effect curves, we pretend then that we understand ratios of effect curves from different types of radiation. Why would any reasonable physicist, or scientist in general, try to analyze the ratio of effects of quite different radiations which depend on so many things: on dose rates, in different ways, and on dose levels, on the presence of oxygen, and in human beings even on such things as modifying agents? I really think that we should make it completely clear that dose is a quantity which we have gotten used to and worked with; at least some of us have developed a feeling for the shortcomings of this quantity. This would be completely wrong, and I would make a strong plea for it, not to substitute it with another wrong quantity, which we have then to start over again to develop the feeling for those shortcomings, and stay with those until we can do something considerably better. For the time being, I think it is a very good and sound recommendation from Bob Katz to say just give the radiation field, the fluence and the spectrum of a photon field or the number of particles per second and so on; because the fluence is the basic input data which you can use then to analyze data and develop new ideas which you can use to interpret the biological effect.

Katz: I must make a statement! Herwig wants to be on record for this meeting for what he says. I want to be on record to strongly support him in everything he has said here.

Varma: What Herwig is saying is probably not entirely correct, because he goes back and uses his parallel processor computers and so on and so forth and produces a track structure and pops up with a microdosimetric distribution. Nobody here is suggesting that you use the absorbed dose. I am saying that we try to find the parameter that helps in understanding biology. The advancement has been to determine energy deposition, statistical distributions of energy deposition in small sites. I am not suggesting that we continue to use absorbed dose as we have in the past. You want to find out if it might be the absorbed dose in small volumes of 10 nanometers or so. It could be the number of ionizations that are produced in that small site; I don't know. What I am suggesting is that since we have that knowledge, we should try to analyze this data in terms of such parameters.

Inokuti: But how do you know if these parameters suffice?

Varma: I assume it must be a better parameter than absorbed dose!

Inokuti: How do you know?

Katz: Analyze it as long as you like, but we do not know that it is a better parameter.

Mechanistic Models

Stanley B. Curtis[a]

Abstract

Several models and theories are reviewed that incorporate the idea of radiation-induced lesions (repairable and/or irreparable) that can be related to molecular lesions in the DNA molecule. Usually the DNA double-strand or chromatin break is suggested as the critical lesion. In the models, the shoulder on the low-LET survival curve is hypothesized as being due to one (or more) of the following three mechanisms: (i) "interaction" of lesions produced by statistically independent particle tracks, (ii) nonlinear (i.e., linear-quadratic) increase in the yield of initial lesions, and (iii) saturation of repair processes at high dose. Comparisons are made between the various approaches. Several significant advances in model development are discussed; in particular, a description of the matrix formulation of the Markov versions of the repair-misrepair (RMR) and lethal-potentially-lethal (LPL) models is given. The more advanced theories have incorporated statistical fluctuations in various aspects of the energy-loss and lesion-formation process. An important direction is the inclusion of physical and chemical processes into the formulations by incorporating relevant track structure theory (Monte Carlo track simulations) and chemical reactions of radiation-induced radicals. At the biological end, identification of repair genes and how they operate, as well as a better understanding of how DNA misjoinings lead to lethal chromosome aberrations, are needed for appropriate inclusion into the theories. More effort is necessary to model the complex end point of radiation-induced carcinogenesis.

Introduction

There are many effects that ionizing radiation can have at the subcellular/molecular level. Those often mentioned are double- and single-strand breaks in the DNA, DNA base damage and DNA-protein crosslinks. Most of the theories/models that have been developed, however, have chosen the double- (and/or single-) DNA strand break as the most likely lesion leading to reproductive cell death and other end points of interest such as mutation and cell transformation. Perhaps because of the large amount of experimental data available using reproductive cell death as an end point under many different physiological conditions and with many different cell types, the majority of models developed have chosen this end point as their focus. It is important to recognize that this end point is crucial, not only in fields such as radiation oncology where selective cell killing is the goal, but also in the field of radiation risk assessment where damaged but surviving cells are critical. Cell killing by radiation will modify the number of cells "at risk" for a transformation event and must be a part of any complete theory of radiation carcinogenesis. This is particularly true in the case of high LET radiation carcinogenesis where the probability is not negligible that one

---

(a) Cell and Molecular Biology Division, Lawrence Berkeley Laboratory, University of California at Berkeley, Berkeley, California

Physical and Chemical Mechanisms in Molecular Radiation Biology
Edited by W.A. Glass and M.N. Varma, Plenum Press, New York, 1991

traversal of a high-LET particle through the nucleus of a cell will kill it. For this reason, and because there are several relatively recent new ideas and developments involving this end point, the present review will emphasize the end point of cell reproductive death.

Good groundwork has been laid by the previous paper (Braby). The background has been covered and many different approaches have been discussed. It should be noted that although the present paper, as the title suggests, deals with mechanisms, the division between mechanistic and phenomenological models is tenuous at best, and several of the approaches discussed here have phenomenological aspects. We will focus on different ideas behind the shape of the dose-response curve (survival curve) as suggested by mechanisms based on effects to the DNA molecule.

Such mechanisms can be divided into three general types: those involving *interaction of lesions,* those requiring a *nonlinear induction of lesions*, and those involving *saturation of repair processes.*

Lesion Interaction Formulations

All interaction formulations have as a basis the idea that lesions or "sublesions" (i.e., sites of damage) can interact with other lesions to produce a lesion that ultimately leads to cell death. Such ideas can be traced originally to the ideas and quantitative description put forth by Lea and Catcheside[1] of pieces of chromosomes misjoining to produce chromosomal aberrations.

Theory of Dual Radiation Action (TDRA)[2,3]

This formulation arose out of a recognition (i) that it might well be important to include the stochastics of the energy-loss process in a theoretical treatment of radiation effects and (ii) that the RBE of neutrons appeared to vary as the reciprocal of the square-root of the neutron dose for many different end points. The basic hypotheses are as follows:

1. Radiation effects in autonomous cells (i.e., cells responding independently of radiation received to other cells) and in particular, reproductive cell death, are due to lesions whose production is proportional to the square of the specific energy, z, deposited in the radiosensitive volume, called here the radiosensitive matrix.

2. These lesions are caused by pairwise interaction of molecular products termed "sublesions," whose production is proportional to z.

The specific energy, z, is a random variable and so the effect (or mean number of lesions) over a population of cells is written

$$\bar{\epsilon} = k \overline{z^2} \tag{1}$$

It can be shown[2] that

$$\overline{z^2} = \overline{z_{1D}} \, D + D^2 \tag{2}$$

where $\overline{z_{1D}}$ is the dose-mean specific energy for a single event. In the 1972 paper, it is called the "energy average" of the event size. The expression for the mean number of lesions becomes

$$\bar{\epsilon} = k(\overline{z_{1D}}\, D + D^2) \tag{3}$$

This form of the model was used to permit the application of microdosimetric data, which provide values of $\overline{z_{1D}}$ experimentally for microscopically defined volumes of different sizes. With the assumption that $\bar{\epsilon}$ was a measure of the mean number of lethal lesions in a cell, the survival was written as the zero class of a Poisson distribution, i.e., the probability that the cell has no lesions,

$$S = e^{-\bar{\epsilon}} \tag{4}$$

Comparison with experimental data showed good agreement if the critical volume had a characteristic size on the order of one or a few micrometers. This interpretation implies a site size and a constant interaction probability of sublesions produced inside the site.

The more generalized version[3] addressed the question of a probability of sublesion interaction dependent on the separation of the sublesions. The formula for $\bar{\epsilon}$ becomes

$$\bar{\epsilon} = k(\xi D + D^2) \tag{5}$$

$$where\ \ \xi = \int_{o}^{\infty} \frac{s(x)\, g(x)\, t(x)\, dx}{4\pi \rho x^2} \Big/ \int_{o}^{\infty} s(x)\, g(x)\, t(x)\, dx \tag{6}$$

Here $s(x)/V$ is the point-pair distribution of distances between points inside the sensitive matrix of volume V where energy transfers can result in sublesions; $g(x)$ is the probability of interaction of two sublesions separated by a distance x; and $t(x)$, the proximity function, is the point-pair distribution of energy-transfer points in a single event (i.e., a passage of an ionizing particle) weighted by the energy transferred. The quantity ρ is the density of the sensitive matrix.

As pointed out by Kellerer and Rossi,[3] two single-strand breaks, each caused by independent electron tracks, cannot be the sublesions leading to the lesion of a double-strand break relevant for cell killing, because at the doses experimentally found to cause the shoulder on the survival curve, the probability is negligible that energy transfers from two statistically independent tracks will each cause a single-strand break on opposite strands of the DNA molecule close enough to produce a double-strand break. More to the point would be two double-strand breaks, representing, say, a chromatin break, interacting to produce a lethal chromosome aberration.

Recently, Brenner[4] has carried the idea a step farther by explicitly assuming that the sublesions are double-strand breaks and the resulting lesions are exchange-type chromosomal aberrations. Also, the survival equation $S = e^{-k\bar{n}}$ with $\bar{n}$ being the mean yield of lesions is only true at low dose and LET, when $e^{-k\bar{n}}$ can be approximated by $1 - k\bar{n}$. Therefore, he considered a cell-by-cell approach to be more appropriate. He simulated passage of radiation tracks through individual cells and has followed the production and interaction of double-strand breaks in time within each cell. Cell viability based on the number of exchange-type aberrations found in the cell was determined at an "appropriate" time. He used 10^4 sec as the time of assay for viability (this is five times his assumed characteristic time for double-strand break repair).

Another assumption in this approach is that one in 1500 ionizations along the track produces a double-strand break. The double-strand breaks were allowed to diffuse stochastically, and interactions were scored if two were found within the "encounter radius," "a," assumed to be 0.5 nm. It was first checked whether either of the two breaks had repaired via a process with characteristic repair time of 2000 sec. The probability for reaction was taken as $p(x,t) = a/x \; erfc \left[(x-a)/\sqrt{4Dt} \right]$, where x is the original separation distance, and Dt was taken as $12 \times 10^4 \; nm^2$. This gave a reasonable fit to the g(x) function of Eq. (6) as deduced from experiment. Of the exchange-type aberrations left at 10^4 sec, one out of two was assumed to be lethal based on symmetry arguments. Good agreement between calculated and experimental survival curves was obtained for V-79 cells exposed to X-rays and ions with LETs from 20 to 170 keV/μm.

Repair-Misrepair Model (RMR)[5]

Initial ("uncommitted") lesions are formed which can subsequently either interact pairwise to form a lethal lesion (quadratic misrepair), or misrepair on their own (linear misrepair), or repair correctly (linear eurepair).

It is convenient (though not essential) to assume that the initial uncommitted lesions are formed linearly with absorbed dose:

$$U_o = \alpha D \tag{7}$$

The kinetic differential equation describing the time-rate of change of the U lesions is

$$\frac{dU}{dt} = -\lambda U(t) - \kappa U^2(t) \tag{8}$$

where the linear term can be divided into two terms, $\lambda \phi \, U(t)$ and $\lambda(1-\phi) \, U(t)$, the eurepair and misrepair terms, respectively. The parameter ϕ gives the fraction of linear repair that is correct.

The solution to this equation is

$$U(t) = U_o e^{-\lambda t} / \left[1 + \frac{U_o}{\epsilon} (1 - e^{-\lambda t}) \right] \tag{9}$$

with initial condition $U(0) = U_0$ and where $\epsilon = \lambda/\kappa$.

Lethal lesions are defined as the sum of the "uncommitted" plus linearly misrepaired plus quadratically misrepaired lesions at time t:

$$N_{lethal}^{(t)} = U(t) + (1 - \phi)\lambda \int_o^t U(t')dt' + \kappa \int_o^t U^2(t')dt' \tag{10}$$

If N_{lethal} is assumed to be the mean number of lethal lesions per cell of a Poisson distribution, the survival expression can be written:

$$S(t) = e^{-N_{lethal}(t)} \tag{11}$$

Making the appropriate substitutions, this yields

$$S(t) = e^{-\alpha D}\left[1 + \frac{\alpha D}{\epsilon}(1 - e^{-\lambda t})\right]^{\epsilon\phi} \tag{12}$$

This is the main result of the model. A subsequent change in parameter designation (δ replacing α and μ replacing $\epsilon\phi$) yields

$$S(t) = e^{-\delta D}\left[1 + \frac{\delta\phi D}{\mu}(1 - e^{-\lambda t})\right]^{\mu} \tag{13}$$

The mechanistic interpretation of this model is that the initial uncommitted lesions are double-strand or chromatin breaks. Quadratic misrepair is due to separate broken strands rejoining ("interacting"), while linear misrepair is due to a single double-strand break misjoining in a way that is incompatible with cell survivability. The initial slope on the survival curve, assuming long repair times, is due to the infidelity of the linear repair process.

The space and time dependence of lesion formation (i.e., LET and dose-rate effects) has also been addressed.[5]

Recently, Albright[6] has formulated lesion repair and misrepair in the RMR model as a Markov process, a discrete sequence of repair steps occurring at random times. This allows dropping the approximations of (i) neglecting the effect of statistical fluctuations in the repair process, and (ii) assuming the final lethal lesion distribution among cells to be Poisson.

Lethal-Potentially-Lethal (LPL) Theory[7,8]

This approach builds on the TDRA and RMR formulations and was originally presented as a unified repair theory[9] joining several of the ideas from these approaches, plus the irreparable-repairable lesion concept found in the cybernetic model.[10]

The theory is based on three hypotheses:

1. Two broad classes of lesions are produced by ionizing radiation: irreparable and repairable. They are distinguishable by the amount of energy that must be deposited locally in order to produce them.

2. Irreparable (or "lethal") lesions are formed linearly with increasing absorbed dose with proportionality constant η_L.

3. Repairable (or "potentially lethal") lesions may be separated into at least two categories. Each is formed linearly with increasing absorbed dose, with proportionality constants η_{PL} and $\eta_{PL'}$. They are distinguishable by their different repair rates. The rapidly repairing lesions, repairing with rate $\epsilon_{PL'}$ ($\sim$ 3 to 6/hr), are not usually expressed because they are normally repaired with high fidelity and so are not experimentally measured in most experiments. The slowly repairing lesions, repairing with rate ϵ_{PL} ($\sim$ 0.5 to 1.0/hr), can be "fixed," i.e., made lethal, at various points throughout the cell cycle, or can "interact" with each other to form an irreparable (lethal) lesion, with rate $2\,\epsilon_{2PL}$ ($\sim$ 0.05 to 0.15/hr). The latter process is called "binary misrepair."

The molecular mechanisms are not specified. The values of the kinetic parameters ϵ_{PL} and ϵ_{PL}', however, are consistent with the repair rates of slowly and rapidly repairing components of double-strand breaks, respectively.

The above hypotheses lead to the following differential equations:

- During irradiation

$$\frac{dn_{PL}(t)}{dt} = \eta_{PL}\dot{D} - \epsilon_{PL}\,n_{PL}(t) - \epsilon_{2PL}\,n_{PL}^2(t) \tag{14}$$

$$\frac{dn_L(t)}{dt} = \eta_L\dot{D} + \epsilon_{2PL}\,n_{PL}^2(t) \tag{15}$$

Here $\dot{D}$ is the dose-rate.

- After irradiation

$$\frac{dn_{PL}(t)}{dt} = -\epsilon_{PL}\,n_{PL}(t) - \epsilon_{2PL}\,n_{PL}^2(t) \tag{16}$$

$$\frac{dn_L(t)}{dt} = \epsilon_{2PL}\,n_{PL}^2(t) \tag{17}$$

If we assume the "high dose-rate approximation," i.e., repair occurring during the irradiation can be neglected, we have just Eqs. (16) and (17) with initial conditions: $n_{PL}(0) = \eta_{PL}D$ and $n_L(0) = \eta_L D$.

The solution of these equations is straightforward.[8] With the assumption that the lesions are distributed among cells in a Poisson distribution and that the number of lethal lesions is the sum of the initially lethal and potentially lethal lesions at time, t_r, the survival can be written as the zeroeth class of the Poisson distribution, i.e., the probability that no lethal lesions exist in the cell:

$$S(t_r) = e^{-n_{PL}(t_r) - n_L(t_r)}$$

$$= e^{-(\eta_L + \eta_{PL})D}\left[1 + \frac{\eta_{PL}D}{\epsilon}\left(1 - e^{-\epsilon_{PL}t_r}\right)\right]^\epsilon \tag{18}$$

where $\epsilon = \epsilon_{PL}/\epsilon_{2PL}$.

It can easily be shown[7] that at low dose, the above equation approximates a linear-quadratic expression:

$$S = e^{-\alpha D - \beta D^2} \tag{19}$$

where

$$\alpha = \eta_L + \eta_{PL} e^{-\epsilon_{PL} t_r} \tag{20}$$

$$\beta = \frac{\eta_{PL}^2}{2\epsilon} \left(1 - e^{-\epsilon_{PL} t_r} \right)^2 \tag{21}$$

and, for low dose-rate and long repair time, t_r,

$$\beta = \frac{\eta_{PL}^2}{2\epsilon} \cdot \frac{2}{(\epsilon_{PL} T)^2} (\epsilon_{PL} T + e^{-\epsilon_{PL} T} - 1) \tag{22}$$

where T is the irradiation time. This is the "dose protraction" factor found in several of the other approaches.

A recent development of the LPL theory has included a Markov formulation[8] for the high dose-rate version patterned after that obtained for the RMR Markov formulation.[6] The conclusion of the reformulation is that for parameter values relevant to experimental survival curves, the error in assuming the Poisson rather than the Markov formulation leads to a 30% decrease in survival at 10 Gy absorbed dose. Such a difference would be very difficult to observe experimentally.

Some of the explicit predictions of this formulation for stationary-phase cells are as follows:

- For low dose-rates, the survival curve will have a constant slope and the shape will be independent of dose-rate.
- For high dose-rates, the survival curve will be independent of dose-rate.
- The initial slope of the delayed plating curve at high dose-rate will equal the slope of the low dose-rate curve (and be equal to η_L).
- The initial slope of the survival curve of a proliferating cell population is dependent on the amount of time available for repair.
- The slope of the survival curve will approach a constant ($\eta_L + \eta_{PL}$) as the dose increases.
- Repair of sublethal damage and repair of the slow component of potentially lethal damage are different manifestations of the repair of the same lesions.
- At low LETs, potentially lethal lesions dominate; at high LETs, lethal lesions dominate.

Nonlinearity of Initial Lesion Yield

The following two approaches assume that the shoulder on the survival curve is the reflection of the nonlinear yield of DNA double-strand breaks.

Critical DNA Target Size Model[11]

In this model, it is hypothesized that DNA double-strand breaks are the critical lesions and that the dose response is nonlinear due to the action of a saturable chemical repair process. Only double-strand breaks occurring within "critical targets" are important, but these initiate recombination events with undamaged sequences, which lead to chromosomal aberrations. The subsequent loss of acentric fragments at mitosis prevents continuity of the genome and leads to cell death by inducing structural changes in the chromatin. Radford cites his own data on the yield of double-strand breaks[12-14] to support his contention that the yield is nonlinear at doses where a shoulder in the survival curve is seen.

One interesting assumption is that there are a number of critical targets of length X_c along the DNA molecule that must remain intact for the cell to survive. This arises from the suggestion coming from experimental data that the number of lethal events is directly proportional to the number of DNA double-strand breaks per unit length of DNA. One problem with this assumption is that X_c appears to be dependent on the radiation type.

The targets are assumed to be sites of chromosome instability: the constitutive or common fragile sites (c-fra), and the proto-oncogenes occurring in "light G-bands" of stained chromosomes. In particular, the c-abl, bcl-1, bcl-2, c-myc, and blym-1 genes are identified as being highly susceptible to radiation-induced breakage.

Since the non-linearity in this model occurs in the DNA double-strand break yield, the molecular mechanism for aberration production is presumed to be the damaged site initiating a recombination event with undamaged sequences. Asymmetrical exchanges would lead to cell death while symmetrical exchanges could lead to transformation. Double-strand breaks occurring in nontarget sequences are assumed to be repaired by a ligation-type mechanism and are not normally lethal to the cell. It is concluded that repair of the vast majority of radiation-induced DNA double-strand breaks is irrelevant to an understanding of cell killing in normal mammalian cells.

Finally, it is hypothesized that cell death is due to the presence of chromosomal aberrations, particularly acentric fragments. The suggestion is that, in the interphase mammalian nucleus, chromosomal DNA is continuous; i.e., the chromosomes are joined together. The existence of a fragment would prohibit the postulated proper interconnection because one chromosome would be missing a telomere. This produces long-range perturbation of chromatin structure and resultant changes in genetic activity, which result in cell death. Such a mechanism is described as "karyotypic discontinuity."

Apparently, these ideas have not been quantified to the extent that a survival expression has been derived or deduced.

Molecular Theory[15]

The unique aspect of the Molecular Theory is that single- and double-strand breaks in the DNA were hypothesized from the beginning as being the important molecular lesions. The survival curve shoulder is suggested to reflect the double-strand break yield curve as a function of absorbed dose. The hypotheses are as follows:

1. Single-strand breaks are produced linearly with dose.

2. Double-strand breaks are produced when two single-strand breaks are produced in close proximity.

3. The production of lethal lesions is proportional to the double-strand break yield.

The result of the formulation is a yield of double-strand breaks having a linear component due to two closely produced single-strand breaks by a single charged particle, and a quadratic component due to two closely produced single-strand breaks by two statistically independent charged particles. The survival equation is written

$$S = e^{-p(\alpha D + \beta D^2)}$$ (23)

As pointed out,[3] in the dose range below 10 Gy, microdosimetric arguments rule out the possibility that two single-strand breaks produced by two statistically independent tracks could yield the experimentally measured number of double-strand breaks. Also, in this range of doses, experimental evidence appears to favor a linear dependence of initial double-strand break yield with absorbed dose.

DNA Fragment Loss and Unrepaired Double-Strand Breaks[16]

This model assumes that cell death is caused by the presence of an unrepaired double-strand break (including DNA fragments) at certain points in the cell cycle. It is concluded that each unrepaired double-strand break becomes a chromosomal aberration.

The survival expression is written:

$$S = \left[e^{-y} \sum_{k=0}^{\infty} (y^k/k!)\, \mu^{(k_1 - 1)} P(T) \right]^N$$ (24)

where N is the number of DNA molecules/nucleus, $y = D/D_{dsb}$ with D_{dsb} = the mean dose for the induction of one double-strand break per molecule, μ is the probability that a DNA fragment will remain in the chromosome, $k_1 = k+1$ for linear DNA and $k_1 = k$ for circular DNA where k is the number of double-strand breaks per molecule, $P(T) = 1$ for $k = 0$, $P(T) = (1 - e^{-T})^k$ for $k > 0$ for repair processes where each double-strand break is repaired independently of other double-strand breaks, and $P(T) = 1 - e^{-T}$ for repair processes for which, if one break is repaired, all breaks on that molecule are repaired. $T = T_{rep}/\tau_{dsp}$ where T_{rep} is the allowable repair time and τ_{dsb} is the mean time for double-strand break repair. In this expression, a sum is taken over k of the product of the (Poisson) probability of inducing k double-strand breaks times the probability that the fragment will remain in the chromosome times the probability that all k breaks in a molecule will be repaired by a time T_{rep}, and the whole expression is raised to the Nth power since there are N DNA molecules per cell nucleus.

A low dose-rate approximation is derived and shows exponential decrease with dose. Repair of potentially lethal damage is addressed, and expressions are given for the (final) number of unrepaired double-strand breaks and the (final) number of prematurely condensed chromatin fragments (PCC fragments). The interpretation is made that the final number of PCC fragments consists of two groups: those separated from DNA molecules soon after irradiation and those separated at the cutoff time for repair. All unrepaired double-strand breaks at the cutoff time become chromosome aberrations. A fraction yields DNA fragments; the rest interact to yield "misrepaired" chromosome aberrations.

The difference between radioresistant and radiosensitive cell lines lies in the relative amount of repair time available for double-strand break repair: long repair times (relative to the characteristic double-strand break repair time) lead to radioresistant cells, and short repair time leads to a high level of "misrepair"-type chromosome aberrations and radiosensitive cells.

Repair Saturation

Many repair saturation models have been presented in various stages of development over the last decade, and several were reviewed by Braby at this conference. Two models that relate to DNA damage and repair are reviewed briefly here.

DNA Repair and Metabolic States[17]

In this approach, it is suggested that cell survival is related to the difference in rates of DNA lesion production and removal as mediated by the rates of metabolic processes required for maintaining cell integrity. It is specifically stated that the presence of residual unrepaired or misrepaired DNA lesions is not required to produce cell death. The probability of cell survival is written:

$$P(S) = 1 - \int_o^t [L(t) - R(t)] P_D(z) \, dt \qquad (25)$$

where
$L(t)$ = number of DNA lesions produced per unit time
$R(t)$ = number of DNA lesions removed per unit time
$P_D(z)$ = probability of death/lesion/unit time given a specific metabolic state, z.

The idea of saturation of repair at high doses is introduced with the assumption that as the dose increases, the number of DNA lesions produced eventually exceeds the number of repair complexes available to remove the damage and the velocity of repair becomes a constant. In this situation, the half-time for damage removal will increase with dose. Experimental data have been presented that purportedly support this idea, but analyses of data sets have not been quantitatively carried out explicitly using the equation above.

Damage Accumulation-Interaction[18]

The last model to be reviewed here is one in which the repair of damage relevant to cell survival is caused by a repair process that saturates at doses as low as 2 Gy. Interaction occurs among the accumulated (repairable) lesions during irradiation, producing irreparable (lethal) lesions. The model is based on three hypotheses:

1. Lesions induced during irradiation either interact at the time of formation or do not interact at all and thus remain repairable if the milieu does not alter DNA conformation.

2. Repair (in V-79 cells) is not affected by postirradiation medium-dependent cell cycle progression or its delay.

3. Repair of repairable damage occurs in cells both with and without irreparable damage but is not detected by survival assays in cells with irreparable damage.

The unique aspect of this model appears to be the assumption that damage can interact only during the irradiation period. Experimental evidence in favor of this assumption is lacking at present.

The lesions suggested as being relevant for this model are breaks in chromatin and DNA which underlie the formation of chromosome aberrations. This model agrees with the preceding model[17] in that the repair process is unsaturated at low dose and is saturated at high dose (but in this model it is claimed that saturation occurs above 2 Gy for V-79 cells irradiated at 2.5 Gy/minute dose-rate).

Explicit survival equations for this model have evidently not been published.

Comparison of the Models/Theories

A comprehensive comparison of the theories and models chosen for review would be too lengthy for inclusion in this chapter. A concise comparison of the mechanism(s) hypothesized for the production of the shoulder on the survival curve is informative, however. Table I presents the mechanisms responsible for the shoulder of a low LET survival curve as envisaged by the developers of the formalisms under discussion. We see that even though DNA is assumed to be the critical target in all formulations and most assume that double-strand breaks or chromatin breaks leading to chromosomal aberrations are the important molecular lesions involved, there is still room for a varied interpretation of the "real" reason for a shouldered survival curve.

Table I. Cause of Shoulder on the Low LET Survival Curve

Theory/ Model	"Interaction" by Lesions Produced from Statistically Independent Tracks	Nonlinear Production of Initial Lesions	Saturation of Repair Processes	Other	Reference
TDRA	√				Kellerer & Rossi 1972, 1978[2,3]
RMR	√				Tobias et al. 1980,[5] Tobias 1985[19]
LPL	√				Curtis, 1986,[7] 1988,[8] 1983[9]
Molecular Theory		√			Chadwick and Leenhouts 1981[15]
Critical DNA Target Size		√	√		Radford et al. 1988[11]
DNA Fragment Loss and Unrepaired Double-Strand Breaks				[Dose-dependent probability of fragment loss from chromosomes]	Ostashevsky 1989[16]
DNA Repair and Metabolic States			√		Wheeler 1987[17]
Damage Accumulation Interaction	√		√		Reddy et al. 1990[18]

One important task for the future is to provide strong experimental evidence for or against at least the three mechanisms mentioned in the table: interaction of lesions, nonlinear yield of initial lesions, and saturation of repair processes. This, of course, assumes that the "lesions" have been identified. At present, the experimental evidence in the dose range below 6 Gy appears to favor interaction of lesions, since it is well established that broken pieces of chromatin/DNA do "misjoin" to form chromosomal aberrations which are lethal to the cell. In this dose range, the evidence is considerably weaker that there is nonlinearity in the production of chromatin/DNA double-strand breaks or that the repair processes are saturating. A definitive paper has just appeared[20] showing that, for the neutral-elution technique of determining double-strand breaks, the nonlinear (e.g., linear-quadratic) dose response of the fraction of eluted DNA (or the logarithm of the fraction retained on the filter) does not necessarily imply nonlinear induction of double-strand breaks with absorbed dose.

Because the DRA, RMR and LPL formulations all hypothesize the same mechanism for the shoulder on the survival curve, it is informative to show explicitly the differences in these approaches. This is presented in Table II. The sublesions envisaged in DRA cannot be lethal individually but need an interaction with another sublesion to become lethal. This is not necessary in either the RMR or LPL approaches. Fixation of individual (sub)lesions can take place. This indeed is assumed to occur in experiments in which repair inhibitors, such as ara-A or hypertonic solution, are added to the cell medium. The difference between the RMR and LPL formulations is mainly that one type of "uncommitted" lesion is postulated in the RMR approach and the initial slope on the survival curve is due to the infidelity of the repair process that causes individual lesions to misrepair (self-misrepair). In the LPL approach, the initially lethal irreparable lesions (requiring more local energy deposition for their formation) are created linearly with dose and cause the initial slope on the survival curve.

Table II. Differences Between DRA, RMR & LPL Models

Model	Cell Survival End Point				
	Self-Repair	Self-Misrepair	Binary Repair	Binary Misrepair	Fixation
DRA	√			√	
RMR	√	√	(√)	√	√
LPL	√			√	√

(√) = present, but has negligible effect on cell survival.

Present and Future Directions

Clearly, this is a very active field with many interesting directions. Already mentioned is the need for definitive evidence in the dose region below 6 Gy for or against interaction of (sub)lesions, nonlinearity in the production of (sub)lesions with absorbed dose, and saturation of processes that repair the (sub)lesions. Several other significant directions for biophysical modeling have been discussed recently.[21] Briefly, it was concluded that in order to develop a complete understanding of the effects of ionizing radiation on a population of proliferating and non-proliferating cells, attention must be paid to (i) the statistical nature of the energy loss process through the concepts of microdosimetry and track structure theory, (ii) the fluctuations of lesion formation from cell to cell throughout the population and (iii) the variation of radiosensitivity throughout the cell cycle when proliferating cells are being considered.

Particularly interesting is the suggestion made recently by a joint DOE/CEC working group to compare various models by applying each to the analysis of the same sets of experimental data of one or more of three specific end points (cell transformation of C3H 10T1/2 cells, chromosomal aberrations and mutations, and interaction of mixed radiations in cell survival experiments). A workshop scheduled in 1991 will bring the analyses together for comparison and evaluation.

Exciting progress is being made in extending the RMR and LPL models to include low dose-rate into the Markov formulations.[22] In fact, this approach is a very general and elegant way of incorporating fluctuations into different theoretical descriptions. Briefly, the approach introduces a matrix formulation in which each matrix element relates to a differential equation describing the rate of change of lesions in a single cell. For example, Eqs. (14-17) above are replaced by individual probabilistic differential equations for the rate of change of the probability $P_n(t)$ of one cell having a given number of potentially lethal lesions, n. We will refer to a cell with n potentially lethal lesions as being in the nth state. The equations in matrix element notation are

$$\frac{dP_n(t)}{dt} = \sum_k M_{nk} P_k(t) \quad n,k = 0,1,... \tag{26}$$

where n is the number of potentially lethal lesions in a cell at time, t, and k denotes the sources [i.e., states that are involved in the movement of lesions into (or, when k=n, out of) the state of n lesions].

The M_{nk} are matrix elements that correspond to transitions from the kth state to the nth state or (when k=n) out of the nth state.

The form of the matrix M is

$$M = \dot{D}/z_F R + A \quad \textit{during irradiation} \tag{27}$$

$$M = A \quad \textit{after irradiation} \tag{28}$$

Here $\dot{D}$ is the dose-rate and z_F is the average specific energy per event which relates the energy deposition (i.e., the dose) in the nucleus to the number of events.

The matrix A contains the transition rates which involve the repair and misrepair processes and are operative both during and after irradiation:

$$A_{nn} = -n\epsilon_{PL} - n(n-1)\,\epsilon_{2PL} \quad \textit{for } k = n \tag{29}$$

$$A_{n,n+1} = (n+1)\,\epsilon_{PL} \quad \textit{for } k = n+1 \tag{30}$$

The biological interpretation of these equations is that the number of cells in the nth state is changing in two ways: first by cells moving out of the state by "correct" repair (n lesions can be repaired, which occurs at a rate ϵ_{PL} per lesion per unit time) and by "incorrect" binary misrepair (there are 1/2 n(n-1) pairwise interactions that can take place which occur at a rate $2\epsilon_{2PL}$ per lesion^2 per unit time), and, secondly, by cells moving into the state from the $(n+1)^{th}$ state with lesions being repaired with a rate ϵ_{PL} per unit time.

The use of just the A matrix (R = 0) yields the results already reported in Curtis[8] for the LPL theory and Albright[6] for the RMR theory. This is valid for dose-rates large enough and exposure times small enough so that repair during the radiation time can be neglected.

For lower dose-rates and long exposure times, lesions are being produced by the radiation and repaired by the cell during the same time period, so both processes must be included in the matrix. Thus, the R matrix elements, which relate to the production of lesions by the radiation, are introduced:

$$R_{nn} = -\mu - \nu \quad where \quad \mu = \sum_k \mu_k \tag{31}$$

$$R_{n,n-k} = \mu_k \tag{32}$$

where μ_k is the probability of producing k potentially lethal lesions in one event, and ν is the probability of producing one or more lethal lesions in one event. This approach expresses the irradiation process as a series of events which, in turn, produce lesions. This appears to be a convenient and conceptually appropriate manner with which to describe the irradiation process, irrespective of the value of the dose-rate.

Written in matrix notation, where dP/dt on the left hand side of the equation is a one-row matrix and P on the right hand side is a one-column matrix:

$$dP(t)/dt = MP(t) \tag{33}$$

the formal solution is

$$P(t_r) = e^{Mt_r}P(T) \tag{34}$$

where

$$P(T) = e^{MT}P(0) \tag{35}$$

and where the matrix M is the appropriate one applying during the radiation [M is taken from Eq. (27) for Eq. (35)] or after the irradiation [M is taken from Eq. (28) for Eq. (34)].

The probability for a cell to survive is the probability of a cell having no potentially lethal or lethal lesions at t_r, the time available for repair after irradiation. It is given by

$$S = P_0(t_r) \tag{36}$$

where t_r is the time post-irradiation after which no repair can take place and the fate of the cell has been determined, and subscript zero indicates no lesions are present.

The overall approach described here provides the capability for calculating statistical distributions of potentially lethal and lethal lesions in a cell population at various times after radiation. Such distributions may ultimately be used to compare theoretical predictions with experimental distributions of "candidate" lesions (e.g., chromatin breaks). Another advantage of this formulation is that various other hypotheses can be accommodated, such as saturable repair, multi-target and multi-hit effects, certain aspects of the dual-radiation-action hypothesis, and so on. Two additions of particular interest are cell-cycle effects and the LET-dependence of the probability of potentially lethal and lethal lesion formation per event.

Finally, end points such as transformation, mutation and tumorigenesis should be addressed within the framework of the statistical fluctuations discussed above. The concept that carcinogenesis does not arise from autonomous cells should be taken seriously. This idea confounds the development of conceptually simple theories, but, if true, it cannot be neglected in our ultimate understanding of radiation-induced carcinogenesis.

Acknowledgments

This work was supported by the U.S. Department of Energy under Contract No. DE-AC03-76SF00098.

References

1. D. E. Lea and D. G. Catcheside. The Mechanism of the Induction by Radiation of Chromosome Aberrations in *Tradescantia, J. Genet.* 44:216-245 (1942).
2. A. M. Kellerer and H. H. Rossi. The Theory of Dual Radiation Action. *Curr. Top Radiat. Res.* Q.8:85-158 (1972).
3. A. M. Kellerer and H. H. Rossi. A Generalized Formulation of Dual Radiation Action, *Radiat. Res.* 75:471-488 (1978).
4. D. J. Brenner. Track Structure, Lesion Development and Cell Survival. *Radiat. Res.* Vol. 124:529-537 (1990).
5. C. A. Tobias, E. A. Blakely, F.Q.H. Ngo, and T.C.H. Yang. The Repair-Misrepair Model of Cell Survival, in: *Radiation Biology and Cancer Research* (R. E. Meyn and H. R. Withers, Eds.), pp. 195-230, Raven, New York (1980).
6. N. Albright. A Markov Formulation of the Repair-Misrepair Model of Cell Survival. *Radiat. Res.* 118:1-20 (1989).
7. S. B. Curtis. Lethal and Potentially Lethal Lesions Induced by Radiation - A Unified Repair Model. *Radiat. Res.* 106:252-270 (1986).
8. S. B. Curtis. The Lethal and Potentially Lethal Model - A Review and Recent Development. In: *Quantitative Mathematical Models in Radiation Biology* (J. Kiefer, Ed.), pp. 137-148, Springer-Verlag, Berlin, Heidelberg (1988).
9. S. B. Curtis. Ideas on the Unification of Radiobiological Theories. In: *Radiation Protection: Proceedings of the Eighth Symposium on Microdosimetry*, Jülich, pp. 527-536, Commission of the European Communities, Luxembourg, 1983.
10. A. Kappos and W. Pohlit. A Cybernetic Model for Radiation Reactions in Living Cells. I. Sparsely Ionizing Radiations; Stationary Cells. *Int. J. Radiat. Biol.* 22:57-65 (1972).
11. I. R. Radford, G. S. Hodgson, and J. P. Matthews. Critical DNA Target Size Model of Ionizing Radiation-Induced Mammalian Cell Death. *Int. J. Radiat. Biol.* 54:63-79 (1988).
12. I. R. Radford. The Level of Induced DNA Double-Strand Breakage Correlates with Cell Killing After X-Irradiation. *Int. J. Radiat. Biol.* 48:45-54 (1985).
13. I. R. Radford. Evidence for a General Relationship Between the Induced Level of DNA Double-Strand Breakage and Cell Killing After X-Irradiation of Mammalian Cells. *Int. J. Radiat. Biol.* 49:611-620 (1986).
14. I. R. Radford. Effect of Radiomodifying Agents on the Ratios of X-ray-Induced Lesions in Cellular DNA: Use in Lethal Lesion Determination. *Int. J. Radiat. Biol.* 49:621-637 (1986).

15. K. H. Chadwick and H. P. Leenhouts. *The Molecular Theory of Radiation Biology*, Springer-Verlag, New York (1981).

16. J. Y. Ostashevsky. A Model Relating Cell Survival to DNA Fragment Loss and Unrepaired Double-Strand Breaks. *Radiat. Res.* 118:437-466 (1989).

17. K. T. Wheeler. A Concept Relating DNA Repair, Metabolic States and Cell Survival After Irradiation. In: *Proceedings of the Eighth International Congress of Radiation Research*, Edinburgh, July 1987, Vol. 2 (E. M. Fielden, J. F. Fowler, J. H. Hendry, D. Scott, Eds.), pp. 325-330, Taylor and Francis, London, 1987.

18. N.M.S. Reddy, P. J. Mayer, and C. S. Lange. The Saturated Repair Kinetics of Chinese Hamster V79 Cells Suggests a Damage Accumulation-Interaction Model of Cell Killing. *Radiat. Res.* 121:304-311 (1990).

19. C. A. Tobias. The Repair-Misrepair Model in Radiobiology: Comparison to Other Models. *Radiat. Res.* 104:S77-S95 (1985).

20. D. Blöcher. Dose Response in Neutral Filter Elution. *Radiat. Res.* 123:176-181 (1990).

21. E. J. Hall, M. Astor, J. Bedford, C. Borek, S. B. Curtis, R.J.M. Fry, C. Geard, T. Hei, J. Mitchell, N. Oleinick, J. Rubin, A. Tu, R. Ulbrich, C. Waldren, and J. Ward. Basic Radiobiology. *Am. J. Clin. Oncol (CCT)* 11(3):220-252 (1988).

22. R. K. Sachs, L. Hlatky, P. Hahnfeldt, and P.-L. Chen. Incorporating Dose-Rate Effects in Markov Radiation Cell-Survival Models. *Radiat. Res.* 124:216-226 (1990).

Discussion

Varma: I really appreciate the review of the models, but these models are on cell survival and killing, and we should consider whether we may want to change. The development of these cell survival models, in the past, has been a milestone in this program. It seems to me that, with respect to what Eric has said in his talk, we may have a particular end point that people can look at that doesn't deal with cell death. It seems to me that the emphasis should be on the development of those models, and I would like you to say something about how that connection is being made, looking at those end points that are, say, relevant to radiation carcinogenesis. I know Aloke has made some progress in that area; he talked about it yesterday, in trying to look at transformation. Should we be continuing to develop models for cell killing?

Curtis: Okay, I will comment on that. First of all, I want to second Les Braby's point that you must include cell survival in a complete theory of carcinogenesis. You need to know how many cells at risk remain alive and so are transformable, particularly for a high-LET radiation. The probability is neither zero nor one that a cell is killed when a high-LET particle goes through the nucleus. I do agree that we should directly address the processes going on in the carcinogenic area, and you mentioned Aloke's work with the C3H/l0Tl/2 cell system. That is a good start. But when I look in this area of transformation, I get somewhat nervous when I consider for instance, data for the 10T1/2 cell. It is a non-diploid cell, as Eric pointed out, and well on the way to transformation already, although in some sense, that may make it simpler because you don't have to worry about as many steps. The fact that there are cells with different numbers of chromosomes within a given cell population may mean that each cell may have quite a different probability for transformation. Also, if you take into account the suggestions of Little, Gould, and Clifton, that radiation produces a generalized non-specific damage that eventually causes a transformation process; one may never be able to identify the specific events that are important that are caused by radiation. That is somewhat frightening from a modeler's point of view.

Hall: I would say that Stan did an outstanding job of summarizing the models, but I too was rather disappointed that there was nothing for transformation or for mutation and noted that in fact it is 19 years since the first dose response curve for trans-formations was published, from our lab as a matter of fact, and it was for a diploid cell line. All the objections you mentioned don't apply. It was for Syrian hamster embryo cells, which were diploid cells where transformation and cell survival could be scored in the same dishes in the same experiments. These data have been available for 19 years and nobody's modeled them yet. That seems to me to be a little bit unfortunate, and I am glad to see that it is going to be remedied in the future. I would strongly suggest that the Syrian hamster embryo data not be neglected. In many ways I think it would be simpler to do than for the 10T1/2 cell system although there are much more data for 10T1/2 cells. But the objections that have been raised for the 10T1/2 cell don't apply to the Syrian hamster embryo. It has been 26 years since the first data were available and 19 since the complete dose response curves were available.

Curtis: That's a point well taken. They are certainly data to look at.

Zaider: I don't think we have to worry too much about modeling survival since, first of all, none of these models is specific for cell killing, and probability of the effect is simply one minus the probability for survival, so there you have your carcinogenesis model. These models are just very general mathematical formulations.

Ward: If you are correct, then we must show that the same types of molecular damage are involved in killing, mutagenesis, and transformation. In addition, if a model includes a repair term, then the repair efficiency and accuracy should be the same for all the biological end points. If we cannot say this, then for different molecular damages different parameters of energy deposition may be important, and different repair processes may be involved.

Varma: I think Marco is saying that you have definitely modeled for cell killing, but this may have correlation with the model for carcinogenesis or double-strand breaks. My question to Stan was, we have a lot of information on double-strand breaks and single-strand breaks. Can you model the single- and double-strand breaks?

Curtis: Aloke, Bill Holley, and John Magee have published a detailed model for that end point (Radiat. Res. 121:161, 1990). As far as transformation is concerned, I thought that we were going to talk about it and about carcinogenesis later on when Moolgavkar brought it up.

Hall: Another point that I wanted to make is that Bob Katz eloquently pleaded that he wanted undigested data, but his concern about digestion only goes to the X-axis. He is perfectly prepared to accept something that is chewed over and digested having to do with the Y-axis, that is, the surviving fraction. Surviving fraction depends on many factors, such as how you stain the cells, how many cells to a colony, and all that, whereas the other endpoints, mutation, and transformation are much less digested.

Curtis: I agree.

Weinstein: I think I understand now what the difficulty was in differentiating mechanistic and phenomenological models. I begin speaking as somebody who is still trying to make out what's what, both in the radiological component and in the cell biology component. What I would expect of a good model is that it allows a challenge of some of the premises that lead to its formulation. We have heard all the trouble that the radiobiologists have to describe the events, the tracks, the energy deposition and whether you should use them at all, and so forth. I wonder if these models tell us anything about solving some of these problems or sorting them out. On the other hand, we just heard a discussion of the key question as far as the cell biology is concerned, "Are the same lesions that produce cell death also the ones that produce any other phenomena that one would decide is important?" If, as Marco said, the models are just mathematical formulations, they have some major shortcomings as far as meeting the requirements of a good model. Another point that Marco made earlier: he demanded of a model that it be falsifiable. In the models you presented, the falsification depends upon comparison with experimental results, a procedure not necessarily lending itself to the immediate identification of false premises.

Curtis: I agree with much of what you say, but I would make two points. First, most of the models I reviewed were specifically developed to address cell killing only, and I do not agree with Marco that they should be considered simply as mathematical formulations usable for other end points as well. Second, some of the models that I presented do make predictions. In fact, several predictions of the LPL model were listed concerning the shapes of survival curves and the types of repair processes, for sublethal and potentially lethal damage, for instance, and whether they are the same or different.

Hall: Don't we already know that some of the assumptions made in some of the models are not correct? For example, the Radford model states that the loss of an acentric fragment leads to cell death, and we know that isn't true. It is not the loss of the acentric fragment. If you have an acentric fragment, then you must have a dicentric chromosome. It is the presence of a dicentric chromosome that kills the cell. Don't we also know that we can lose a large deletion and that doesn't necessarily lead to cell death?

Curtis: That is immediate cell death. But a lot of times the cell dies after several divisions and we don't know quite what it is that causes death there. A large deletion might be enough.

Hall: But we know of established tumor cell lines that have whole chromosomes missing, half an arm missing, that live for years. You know that a large deletion does not necessarily kill them.

Varma: Eric, related to that question, talking about the cell killing, please talk about the gross rearrangement of the chromosome itself. That could also lead to the cell death.

In modeling, is there any indication or any possibility of relating all of that information to the activation of the oncogene?

Curtis: You are making an assumption that the activation of an oncogene is the most important process. I'm not sure that everyone would agree with that. As far as radiation is concerned, people have said that the loss of a suppressor gene may be the most important.

Varma: My question is that with all this knowledge, would we be able to put that in proper perspective to talk about the probability of whatever end point you want to assume, rearrange the suppressor gene, rearrangement of chromosomes, etc? Aloke, can you do it now?

Chatterjee: Yes, in principle you can do this. But we need more experiments where we find the point mutations or the transformations at very low doses.

Varma: There are a lot of data from Berkeley on PCC, premature chromosome condensation, experiments. Why can't they be used to model?

Ward: Some aspects of this are covered in my presentation. To move genes around in DNA, the DNA must be cut. Gene movements would be initiated by DSBs where the ends can separate and subsequently rejoin with the wrong partner. This is something Dr. Curtis has discussed—misrepair. It should be pointed out that even if the DSB end does join with the correct partner, there may still be a point mutation at that site.

Christophorou: The description of the models was a very interesting one. Wouldn't it be nice if you could perhaps give an assessment of the assumptions that went into the models? As data have been accumulating in time, we can find out that this model really is not so good anymore, and that model is still okay.

Curtis: I think that is a very important thing to do. I tried to indicate this as I went along. But I think it is quite presumptuous to say that this model is not valuable or that these assumptions are wrong. I did not feel that this was the place to do that. It is only a personal point of view. I don't think there is a consensus in the field.

Kasha: Even today with so much time and so many models to compare, there isn't consensus on some things which are inoperative?

Curtis: There is not a consensus. For instance, this idea of the initial yield of strand breaks being linear-quadratic. For many years people thought that it was linear-quadratic. Lately, the data have come out to say, in the dose regions that we are interested in, the yield is linear, and Radford came up with neutral elution experiments to show that it is not linear. Even in that dose range, it is linear-quadratic. In my manuscript, I point to a paper that has just come out in *Radiation Research* by Detlef Blöcher (Radiat. Res. 123:176, 1990). He points out that, if you get a yield in a neutral elution experiment that is linear-quadratic, it does not necessarily mean that the initial induction of double-strand breaks was linear-quadratic. He has come up with a very nice theory of how you get the yield, and you can introduce an initial linear induction of double-strand breaks and still get a linear-quadratic yield. But that is still in a controversial state in the field.

Geacintov: Why are there no models which include the effect of replication stops or other forms of damage which inhibit or cause errors in DNA replication, without the need of a double- or single-strand break?

Curtis: I don't see any models like that in the literature. There might be some.

Geacintov: But it is known that this is happening. If you have a DNA lesion, you may have a replication stop, and I was just wondering if anybody had even considered that. It would be interesting to include such processes.

Curtis: You say that that in itself could lead to cell killing?

Geacintov: If you arrest DNA synthesis, the cell can't replicate.

Ward: Isn't it possible that those lesions could be excised at some time and the cell can then replicate their DNA? You have to predict that those lesions will be very long-lived.

Chatterjee: I would like to go back to your question on how ready we are for transformation models. There are some difficulties that we are experiencing and I thought I would share them with this group, so that with developments in molecular biology or techniques we can focus on those problems. One of the issues involving large deletion of suppressor genes is that we don't know how many suppressor genes there are. That is one of the biggest problems I am facing in the modeling. The second problem is that of gene size. Let's say we have 50 or 55 proto-oncogenes, and maybe all of them are involved or maybe some of them may be involved. We are not sure whether the damages have to be on the introns or the exons. That is the second question with which we have problems. The third question concerns what point mutations are at the proto-oncogenes. We need more experiments at lower dose levels, perhaps where point mutations seem to be important.

Ward: Just one comment along these lines from the Greg Freyer experiment described by Dr. Hall. That result cannot be the absence of the suppressor since DNA is transfected into a cell to get the effect. You cannot transfect the absence of something!

Varma: That is a good point.

Hall: It's got to be a dominant-acting gene.

Braby: The business of blocking DNA synthesis is likely to show up as a division delay, and there have been some attempts at modeling division delay processes. The problem is that it requires very careful control of cell cycle times, and so forth, in order to have a population that behaves as *in vivo* cells. There is normally too much biological variability to test even simple models. Division delay is a combination of several different processes which occur at different times in the cell cycle. We investigated one of those processes, mitotic delay, which can be measured very precisely using mitotic selection. Delays caused by processes, like delayed DNA synthesis, which occur earlier in the cell cycle, are harder to deal with because measurements will include the effects of later processes as well.

Carcinogenesis Models: An Overview[a]

Suresh H. Moolgavkar[b]

Abstract

Biologically based mathematical models of the process of carcinogenesis are not only an essential part of a rational approach to quantitative cancer risk assessment, but also raise fundamental questions about the nature of the events leading to malignancy. In this paper two such models are reviewed. The first is the multistage model proposed by Armitage and Doll in the 1950s. The larger part of the paper is devoted to a discussion of the two-mutation model proposed by Moolgavkar and colleagues. This model is a generalization of the idea of recessive oncogenesis proposed by Knudson, and has been shown to be consistent with a large body of epidemiological and experimental data. The usefulness of the model is illustrated by analysis of a large experimental data set in which rats exposed to radon develop malignant lung tumors.

Introduction

A comprehensive model for carcinogenesis in response to exposure to an environmental hazard would consist of at least two major components or submodels. The first would relate environmental exposure to tissue dose. Once tissue dose had been determined, the second submodel would relate tissue dose to the probability of tumors of that tissue, and would itself consist of two components. The first of these would model the interactions between the environmental agent in the tissue and macromolecules, such as DNA, that are relevant to carcinogenesis; the second and final component would relate alterations in the macromolecules to the end point of interest, namely the appearance of malignant tumors.

As an example, consider malignant lung tumors arising as a result of exposure to radon in the environment. Within the framework of the scheme outlined above, the first submodel would relate exposure to the pattern of distribution of dose of radon-daughters to the individual cells of the lung epithelium. The second submodel would relate the appearance of malignant lung tumors to the distribution of tissue dose. The first component would model the interaction of alpha particles with the tissue leading ultimately to changes in critical cellular targets; the second component would relate these changes to the appearance of malignant tumors.

In this paper, I shall be focusing attention only on the last part of this comprehensive model; that is, I shall be talking about models that relate fundamental cellular processes, such as mutations, to the incidence of malignant tumors in the

(a) This paper will also appear in the Proceedings of the 29th Hanford Symposium on Health and the Environment, "Indoor Radon and Lung Cancer: Reality or Myth?" (F. T. Cross, editor), Battelle Press

(b) Fred Hutchinson Cancer Research Center, Seattle, Washington

Physical and Chemical Mechanisms in Molecular Radiation Biology
Edited by W.A. Glass and M.N. Varma, Plenum Press, New York, 1991

387

tissues of interest. Because detailed information relating exposure and tissue and molecular dosimetry is not usually available, the parameters of the model are treated as functions of exposure.

I shall discuss two carcinogenesis models here. The first of these, due to Armitage and Doll, was first proposed in the 1950s[1] and has been extensively used for the analysis of epidemiologic data. It continues to be used by regulatory agencies for risk assessment. The second model I shall consider is a generalization of the idea of recessive oncogenesis according to which inactivation of homologous tumor suppressor genes leads to cancer.[2,3] Because it explicitly considers cell kinetics, and can therefore be used to model the effects of non-genotoxic carcinogens, this model is gradually gaining popularity over the Armitage-Doll model. Thus, I will devote a major part of this talk to a discussion of this two-mutation model and, in particular, I will illustrate its usefulness by analysis of an experiment in which rats exposed to radon developed malignant lung tumors. One of the main issues addressed in that experiment was the effect of fractionation of a given total exposure. The two-mutation model offers a biological explanation of the increased lifetime probability of tumor with fractionation of exposure.

The Armitage-Doll Model

This model was proposed in the 1950s to explain the observation that the age-specific incidence of many human adult carcinomas increases roughly with a power of age. Armitage and Doll[1] postulated that carcinogenesis was a multistage process during which a normal susceptible cell became a malignant cell via a sequence of intermediate stages, with the sojourn time distribution in each stage being exponential. The model may be represented as follows:

$$E_0 \xrightarrow{\lambda_0} E_1 \xrightarrow{\lambda_1} \dots \xrightarrow{\lambda_{n-1}} E_n$$

where E_0 is the normal cell, E_n is the malignant cell, and λ_i is the parameter of the exponential waiting time distribution in stage i.

Suppose now that cells are independent, and let p(t) be the probability that any given cell is transformed by time t. Then the age-specific incidence rate per 100,000 population is given by h(t) x 100,000, and h(t) the hazard function can be seen to be

$$h(t) = N p'(t)/(1 - p(t))$$

where N is the total number of susceptible cells in the tissue and p'(t) is the derivative of p(t) with respect to t. Since at the level of a cell, cancer is a very rare disease, p(t) is very close to zero, and thus

$$h(t) \approx N p'(t)$$

Now, p'(t) can be expanded in a Taylor series, and if only the first non-zero term is retained we obtain

$$h(t) \approx \frac{N \lambda_0 \lambda_1 \cdots \lambda_{n-1} t^{n-1}}{(n-1)!} \tag{1}$$

Thus, with the two approximations made above, h(t) increases as age raised to a power that is one less than the number of stages required for malignant transformation.

The Armitage-Doll model continues to be used for risk assessment by regulatory agencies. Unfortunately, the approximate form of the model (Eq. 1) is used, and this may not be entirely appropriate as discussed in a previous publication.[4]

A serious deficiency of the Armitage-Doll model is that it does not take explicit account of cell division and differentiation, although these are clearly important in carcinogenesis, and, indeed, may be the key to promotion.[5,6] In the next section, I discuss a two-mutation model that incorporates cell kinetics and has been shown to be consistent with a large body of epidemiologic and experimental data.

The Two-Mutation Model

This model is a generalization of the concept of recessive oncogenesis first advanced by Knudson for embryonal tumors in the 1970s.[2] It is instructive to recapitulate briefly Knudson's model for retinoblastoma.

Retinoblastoma is a rare malignant tumor of the retina in children below the age of 10 years. It occurs in two forms. One is sporadic; the other appears to cluster in families, and on pedigree analysis appears to segregate in an autosomal dominant fashion. Inheritance of the gene for retinoblastoma is the strongest known risk factor for cancer in humans. Thus, while the probability of sporadic retinoblastoma is approximately three per 100,000 children, a gene carrier develops on average three to four tumors (in both eyes combined). Thus, at the level of the susceptible cell, the retinoblast, the gene for retinoblastoma increases the risk some 100,000-fold.

In developing his model for retinoblastoma, Knudson noted two crucial facts. First, inheritance of the gene is not sufficient for malignant transformation because although every cell in the retina carries the gene, only three or four tumors arise in the susceptible individual, which indicates that at least one other event is necessary. Second, cytogenetic analysis reveals that in many instances of hereditary retinoblastoma, a critical segment of Chromosome 13 is deleted. Thus, in contrast to oncogenes, it is the inappropriate inactivation of these genes that leads to malignancy. Such genes are called "antioncogenes" or "tumor suppressor genes."

Based on these observations, Knudson postulated that retinoblastoma resulted from homozygous loss of gene function at a specific locus (the retinoblastoma locus) on Chromosome 13. The first inactivation event was somatic in the sporadic cases, whereas it had occurred in the germ cell of an ancestor and had been inherited in the hereditary cases. The second inactivation event was always somatic. In the past few years, spectacular advances in the technology of molecular biology have made it possible to demonstrate that the salient features of Knudson's model for retinoblastoma are correct. The gene for retinoblastoma has been cloned.[7]

In 1973 Comings proposed a genetic regulatory schema incorporating the oncogenes and antioncogenes.[8] According to this schema, all cells contain genes (oncogenes) capable of coding for transforming factors that can release the cell from normal growth constraints. The oncogenes are expressed during histogenesis and tissue renewal; their expression is controlled by (diploid) antioncogenes (tumor suppressor genes). Malignant transformation of a cell occurs when the oncogenes are turned on to inappropriately high levels. For most human tumors, this occurs with inactivation of the appropriate pair of antioncogenes. However, oncogenes could be directly activated by chromosomal rearrangement to bring an oncogene adjacent to a promoter site, as happens, for example, with Burkitt's lymphoma, in which the c-myc gene on Chromosome 8 is moved adjacent to one of the immunoglobulin loci on Chromosomes 2, 14, or 22. In view of the Comings' schema, it is interesting to note that the retinoblastoma gene product has been reported to turn off the transcription of the c-fos oncogene.

The schema briefly described above is clearly an oversimplification of the complex process of carcinogenesis. However, it turns out that, providing tissue growth kinetics are taken into account, a model for carcinogenesis postulating two rare and heritable events on the pathway to malignant transformation is consistent with the data from animal experiments and human epidemiology.

The specific model that I am discussing is shown in Fig. 1. It assumes that cancers are clonal and that normal susceptible (stem) cells become malignant via an intermediate stage in two rare and heritable (at the level of the cell) steps. In the parlance of chemical carcinogenesis, the first mutation may be identified with initiation; the clonal expansion of intermediate (or initiated) cells may be identified with promotion; and the second mutation may be identified with conversion. In this model, the salient feature of promotion is the clonal expansion of intermediate cells. The net effect of this is to increase the number of cells with the first mutation, thus increasing the probability that one of them will sustain the second mutation and become malignant. This explains the great efficiency of promotion in bringing about malignant transformation. I refer the interested reader to recent publications for more details.[9,10]

For a mathematical development of the model, a few more assumptions are necessary: the process by which intermediate cells arise from normal cells is non-homogeneous Poisson with intensity function $v(s)X(s)$ where $v(s)$ is the first mutation rate and $X(s)$ is the number of normal susceptible cells at age s. Once intermediate cells are generated, they develop into clones with cell division rate $\alpha(s)$ and cell death (or differentiation) rate $\beta(s)$. During cell division there is a small probability that an intermediate cell will sustain the second mutation and become malignant. The second mutation rate is denoted by $\mu(s)$. All the parameters of the model can be made functions of dose of agent of interest. The reader is again referred to a recent publication for details.[10]

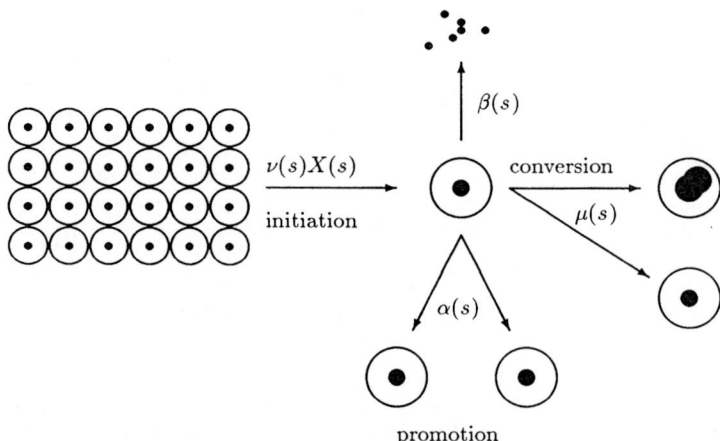

Figure 1. Two-mutation model for carcinogenesis.

Initiation and Promotion

As mentioned above, initiation can be identified with the first mutation, promotion with the clonal expansion of intermediate cells, and conversion with the second mutation. Some implications of this model are as follows. Clearly the model provides a framework for the so-called I-P-I (initiation, promotion, initiation protocol). Thus, the model predicts that such a protocol should lead to a higher yield of malignant tumors than an initiation promotion protocol, which should lead to a high yield of benign (intermediate) lesions. This prediction of the model has been verified in two animal systems: the mouse skin[11] and the rat liver.[12] An I-P-I experiment with radon as the initiator, cigarette smoke as the promoter, and lung tumors in rats as the end point is currently being conducted by Dr. F. T. Cross at Battelle, Pacific Northwest Laboratories. This experiment should shed light on the interaction of radon and cigarette smoke in producing lung tumors.

One mathematical consequence for a population undergoing a birth-death process, such as the population of intermediate cells in the two-mutation model, is that there is a non-zero probability of extinction, i.e., of the entire population dying out. Thus, when an initiated cell is created by mutation from a normal cell, there is a non-zero probability that this cell will die without giving rise to an intermediate clone. In fact, when α and β are constant, the (asymptotic) probability of this event is exactly β/α when $\alpha > \beta$ and is 1 when $\beta \leq \alpha$. Such considerations may apply to malignant tumors as well. For example, it is well known that several hundred thousand malignant cells may have to be injected into a nude mouse before a malignant tumor develops. Thus, the vast majority of malignant cells presumably die without giving rise to tumors. This implies that even without defensive mechanisms, such as immune surveillance, some fraction of malignant cells in an intact animal never gives rise to tumors.

Promotion is viewed rather differently within the framework of the Armitage-Doll multistage model. An initiator is often viewed as an "early" stage carcinogen, a promoter as a "late" stage carcinogen. This view of initiation and promotion does not appear to be consistent with the evidence that clonal expansion of initiated cells is the key to promotion.

Recently, Vogelstein and colleagues have studied the genetic alterations in colon cancers and proposed a multistage model for the genesis of these neoplasms.[13] Their schema is depicted in the upper panel of Fig. 2. I believe that Vogelstein's data are equally consistent with the schema shown in the lower panel of Fig. 2, i.e., with a two-mutation model. Specifically, there are at least two forms of dominantly inherited colon cancer, one associated with polyposis, the other not. A model that is consistent with Vogelstein's data postulates that colon cancer results from homozygous loss of gene function at either the polyposis locus (known to be on Chromosome 5) or the as-yet-unmapped nonpolyposis colon cancer locus. The other genetic events, such as the ras gene mutation, would in this scenario lie on the branch point in the schema and confer a growth advantage on intermediate (initiated) cells. Thus, these mutations would not be necessary for malignant transformation, but would greatly increase its efficiency by increasing the population of target intermediate cells. Such mutations could therefore be called "promoter" mutations. [Please see note added in proof, end of paper.]

An Application: Radon and Lung Tumors in Rats

Several epidemiologic and experimental data sets have been analyzed within the framework of the two-mutation model. With epidemiologic data the end point of interest is usually the incidence of malignant tumor, and the mathematical quantity that

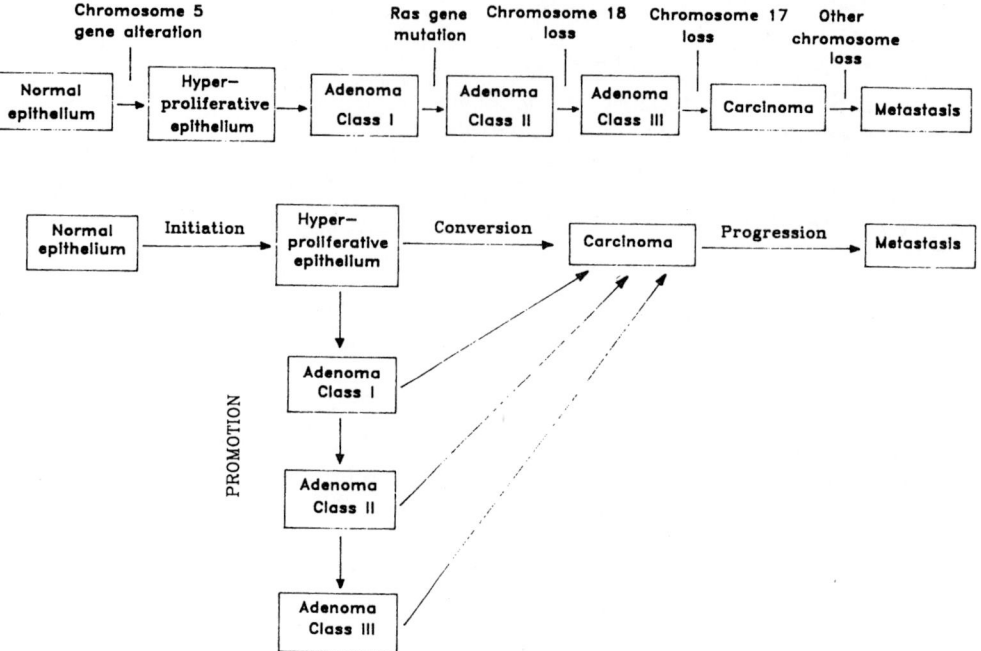

Figure 2. Two schemata for carcinoma of the colon. The upper schema is due to Vogelstein [see Stanbridge[13]]. The lower schema represents an interpretation of Vogelstein's data within the framework of the two-mutation model. In this interpretation two mutations are necessary (in the sense that repair of any one of these mutations will lead to reversal of the phenotype of malignancy) for malignant transformation. The other genetic events increase the efficiency of the second necessary mutation by increasing the population of initiated target cells (promotion).

needs to be derived from the model is the hazard or incidence function. With experimental data when malignant tumors are of primary interest, the hazard function is again the relevant mathematical quantity. However, in some experimental situations, for example the rodent liver model for chemical carcinogenesis, good quantitative data are available on the number of premalignant foci and their size distribution. The relevant mathematical quantities for the analysis of such data can also be derived from the model.

Here I will briefly describe the analysis of an experiment in which rats exposed to radon developed malignant lung tumors. The experiment that is analyzed here was conducted by Dr. Cross at Battelle. One of the main goals of the analysis was to study the effects of total exposure to radon and the exposure rate on the incidence of malignant lung tumors in rats. In particular, there are reports in the literature that fractionation of a given total exposure to high LET radiation leads to an increased lifetime probability of tumor. Thus, the main goal of the analysis was to confirm this result and to seek an explanation for it in biological terms.

Seventeen hundred and ninety-seven rats were subjected to different total lifetime exposures to radon. Within each total exposure group, there were several exposure rate regimens. The immediate goal of the statistical analysis was to estimate the parameters of the two-mutation model as functions of the exposure rate of radon. We constructed the likelihood of the data and estimated the parameters by maximizing the likelihood. The technical details of the procedure are provided in a recent paper.[14]

Table I. Observed and expected number of tumors generated by model. Exposure refers to total of radon daughters received by the animals. Each group of animals was exposed to a particular exposure rate regimen, d. Thus, for example, the exposure rate regimen in group 6 ranges from 150 to 180 working level months per week. For each group we show, within each exposure rate and total exposure category (from top to bottom), the number of animals in the experiment, the number of malignant tumors observed, and the expected number of malignant tumors generated by the model.

Group expos:			0	320	640	1280	2500	5000	10000 WLM	Total
1	d	5 WLM/w	32	128	0	0	0	0	0	160
		Obs.	0	16	0	0	0	0	0	16
		Expec.	0.83	18.35	0	0	0	0	0	19.18
2	d	50 WLM/w	96	127	64	32	32	32	0	383
		Obs.	1	8	17	17	22	25	0	90
		Expec.	2.29	14.89	14.08	13.91	19.68	22.11	0	86.96
3	d	500 WLM/w	81	169	105	74	72	71	76	648
		Obs.	0	12	5	11	20	27	43	118
		Expec.	1.34	7.14	7.24	9.03	18.95	31.04	47.37	112.12
4	d	260 WLM/w	64	0	0	0	0	128	0	192
		Obs.	1	0	0	0	0	68	0	69
		Expec.	1.47	0	0	0	0	65.59	0	67.05

Group expos:			0	20	40	80	160	320	640 WLM	Total
5	d	50 WLM/w	18	18	18	18	18	18	18	126
		Obs.	0	0	0	0	0	0	0	0
		Expec.	0.08	0.10	0.12	0.16	0.23	0.46	0.97	2.12
6	d	150-180 WLM/w	0	0	0	0	0	96	0	96
		Obs.	0	0	0	0	0	8	0	8
		Expec.	0	0	0	0	0	6.55	0	6.55
7	d	50-170 WLM/w	32	0	0	0	0	0	160	192
		Obs.	0	0	0	0	0	0	25	25
		Expec.	0.52	0	0	0	0	0	19.98	20.50

Table I shows the results of fitting the model to the data and also summarizes the experimental protocol. As can be seen from that table, the model describes the data quite well.

The parameter estimates indicate that the first mutation rate is strongly increased by exposure to radon. In contrast, the second mutation is only slightly elevated, and in fact this increase is not statistically significant at the 0.05 level. Surprisingly, exposure to radon appears to have a strong positive effect on intermediate cell kinetics. A strong increase in the net growth rate of intermediate cells, i.e., in α-β, is estimated during radon administration. Thus, radon appears to be both an initiator and a promoter. The initiating activity of radon is not surprising in view of its mutagenic properties. However, why should radon be a promoter? On this question, it should be noted that radon was administered along with uranium ore dust, and it is not possible to determine in this experiment whether it was the radon or the chronic irritation of the uranium ore dust that was responsible for the promotion.

How could it be that radon affects the two-mutation rates so differently? The nature of the two "mutational" events may be quite different, and radon may be quite efficient in producing one and not the other. For example, high LET radiation is well known to cause deletions. If the first mutational event is a deletion, then the second event may have to be a point mutation because a second (radon-induced or spontaneous) deletion may prove to be lethal to the cell.

Figure 3 shows that fractionation of exposure increases the lifetime probability of tumor. Within the framework of the model an intuitive explanation of the efficiency

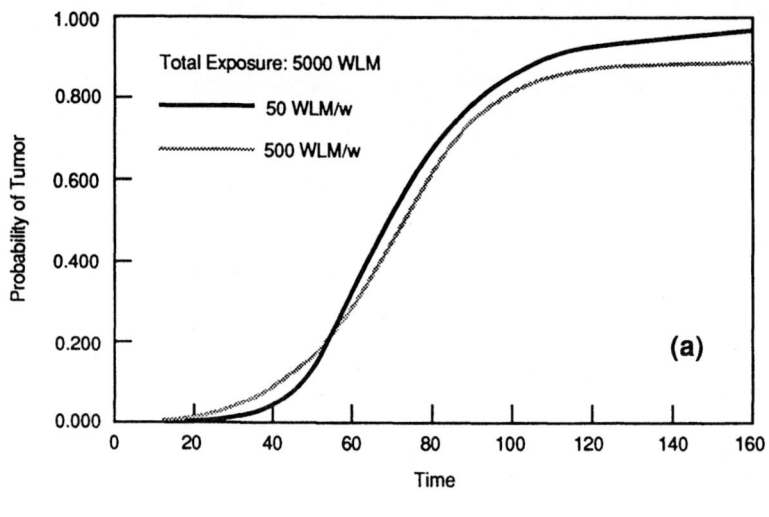

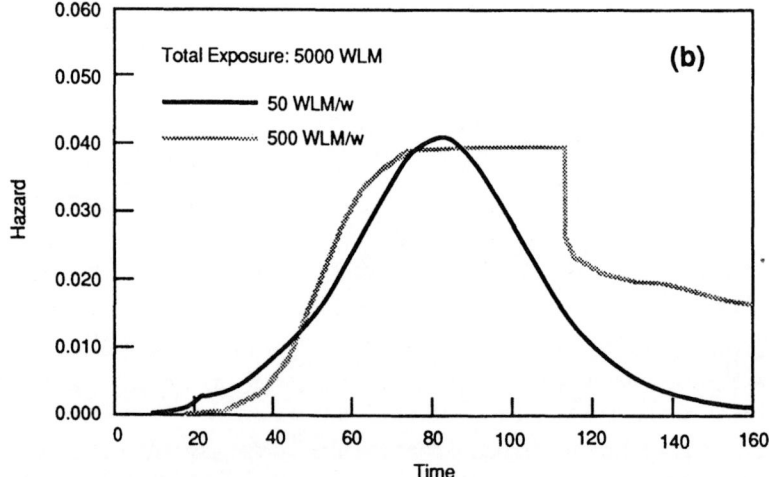

Figure 3. (a) Probabilities of tumor and (b) hazard functions associated with a total exposure to radon daughters of 5000 working level months with two exposure rate regimens, 50 WLM/w and 500 WLM/w. Note that fractionation of exposure leads to a higher lifetime probability of tumor.

of fractionation is as follows. During exposure to radon, there is an increase in the pool of intermediate cells due to direct mutation of the normal cells. These cells in turn grow into clones, the sizes of which depend on continued exposure to radon (or uranium dust). The increase in the number of intermediate cells, of course, increases the probability of a malignant tumor. Now, for a given total exposure to radon, a low exposure rate over a prolonged period allows more time for exposure-induced proliferation of intermediate cells. Further, the estimated mutation rates $\nu(d)$ and $\mu(d)$ are sublinear functions of exposure rates. As a consequence, a unit exposure is less efficient at producing mutations at high exposure rates. Thus, the probability of malignant tumor depends in a complicated way on the rate of mutation, the rate of intermediate cell proliferation, and the length of time that exposure-induced proliferation of intermediate cells goes on. For a more detailed discussion see Moolgavkar et al.[14]

Concluding Remarks and Directions for Future Research

A dose-response model based on biological considerations is an important component of a rational approach to quantitative cancer risk assessment. Such a model also serves to focus attention on the sorts of data that are needed for risk assessment. Not too surprisingly, what is needed is quantitative information on fundamental biological questions. How does the number of stem cells of a tissue evolve with age? What are the rates of stem cell division and differentiation? What are the locus-specific mutation rates? How do these rates change in response to administration of the agent of interest?

If there are intermediate stages on the pathway to malignancy, it is clearly of interest to identify and characterize these. In particular, the two-mutation model discussed here postulates one intermediate stage. Are there proliferative epithelial lesions that could be identified with this intermediate stage? If so, can these lesions be quantified; i.e., can their number and size distribution be determined? In one animal model system for chemical carcinogenesis, the rodent liver, a distinguishing feature is the appearance of phenotypically altered clonal populations of cells that are characterized by changes in the expression of a variety of biochemical markers, and at least some of which are believed to be precursor lesions for malignant tumors. Considerable effort has been expended at quantifying the number and size distribution of these enzyme-altered foci. These data are considerably more useful for modeling than data on incidence of malignant tumor alone, such as the data on radon-induced lung tumors in rats. Could the techniques used in rodent hepatocarcinogenesis experiments be modified in order to pick up putative intermediate lesions in the rodent lung? If this could be done, the quantitative dependence of the parameters of the two-mutation model on exposure to radon could be made much more precise.

All models make simplifying assumptions. The two-mutation model discussed here is no exception; it makes both biological and mathematical simplifications. These and other limitations of the model are discussed in detail in a recent paper.[10] Thus, although the model focuses attention on the types of data needed for a rational approach to risk assessment and raises interesting questions, much more work needs to be done before quantitative estimates derived from the model can be accepted with confidence.

Note added in proof: Recent analyses of colon cancer incidence in the general population and among polyposis subjects indicate that the data are consistent with either a two- or a three-mutation model. However, a three-mutation model yields more realistic estimates of mutation rates and appears to be more consistent with the laboratory data (Moolgavkar and Luebeck, submitted for publication).

Acknowledgment

This work was supported by the U.S. Department of Energy under Contract No. DE-FG06-88GR60657.

References

1. Armitage, P. and R. Doll. The Age Distribution of Cancer and a Multistage Theory of Carcinogenesis. *Br. J.* 8:1-12 (1954).
2. Knudson, A. G. Mutation and Cancer: Statistical Study of Retinoblastoma. *Proc. Natl. Acad. of Sciences U.S.A.* 68:820-823 (1971).
3. Knudson, A. G. Hereditary Cancer, Oncogenes and Antioncogenes. *Cancer Res.* 45:1437-1443 (1985).
4. Moolgavkar, S. H. and A. Dewanji. Biologically-Based Models for Cancer Risk Assessment: A Cautionary Note. *Risk Analysis* 8:5-6 (1988).
5. Ames, B. N. and L. S. Gold. Too Many Rodent Carcinogens: Mitogenesis Increases Mutagenesis. *Science* 249:970-971 (1990).
6. Cohen, S. M. and L. B. Ellwein. Cell Proliferation in Carcinogenesis. *Science* 249:1007-1011 (1990).
7. Friend, S. H., R. Bernards, S. Rogelj, R. A. Weinberg, J. M. Rapaport, D. M. Albert, and T. P. Dryja. A Human DNA Segment with Properties of the Gene That Predisposes to Retinoblastoma and Osteosarcoma. *Nature* 323:643-646 (1986).
8. Comings, D. E. A General Theory of Carcinogenesis. *Proc. Natl. Acad. of Science U.S.A.* 70:3324-3328 (1973).
9. Moolgavkar, S. H. and A. G. Knudson. Mutation and Cancer: A Model for Human Carcinogenesis. *JNCI* 67:15-23 (1981).
10. Moolgavkar, S. H. and A. G. Knudson. Two-Event Model for Carcinogenesis: Biological, Mathematical and Statistical Considerations. *Risk Analysis* 10:323-341 (1990).
11. Hennings, H., R. Shores, M. L. Wenk, E. F. Spangler, R. Tarone, and S. H. Yuspa. Malignant Conversion of Mouse Skin Tumors Is Increased by Tumor Initiators and Unaffected by Tumor Promoters. *Nature* 304:67-69 (1983).
12. Scherer, E., A. W. Feringa, and P. Emmelot. Induction of Neoplastic Foci Within Islands of Precancerous Liver Cells in the Rat. In *Models, Mechanism and Etiology of Tumor Promotion*, M. Borzsonyi, N. E. Day, K. Lapis, and H. Yamasaki, eds. IARC Scientific Publications 56, Lyon, France (1984).
13. Stanbridge, E. J. Identifying Tumor Suppressor Genes in Human Colorectal Cancer. *Science* 247:12-13 (1990).
14. Moolgavkar, S. H., F. T. Cross, E. G. Luebeck, and G. E. Dagle. A Two-Mutation Model for Radon-Induced Lung Tumors in Rats. *Radiat. Res.* 121:28-37 (1990).

Discussion

Varma: What is inappropriate activation?

Moolgavkar: One of the mechanisms that has been suggested for carcinogenesis is the inappropriate activations of an oncogene that has been suppressed or the expression of an oncogene that has been changed in some way.

Varma: How do you define a genomic event?

Moolgavkar: They could be deletions; they could be mutations; they could be any alteration of the DNA.

Varma: But all those processes have different probabilities, so when you do your modeling are you just combining those different probabilities?

Moolgavkar: You are right. All I can get from fitting data of this kind is that I can estimate the probability that these events occur. I don't know what these events are. They are on the same time scale, I would think, in terms of number of such events or cell division, spontaneously maybe between 10^{-6} and to 10^{-7} per cell division. So that is a part of the model where it is most useful to have input from the kinds of things we have been talking about the last few days.

Zaider: Can your models, the Armitage-Doll model for example, explain all those slopes on your log-log plot for a variety of cancers?

Moolgavkar: Oh yes, what I should say is that with the Armitage-Doll model, in fact, the slope depends upon the number of stages of the model, and also one can show rigorously that the hazard is a monotonically increasing function which approaches an asymptote. The Armitage-Doll model cannot explain the age-specific incidence rates of tumors that have peaked incidences, like the childhood tumors. But the two-mutation model, because it takes into account the cell differentiation kinetics can do so. The difference is that in the Armitage-Doll model, the shape depends upon the number of stages involved in the transformation. For the two-mutation model, the shape of the age-specific incidents depends entirely upon the kinetics of tissue growth and development. For example, retinoblastoma does not occur after the age of 5 or 7 years because there are no retinoblasts; they have all differentiated; they are gone.

Zaider: Since you mentioned this, I thought that one of the strengths of this model was to explain those cancers that appear early and those that appear late by affecting selectively certain agents, initial stages or late stages, and that was their explanation of why certain cancers are prevalent in childhood and others at other ages. The other questions is; can you show us when you have an actual dose response curve? That's what I would really like to see.

Moolgavkar: Do you mean dose mutation rate?

Zaider: No, I mean hazard rates, say for the radon workers as a functional dose because that tells us if there are two events or....

Moolgavkar: It does not tell you if there are two events or not because two events can explain the function clearly.

Zaider: You told us that there are two events in retinoblastoma.

Moolgavkar: Yes, two or more events...you can't conclude from the doses response curves how many events there are, not for carcinogenesis. Because all bets are off, you know, this simple power law where you have chemicals or whatever interacting with macromolecules in the cell. That is valid only to a certain extent because cell kinetics play an important role and then all bets are off. I'm not sure about your dose response curve..., I could do it, but as a function of what age, it is a continuous function of age.

Zaider: What I thought would be a good test of the theory that you showed us in the beginning is if you plot hazard rate versus time for people to inherit one oncogene from the parents, one antioncogene; then if you cross the same hazard rate for people who did not inherit that, you should get a linear curve and a quadratic curve.

Moolgavkar: Well, that is approximately true, and that has been done for retinoblastoma. In fact, that was the original observation that led Al to his first paper in 1971. Here he proposes this two-mutational model for retinoblastoma, a statistical study of retinoblastoma.

Hall: Is this study with the radon and the lung cancer in rats, and the conclusion that there is an inverse dose rate effect, has that been published?

Moolgavkar: Yes, it was published in *Radiation Research* in January, 1990.

Kasha: My question is on the molecular basis of the abnormality of the oncogene and also neoplastic tissue. If you could have the base sequence, will you find oncogenes recognizable in base sequence, or would the triplet code be simply normal and not recognizable? If you could do the amino acid sequencing of neoplastic tissue, would that appear abnormal?

Moolgavkar: Oh, I suppose so. It depends upon whether a mutated oncogene was responsible for that transformation or not; besides, if you look at tumors, there are many things that happen by the time you get a tumor. Genetic instability is a hallmark for cancer, and it is hard to know which ones are responsible for the malignant transformation.

Kasha: How are genetic changes recognized? That's what I'm asking. What is the molecular marker? In Eric Hall's pictures, we see clustering of cells, and we see different changes in the different changes in structure. What is pressing the amino acid sequencing of those?

Moolgavkar: I am not an expert in this field. Harel could probably answer that question better.

Weinstein: Maybe I should clarify this. In most cases, if you were to sequence the genome, you might find all the genes being identical in the normal cell and in a cell in which the process has been activated. In many instances, the question is one of regulation of gene expression because not all the genes are active in any given cell, so that identifying the culprit for cell transformation would also require knowing which genes are being translated in that particular cell. If, however, the lesion is in a gene, which, in the affected cell, it is now defective in some way, or missing, or produces a defective gene crossing, then this could be observed from sequencing. In the case of mutation, the genomic sequence would produce a gene product that would be somewhat defective. Here again, there is an enormous probability question because not every defect in the gene product will actually affect the function. One can have proteins that have been modified, and still function very well. So, there is no simple, single marker that will allow you to correlate, one to one, what you would see in the genome, or even among gene products, and what one would expect from the fate of the cell. In most cases we are reduced to observing what happens and saying, "This is a phenotype of a transformed, irregular cell."

Braby: Your model provides a dose rate effect, essentially, but the cell effect models that we were talking about earlier also deal with dose rate effect on a different time scale on a matter of hours instead of months. Is there any possibility of combining those two effects to determine whether or not the dose rate effect at the cellular level, for example, plays any role in the overall dose rate effects for tumor production?

Moolgavkar: I would hope so. I think that when you are talking about models in such different time scales, it is better to model them in tandem, taking the output of one model and making it the input of the second one rather than having a comprehensive model. We could take the kind of output that you could provide us and use that as a kind of dose response for, let's say, for the mutation rate and see if it is consistent with what you find at the other end.

Curtis: I have a question about the inverse dose rate effect. You said that you explained the inverse dose rate effect by assuming that you had an increased proliferation rate due to the presence of the uranium dust. Now, wouldn't you also get an inverse dose rate effect if you didn't have an increase in the proliferation rate, simply because, over time for the low-dose-rate experiment, you have an increase due to proliferation of the cells that have these mutations, so that the second event would have a higher probability of occurring simply because there are more cells?

Moolgavkar: Let's look at the simplest case. Let's assume that the mutation rate is linear as a function of dose. Then, if you take 10 times the dose rate, you get 10 times as many cells right in the beginning. With fractionation of dose, you get the same number of cells because the effect is linear. You get the same number of cells over a much longer period of time. Ultimately you end up with the same number of cells that are initiated from normal ones, and all of them have the capability of growing because the cell division rates are not affected. So you need that to get the inverse dose effect.

Kopelman: The Armitage-Doll model, which is based on the waiting time distribution, is a tremendously robust type of model, and there are many waiting time distribution models that have been so successful in physics and chemistry. Some of those are in your book. Even though the actual mechanism is totally different in those different situations, this is a kind of general warning because people often used to assume that if you have two phenomena and they have the same, you can explain it in the same model. The mechanism must be similar--and the mechanism has turned out to be totally different. I had a specific question about the use here. There is an order of events.

Moolgavkar: You mean, does Event 1 have to be different from Event 2? Is that what you are asking?

Kopelman: No, they are ordered. Does Event 2 have to come after Event 1?

Moolgavkar: Yes. Another way of putting it is that the first and second mutation rates do not appear in the mathematical formula in a symmetric fashion. You can't exchange them.

Kopelman: That's the question. Even though you have an exponential waiting time.

Moolgavkar: Yes.

Molecular Radiation Biology

DNA Damage and Repair

John F. Ward[a]

Abstract

Some of the background to radiation chemical studies of DNA damage is presented, followed by a review of measurements of such damage and its repair in mammalian cells. While most effort has been placed on the measurement of radiation-induced strand breaks (because assays can be used in the biologically relevant dose range), the radiation-altered bases are less studied. Attempts have been made to devise assays for base-damage measurement in cellular DNA after irradiation; the problems of using such assays at reasonable radiation doses are discussed. The alternate approach to measuring yields and chemical identities of damages in cells is extrapolation from model systems. The limitations of extrapolation are considered in the context of the intracellular structures in which DNA exists and the problems in predicting mechanisms of intracellular damage induction. The complexities of damages that could be lethal are considered; "double-strand break" is the generic term for a wide variety of damages, each of which is produced by a similar mechanism. Many of the damages caused by multiple radicals have the potential to be lethal. The current work of the author is outlined along with his attempts to throw light on the topics described above. Some potentially fruitful directions for future research are suggested. These would help to build bridges between studies of the physics of energy deposition and the chemistry of cellular radiation damage and to provide a comprehensive basis for the prediction of the biological effects consequent to the deposition of radiation energy.

Introduction

The initially observed subcellular effects of ionizing radiation were, of course, the induction of mutagenesis and of chromosome aberrations. Consequently, the genetic material was implicated early as the cellular target for radiation damage. Early studies of damage to DNA, per se, were of changes of viscosity and of streaming birefringence,[1] physical changes that could be assigned to depolymerization of the macromolecule. Chemical changes induced in DNA by irradiation were reported at a Brookhaven National Laboratory Symposium; these included the release of inorganic phosphate and of ammonia (after high-radiation doses).[2] Later, specific radiation products were identified, such as thymine hydroxyhydroperoxide.[3]

Studies of quantitative changes produced in DNA irradiated in dilute solution[4] indicated that

- the major radical capable of destroying DNA was the OH species
- all four bases were destroyed to a similar extent in the macromolecule

(a) Division of Radiation Biology, Department of Radiology, School of Medicine, University of California at San Diego, La Jolla, California

Physical and Chemical Mechanisms in Molecular Radiation Biology
Edited by W.A. Glass and M.N. Varma, Plenum Press, New York, 1991

- the OH reacted with the base moieties four times more frequently than it reacted with the deoxyribose.

Subsequent studies, particularly by the French group,[5] helped to establish the identities of the plethora of products produced from the DNA bases, while the Mülheim group worked out many of the sugar damage mechanisms.[6]

Studies of DNA damage within cells began with the introduction of the technique of alkaline sucrose gradient centrifugation and the consequent observation of DNA single-strand break production and repair in bacteria by McGrath and Williams[7] of Oak Ridge National Laboratory, and later observation by Lett et al.[8] of single-strand break in mammalian cells. Corry and Cole[9] measured the production of DNA double-strand breaks and their repair in mammalian cells using neutral sucrose gradient centrifugation.

At first the evidence that cellular DNA was the target for radiation-induced damage was strong only for bacteria,[10,11] but evidence in mammalian cells was not so conclusive, although strong arguments were made in this direction. More recently,[12,13] the evidence for the importance of DNA damage in killing mammalian cells has been more clearly defined, although there are also data that show low-dose sensitivity of mammalian cell membranes.[14]

Damages Introduced in DNA by Ionizing Radiation in the Cell

Other than strand-break measurements, there have been few studies of DNA damages induced in mammalian cells. The problems in measuring other types of damage lie in the sensitivity required at biologically relevant doses. The sensitivity of the strand-break assays relies on biophysical measurements of changes in macro-molecular properties of the DNA; no such measurable change is induced by other types of damage except DNA-protein cross-links (DPC).[15]

Compared to ionizing radiation, work with other DNA-damaging agents is easier because of the large numbers of agent-induced DNA damages needed to kill the cell.[13] For ultraviolet light, for instance, 400,000 thymine dimers are induced at a mean lethal dose. For ionizing radiation, on the other hand, only 40 double-strand breaks and 1,000 single-strand breaks are present after a lethal dose. At the same radiation dose, the yield of total base damages would be expected to be about five times the yield of single-strand breaks, i.e., 5,000 per 2×10^{10} bases, or $2.5 \times 10^{-5}\%$ of the bases altered. These damages would be distributed among all four bases and among the large variety of altered base identities possible. This is a major challenge to any assay procedure.

Several approaches to the ultrasensitive analysis of base damages have been used. These include immunoassays,[16-18] gas chromatography-mass spectrometry,[19] post-labeling,[20-22] radiochemical,[23] and damage-specific endonuclease.[24] As yet these assays still do not approach the required radiosensitivity for measurement of damage at reasonable radiation doses, as is illustrated in the following case.

Several years ago, we used the RIA approach and achieved a sensitivity of 4 fmole of thymine glycol[16] and 40 fmole of 8 hydroxyadenine.[25] However, in the latter case, where a valid measurement could be made, undamaged adenine cross-reacted at a concentration that was 10^5 higher. In a mammalian cell, there are 10^{10} bases, so achieving the same signal from the product as from the starting material would necessitate the formation of $10^{10}/(4 \times 10^5) = 2.5 \times 10^4$ 8-hydroxyadenines. The yield of 8-hydroxyadenine is about one-fourth of that of the single-strand break, indicating that a dose of 100 Gray would be needed to detect this product under optimum conditions. Of course, if repair of the damage is to be followed by measuring the residual altered base in macromolecular DNA, greater initial doses of radiation would be required.

While we were using this approach to measure base damage types, we realized that single damages in DNA are not a significant cause of cell death. We found that producing individual damages in DNA by treating Chinese hamster cells with hydrogen peroxide was not an effective way to kill the cells.[26] While we only measured DNA single-strand breaks in that work, it is clear that single-base damages must be accompanying these lesions[27] and that these also are not lethal. Therefore, we discontinued attempts to increase the sensitivity/selectivity of our immunoassays.

In contrast to the paucity of base damage measurements, there are many publications in which repair of strand breaks has been measured. Since assays of this type of damage were usable in the dose region at which cell killing was measured, these were—and are still—the lesions most frequently studied. The early studies of strand breaks used procedures based on good biophysical justifications, i.e., velocity sedimentation through sucrose gradients. Recently newer, more sensitive techniques have been employed, such as alkaline and neutral elution, alkaline unwinding, and pulse field gel electrophoresis. The biophysical principles underlining the latter techniques are not yet fully understood.

The yields of double-strand breaks formed by irradiation of mammalian cells have recently been evaluated.[28] The literature was reviewed and the data considered in light of the mechanisms of energy deposition and the possible secondary reactions of the radicals formed on DNA. It was concluded that the yield of this double-strand break damage is linear with radiation dose, in contrast to suggestions raised from the interpretation of neutral elution data.[29] Secondly, the yield of double-strand breaks per mass of DNA and per dose is constant from cell line to cell line if irradiation is carried out under identical conditions. Thus variations in radiosensitivity among cell types must be assigned to differences in repair capability.[28]

As has been mentioned, measurement of strand breaks gives no information about the chemical nature of the strand break termini that are produced. The latter are what the cellular enzymes must handle in their attempts to repair the damage. This problem has been approached by one group[30] at admittedly high radiation doses. To make advancements in specific lesion identification from cellular measurements is a major problem because of the sensitivity required at biologically significant dose levels.

The alternative to actually measuring the yields of damages in mammalian cells is to extrapolate from data obtained in model systems. Such extrapolations are not straightforward for a variety of reasons, some of which are discussed below.

Problems in Extrapolation

The place at which radiation physics, radiation chemistry, and radiation biology meet is at the mechanisms of production of damage and the identities of damages produced in DNA within the mammalian cell. Physicists make predictions based on energy deposition patterns; chemists attempt to identify damage types and yields; and biologists consider the cell's abilities to handle the so-far-uncertainly characterized damages. As mentioned above, the actual information about DNA damage in the mammalian cell is limited. Scientists from each of the three disciplines make extrapolations into this area from what is known in their disciplines. At a meeting in San Miniato, Italy (April 1990), jointly sponsored by DOE, NIH, NATO, and CEC, the early effects of radiation on DNA comprised the discussion topic.[31]   While the majority of the presentations were of data from model systems, a concerted effort was made to extrapolate these findings to cellular DNA.

Large amounts of chemical data have accumulated from the radiology of dilute solutions, including time-resolved studies, and from electron spin resonance examinations of irradiated crystals and ices.   The question arises: How can these be used to predict damage in the mammalian cell?   First it must be remembered that in the cell the DNA does not exist either dry or in dilute solution.   It exists in chromatin and is damaged by energy deposited directly in it, by reaction with ionized molecules close to it (including possible attack by H_2O^+), and by OH radicals produced from surrounding water diffusing a limited distance (20nm) from their initial point of origin (for a review, see Ref. 13).

At the San Miniato meeting[31] a consensus was reached that two types of damage could be distinguished: that which can be scavenged by OH radical scavengers and that which cannot.   It was argued that any attempt to further classify damage was not useful.   Major problems exist in extrapolating from solid systems, frozen or single-crystal, to the cell.   Studies of these systems have the great advantage of being able to assign radical structures; the drawbacks of the systems lie in deciding the subsequent reactions of these radicals to form stable products.   In the solid state, reaction is limited to neighboring molecules, whereas in the cell these radicals would react in a liquid water environment with other molecules, notably oxygen and water.

Model systems in solution have clearly described the reactions of OH radicals with nucleic acid constituents; rate constants for reaction and sites of attack are known. Mechanisms have been described for base damaging and for strand breaking (reviewed by C. von Sonntag at this meeting).   The extrapolation of these data to cellular DNA presents two questions: (i) What fraction of the attack occurs on the sugar moieties and what on the bases?   (ii) What are the subsequent reactions of the initially formed macromolecular radicals?   The former is important in establishing strand-break mechanisms; there is clear evidence in polyuridylic acid that base radicals can transfer the radical to the sugar moiety initiating a strand break,[32] but it is unclear whether such a radical transfer occurs in DNA or inside the cell.   The reader is referred to the papers presented at the San Miniato meeting for more information on this point.

In addition to these difficulties of extrapolation, two mechanisms of DNA double-strand break formation in solution have been presented, each of which could occur in the mammalian cell and each with different consequences.[33]

The Lethal Lesions

There is much evidence that points to the DNA double-strand break as the lethal lesion (reviewed in Refs. 12 and 13).   When a double-strand break is formed in a cell, the possibility exists that the two ends of the break may move apart and become enzymatically joined with one of the termini of a different double-strand break—this constitutes "misrepair." On the other hand, each of the constituent single-strand breaks occurs by loss of a deoxynucleoside moiety;[34] thus, even if joining occurred with the correct partner, loss of biological information leading to a "point mutation" could result.[35]

Double-strand breaks are measured in cells as a reduction in double-stranded molecular weight without consideration of the structure(s) of the damage induced. Their mechanisms of formation by free radical processes[33] indicate that "double-strand break" is a generic term that covers a variety of damages; the possible components of this class of lesions deserve closer examination.   If it is granted that the proposed mechanism for double-strand break formation as a consequence of multiple radical attack can cause a Locally Multiply Damaged Site (LMDS),[36] then there are several corollaries.

- Since there is a range in the sizes of energy deposition "events," there will be a comparable range in the number of radicals produced per "event." Consequently as the amount of the energy deposited per event increases, then the numbers of damages per LMDS will increase. (The number of radicals from each event that react with DNA will depend on the proximity of the individual radicals to the DNA.)

- Damages within an LMDS would be on bases and sugars, with the ratio of attacks on these constituents being 2.7 base damages per sugar damage.[35] This calculation takes into account the contributions of damages produced by scavengeable and non-scavengeable mechanisms occurring in a 65/35 ratio.

- The distributions of these radical attacks at the dimension of nanometers is not known, but, from the size of the energy deposition events and the known migration distance of the OH free radical, a single LMDS could spread across tens of base pairs.

For any specific type of radiation, the distributions described in A, B and C would be independent of the radiation dose.

We have previously shown[37] that of the LMDSs produced by two damaging species the actual fraction of these damages which produce double-strand breaks will be small. The percentage of LMDSs which result in double-strand breaks from different numbers of radical damages per site can be calculated and is shown in Table I below. This assumes that the ratio of base to sugar damage is 2.7 and that only attack on the sugar leads to a strand break.

Table I. Composition of Locally Multiply Damaged Sites

No. of Radicals on DNA	Percentage of Locally Multiply Damaged Sites	
	a) as Double-Strand Breaks	b) with Only Base Damages
2	3.6	53
3	9.0	39
4	16	28
5	24	21
10	57	4
14	75	1.2

The yield of double-strand breaks from an energy deposition event is dependent almost on the square of the number of radicals reacting (at fewer than five radicals per site). This finding is in agreement with the proposition that track ends (containing multiple radicals) are more important in biological damage production than the more common spur type of event.[38] It also could explain the greater efficacy of soft X-rays, which deposit their energy in events of hundreds of eV, i.e., enough for the formation of multiple radicals.[39]

We have pointed out earlier the problems facing a cell in repairing LMDS containing base damages:[13] In order to remove a base-damaged site, the repair enzymes cut the DNA backbone. If such a cut results in the formation of a double-strand break (e.g., in conjunction with a single-strand break on the other strand), then the consequences can be the same as if a double-strand break had been initially induced. This possibility would increase as the number of damages per LMDS increases, i.e., more base damages per site. However it should be pointed out that some damage-specific enzymes do not react as efficiently on single-stranded regions of DNA (i.e., opposite a single-strand break) as on double-stranded DNA; indeed, some prefer supercoiled DNA.[40] The cell may, therefore, have a built-in self defense against removal of base damage from DNA when other damage is in the vicinity.

The assays of double-strand breaks are not sensitive to the structure of the damage. Because of the mechanism(s) by which they are produced, double-strand breaks will consist of the constituent single-strand breaks separated on opposite strand by a variety of distances; up to 20 base pairs for low LET radiation.[13] This factor will be of importance when the repair of these breaks is considered (see below).

DNA Repair

The majority of DNA repair studies carried out following ionizing radiation damage focus on measurements of strand breaks. Some studies of base damages have been accomplished. In one study using the radiochemical assay, Mattern et al.[41] measured the rate of removal of oxidized thymine damage from irradiated rodent and human cells. They found, after a high dose of radiation (2500 Gray), that both cell types removed the damage at about the same rate; 90% removed from macromolecular DNA in 15 minutes.

Paterson et al.,[42] attempting to probe the reason for the radiosensitivity of *ataxia telangiectasia* cells, assayed base removal using a gamma-specific endonuclease. Their data indicated that these cells (relative to normal cells) were deficient in removing endonuclease-sensitive sites introduced by 500 Gray of gamma irradiation under hypoxia. These studies have not been confirmed; an alternative reason for this cell line's radiosensitivity would be its inability to shut down DNA synthesis following a dose of radiation (i.e., taking time out for repair).[43]

In contrast to the paucity of base damage repair measurements made, many publications report the measurement of strand break repair. The early efforts using velocity sedimentation measurements have been for the most part superceded by more sensitive techniques, such as alkaline and neutral elution, alkaline unwinding, and pulse field gel electrophoresis.

Bryant has pointed out that not all methods of measurement give the same results;[44] centrifugation gives a half time for repair of DNA double-strand break of 1 to 2 hours (in Ehrlich ascites tumor cells),[45] while elution has a component with a shorter time period. Our own data[a] for measuring double-strand break repair by neutral elution indicate a fast-repairing component (60% of total) with a half-life of repair of 6 minutes. In that work we suggest that these fast-repairing double-strand breaks reflect resealing of the lesions having the constituent single-strand breaks sufficiently separated that they are repaired as single-strand breaks. These types of double-strand break may not be visualized in neutral centrifugation, since it is traditionally carried out at a high ionic strength (μ). If the constituent single-strand

---

(a) J. F. Ward, C. L. Limoli, and P. M. Calabro-Jones. An Examination of the Repair Saturation Hypothesis for Describing Shouldered Survival Curves. *Radiat. Res.* (in press).

breaks of the double-strand breaks are separated by a distance of several base pairs, then the high μ would prevent the local denaturation of the hydrogen bonds holding the strands together. In addition, the forces that may tend to pull apart such potential breaks may be weaker for centrifugation than for elution.

Since we assign fast-repairing breaks to widely separated single-strand breaks of a double-strand break, we consider such damages to be of no consequence biologically; they can be readily and accurately repaired using the other strand as a template.

The repair of double-strand breaks can be inhibited by a variety of procedures; some of these affect cell survival while others do not. For instance Bryant and Blöcher[46] inhibited strand break repair using 9-beta-arabinofuranosyladenine and also sensitized the cells to irradiation. We were also able to sensitize cells using a variety of strand break repair inhibitors,[47] but slowing the repair by reducing the post-irradiation temperature did not radiosensitize the cells.[a]

There is good evidence that the variation of cell radiosensitivity through the cycle can be assigned to the efficiency of cell-cycle-dependent DNA double-strand break repair efficiency.[48,49]

Current Interests

The author's current interests lie in relating radiation-induced molecular damages to DNA to biological end points. Two major directions are being pursued: Studying mechanisms of production of molecular damage to DNA within mammalian cells and its modulation and repair; and validating and using extracellular molecular systems to study DNA damage from which extrapolations can be made to DNA in the intracellular environment.

For a model system, we have chosen to use SV40 DNA complexed with histones, i.e., in the form of chromatin. These "minichromosomes" share many of the characteristics of the DNA of a mammalian cell,[50] including conformation and protein "masking." This system has other advantages: it is easily manipulated; it is relatively easy to use to measure the important types of DNA damage; it can be used in the presence of additives whose radiation chemical behavior can be controlled; and, in the future, it may be usable to monitor repair in cells.

We have used this model to probe the mechanisms of formation of DNA single-strand breaks and double-strand breaks by chemically produced free radicals and by gamma and alpha radiation, in the presence and absence of scavengers of OH radicals. This has been accomplished in the SV40 DNA in the naked state and in chromatin.[51]

Future work along these lines will be aimed at continued studies of LMDS from alpha and gamma radiation. Our work has indicated that there is not a major difference in the yield of DNA damages produced by alpha particles compared to gamma-radiation in a simulated cellular environment. Consequently, to account for the large RBE of the former, we propose that the quality of the damage should be studied. For this we would use direct measurements of oligonucleotide fragments released by the radiations, probably by using ^{32}P end labeling to attempt to find evidence of the distribution on a nanometer scale of damaging species production. The findings may also enable us to determine whether a signature of the radiation damage is shown in the structure of the released oligonucleotides.

In addition to measuring the immediately released materials, we would study the additional fragments that are released by treatment with enzymes such as S_1 nuclease (acting on "bulky lesions")[52] and/or damage-specific endonucleases (provided by Dr. S. S. Wallace). The findings from these studies will be used to extend the studies to an intracellular target DNA.

Our cellular studies include measurements of damage by elution procedures (we hope to use pulse field gel electrophoresis in the future), their modulation in the presence of sensitizers such as 5-bromodeoxyuridine,[53] and protectors such as WR2721.[54] The data are consistent with the concepts of damage being caused by OH radicals and non-scavengeable damage.[37] The latter includes direct ionization of DNA and reactions of possibly H_2O^+ or non-scavengeable OH radicals. Studies of scavenging of BU damage precursors[a] indicate the involvement of hydrated electrons as a cause of increased single-strand breaks, double-strand breaks and cell killing, as predicted earlier.[55,56] The latter are also additional support for the LMDS concept of damage induction.

In our studies of repair,[b] we have shown that, while post-irradiation temperature of holding has a marked effect on rates of repair of DNA damage in cells, it does not affect cell survival. The saturation of repair hypothesis for shouldered survival curves suggests that slower repair permits damage to be fixed as irreparable. The absence of a temperature effect on cell survival suggests that damage repair and damage fixation are the same process, probably misrejoining of double-strand break ends.

We have begun exploring the use of other means to induced DNA double-strand breaks in cells with a view to testing the above conclusion. Methods to be considered include the use of bleomycin[57] or neocarcinostatin[58] or dyes + light (C. L. Limoli and J. F. Ward, unpublished observation). These damaging agents will be probed for their interaction with gamma-radiation damage to see whether the misrejoining is a feasible mechanism. Such studies of synergism will also enable us to probe whether these chemically relatively simple double-strand breaks are as efficient in cell killing as gamma-radiation-induced double-strand breaks or whether more complex damages such as those which would be produced by track ends are required.[38]

Future Directions

Molecular

Radiation physicists currently model the biological effects of ionizing radiation in terms of energy deposition. It is clear that between the energy deposition and the biological end point, a series of steps must be accomplished. This was pointed out many years ago as a criticism of dual action theory by J. W. Boag:

> The complex radiation chemistry which follows the initial energy deposition and which can be modified by various added solutes, together with the biochemical damage and repair mechanisms about which more than a little is already known, must all be compressed into 'the probability that 2 sublesions shall combine to form a lesion within a sensitive site.'[59]

While we have improved in our understanding of repair mechanisms, progress has been limited in the measurement and understanding of the complex radiation chemistry. It is unlikely that the significant chemistry within the cell will be approachable by conventional pulse radiolysis techniques (although the gas explosion technique has been used to advantage by B. D. Michael and his co-workers, e.g., Ref. 60). Thus, to understand the relevant chemistry, we are in the position of radiation chemists before the advent of pulse radiolysis, when reaction mechanisms were worked out from yields and identities of radiation products.

(a) L. L. Ling, C. M. Webb, J. F. Ward, and J. Aguillera. Hydrated Electrons are Involved in the Radiosensitization of Bromodeoxyuridine Substituted Cells (in preparation).

(b) J. F. Ward, C. L. Limoli, and P. M. Calabro-Jones. An Examination of the Repair Saturation Hypothesis for Describing Shouldered Survival Curves. *Radiat. Res.* (in press).

Accomplishing our goal requires a description of the series of steps between the physics of energy deposition and the cellular consequences. Within each step of this progression, there exists a variety of possibilities.

- Reactions of radicals with DNA: The packaging of DNA is not uniform, being associated with histone proteins in nucleosomes with internucleosomal regions. At higher levels of packaging, there is attachment to other matrix proteins, etc. Thus, the access of radicals to all sites of the DNA will not be uniform.

- Varieties of damages: As mentioned there will be different numbers of damages per LMDS, i.e., containing a range of different base damages and strand breaks. These distributions will bear a marked dependence on the quality of the radiation used.

- Repair mechanisms: The varieties of damages will present different repair challenges to the cell, proximities of damages, heterogeneity of damages involved, etc.

- Cell cycle: There will not be major differences in the yields and types of damage produced in DNA as a function of position of the cell in the cell cycle (or indeed in different chromatin structures).[28] However, repair varies throughout the cell cycle and also in active versus inactive chromatin: If an asynchronous population of cells is irradiated then different repair capacities exist from cell to cell. There is also the phenomenon of radiation-induced interruption of cell cycle progression. With these variables the assignment of relatively simple two- or three-component models would seem to be insufficient.

To improve modeling capabilities, I suggest that attempts be made to analyze the types of damage induced in the accepted cellular target, DNA, as a function of radiation quality. Model systems could be used, if sufficiently validated; and selected radiations, gamma-radiation, soft X-rays, particle beams, etc., should be compared for the damage they induce. It is clear that the important damage is not of singly altered bases, or of single-strand breaks or of all double-strand breaks. It is also clear that the initial damage of a wide variety of types would be randomly induced. The variety in identity and the randomness of the site of induction of potentially significant damage places demands on the investigator wishing to devise relevant assays. Clearly, molecular biological techniques can be used, but not in the manner in which they are usually employed.

The existence of this damage produced as a consequence of several radicals close together on the DNA was predicted (and was assigned the name "locally multiply damaged sites," or LMDS), as was the increasing complexity of the damage and importance with increasing LET of the radiation.[35] Experiments to define the molecular nature of such damage were begun two years ago using a model SV40 chromatin system. The system is now well characterized; excellent preparations of starting material can now be obtained; the yields of the more easily measured products have been made; and we are beginning to make an approach to measuring the complexities of LMDS after gamma- and alpha-particle irradiation processes. Clearly, when developed the system could also be used to examine the damages for other particles and for soft X-rays.

Other Damaging Agents

There are other ways to induce DNA double-strand breaks in cells, using chemicals and also dyes combined with light. These treatments may not cause complex LMDSs such as those induced by radiation but should form simpler double-strand breaks with the distance between damages being relatively fixed and dependent on the agent used.

If, indeed, simple double-strand breaks are lethal events, then synergism between these methods of damage induction and radiation should be examined for cell survival, mutagenesis, and transformation.

Premature Chromosome Condensation and Chromosomal Aberrations

Some double-strand breaks produced by ionizing radiation may not have the potential of becoming biologically significant. The yield of double-strand breaks is about 40 per cell per Gray; the number of breaks measured by premature chromosome condensation is 7 per cell per Gray; and the yield of chromosome aberrations is 1 per cell per Gray. The reason for these differences in yield is clearly cellular repair of damage.

The distance separating the constituent single-strand breaks may be sufficiently great that each break can be repaired as if it were an individual break, i.e., using the opposite strand as a template for accurate repair. Such double-strand breaks would not be expected to be biologically important. The double-strand breaks expected to be biologically significant would occur when the constituent single-strand breaks are sufficiently close together on opposite strands to either (i) permit separation of the ends, or (ii) where repair can not take place using undamaged DNA sequence on the other strand and is forced on a damaged template.

The general methods for measuring DNA double-strand breaks are sensitive to molecular size (weight/length) and sometimes to conformation (e.g., supercoiling, double double strands in newly replicated DNA) but are not a measure of the characteristics of the end groups produced by radiation.

This indicates that the PCCs and chromosomal aberrations are a more appropriate measure of biologically significant damage; however, these assays do not give the yield of such damage immediately after irradiation. If the chemical nature of the significant initial lesion could be determined, an assay could be developed for use in cellular studies.

Mutagenesis

A radiation-induced mutation represents a "biological end point" whereby the consequences of irradiation could be assigned to a specific change in DNA sequence. Current information in this area indicates that there are two types of mutations detectable: point mutations and total deletion. It is not known whether these two different mutagenic types are caused by the same type of radiation damage. (It is possible, of course, to hypothesize mechanisms whereby both mutation types arise from DNA double-strand breaks.) It would be of interest to define the molecular events which precede these changes by varying the radiation conditions—hence, if the DNA double-strand break misrejoining hypothesis[a] as a function of dose is valid, it might be expected that the ratio of point to total deletion mutations would decrease with increasing dose. As yet there are no data on the distributions of these mutations as a function of dose. The situation is unfortunately complicated by the variations among cell lines of the yield and composition of the product mutants.[61] An urgent need exists to develop better assay systems for mutagenesis in cells of human origin.

(a) J. F. Ward, C. L. Limoli, and P. M. Calabro-Jones. An Examination of the Repair Saturation Hypothesis for Describing Shouldered Survival Curves. *Radiat. Res.* (in press).

Acknowledgments

Work in the author's laboratory is supported by grants from the U.S. Department of Energy under Contract #DE-FG0388ER60660 and from the U.S. Public Health Service: CA 26279 and CA 46295.

References

1. J.A.V. Butler and B. E. Conway. The Action of Ionizing Radiations and of Radiomimetic Substances on Deoxyribonucleic Acid. Part II. The Effect of Oxygen on the Degradation of the Nucleic Acid by X-Rays. *J. Chem. Soc.* 3418-3421 (1950).
2. G. Scholes and J. Weiss. Chemical Action of X-rays on Nucleic Acids and Related Substances in Aqueous Systems. *Exp. Cell Res. Suppl.* 2:219 (1952).
3. M. Daniels, G. Scholes, and J. Weiss. Chemical Action of Ionizing Radiations in Solution. Part XVII Degradation of Deoxyribonucleic Acid in Aqueous Solution by Irradiation with X-Rays. *J. Chem. Soc.* 38:226-232 (1957).
4. G. Scholes, J. F. Ward and J. Weiss. Mechanisms of the Radiation Induced Degradation of Nucleic Acids. *J. Mol. Biol.* 2:379-391 (1960).
5. R. Téoule and J. Cadet. Radiation-Induced Degradation of the Base Components in DNA and Related Substances-Final Products. In *Effects of Ionizing Radiation on DNA.* J. Hütterman, W. Köhnlein, R. Téoule and A. J. Bertinchamps, eds., pp. 171-203. Springer, Berlin (1978).
6. C. von Sonntag. Carbohydrate Radicals: from Ethylene Glycol to DNA-Strand Breakage. *Int. J. Radiat. Biol.* 46:507-519 (1984).
7. R. A. McGrath and R. W. Williams. Reconstruction in vivo of Irradiated Escherichia Coli Deoxyribonucleic Acid: The Rejoining of Broken Pieces. *Nature* 212:534-535 (1966).
8. J. T. Lett, I. Caldwell, C. J. Dean and P. Alexander. Rejoining of X-ray Induced Breaks in the DNA of Leukaemia Cells. *Nature* 214:790-792 (1967).
9. P. M. Corry and A. Cole. Double-Strand Break Rejoining in Mammalian DNA. *Nature New Biology* 245:100-101 (1973).
10. F. Hutchinson. The Molecular Basis for Radiation Effects on Cells. *Cancer Res.* 26:2045-2052 (1966).
11. S. Okada. DNA as Target Molecule Responsible for Cell Killing. In *Radiation Biochemistry: Cells.* Chapter III, pp. 103-147. Academic Press, New York (1970).
12. R. B. Painter. The Role of DNA Damage and Repair in Cell Killing Induced by Ionizing Radiation. In *Radiation Biology in Cancer Research,* R. E. Meyn and H. R. Withers, eds., pp. 59-68. Raven Press, New York (1979).
13. J. F. Ward. DNA Damage Produced by Ionizing Radiation in Mammalian Cells: Identities, Mechanisms of Formation and Repairability. *Prog. in Nucleic Acid Res. and Mol. Biol.* 35:95-125 (1988).
14. Z. Somosy, T. Kubasova and G. J. Köteles. The Effects of Low Doses of Ionizing Radiation upon the Micromorphology and Functional State of the Cell Surface. *Scanning Microsc.* 1:1267-1278 (1987).
15. S. M. Chiu, L. R. Friedman, N. M. Sokany, L. Y. Xue and N. L. Oleinick. Nuclear Matrix Proteins are Crosslinked to Transcriptionally Active Gene Sequences by Ionizing Radiation. *Radiat. Res.* 107:24-38 (1986).
16. G. J. West, I.W.-L. West and J. F. Ward. Radioimmunoassay of a Thymine Glycol. *Radiat. Res.* 90:595-608 (1982).
17. S. A. Leadon and P. C. Hanawalt. Monoclonal Antibody to DNA Containing Thymine Glycol. *Mutat. Res.* 112:191-200 (1983).
18. K. Hubbard, H. Huang, M. F. Laspia, H. Ide, H.B.F. Erlanger and S. S. Wallace. Immunochemical Quantitation of Thymine Glycol in Oxidized and X-irradiated DNA. *Radiat. Res.* 118:257-268 (1989).
19. A. F. Fuciarelli, B. J. Wegher, E. Gajewski, M. Dizdaroglu and W. F. Blakely. Quantitative Measurements of Radiation-Induced Base Products in DNA Using Gas Chromatography-Mass Spectrometry. *Radiat. Res.* 119:219-231 (1989).
20. J. F. Mouret, F. Odin, M. Polverelli and J. Cadet. ^{32}P Postlabelling Measurement of Adenine-N-1-oxide in Cellular DNA Exposed to Hydrogen Peroxide. *Chem. Res. Toxicol.* 3:102-110 (1990).
21. M. Weinfeld, M. Liuzzi and M. C. Paterson. Response of Phage T4 Polynucleotide Kinase Towards Dinucleotides Containing Apurinic Sites: Design of a ^{32}P-Postlabelling Assay for Apurinic Sites in DNA. *Biochemistry* 29:1737-1743 (1990).

22. M. Sharma, H. C. Box and D. J. Kelman. Fluorescence Postlabelling Assay of Thymidine Glycol Monophosphate in X-irradiated Calf Thymus DNA. *Chemico-Biol. Interact.* 764:107-117 (1990).

23. P. V. Hariharan and P. A. Cerutti. Formation and Repair of Gamma-Ray Induced Thymine Damage in Micrococcus Radiodurans. *J. Mol. Biol.* 66:65-81 (1972).

24. M. C. Paterson and R. B. Setlow. Endonucleolytic Activity from Micrococcus Luteus that Acts on γ-Ray-Induced Damage in Plasmid DNA of Escherichia Coli Minicells. *Proc. Nat. Acad. Sci.* (USA) 72:1997-2001 (1975).

25. G. J. West, I.W.-L. West and J. F. Ward. Radioimmunoassay of 7,8-Dihydro-8-Oxoadenine (8-hydroxyadenine). *Int. J. Radiat. Biol.* 42:481-490 (1982).

26. J. F. Ward, W. F. Blakely and E. I. Joner. Mammalian Cells are not Killed by DNA Single-Strand Breaks Caused by Hydroxyl Radicals from Hydrogen Peroxide. *Radiat. Res.* 104:383-393 (1985).

27. W. F. Blakely, A. F. Fuciarelli, B. J. Wegher and M. Dizdaroglu. Hydrogen Peroxide-Induced Base Damage in Deoxyribonucleic Acid. *Radiat. Res.* 121:338-343 (1990).

28. J. F. Ward. The Yield of DNA Double-Strand Breaks Produced by Ionizing Radiation. *Int. J. Radiat. Biol.* 57:1141-1150 (1990).

29. I. A. Radford. Evidence for a General Relationship Between the Induced Level of DNA Double-Strand Breakage and Cell Killing After X-Irradiation of Mammalian Cells. *Int. J. Radiat. Biol.* 49:611-620 (1986).

30. T. Coquerelle, A. Bopp, B. Kessler and U. Hagen. Strand Breaks and 5' End Groups in DNA of Irradiated Thymocytes. *Int. J. Radiat. Biol.* 24:397-404 (1973).

31. M. Fielden and P. O'Neill, eds. *Early Effects of Radiation on DNA.* NATO Advanced Research Workshop Series. Springer Verlag, New York (1991).

32. D. Schulte-Frohlinde and E. Bothe. Identification of a Major Pathway of Strand Break Formation in Poly U Induced by OH Radicals in the Presence of Oxygen. *Zeitshrift für Naturforschung* 39c:315-319 (1984).

33. J. F.Ward. DNA Lesions Produced by Ionizing Radiation: Locally Multiply Damaged Sites. In *Ionizing Radiation Damage to DNA: Molecular Aspects,* S. S. Wallace and R. B. Painter, eds. Wiley-Liss, New York (1990).

34. J. F. Ward and I. Kuo. Deoxynucleotides - Models for Studying Mechanisms of Strand Breakage in DNA. II Thymidine 3'5'- Diphosphate. *Int. J. Radiat. Biol.* 23:543-557 (1973).

35. J. F. Ward. Biochemistry of DNA Lesions. *Radiat. Res.* 104:S103-S111 (1985).

36. J. F. Ward. Some Biochemical Consequences of the Spatial Distribution of Ionizing Radiation Produced Free Radicals. *Radiat. Res.* 86:185-195 (1981).

37. J. F. Ward. Mechanisms of Radiation Action on DNA in Model Systems - Their Relevance to Cellular DNA. In *The Early Effects of Radiation on DNA,* E. M. Fielden and P. O'Neill, eds. NATO Advanced Research Workshop. Springer-Verlag, New York (in press).

38. D. T. Goodhead and H. Nikjoo. Track Structure Analysis of Ultrasoft X-Rays Compared to High- and Low-LET Radiations. *Int. J. Radiat. Biol.* 55:513-529 (1989).

39. M. R. Raju, S. G. Carpenter, J. J. Chmielewski, M. E. Scillaci, M. E. Wilder, J. P. Freyer, N. F. Johnson, P. L. Schor, J. Sebring and D. T. Goodhead. Radiobiology of Ultrasoft X-Rays. I. Cultured Hamster Cells (V79). *Radiat. Res.* 110:396-412 (1987).

40. S. S. Wallace. AP Endonucleases and DNA Glycosylases that Recognize Oxidative DNA Damage. *Environ. and Mol. Mutagen.* 12:431-477 (1988).

41. M. R. Mattern, P. V. Hariharan and P. A. Cerutti. Selective Excision of Gamma Ray Damaged Thymine Products from the DNA of Cultured Mammalian Cells. *Biochim. et Biophys. Acta* 395:48-55 (1975).

42. M. C. Paterson, B. P. Smith, P.H.M. Lohman, A. K. Anderson and L. Fishman. Defective Excision Repair of γ-Ray Damaged in Human (Ataxia Telangiectasia) Fibroblasts. *Nature* (London) 260:444-447 (1976).

43. R. B. Painter and B. R. Young. Radiosensitivity in Ataxia Telangiectasia: A New Explanation. *Proc. Nat. Acad. Sci. (USA)* 77:7315-7317 (1980).

44. P. E. Bryant. Kinetics and Mechanism of DNA-Strand Break Repair. *J. Cancer Res. and Clinic Onc.,* 116:843. Suppl. Part II, Abst. 05.16.05 (1990).

45. D. Blöcher, M. Nüsse, and P. E. Bryant. Kinetics of Double-Strand Break Repair in the DNA of X-Irradiated Synchronized Mammalian Cells. *Int. J. Radiat. Biol.* 43:579-584 (1983).

46. P. E. Bryant and D. Blöcher. The Effects of 9-β-D-Arabino-furanosyladenine on the Repair of DNA-Strand Breaks in X-Irradiated Ehrlich Ascites Tumor Cells. *Int J. Radiat. Biol.* 42:385-394 (1982).

47. J. F. Ward, E. I. Joner, and W. F. Blakely. Effects of Inhibitors of DNA-Strand Break Repair on HeLa Cell Radiosensitivity. *Cancer Res.* 44:59-63 (1984).

48. A. Giaccia, R. Weinstein, J. Hu, and T. D. Stamato. Cell Cycle-Dependent Repair of Double-Strand Breaks in a γ-Ray Sensitive Chinese Hamster Cell. *Somatic Cell and Mol. Genet.* 11:485-491 (1985).

49. G. E. Iliakis and T. Okayasu. Radiosensitivity Throughout the Cell Cycle and Repair of Potentially Lethal Damage and DNA Double-Strand Breaks in an X-Ray-Sensitive CHO Mutant. *Int. J. Radiat. Biol.* 57:1195-1211 (1990).

50. P. Oudet, E. Weiss, and E. Regnier. *Preparation of Simian Virus 40 Minichromosomes, Methods in Enzymology: Nucleosomes.* P. M. Wasserman and R. D. Kornberg, eds. Academic Press 170:14-52 (1989).

51. P. M. Calabro-Jones, D. T. Lai, J. R. Milligan, T. R. Nelson, and J. F. Ward. Alpha Particle Irradiation of DNA. *38th Annual Radiat. Res. Soc. Mtg.*, New Orleans, Abstract Ei4 (1990).

52. H. Martin-Bertram, P. Hartl and C. Winkler. Unpaired Bases in Phage DNA after Gamma-Irradiation In-Situ and In-Vivo. *Radiat. Environ. Biophys.* 23:95-105 (1984).

53. L. L. Ling and J. F. Ward. Radiosensitization of Chinese Hamster Cells by Bromodeoxyuridine Substitution of Thymidine: Enhancement of Radiation-Induced Toxicity and DNA-Strand Break Production by Monofilar and Bifilar Substitution. *Radiat. Res.* 121:76-83 (1990).

54. P. M. Calabro-Jones, J. Aguilera, J. F. Ward, G. D. Smoluk, and R. C. Fahey. Radioprotection of Cells in Culture by WR-2721 and WR-1065: Dependence of Protection upon Intracellular WR-1065. *Cancer Res.* 48:3634-3640 (1988).

55. J. D. Zimbrick, J. F. Ward and L. S. Myers, Jr. Studies on the Chemical Basis of Cellular Radiosensitization by 5-Bromouracil Substitution of DNA. I. Pulse and Steady-State Radiolysis of 5-Bromouracil and Thymine. *Int. J. Radiat. Biol.* 16:502-523 (1969).

56. J. D. Zimbrick, J. F. Ward and L. S. Myers, Jr. Studies on the Chemical Basis of Cellular Radiosensitization by 5-Bromouracil Substitution of DNA. II. Pulse and Steady-State Radiolysis of Bromouracil Substituted and Unsubstituted DNA. *Int. J. Radiat. Biol.* 16:524-534 (1969).

57. J. W. Lown. Molecular Mechanisms of Action of Anticancer Agents Involving Free Radical Intermediates. *Adv. Free Radical Biol. Med.* 1:225-264 (1985).

58. I. H. Goldberg. Atypical Abasic Sites Generated by Neocarzinostatin at Sequence Specific Cytidylate Residues in Oligodeoxynucleotides. *Biochemistry* 27:4331-4340 (1988).

59. J. W. Boag. Summing up the Seminar. *Int. J. Radiat. Oncol. Biol. Phys.* 5:1131-1134 (1979).

60. D. Frankenberg, B. D. Michael, M. Frankenberger-Schwarger and R. Harbich. Fast Kinetics of the Oxygen Effect for DNA Double-Strand Breakage in Cell Killing in Irradiated Yeast. *Int. J. Radiat. Biol.* 57:485-501 (1990).

61. J. B. Little. Low Dose Radiation Effects: Interactions and Synergism. *Health Phys.* 59:49-56 (1990).

Discussion

Zaider: What kind of energy do you need to produce one of these double-strand breaks?

Ward: To produce an OH radical you need 17 eV, to produce two OH radicals you need 34 eV. But to produce a double-strand break, each ·OH must react with a deoxyribose. The probability is only 4 percent that this will happen. If 200 eV is deposited, then 70 percent of such depositions give a double-strand break. This is into the range of soft X-rays.

Zaider: Over what distance are these six or ten damaged sites distributed?

Ward: It is possible that they will be 20 bases apart. It is also possible that they are close together; nanometer dosimetry should tell us this.

Paretzke: It's no problem to have three electron tracks or two ionizations within 5 angstroms each electron with low energy.

Varma: I think you are really begging the question; 17 eV is an average energy. It is not energy deposited in 10 nanometers or 5 nanometers; it's the average energy of 17 which produces one OH radical. In a linear way 34 eV will produce two, and only 4 percent of those would produce double-strand breaks. We don't know where that energy is deposited; could you comment on the significance of assuming average energy for this end point?

Ward: We are dealing with a sequence of probabilities here: First, the possibility that the energy deposition is close enough to cause damage; second, that enough energy is deposited to cause a double-strand break; and third, that the reactive species react with the deoxyribose moieties. The situation should therefore be described by products of probabilities. The statement made is a simplification; if you have enough energy to produce two OH radicals, then you could form a double-strand break.

Von Sonntag: Dr. Chatterjee has some calculations which he might quote, which indicates that with a scavenger not many double-strand breaks were formed by the two OH radical mechanisms.

Ward: In a mammalian cell, if you have one OH radical that is producing a damaged site in DNA, it has been shown that this radical has to have been produced within 20 angstroms or so of the DNA. The energy deposition event from which this ·OH comes may have had enough energy to produce more than one radical; then, you would have had OH radicals produced within the vicinity of the DNA. Thus with every single-strand break in a cell there is a probability of other radicals close by.

Chatterjee: What Dr. von Sonntag is saying is that we talked about calculations. Calculations show that when tracks form at random around DNA, the number of tracks very close to the DNA is very small. From those tracks very close to the DNA, ·OH are formed very close to the DNA. Many of them go to the bases. We assume that base damage by OH radical does not lead to strand breaks. If this is correct, then we do not find two hydroxyl radicals creating double-strand breaks, if you define double-strand breaks as two breaks on opposite strands within 10 base pairs. So we do not find enough probability to cause double-strand breaks.

Ward: Our calculations indicate that for the damages where two OH radicals attack DNA, only 4 percent would form double-strand breaks. Let's take another point. If a spur is formed so that energy is deposited in DNA, this will cause direction ionization. Thus, where the direct ionization of the DNA occurs there has to be a spur close, with the consequent radicals, so you can have double-strand breaks formed by a direct ionization and an OH radical.

Chatterjee: How many tracks go to the DNA, can you estimate?

Ward: If 35 percent of the damage to the DNA is by direct effect, then 35 percent of the spur (plus blob) energy causing DNA damage is deposited in the DNA itself.

Chatterjee: No, 35 percent of the energy in the whole cell for killing, not in the DNA itself.

Ward: Thirty-five percent of the damage to DNA is nonscavengeable by radical scavengers.

Chatterjee: OH radicals produced very close to the DNA may not be scavengeable.

Ward: If it is so close to the DNA, then that means that the energy deposition is also close to the DNA!

Chatterjee: I'd like to mention other experimental results from Krisch [Radiat. Res. Soc. Mtg., Abstr. Ei23 (1990)]. They have done experiments as a variation of tris scavenger. They find the ratio of double-strand breaks to single-strand breaks constant, as a function of tris concentration. That tells you that a single OH radical is creating a double-strand break.

Ward: According to Krisch's abstract, the ratio of single-strand to double-strand breaks at lower scavenger is 200:1, while above 500mM glycerol it decreases to 15-20:1. The single ·OH mechanism (0.5 percent of single-strand breaks giving double-strand breaks) predominates at low scavenger concentration, but as it approaches the cellular scavenging capacity the two ·OH mechanisms become more apparent.

Katz: There are some questions of consistency. You are saying that a single electron can create a double-strand break. The OH radical comes from a single electron; 2 ·OH is from one electron. Is that right? So one electron passing through can make a double-strand break. That would mean that high LET radiation would be less efficient than low LET radiation for double-strand breaks.

Chatterjee: Which it is for ·OH.

Katz: Is high LET radiation less efficient than low LET radiation? This is required if a single electron can produce a double-strand break.

Ward: The yield of double-strand breaks from alpha particles equals the yield of double-strand breaks of gamma radiation in conditions where the DNA is in intra-cellular radical scavenging conditions.

Ward: Can you do the calculations with dose from 5MeV alpha particles traveling through 40 microns of water?

Katz: Yes.

Ward: How do you relate that to gamma rays when the electrons are from the gamma ray?

Katz: The calculation is based on Butts and Katz [*Radiat. Res.* 30:855 (1967)]. We would want to know the D_{37} dose for gamma rays--then we would calculate the response to alpha particles by track theory. This is straightforward to a one-hit response.

Moolgavkar: A point of clarification; would you put up that slide where you have the different double-strand breaks? The one with sticky ends and the one with semi-blunt ends. The strand breaks are well separated; did you say that kind of damage is easily repaired by itself? And the second one: you had strand breaks close together. Would you explain again how repair takes place?

Ward: This is the data from the literature. When you have a radiation-damaged site, the cell inserts three or four nucleotides so that when the constituent single-strand breaks are well separated they can be easily repaired, but when the two single-strand breaks are close together there are problems in repairing with an intact template.

Moolgavkar: So, it excises those other nucleotides first and then fills the gap.

Ward: Correct. If you look at the repair of DNA double-strand breaks in Chinese hamster V-79 cells, there is a very fast component which has a 6-minute half-life. There is also a slow component with a half life of more than one hour. What we think is that this fast component repairing is of double-strand breaks in which the two breaks are well separated; the ones that are slowly repairing are those in which the breaks are directly opposite, or closely opposite, which the cell has problems putting back together.

Weinstein: We are talking again about the lethal damage. Can this discussion be generalized to other types of damage and impairments of biological processes?

Ward: There is not so much data for these other impairments; I suppose you mean mutagenesis and transformation. We have done experiments looking at mutation after hydrogen peroxide treatment. You can put a high concentration of hydrogen peroxide into a cell at 0°C (causing hundreds of Grays' worth of single-strand breaks) without the formation of mutations. To us this suggests that singly damaged sites do not cause mutagenesis.

Kopelman: I'd like clarification. Did I understand right that the DNA separates into single strands so that it can be repaired?

Ward: No, it repairs while in the double-strand state.

Kopelman: Is there a possibility of repair when the ends of a double-strand break separate?

Ward: In some of the models that Dr. Curtis described, there are suggestions that some double-strand break ends separate and then can rejoin later. There are two factors to consider: (i) the probability of the two ends correctly matching up again among the total mass of DNA, and (ii) loops of DNA being anchored to scaffold proteins, possibly preventing the ends from separating too far.

Menzel: I appreciate very much the clear presentation of your paper and I believe that your explanation of the radiation quality effect appears to be plausible. Can you explain the dose rate effect in a similar way in view of the relatively low frequencies of the free ends?

Ward: Cells are capable of "repairing" these damages. They can rejoin double-strand breaks at 24°C. Split dose recovery occurs at 24°C. At a low dose rate, the breaks are introduced slowly, and the cells may repair them more accurately at low doses when there are few present. At low dose rates there will be no built-up concentration of damage.

Menzel: So survival is a function of the cell as a whole as a repairing organism.

von Sonntag: Immediately after irradiation would a slight increase in double-strand break yield be expected because of the excision of base damage?

Ward: If the rate of incision at base-damaged sites is slower than the rate of resealing, then the incision would not show up above the resealing. However, incision would show up as slower resealing.

Chatterjee: Would you comment on the situation where the DNA is wrapped around histones? Most of this discussions could be applied to linker DNA, but what about the chromatin-wrapped DNA. Would you comment about base damage, the different chromatin breaks, the number of single-strand breaks and the number of double-strand breaks?

Ward: This is something to which you referred in your talk, breaks which occur between the chromatin beads and those which occur on the nucleosome. If a double-strand break occurs in a linker region, the DNA ends may be free to move away from each other; however, a double-strand break occurring in the DNA which is salt-linked to the positively charged histone may be held together and more readily repaired.

Chatterjee: Does histone H_1 play a role in terms of DNA protein cross link formation?

Ward: The recent papers on DNA protein cross links (Oleinick, Cress), show no DNA protein cross links with histones. It is the scaffold proteins which are involved. This could indicate that the histones are not so close to the DNA.

Chatterjee: Dr. Curtis' model considers that the breaks that are produced on the chromatin core are repaired much more easily than those on the linker DNA.

Ward: There is a possibility that the rate is less for internucleosomal break rejoining. This may be difficult to prove experimentally.

Varma: How fast can you measure this repair?

Ward: In this work the damage is put into the cells at 0°C, and then the cells are moved to the temperature under study. At various times samples are taken and rapidly cooled (no repair takes place below 10°C). Therefore time is not a problem since the repair can be stopped. At intermediate temperatures repair is slowed over the immediate post-irradiation time period (when "sub-lethal damage repair" takes place). If repair is in competition with damage fixation, then this change of

temperature might be expected to change the relative rates of these two processes. The ratio of the two reactions at 37°C should be different at lower temperatures if the activation energy of the reactions differs. However, with 3 hours holding after irradiation at a series of temperatures from 10° to 37°C, there are no effects on cell survival.

Weinstein: I was wondering about a slide in which you showed the number of strand breaks that are required to kill cells from radiation compared to the number of breaks from chemical means. If in models, for example, one quantitates and tries to relate radiation to cell killing and yet there is an intermediator, some translation function that translates number of breaks to cell killing, how then can the model be based mechanistically on breaks in the quantitative aspect?

Ward: You are talking about single-strand breaks. However for the ionizing radiation, the number of double-strand breaks required to kill a cell is 40, and for bleomycin the number of double-strand breaks to kill a cell is 30.

Weinstein: And the other agents; we don't know?

Ward: We don't find double-strand breaks in any of the other situations. This is why we will use agents such as bleomycin, to see if the number of double-strand breaks per kill is constant. We will try to see whether the quality of the double-strand breaks has any effect on the lethality of the break.

Zaider: How do you explain that you need only 1,000 single-strand breaks from ionizing radiation to kill a cell?

Ward: It is not the single-strand breaks that cause the death. It is the double-strand breaks.

von Sonntag: Perhaps you might indicate a mechanism that would explain single-strand breakage by the hydrogen peroxide in the cells.

Ward: The cells are treated with hydrogen peroxide; we rely on adventitious metal ions bound to the DNA. The hydrogen peroxide reacts with either cuprous or ferrous, for instance, producing OH radicals which then go on to react to the DNA. We have shown that OH radicals are involved by using OH radical scavengers; however, there is a certain yield of damage which is not removable by OH radical scavengers. There may be ferryl species involved in causing breaks.

Varma: What is the accepted definition of a single-strand break?

Ward: The discontinuity in one strand.

Rossi: If you say it takes 1,000 single-strand breaks and 40 double-strand breaks with ionizing radiation, do you mean any ionizing radiation or X-rays?

Ward: This is for low LET irradiation.

Rossi: You don't know what it is for neutrons do you?

Ward: No.

Varma: You know for alpha particles, though.

Ward: The yield of breaks for alpha particles is about the same yield per Gray as for ^{137}Cs gamma radiation.

Moolgavkar: That list of experiments you showed relating damage to survival; were they all done with the same cells?

Ward: No, they were not done with the same cells.

Braby: You said that the yield of double-strand breaks per Gray for alpha particles is about the same as for gamma radiation. What are the ratios of double-strand breaks to single-strand breaks for the two radiations?

Ward: The values that I use are for SV40 DNA irradiated in a simulated cellular environment, using 0.1 M DMSO as an ·OH scavenger. The ratio of double-strand breaks to single-strand breaks for gamma radiation is 11, and for alpha particles it is 29 (manuscript in preparation).

Varma: These single-strand breaks are repairing very fast.

Ward: They have a half life of 3 to 4 minutes.

Varma: The loss of bases from cellular DNA is occurring at a high rate at 37°C. [Lindahl and Nyberg, *Biochemistry* 11:3610 (1972)]. About 600 purines are lost per cell per hour, normally. If the number being repaired is that high, why do we see all these single-strand breaks?

Ward: Because the cells are irradiated at 0°C where repair does not take place. The number you quote is for loss of purines leaving apurinic sites--not strand breaks. At current background radiation dose rates (1×1^{-3} Gray per year), a mammalian cell would see a double-strand break once every 25 years!

Varma: Is that the normal frequency of the double-strand break? What is the number for single-strand breaks?

Ward: Twenty-five times higher—one per year.

Rossi: I would like to go back to the point I made before. I think it is extremely important to know whether or not the result would be the same for neutrons.

Paretzke: This can be answered; if there are the same number of double-strand breaks in DNA, 40 per D_0 from low LET radiation and from alpha particles. The D_{37} for alpha particles is a fraction of that for gamma radiation, so it takes only two to four double-strand breaks to inactivate the cell when they are produced by high LET particles. Of course, they are more difficult to repair, as you showed in your other slide.

Ward: It should be remembered that the yields of the alpha particles are for the model system.

Varma: But from a mechanistic point of view, it is very important to recognize that the number of tracks is very different; yet, you get the same number of double-strand breaks.

Ward: This is probably because the quality of the double-strand breaks is different.

Varma: In the experiment of Meyrick Peak from Argonne National Laboratory [Peak et al. *Int. J. Radiat. Biol.* 55:761 (1989)] that we discussed yesterday, he claims that the qualitative picture of double-strand breaks is different with fission spectrum neutrons at Argonne than with gamma radiation. Are there two types of double-strand break, by high and low LET radiation?

Ward: The sort of damage from neutron radiation would be expected to be more complex than that from gamma radiation. If the neutron deposits its energy sufficiently close to the DNA, there may be multiple base damages in the vicinity of the double-strand break, possibly between the constituent single-strand breaks on opposite strands.

Varma: Do we have the technique to distinguish these multiply damaged sites?

Ward: In the work reported by Dr. Peak, neutrons produce a higher proportion of nonrejoining single-strand breaks. The bluntness of the ends is inferred from the cell's inability to repair these breaks. This nonrejoinability has been reported as a general characteristic of high-LET-induced breaks [Ritter et al. *Nature*, 266:653 (1977)].

von Sonntag: I would like to mention work of Dr. Martin-Bertram [*Radiat. Environ. Biophys.* 23:95 (1984)]. She irradiated bacteriophages and observed the formation of "bulky" lesions, in which bases no longer matched. These were produced by gamma rays. To what extent do they play a role?

Ward: My interpretation of her data is that what she might be seeing in addition to mismatched or non-hydrogen-bonded bases is two strand breaks which occur on the same strand, deleting a short patch of deoxynucleotides. When they treated the DNA, which had been irradiated in the phage head with S_1 nuclease, the yield of single-strand breaks increased.

von Sonntag: Would the short single-stranded region leave the complementary strand?

Ward: If it were short enough, yes.

Inokuti: Please explain the nature of the observation.

Ward: The observation is that if the DNA was irradiated in solution, it is not susceptible to S_1 nuclease. This enzyme specifically cleaves single-strand regions of DNA which are more than three deoxynucleotides long. If the DNA was irradiated in a situation where it was packaged in a phage head or in yeast (Andrews et al. *Int. J. Radiat. Biol.* 45:497 (1984); Geigl and Eckardt-Schupp, *Molec. Microbiol.* 4:801 (1990)], the effect was as follows. When the DNA is assayed for strand breaks, a certain yield is obtained; if the same DNA is treated with an enzyme that specifically cuts single-stranded DNA, an increase in strand breaks yield is observed. My interpretation is that irradiation induces deletions of short stretches of single-stranded DNA leaving a single-stranded region susceptible to the enzyme.

Structure-Function Relations in Radiation Damaged DNA

Roman Osman, Karol Miaskiewicz, and Harel Weinstein[a]

Introduction

The most important biological effects of exposure to ionizing radiation can be related to a variety of changes in cell function. Some of these changes can produce cell death, but others lead to less final deleterious effects such as carcinogenesis or altered cell function as a result of energy deposition in the biological system. All the changes in cell function can be linked to DNA damage, with the double-strand break and the radiation-induced mutations causing most of the lethal damage. Increasingly more accurate and direct measurements in radiation dosimetry, as discussed at this Conference, and the understanding provided by the theories and formulations of condensed matter physics, also presented and discussed at great length at this meeting, have offered important insight into the parameters and measurable outcomes of exposure to ionizing radiation. These are enhanced by findings, such as those presented at this Conference by Clemens von Sonntag, that emerge from radiochemistry measurements *in vitro* of chemical changes produced by radiation exposure. Also as described at this conference by Aloke Chatterjee and Herwig Paretzke, computational simulations based on Monte Carlo algorithms have been developed to explore the parameters of the energy deposition processes and their consequences in models of the biological systems. But a clear, mechanistic link between the physical processes and the biological consequences remains somewhat elusive, suggesting that it will be necessary to know and understand first the sequence of events that lead from energy deposition of radiation in condensed matter to the biophysical and biochemical processes that occur at the level of cellular DNA.

DNA is one of the most important targets that are damaged as a result of the chemical reactions caused by ionizing radiation. Irreparable damage produced in DNA usually leads to cell death. It is well established that ionizing radiation can create lesions in DNA either by a *direct* mechanism in which the high-energy radiation ionizes the DNA molecule, or by an *indirect* mechanism in which hydrated electrons (e_{aq}^-), hydrogen radicals ($H\cdot$), and hydroxyl radicals ($OH\cdot$), that are formed by the interaction of the ionizing radiation with water, react with DNA. The various lesions induced in DNA through such mechanisms have been demonstrated as specific molecular events [for a review see [1]]. For example, most of the primary damage, direct and indirect, induced by ionizing radiation is centered on the bases of DNA. Direct ionization of DNA produces radical cations, usually $Pu^{+}\cdot$, which upon deprotonation can yield the radical $Pu\cdot(-H)$. Electron capture produces radical anions, usually $Py^-\cdot$, which upon protonation yield the radical $PyH\cdot$. The protonation on heteroatoms (N

(a) Department of Physiology and Biophysics, Mount Sinai School of Medicine of the City University of New York, New York

Physical and Chemical Mechanisms in Molecular Radiation Biology
Edited by W.A. Glass and M.N. Varma, Plenum Press, New York, 1991

423

or O) is rapid and kinetically controlled, but the protonation on carbon yields a product that is much more stable thermodynamically.[2,3] Similar products can be obtained by additions of H· to pyrimidines, which result in carbon-centered radicals with the concomitant reduction of the C_5-C_6 double bond. Another major primary damage to DNA bases is the addition of OH· to double bonds of the bases. The addition of OH· to pyrimidines has been studied extensively[1] and recently new progress has been also made in the studies of OH· addition to purines.[4] Base damage can lead to a lesion in DNA, which can be demonstrated as the release of bases from irradiated DNA.[1,5,6] However, there is increasing evidence that cell death can be related to unrepaired double-strand breaks,[7-12] which can be induced only through damage generated in the deoxyribose portion of the DNA.

One way of producing a strand break through a lesion in the deoxyribose portion of DNA is through an *indirect* mechanism of hydrogen abstraction by an OH·. Such a process will produce a carbon-centered radical on the sugar which can undergo a β-elimination to produce a strand break. Alternatively, a damaged base, produced in one of the processes described above, can abstract a hydrogen from a neighboring sugar moiety, transferring the radical to the sugar and ultimately leading to a strand break through a β-elimination process. These alternatives are discussed in Dr. von Sonntag's contribution to this volume in relation to the currently accepted mechanistic hypotheses for the production of lethal lesions in DNA.[13]

At the present time, when the methods of molecular biology and cellular physiology make possible the investigation of biological processes from a mechanistic viewpoint at a discrete molecular level, it becomes reasonable to seek an understanding of the chemical nature and structural consequences of radiation damage in DNA. These constitute the basis for understanding the molecular mechanisms that underlie the biological consequences of exposure to ionizing radiation. Such an approach to elucidating the mechanisms for the biological effects of ionizing radiation would take advantage of the newly obtained mechanistic clues offered by findings from molecular biology, and the structural clues offered by the information that structural biology has derived from X-ray crystallography, NMR spectroscopy, and the new generation of electron microscopies—examples for all of which were also presented at the conference. Our own work follows this new avenue of research made possible by the current explosive progress in these disciplines of molecular research, and integrates the experimental results and theoretical models with the use of the conceptual and computational tools offered by theoretical chemistry and molecular biophysics. Some of the recent findings and new concepts, as well as the models they suggest for further study, are described and briefly illustrated below.

We begin with an illustration of some of the approaches and results from the study of the relevant molecular processes in terms of explicit molecular mechanisms. These mechanisms describe the conformational changes that radiation can produce in DNA, and that can be considered—in relation to impaired repair mechanisms and regulation of gene expression—to lead to altered biological function. The understanding we seek of the chemical reactivities and structural changes in DNA at the molecular level of detail should allow us to establish a causal relationship between the local lesions—ionizations, base modifications, base loss, and strand breaks—and the global changes in conformational properties which may lead to the impairment of regulation of gene expression. As a result of such an impairment, new gene products can be introduced into the cell, as discussed at this conference by Eric Hall, and repair mechanisms may be activated or altered, as discussed by John Ward. The repair mechanisms, whether original or altered, have only limited fidelity and could thus transform the local lesions in DNA either into lethal lesions (e.g., double-strand breaks) or into inheritable damage in the form of irreparable mutations. The lethal lesions are conducive to the cell death end point that is so frequently observed. Mutations, on the other hand, especially if they are not repaired, can persist and lead to an altered biological activity which may result in the induction of the carcinogenesis discussed at this conference by Suresh Moolgavkar. The feasibility and consequences of such mechanistic details require careful exploration if the process of DNA damage is to be understood at the molecular level. To this end we have also studied by computational simulations some of the chemical reactions responsible for the local damage, in addition to the investigation of the structure-function relations in damaged, conformationally altered DNA.[14-19]

We conclude with proposed directions for combined theoretical and experimental investigations of the mechanistic details of processes that have emerged thus far as the key elements in the evoked biological consequences of cell exposure to ionizing radiation.

Notes on Methodology

The theoretical constructs and computational approaches used to obtain the results summarized below are based on the methods of theoretical chemistry that are very well suited to address the properties of the reactive species produced by radiation and the mechanisms that lead to global conformational changes. With the use of quantum chemistry, it is possible to characterize the properties of reactive radicals and to simulate the processes of bond breaking as models of strand breaks. With the methods of molecular dynamics, it is possible to explore the conformational properties of damaged DNA in solution as well as the interactions with regulatory proteins as models of the impairment of the regulation of gene expression. All these methods provide insight into the most likely pathways, the energetic costs, and the structural details of the molecular processes that connect the local damage induced in DNA to the global change in DNA structure that is reflected in DNA function.

The successful use of computational methods developed in theoretical chemistry—including quantum chemistry, molecular dynamics, and molecular mechanics—in the investigation of biological structure and mechanisms is well documented in the literature that includes extensive reviews of the strengths and limitations of these methods and of their performance in computational simulations.[20-30] These methods have also been applied to the study of the electronic structure of DNA components, conformational properties of short strands of DNA, and to the electrostatic and hydration properties of DNA [see references above, also[31] and references therein]. With this wealth of reviews, and specific details given in our own publications,[14-19,32-43] we will not attempt here a detailed account of the methods of theoretical chemistry that were used in the study. Based on these well-defined methods, the following protocols were used for specific components of our quantum chemical calculations:

Structure optimizations by energy minimization criteria, which include ground-state and transition-state structures, were carried out with the gradient methods available in the standard packages of programs for quantum chemistry, GAUSSIAN 86 and GAUSSIAN 88, using the Unrestricted Hartree Fock (UHF), the Restricted Open Shell Hartree Fock (ROHF), and the General Valence Bond (GVB) formalisms with the high-quality (split-valence or better) basis sets built into the program.

Electron correlation effects were calculated with the Moller-Plesset Perturbation (MPn), Configuration Interaction (CI), and Multi-Configurational Self Consistent Field (MCSCF) methods also available in the GAUSSIAN packages.

Specific details of the molecular dynamics simulations are given below, in the description of the first results obtained for a DNA molecule incorporating radiation-induced damage.

Global Conformational Properties of Radiation-Damaged DNA and Its Relation to Function

Mechanistic Hypotheses

Specific protein-DNA interactions are the key elements in the regulation of gene expression. Recently, several molecular structures of protein-DNA complexes have been elucidated with the experimental methods of X-ray crystallography and two-dimensional (2-D) and three-dimensional (3-D) NMR. Significant structural modifications were observed in the crystalline complexes of DNA with repressor proteins,[44-50] as illustrated in Fig. 1. Changes in the conformation of DNA, especially distal bends induced in DNA by interactions with proteins involved in the regulation of gene expression, have been shown to be essential for function.[51-53] Notably, appropriate bends were shown to be conducive to function whether they were induced by the authentic proteins or by other structures that have the same effect on conformation[54]. These novel findings offered by elucidation of DNA function with the methods of molecular biology, and of DNA conformation with methods of structural biology, point to another mode in which radiation damage can impair cell function. Thus, conformational changes could be produced by local radiation damage and propagated into the macromolecular structure, as suggested by our calculations (see below). These conformational changes can interfere with cellular processes that depend on protein-DNA interactions and are sensitive to the conformation of the protein-binding region of DNA. The processes of DNA repair as well as DNA transcription fall into this category of mechanisms susceptible to impairment by modification of DNA conformational properties, because they involve protein-DNA complexes. Properties of the DNA molecule affect both the specificity and the affinity of DNA-protein complexes. Yet these mechanisms of transcription and of repair are also key factors in transforming radiation-induced lesions into the well-known biological consequences of cell death or carcinogenesis. To understand the mechanisms leading from radiation damage to these biological consequences, it thus becomes essential to elucidate the relation between local damage to DNA and conformational changes induced in the macromolecular structure by the presence of a local lesion.

Conformational changes are also essential elements in the mechanistic sequence of events that incorporates experimental results on damage to single- and double-strand DNA into a working hypothesis for the link between radiation damage to DNA and long term (inheritable) cellular damage:

Step 1: The *primary damage* to DNA, in the form of base ionization or OH· addition to bases, causes a conformational change in DNA by modifying the interactions that are important to maintain the structure of intact DNA. Thus, addition of OH· to thymine changes the structure of DNA because the C_5 of thymine is now a tetrahedral center and the additional OH group on C_5 brings about new interactions with the neighboring atoms that are transmitted into conformational changes (see below). Ionizations of bases will lead to charge shifts or to deprotonations that will change the base pair interactions and will probably weaken the stabilizing forces that hold the double-strand DNA in its intact form.[55]

Step 2: The conformational changes lead to the conversion of *primary damage* to *secondary damage*. Loss of hydrogen-bonding or stacking interactions will lead to DNA bending which can then facilitate further unwinding of DNA.[55] Once the two strands

Osman, Miaskiewicz, and Weinstein

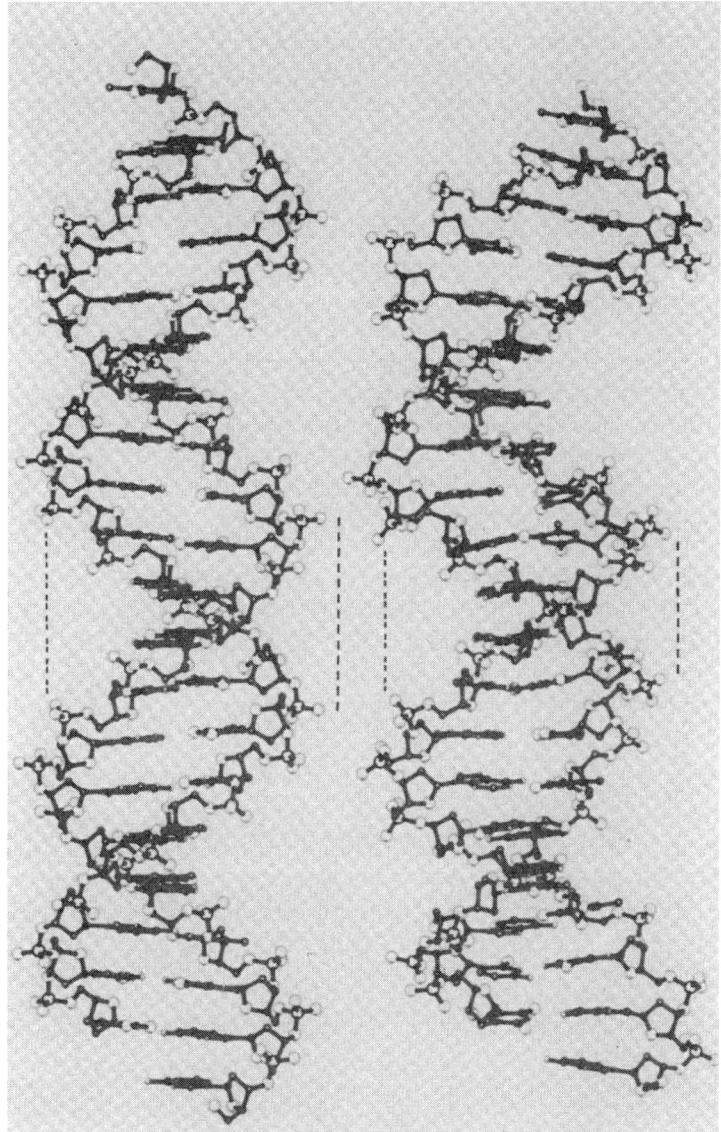

Figure 1. A comparison between canonical B-DNA (left) and the DNA operator of phage 434 repressor (right)[46]. The operator has been separated from its complex with the repressor protein and redrawn in isolation. The distortion of the operator is highlighted by the vertical lines on the left and right of each helix, which denote the reduced opening of the major groove and the increased opening of the minor groove in the DNA operator, respectively.

of DNA are unwound, the behavior of each strand could locally resemble that of a single-strand DNA and the transfer of a base radical to the sugar becomes possible. Sensitivity of irradiated DNA to single-strand-specific nucleases is in full accord with such a hypothesis.[56,57] The formation of sugar radicals should lead to strand breaks. Strand-broken DNA must exhibit different conformational properties than intact DNA because the continuity of one of the strands has been disrupted, endowing the DNA with a much greater flexibility. Whether such a strand break leads to a lethal consequence or not could depend on the presence of another damaged site in the vicinity according to the hypothesis of Locally Multiply Damaged Sites (LMDS).[13] Alternatively, the primary damage formed on the base can lead to base loss[58-60] or to base destruction and its conversion to a non-coding residue.[61] Again, the survival of such a damage depends on its being in an LMDS area; otherwise, it can be excised by a glycosylase/apurinic/apyrimidinic (AP)-endonuclease and ultimately repaired.[62-64] The AP sites are non-instructional, and their genetic toxicity is well documented.[65] However, the repair mechanism, upon an encounter of an AP site, will insert an adenine nucleotide at a high preference over any other nucleotide.[66] Thus, the AP sites generated through radiation damage to bases can be highly mutagenic; indeed, this is supported by experimental findings.[67]

The impact of AP sites on the properties of DNA is not well understood. It is very likely that AP sites cause conformational changes in DNA (see above) but there is little doubt that the electrostatic recognition properties of DNA must change upon loss of a base from its sequence. To what extent the change in the recognition properties and the change in the conformation of DNA trigger the AP-endonuclease to engage in excision of the AP site is not known. As discussed further below, in the section on Mechanistic Conclusions, molecular dynamics simulations of the conformational behavior of DNA with *secondary damage*, e.g., AP sites, and the analysis of the electrostatic and steric properties of the average structure that results from the simulations, make it possible to identify the salient differences between radiation-damaged and intact DNA. The results should lead to specific hypotheses concerning the molecular mechanisms of recognition of modified DNA by proteins regulating gene expression, and of DNA damage by the repair enzymes. These hypotheses connect the findings at the discrete molecular level to biological mechanisms that are relevant to the processes of inheritable and/or lethal cell damage.

Molecular Dynamics Simulation of Radiation-Damaged DNA

In preliminary studies, we have conducted a 150 ps molecular dynamics simulation of a dodecamer of DNA in which one of the thymines was replaced with a 5-hydroxy-6-thymidyl radical. This species is the main product of the attack of an $OH\cdot$ produced by energy deposition in aqueous media. The characteristics of the reaction and of the chemical products were first elucidated in detail from the results of quantum chemical calculations of the reaction mechanism and the energetics of the products. These results are briefly summarized below as a basis for the characterization of the molecular properties and the formation of the damaged DNA segment used in the computational simulation of the conformational changes produced by the local damage.

DNA Base Radicals Formed by $OH\cdot$ Addition. The significance of addition of $OH\cdot$ to the C_5-C_6 double bond of uracil and thymine stems from the reactivity of the produced base radicals that are formed as a result. Such radicals of uracil have been found to abstract a hydrogen atom from the neighboring sugar to produce sugar radicals.[68,69] The subsequent reactions of the sugar radicals may lead to lethal DNA strand breaks. The $OH\cdot$ adducts to bases may also induce other mutagenic or carcinogenic damage. Experimental evidence shows that the addition to C_5 of uracil is 4.6 times more prevalent than the addition to C_6.[70] The addition to C_5 of thymine is only two times preferred over the addition to C_6. But the results from $OH\cdot$ addition do not agree with the relative stabilities of these radicals inferred from other experiments. For example, analysis of the yields of the products of water addition to the radical cation formed by $SO_4^-\cdot$ oxidation of thymine indicates that the OH(5) 6-yl radical is not formed at all, whereas the OH(6) 5-yl radical is formed in 70% yield. The rest is the allylic radical formed from the deprotonation of the C_5-methyl.[71] Similar results are obtained from irradiation of frozen solution of thymidine.[72]

Unfortunately, such an analysis cannot be conducted with uracil (or 1,3-dimethyl uracil) because it undergoes a chain reaction which prevents the system from reaching equilibrium.[73] Based on quantum chemical calculations of the radical cation of uracil, Schuchmann et al.[73] speculated that the OH(6) 5-yl radical should be the more stable product. The question that is raised by these results is whether the addition of OH· to the C_5-C_6 double bond reflects the relative stabilities of the products or the relative energies of the transition states.

Energetics of OH· addition - The energetics of OH· addition to thymine and uracil are summarized in Table I below. The results demonstrate that the most stable structure for thymine is the OH(6)-axial 5-yl radical. For uracil, the OH(5) 6-yl radical is preferred by a small amount at the quantum chemical level of calculation. Interestingly, for both thymine and uracil the isomer with the OH(6) in an equatorial conformation was found to be unstable, and during the optimization procedure it always converted to the structure with the OH(6) in an axial conformation. In contradistinction to the 5-yl radical, the OH group on C_5 in the 6-yl radical can be in either the axial or the equatorial position; the latter is energetically preferred. The likely reasons for these preferences are discussed below in the section concerning the structures of the radicals.

The results from the quantum chemical calculations show that the preferred products of the addition of OH· are the OH(5) 6-yl radical of uracil and the OH(6) 5-yl radical of thymine. However, it should be kept in mind that the *ab initio* calculations refer to the molecules in gas phase, while the experiments have been done in aqueous solution. In an attempt to model the situation in solution, the hydration energies, E_{solv}, have been calculated with the method developed by Rashin based on

Table I. The Energetics of OH· Addition to Thymine and Uracil

	TE(a.u.)[a] PMP2/6-31G*	E_{solv} (Kcal mol^{-1})	Relative Energy (Kcal mol^{-1}) PMP2/6-31G*	PMP2+E_{solv}
Thymine				
OH(6)-ax	-528.373579	-19.80	0.00	0.00
OH(5)-ax	-528.366487	-22.09	4.45	2.16
OH(5)-eq	-528.368527	-19.98	3.17	2.99
Uracil				
OH(6)-ax	-489.193098	-21.87	0.00	0.00
OH(5)-ax	-489.191481	-21.85	1.01	1.03
OH(5)-eq	-489.193682	-19.65	-0.37	1.89

[a] PMP2/6-31G* stands for Projected Moller Plesset perturbation calculation to second order with the basis set specified after the slash.

the Born model of solvation by a continuum dielectric.[74] The hydration energies and the corrected Projected Moller Plesset perturbation (PMP2) relative energies which include hydration are presented in Table I. They show that the OH group in the axial position at C_5 is slightly better hydrated than the OH group in the equatorial position. Consequently, structures with OH(5) axial become more stable in solution than those with OH(5) equatorial. This situation is the opposite of that in gas phase, probably due to the higher accessibility of water to an axial OH group. The hydration in thymine radicals favors OH(5)-axial over the other two structures. As a result of the inclusion of E_{solv} the OH(6) 5-yl radical of thymine is more stable than the OH(5) 6-yl radical by 2 Kcal mol^{-1}. This is in agreement with the experimental results showing that the OH(5) 6-yl radical of thymine was not formed as a consequence of the interaction of water with the radical cation.[71] Similarly, the OH(6) 5-yl radical of uracil is more stable than the OH(5) 6-yl radical by 1 Kcal mol^{-1}, which is in agreement with the speculation of Schuchmann et al.[73]

Taken together, these results indicate that the distribution of radicals obtained as products of the addition of OH· must be kinetically controlled because the relative energies of the product radicals prefer the radicals with the OH group on C_6, whereas C_5 is preferred in the addition reaction. This is consistent with the observed high rates of addition of OH radicals to pyrimidines, suggesting that the transition states for the addition will occur early along the reaction coordinate and will resemble the reactants.

Structures of the base-radicals - The structures of base radicals formed from the addition of OH· to thymine and uracil radicals are qualitatively the same. As pointed out above, structures with OH(6) in an equatorial position do not exist for either uracil or thymine. This can be understood from simple sterical considerations. An axial position for a large substituent such as the OH group should be favored in the pyrimidine ring because the unfavorable eclipsing interactions at the equatorial plane will be eliminated. The structure of uracil with an OH(6) in an axial orientation is shown in Fig. 2.

In view of this explanation, it is surprising to find that when the OH group is on C_5 both the axial and the equatorial isomers exist. Moreover, in spite of the steric eclipsing in the equatorial plane, the OH(5)-equatorial isomer is the more stable one (see Table I). The structures of the OH(5) 6-yl radicals of uracil shown in Fig. 3 suggest that the reason for different structural preference at C_5 and C_6 lies in the

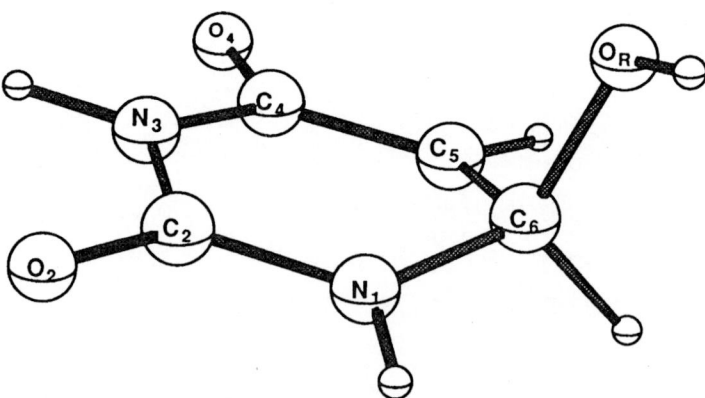

Figure 2. Structural model of 6-hydroxy-5-uracilyl radical. The conformation of the OH group is axial.

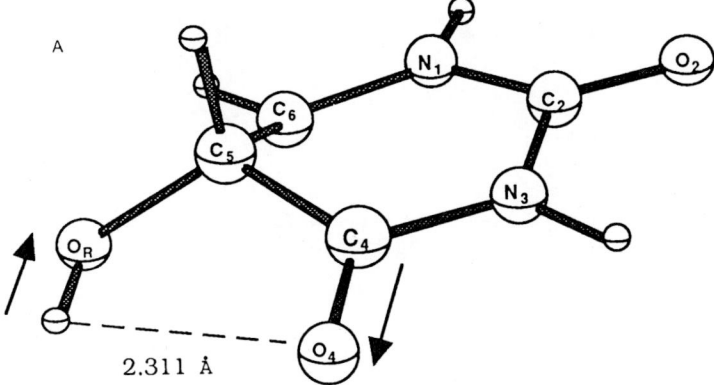

Figure 3A. Structural model of 5-hydroxy-6-uracilyl radical with equatorial hydroxyl. Arrows indicate the alignment of the dipoles and the possible hydrogen bonding distance between O_4 and the hydroxyl hydrogen.

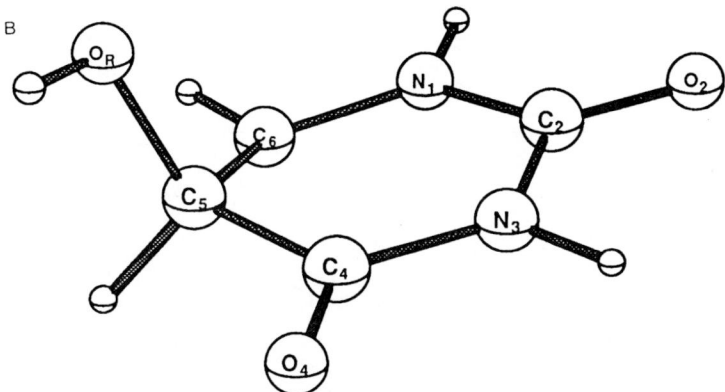

Figure 3B. Structural model of 5-hydroxy-6-uracilyl radical with axial hydroxyl.

electrostatic interaction between the O_R-H group and the C_4-O_4 carbonyl group. Both groups lie almost ideally in one plane with antiparallel orientation of the dipoles of these two bonds, which contributes to the electrostatic stabilization of the equatorial isomer. Such a stabilization is impossible when OH(6) is equatorial because the O-H and N_1-H bonds will be in a highly unfavorable parallel orientation of bond dipoles.

An important property of the uracil radicals is the conformation adopted by the pyrimidine ring. It is important to remember that these radicals are formed in DNA and their conformation may disturb the local structure of DNA (see results from molecular dynamics simulations, below). The schematic representations in Fig. 4 demonstrate that the pyrimidine ring is not planar. The ring adopts an envelope conformation, with the C_6 lying out of plane for the axial 6-hydroxy-5-uracilyl and 5-hydroxy-6-uracilyl radicals and a half-chair conformation for the equatorial 5-hydroxy-6-uracilyl radical. The conformational difference between the half-chair of the equatorial 5-hydroxy-6-uracil radical and the envelope of the two other structures is likely due to the dipole-dipole bond interaction described above. This interaction forces the C_5 atom out of plane to align the dipoles of the OH and CO bonds parallel to each other. The ring conformation of the axial 5-hydroxy-6-uracilyl radical is closest

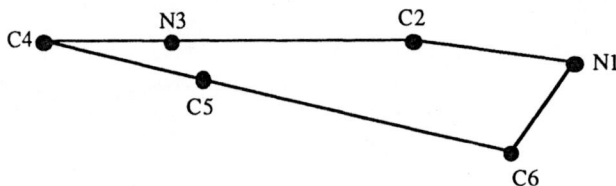

Figure 4A. Conformation of axial 6-hydroxy-5-uracilyl radical. The plane defined by $C_4N_3C_2$ is perpendicular to the plane of the figure.

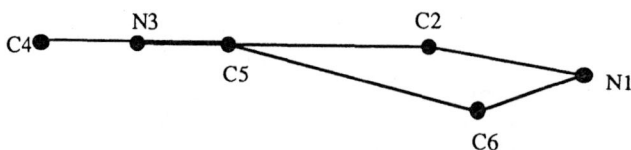

Figure 4B. Conformation of axial 5-hydroxy-6-uracilyl radical. The plane defined by $C_4N_3C_2$ is perpendicular to the plane of the figure.

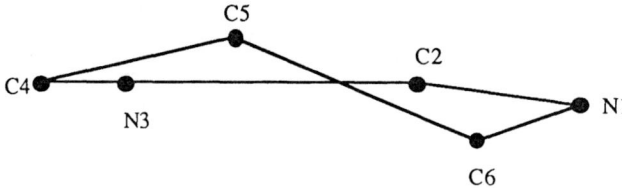

Figure 4C. Conformation of equatorial 5-hydroxy-6-uracilyl radical. The plane defined by $C_4N_3C_2$ is perpendicular to the plane of the figure.

to that of the original uracil. The C_5, C_4, N_3 and C_2 atoms lie in one plane and the out-of-plane deviation of N_1 is quite small. In the two other structures, perturbations of the initial planar conformation of the ring are much larger.

As shown in Fig. 4, the structure of the radical center is very different for OH(5) and OH(6) adducts. Thus, the C_5-yl radicals (i.e., OH at C_6) are almost planar, whereas the C_6-yl radicals (i.e., OH at C_5) have a very distinct pyramidalization of the pyrimidine ring. The dihedral angle for C_6-yl is 170°, but for C_5-yl it is only 152-155°. These differences in the conformations of the ring suggest that the electronic structure of these two radicals must be different. In the C_6-yl radicals, the unpaired electron occupies an orbital, which is closer to a *sp*-type hybrid than to a π-type orbital. In the C_5-yl radical, the interaction of the unpaired electron with the π-system of neighboring C_4-O_4 carbonyl group delocalizes it and causes the structure of the radical to be planar. These conclusions are supported by the spin distribution obtained from the analysis of the wave function (not shown), which indicate that both radicals, but especially the C_6-yl, are well localized on the respective carbon atoms. In the C_5-yl radical, a significant spin density is also observed on the O_4 atom. This confirms the explanation given above for the planarity of the C_5-yl radical as being due to the interaction of the unpaired electron with the π-system of the C_4-O_4 carbonyl bond. For the C_6-yl radicals, the only meaningful contribution to spin density, besides that on the C_6, comes from the N_1 atom, but it is still much smaller than that from the O_4 atom in the C_5-yl radical.

In summary, the stability of the radicals formed by the addition of OH· to thymine and uracil indicate that the addition process is kinetically controlled. Thus, the

Osman, Miaskiewicz, and Weinstein

addition to C_5 is preferred because of a lower barrier rather than due to the stability of the final product. The 6-hydroxy adducts are more stable than the 5-hydroxy adducts, which is in agreement with the relative stability obtained from the hydration of the radical cations. The resulting structures of the radicals are distorted, but while the 6-hydroxy-C_5-yl radical is almost planar, the 5-hydroxy-C_6-yl radical is more distorted around the atom on which the radical is localized. This structural difference is explained by the different spin distribution and its delocalization into the C_4-O_4 carbonyl when the radical is centered on C_5. These results are important for the understanding of the molecular parameters that will determine the consequences of the initial damage, and its translation into secondary damage and ultimately to the global changes in the conformation of DNA.

Calculation of the Conformational Modification of DNA by the Base Radical. For the preliminary studies, the simulated system consisted only of the DNA fragment and counterions placed as described below. Dielectric effects are represented only by a distance-dependent permittivity constant ($\epsilon = r$). A simulation with similar constraints and parameters was recently published for the intact DNA.[75]

The following procedure was used for the simulation:

1. The crystallographic coordinates of the duplex dodecamer d(CGCGAATTCGCG) obtained by Drew et al.[76] were taken from the Brookhaven Protein Data Bank.

2. Na$^+$ ions were placed initially at positions bisecting the O-P-O angle, at a distance of 5.0 Å from the phosphorus.

3. The thymine in Position 7 was replaced with an equatorial 5-hydroxy-6-thymidyl radical representing the lesion. The geometry of the modified thymine was taken from the *ab initio* calculations described above.

4. New parameters for use with the AMBER program package[77] were derived as described recently.[19] A new molecular fragment representing the damaged site was added to the AMBER library of fragments. The charges for that fragment were fitted with the program QUEST within the AMBER package to reproduce the electrostatic potential calculated from the *ab initio* wave function in several layers that surround the molecule. New atom types were defined to enable the definition of new force field parameters, which were optimized to reproduce the *ab initio* optimized structure.

5. The energy of the new dodecamer incorporating the radiation-damaged thymine was minimized while keeping the geometry of the modified residue frozen (in its geometry calculated quantum mechanically) during the minimization and during the subsequent dynamics simulation runs (Belly optimization).

6. After minimization the structure was heated to 300°K in steps of 20°K at time intervals of 0.15 ps. The total heating time was 3.0 ps.

7. At the end of the heating period, the velocities were reassigned from the Maxwell-Boltzmann distribution, and a 1.2 ps dynamics simulation at 300°K was performed while rescaling the velocities at the end of every 0.1 ps.

8. The molecular dynamics simulation was continued for 150 ps, yielding a total simulation time of 154.2 ps.

9. The structures from the last 50 ps were averaged and minimized for further analysis.

As can be seen from Fig. 5, the total energy stabilizes quickly and the temperature remains constant throughout the simulation. The maximum excursion of the temperature during the simulation does not exceed 1°C.

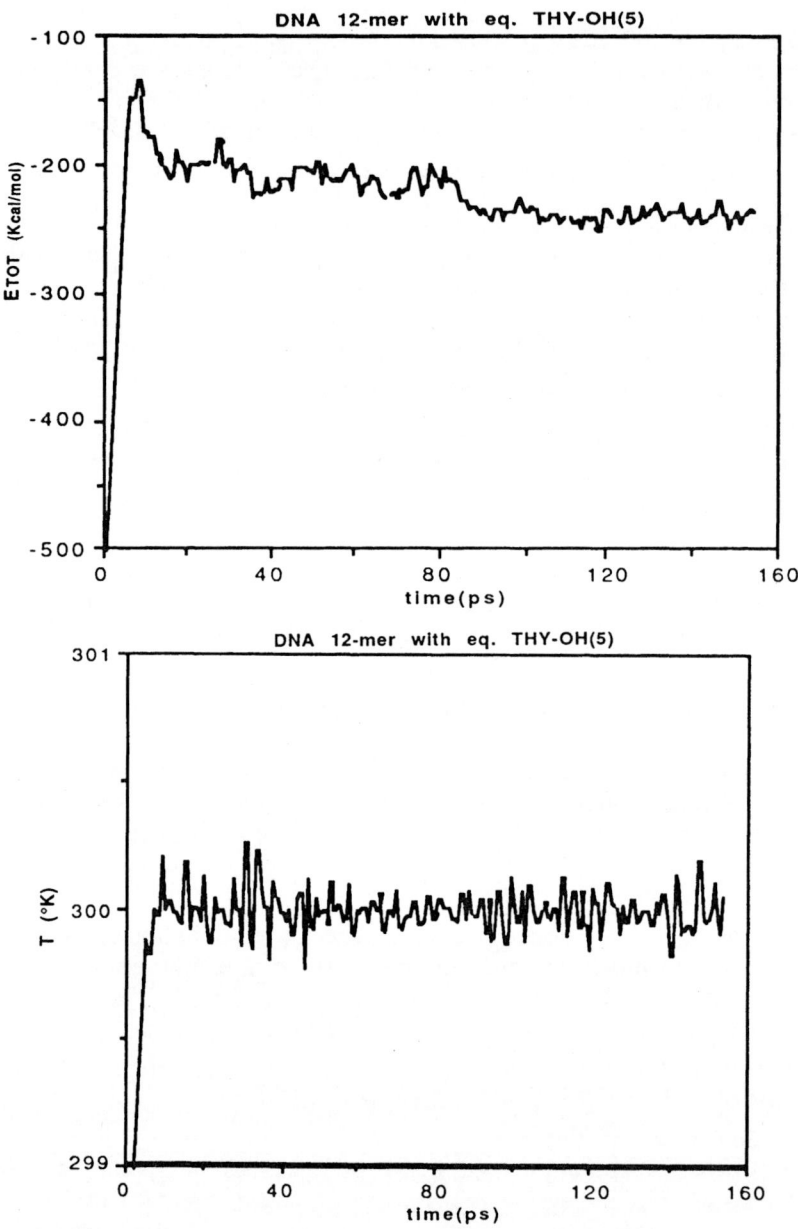

Figure 5. Averaged total energy and temperature during the molecular dynamics simulation of the radiation-damaged dodecamer of DNA.

Osman, Miaskiewicz, and Weinstein

A comparison between the averaged structure and a standard canonical B-DNA shows the extent of conformational change that the radiation-damaged DNA underwent in the course of the dynamics simulation. In Fig. 6, the damaged DNA is compared to the canonical B-DNA; the two molecules are presented side by side for comparison. Clearly, the conformational change in the global structure of the dodecamer changes the major characteristics of this structure. The major groove is reduced and the minor groove is somewhat opened. Most impressive is the global kink that appears in the structure of the damaged dodecamer, because it was induced in part by the presence of the altered thymine.

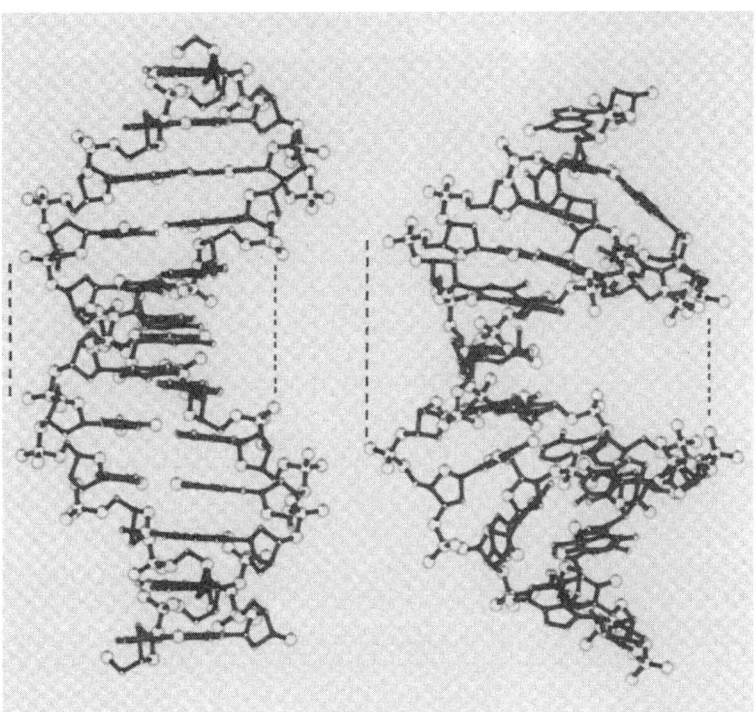

Figure 6. A comparison between canonical B-DNA (left) and a structure of the same sequence in which the thymine in position seven was replaced by equatorial 5-hydroxy-6-thymidyl radical representing a radiation damaged thymine (right). The structure of the damaged DNA is an averaged structure obtained after 150 ps of molecular dynamics simulation at 300°K (see text). Note that the distortion of the radiation damaged DNA is distributed over the entire structure. Vertical lines on the left and right of the helix highlight the changes in the span of the major and minor grooves compared to the canonical B-DNA.

A more detailed analysis of the conformational changes in the locally damaged DNA was performed with the program CURVES and is presented in a graphical form for each base pair in Figs. 7A-7E.

The puckering of the sugars also undergoes a significant change. The strand incorporating the 5-hydroxy-6-thymidyl radical exhibits a significant change in the puckering of the deoxyribose of the cytosine C9, which is removed two bases away from the point of damage. Its sugar is found in the O_1'-*endo* conformation, which is actually a relative maximum in the pseudorotation curve. In the complementary strand, two nucleotides are distorted: adenine A17 and guanine G14. The former is paired with thymine T8, and its sugar assumes a conformation of O_1'-*endo*; the latter is paired with cytosine C11, and its sugar is distorted into an unusual conformation of C_4'-*endo*, which is near the higher maximum in the pseudorotation curve.

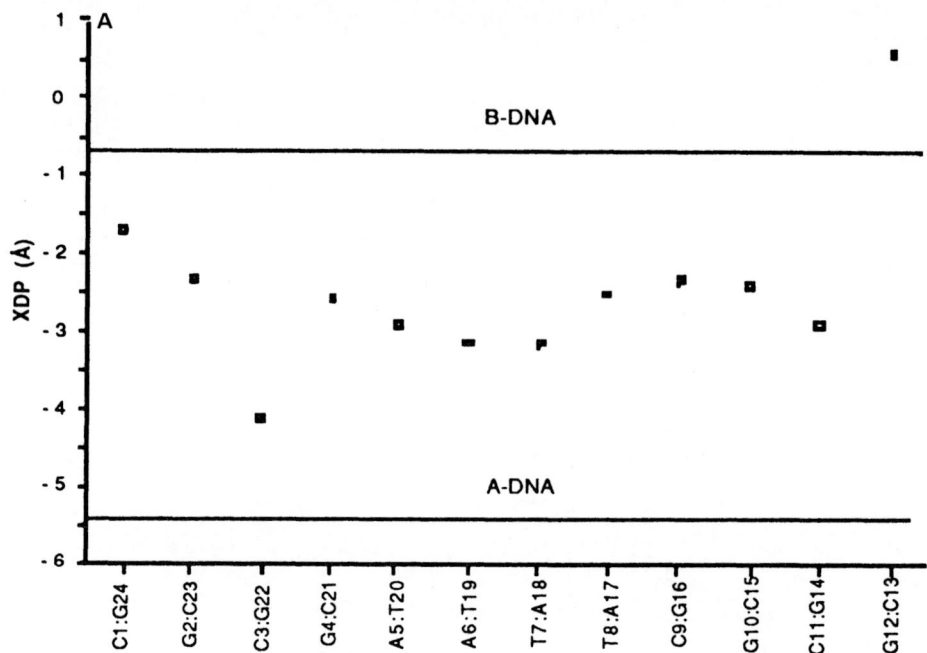

Figure 7A. The axis base-pair parameter XDP as a function of the base pair along the strand of the dodecamer. XDP describes the perpendicular displacement of the base pair from the axis. Solid lines describe the value of XDP in standard A- and B-DNA. The axis base-pair parameter, XDP (X-displacement) shows that the structure of the dodecamer changed from its initial B-DNA form into a form that on the average is shifted towards that of A-DNA.

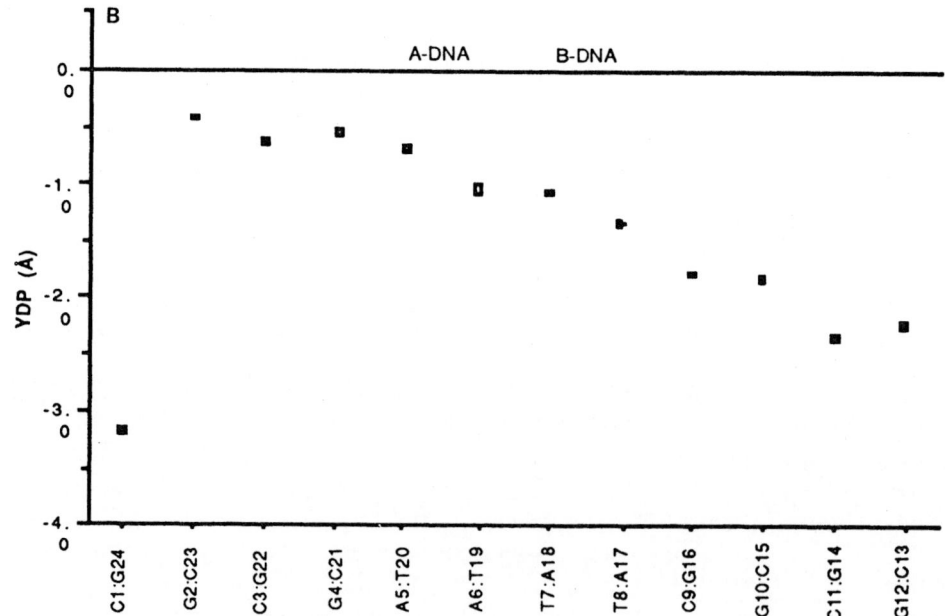

Figure 7B. The axis base-pair parameter YDP as a function of the base pair along the strand of the dodecamer. YDP describes the displacement of the base pair perpendicular to the axis. YDP shows that many base pairs are moved below their expected positions. This is consistent with an 8% contraction of the length of the helix.

Osman, Miaskiewicz, and Weinstein

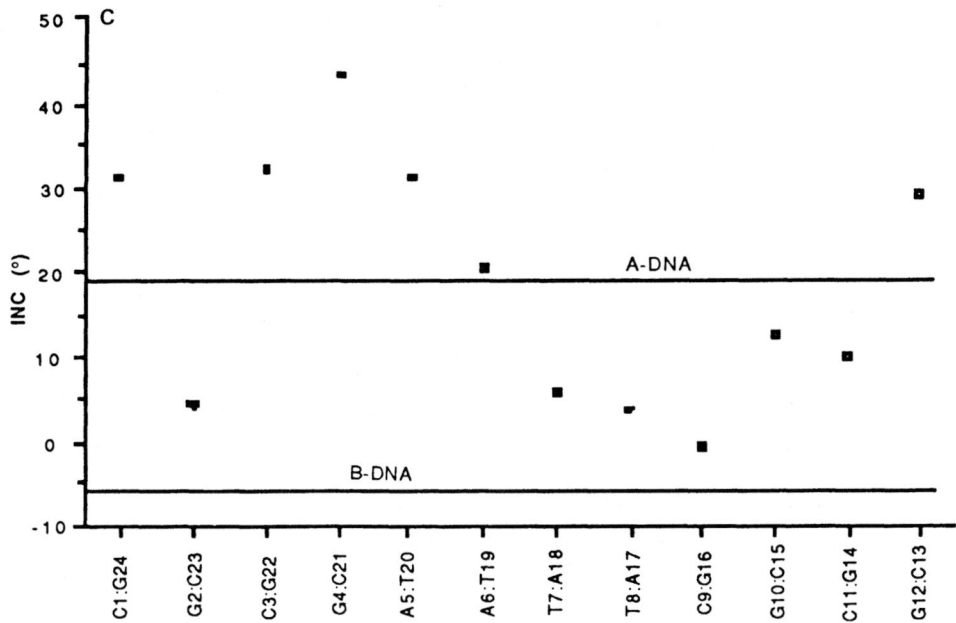

Figure 7C. The axis base-pair parameter INC as a function of the base pair along the strand of the dodecamer. INC describes the inclination of the base pair with respect to the axis. The parameter INC shows an oscillatory behavior which cannot be classified as either A- or B-DNA.

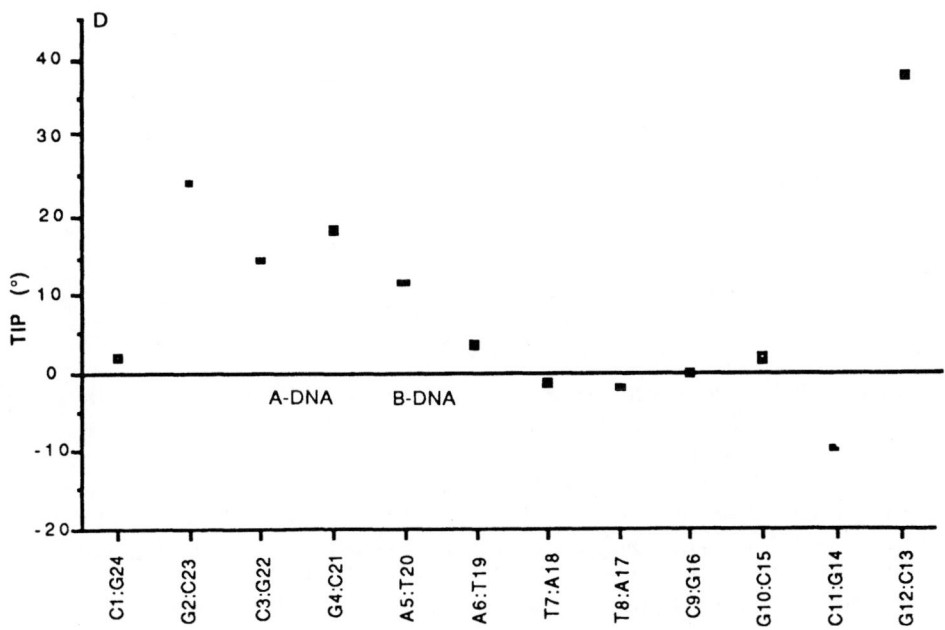

Figure 7D. The axis base-pair parameter TIP as a function of the base pair along the strand of the dodecamer. The TIP parameter shows that the distortion is not uniform; the sequence C1:G24 through A5:T20 is more affected than the rest of the dodecamer.

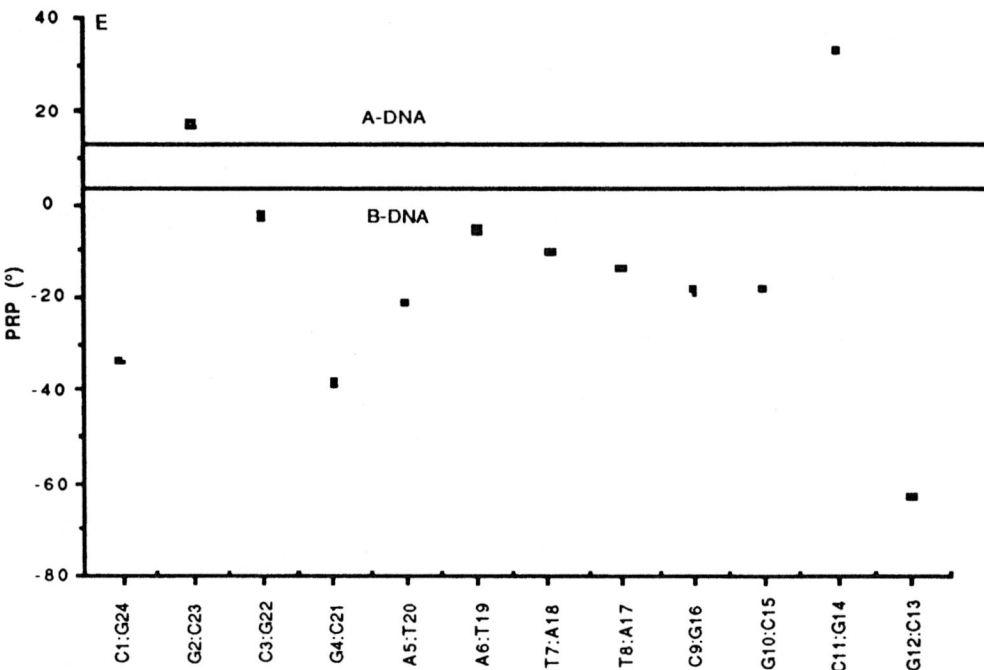

Figure 7E. The intra-base-pair parameter PRP as a function of the base pair along the strand of the dodecamer. PRP is the propeller twist of one base with respect to the other. PRP is negative and significantly large for most of the sequence. The other intra-base-pair parameters show small degrees of distortion that are mostly concentrated in the C1 through A5 sequence.

Clearly, the introduction of a base damaged by radiation produced *local distortions*, expressed in the unusual puckerings of the sugars, and *global distortions* exhibited by the shortening of the helix and the changes in the structure of the major and minor grooves. Further analysis is needed to relate these structural findings to the distortions observed in the crystal structure of DNA complexes with proteins and intercalating agents (e.g., see Fig. 1). Such comparisons will become useful in an attempt to elucidate the relation of DNA distortions to possible molecular events that transform primary to secondary damage and ultimately exhibit themselves in the altered biological properties of the damaged DNA.

Mechanistic Conclusions and Future Directions

Our calculations have shown the profound conformational changes induced in a macromolecular segment of DNA by the formation of base radicals following addition of OH· species to thymine. These changes illustrate the significance of even small structural modifications, such as the introduction of a hydroxyl group on thymine, for the structural integrity of DNA. Further analysis is required before the forces that lead to the conformational changes can be unequivocally identified. The parts of the simulations performed *in vacuo* with a distance-dependent dielectric to simulate the effect of the bulk water must be replaced with full dynamics simulations of DNA in the presence of a microscopic description of the solvent water, as illustrated in Fig. 8. More-accurate energetic results can be expected from these technological improvements, but the qualitative structural result—that the conformational changes are reminiscent of the bending of DNA by the interaction with proteins—is not likely to be

438   *Osman, Miaskiewicz, and Weinstein*

affected. The quantitative parameters describing the degree and feasibility of the structural changes will become more reliable as a result of the improved representation of the molecular systems to be simulated. But the current results are already intriguing because they indicate how long-range conformational changes induced by local lesions can be responsible for significant impairment of DNA function and for permanent, inheritable damage to the cellular genome.

The mechanistic hypotheses presented here for the relationship between the structural changes and the biological consequences involve mechanisms that are beginning to yield to investigation at the molecular level. Of particular importance, for the integration of these investigations into a comprehensive picture of the mechanisms

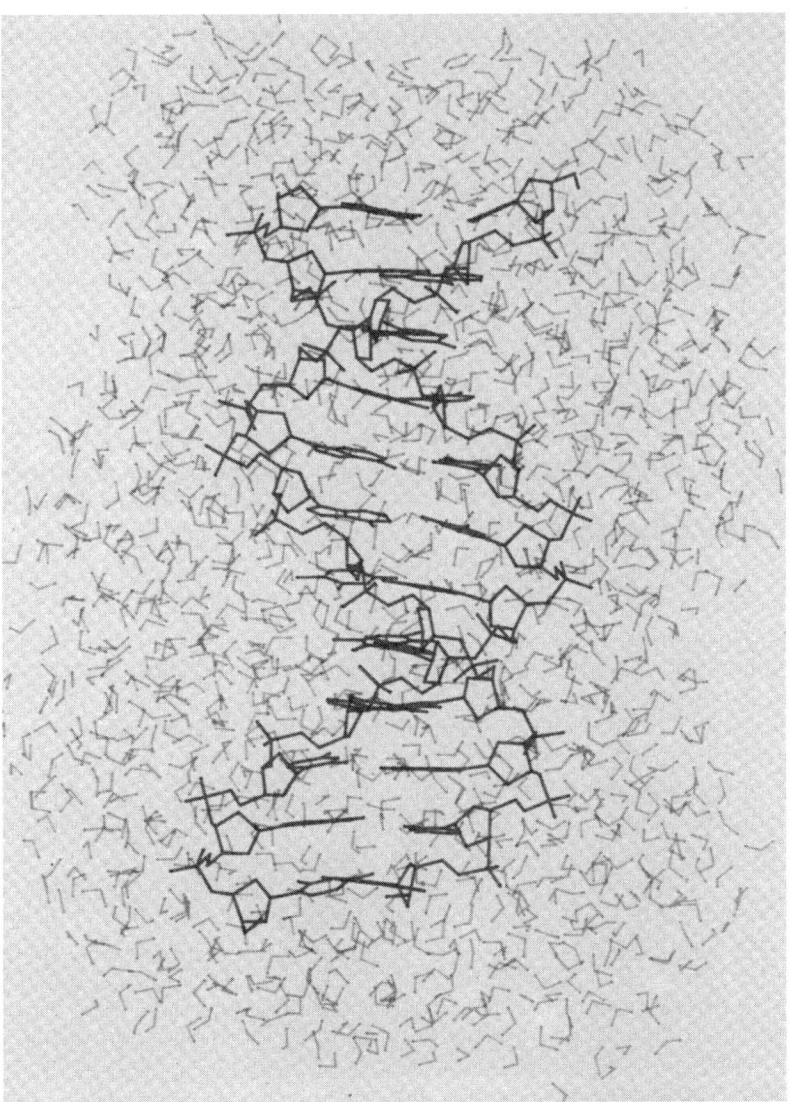

Figure 8. A snapshot from molecular dynamics simulation of a DNA dodecamer with the same sequence as in Fig. 6 (see text) in solvent bath consisting of 1437 water molecules and 22 counterions. The system is shown after the first 4.5 ps of simulation at 300°K.

by which the biological consequences of cell exposure to ionizing radiation are evoked, is the accumulating understanding regarding the molecular details of the lesioning processes and their structural consequences, as summarized briefly below.

The Relation of Local Lesions to DNA Strand Breaks

There is little doubt today that the majority of radiation-induced strand breaks in polynucleotides is caused by base radicals that attack the sugar moiety.[78-82] It appears that the same mechanisms that lead to strand breaks are also responsible for base release.[58-60] In poly-U, where most of the primary radiation damage is on the base, a specific sugar radical has been identified as localized on C_2',[68,69] thus establishing the mechanism by which both the strand break and the base release take place in this polynucleotide. Several issues, however, remain unexplained. Strand breaks in poly-U show a dependence on pH, in contrast to β-cleavage in sugars, which lack a dependence on pH.[83] This difference suggests that the rate-determining step may be in the conversion of a primary uracil-OH radical to another form, whose identity is not known but which is a more powerful H-abstractor. For example, the radical cations of pyrimidines and related structures that can be formed from them have been suggested[84] as possible candidates for H-abstraction in addition to the radicals formed by the addition of OH· to pyrimidines. These radical cations obtained in aqueous solutions are perhaps the best models of *direct* ionizations of bases in DNA. Nevertheless, the high reactivity of these species reflected in their short lifetime (a few μsec) presents a difficult experimental challenge when the characterization of their properties is attempted. Theoretical methods, on the other hand, provide ideal tools for the characterization of the properties of radical cations. Thus, theoretical investigations of the properties and reactivities of the radical cations of pyrimidines should yield an answer to the question whether radical cations of pyrimidines could be responsible for H-abstraction from the sugar. Since radical cations are the initial products of the *direct* effect of ionizing radiation, the answers from such studies should provide a better understanding of the involvement of radical cations in the primary damage to DNA induced by ionizing radiation.

Radical cations of purines have also been investigated in detail. A recent review[4] summarized the current knowledge of their properties. Guanine is well known to be the most easily ionized base,[1,85,86] a fact that is deduced from experimental measurements and is supported by MO calculations.[87-90] This property of guanine led to the suggestion that the radical cation of guanine might be a precursor for strand breaks in DNA.[91] However, the deprotonation of the radical cation of guanine is so rapid[92] that it is unlikely that the radical cation will be able to abstract a hydrogen from a neighboring sugar. This conclusion is also supported by the low sensitivity of poly-G and poly-dG to radiation-induced strand breaks (see above). The radical cation of adenine has similar properties; i.e., it deprotonates rapidly,[93] and it is a strong oxidant.[94-96] But while the reduction in the pK_a of guanine upon ionization is only 5.5 units, the ionization of adenine reduces the pK_a by >13 units.[93] Undoubtedly, the deprotonation reactions of the radical cations of purines in aqueous solutions are important to the understanding of their properties, but the proton affinities *inside the DNA* are of major importance in relation to potential damage to DNA. These properties are difficult to measure in experimental settings, but they can be easily calculated with theoretical methods, as was shown for a variety of macromolecules.[37,97-100] The importance of these properties is related to the fact that in DNA the complementary base will be affected by changes in pK_a due to ionization of one of the bases. For example, the pK_a of cytosine is 4.45 and the pK_a of the ionized guanine is reduced from 9.4 to 3.9. Thus, upon ionization of guanine a proton will be transferred to cytosine, leading to a separation of charge and spin. Similarly, ionization of adenine makes it a strong acid ($pK_a < 1$), and the complementary thymine will be able to receive the proton from adenine. The redistribution of charge due to proton transfer will affect the stacking interactions between the bases. Furthermore, the ionized DNA (whether from guanine or from adenine) can release the proton into the solution, thereby changing the charge distribution in DNA and also destroying some of the hydrogen bonds between the base pairs. Such changes in the major forces that are responsible for the integrity of the *double-strand* DNA will have important consequences for the stability and flexibility of DNA.

Local unwinding, especially of the kind produced by local lesions, can facilitate strand breaks by exposing segments of essentially single-strand DNA to the reactive radicals produced by energy deposition in the cellular medium. The susceptibility of several polynucleotides and single-strand DNA to radiation induced strand breakage has been recently investigated,[101,102] showing a differential distribution of sensitivities of polynucleotides to radiation. Poly-U and poly-C show a high degree of strand breakage, which is contrasted by a much lower susceptibility of strand breakage in poly-dU and poly-dC. On the other hand, it is poly-dA that shows a high sensitivity to radiation-induced strand breakage, whereas poly-A, poly-dG and poly-G show a low sensitivity. One possible explanation for this observation is that the respective base radicals or radical cations, which act as the H-abstractors, are responsible for these differences. A complementary hypothesis is that the conformational differences between these polynucleotides are responsible, at least in part, for this behavior. Clearly, the properties and reactivities of base radicals and radical cations need to be characterized in order to establish whether the difference in the observed susceptibilities can be explained by the different properties of these primary sites of radiation damage in DNA.

The charge redistribution produced by ionization may affect not only the conformational properties of DNA but also the electrostatic recognition elements that the DNA presents to other molecules (e.g., regulatory proteins and repair enzymes) that interact with it. These details must be explored with the methods of theoretical chemistry because the mechanistic link between ionizing energy deposition, radiation damage, and observable biological consequences depends on the interrelated changes in molecular structure and intermolecular interaction properties of cellular DNA, as discussed above.

The Importance of DNA Conformation in Mechanisms of Radical Transfer to Sugars

The wealth of structural information about double-strand DNA is well documented.[103] However, the results presented here concerning the conformational changes induced in DNA are important not only for the mechanistic hypothesis elaborated here on this basis, but also because it is clear that the formation of radicals on the base cannot in itself produce a strand break. The necessary step is the transfer of the radical from the base to the sugar. This can be accomplished by a hydrogen abstraction from the sugar by the radical on the base. A special three-dimensional arrangement of the base with respect to the sugar must be achieved first to allow the radical on the base to approach a hydrogen on the sugar within a reactive distance. It is important to keep in mind that the structural constraints in the intact DNA are likely to be modified significantly by the conformational changes induced in damaged DNA, as discussed above. The modified structure could enable an otherwise less-probable H radical transfer.

The demonstration of the mechanism of H-abstraction from ribose by uracil radical in poly-U[68,69] established the likelihood of such a mechanism and its relevance to radiation damage in DNA. Similar strand-breaking phenomena are observed in other polynucleotides as well as in single-strand DNA (see above). However, the nonuniform nature of such damage indicates that different polynucleotides exhibit differential sensitivity to the transfer of base radicals to a sugar radical, which in turn causes a strand break. There are two possible sources for this difference: one is the difference in reactivity of different base radicals, determining their ability to abstract a hydrogen from a sugar (either ribose or deoxyribose), while the other is related to the different structures and conformational properties of various homo-polynucleotides and single-strand DNA.

An important mechanistic detail concerns the steric constraints imposed by the macromolecular structure of DNA on the H-abstraction process between base and the sugar. The details of the constraints imposed by the reactivity properties of the reactants were outlined in the simulation of the mechanism of H-abstraction we have recently completed (Osman et al., in preparation), in which it was observed that the H-abstractor has to approach the H-donor to an approximate distance of 2.5-2.6 Å in order for the reaction to proceed at the optimal rate. From the structure of various polynucleotides,[31] it is known that the helical step size from one nucleotide to another is *ca.* 3.0 Å and the distance to the sugars lies approximately in the range of 3.5-5.0 Å. Thus, it is clear that for the base radical to be able to abstract a hydrogen from a sugar, the polynucleotide has to undergo a specific conformational change that will bring the two partners of the H-abstraction reaction to an appropriate reactive distance. In this context the *reactive distance* is specified as the distance between the atom on which the radical is centered in the base and a carbon atom with a detachable hydrogen. The ability of a base radical to abstract a hydrogen depends on the thermochemical balance between the reactants (i.e., base radical + sugar) and the products (i.e., the reduced base and the sugar radical). Such thermochemical balances are difficult to obtain experimentally because heats of formation of many radicals are not known, but quantum chemical calculations can provide very reliable thermochemical results. The energetic conditions required for the process to occur should be defined in terms of the identity of the base radicals that are sufficiently reactive to abstract a hydrogen from a sugar, as well as those of the hydrogens that are susceptible enough to be abstracted. The structural constraints imposed by the DNA macromolecule on the H-abstraction process are thus superimposed on the thermochemical balance.

Experimental validation of such a mechanistic conjecture based on our findings to date is clearly required. The best-studied processes of radical transfer from base to sugar are in polynucleotides. Of special interest in this context are the differential sensitivities to strand breaks of polyribonucleotides and polydeoxyribonucleotides. For example, poly-U is very susceptible to radiation-induced strand breakage, whereas poly-dU shows a much smaller sensitivity. It has been suggested that the mechanism for such a process in poly-U is in the abstraction of H_2' from ribose by the uracil radical. An important question is why the same process does not occur in poly-dU. A possible reason may be in the relative difficulty of either abstracting the H_2' from deoxyribose or producing a strand break from the C_2'-radical of deoxyribose. Alternatively, it is possible that the conformation of poly-dU is quite different from poly-U. These alternatives can be explored with various polynucleotides using molecular dynamics methods in a manner similar to the studies reported herein, and with the inclusion of a discrete molecular representation of the aqueous solvent.

Strand Breaks and the Conformation of DNA

The discontinuity of the backbone introduced by a strand break must have an effect on the conformational properties of DNA that is even larger than the effect we demonstrated for the modified thymine. The forces that hold the double-strand, undamaged DNA in the helical conformation are out of balance when the strand is broken, and the DNA molecule is likely to depart significantly from its original conformation. The expected conformational changes in strand-broken DNA are presently unknown, but should be significant in view of the findings we presented that even as small a damage as the addition of OH· to thymine has profound effects on DNA conformation. The conformational properties of models of DNA with a strand break and with AP sites require investigation with the same approaches as presented here for the double-strand molecule. Such studies should identify possible geometries of damaged DNA that may reside in cells for a considerable time and therefore can become the substrates of misrepair leading to mutations. The combined molecular mechanics and molecular dynamics simulations of damaged DNA can thus be expected to provide a link between the initial radiation damage and the molecular changes in the global structure of DNA. By following the underlying mechanisms at the molecular level, we propose that developments in future years could yield an integrated representations of radiation damage to biological systems, linking energy deposition processes to health effects.

Acknowledgments

The authors are greatly indebted to their many collaborators who have made possible the work described in the original reports reviewed here. This work was supported in part by a grant from the U.S. Department of Energy (DOE), USDOE DE-FG02-88ER60675. H. Weinstein is recipient of a NIDA Research Scientist Award DA-00060. Computations were performed at the National MFE Computer Center of the DOE, on the supercomputer systems at the Pittsburgh Supercomputer Center, and the Cornell National Supercomputer Facility—both of which are sponsored by the National Science Foundation—as well as at the Advanced Scientific Computing Laboratory at the Frederick Cancer Research Facility of the National Cancer Institute (Laboratory for Mathematical Biology). The authors gratefully acknowledge these resources and the generous allocations of computer time at the University Computer Center of the City University of New York.

References

1. C. von Sonntag. *The Chemical Basis of Radiation Biology*. Taylor & Francis, London (1987).
2. S. Das, D. J. Deeble, M.-N. Schuchmann and C. von Sonntag. Pulse Radiolytic Studies on Uracil and Uracil Derivatives. Protonation of Their Electron Adducts at Oxygen and Carbon. *Int. J. Radiat. Biol.* 46:7-9 (1984).
3. M.C.R. Symons. ESR Spectra for Protonated Thymine and Cytidine Radical Anions: Their Relevance to Irradiated DNA. *Int. J. Radiat. Biol.* 58:93-96 (1990).
4. S. Steenken. Purine Bases, Nucleosides, and Nucleotides: Aqueous Solution Redox Chemistry and Transformation Reactions of their Radical Cations End e⁻ and OH Adducts. *Chem. Rev.* 89:503-520 (1989).
5. G. Scholes, G. Stein and J. Weiss. Action of X-rays on Nucleic Acids. *Nature (Lond.)* 164:709-710 (1949).
6. J. F. Ward and I. Kuo. Strand Breaks, Base Release and Postirradiation Changes in DNA γ-irradiated in Dilute O_2-saturated Aqueous Solution. *Radiat. Res.* 66:485-498 (1976).
7. H. P. Leenhouts and K. H. Chadwick. The Crucial Role of DNA Double-Strand Breaks in Cellular Radiobiological Effects. *Adv. Radiat. Biol.* 7:55-101 (1978).
8. D. Frankenberg, M. Frankenberg-Schwager, D. Blocher and R. Harbich. Evidence for DNA Double-Strand Breaks as the Critical Lesions in Yeast Cells Irradiated with Sparsely or Densely Ionizing Radiation Under Oxic or Anoxic Conditions. *Radiat. Res.* 88:524-532 (1981).
9. S. E. Bresler, L. A. Noskin and A. V. Suslov. Induction by Gamma Irradaition of Double-Strand Breaks of *Escherichia Coli* Chromosomes and Their Role in Cell Lethality. *Biophys. J.* 45:749-754 (1984).
10. I. R. Radford. The Level of Induced DNA Double-Strand Breakage Correlates with Cell Killing After X-irradiation. *Int. J. Radiat. Biol.* 48:45-54 (1985).

11. C. J. Roberts and P. D. Holt. Induction of Chromosome Abberation and Cell Killing in Syrian Hamster Fibroblasts by γ-rays, X-rays and Neutrons. *Int. J. Radiat. Biol.* 48:927-939 (1985).

12. D. C. Lloyd. Comment on the Paper by Roberts and Holt. *Int. J. Radiat. Biol.* 48:940-942 (1985).

13. J. F. Ward. DNA Damage Produced by Ionizing Radiation in Mammalian Cells: Identities, Mechanisms of Formation, and Reparability. *Prog. Nuc. Acid Res. Mol. Biol.* 35:95-125 (1988).

14. R. Osman, W. J. Clark, A. P. Mazurek, and H. Weinstein. Theoretical Studies of Molecular Mechanisms of DNA Damage Induced by Hydroxyl Radicals. *Free Rad. Res. Comms.* 6:131-132 (1989).

15. L. Pardo, A. P. Mazurek and R. Osman. Computational Models for Proton Transfer in Biological Systems. *Int. J. Quantum Chem.* 37:701-711 (1989).

16. L. Pardo, R. Osman, J. Banfelder, A. P. Mazurek, and H. Weinstein. Molecular Mechanisms of Radiation Induced DNA Damage: H-abstraction and β-cleavage. *Free Rad. Res. Comms.* 12:461-463 (1991).

17. H. Weinstein and R. Osman. Molecular Biophysics of Specificity and Function in Enzymes, Receptors and Calcium Binding Proteins. In *Theoretical Biochemistry and Molecular Biophysics: A Comprehensive Survey,* D. L. Beveridge and R. Lavery, ed., pp. 275-289. Adenine Press, New York (1990).

18. H. Weinstein and R. Osman. On the Structural and Mechanistic Basis of Function, Classification and Ligand Design for 5-HT Receptors. *Neuropsychopharmacol.* 3(5-6):397-409 (1990).

19. H. Weinstein, R. Osman, G. A. Mercier, A. P. Mazurek, L. Pardo, and L. A. Rubenstein. Theory and Computation of Molecular Mechanisms in Biological Processes: Radiation-Induced Damage to DNA and Neurotransmitter Receptor Function. In *Computer Assisted Analysis and Modeling on the IBM 3090,* H. U. Brown, ed., pp. 629-673. MIT Press, Boston (1990).

20. H. Weinstein and J. P. Green. Quantum Chemistry in Biomedical Sciences. *Ann. N.Y. Acad. Sci., 367* (1981).

21. E. Clementi, ed. *Modern Techniques in Computational Chemistry: MOTECC-89.* ESCOM, The Netherlands (1989).

22. E. Clementi, ed. *Modern Techniques in Computational Chemistry: MOTECC-90.* ESCOM, The Netherlands (1990).

23. B. Venkataraghavan and R. J. Feldmann. Macromolecular Structure and Specificity: Computer-Assisted Modeling and Applications. *Ann. N.Y. Acad. Sci.* 439 (1985).

24. E. Clementi and R. H. Sarma. *Structure and Dynamics: Nucleic Acids and Proteins.* Adenine Press, New York (1983).

25. E. Clementi, G. Corongiu, and R. H. Sarma. *Structure and Motion: Membranes, Nucleic Acids and Proteins.* Adenine Press, Guilderland, New York (1985).

26. C. L. Brooks, M. Karplus, and B. M. Pettitt. *Proteins: A Theoretical Perspective of Dynamics, Structure, and their Thermodynamics*; John Wiley & Sons: New York (1988).

27. J. A. McCammon and S. C. Harvey. *Dynamics of Proteins and Nucleic Acids.* Cambridge University Press, New York (1987).

28. W. G. Richards. *Computer-Aided Molecular Design.* IBC Technical Services, Ltd., London (1989).

29. W. F. van Gunsteren and P. K. Weiner. *Computer Simulation of Biomolecular Systems.* ESCOM, Leiden, The Netherlands (1989).

30. U. C. Singh and P. A. Kollman. An Approach to Computing Electrostatic Charges for Molecules. *J. Comp. Chem.* 5:129-145 (1984).

31. W. Saenger. *Principles of Nucleic Acid Structure.* Springer-Verlag, New York (1983).

32. R. Osman, K. Namboodiri, H. Weinstein, and J. R. Rabinowitz. Reactivities of Acrylic and Methacrylic Acids in a Nucleophilic Addition Model of Their Biological Activities. *J. Amer. Chem. Soc.* 110:1701-1707 (1988).

33. M. T. Carrol, J. R. Cheeseman, R. Osman, and H. Weinstein. Nucleophilic Addition to Activated Double Bonds: Predictions of Reactivity from the Laplacian of Charge Densities. *J. Phys. Chem.* 93:5120-5123 (1989).

34. J. P. Dijkman, R. Osman, and H. Weinstein. A Theoretical Study of the Effect of Primary and Secondary Structure Elements on the Proton Transfer in Papain. *Int. J. Quantum Chem.* 35:241-252 (1989).

35. K. Hori, J. N. Kushick, and H. Weinstein. Structural and Energetic Parameters of Ca^{2+} Binding to Peptides and Proteins. *Biopolymers* 27:1865-1886 (1988).

36. G. A. Mercier, R. Osman, and H. Weinstein. Role of Primary and Secondary Protein Structure in Neurotransmitter Activation Mechanisms. *Protein Eng.* 2:261-270 (1988).

37. G. A. Mercier, J. P. Dijkman, R. Osman, and H. Weinstein. Effects of Macromolecular Environments on Proton Transfer Processes: The Calculation of Polarization. In *Quantum Chemistry; Basic Aspects, Actual Trends*, R. Carbo, ed., pp. 577-596. Elsevier Scientific Publ., Amsterdam (1989).

38. G. A. Mercier, R. Osman, and H. Weinstein. A Molecular Theoretical Model of Recognition and Activation of a 5-HT Receptor. In *Computer Assisted Modeling of Receptor - Ligand Interactions*, R. Rein and A. Golombek, ed., pp. 399-410. Alan R. Liss Publ., New York (1989).

39. L. Pardo, A. P. Mazurek, R. Osman, and H. Weinstein. Theoretical Studies of the Activation Mechanism of Histamine H_2-Receptors. Dimaprit and the Receptor Model. *Int. J. Quantum Chem.* QBS16:281-290 (1989).

40. H. Weinstein, R. Osman, and G. Mercier. Recognition and Activation of a 5-HT Receptor by Hallucinogens and Indole Derivatives. In *NIDA Research Monograph 90, Proc. 50th Ann. Meet. Problems of Drug Dependence*, L. S. Harris, ed., pp. 243-255. US-HHS, NIDA, Maryland (1988).

41. H. Weinstein and R. Osman. *Interaction Mechanisms at Biological Targets: Implications for Design of Serotonin Receptor Ligands*. In *Computer-Aided Molecular Design*, W. G. Richards, ed., pp. 105-108. IBC Technical Services, London (1989).

42. H. Weinstein and R. Osman. Simulations of Ligand-Receptor Interactions as Guides for Design. In *Frontiers in Drug Research, Alfred Benzon Symposium 28*, B. Jensen, F. S. Jorgenson, and H. Kofod, ed., pp. 169-182. Munksgaard, Copenhagen (1989).

43. H. Weinstein, K. Hori, J. N. Kushick, F. Sussman, and A. Factor. Computer Simulation Studies of Structure-Function Relations in Calcium-Binding Proteins. In *Proc. 4th Intl. Conf. Supercomputing Intl.*, L. P. Kartashev and S. I. Kartashev, ed., pp. 106-108. Supercomputing Inst. Inc. (1989).

44. T. A. Steitz. Structural Studies of Protein-Nucleic Acid Interaction: the Sources of Sequence-specific Binding. *Quarterly Reviews of Biophys.* 23:205-280 (1990).

45. Y. Kim, J. C. Grable, and R. Love. Refinement of Eco RI Endonuclease Crystal structure. *Science* 249:1307-1310 (1990).

46. A. K. Aggarwal, D. W. Rodgers, M. Drottar, M. Ptashne, and C. Harrison. Recognition of a DNA Operator by the Repressor of Phage 434: A View at High Resolution. *Science* 242:899-907 (1988).

47. S. R. Jordan and C. O. Pabo. Structure of the Lambda Complex at 2.5 A Resolution: Details of the Repressor-Operator Interactions. *Science* 242:893-899 (1988).

48. J. A. McClarin, C. A. Frederick, B. C. Wang, P. Greene, H. W. Boyer, J. Grable, and J. M. Rosenberg. Structure of the DNA-Eco RI Endonuclease Recognition Complex at 3 A Resolution. *Science* 234:1526-1541 (1986).

49. C. O. Pabo, A. K. Aggarwal, S. R. Jordan, L. J. Beamer, U. R. Obeysekare, and S. C. Harrison. Conserved Residues Make Similar Contacts in Two Repressor-Operator Complexes. *Science* 247:1208-1214 (1990).

50. C. Wolberger, Y. Dong, M. Ptashne, and S. C. Harrison. Structure of a Phage 434 Cro/DNA Complex. *Nature* 335:789-795 (1988).

51. M. Ptashne. Gene Regulation by Proteins Acting Nearby and at a Distance. *Nature* 322:697-701 (1986).

52. M. Ptashne. How Eukaryotic Transcriptional Activators Work. *Nature* 335:683-689 (1988).

53. O. K. Snyder, J. F. Thompson, and A. Landy. Phasing of Protein-Induced DNA Bends in a Recombination Complex. *Nature* 341:255-258 (1989).

54. S. D. Goodman and H. A. Nash. Functional Replacement of a Protein-Induced Bend in a DNA Recombination Site. *Nature* 341:251-254 (1989).

55. J. Ramstein and R. Lavery. Energetic Coupling Between DNA Bending and Base Pair Opening. *Proc. Natl. Acad. Sci. USA* 85:7231-7235 (1988).

56. H. Martin-Bertram, P. Hartl, and C. Winkler. Unpaired Bases in Phage DNA After Gamma-Irradiation In-situ and In-vitro. *Radiat. Environ. Biophys.* 23:95 (1984).

57. J. Andrews, H. Martin-Bertram, and U. Hagen. S1 Nuclease-Sensitive Sites in Yeast DNA: An Assay for Radiation-Induced Base Damage. *Int. J. Radiat. Biol.* 45:497 (1984).

58. D. Deeble, D. Schultz, and C. von Sonntag. Reactions of OH Radicals with Poly (U) in Deoxygenated Solutions: Sites of OH Radical Attack and the Kinetics of Base Release. *Int. J. Radiat. Biol.* 49:915 (1986).

59. D. J. Deeble and C. von Sonntag. γ-Radiolysis of Poly(U) in Aqueous Solution. The Role of Primary Sugar and Base Radicals in the Release of Undamaged Uracil. *Int. J. Radiat. Biol.* 46:247-260 (1984).

60. D. J. Deeble and C. von Sonntag. Radiolysis of Poly(U) in Oxygenated Solutions. *Int. J. Radiat. Biol.* 49:927-936 (1986).

61. C. R. Paul, J. C. Wallace, J. L. Alderfer, and H. C. Box. Radiation Chemistry of d(TpApCpG) in Oxygeneated Solution. *Int. J. Radiat. Biol.* 54:403-415 (1988).

62. T. Lindahl. DNA Glycosylases, Endonucleases, and Base-Excision Repair. *Prog. Nucl. Acids Res. Mol. Biol.* 22:135-190 (1979).

63. G. Teebor, R. Boorstein, and J. Cadet. Repairability of Oxidative Free Radical-Mediated Damage to DNA: A Review. *Int. J. Radiat. Biol.* 54:131-150 (1988).

64. W. A. Deutsch and S. Linn. DNA Binding Activity from Cultured Human Fibroblasts That is Specific for Partially Depurinated DNA and That Inserts Purines into Apurinic Sites. *Proc. Natl. Acad. Sci. USA* 76:141-146 (1979).

65. L. A. Loeb and B. D. Preston. Mutagenesis by Apurinic/Apyrimidinic Sites. *Ann. Rev. Genetics* 20:201-230 (1986).

66. D. Sagher and B. Strauss. Insertion of Nucleotides Opposite Apurinic/Apyrimidinic Sites in Deoxyribonucleic Acid During *in vitro* Synthesis: Uniqueness of Adenine Nucleotides. *Biochemistry* 22:4518-4526 (1983).

67. A. Gentil, A. Margot, and A. Sarasin. Apurinic Sites Cause Mutations in Simian Virus 40. *Mutat. Res.* 129:141-147 (1984).

68. K. Hildenbrand and D. Schulte-Frohlinde. E.S.R. Studies on the Mechanism of Hydroxyl Radical Induced Strand Breakage of Polyuridilic Acid. *Int. J. Radiat. Biol.* 55:725-738 (1989).

69. K. Hildenbrand and D. Schulte-Frohlinde. ESR Studies on the Mechanism of OH-Induced Strand Breakage in Poly(U). *Free Radical Res. Commun.* 6:137-138 (1989).

70. S. Fujita and S. Steenken. Pattern of OH Radical Addition to Uracil and Methyl- and Carboxyl-Substituted Uracils. Electron Transfer of OH Adducts with N,N,N',N'-Tetramethyl-p-phenylenediamine and Tetranitromethane. *J. Am. Chem. Soc.* 103:2540-2545 (1981).

71. C. von Sonntag. The Chemistry of Free-Radical-Mediated DNA Damage, This Proceedings.

72. A. A. Shaw and J. Cadet. Radical Combination Processes Under the Direct Effects of Gamma Radiation on Thymidine. *J. Chem. Soc., Perkin Trans. 2,* in press.

73. H.-P. Schuchmann, D. J. Deeble, G. Olbrich, and C. von Sonntag. The SO_4^--Induced Chain Reaction of 1,3-dimethyluracil with Peroxidosulfate. *Int. J. Radiat. Biol.* 51:441-453 (1987).

74. A. Rashin. Hydration Phenomena, Classical Electrostatics, and the Boundary Element Method. *J. Phys. Chem.* 94:1725-1733 (1990).

75. J. Srinivasan, J. M. Withka, and D. L. Beveridge. Molecular Dynamics of an *in vacuo* Model of Duplex d(CGCGAATTCGCG) in the B-form Based on the Amber 3.0 Force Field. *Biophys. J.* 58:533-547 (1990).

76. H. R. Drew, R. M. Wing, T. Takano, S. Broka, S. Tanaka, K. Itakura, and R. E. Dickerson. Structure of a B-DNA Dodecamer: Conformation and Dynamics. *Proc. Nat. Acad. Sci (USA)* 78:2179-2183 (1981).

77. U. C. Singh, P. K. Weiner, J. Caldwell, and P. A. Kollman. *AMBER 3.0.* University of California, San Francisco, California (1986).

78. D.G.E. Lemaire, E. Bothe, and D. Schulte-Frohlinde. Yields of Radiation-Induced Main Chain Scission of Poly U in Aqueous Solution: Strand Break Formation Via Base Radicals. *Int. J. Radiat. Biol.* 45:351-358 (1984).

79. D. Schulte-Frohlinde and E. Bothe. Identification of a Major Pathway of Strand Break Formation in Poly-U Induced by OH Radicals in Presence of Oxygen. *Z. Naturforsch.* 39c:315-319 (1984).

80. E. Bothe, G. Behrens, E. Bohm, B. Sethuram, and D. Schulte-Frohlinde. Hydroxyl Radical Induced Strand Break Formation of Poly(U) in the Presence of Oxygen. Comparison of the Rates as Determined by Conductivity, ESR and Rapid-Mix Experiments with a Thiol. *Int. J. Radiat. Biol.* 49:57-66 (1986).

81. D.G.E. Lemaire, E. Bothe, and D. Schulte-Frohlinde. Hydroxyl Radical Induced Strand Break Formation of Poly(U) in Anoxic Solution. Effect of Dithiothreitol and Tetranitromethane. *Int. J. Radiat. Biol.* 51:319-330 (1987).

82. E. Bothe, M. Adinarayana, and D. Schulte-Frohlinde. Rate and Yield of OH-Induced Strand Break Formation of Polynucleotides and DNA. *Free Radical Res. Commun.* 6:139 (1989).

83. E. Bothe and D. Schulte-Frohlinde. Release of K^+ and H^+ from Poly U in Aqueous Solution upon γ and Electron Irradiation. Rate of Strand Break Formation in Poly U. *Z. Naturforsch.* 37c:1191-1204 (1982).

84. D. J. Deeble, M. N. Schuchman, S. Steenken, and C. von Sonntag. Direct Evidence for the Formation of Thymine Radical Cations from the Reaction of SO_4^- with Thymine Derivatives: A Pulse Radiolysis Study with Optical and Conductance Detection. *J. Phys. Chem.* 94:8186-8192 (1990).

85. A. J. Bertinchamps, J. Huttermann, W. Kohnlein and R. Teoule. *Effects of Ionizing Radiation on DNA.* Springer, Berlin (1978).

86. G. Scholes. The Radiation Chemistry of Pyrimidines, Purines and Related Substances. In *Photochemistry and Photobiology of Nucleic Acids*, S. Y. Wang, ed., p. 521. Academic Press, New York (1976).
87. B. Rakvin and J. N. Herak. Localization of Radiation Energy in DNA. *Radiat. Phys. Chem.* 22:1043 (1983).
88. B. Pullman and A. Pullman. *Quantum Biochemistry*. Interscience Publishers, New York (1963).
89. N. Bodor, M.J.S. Dewar, and A. J. Harget. Ground States of Conjugated Molecules. XIX. Tautomerism of Heteroaromatic Hydroxy and Amino Derivatives and Nucleotide Bases. *J. Am. Chem. Soc.* 92:2929 (1970).
90. H. Berthod, C. Giessner-Prettre, and A. Pullman. Theoretical Study of the Electronic Properties of the Purine and Pyrimidine Components of the Nucleic Acids. *Theor. Chim. Acta* 5:53 (1966).
91. P. J. Boon, P. M. Cullis, M. C. R. Symons, and B. W. Wren. Effects of Ionizing Radiation on Deoxyribonucleic Acid and Related Systems. Part I. The Role of Oxygen. *J. Chem. Soc. Perkin Trans.* II:1393-1399 (1984).
92. L. P. Candeias and S. Steenken. Structure and Acid-Base Properties of One-Electron-Oxidized Deoxyguanosine, Guanosine, and 1-methylguanosine. *J. Am. Chem. Soc.* 111:1094-1099 (1989).
93. A.J.S.C. Vieira and S. Steenken. Pattern of OH Radical Reaction with 6- and 9-Substituted Purines. Effect of Substituents on the Rates and Activation Parameters of the Unimolecular Transformation Reactions of Two Isomeric OH Adducts. *J. Phys. Chem.* 91:4138-4144 (1987).
94. P. O'Neill. Pulse Radiolytic Study of the Interaction of Thiols and Ascorbate with OH Adducts of dGMP and dG: Implications for DNA Repair Processes. *Radiat. Res.* 96:198 (1983).
95. P. O'Neill, P. W. Chapman, and D. G. Papworth. Repair of Hydroxyl Radical Damage of dA by Antioxidants. *Life Chem. Rep.* 3:62 (1985).
96. P. O'Neill and S. E. Davies. A Pulse Radiolytic Study of the Interaction of Nitroxyls with Free-radical Adducts of Purines: Consequences for Radiosensitization. *Int. J. Radiat. Biol.* 49:937 (1986).
97. M. K. Gilson and B. H. Honig. Calculation of Electrostatic Potentials in an Enzyme Active Site. *Nature* 330:84-86 (1987).
98. M.J.E. Sternberg, F.R.F. Hayes, A. J. Russell, P. G. Thomas, and A. R. Fersht. Prediction of Electrostatic Effects of Engineering of Protein Charges. *Nature* 330:86-88 (1987).
99. K. Soman, A. S. Yang, B. Honig, and R. Fletterick. Electrical Potentials in Trypsin Isozymes. *Biochemistry* 28:9918-9926 (1989).
100. E. L. Mehler and G. Eichele. Electrostatic Effects in Water-Accessible Regions of Proteins. *Biochemistry* 23:3887-3891 (1984).
101. L. R. Karam, M. Dizdaroglu, and M. G. Simic. Intramolecular H Atom Abstraction from the Sugar Moiety by Thymine Radicals in Oligo- and Polydeoxynucleotides. *Radiat. Res.* 116:210-216 (1988).
102. M. Adinarayana, E. Bothe, and D. Schulte-Frohlinde. Hydroxyl Radical-Induced Strand Break Formation in Single-Stranded Polynucleotides and Single-Stranded DNA in Aqueous Solution as Measured by Light Scattering and by Conductivity. *Int. J. Radiat. Biol.* 54:723-737 (1988).
103. O. Kennard and W. N. Hunter. Oligonucleotide Structure: A Decade of Results from Single Crystal X-ray Diffraction Studies. *Quarterly Rev. Biophys.* 22:327-379 (1989).

Discussion

Ward: The radicals are very transient. They are gone in less than a millisecond to form other products.

Weinstein: A millisecond is a long time for the type of reactions we consider, as well as for the conformational changes we simulate. That's why I emphasize that at least in the simulation the structure begins to change in the first 10 picoseconds.

Ward: If this initial change in structure is to affect DNA protein interactions, what is going to happen to the protein in that millisecond? It isn't going to move away in a millisecond.

Weinstein: Protein-DNA interactions can be affected in several ways: One is, of course, by chemical reaction at the site, producing cross-linking which would fix the proteins with the DNA. Another is to interfere with the formation of the complex and affect the process regulating gene expression. You may be thinking of longer time frames because the process takes longer in the experiments due to the nature of the observations. If you look at how long it takes for the repressor protein to act in an *in vitro* expression system, you may come up with a long time. But the interaction itself, the recognition process when the protein is there, is extremely fast. Dissociation of the protein-DNA complex has already occurred long before that point. If, in addition to this, a local strand break is formed in conjunction with the conformational change induced by the local lesion, it will further weaken the type of interactions that will occur between that operator and the protein. The same holds for the interaction of repair enzymes with DNA.

Ward: By the same token, if the protein does move away then there is a recognition site, a distortion of DNA which might attract repair mechanisms.

Weinstein: Yes, but the conformational change may also prevent the repair.

Ward: What is the feasibility that a wrong message goes out from the site that is deleterious to the cell, before repair takes place? Does the distortion of the protein make the mistake? What is the time domain of all this?

Weinstein: A reliable answer to this question requires much more study and more simulations than we have done so far. I can tell you this much. The experiments that could probe the feasibility of the hypothesis we present are conceptually easy. It is probably feasible experimentally to see if any change in the regulation of DNA *in vitro* is detected after such an irradiation. If they are, then the exact components of the mechanism become the focus of further study.

Varma: You have raised an important question. You talk about ten picoseconds bending of the DNA taking place, that it affects some structures or some conformation that changes the DNA somewhere. Do the changes in conformation that you produce in 10 picoseconds really represent changes in the eventual protein regulation?

Weinstein: At this stage of the game I have no reason to disbelieve the model we presented. Any hypothesis is there to be addressed theoretically and experimentally and to be proved correct or incorrect; but based on what I have shown you from experimental data, I have no reason to doubt the importance of the bending in the DNA structure that we observed, because of the relationship between what we calculate, and what we see in the crystal structures of protein-DNA complexes.

Moolgavkar: You want a change that is meaningful for carcinogenesis, you want a change in the cell which is inherited by daughter cells. How do you see these conformational changes playing a role in bringing about a permanent change which is going to be inherited by the daughters? That seems to be the crucial question.

Weinstein: Well, I don't know if that is the crucial question because to me what I described constitutes the initiation of the sequence of events. It could be the first trigger for the generation of the first oncogene products, for example.

Moolgavkar: If the change is transient, it is not going to matter. It has to be inherited by the daughters.

Weinstein: Well, if the changes are made permanent in some way, that is, if a break really occurs in that place, then you can have a permanent change that will go on further.

Moolgavkar: So, you are talking about a quick reaction that would ultimately bring about the sort of things that John was talking about yesterday.

Weinstein: Yes, or the sort of things that initiate them.

Chatterjee: From the slides you showed on the deformation of structures, I am not convinced that there will be a significant effect with respect to hydroxyl radical attack on various sites. By that I mean that the coordinates of undamaged sites will not be significantly distorted to make a difference in the $\cdot OH$ reactivity, for example. These radicals are migrating over a characteristic diffusion length of 3 or 4 nanometers, which seems to me quite large compared to the extent of deformation you demonstrated.

Weinstein: I know that you maintain that the local structure, and changes in it produced by preceding lesions, may not be important, but I cannot agree based on the results from our study at a molecular level. When we do these calculations of the interaction between the OH radical and the molecule to obtain hydrogen abstraction, for example, there is an energy balance that determines what can happen. If you don't have the 10, 15, or 20 kilocalories locally, or if it should take more than that for the hydroxyl to penetrate close to the site of H-abstraction because the site is now blocked by a conformational change due to a preceding lesion, then the reaction may not occur at that site. The Monte Carlo simulation based on an intact DNA structure would not pick this up. Moreover, the converse is also true, and sites that appeared inaccessible in intact DNA become dynamically accessible. So if you could connect the requirements for some kind of site specificity with the change in the energy profile, in the energy surface, then the simulations would become more realistic, and I find it difficult to see how we cannot agree on theoretical grounds.

Chatterjee: I have no problem in agreeing with you on theoretical grounds. My question is, how quantitatively significant is the effect of distortion?

Weinstein: I can promise you one thing. I will now go back, and for each of the structures that we have, I will recalculate the accessibility of several key sites, but it seems to me that both because of the changes in the electronic structure and the change in the steric constraints, there will be a change in susceptibility. I will not fly in the face of experimental observations if they show that there is no change, and no effect, but I will only say this: From the type of energy and structure calculations done with quantum mechanics and from the molecular dynamics simulations that we do, we have the opportunity to talk about the probability of specific changes, for example that it is this part of the DNA or this operator site that is affected. This is an opportunity that we have not really explored, because if you look at cell death as the only end point, or if you only look at double-strand breaks and everything falls to pieces, then there really is no discussion of the details of biological mechanisms of regulation that can be damaged by the effects of energy deposition from radiation exposure. But when you discuss lesions that allow the cell to continue living but producing gene products that have nothing to do with what that cell needs to produce, then you have to talk about specific loci and interactions, and I'm saying that these specificities may be determined by the changes in the structure.

Chatterjee: That I don't have a problem with. The ·OH reactivities; that's what I have a problem with.

Weinstein: I will agree only if you say that this is irrelevant if one measures non-specific numbers of hydroxyl radical-caused strand breaks. If you limit yourself to that, I will agree with you.

Varma: Maybe you can discuss that with Aloke later. You talk about the non-local effects of DNA. Would you define for us what you mean by that?

Weinstein: What we want to look at is not only the changes in the base that has been attacked, but also at changes in electronic structure and in conformation down any number of base pairs. From the experimental information we have in the crystals of protein-DNA complexes, there is no reason to believe that the distortion stops at the site of interaction, because you see that it is greatest at the ends of the segment. The local lesion and the distortion effect, and the rearrangement effect, and therefore the effects on the interaction with the other proteins at other sites, will be propagated to rather large distances away from the point of the lesion. That is what I am talking about.

Menzel: I would like to recall the fact that the radiation-induced observable biological effect is a very rare event. The ratio of the frequencies of primary events to final events is extremely large. The explanation of mechanisms must be related to the identification of relevant rare processes.

Weinstein: You really have two issues here. One is Suresh's original point, which is an important point that we have already discussed. The other issue that you mentioned has to do with the probability of the event occurring and affecting a part of DNA that is functional and involved in protein-DNA interactions. If you consider the entire genome being exposed, and if you consider the relatively small number of sites at which you have these protein-DNA interactions that are relevant to regulation, you already are in a region of low probability. I will thus not be surprised to find a low probability of this mechanism because the probability of making a structural change that will have an effect on the regulation of gene expression is already lowered enormously by the sparseness of these sites in the entire genome.

Rossi: When you talk about the effects of radiation on the molecule, are you thinking in terms of individual ionizations?

Rossi: Well, if you are, that means the effects you are talking about do not depend on radiation quality.

Weinstein: Probably; I will take your word for it.

Varma: It could depend on radiation quality because what he is saying is that if you have a higher LET radiation you might have two ionizations coming close together. So you could have an LET dependence.

Ward: In sunlight many thymine dimers are formed in skin cells. Thymine dimers cause distortions; however, the incidence of skin cancer is not great compared to the numbers of dimers produced and the number of cells exposed. Compared to background radiation, low doses of radiation, there are many fewer lesions per transformation.

Weinstein: The fact that you do get local distortions in DNA structure does not contradict the importance of the mechanistic hypothesis I presented, for the same reason that I mentioned before in answer to Dr. Menzel. The probability of any one distortion's being at a regulatory site like this is extraordinarily small, but when it happens it could lead to the biological consequences I described.

Ward: But if there are orders of magnitude more from ultraviolet light than you get from any low dose of radiation, then there should be a higher probability of the effects occurring from ultraviolet.

Weinstein: So your argument would be that this is easily repaired.

Ward: Yes.

Paretzke: I would like to add another argument to this. Most of the radiation effects which are of interest to us, as mutation or transformation, are sensitively dependent on the energy density, the density of ionizations, the effect of 10, 20 or even higher number of ions. That number of ionizations produced within the cell at different densities affects the outcome we are interested in as significant, and as long as you can take this into consideration, it might indicate that this effect might not play a major role.

Weinstein: It might, but it also might not. As far as I know, these statistics are based on the gross effects, such as cell death, as end points of the observations. In that case, I can see why it would be independent of the detailed mechanisms I described, and it would be dependent on those densities you mention. But I don't know that you can say that all the other phenomena of interest in a cell are also dependent in the same way. I would like to see the evidence that we have discriminated among the components of these phenomena, so that what we consider cell transformation is of the same kind and follows the same mechanism as the other, more radical end points. I would like to know if they behave similarly as a function of density. I would pose it to you that they probably do not.

von Sonntag: At this point, I would like to pose a question: What is the probability of the transformation of the cell compared to the cell killing? That's very low, isn't it?

Varma: About 10^{-5}.

Inokuti: You see, you report an awfully short time of 10 picoseconds for the time. How do you get that number?

Weinstein: From the simulation. What I said was that in our simulation the bend begins to form in the first 10 picoseconds.

Inokuti: What kind of simulations do you do?

Weinstein: These are molecular dynamic simulations, based on the AMBER force field with new parameters for the base radical obtained from quantum chemical calculations of the electronic structures, and constructed so as to be compatible with the AMBER formalism.

Inokuti: You also presume a temperature.

Weinstein: Yes, of course, we do the simulations at 300 degrees. We could go to any temperature.

Rossi: To continue my point, for most of the biological effects of radiation at least two ionizations (or energy transfers) are required, and single ionizations must be regarded as ineffectual.

Weinstein: That is an important point.

Christophorou: How unique is the structure you obtain?

Weinstein: It probably is not. The computations are exorbitantly expensive and thus, even over 600 picoseconds, limited in the amount of time over which the structure is simulated. Even if the average structure is stable, it is still only one of the possible structures, and I don't know how unique the structure is in the crystal. The only thing that we may have some confidence in is the overall nature of the structure. That is all I was talking about; that is the difference between the B-DNA and the structure that appears in the complexes and in our atomic simulations. If I were to emphasize the atomic details of one particular structure, or to put constraints on that structure and build a hypothesis based on the specific geometry of that structure, then you have the right to ask me how unique it is. At this point, I am saying only that there is enough change in the structure to produce the kind of distortion that will either help protein binding or cause dissociation and thus interfere with regulation of translation or with repair.

Christophorou: You don't even know if that is the structure with the minimum energy?

Weinstein: No, not necessarily the global minimum. In a dynamic simulation, one converges to an equilibrium structure that is a local minimum in an enormous potential surface. How close you are, then, to the absolute minimum is an open question in these types of calculations because of the enormous number of degrees of freedom. But if the computations are done correctly, the structure is usually in excellent agreement with the experiment. Molecular dynamics compares itself to NMR and to crystallography all the time. The result is a minimum which very often coincides with the minimum of average structures observed experimentally.

Varma: These are all just self-consistent field calculations?

Weinstein: No, the quantum chemical self-consistent field calculations with corrections for election correlation were used to find the electronic structures of the bases, the sugars, and the radicals produced by radiation, and to calculate their interactions. This is how we obtained the changes in the electronic structure of the base and produced new parameters for the molecular dynamics simulations. The molecular dynamics calculations give us the change in the macromolecular structure that is caused by the local lesion in DNA and is propagated to sites distal from the lesioned base or the break in the strand.

Chemical, Molecular Biology, and Genetic Techniques for Correlating DNA Base Damage Induced by Ionizing Radiation with Biological End Points

Nicholas E. Geacintov[a] and Charles E. Swenberg[b]

Abstract

The types of DNA base damage induced by ionizing radiation, and also relevant model system investigations on replication and mutagenesis, are reviewed in this paper. Recent advances in DNA synthesis technology and site-directed mutagenesis suggest that these methods can be profitably utilized to correlate specific types of DNA base damage with selected biological end points. A deeper insight can be obtained into the molecular origins of mutations, and the effects of base sequence surrounding the lesions on the nature and types of mutations.

Introduction

The effects of ionizing radiation on cells can result in cytotoxic, carcinogenic and mutagenic cellular responses. Numerous studies have demonstrated that damage to DNA is of critical importance in these processes. In general, genotoxic agents, e.g., chemical carcinogens and ionizing radiation, generate activated chemical species which can react with nucleic acids. The genetic alterations induced can range from chromosomal changes to a simple nucleotide modification. For obvious reasons, point mutations, or mutations involving a few bases have been studied in more detail than large genomic rearrangements. Several types of known radiation-induced damage are known: base damage, single-strand and double-strand breaks, apurinic and apyrimidinic (AP) sites, alkali-labile sites, cross-links between DNA strands, and cross-links between DNA and other biological macromolecules.[1,2] A major task in the field of radiation biology is to assess the relative importance of each DNA alteration, and to associate a particular form of damage with a specific biological response. The complexity of this problem is evidenced by noting that several thousand DNA lesions are produced in a single cell irradiated with one Gray.[3]

Advances in the field of molecular biology within the last 10 years, and the feasibility of synthesizing oligodeoxynucleotide sequences of defined base composition and sequence, have opened the possibility of exploring the contribution of a particular DNA lesion to the biological end points produced by ionizing radiation. Considerable information on the nature of DNA lesions generated by ionizing radiation is now available. Many of these altered bases can be incorporated into short DNA fragments,

---

(a) Chemistry Department and Radiation and Solid State Laboratory, New York University, New York
(b) Radiation Biochemistry Department, Armed Forces Radiobiology Research Institute, Bethesda, Maryland

Physical and Chemical Mechanisms in Molecular Radiation Biology
Edited by W.A. Glass and M.N. Varma, Plenum Press, New York, 1991

453

and subsequently these modified oligonucleotides can be built into plasmids or phage DNA, which can then be transfected into cells to monitor survival and mutation frequencies.[4] Since mutagenesis due to ionizing radiation is an infrequent event as compared to lethal hits, any lesion—even one produced in low yields—must be identified and its biological effects evaluated. Once the chemical, structural, and biological characteristics of a DNA lesion have been established, detailed molecular models of the lesion may help to eventually elucidate how repair enzymes recognize DNA damage sites and implement repair. In addition, the influence of these lesions on critical events at the replication forks might also be clarified.

In this paper, we review the present knowledge of DNA base damage induced by ionizing radiation, and the currently available information on mutation spectra obtained by base sequencing methods. We describe several applications of modern research techniques involving short deoxyoligonucleotides specifically modified at selected sites for characterizing various DNA lesions, and also the biological end points associated with these lesions. Our conclusion is that new and deeper insights into the molecular bases of mutagenesis induced by ionizing radiation can be achieved by selective use of such methods in this field of research.

Damage Induced by Ionizing Radiation

In this section the documented types of DNA damage are summarized and discussed.

Strand Breaks and Damage to Sugar Moieties

The occurrence of strand breaks and their biological significance has been discussed by many researchers[1-3,5] and so will not be treated in detail here. Although single-strand breaks are the most prominent lesions caused by ionizing radiation, there is abundant evidence that these lesions are rapidly repaired in most cells.[6-8] It has been suggested that DNA double-strand breaks may constitute the critical lesions associated with cellular lethality; presumably, even a single unrepaired double-strand break is lethal.[9,10]

The types of damage experienced by the 2′-deoxyribose groups in irradiated DNA have been discussed by Ward[3] and von Sonntag.[1] Radiation damage to the sugar moieties results mostly in strand breaks with both 3′ and 5′ monophosphate groups, and 3′-phosphoglycolate at the termini.[11] Also 8-5′ cyclization of the sugar with guanine and adenine is observed. Strand scission during gamma-irradiation occurs uniformly at all nucleotide sites and is thus independent of sequence.[11] However, the occurrence of alkali-labile sites, which can be revealed after the irradiation, is sequence-dependent.[12]

DNA Base Damage

The damage to nucleic acids caused by radiation has been studied at different levels of complexity. Téoule and Cadet[13] have reviewed their extensive work on radiation-induced damage of the individual DNA bases. The radiation chemistry of the dinucleotides d(CpG) and d(GpC) in N_2O and O_2 atmospheres,[14] d(ApT) and d(TpA) in N_2O atmosphere,[15] and O_2 atmosphere,[16] subjected to X-ray irradiation, have been studied by high-performance liquid chromotography and nuclear magnetic resonance. The degradation of the DNA bases in the oligonucleotide d(TpApCpG) exposed to X-rays in an oxygen atmosphere was investigated by Paul et al.[17] Of interest also are the studies of Henner et al.[11] and Duplaa and Téoule[12] on longer oligonucleotides exposed to gamma-irradiation.

During irradiation, a fraction of modified and unmodified bases is released from the DNA backbone, leaving behind an apurinic or apyrimidinic site. Base damage caused by ionizing radiation and not immediately released from DNA can be excised as free bases by the action of DNA glycosylases[18,19] through their catalysis of hydrolysis of the N-glycosylic bonds. The chemical nature of the unmodified and modified free base products can be identified by standard analytical methods and by comparisons of their properties with those of authentic standards.[20] The chemical nature of the modified bases remaining on the DNA backbone is usually determined after mild acid hydrolysis, which releases the products into solution. There is some concern that the very nature of these altered bases can be further changed by even such mild hydrolysis treatments, or that some bases will resist degradation and thus remain unidentified. The use of short oligonucleotides offers the possibility of circumventing these potential complications by identifying the nature of the base alterations without first releasing them through hydrolysis. Nevertheless, based on only one example of such an approach,[17] the major products which have been identified in this manner following x-irradiation of the sequence d(TpApCpG) in aerated solution are the same as those previously listed by Téoule.[20]

The types of chemical DNA damage and DNA base alterations induced by ionizing radiation have been most ably summarized by Téoule[20] and von Sonntag.[1] In this paper, we are mostly interested in those lesions and chemically altered bases which remain attached to the DNA after irradiation; such lesions may remain unrepaired or be subject to error-prone repair, and thus might give rise to mutations. In the following sections, we have omitted discussion of any mechanistic details of how base alterations occur, or of the relative proportions of the different defects produced by different kinds of irradiation and under different experimental conditions.

Téoule[20] counted 29 known and different modified DNA base and sugar residues, but there may be as many as 100 different radiation products induced in DNA by ionizing radiation.[2] Taking into account that even minor radiation products may be biologically significant, as well as the possible effects of DNA sequence in which the lesions reside, it is evident that correlations between DNA base damage and specific biological end-points represent problems of enormous complexity. Nevertheless, initiatives along these lines could begin with some of the better characterized products listed below; the choice of altered DNA bases for further investigation should also depend on results of studies of mutation spectra induced by ionizing radiation, as discussed below.

Pyrimidines

In general, pyrimidines are more susceptible to radiation damage than purines.[14] The types of products arising from radiation damage to pyrimidines, especially thymine, have been well characterized.[20] The major products derived from thymine in DNA originate from attacks of hydroxyl radicals at the 5 and 6 positions. In the presence of oxygen, many of these products are unstable, and formamido degradation products arise predominantly.

Thymine Products. The known thymine products include the following:

I.   5,6-dihydroxy-5,6-dihydrothymine, or thymine glycol (t′); there are four stereoisomers of t′: (cis (5R,6S), cis (5S,6R), trans (5S,6S) and trans (5R,6R), which have been characterized.[15]

II. 6-hydroxy-5,6-dihydrothymine; two cis and two trans isomers have been identified.[15]

III. 5-hydroxy-5,6-dihydrothymine.

IV. Formamido and urea derivatives.[15,16,20,21]
V. 5,6-dihydrothymine.
VI. 5-hydroxymethyl uracil.[20,22]
VII. 5-hydroxy-5-methylhydantoin.
VIII. *Cis-syn* thymine dimers.[23,24]

The structures of some of these products listed above are depicted in Fig. 1.

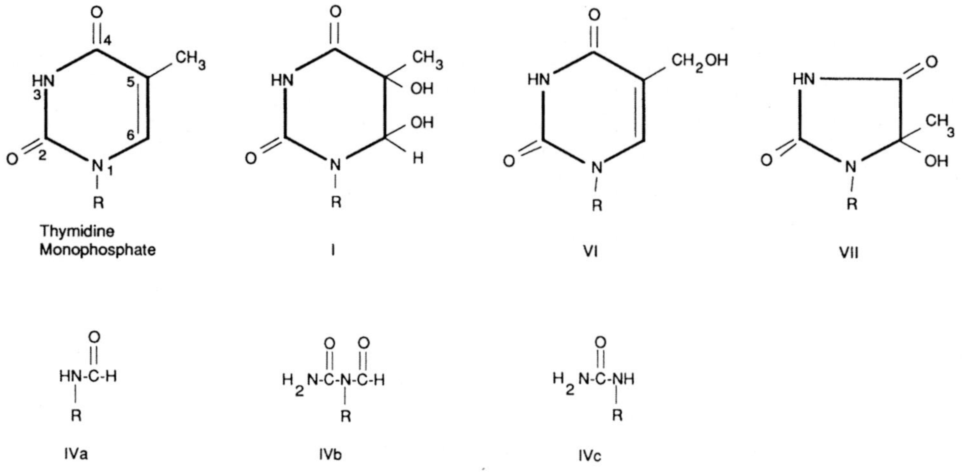

Figure 1

Base-pairing in DNA occurs via the 3 and 4 positions of the thymine base residues; modification of thymine at the 5 and 6 positions (products I - III) is not expected, by itself, to inhibit such base-pairing. However, these modifications may lead to significant local structural alterations in the DNA molecules, which can affect the normal functioning of the polymerases (see below).

The formation of urea, formamide, and formamido products occurs when irradiation is carried out in the presence of oxygen.[13,17,21] The *cis-syn* thymine dimer is formed under the influence of ultraviolet light (below 300 nm); however, such dimers have also been detected in cellular DNA exposed to ionizing radiation, an effect which is attributed to Cerenkov radiation.[1]

Cytosine Products. During irradiation of DNA in aerated solution, monomeric cytosine and some radiation products are spontaneously released into the solution.

The radiation products that remain attached to the DNA backbone tend to be unstable.[20] The following DNA-cytosine products have been recognized:[20]

IX. 5-hydroxycytosine.

X. 5,6-dihydroxy-5,6-dihydrouracil, or uracil glycols (*cis* and *trans*).
XI. 3-carbamoyl-4,5-dihydroxy-2-oxo-imidazolidine, two *trans* isomers.[14,21]
XII. 3-carbamoyl-4-hydroxy-hydantoin.

The structures of some of these products are shown in Fig. 2. In deaerated solutions, the deaminated products of cytosine (X) dominate, whereas in the presence of oxygen, products XI are dominant.[14,17,21]

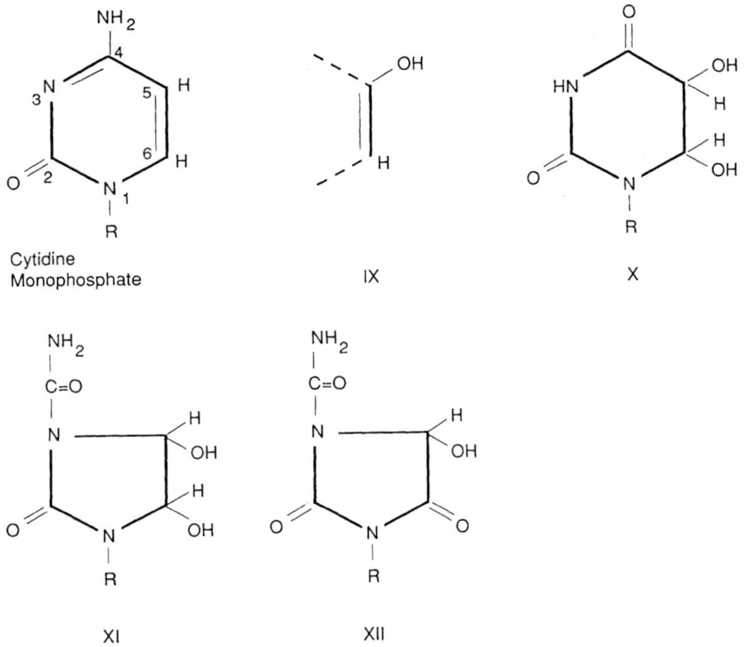

Figure 2

Purines

The following major products have been described in the literature (reviewed by Téoule):[20]

  XIII. 8-hydroxyguanine
  XIV. 2,4-diamino-5-formamido-pyrimidin-6-one
   XV. 8,5′-cycloguanine residue.
  XVI. 8-oxo-7,8-dihydroadenine (often called 8-hydroxyadenine, which represents the less abundant enol form).
 XVII. 4,6-diamino-5-formamido-pyrimidine
XVIII. 8,5′-cycloadenine residue.

The structures of these products are shown in Fig. 3. Products XVI and XVII are partially liberated from DNA during irradiation, but a significant fraction remains attached to the DNA backbone. Product XIV is the major degradation product of guanine in aerated aqueous solutions. Product XIII is formed both in aerated and in

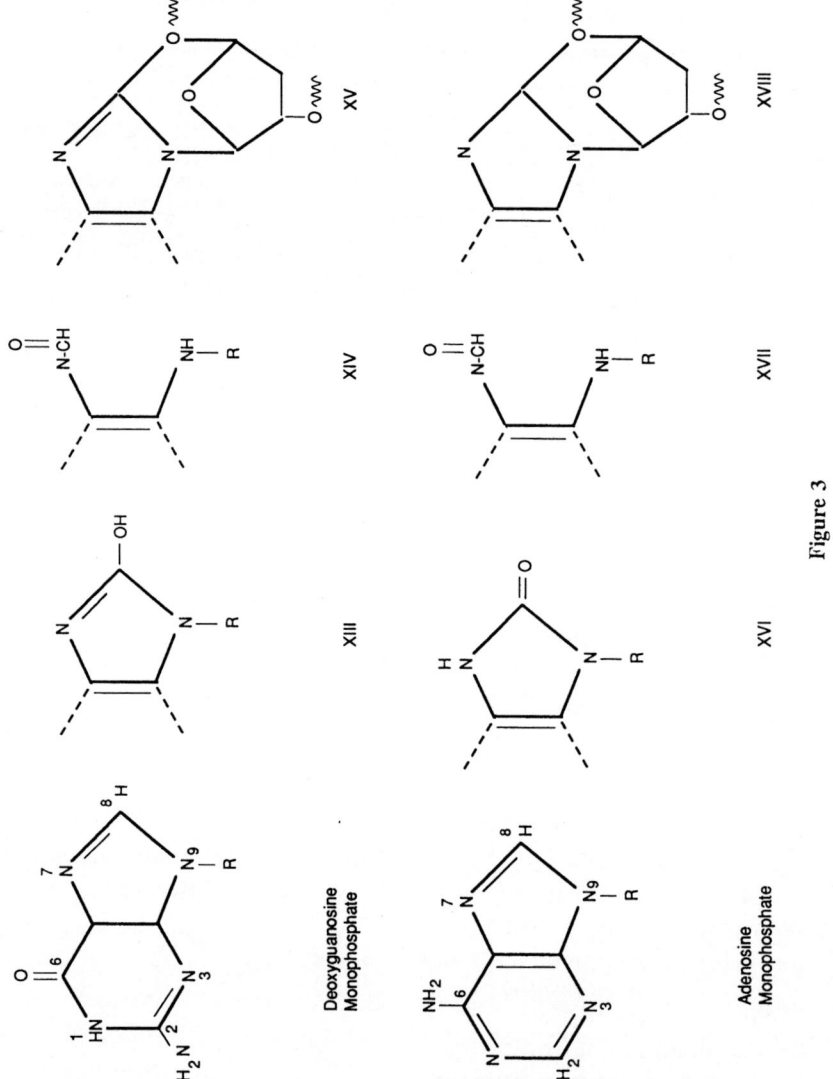

Figure 3

Deoxyguanosine
Monophosphate

Adenosine
Monophosphate

oxygen-free solutions;[20,25] it is also the major radiation product of guanine in the sequence d(TpApCpG) in oxygenated aqueous solutions subjected to X-rays. Interestingly, 8-hydroxyguanine is a major product of the irradiated dinucleotide d(GpC), but not of its sequence isomer d(CpG).[14] Analogous observations were reported for the sequence isomers d(ApT) and d(TpA), since only the former yielded the analogous purine radiation product XVI.[15] The formation of different damaged DNA bases may thus depend on the nature and sequence of the surrounding bases. Furthermore, the extent of local distortion in the native DNA structure, as well as the enzyme repair probabilities and the mutation frequencies due to unrepaired lesions, may also depend on the nature of the DNA sequences in which a given radiation-damaged base is located.

Apurinic/Apyrimidinic Sites

Apurinic/apyrimidinic (AP) sites are common forms of DNA damage and are continuously generated in normal cells as a result of spontaneous depurination, estimated to occur at a rate of 10^4 depurinations per mammalian cell per day;[26] They also result from the protective action of DNA N-glycosylases which remove damaged or unusual bases from DNA.[27] The AP sites are non-instructional; that is, they are devoid of information for Watson-Crick base pairing; however, there is a substantial body of experimental evidence for an unexpected specificity for the insertion of adenine nucleotides opposite such AP sites.[28] The mechanisms involved in this nucleotide selectivity are not understood. Since most depurination events *in vivo* arise from the spontaneous loss of guanine,[26] the preference for adenine insertion opposite AP sites implies that G·C → A·T transversions are the major spontaneous mutations; A·T → T·A transversions are also observed to a lesser extent.[29] The role of AP sites in genetic toxicology has been reviewed by Loeb and Preston.[30]

AP sites can be introduced into DNA by heating (to approximately 70°C) under acidic conditions or by using uracil-containing DNA templates. Because of the low stability of AP, it has not yet been possible to incorporate such sites into oligonucleotides for site-directed mutagenesis studies. Other spontaneously generated DNA lesions, such as those resulting from the deamination of cytosine, are also observed, although at a rate about 1000 times lower than that of the formation of AP sites.[31]

Biological Effects Associated with DNA Base Damage

DNA damage produced by ionizing radiation can inhibit DNA replication and thus lead to cell death. When cells, on the average, sustain one single lethal event which requires a dose of approximately one Gy,[32] about 1000 single-strand breaks, 40 double-strand breaks, 150 DNA-protein cross-links, and about 500 DNA base-damage sites are generated.[3] Apparently, most or nearly all (except the lethal lesion) of this damage is repaired, or somehow does not affect cell survival. DNA damage which is not repaired, or which is repaired by error-prone mechanisms, can lead to mutations. The molecular bases of these events are poorly understood. However, as our knowledge of the specific chemical nature of DNA damage increases,[1-3,20] it should be feasible to correlate specific mutational events with the forms of DNA base damage which give rise to lethal events and to mutations.

AP Sites

Damage produced by ionizing radiation, through the action of glycosylases, can result in the formation of AP sites.[33] Multiple mechanisms for the repair of such lesions exist: AP endonucleases which specifically recognize abasic sites, and the enzyme insertases which insert the appropriate purine into apurinic sites in duplex

DNA.[34] The effects of AP sites on mutagenesis and cell survival has been investigated in both mammalian and prokaryotic cells by transfecting cells with exogenous probes bearing AP sites. Gentil et al.,[35] using monkey kidney cells transfected with modified SV40, have demonstrated that the AP sites are highly mutagenic. In addition, cell survival is significantly lowered, confirming the results of Schaaper and Loeb[36] concerning the effects of AP sites in single-strand φX174 DNA—which strongly decrease the survival of the progeny in *E. coli.*

In prokaryotes, the SOS response is required for the expression of mutations induced by apurinic sites since the mutagenic response is not observed in ultraviolet-irradiated *E. coli* strains deficient in recA, recF or umvC genes.[37] In cases where bypass DNA synthesis occurs, purines were found to be inserted opposite unrepaired AP sites more readily than were pyrimidines. Although AP sites generally block DNA synthesis, the termination of synthesis is a function of the polymerase and the nucleotide sequence. Experiments indicate that T4 DNA polymerase terminates chain elongation one nucleotide before the AP sites, whereas DNA polymerase I, AMV inverse transcriptase, and DNA polymerase α terminate synthesis either before or at the AP lesion, depending on the particular base sequence. In the case of AP lesions which are not repaired *in vivo*, depurinations should most likely lead to transversions (e.g.,G → T), whereas depyrimidination should induce transitions.[28]

Mutation Spectra

Advances in genetics and molecular biology, coupled with the availability of efficient techniques for sequencing DNA, allow an identification of the DNA base sequences which accompany the observed mutations. Within the last few years, some interesting studies along these lines have been published.

Ayaki et al.[38] exposed M13 mp10 single-stranded phage DNA to ^{60}Co gamma rays, and the irradiated phage DNA was subsequently transfected into *E. coli*. The M13 mp10 DNA contains the *lac* regulatory genes and part of the *lac Z* structural gene of *E. coli* (SOS-noninduced). Mutant DNA was isolated from 15 clones, and ≈ 200 base fragments within the *lac* region were sequenced using the chain termination method of Sanger et al.[39] In order to generate a number of mutants significantly above the spontaneous, or background level, radiation doses of the order of 100 Gy were used. At 200 Gy, the survival probability was 0.1%, while the mutation frequency was only five times greater than that of background. Thus, lethal events greatly dominate mutational events.

Nearly 90% of mutations observed by Ayaki et al. involved single-base substitutions, with transversions and transitions occurring with about equal frequencies.[38] Among the substitutions, mutations at cytosine bases were most frequent (7/15) and C·G → T·A transitions were dominant; the authors speculated that C → T transitions may result from the gamma radiation-induced product X, 5,6-dihydroxy-5,6-dihydrouracil, which has the same hydrogen-bonding capacity as thymine. There were two G·C → T·A transversions, involving a G → T change. These two types of mutations include 9/15, or 60% of the total number of observed mutations, which involve misincorporation of dAMP opposite damaged dC or dG sites on the template strand during the first round of replication. Since AP sites also preferentially involve the incorporation of dAMP opposite the abasic sites, particularly in SOS-induced cells (see above), these results suggest that the formation of AP sites by ionizing radiation may be also quite important.[38] Such abasic sites can be formed by repair of DNA by glycosylases[20] or by destabilization of the N-glycosidic bond due to DNA base damage.[1] However, such a response usually requires the induction of SOS repair,[30] which was not the case in the Ayaki et al. experiments.[38] Hoebee et al.[40]

speculated that SOS-response might be induced by the relatively heavy doses of ionizing radiation used in experiments of this type, thus triggering the error-prone SOS repair system (see also Ref. 43).

In addition, only one other transition (T·A → C·G) occurring at dT sites was observed by Ayaki et al., and one frameshift mutation. The other six mutations (40%) involved the following transversions:[38]

(i) G·C → C·G (one), possibly involving 8-hydroxyguanine (XIII).[40,41]

(ii) A·T → C·G (one).

(iii) C·G → C·A (one) and C·G → A·T (two).

Grosovsky et al.[42] characterized various mutations induced by gamma rays at the endogenous adenine phosphoribosyltransferase (*aprt*) locus of Chinese hamster ovary cells at the DNA sequence level. Only 16% of the mutants involved major genomic rearrangements, while 84% involved point mutations (base substitutions, small deletions, insertions, and frameshift mutations). A total of 16 mutants were sequenced, and 11 base substitutions and 5 frameshift deletions (1-18 base pairs) were found. Out of the 11 base substitutions, one involved C (C → A base substitution). Four involved G: G → A (three) and G → C (one). Three involved T: T → G (two) and T → C (one). Three involved A: A → T (two), and A → C (one). Thus, four of these point mutations were the transitions G·C → A·T (three) and A·T → G·C (one), and seven were transversions T·A → G·C (three), A·T → T·A (two), C·G → A·T (one), and G·C → C·G (one). Grosovsky et al. concluded that the distribution of mutations in the *aprt* locus sequenced is practically random, in strong contrast to the occurrence of spontaneous mutations. Two types of mechanisms, sequence-directed events and random breakage, may be involved in the gamma-ray-induced deletions, in which slippage and misalignment of damaged DNA sequences may occur.[42]

Ayaki et al. and Grosovsky et al. observed a total of seven different mutations; however, only four of these were observed by both groups (see Table I).

Tindall et al.[43] studied mutations in the cI (repressor) gene of lambda phage, either as free phage or prophage inserted in the genomes of lysogens, induced by gamma radiation. The host cells were either pre-induced or non-induced for SOS response. In SOS-induced cells, 6 of the 41 base substitutions (there were an additional 7 frameshift mutations) involved G·C → A·T transitions, 14 were G·C → T·A transversions, and there was one A·T → C·G transversion. An analysis of their mutation spectrum [Fig. 3 of Ref. (43)] suggests that 22/41 (about 54%) of the observed point mutations appear to have involved misincorporation of dAMP opposite the damaged base site during the first round of replication. Overall, in 40/41 of the observed base mutations, A·T base pairs replaced either G·C, C·G or T·A base pairs. The authors suggest that the preferential mutations resulting in T·A or A·T base pair substitutions may involve non-informational base damage such as AP sites, which induce mutations via SOS repair-dependent mechanisms. It is noteworthy that Schaaper et al.[29] observed a similar mutational specificity in depurinated single-stranded M13 phage DNA.

In SOS non-induced host cells, there were fewer point mutations (18 instead of 41 in SOS-induced cells): 13 of these involved G·C → A·T transitions, there were 2 G·C → T·A transversions, and 3 A·T → G·C transversions. This latter mutation was not observed in the SOS-induced cells. Three frameshift mutations were also observed. The G·C → A·T transitions may result from a deaminated radiation product of cytosine which mispairs with adenine during replication.[43]

Table I. Some Published Examples of Single Base Substitution Mutation Spectra Resulting from Gamma Irradiation

Mutations	Base Change	Fraction of B.P. Substitutions	Remarks	Mechanism
(TS) G·C → A·T	G → A	3/11 (G)		
		9/40 (T)	+SOS	
		5/17 (T)	-SOS	
(TS) C·G → T·A	C → T	7/15 (A)		AP? X?
		10/40 (T)	+SOS	
		6/17 (T)	-SOS	
		1/19 (H88)	+O$_2$	X?
(TS) T·A → C·G	T → C	1/15 (A)		
		1/11 (G)		
(TS) A·T → G·C	A → G	1/17 (T)	-SOS	
(TV) G·C → T·A	G → T	2/15 (A)		AP?
		8/40 (T)	+SOS	
		2/17 (T)	-SOS	
		2/19 (H88)	+O$_2$	
(TV) G·C → C·G	G → C	1/15 (A)		
		1/11 (G)		
		1/19 (H88)	+O$_2$	
(TV) C·G → G·C	C → G	1/15 (A)		
		15/19 (H88)	+O$_2$	XIII?
		15/20 (H89)	+N$_2$O	XIII?
(TV) C·G → A·T	C → A	2/15 (A)		
		1/11 (G)		
		6/40 (T)	+SOS	AP?
		1/17 (T)	-SOS	
		5/20 (H89)	+N$_2$O	AP? XIII?
(TV) A·T → C·G	A → C	1/15 (A)		
		1/11 (G)		
		1/40 (T)	+SOS	
(TV) T·A → G·C	T → G	2/11 (G)		
(TV) A·T → T·A	A → T	2/11 (G)		
		4/40 (T)	+SOS	
(TV) T·A → A·T	T → A	2/40 (T)	+SOS	
		2/17 (T)	-SOS	

Product X: 5,6-dihydroxy-5,6-dihydrouracil; XIII: 8-hydroxyguanine.
AP: apurinic/apyrimidinic sites.
TS: transitions; TV: transversions.

Geacintov and Swenberg

References to Table I:

(A):   H. Ayaki, K. Higo and O. Yamamoto. *Nucleic Acids Res.* 14:5013-5018 (1986).

(G):   A. J. Grosovsky, J. G. deBoer, P. J. deJong, E. A. Drobetsky and B. W. Glickman. *Proc. Natl. Acad. Sci.* 85:185-188 (1988).

(T):   K. R. Tindall, J. Stein and F. Hutchinson. *Genetics* 118:551-560 (1988).

(H88):   B. Hoebee, J. Brouwer, P. v. d. Putte, H. Loman and J. Retèl. *Nucleic Acids Res.* 16:8147-8156 (1988).

(H89):   B. Hoebee, M.A. v.d. Ende, J. Brouwer, P.v.d. Putte, H. Loman and J. Retèl. *Int. J. Rad. Biol.* 56:401-411 (1989).

Hoebee and coworkers[40,41] studied the nature of the mutations in double-stranded M13 mp10 bacteriophage DNA with an in-frame insert of target sequence of 144 base pairs in the region of the *lacZα* gene. The DNA was exposed to ^{60}Co gamma irradiation either in oxygenated aqueous solutions or anoxic solutions (purged with N_2O), and then transfected into *E. coli* cells. Following irradiation doses of several Gy, the survival rate was about 10%, and the mutation frequency exceeded the spontaneous frequency by a factor of about 7. Out of 23 mutations sequenced in the 144-base-pair insert, 19 were base substitutions and 4 were frameshift mutations. A strikingly large fraction (15 out of 19, or 79%) of the mutants involved C → G base changes (C·G → G·C transversions), and the other four mutants exhibited either G → C, C → T, or G → T base changes. Hoebee et al.[41] noted that the preponderance of G → C transversions is consistent with the presence of the 8-hydroxyguanine (XIII) radiation product which, in the enol form, might base-pair with guanine in the enol/imino form. However, Kuchino et al.,[44] using an *in vitro E. coli* DNA synthesis polymerase I system, showed that XIII can indeed cause mutations, but that there was no preference for the incorporation of G opposite the 8-hydroxyguanine radiation product. While it is possible that *in vivo* replication in *E. coli* cells may be quite different from that of the *in vitro* system of Kuchino et al.,[44] Hoebee et al.[41] noted that the C·G → G·C transversions may also result from specific processing of the lesion XIII in *E. coli*, or might be due to a still-different radiation lesion. The C·G → T·A transition observed by Hoebee et al.[41] was the dominant mutation observed with single-stranded DNA in the mutation experiments of Ayaki et al.[38] The fact that this transition was observed in only one out of 19 mutations in double-stranded DNA suggests that the putative radiation product uracil glycol (X), which can lead to a mutation of this type, is not formed efficiently in double-stranded DNA, or may be repaired in *E. coli*.[41] The other two types of mutations, C·G → T·A and G·C → T·A,[41] were also observed in the experiments of Ayaki et al.[38] Another interesting observation in the experiments of Hoebee et al.[41] is that about 60% of base pair substitutions were concentrated at three different positions, all in the vicinity of (TGCT)·(ACGA) sequence. This suggests that gamma-radiation-induced mutations are somehow favored near this sequence, for reasons which remain unknown.

Irradiation of the same system under anoxic conditions (N_2O), also preferentially gives rise to C·G → G·C transversions (15 out of 20) within the same 144 base-pair target sequence; the other five base pair substitutions were C·G → A·T transversions.[40] Eight frameshift mutations were counted when the irradiation was carried out under anoxic conditions. Thus, most of the types of mutations observed in the presence and absence of oxygen are the same. A noticeable difference is the occurrence of -1 base pair deletions which are found in the anoxic mutation spectrum, but not in the presence of oxygen; these deletions were attributed to the results of DNA damage induced by H radicals under anoxic conditions, or to the biological repair of the resulting damage.

The types of base-substitution mutations observed as a result of gamma irradiation, the relative frequency of occurrence of each base substitution, and some of the suggested molecular origins of these mutations are summarized in Table I. It is evident that every one of the 12 possible base substitutions involving the four bases C, T, A, and G have been observed, although with different frequencies. The molecular origins of these mutations have not yet been established. Only a few of the known radiation products, such as 8-hydroxyguanine, the uracil glycols, and putative abasic sites, have been implicated so far. It is clear that much more research will be necessary in order to reach a deeper understanding of the molecular basis underlying these phenomena.

Recent developments suggest that new approaches for solving these complex problems should be explored. These new approaches involve studies of well defined and characterized model systems based on the synthesis of chemically defined single-stranded oligonucleotides containing a single lesion. These oligonucleotides can then be characterized physico-chemically, and site-directed mutagenesis,[4] *in vitro* polymerase stop, and other experiments can be used to assess the biochemical and biological characteristics of each defined lesion. Some extremely interesting examples along these lines of research have already been published, and are briefly discussed below.

Synthesis of Site-Specific Oligodeoxynucleotides of Defined Sequence with A Single Lesion

There are currently three methods for synthesizing oligonucleotide sequences with a specifically altered base. The first method involves total synthesis of the modified DNA segments.[45] Whenever possible, this should be the method of choice. The advantages are high yield, precise adduct placement, and high purity. Unfortunately, this method is not feasible for many adducts because of their instability under the conditions of chemical synthesis.

The second method involves direct modification of one particular base in an existing oligonucleotide. This can be achieved by treatment of the oligonucleotide with a chemically reactive form of a carcinogen, or by irradiating a DNA segment with ultraviolet light containing two adjacent thymine residues to form thymine dimers, or by producing an altered thymine base by oxidation to thymine glycol with osmium tetroxide.[46] While this approach is more simple than the total-synthesis method, a complicating feature is that this method generally produces a heterogeneous set of products which must be separated from one another, and extensive purification of the desired product is usually necessary. Thus, contamination is an ever-present danger and may lead to incorrect conclusions regarding the biological end points of a given lesion.

The third method involves enzymatic synthesis in which DNA polymerase and ligase are employed.[47,48] Although the synthesis conditions are mild, the method is of limited utility because of the small yields and types of adducts that can be incorporated.[4]

A large number of different kinds of oligonucleotides with certain defined bases, altered either chemically or via a covalent bond with an aromatic or aliphatic mutagenic moiety, have been synthesized and studied until now. These include oligonucleotides containing alkylated DNA bases, e.g. O^6-alkylguanines,[49] aromatic amine adducts such as 2-aminofluorene,[50] ultraviolet and psoralen photoadducts,[51] and oxidized bases such as thymine glycol.[52] Examples which are relevant to radiation chemistry are thymine glycol[53] and product XVI[54] incorporated into defined oligonucleotides.

In this paper, we restrict our attention to the oligonucleotides derived from thymine glycol (I), and those relevant in ultraviolet radiation of DNA (Fig. 4). These adducts have been studied extensively; thus, the value of the methods used has been demonstrated. Photo-pyrimidine dimers are the most frequent type of DNA damage produced by ultraviolet light; failure to repair such lesions has been correlated with the genetic disease *xeroderma pigmentosum* and skin cancer.[55] Apurinic/apyrimidinic (AP) sites represent one of the most common forms of DNA damage, and exhibit cytotoxic and mutagenic effects.[56] Release of damaged base residues by DNA glycosylases[57] results in AP site formation.

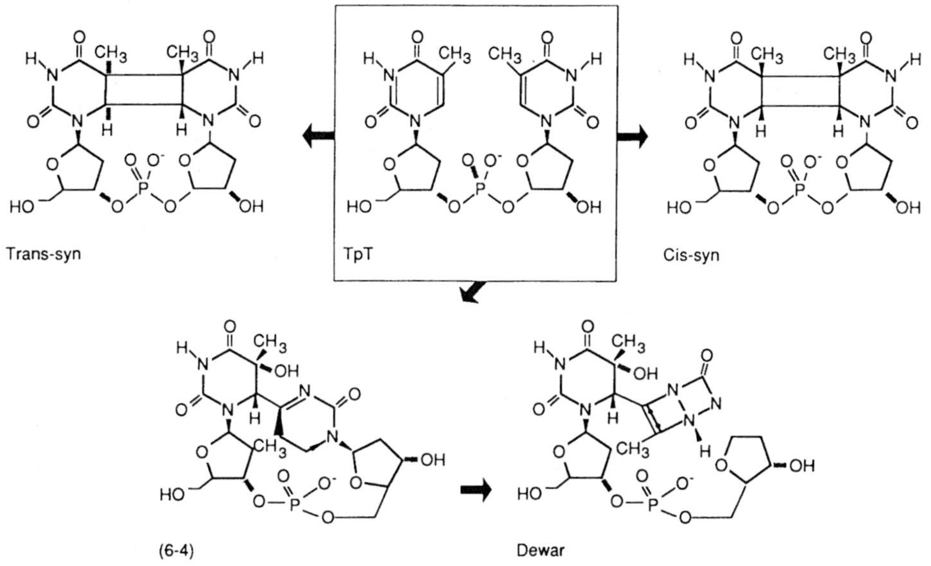

Figure 4

Some Applications of Techniques Involving Oligonucleotides with Site-Specifically Incorporated DNA Base Lesions

Thymine Glycol (t´)

Thymine glycol (t´) is one of the major products of ionizing radiation[20] and cellular oxidation.[58] In order to assess its biological effects *in vivo* and *in vitro*, t´ has been incorporated into oligonucleotides of defined sequences at specific positions within these short DNA molecules.[52,59,60] Total synthesis of this adduct is difficult due to the susceptibility of t´ to rapid degradation to urea and other products under alkaline conditions.[59] For this reason, methods involving the modification of existing oligonucleotides were used; well-defined oligomers containing a single thymine base were exposed to osmium tetroxide,[46] in the presence of aqueous pyridine,[61] the latter being added primarily to accelerate the rate of oxidation. Under these conditions, *cis*-t´ is the major product, although both diastereomeric forms are produced in quantities not yet established. Clark and Beardsley[60] used this procedure for two different 18-mers containing a single thymine base located at position 16 from the 3´-end. To determine whether *cis*-t´ introduces sufficient distortions into the helix to alter DNA polymerase activity, each oligomer was annealed with a 14-mer primer complementary to the lesion-containing template, and ^{32}P-labeled at its 5´-end by standard methods.[62] The DNA synthesis reaction was performed in the presence of different 2´-deoxynucleotide triphosphates (dNTP) with different polymerases, and was analyzed by denaturing polyacrylamide gel electrophoresis. In the case of the Klenow fragment (polymerase I), T4 DNA polymerase, and (human) polymerase a_2, synthesis was arrested at the site of the lesion in the template strand, although in certain cases DNA polymerase was able to bypass the lesion.[63] In contrast, the *cis-syn* thymine dimer is bypassed with a frequency of 10 to 20% by *E. coli* DNA polymerase I and the Klenow fragment (polymerase I) *in vitro*[64] (see below).

The DNA synthesis pattern in the case of the t´-containing template and in the presence of AMV reverse transcriptase was complex, and in some cases synthesis beyond the lesion did occur. Comparison of the denaturing polyacrylamide gel bands produced in the presence and absence of dATP showed that t´ base-paired almost exclusively with the correct nucleotide. Clark et al.,[65] using coordinates of the *cis*-t´ determined by X-ray crystallography[66] and energy minimization techniques, showed that the structural alteration produced by t´ are considerably less than those imposed by thymine dimers, and that there is unfavorable steric overlap between the 5´ base and the (axial) methyl group of glycol. Presumably the increase in tilt angle of the 5´-template base distorts the geometry sufficiently so as to preclude the stable addition of the next nucleotide beyond the lesion. Measurements also indicate an enhanced susceptibility of the adenine-t´ pair to hydrolysis by 3´-5´-exonucleases as compared with the normal A·T base pairs. In effect, the enhanced turnover further contributes to the inhibitory effect of thymine glycol in DNA synthesis. A change in base immediately 5´ to the lesion site (A → C) did not significantly alter these results.

Why some DNA polymerases such as AMV reverse transcriptase are partially able to continue synthesis beyond the t´ lesion is not known, although one could speculate as did Clark and Beardsley[60] that one of the *cis* diastereomers, by producing a slightly different distortion in the DNA helix, impedes the enzyme. The conclusion inferred from this *in vitro* study is that thymine glycol is unlikely to be a potent mutagenic lesion *in vivo* and that t´ arrests DNA synthesis.

The biological effects of t´ were studied by Basu et al.,[53] who investigated its effect on replication and mutagenesis *in vivo* by employing modified phage M13mp19-*NheI* DNA transfected into several different *E. coli* strains. Scoring for mutants was facilitated from observing phenotypical changes in the clones from assaying in the presence of 5-bromo-4-chloro-3-indoyl ß-D-galactopyranoside, provided that the site-specific alteration was positioned in the amber codon (GCTAGC) within the hexanucleotide insert. Specifically, a modified 6-mer d(GCt´AGC) was produced either by oxidation of d(GCTAGC) with $KMnO_4$ or OsO_4. After purification, the modified hexanucleotide was inserted into the M13mp19 DNA at the unique *SmaI* site. The technique is quite general and can be utilized to generate other site-specific adducts for *in vivo* studies of mutagenicities. Several different *E. coli* strains with different enzyme repair capacities were transfected with either single-stranded t´-M13mp19-*NheI* [either on the (+) or the (-) strand at the same position], or with double-stranded t´-M13mp19-*NheI* DNA. Scoring for mutations in the clones was performed phenotypically; M13-*NheI*-containing *E. coli* produced light-blue plaques in supE strains in the presence of 5-bromo-4-chloro-3-indoyl ß-D-galactopyranoside, whereas point mutations at the unique *NheI* site generated dark-blue plaques. Such dark-blue plaques, independent of t´, are also produced by cells with deleted *NheI* sequences in the M13-*NheI* phages. The mutation frequency of t´ was 0.2-0.4% in single-stranded DNA, whereas t´ was not detectably mutagenic in the case of the modified double-stranded DNA. The latter result suggested either that t´ blocked replication of the strand, a result consistent with the conclusions of Clark and Beardsley[60] discussed above, or that t′ was efficiently repaired. The mutation frequency was SOS-independent, since strain DE1018, having an impaired SOS capability, exhibited the same mutation frequency as SOS-competent cells (strain MM294A).

When the t′-induced mutant phages were sequenced, only t′·A → C·G mutations were found, suggesting that t′ can mispair with guanine during replication. Interestingly, molecular modeling studies can be used to rationalize the preference for the t′ → C mutation; it was shown that t′ is displaced by ≈ 0.5 Å further than thymine toward the major groove, which is expected to facilitate base-pairing of t′ with guanine. Results of molecular mechanics calculations suggest that base-pairing t′·A is favored over t′·G pairing; however, the t′·G interaction was calculated to be approximately 1.2 Kcal/mol more favorable than T·G mispairing.[53]

An untested prediction of the model calculation is that the 5′-flanking base of t′ has little influence on the pairing interaction, whereas the nucleotide on the 3′-side of t′ influences the extent of base wobbling at t′. The degree of displacement of t′ was predicted to be markedly dependent on the nature of the 3′-flanking base; this displacement increased in the order G < A < T ≈ C. These predictions are consistent with the known experimental result that t′ polymerases bypass t′ only in sequences with 3′-purines.[63,65] Therefore, there may be an effect of base sequence on mutation frequencies, a prediction which can be verified experimentally with modified single-stranded t′-M13mp19-NheI.[53]

Thymine Dimer Photoproducts

In the presence of 254 nm light,[67] the major photoproducts *cis-syn* cyclobutane and (6-4) dipyrimidine dimers (Fig. 4) are formed in double-stranded DNA. Both photoproducts have been implicated in ultraviolet light-induced mutagenesis and carcinogenesis.[68-70] In the presence of 313 nm irradiation, the (6-4) dipyrimidine dimer is converted to the Dewar pyrimidinone (Fig. 4), a reaction which is reversed in the presence of 240 nm irradiation.[71] The possibility that the photoproduct of the (6-4) dipyrimidine dimer also contributes significantly to the biological response of UV-irradiated cells should be considered.[72] A detailed discussion of the spectroscopy, conformational, and molecular modeling analysis of the Dewar pyrimidinone has been provided by Taylor et al.[73] A *trans-cis* dimer (Fig. 4) is also formed in very low yield; little is known about its mutagenic properties, however.[74]

Results of 2-D NMR studies of the oligonucleotide duplex d(GCGTTGCG).d(CGCAACGC) of Kemmink et al.[51] show that there are small differences in both the ¹H chemical shifts of the normal and *cis-syn* thymine dimerized octamer duplexes and the NOE cross-peak intensities. This data suggests that the B-DNA double-helical structure is perturbed only in the thymine dimer region and that these distortions are small.

Model calculations by Rao et al.[75] for a dodecamer of DNA indicated little distortion in the B-DNA structure, a result consistent with NMR data though not in agreement with the results of molecular modeling studies of Pearlman et al.[76] These authors concluded that DNA is bent at the site of the thymine dimer by approximately 27°, and that significant conformational distortion occurs on the strand complementary to the one bearing the thymine dimer lesion. Experimental support for thymine dimer-induced bending comes from one- and two-dimensional gel electrophoresis measurements on DNA fragments containing thymine dimers.[77] Photodimer-containing DNA fragments migrated more slowly than their undamaged counterparts. Exposure of the samples to photoreactivating light not only annealed the dimer lesions, but caused the anomalous electrophoretic DNA migration to disappear. It is difficult to reconcile these results with current 2-D NMR data[51] and the results of molecular mechanics calculations,[78] which indicate that the structural changes are limited to a few base pairs of the photodimer without inducing significant bending of the DNA backbone. A first step towards the resolution of these conflicting interpretations necessitates a calculation of the NMR parameters for the different theoretical models; it may well be that bending is consistent with the 2-D NMR data.

Taylor and O'Day[64] performed *in vitro* polymerase replication experiments with *cis-syn* thymine dimers incorporated site-specifically into a 41-mer oligonucleotide similar to the one employed in the studies of thymine glycol by Clark and Beardsley.[60] Incorporation of the *cis-syn* thymine dimer into the oligonucleotide was accomplished via phosphoramidite-based solid phase DNA synthesis.[79]

Incubation of the 41-mer template with a $5'$-^{32}P end-labeled 15-mer, with the four dNTP's at different 15-mer and polymerase concentrations, showed that DNA synthesis was viable and that the *cis-syn* thymine dimer was bypassed by the polymerase 10% to 20% of the time. Both polymerase I, the best characterized DNA polymerase of *E. coli*, and the Klenow fragment were able to synthesize beyond the thymine dimer lesion, in contrast to previous reports that pyrimidine dimers block polymerase I DNA synthesis.[80,81] These differences may have been due to possibly inhomogeneous templates used in the older studies.[64] It is noteworthy that polymerase I has also been shown to bypass a site-specific aminofluorene adduct.[82,83] For DNA synthesis beyond the dimer, the frequency of different nucleotides incorporated at the lesion site decreased in the following order: pdA > pdT ≈ pdG > pdC. The high rate of incorporation of pdA could be related to the "A-rule," which states that pdA is preferentially inserted opposite non-instructional lesions by polymerases.[81] The observation that there is a twenty-fold increase in pdA incorporation over that of pdG opposite the $3'$-T of the dimer in bypass products, as compared with halted DNA synthesis, supports the notion that an additional process is operative; presumably there is a preferential DNA elongation in the presence of correct base pairing at the lesion site.[84]

Banerjee et al.[85] synthesized an 11-mer oligonucleotide (dGCAAGTTGGAG) which was photochemically modified to produce the TT cyclobutane dimer. The *cis-syn* oligonucleotide was separated from other products and purified extensively. This modified oligonucleotide, containing the unique *cis-syn* thymine dimer, was subsequently inserted into a single-stranded M13mp7 hybrid phage vector, which was then transfected into ultraviolet-irradiated A6 mutants of *E. coli* for determination of survival rate, mutation frequency, and mechanisms of mutation. In host cells which were not induced for SOS response, very few progeny plaques were observed, implying that the *cis-syn* thymine dimer blocks replication with an efficiency > 99.5%. As noted above, this lesion is less effective in blocking replication *in vitro* (polymerase I and the Klenow fragment);[64] this difference can probably be attributed to the fact that additional polymerases are active *in vivo*. As expected for *cis-syn* dimers, the inhibition of DNA synthesis in vivo was abolished upon photoreactivation with *E. coli* photolyase.

In host cells in which the ultraviolet response had been photoinduced with ultraviolet irradiation, transfection efficiencies were 50 to 100 times higher and the phage survival levels were 30% of control values. A total of 529 progeny phages were sequenced in order to assess the effects of the *cis-syn* dimer on replication and mutagenesis. The normal sequence was found in 93% of the progeny, suggesting efficient and accurate replication past the lesion. The mutation frequency was 7%, and all of the mutations involved single base substitution. Slightly more than half of the mutations were T → A transversions, while the remainder were T → C transitions. There were no T → G transversions in evidence. About 90% of the single base substitutions were targeted at the 3′ thymine of the lesion, which is the first of the two modified thymines encountered by polymerase. Interestingly, these results suggest that a single and chemically defined lesion can give rise to more than one type of mutation.

Finally we note that the T·A → A·T transversion and T·A → C·G transition seen in the site-directed mutagenesis studies of Banerjee et al.[85] is also observed (besides many other mutations) in the ultraviolet mutation spectra in *E. coli*.[68]

Summary and Conclusions

The limited number of studies on mutation spectra reported thus far suggest that AP sites, 8-hydroxyguanine, and uracil glycols are DNA lesions which could give rise to some of the observed mutational base substitutions. These and other studies suggest that thymine glycols arrest DNA synthesis, but do not appear to give rise to mutations, perhaps because of efficient repair or low rates of survival. Site-directed mutagenesis experiments using modified single-stranded DNA show that single t′·A → C·G mutations are observed, whereas t′ incorporated site-specifically into double-stranded DNA was not noticeably mutagenic.[53] Analogous experiments with *cis-syn* thymine dimers indicate that one single lesion can give rise to several different base-substitution mutations.[85] There are indications that DNA damage induced by ionizing radiation, as well as the nature of mutations, may be in some cases a function of base sequence.

The recent results summarized here suggest that experiments with synthetic DNA sequences containing a single, well-defined lesion should be extremely useful for clarifying relationships between specific types of damage with biological end points such as survival, mutation spectra, and DNA polymerase inhibition. For instance, the question whether 8-hydroxyguanine is mutagenic or not[40,41] could be clarified by site-specific mutagenesis techniques.[4] The elegant new genetic and molecular biology techniques developed within the last 10 to 15 years should provide new insights into the molecular mechanisms underlying the mutational effects of ionizing radiation; however, the large number of known DNA lesions, possible effects of base sequence which multiply the number of variables, and the labor-intensive techniques associated with synthesizing site-specific modified DNA sequences, indicate that progress in this field will nevertheless be difficult.

Note added in proof: It has been demonstrated that the (5R, 6S) and (5S, 6R) *cis* isomers of 5,6-dihydroxy-5,6-dihydrothymidine are formed in equal amounts in gamma-irradiated DNA [G. Teebor, A. Cummings, K. Frenkel, A. Shaw, L. Voituriez, and J. Cadet. Quantitative Measurement of the Diastereoisomers of *cis*-Thymidine Glycol in Gamma-Irradiated DNA. *Free Rad. Res. Comms.* 2:303-309 (1987)]. The differences in the biological effects of these two diastereoisomers have not yet been established.

Acknowledgment

The writing of this review was supported by the U.S. Department of Energy, Grant DE-FGO2-86ER60405.

References

1. C. von Sonntag. *The Chemical Basis of Radiation Biology.* Taylor & Francis, London (1987).
2. F. Hutchinson. Chemical Changes in DNA Induced by Ionizing Radiation. *Progr. Nucleic Acid Res. Mol. Biol.* 32:115-154 (1985).
3. J. F. Ward. DNA Damage Produced by Ionizing Radiation in Mammalian Cells: Identities, Mechanisms of Formation and Repairability. *Progr. Nucleic Acids Res. Mol. Biol.* 35:95-125 (1988).

4. A. K. Basu and J. M. Essigman. Site-Specifically Modified Oligonucleotides as Probes for the Structural and Biological Effects of DNA-Damaging Agents. *Chem. Res. Toxicol.* 1:1-18 (1988).

5. D. Schulte-Frohlinde. Comparisons of Mechanisms for DNA Strand Break Formation by the Direct and Indirect Effect of Radiation. In *Mechanisms of DNA Damage and Repair*, M.G. Simic, L. Grossman and A.C. Upton, eds., pp. 19-27. Plenum Press, New York (1986).

6. I. R. Radford. Effect of Cell Cycle Position and Dose on the Kinetics of DNA Double-strand Breakage Repair in X-irradiated Chinese Hamster Cells. *Int. J. Radiat. Biol.* 52:555-563 (1987).

7. P. Wlodek and W. H. Hittelman. The Repair of Double-Strand DNA Breaks Correlates with Radiosensitivity of L5178 Y-S and L51178 Y-R Cells. *Radiat. Res.* 112:146-155 (1987).

8. C. K. Hill, J. Holland, C.M.C. Liu, E. M. Buess, J. G. Peak and M. J. Peak. Human Epithelial Teratocarcinoma Cells (P_3): Radiobiological Characterization, DNA Damage, and Comparison with Other Rodent and Human Cell Lines. *Radiat. Res.* 113:278-288 (1988).

9. D. Blocher and W. Polit. DNA Strand Breaks in Ehrlich Ascite Tumor Cells at Low Doses of X-rays. II. Can Cell Death be Attributed to Double-strand Breaks? *Int. J. Radiat. Biol.* 42:329-338 (1982).

10. I. R. Radford. The Level of Induced DNA Double-Strand Breakage Correlates with Cell Killing After X-irradiation. *Int. J. Radiat. Biol.* 48:45-54 (1985).

11. W. D. Henner, S. M. Grunberg and W. A. Haseltine. Sites and Structure of Gamma-Radiation-Induced DNA Strand Breaks. *J. Biol. Chem.* 257:11750-11754 (1982).

12. A. M. Duplaa and R. Téoule. Sites of Gamma Radiation-Induced DNA Strand Breaks After Alkali Treatment. *Int. J. Radiat. Biol.* 48:19-32 (1985).

13. R. Téoule and J. Cadet. Radio-Induced Degradation of the Base Component in DNA and Related Substances. In *Effects of Ionizing Radiation in DNA*, J. Huttermann, W. Kohnlein, R. Téoule and A. Bertinchamps, eds., pp. 171-202. Springer Verlag, Berlin (1978).

14. C. R. Paul, A. V. Arkali, J. C. Wallace, J. McReynolds and H. C. Box. Radiation Chemistry of 2′-Deoxycytidylyl-(3′-5′)-2′-Deoxyguanosine and its Sequence Isomer in N_2O and O_2-Saturated Solutions. *Radiat. Chem.* 112:466-477 (1987).

15. C. A. Belfi, A. V. Arkali, C. R. Paul and H. C. Box. Radiation Chemistry of a Dinucleoside Monophosphate and its Sequence Isomer. *Radiat. Res.* 106:17-30 (1986).

16. C. R. Paul, C. A. Belfi, A. V. Arkali and H. C. Box. Radiation Damage to Dinucleoside Monophosphates: Mediated Versus Direct Damage. *Int. J. Radiat. Biol.* 51:103-114 (1987).

17. C. R. Paul, J. C. Wallace, J. L. Alderfer and H. C. Box. Radiation Chemistry of d(TpApCpG) in Oxygenated Solution. *Int. J. Radiat. Biol.* 54:403-415 (1988).

18. P. C. Hanawalt, P. K. Cooper, A. K. Ganesan and C. A. Smith. DNA Repair in Bacteria and Mammalian Cells. *Ann. Rev. Biochem.* 48:783-836 (1979).

19. T. Lindahl. DNA Repair Enzymes. *Ann. Rev. Biochem.* 51:61-87 (1982).

20. R. Téoule. Review: Radiation-Induced DNA Damage and Its Repair. *Int. J. Radiat. Biol.* 51:573-589 (1987).

21. A. V. Arkali, J. L. Alderfer, C. R. Paul, C. A. Belfi, and H. C. Box. Characterization of Radiation and Autoxidation-Initiated Damage in DNA Model Compounds. *Radiat. Phys. Chem.* 32:511-517 (1988).

22. K. Frenkel, A. Cummings, J. Solomon, J. Cadet, J. J. Steinberg and G. W. Teebor. Quantitative Determination of the 5-(hydroxymethyl) Uracil Moiety in the DNA of Gamma-Irradiated Cells. *Biochemistry* 24:4527-4533 (1985).

23. T. L. Morgan, J. L. Redpath and J. F. Ward. Pyrimidine Dimer Induction in E. coli DNA by Cerenkov Emission Associated with High Energy X-irradiation. *Int. J. Radiat. Biol.* 46:443-449 (1984).

24. V. V. Duba, V. A. Pitkevich, N. G. Selyowa, I. V. Petrova and M. N. Myasnik. The Formation of Photoreactivable Damage by Direct Excitation of DNA in X-Irradiated E. coli Cells. *Int. J. Radiat. Biol.* 47:49-56 (1985).

25. M. Dizdaroglu. Formation of an 8-hydroxyguanine Moiety in Deoxyribonucleic Acid on Gamma-Irradiation in Aqueous Solution. *Biochemistry* 24:4476-4481 (1985).

26. T. Lindahl and N. Nyberg. Rate of Depurination of Native Deoxyribonucleic Acid. *Biochemistry* 11:3610-3618 (1972).

27. T. Lindahl. DNA Glycosylases, Endonucleases, and Base-Excision Repair. In *Proc. Nucleic Acids Res. and Mol. Biol.* 22:135-190 (1979).

28. D. Sagher and B. Strauss. Insertion of Nucleotides Opposite Apurinic/Apyrimidinic Sites in Deoxyribonucleic Acid During in vitro Synthesis: Uniqueness of Adeneine Nucleotides. *Biochemistry* 22:4518-4526 (1983).

29. R. M. Schaaper, T. A. Kunkel and L. A. Loeb. Infidelity of DNA Synthesis Associated with Bypass of Apurinic Sites. In *Proc. Natl. Acad. Sci. USA* 80:487-491 (1983).

Geacintov and Swenberg

30. L. A. Loeb and B. D. Preston. Mutagenesis by Apurinic/Apyrimidinic Sites. *Ann. Rev. Genetics* 20:201-230 (1986).

31. J. W. Drake and R. H. Baltz. The Biochemistry of Mutagenesis. *Ann. Rev. Biochem.* 45:11-38 (1976).

32. U. Hagen. Genetische Wirkungen Kleiner Strahlendosen. *Naturwissen.* 74:3-11 (1987).

33. G. Teebor, R. Boorstein and J. Cadet. Repairability of Oxidative Free Radical-Mediated Damage to DNA: A Review. *Int. J. Radiat. Biol.* 54:131-150 (1988).

34. W. A. Deutsch and S. Linn. DNA Binding Activity from Cultured Human Fibroblasts that is Specific for Partially Depurinated DNA and that Inserts Purines into Apurinic Sites. *Proc. Natl. Acad. Sci. USA* 76:141-146 (1979).

35. A. Gentil, A. Margot and A. Sarasin. Apurinic Sites Cause Mutations in Simian Virus 40. *Mutat. Res.* 129:141-147 (1984).

36. R. M. Schaaper and L. A. Loeb. Depurination Causes Mutation in SOS-induced Cells. *Proc. Natl. Acad. Sci.* 78:1773-1777 (1981).

37. R. M. Schaaper, B. W. Glickman and L. A. Loeb. Mutagenesis Resulting from Depurination is an SOS Process. *Mutat. Res.* 106:1-9 (1982).

38. H. Ayaki, K. Higo and O. Yamamoto. Specificity of Ionizing Radiation-Induced Mutagenesis in the lac Region of Single-Stranded Phage M13 mp10 DNA. *Nucleic Acids Res.* 14:5013-5018 (1986).

39. F. Sanger, S. Nicklen and A. R. Coulsen. DNA Sequencing with Chain-Terminating Inhibitors. *Proc. Natl. Acad. Sci. USA* 74:5463-5467 (1977).

40. B. Hoebee, M.A. v.d. Ende, J. Brouwer, P. v.d. Putte, H. Loman and J. Retèl. Mutations Induced by ^{60}Co Gamma Irradiation in Double-Stranded Bacteriophage DNA in Nitrous-Oxide Saturated Solutions are Characterized by a High Specificity. *Int. J. Rad. Biol.* 56:401-411 (1989).

41. B. Hoebee, J. Brouwer, P. v.d. Putte, H. Loman and J. Retèl. ^{60}Co Gamma-Rays Induce Predominantly C/G to G/C Transversions in Double-Stranded M13 DNA. *Nucleic Acids Res.* 16:8147-8156 (1988).

42. A. J. Grosovsky, J. G. deBoer, P. J. deJong, E. A. Drobetsky and B. W. Glickman. Base Substitutions, Frameshifts, and Small Deletions Constitute Ionizing Radiation-Induced Point Mutations in Mammalian Cells. In *Proc. Natl. Acad. Sci.* 85:185-188 (1988).

43. K. R. Tindall, J. Stein and F. Hutchinson. Changes in DNA Base Sequence Induced by Gamma-Ray Mutagenesis of Lambda Phage and Prophage. *Genetics* 118:551-560 (1988).

44. Y. Kuchino, F. Mori, H. Kasai, H. Inoue, S. Iwai, K. Miura, E. Ohtsuka and S. Nishimura. Misreading of DNA Templates Containing 8-hydroxydeoxyguanosine at the Modified Base and at Adjacent Residues. *Nature* 327:77-79 (1987).

45. M. J. Gait, Ed. *Oligonucleotide Synthesis: A Practical Approach.* IRL Press, Washington, D.C. (1984).

46. M. Beer, S. Stern, D. Carmalt and K. H. Muhlhenrich. Determination of Base Sequence in Nucleic Acids with the Electron Microscope. V. The Thymine-Specific Reactions of Osmium Tetroxide with Deoxyribonucleic Acid and its Components. *Biochemistry* 5:2283-2288 (1966).

47. G. C. Walker and O. C. Uhlenbeck. Stepwise Enzymatic Oligonucleotide Synthesis Including Modified Nucleotides. *Biochemistry* 14:817-824 (1975).

48. C. A. Brennan and R. I. Gumport. T4 RNA Ligase Catalyzed Synthesis of Base-Analogue-Containing Oligodeoxyribonucleotides and Characterization of Their Thermal Stabilities. *Nucleic Acids Res.* 13:8665-8684 (1985).

49. A. Lovelace. Possible Relevance of O-6 Alkylation of Deoxyguanine to the Mutagenicity and Carcinogenicity of Nitrosamines and Nitroamides. *Nature* (London) 233:206-207 (1969).

50. P. K. Gupta, D. L. Johnson, T. M. Reid, M. S. Lee, L. J. Romano and C. M. King. Mutagenesis by Single Site-Specific Arylamine-DNA Adducts. *J. Biol. Chem.* 264:20120-20130 (1989).

51. J. Kemmink, R. Boelens, T.M.G. Koning, R. Kaptein, G.A. van der Marel and J.H. van Boom. Conformational Changes in the Oligonucleotide Duplex d(GCGTTGCG).d(CGCAACGC) Induced by Formation of a cis-syn Thymine Dimer. *Eur. J. Chem.* 162:37-43 (1987).

52. P. Rouet and J. M. Essigmann. Possible Role of Thymine Glycol in the Selective Inhibition of DNA Synthesis in Oxidized DNA Templates. *Cancer Res.* 45:6113-6118 (1985).

53. A. K. Basu, E. L. Loechler, S. A. Leadon and J. M. Essigmann. Genetic Effects of Thymine Glycol: Site-Directed Mutagenesis and Molecular Modeling Studies. *Proc. Natl. Acad. Sci. USA* 86:7677-7681 (1989).

54. A. M. Duplaa, A. Guy and R. Téoule. Measurement of the Radiosensitivity of a Radiation-Induced Adenine Defect in a Single Stranded DNA Chain. *Radiat. Environ. Biophys.* 28:169-176 (1989).

55. R. B. Setlow. Repair Deficient Human Disorders and Cancer. *Nature* (London) 271:713-717 (1978).

56. L. A. Loeb. Apurinic Sites as Mutagenic Intermediates. *Cell* 40:483-484 (1985).

57. T. Lindahl. DNA Glycosylases, Endonucleases for Apurine/Apyrimidine Sites and Base-Excision Repair. *Progr. Nucleic Acids Res. Molec. Biol.* 22:135-192 (1979).

58. K. Frenkel, K. Chrzan, W. Troll, G. W. Teebor and J. J. Steinberg. Radiation-Like Modification of Bases in DNA Exposed to Tumor Promoter-Activated Polymorphonuclear Leukocytes. *Cancer Res.* 46:5533-5540 (1986).

59. H. Ide, Y. W. Kow and S. S. Wallace. Thymine Glycols and Urea Residue on M13 DNA Constitutes Replicative Block in vitro. *Nucleic Acids Res.* 13:8035-8052 (1985).

60. M. Clark and G. Beardsley. Functional Effects of cis-Thymine Glycol Lesions on DNA Synthesis in vitro. *Biochemistry* 26:5398-5403 (1987).

61. T. Friedmann and D. M. Brown. Base-Specific Reactions Useful for DNA Sequencing: Methylene Blue - Sensitized Photo-Oxidation of Guanine and Osmium Tetraoxide Modification of Thymine. *Nucleic Acids Res.* 5:615-622 (1978).

62. A. M. Maxam and W. Gilbert. Sequencing End-Labeled DNA with Base-Specific Chemical Cleavages. *Methods Enzymol.* 65:499-560 (1980).

63. R. C. Hayes and J. E. LeClerc. Sequence Dependence for Bypass of Thymine Glycols in DNA by DNA Polymerase I. *Nucleic Acids Res.* 14:1045-1061 (1986).

64. J. S. Taylor and C. L. O'Day. Cis-syn Thymine Dimer Are not Absolute Blocks to Replication by DNA Polymerase I of Escherichia coli in vitro. *Biochemistry* 29:1624-1632 (1990).

65. J. M. Clark, N. Pattaberaman, W. Taurs and G. P. Beardsley. Modeling and Molecular Mechanical Studies of the cis-Thymine Glycol Radiation Damage Lesion in DNA. *Biochemistry* 26:5404-5409 (1987).

66. J. L. Flippen. The Crystal and Molecular Structures of Reaction Products from Gamma-Irradiation of Thymine and Cytosine:cis-thymine Glycol, $C5H_8N_2O_4$, and Trans-1-carbamoyl-imidazilidone-4,5-diol, $C_4H_7N_3O_4$. *Acta Cryst. Chem. Cryst. Chem.* B29:1756-1762 (1973).

67. S. Y. Wang, Ed. Photochemistry and Photobiology of Nucleic Acids, Vols. 1 and 2. Academic Press, New York (1976).

68. F. Hutchinson. A Review of Some Topics Concerning Mutagenesis by Ultraviolet Light. *Photochem. Photobiol.* 45:897-903 (1987).

69. D. E. Brash. UV Mutagenic Photoproducts in Escherichia coli and Human Cells: A Molecular Genetics Perspective on Human Skin Cancer. *Photochem. Photobiol.* 48:59-66 (1988).

70. W. A. Franklin and W. A. Haseltine. The Role of the (6-4) Photoproduct in Ultraviolet Light-Induced Transition Mutations in E. coli. *Mutat. Res.* 165:1-7 (1986).

71. H. E. Johns, M. L. Pearson, J. C. LeBlanc and C. W. Helleiner. The Ultraviolet Photochemistry of Thymidylyl-(3′--5′) Thymidine. *J. Mol. Biol.* 9:503-524 (1964).

72. J. S. Taylor, H.-F. Lu and J. J. Kotyk. Quantitative Conversion of the (6-4) Photoproduct of TpdC to its Dewar Valence Isomer Upon Exposure to Simulated Sunlight. *Photochem. Photobiol.* 51:161-167 (1990).

73. J. S. Taylor, D. S. Garrett and M. P. Cohrs. Solution-State Structure of the Dewar Pyrimidinone Photoproduct of Thymidylyl-(3′-5′)-Thymidine. *Biochemistry* 27:7206-7215 (1988).

74. J. S. Taylor and I. R. Brockie. Synthesis of a Trans-syn Thymine Building Block. Solid Phase Synthesis of CGTAT[t,s]TATGC. *Nucleic Acids Res.* 16:5123-5136 (1988).

75. N. Rao, J. W. Keepers and P. Kollman. The Structure of d(CGCGAATTCGCG): The Incorporation of a Thymine Photodimer into a B-DNA Helix. *Nucleic Acids Res.* 12:4789-4807 (1984).

76. D. A. Pearlman, S. R. Holbrook, D. H. Pirkle and S. H. Kim. Molecular Models for DNA Damaged by Photoreaction. *Science* 227:1304-1308 (1985).

77. I. Husain, J. Griffith and A. Sancar. Thymine Dimers Bend DNA. *Proc. Natl. Acad. Sci. USA* 85:2558-2562 (1988).

78. G. Raghunathan, T. Kieber-Emmons, R. Rein and J. L. Alderfer. Conformational Features of DNA-Containing cis-syn Photodimer. *J. Biomol. Struct. Dyn.* 7:899-913 (1990).

79. J. S. Taylor, I. R. Brockie and C. L. O'Day. A Building Block for the Sequence-Specific Introduction of cis-syn Thymine Dimers into Oligonucleotides. Solid-Phase Synthesis of TpT[c,s]pTpT. *J. Am. Chem. Soc.* 109:6735-6742 (1987).

80. G. L. Chan, P. W. Doetsch and W. A. Haseltine. Cyclobutane Pyrimidine Dimers and (6-4) Photoproducts Block Polymerization by DNA Polymerase I. *Biochemistry* 24: 5723-5728 (1985).

81. K. L. Larson and B. S. Strauss. Influence of Template Standedness on in vitro Replication of Mutagen-Damaged DNA. *Biochemistry* 20:2471-2479 (1987).

82. D. O'Connor and G. Stöhrer. Site-Specifically Modified Oligodeoxyribonucleotides as Templates for Eschericia coli DNA Polymerase I. *Proc. Natl. Acad. Sci. USA* 82:2325-2329 (1985).

83. M. L. Michaels, D. L. Johnson, J. M. Reed, C. M. King and L. J. Romano. Evidence for in vitro Translesion DNA Synthesis Past a Site-Specific Aminofluorene Adduct. *J. Biol. Chem.* 262:14648-14654 (1987).
84. R. D. Kuchta, V. Mizrahi, P. A. Benkovic, K. A. Johnson and S. J. Benkovic. Kinetic Mechanism of DNA Polymerase I (Klenow). *Biochemistry* 26:8410-8417 (1987).
85. S. K. Banerjee, R. B. Christensen, C. W. Lawrence and J. E. LeClerk. Frequency and Spectrum of Mutations Produced by a Single cis-syn Thymine-Thymine Cyclobutane Dimer in a Single-Stranded Vector. *Proc. Natl. Acad. Sci. USA* 85:8141-8145 (1988).

Discussion

Weinstein: Nick, in the *lac* gene, you seem to have 15 mutations, but they were localized. They were not randomly scattered. They were localized to maybe 4 loci. What do you make of that?

Geacintov: The question is whether mutations are favored in certain base sequences. Hoebee et al. (Reference 41) observed that 60 percent of the base substitutions occur near (TGCT)•(AGCA) sequences in a 144 base-pair in-frame target sequence of the *lacZ* gene in double-stranded M13 DNA. However, such base sequence specificities were not found in (i) the ionizing radiation-induced mutation spectra in the *lacI* gene of *E. coli* (Glickman, B.W., Rietveld, K., and Aaron, C.S. (1980) Gamma-Ray Induced Mutational Spectrum in the *lacI* Gene of *E. coli*. *Mutation Res.* 69:1-12), (ii) in the *arpt* gene of Chinese hamster ovary cells (Grosovsky et al., Reference 42), and (iii) in irradiated bacteriophage and prophage lambda DNA (Tindall et al., Reference 43). Thus, the evidence for the existence of base-sequence effects is mixed. In principle, however, such effects may be important. More experimental evidence is needed along these lines.

Ward: You briefly mentioned there must be competition between repair or removal of these lesions and their becoming mutagenic. If mammalian cells are irradiated, DNA synthesis stops, there is no replicon initiation. The cell takes time to fix itself before it progresses. In the Basu et al. experiment, did they check to see if the glycol was still there after the challenge? Had the cell removed the thymine glycol and that is why they didn't see an effect?

Geacintov: In the site-directed mutagenesis experiments of Basu et al. (Reference 53), the oligonucleotide containing the thymine glycol lesion did not reduce the viability of *E. coli* cells, implying efficient bypass of the lesion by polymerases and/or repair. The lack of mutations observed when double-stranded modified DNA was used was indeed attributed to repair. Endonuclease III treatment of the DNA prior to transformation did decrease the mutation frequency. The mutations which were observed by Basu et al. were due to unrepaired thymine glycol (Reference 53). Experiments monitoring the survival of this lesion were not specifically performed in that work.

von Sonntag: But these oxidants have also a 5 to 10 percent formation of other products. You don't know whether these 5 to 10 percent other products have been the cause of this destruction.

Geacintov: You have a good point, of course. One aspect of this type of work is that there are two methods for generating specifically modified oligonucleotides: (i) total synthesis in which a base is first modified and then the other bases are added on the 5′ and 3′ sides of the modified base to form the oligonucleotide, and (ii) synthesis of an oligonucleotide first, and then specifically modifying one of the bases in this oligonucleotide. Basu et al. (53) used the latter method to generate the thymine glycol-containing oligonucleotide. These workers went to great lengths to purify the modified oligonucleotide and to demonstrate that they indeed made a thymine glycol-containing oligonucleotide.

Varma: You just said there was not enough data on mutations at different LETs. Does the mutation's frequency or the mutation spectrum change with LET?

Chatterjee: These radiation qualities have been studied and definitely show that high LET energy particles give rise to higher mutation frequencies over and above the low LET.

Hall: This question is going to show my ignorance, but do we, in fact, have information where you can see the balance between deletions and point mutations as a function of dose or as a function of LET? Is that information available?

Geacintov: No, it's not. In the examples which I have cited, deletions were observed, sometimes base pair deletions or base pair insertions, in some cases 10 base pair duplications, and in some cases a very large deletion. None of these effects have been studied as a function of dose or as a function of LET. Such studies would be very interesting!

Varma: Do you require very high doses like John was saying to get observable quantities? Hundreds of kilorad dose to see one? What kind of dose do you have?

Geacintov: To observe mutation spectra you need only 1 Gray or a couple of Grays.

Exciton Microscopy and Reaction Kinetics in Restricted Spaces

Raoul Kopelman[a]

Abstract

We describe the development of a new biologically non-invasive ultraresolution light microscopy, based on combining the energy transfer "spectral ruler" method with the micro-movement technology employed in scanning tunneling microscopy (STM). We use near-field scanning optical microscopy, with micropipettes containing crystals of energy packaging donor molecules in the tips that can have apertures below 5 nm. The excitation of these tips extends near field microscopy well beyond the 50 nm limit. The theoretical resolution limit for this spectrally sensitive light microscopy is well below 1 nm. Exciton microscopy is ideally suited for kinetic studies that are spatially resolved on the molecular scale, i.e., at a single molecule site. Moreover, the successful operation of the scanning exciton tip depends on an understanding of reaction kinetics in restricted spaces. In contrast to the many recent reviews on scanning tip microscopies, there is no adequate review of the recent revolutionary developments in the area of reaction kinetics in confined geometries. We thus attempt such a review in this paper. Reactions in restricted spaces rarely get stirred vigorously by convection and are thus often controlled by diffusion. Furthermore, the compactness of the Brownian motion leads to both anomalous diffusion and anomalous reaction kinetics. Elementary binary reactions of the type A + A → Products, A + B → Products and A + C → C + Products are discussed theoretically for both batch and steady-state conditions. The anomalous reaction orders and time exponents (for the rate coefficients) are discussed for various situations. Global and local rate laws are related to particle distribution functions. Only Poissonian distributions guarantee the classical rate laws. Reactant self-organization leads to interesting new phenomena. These are demonstrated by theory, simulations, and experiments. The correlation length of reactant production affects the self-ordering length-scale. These effects are demonstrated experimentally, including the stability of reactant segregation observed in chemical reactions in one-dimensional spaces, e.g., capillaries and microcapillaries. The gap between the reactant A (cation) and B (anion) actually increases in time, and extends over millimeters. Excellent agreement is found among theory, simulation, and experiment for the various scaling exponents.

Introduction to Reaction Kinetics in Restricted Dimensions

Chemical kinetics in low dimensions usually involve active sites or surfaces which significantly lower the activation energy barriers. At the same time, there is essentially no convective stirring in low dimensions and the diffusion process is less efficient due to the compactness[1] of the Brownian motion. As a result, low-dimensional reactions are very often diffusion limited. Furthermore, the so-called "compact" Brownian motion leads to anomalous diffusion[2] and to anomalous rate laws.[3] Finally, these

---

(a) Department of Chemistry, University of Michigan, Ann Arbor, Michigan

Physical and Chemical Mechanisms in Molecular Radiation Biology
Edited by W.A. Glass and M.N. Varma, Plenum Press, New York, 1991

anomalous macroscopic rate laws are a manifestation of the unexpected entropic self-ordering of reacting molecules, i.e., a dramatic self-organization on a microscopic or mesoscopic (and sometimes macroscopic) length scale.[4]

Practically every heterogeneous reaction is a low-dimensional reaction, whether it occurs on a surface, inside a pore, at a micelle, or elsewhere. In addition, some homogeneous reactions, such as solid-state reactions, occur preferentially in low-dimensional environments, e.g., domain boundaries, defect sites, surface steps or radiation tracks. Furthermore, reactions such as polymerization, gelation and cellular growth have low-dimensional dynamics. We also note that, mechanistically, these low-dimensional reactions may involve molecules, radicals, atoms, electrons, holes, excitons, or solitons.

Experimentally, simple elementary reactions that have been observed to follow strange rate laws range from solution photodimerization inside porous polymeric membranes[5,6] to exciton fusion in molecular alloys.[3] Further evidence for such anomalies comes from excimer kinetics in polymer blends,[7-9] oxygen quenching in porous media,[10] energy transfer in photosynthetic systems,[11] and neuron-gating kinetics in sodium channels.[12] There also appear to be many other instances where anomalous rate laws have been fitted (unnecessarily) by complex mechanisms.

There are three major categories of simple, elementary, bimolecular mechanisms:

$$A + A \rightarrow \text{Products} \tag{1}$$

$$A + B \rightarrow \text{Products} \tag{2}$$

$$A + C \rightarrow C + Products \tag{3}$$

We assume, here, that the products are inert or leave the reaction medium (e.g., desorb). We also assume that in reaction (2) there is no A + A or B + B reaction mechanism. We note that in reaction (3) C is a catalyst or catalytic site. Reactions (1) and (2) are described by the textbook bimolecular rate equations and reaction (3) by the pseudo-monomolecular equation. We also note that reaction (1) is "second order" in A concentration while reaction (2) is "second order" overall, i.e., first order in A and first order in B. Reaction (3) is actually first order in both A and C (if C is varied from case to case) and thus also "second order" overall, while being pseudo-first order in a given case (where only A varies in time). It turns out that these three categories (1-3) give rise to extremely different rate laws for low dimensions. Furthermore, within each category there arise various sub-categories with drastically different rate laws, depending on distinctions such as dimensionality, initial conditions, source (input) conditions and other subtle constraints which do not affect the textbook rate laws. Even the achievement of a steady state is no longer guaranteed by a steady input of particles. The new "zoo" of rate laws is demonstrated by a few examples of reactions constrained to one dimension (Table I). While some early theoretical work and some early experiments may have indicated rate law anomalies,[13,14] it was the development of computer simulations that became the major force behind the recent developments in this area.

Table I. Reaction Orders X in One Dimension

Reaction	Steady State	$t \to \infty$
Random $A + A \to A_2 \uparrow$**	3	3
$\downarrow A_2 \to A + A \to A_2 \uparrow$	2	3
Classical bimolecular	2	2
Random $A + B \to AB \uparrow$	4*	5
$\downarrow AB \to A + B \to AB \uparrow$	2	3
If $A + A \to A_2 \uparrow$ and $B + B \to B_2 \uparrow$	3	3
Examples: hydrogen on Pt Wire		
$\downarrow H_2 \to H + H \to H_2 \uparrow$	2	3
$\downarrow D_2 \to D + D \to D_2 \uparrow$	2	3
$H + D \to HD \uparrow$	3	3

* For hard-sphere particles and vertical reactions; otherwise X = 2.
** ↑ means that the species leaves the system; ↓ means that the species enters the system.

The importance of short-time fluctuations has already been pointed out by Einstein in his 1905 papers on Brownian motion. However, the possibility of drastic long-time fluctuations was probably first pointed out by the famous Soviet chemist, physicist, and astrophysicist, Zeldovich.[15] Similar long-time fluctuation phenomena have been pointed out in the context of solid-state kinetics.[16] It is the occurrence of such long-time and long-range fluctuations that leads to reactant self-ordering and to the strange new rate laws for elementary reaction mechanisms.

We first review the theory and simulations from the standpoint of an experimentalist, emphasizing the nature and form of the novel laws. We also attempt to give physicochemical rationalizations for, and insights into, this new paradigm. We then describe some selected experiments in order to demonstrate these laws and the consequences and opportunities which they create.

Theory and Simulations of Restricted Reaction Kinetics

Batch Reactions

A "batch reaction" is one where the reaction is suddenly turned on at time zero (hence often called "big bang"). This may occur because of a sudden introduction of matter, radiation, or heat. These batch reactions were the first to be treated with respect to the long-time, long-range "fluctuations" that give rise to the anomalous asymptotic (long-time) behavior. For instance,[17,18] the "trapping reaction" and the "target reaction" are both special cases of reaction (3):

$$A + C \rightarrow C \qquad \text{(3a)}$$

In the trapping reaction, C is fixed and A moves, while in the target (quenching) reaction, A is fixed and C is mobile. Classically, one expects the survival function C_A (A concentration) to decrease exponentially in time:

$$C_A = C_A^\circ \exp(-kt) \qquad \text{(4)}$$

where C_A° is the "big bang" (t = 0) reactant concentration and k the rate constant. In the Smoluchowski treatment of diffusion-limited reactions,[17] $k \sim D$ where D is the diffusion constant. However, this is only valid in three (or higher) dimensions. For instance, the Smoluchowski approach in one dimension leads to a different asymptotic solution:

$$C_A \sim \exp(-k't^{1/2}) \qquad \text{(5)}$$

Furthermore, taking fully into account the long-range fluctuations leads to[16]

$$C_A \sim \exp(-k''t^{1/3}) \qquad \text{(6)}$$

Many such new relations have been derived[16,18] for low dimensions (including fractal dimensions).

Replacing Eqs. (1) to (3) by

$$A + i = P \qquad i = A,B,C \qquad \text{(7)}$$

where P is any resulting products (including A and/or i), a simple way of describing many of the results is to consider the differential rate equation (rather than the integrated forms, involving the "survival function" C_A):

$$- (dC_A/dt) = k(t)C_A C_i \qquad i = A,B,C \qquad \text{(8)}$$

The non-classical results can now be written (for long times, $t \rightarrow \infty$)

$$K(t) \sim t^{-h} \qquad h < 1 \qquad \text{(9)}$$

with the special case h = 0 for the classical (textbook) case. In particular, for low dimensions (d < 2), one finds[19] for certain A + A reactions:

$$h = 1 - d_s/2 \qquad \text{(10)}$$

while for the same $A + B \rightarrow 0$ reactions,[20]

$$h = 1 - d_s/4 \qquad\qquad (11)$$

whenever the spectral (fracton)[21] dimension (d_s) is $d_s < 2$.

Another simple way of describing many of the results, and particularly those for steady-state reactions, is

$$-dC_A/dt = kC_A^y C_i^z \qquad\qquad (12)$$

where k is a time-independent constant and the partial orders y and z may be non-integers, and the same is true for the overall order

$$x = y + z \qquad\qquad (13)$$

We emphasize that all this is for an elementary bimolecular reaction, where the molecularity of the reaction is described by Eq. (8). Obviously, by the classical (textbook) reaction laws, x, y and z are all integers (y = 1, z = 1, x = 2). For instance, for A + A reactions on a critical percolation cluster, x = 5/2; for a Sierpinski gasket it is x = 2.47, and, in general, it is $x = 1 + 2/d_s$. Similarly, this empirical form [Eqs. (12) and (13)] accounts for the strange results listed in Table I.

In the last few years, there has been a bewildering explosion of new results. It would be impossible to list them all, let alone point out problems and contradictions. We list a few papers that include partial reviews[4,22,23] and give below what we consider to be the main problems of interest.

The beauty of classical kinetics lies in the universality of its formulation. The functional form of the rate laws does not depend on the dimension or dimensionality of the medium, initial conditions (for batch reactions), details of source (for steady-state reactions) nature of motion (ballistic, diffusion), relative mobility and reaction of the components, etc. Such factors only affect the values of the parameters. It is also implied that global and local rate laws are the same and that the particle distribution functions depend only on the concentrations, but not on whether the ensemble is at equilibrium or far from it, or at steady state or far from it.

When deviations from classical behavior are studied, it is important to distinguish the various factors behind such deviations and the interactions among them. Often it is only with a combination of two such factors that one finds deviations from the classical picture, e.g., with a combination of low dimensionality and a random source term (but not with low dimensionality and a "geminate" source term).

Non-classical diffusion-controlled reaction kinetics does retain scaling and universality relations.[24-26] However, they differ in detail for different kinds of reactions (A + A versus A + B), for different modes of reactions (steady-state versus batch), for different conditions (source term structure, initial distributions), for different conservation laws (exact versus statistical equality of reactant species), etc. To get a better insight into the class structure of such scaling laws is one problem of interest to us.

Global rate laws have been of much interest. They indeed were the first targets of modern investigations. However, it has become apparent that there are instances where precise global laws do not exist. Can they be substituted by local laws? Can one always define a local law? What is the relation between a local and a global law? This is another problem of interest to us.

Distribution functions have been the linch pins of statistical mechanics. They have also been recognized as the basic information required for the qualitative and quantitative understanding of reaction kinetics.[4,27] Probably the most stringent test of any kinetic theory is the goodness of the distribution functions predicted by it. Computer simulations are adept at testing distribution functions with almost as much ease as testing rate laws.[27,28] Examples[29] are given in Figs. 1 and 2. The distribution functions can also serve as a bridge between local and global rate laws (to the extent that they exist). Furthermore, on an empirical basis, computer simulations of distribution functions show trends and clues that may be useful for the design of improved theories. Such trends may also be of much aid to experimentalists, providing new qualitative or intuitive model pictures. For instance, the build-up of depletion zones around reactants or catalytic sites depends on specific parameters such as medium dimensionality and reactant mobility.[30,31] Recently there has been much progress in the area of analytical models, for diffusion-controlled reaction kinetics.[32,33]

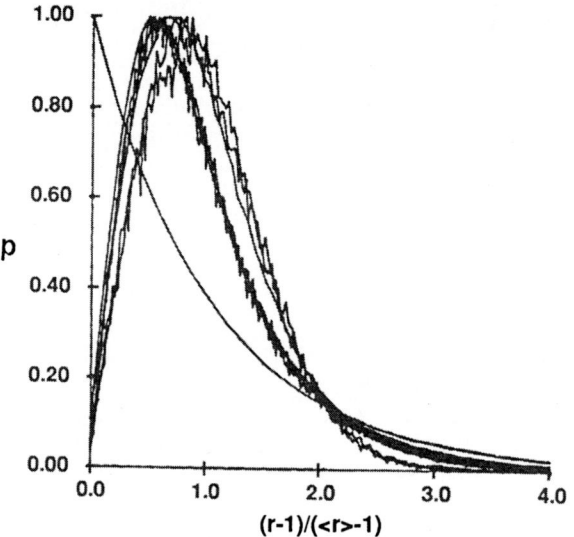

Figure 1. The normalized distribution of nearest-neighbor distances r for 1-dimensional kinetics.[29] The normalization of r is done by calculating the quantity $(r - 1)/(<r> - 1)$, where $<r>$ is the average nearest-neighbor distance for all particles at any given time. Lattice size is $L = 10,000$ sites; initial density is $\rho_o = 0.05$. The exponential single curve is the initial ($t = 0$) distribution, the two groups of curves are for the $A + A \rightarrow 0$ reaction (left) and $A + A \rightarrow A$ reaction (right); in each group the distributions are shown for different reaction times: $t = 100, 1000,$ and 2000 steps. All curves are averages of 5000 runs ($A + A \rightarrow 0$) and 1000 runs ($A + A \rightarrow A$).

A + A Reactions. This is apparently the simplest and best understood class of reaction. Somewhat surprisingly, the coagulation reaction (where A designates any particle)

$$A + A \rightarrow A \qquad (14)$$

is better understood than the annihilation reaction,

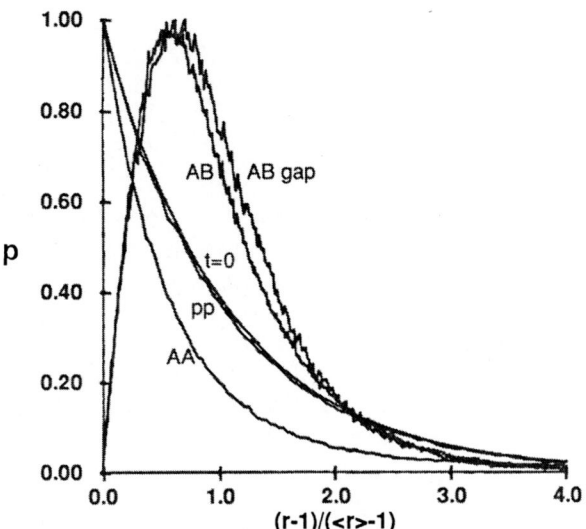

Figure 2. NNDD (nearest-neighbor distance distributions) for one-dimensional A + B → 0 reactions.[29] From the left, the three exponential-type curves are NNDD for t = 1000 like-like (AA), for t = 1000 particle-particle (PP), and for t = 0 (totally overlapped for all cases, i.e., AA, BB and PP), respectively. The two curves with a well-developed maximum are t = 1000 like-unlike (AB) NNDD (left) and like-unlike (AB) gap distribution (right). All NNDD curves are averages of 2000 runs and the gap curve is that of 5000 runs. Total number of lattice sites is 10,000 and the initial (t = 0) concentration is 0.05 for A and 0.05 for B (0.1 for P).

$$A + A \rightarrow O \qquad (15)$$

where 0 designates a reactant that is inert or leaves the system. Even more surprising are the drastic differences between these two, in particular the difference in the particle distribution functions. This difference first became apparent from our simulations on interparticle distribution functions.[4,28] It was elegantly proven in the recent analytical work of Doering and Ben Avraham.[32] Specifically, in one dimension, the normalized interparticle distance distribution probability P(r) was found to have the form

$$P(r) = (\pi/2)Cr \, \exp\left[-(\pi/2)C^2r^2/2\right] \qquad (16)$$

for long times, where C = C(t) is the instantaneous density and r is the interparticle distance. (This is shown in Fig. 1 together with simulation results.) The first important observation is that this distribution differs drastically from the Poisson distribution (in one dimension):

$$P(r) = C \exp(-Cr) \qquad (17)$$

The "maximum" of the Poisson distribution is at r = 0. In contrast, the kinetic distribution [Eq. (16)] has a minimum there [with P(0) = 0]. The maximum of the kinetic distribution is at $r = (2/\pi)^{1/2} C^{-1}$. Obviously this maximum increases with time, i.e., as the density decreases (this is obviously not so for the Poissonian distribution).

Furthermore, for large values of r, the kinetic distribution has a Gaussian tail (while the Poissonian has an exponential one). We also note that for small values of r, the kinetic distribution increases linearly with r, while the Poissonian one decreases exponentially. As the reaction kinetics is mainly determined by the small r range, the resulting rate law is cubic in C, while with a Poissonian distribution it is quadratic in C (the classical result). This is consistent with the older Torney and McConnell[34] result (with the Anacker et al.[35] transformation).

We now turn to the annihilation reaction (A + A → 0). There is no analytical result, so far, for one dimension. However, there is an asymptotic analytical result,[36] for r → ∞: The tail of the distribution has an exponential form, rather than a Gaussian form. (This is also seen from the simulation results in Fig. 1.) This is a major distinction. However, the small r part of the distribution is linear, as in the coagulation (A + A → A) case, thus giving a similar cubic rate law. On the other hand, the central part (intermediate r) already differs from the coagulation case, and the maximum falls at about 0.4 C^{-1} (half the value of the coagulation case). A rough approximation[28,29] is

$$P(r) \sim Cr \exp[-\alpha Cr] \tag{18}$$

where α is a coefficient that has not been determined from simulations, due to the poor quality of the functional fit.[29]

A + B → 0 Reactions. We imply that there occur no A + A or B + B reactions (otherwise this case reverts to the A + A → 0 reaction as A and B become kinetically indistinguishable). This is the Zeldovich[15] case, which essentially was the start of modern diffusion-reaction theory. This model assumes a random "big bang" (t = 0) distribution, with random fluctuations. Globally $C_A = C_B = C/2$ at t = 0 and at every other instant in time. However, locally, $\Delta C = C_A - C_B$ fluctuates around zero at t = 0. Zeldovich's formalism shows that $2|\Delta C|/C$ increases in time and becomes of the order of unity at t → ∞, provided that the dimension of the reaction space is d < 4. This means that a segregation of reactants proceeds in time. A very small segregation $(2|\Delta C|/C << 1)$ at t = 0 turns into a large one (~1) at t → ∞. Concomitantly, the classical survival probability

$$C = 2C_A = 2C_B \sim t^{-1} \qquad t \to \infty \tag{19}$$

is valid only for high dimensions (d > 4) but is replaced by

$$C \sim t^{-d/4} \qquad t \to \infty, d < 4 \tag{20}$$

for all "practical" dimensions (d ≤ 3). This formalism was generalized somewhat (and proved numerically) by Toussaint and Wilczek,[37] and generalized to fractals by Kang and Redner[20] (replacing d by d_s, the "spectral" dimension).

In the above discussions, it was implied that A and B have identical mobilities. The other extreme case is where one of them (say B) is immobile ("sitters"). This case, as well as the intermediate cases, has been studied by Zumofen et al.[22,38] with some interesting results. It is intuitively obvious that complete spatial segregation is harder to achieve and will take longer. Quantitatively, it is not yet clear how the particle distributions depend on the relative mobilities of the two species.

Toussaint and Wilczek[37] have also argued that for a correlated t = 0 distribution (e.g., geminate AB pairs), the asymptotic rate law is different:

$$C \sim t^{-(d/2)} \qquad d \leq 2, t \to \infty \qquad \textbf{(21)}$$

i.e., the same as for A + A reactions,[3,20] and that there is no segregation. It turns out[39] that the situation is more complex and Eq. (21) is only correct for strictly geminate conditions. For correlated but non-geminate initial conditions (where the correlation length is much larger than the particle size):

$$C \sim t^{-(d+2)/4} \qquad d \leq 2, t \to \infty \qquad \textbf{(21a)}$$

$$C \sim t^{-1} \qquad d \geq 2, t \to \infty \qquad \textbf{(21b)}$$

Finally we note that there are no published experiments showing self-organization or anomalous kinetics for A + B reactions.

A + C → C Reactions. The early work for this case has centered around "trapping" reactions, i.e., where A diffuses and C is fixed in space. This work also originated in the Soviet Union and is summarized, in part, in the monograph of Agranovich and Galanin.[16] The typical rate law in one dimension was given in Eq. (6) and was rationalized by a long-time distribution of particles away from the traps. The first exact (analytical) work on particle distributions appears to be that by Weiss et al.[30] It is for a single trap C in three- and one-dimensional continua of particle A (with density C_A). Again the nearest neighbor (C-A) distance distribution is far from random in one dimension (while essentially random in d = 3). Actually, the distribution function is quite similar to that for A + A → A (Eq. 16) in one dimension. Similarly, also, the global rate law is, in one dimension,

$$|dC_A/dt| \sim t^{-1/2}C_A \qquad \textbf{(22)}$$

while in three dimensions the classical result is recovered:

$$|dC_A/dt| \sim C_A \qquad \textbf{(23)}$$

This is again similar to the A + A case. We note that Eqs. (22) and (23) are the first exact results for the original Smoluchowski problem[17] of accretion of molecules by a colloidal particle (or a crystallization seed). Smoluchowski's solution for the three-dimensional case is indeed equivalent to Eq. (23).

Self-Ordering. In each of the three cases above, one finds deviations from uniformly random particle distributions, deviations that increase with time. Often only a few time "steps" (in the simulation or the discrete model) are required to attain a drastically different (e.g., "Wigner-like"[4,28]) distribution. This spontaneous self-ordering, based totally on diffusion and reaction (but no long-range forces) is of major interest. There are similarities[29] in the "patterns" of different reaction classes (A + A versus A + B) and dissimilarities[4,29] in those of very similar reactions (A + A → 0 versus A + A → A). Some of the patterns take much longer to form than others. Dimensionality and initial conditions are also of prime importance. The investigation of such patterns requires some new tools (e.g., order parameters). For instance, we have defined[27,40] about a dozen different "segregation parameters." The relative usefulness of such order parameters in different situations is also of some importance. Particle distributions or partial particle distributions, as well as quantities derived from them, can also be used as such tools.

A + B → C Where A and B Are Segregated Throughout the Reaction. Recently, analytical work on an A + B → C type reaction-diffusion process in an effective one-dimensional system has been done by Galfi and Racz.[41] The reactant A of constant density a_0 and B of constant density b_0 are initially separated. They meet at time 0, forming a single reaction boundary, which makes the system effectively one-dimensional. The motion of the reaction front with time is shown in Fig. 3. This model is similar to that of Weiss et al.[30] for A + C → C where A is a one-dimensional continuous solute and C is a single trap. However, here both A and B are continuous (and A and B are like traps to each other). The results from the set of the reaction-diffusion equations for A and B, which are valid in the long-time limit, show that x_f (the position of the center of reaction front) scales with time as $x_f \sim t^{1/2}$, while w (the width of the reaction front) scales as $w \sim t^{1/6}$ and the production rate rf (at $x = x_f$) scales as $t^{-2/3}$. We can also find that the global reaction rate R scales as $t^{-1/2}$; i.e.,

$$R = dC/dt \sim t^{-1/2}$$

This is similar to the results of Weiss et al.'s work for A + C → C [Eq. (22) for dA/dt].

Steady-Source Reactions

Steady-source reactions, where the reactants are created continuously, are of prime importance in biology, ecology, geology, technology and many other areas.[4] They also give a wider range of phenomena due to three factors: 1) the riches of source-term structures (see below); 2) the inclusion of strange topologies (e.g., "dust" or "fractal dust"), which would not support batch reactions for long; 3) ordering phenomena which for batch reactions are only achievable at extremely long times, i.e., at vanishing densities, but which, under steady-source conditions, may develop even at high densities.

Source-Term Structures. The most common source term is uniformly random in both space and time. One of the surprising results has been the fact that the mere presence of such a source, combined only with diffusion and reaction, may lead to highly self-ordered systems.[4,42] Another common source produces particles correlated in space and/or time. For instance, when H_2 particles land on a platinum surface they dissociate into pairs of atoms; or, when electrons and holes are produced by photons hitting a photovoltaic cell, they are produced geminately. Alternatively, the source may be a hot beam of dissociated AB molecules, landing as pairs of A and B atoms on a surface, correlated in time but not in space. The particles may be able to react as they land on top of other particles (vertical reaction), or they may not, depending on the chemistry or physics of the situation. If they are not able to react, they may be discarded or they may land in the immediate vicinity, depending again on the particular reaction. Sources with vertical reaction may produce particle distributions that differ drastically from those produced by sources with no vertical reaction. Also, the atoms on a surface may occasionally desorb, and the excitons

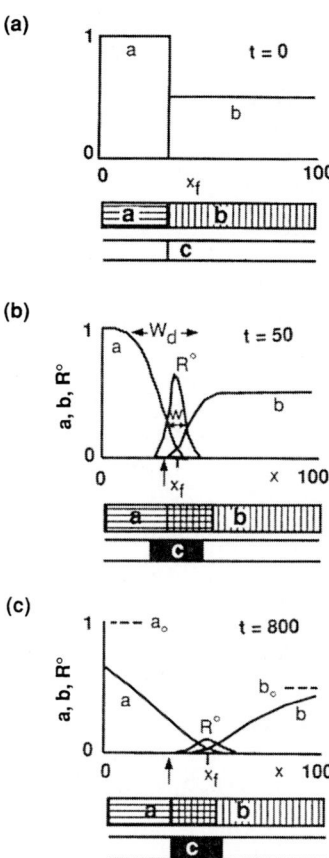

Figure 3. The motion of the reaction front with time. Length (x), time (t), the densities (a,b) of the reagents (A,B), and the magnified production rate (R* = 100R = 100kab) of C are scaled to be all dimensionless. As an initial condition, we used a = 1, b = 0 for x < 30; and a = 0, b = 0.5 for x > 30 (the t = 0 position of the center of the front, x_f is shown by an arrow). The top figures of (b) and (c) are from Ref.(41).

produced in a solid have a natural decay time, independent of the annihilation or fusion reaction. Such a "monomolecular" (first-order) decay process adds another random (negative) contribution to the source terms. All such possible source structures have significant effects on the outcome, in addition to the well-known source-term distinctions, such as an unevenness in A and B production terms (alternatively viewed as a strong A source and a weaker B source). Furthermore, initial conditions may vary, such as an initially empty lattice (the usual case), versus a fully or partially occupied one. We also note that the source structure usually determines the conservation laws. For instance, a time-correlated, pairwise production of A and B particles (in an A + B → 0 reaction with no decay) will guarantee an exact global conservation of the equality $C_A = C_B$. On the other hand, a geminate production of A and B will guarantee both an exact global and an exact local conservation of $C_A = C_B$ (i.e., no segregation). However, a uniformly random source term gives only a statistical global conservation of $C_A = C_B$, and, under certain circumstances (e.g., no desorption) this may lead to long-time realizations with $C_A >> C_B$ or vice versa.

An obvious question of interest is whether a given source term will produce a steady-state system. The answer may depend on a combination of source structure and medium dimensionality, or a combination of other conditions. There have been several recent surprises in this respect,[23-26,43] which lends importance to future work aimed at clarifying this issue. If a steady state is achieved, the next question is whether the particle distribution is uniformly random or ordered. The self-ordering process itself may either hinder or aid the generation of a steady state. The self-ordering is, therefore, an interesting related question.

Certain source terms may generate a steady state that mimics an equilibrium state. For instance, strictly geminate pair creation, with an annihilation reaction (A + A → 0 or A + B → 0), is expected to set up a steady state that mimics equilibrium. We expect equilibrium states to be totally (randomly) disordered (by the second law of thermodynamics). Such source terms and steady states are thus of special interest. However, the steady states which are self-ordered (i.e., far from equilibrium) are of even more interest. It is they that are responsible for the anomalous reaction rate laws. They may even lead to much speculated-upon phenomena, such as biomorpho-genesis,[23,24] which are in apparent (though not real) contradiction with the second law.

A + A Reactions. Analytical solutions exist so far only for a limited number of cases and are restricted to one dimension. For the A + A → 0 reaction, Racz[44] has solved "one and a half" cases. For the geminate landing case, the solution is complete: For the steady-state reaction, the rate law is classical (second order) and the particle distribution is uniformly random. This is actually expected for all simple geminate landing cases, because, as pointed out above, such steady states map onto equilibrium states,

$$A + A \leftrightarrow A_2 \tag{24}$$

and for equilibrium states we expect this classical behavior. On the other hand, for uniformly random landing, where a steady state is also achieved, the rate law is third order in density. Furthermore, the particle distribution is non-random. However, Racz[44] did not derive a functional form for such a distribution.

A recent paper by Doering and Ben-Avraham[33] gives a similar derivation for the one-dimensional A + A → A problem. Again a steady state is found for the random (and geminate?) source terms. The rate law (for the random source) is again third order in density, and the particle distribution is non-random. In addition, the interparticle distance distribution was derived specifically (in terms of Airy functions). Simulation-based results were obtained for all dimensions (or spectral dimensions, for fractals), both higher and lower than unity. Eqs. (12) and (13) actually give, for irreversible A + A reactions,

$$-dC_A/dt = kC_A^x + R \tag{12a}$$

where R is the source term. As mentioned above, for these A + A reactions,

$$X = 1 + 2/d_s \qquad d_s < 2 \tag{25a}$$

$$X = 2 \qquad d_s > 2 \tag{25b}$$

while for the marginal (critical dimension d = 2), one indeed obtains X = 2 but a logarithmic correction appears:

$$-dC_A/dt = kC_A^2/\ln C_A + R \qquad (26)$$

A more rigorous theoretical argument for the form of the last three equations has been derived recently,[26] also including particle correlation functions as well as higher-order terms for higher densities and modifications for source term correlations.

A + B Reactions. This class of reactions is the most abundant, important, intriguing, and challenging. The first demonstration that a completely segregated steady state can be achieved with a uniformly random source term[42] was an intellectual surprise. While the simulation demonstrations were performed on fractal lattices (Sierpinski gasket[45]) and critical percolation clusters,[46] it turns out[27,41] that this phenomenon also occurs in one- and two-dimensional media, and, theoretically, even in three-dimensional media (however, only for infinite systems).[47] In addition, it has been shown[42] that segregation occurs also for systems made of walkers (A) and "sitters" (B), although the distributions and the rate laws may differ. However, the most interesting aspect of steady-source reactions is the role of the source structure and statistics. The most subtle differences in the nature of the source may cause very dramatic differences in particle distributions and rate laws. An illustration of this is given in Table II [Table I of Clement et al.[25]]. We see here how the segregation

Table II. Scaling Behavior of Λ, Typical size of segregated domains, as a function of the Euclidean dimension d and various source conditions. Cases (a) and (b) are, respectively, random and correlated landing (δ is the separation between landing pairs) with conservation of the number of A's and B's on the system. For cases (c), the conservation condition is removed and a first-order decay mechanism is included (with a constant K). The system size is L, the size of a particle is a, and the external rate of arrival of particles is R.

Conservation Conditions	Source Case		d = 1	d = 2	d = 3
(a) Strict	random landing (random δ)	K = 0	L/6	a lnL/a	3a
(b) Strict	correlated landing (fixed δ)	K = 0	δ	a lnδ/a	2a
(c) Statistical	independent landing				
	Non-vertical annihilation	K = 0	saturation*	saturation	saturation
	Vertical annihilation	K = 0	ξ**	a lnξ/a	3a
	Symmetric desorption	K $\neq$0	ξ**	a lnξ/a	3a

* Saturation means that only one of the species occupies the whole system.

** For vertical annihilation: $\xi = \sqrt{D/a^d R}$

For symmetric desorption: $\xi = \sqrt{D/K}$

length Λ is affected by the source conservation laws (strict versus statistical), source correlation length (δ) and nature (random versus fixed), by the presence of vertical annihilation and by the presence of a "negative" source term, such as desorption or natural decay (with first order rate constant K). The reaction-diffusion parameters that enter the quantitative expression are the diffusion constant (D), the source rate (R), the particle size (a), the lattice size (L), and of course the dimensionality (d).

A + C Reactions. It had already been shown a decade ago (Agranovich and Galanin[16]) that for a trapping reaction $A + T \to T$, on a one-dimensional lattice, with a steady, uniformly random source of A particles, the trapping rate is linear in A density but quadratic in trap density. This is true for both random and even (superlattice) trap distributions (although the rate coefficients differ):

$$Rate \sim C_A C_B^2 \tag{27}$$

This result has also been derived via a somewhat different approach by Lopez-Peacock and Keiser.[48] This non-classical rate law is again due to a non-uniform A particle distribution.

Based on preliminary simulations[49] and theoretical work,[31] the last equation can be generalized in the following way:

$$Rate \sim C_A C_B^{2/ds} \qquad d_s < 2 \tag{28a}$$

$$Rate \sim C_A C_B \qquad d_s > 2 \tag{28b}$$

where d_s is an effective spectral dimension and the marginal dimensionality is again two (at $d = 2$ there is probably again a logarithmic correction to Eq. 28b).

Experiments of Restricted Reaction Kinetics

Examples of chemical reactions, photochemistry and photophysics have been selected in order to illustrate the peculiarities of diffusion-controlled reaction kinetics in low dimensions.

Photochemistry in Pores Inside Membranes

The photodimerization of anthracene in solution was probably the first well-studied diffusion-limited reaction.[50] The bimolecular reaction $A + A \to A_2$ (A = anthracene) followed the classical kinetic laws. A similar reaction provided the first study of fractal chemical kinetics:

$$N^* + N^* \to N_2^{**} \to N + N + h\nu \tag{29}$$

Here N* is a naphthalene molecule excited to its first triplet state and N_2^{**} is the transitory dimer in the excited first singlet state.[5,6] The experiment is carried out in a solution embedded in various porous membranes (nylon, acetate, paper; see Fig. 4). The naphthalene molecules diffuse inside the solvent inside the pores.

Figure 5 shows the pores of some of the channel-pore membranes used as "test tubes." These cylindrical tubes are 6000 nm long and have radii ranging from 1000 to

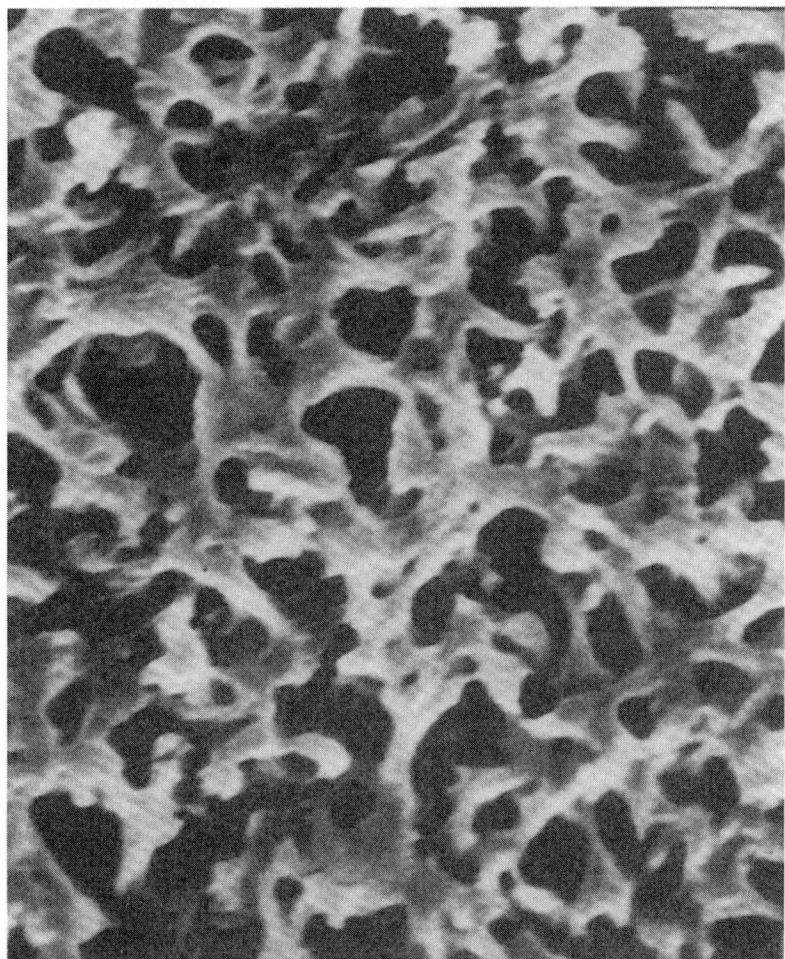

Figure 4. Porous nylon membrane (Gelman Science, Inc., Ann Arbor, Michigan). Magnification about 10^4.

10 nm. Typical molecular diffusion lengths in these experiments are on the order of 100 nm. We thus expect to see a range of kinetic behaviors, from three-dimensional to one-dimensional, depending on the molecular diffusion parameters. Indeed,[5] the heterogeneity exponent h [see Eq. (9)] ranges from h = 0.07 for 1000-nm pores to h = 0.41 for 25-nm pores. This is in good agreement with the theoretical limits; i.e., h = 0 for three-dimensional media and h = 1/2 for one-dimensional media. Moreover, for an intermediate pore radius (100 nm), one observes a cross-over in time, from h = 0.06 in the millisecond regime to h = 0.48 in the sub-second regime.

Earlier experiments,[6] carried out in ordinary porous membranes made of nylon (Fig. 4), acetate, or paper, give h values in the range of 0.3 to 0.4. This can be interpreted simply as some statistical averaging over various pore domains ranging between one-dimensional and three-dimensional topologies. Alternatively, one can ascribe to such membranes "effective" spectral dimensions d_s (see Eq. 10), with values between 1.2 and 1.6.

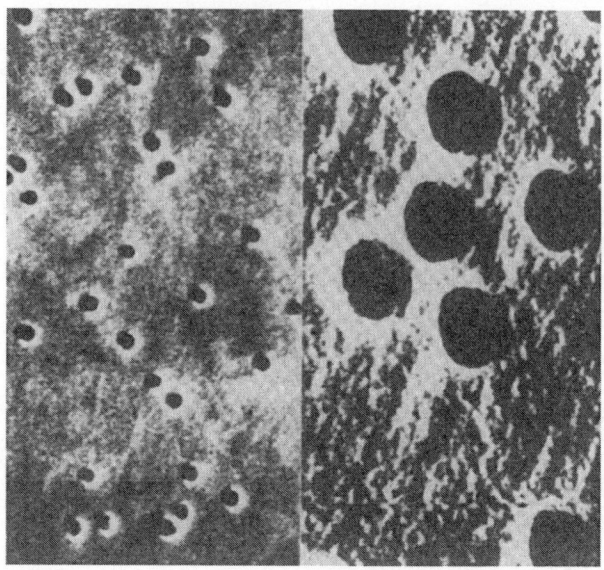

Figure 5. Polycarbonate channel-pore membranes (Nuclepore, Inc., Pleasanton, California). Enlargement about X10^4. Channel length is 6 μm and the radius of the channels on the left is 1 μm while that on the right is 0.1 μm.

Photophysics In Polymer Blends

Excitation kinetics in molecular aggregates and materials is of biological interest (e.g., photosynthesis and neuron transmission),[11,12] morphological interest (e.g., studies of polymeric materials),[7-9] and potential technological interest (nano-devices).[51] Both A + A and A + C type reactions have been studied under tightly controlled conditions using an extremely dilute blend of P1VN (poly-1-vinyl-naphthalene) in PMMA (poly-methyl-methacrylate). The electronic excitation is confined to the naphthalene moieties of the P1VN. Triplet excitations are utilized because of their limited range of energy transfer (4 to 8 Å). In the absence of neighboring P1VN chains, the excitation transport is thus limited to a single chain and reflects its one-dimensional topology. We therefore limit ourselves to the very dilute blends (about 0.01% P1VN or less) where the P1VN chains are effectively isolated from each other.[9]

A + A → Product Reaction (Homofusion). With short pulse laser excitations,[9] the observed phenomenon is that of triplet-triplet fusion ("annihilation"):

$$T + T \rightarrow S \tag{30}$$

The singlet excitations (S) have a natural lifetime of about 100 ns and are monitored via the resultant fluorescence (ultraviolet), while the triplet excitations (T) have a natural lifetime of about 3 seconds and are monitored via their phosphorescence (green).

For the most dilute samples measured (e.g., 0.05%), the heterogeneity exponent was found to be about 0.4, thus approaching the theoretical expectation, h = 1/2, for one-dimensional confinement of the reaction kinetics. On the other hand, for pure

P1VN material (100%), the experimental result is h = 0.02±0.02, in good agreement with the expectation for three-dimensional (classical) systems (h = 0).

A + C → C + Products Reaction (Heterofusion). Under prolonged ultraviolet excitation of P1VN blends, excimers are produced,[7-9] both in the singlet and the triplet states. The excimers are essentially defect sites with trapped excitations. The mechanism of triplet-triplet annihilation can now be written, to a good approximation[7,8] as

$$T_F + T_E \rightarrow T_E \tag{31}$$

where T_F is the free triplet excitation and T_E the trapped excimer excitation. The excimer sites are thus "catalytic" sites where the free excitations are "trapped" and consumed in an annihilative reaction.

From Eqs. (8) to (10) we expect the heterogeneity exponent h = 1/2 for d = 1 and h = 0 for d = 3, in agreement with the exact analytical results[30] for binary reaction kinetics [Eqs. (22) and (23)]. Measurements have been conducted for very dilute blends[9] (0.01% and 0.005%); the average result is h = 0.50±0.03, as expected for truly one-dimensional reaction topologies.

Bimolecular Reaction in Capillary with Gel

The chemical reactions were performed in a rectangular or a square glass capillary. The binary complex reaction

$$Cu^{++} + disodium\ ethyl\ bis(5-tetrazolylazo)acetate\ trihydrate \tag{32}$$

$$\rightarrow (1:1\ complex)$$

was carried out in a gel inside the glass tube, which allowed efficient diffusion but little or no convection. At time 0, there is a sharp boundary between A and B, and that is where product C is being formed (see Fig. 3). The absorbance of the product C was monitored at every fixed period by moving a detector (PMT) and a light source (green LED), which are fixed on a stepping motor, along the reaction front domain, while the reactor is fixed in time.

The results are summarized in Table III below. These results agree well with the theoretical expectations (α = 1/2, β = 1/6, δ = 1/2) given in the section "A + B → C Where A and B Are Segregated Throughout the Reaction" for the theory and simulation of batch reactions.

Scanning Optical Nanoscopy and Exciton Microscopy

The near-field scanning optical microscope (NSOM), developed by Lewis and co-workers, has enabled researchers to optically examine a variety of specimens without being limited in resolution to one-half a wavelength of light.[52-54] All NSOM imaging is based on the fact that as an electromagnetic wave emerges from an aperture, it is at first highly collimated to the aperture dimension. It is only after the wave has propogated a finite distance from the aperture that the diffraction which

Table III. Time Exponents from Experiments

Reactor Size	C_{Cu}	C_{tetra}	α	β	δ
4x2 mm	10^{-3} M	5×10^{-5} M	0.48	0.16	0.53
	10^{-3} M	7.5×10^{-5} M	0.53, 0.48	0.15, 0.20	0.51, 0.54
	2×10^{-3} M	6×10^{-5} M	0.52	0.16	0.53
2x2 mm	2×10^{-3} M	6×10^{-5} M	0.60	0.22	0.52

*   Cu is Cu^{++} and tetra is disodium ethyl bis(5-tetrazolylazo)acetate trihydrate, where the solvent is 0.15% agarose/water. C_{Cu} and C_{tetra} are t=0 concentrations. Here $x_f \sim t^\alpha$, $w \sim t^\beta$, and $R \sim t^{-\delta}$.

limits classical optical imaging takes effect. In the near-field region, a beam of light exists that is largely independent of the wavelength and determined solely by the size and shape of the aperture.[55]

In a typical NSOM system, a sub-micron aperture is fabricated and brought to within several hundred angstroms of the surface to be imaged. The aperture is then illuminated. The spot of light that emerges is scanned across the surface, generating a picture one point at a time. A feedback mechanism is incorporated to maintain a fairly constant distance between the aperture and sample, as the size of the near-field spot is strongly dependent on the distance from the aperture. The system can be operated in a transmission or scattering mode; alternatively, with the same resolution obtained the sample can be illuminated and the near-field collimation effect used to collect the light in the aperture.

The most successful results to date have been obtained using a metal-coated glass micropipette as an aperture.[53] Pipettes can readily be produced with inner diameters (at the tip) of several hundred angstroms.[54] Their tapered profile is almost ideal for probing recessed regions of non-planar surfaces. The metallic coating serves both to confine the light to the dimension of the pipette and as a conducting probe where tunneling feedback is employed. The details of pipette fabrication have been described elsewhere.[54]

The current resolution limitation of the NSOM technique derives from the less-than-ideal characteristics of the aperture. There are no propagating electromagnetic modes in a subwavelength cylindrical metallic waveguide such as a metal-coated pipette. The least attenuated mode for a round aperture has been found to be the TE_{11} mode, for which the energy decays at a rate of

$$E = E_o \exp\left(-2 \times 1.81 \, L/a\right) \tag{33}$$

where L is the length of the aperture and a is the radius.[55] With a sufficiently rapid tapering of the pipette, however, this evanescent region can be kept short enough to obtain a fairly large throughput of light. What ultimately limits the resolution is the finite conductivity of the metallic coating around the pipette. The electromagnetic wave penetrates the coating and decays within it at a finite rate as well, given by

$$E = E_o \exp(-d/\delta) \qquad (34)$$

where d is the depth of the penetration and δ is the extinction length of the metal. When the attenuation due to the waveguide effect exceeds the attenuation in the metal, the contrast between the aperture and surrounding medium becomes insufficient for super-resolution applications. The metal with the largest opacity in the visible region is aluminum, for which $\delta = 65$ Å when the wavelength (lambda) is 5000 Å. This yields a minimum usable aperture of about 500 Å. In practice, multilayered coatings are necessary to obtain good adhesion, and a practical aperture should be expected to perform even worse.

The solution to this problem is to use the energy-packaging capabilities of certain materials to circumvent the boundary problem of the edge of the aperture. According to fundamental understanding of energy propagation in materials, excitation can be confined to molecular and atomic dimensions under appropriate conditions.[56] By growing a suitable crystal within the submicron confines of a pipette, energy can be guided directly to the aperture at the tip instead of being allowed to propagate freely in the form of an electromagnetic wave. Such a material can be excited through an electrical or radiative process to produce an abundance of excitons that allows light to be effectively propagated through the bottleneck created by the subwavelength dimensions of the tip near the aperture. The excitons can be excited directly at the tip, or they can be generated within the bulk of the material and allowed to diffuse to the tip via an excitonic (electric dipole, Forster) transfer.[56,57] In either case, in a suitable material these excitons will then undergo a radiative decay, which produces a tiny source of light at the very tip of the pipette.

The excitonic throughput is basically independent of the wavelength and is a linear function of the cross-section of the aperture. However, due to singlet-singlet exciton-exciton annihilation [analogous to Eq. (30)], it is not a linear function of excitation intensity. Here the restricted dimension of the excitonic medium plays a major role. According to Eqs. (25a) and (25b), for effective dimensionalities below 2, the annihilation reaction order X is higher than 2 (see also Table I). As the excitation density, per site, is much less than unity, a higher power X means a significantly lower annihilation rate [Eq. (12a)]. The exciton distribution is closer to that of Eq. (16) than that of Eq. (17). This kinetic consideration is highly relevant to the successful operation of the exciton tip.

Experimental Procedure

To demonstrate the feasibility and usefulness of the method, we chose to work crystals of molecular anthracene. While not necessarily the best material available for this purpose, anthracene is easy to work with. Moreover, its electrical and radiative properties have been extensively characterized.[56]

The crystal of anthracene was grown inside the very tip of a pipette from a benzene solution. The pipette was held vertically, pointing downwards, while a tiny drop of solution was injected into the top. The strong capillary action immediately pulled the liquid down into the tip. The benzene was then allowed to evaporate out from the top, with the anthracene crystal precipitating inside the tip. By varying solution concentration, the size of the deposited crystal could be accurately controlled.

The source of excitation for our experiments was the 3638 line of an argon ion laser. Anthracene exhibits a very strong fluorescence in the blue with a quantum efficiency approaching unity when illuminated in the near ultraviolet. The crystal can be illuminated either by directing light through the pipette (as in the standard aperturing method) or by having an external beam incident on the crystal at the tip of the pipette. With the second method, large amounts of energy can be brought to bear directly on the spot where the illumination is desired, with the upper limit imposed only by the onset of photochemical bleaching. A fairly large (~1 micron) crystal was grown to extend past the tip of the pipette and was illuminated head on with several milliwatts at 3638 Å. The spot of light produced was clearly visible to the naked eye under room lighting and is shown here magnified 100 times (Figs. 6-8).

Figure 6.   Micropipette exciton light sources excited with 5mW of 363.8 nm of ultraviolet light from an argon ion laser. Tip only shown, magnification about 1000, ID = 150 nm.

To test the imaging capability of the light source, a small crystal was grown inside a half-micron pipette. In order to obtain high-resolution images, it was necessary for the anthracene to be confined to the inside of the pipette as much as possible so that the emerging beam remains collimated. Any excess material could readily be removed from the outer surface with a small drop of benzene. The test sample was a 2-micron aluminum grating. The imaging was done in the transmission mode with the excitory beam being directed through the pipette. A filter, placed behind the grating to prevent any stray ultraviolet light from reaching the photomultiplier, confined the signal to the crystal fluorescence. The lifetime of the singlet state in anthracene, responsible for most of the luminescence, was sufficiently short (about 10 nanoseconds) to allow the beam to be chopped and a lock-in detection scheme to be employed.

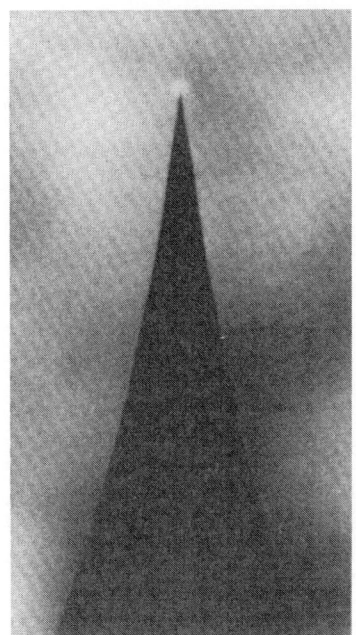

Figure 7. Same as Fig. 6, ID = 100 nm.

Figure 8. Same as Fig. 6, ID = 60 nm.

Results and Discussion

The result of a single line scan 1 micron long and a complete two-dimensional scan is shown in Fig. 9. The image is unprocessed, and the resolution could be further improved by deconvoluting the image with the well-defined shape of the pipette aperture.

Figure 9.   XY Scan of aluminum grating with dust particle (20 min. scan, 20 Å steps, magnification about 10,000). The vertical (Z) axis designates photon transmission.

For the smallest pipette experiment calibrated so far (ID 60 nm), the improvement in light transmission via the excitonic process is at least 150%. Extrapolating to smaller aperture dimensions, the improvement at 6 nm should be about a factor of 10^{13} and the throughput about 10^4 photons/sec. Nanometer scanning optical microscopy has thus become a reality.

References

1. P. G. de Gennes.   Kinetics of Diffusion-Controlled Processes in Dense Polymer Systems. 1. Nonentangled Regimes. *J. Chem. Phys.* 76:3316 (1982).
2. S. Havlin and D. Ben-Avraham. Diffusion in Disordered Media. *Adv. Phys.* 36:695 (1987); J. Rudnick and G. Gaspari. *Science* 237:384 (1987).
3. P. Argyrakis and R. Kopelman. Diffusive and Percolative Lattice Migration: Excitons. *J. Chem. Phys.* 72:3053 (1980);
    P. W. Klymko and R. Kopelman. Fractal Reaction Kinetics: Exciton Fusion on Clusters. *J. Phys. Chem.* 87:4565 (1983).
4. R. Kopelman. Fractal Reactor Kinetics. *Science* 241:1620-1626 (1988).
5. J. Prasad and R. Kopelman. Molecular Reaction Kinetics Inside Channel Pores: Delayed Fluorescence of Naphthalene in Methanol. *Chem. Phys. Lett.* 157:535 (1989).
6. J. Prasad and R. Kopelman. Fractal-Like Molecular Reaction Kinetics: Solute Photochemistry in Porous Membranes. *J. Phys. Chem.* 91:265 (1987).
7. C-S. Li and R. Kopelman. Fractal-Like Triplet Excitation Kinetics and Chromophore Morphology in Blends of Poly(1-vinylnaphthalene) and Poly-(methylmethacrylate). *Macromolecules* 23:2223-2231 (1990).
8. C-S. Li and R. Kopelman. Photophysical Degradation of Poly(1-vinylnaphthalene) and Poly(methylmethacrylate) Blends. *J. Phys. Chem.* 94:2135-2140 (1990).

9. Z-Y. Shi, C. S. Li, and R. Kopelman. Excimer and Exciton Fusion of Blends and Molecularly Doped Polymers--A New Morphological Tool. *Polymer Based Molecular Composites*, J. E. Mark and D. W. Schaefer, eds. Proceedings of Materials Research Society 171:245-254, Pittsburgh (1990).

10. M. Drake. Private communication.

11. E. B. Fauman and R. Kopelman. Excitons in Molecular Aggregates and a Hypothesis on the Sodium Channel Gating. Comments *Mol. Cell. Biophy.* 6:47 (1989).

12. L. A. Dissado. Ion Transport Through Nerves and Tissues. Comments *Mol. Cell. Biophysics* 4(3):143-169 (1987).

13. J. Hoshen and R. Kopelman. Exciton Percolation I. Migration Dynamics. *J. Chem. Phys.* 65:2817 (1976).

14. P. Klymko and R. Kopelman. Heterogeneous Exciton Kinetics: Triplet Naphthalene Homofusion in an Isotopic Mixed Crystal. *J. Phys. Chem.* 86:3686 (1982).

15. A. A. Ovchinikov and Ya. B. Zeldovich. Role of Density Fluctuations in Biomolecular Reaction Kinetics. *Chem. Phys.* 28:215 (1978), and references cited therein;
V. N. Kuzovkov and E. A. Kotomin. Many-Particle Defects in Accumulation Kinetics of Frenkel Defects in Crystal. *J. Phys. Chem.*, Solid-State Phys. 17:2283-2292 (1984).

16. V. M. Agranovich and M. D. Galanim. *Electronic Excitation Energy Transfer in Condensed Matter*, North Holland, Amsterdam, 1982.

17. M. V. Smoluchowski. Mathematical Theory of the Kinetics of the Coagulation of Colloidal Solutions. *Z Phys. Chem.* 92:129-168 (1917);
J. Kelzer. Nonequilibrium Statistical Thermodynamics and the Effect of Diffusion on Chemical Reaction Rates. *J. Phys. Chem.* 86:5052-5067 (1982), and references cited therein.

18. J. Klafter, A. Blumen, and G. Zumofen. Fractal Behavior in Trapping and Reaction: A Random Walk Study. *J. Stat. Phys.* 36:561-577 (1984).

19. R. Kopelman. Rate Processes on Fractals: Theory, Simulation and Experiments. *J. Stat. Phys.* 42:185 (1986).

20. K. Kang and S. Redner. Scaling Approach for the Kinetics of Recombination Processes. *Phys. Rev. Lett.* 52:955-958 (1984).

21. S. Alexander and R. Orbach. Density of States of Fractals: Fractions. *J. Phys. Lett.* (France) 43:L625-L631 (1982).

22. A. Blumen, J. Klafter, and G. Zumofen. In I. Zschokke (Ed.), *Optical Spectroscopy of Glasses*, p. 199, Reidel, Dordrecht (1986).

23. K. Lindenberg, B. J. West, and R. Kopelman. Diffusion Limited A + B → Reaction: Spontaneous Segregation. *Proceedings of Conference on Noise and Chaos in Nonlinear Dynamical Systems*, F. Moss, L. Lugiato, and W. Schleich, eds., pp. 142-171. Cambridge University Press (1990).

24. E. Clement, L. M. Sander, and R. Kopelman. Source Term and Excluded Volume Effects on the Diffusion Controlled A + B → 0 Reaction in One-Dimension: Rate Laws and Particle Distributions. *Phys. Rev. A* 39:6455-6465 (1989).

25. E. Clement, L. M. Sander, and R. Kopelman. Steady State Diffusion Controlled A + B → 0 Reactions in Two and Three Dimensions: Rate Laws and Particle Distributions. *Phys. Rev. A* 39:6466-6471 (1989).

26. E. Clement, L. M. Sander, and R. Kopelman. Steady State Diffusion Controlled A + A → 0 Reaction in Euclidean and Fractal Dimensions: Rate Laws and Particle Self-Ordering. *Phys. Rev. A* 39:6472-6477 (1989).

27. L. A. Harmon, L. Li, L. W. Anacker, and R. Kopelman. Segregation Measures for Diffusion-Controlled A + B Reactions. *Chem. Phys. Lett.* 163:463-468 (1989).

28. S. J. Parus and R. Kopelman. Self-Ordering in Diffusion Controlled Reactions: Exciton Fusion Experiments and Simulations on Naphthalene Powder, Percolation Clusters and Impregnated Porous Silica. *Phys. Rev. B* 39:889, (1989).

29. P. Argyrakis and R. Kopelman. Nearest-Neighbor Distance Distribution and Self-Ordering in Diffusion Controlled Reactions: I. A + A Simulations. *Phys. Rev. A* 41:2114-2120 (1990);
P. Argyrakis and R. Kopelman. Nearest-Neighbor Distance Distribution and Self-Ordering in Diffusion Controlled Reactions: II. A + B Simulations. *Phys. Rev. A* 41:2121-2126 (1990).

30. G. Weiss, R. Kopelman, and S. Havlin. Density of Nearest-Neighbor Distances in Diffusion-Controlled Reactions at a Single Trap. *Phys. Rev. A* 39:466 (1989).

31. E. Clement, R. Kopelman, and L. M. Sander. Trapping Reaction in Low Dimensions: Steady-State Self-Organization. *Europhys. Lett.* 11:707-712 (1990).

32. C. R. Doering and D. Ben-Avraham. Interparticle Distribution Functions and Rate Equations for Diffusion-Limited Reactions. *Phys. Rev. A* 38:3035-3042 (1988).

33. C. R. Doering and D. Ben-Avraham. Diffusion-Limited Coagulation in the Presence of Particle Input: Exact Results in One Dimension. *Phys. Rev. Lett.* 62:2563-2566 (1989).

34. D. C. Torney and H. M. McConnell. *J. Phys. Chem.* 87:1941-1951 (1983).

35. L. W. Anacker, R. P. Parson, and R. Kopelman. Diffusion-Controlled Reaction Kinetics on Fractal and Euclidean Lattices: Transient and Steady-State Annihilation. *J. Phys. Chem.* 89:4758 (1985).

36. M. Bramson and D. Griffeath. Clustering and Dispersion Rates for Some Interacting Particle Systems on ZZ. *Ann. Probab.* 8:183-213 (1980).

37. D. Toussaint and F. Wilczek. Particle-Antiparticle Annihilation in Diffusive Motion. *J. Chem. Phys.* 78:2642-2647 (1983).

38. G. Zumofen, A. Blumen, and J. Klafter. Concentration Fluctuations in Reaction Kinetics. *J. Chem. Phys.* 82:3198-3206 (1985).

39. K. Lindenberg, B. J. West, and R. Kopelman. Diffusion Limited A + B → 0 Reaction: Correlated Initial Condition. *Phys. Rev A* 42:890-894 (1990).

40. L. Li. *Diffusion-Limited Binary Reactions in Low Dimensional Systems: Monte Carlo Simulations.* Ph.D. Thesis, University of Michigan, Ann Arbor, Michigan (1989).

41. L. Galfi and Z. Racz. Properties of the Reaction Front in an A + B → C Type Reaction-Diffusion Process. *Phys. Rev. A* 38:3151-3154 (1988).

42. L. W. Anacker and R. Kopelman. Steady-State Chemical Kinetics on Fractals: Segregation of Reactants. *Phys. Rev. Lett.* 58:289 (1987);
    L. W. Anacker and R. Kopelman. Steady-State Chemical Kinetics on Fractals: Geminate and Nongeminate Generation of Reactants. *J. Phys. Chem.* 91:5555 (1987).

43. K. Lindenberg, B. J. West, and R. Kopelman. Steady-State Segregation in Diffusion-Limited Reactions. *Phys. Rev. Lett.* 60:1777 (1988).

44. Z. Racz. Diffusion-Controlled Annihilation in the Presence of Particle Sources: Exact Results in One Dimension. *Phys. Rev. Lett.* 55:1707-1710 (1985).

45. B. B. Mandelbrot. *The Fractal Geometry of Nature,* Freeman, San Francisco, 1983.

46. J. S. Newhouse and R. Kopelman. Reaction Kinetics on Clusters and Islands. *J. Chem. Phys.* 85, 6804 (1986);
    J. S. Newhouse and R. Kopelman. Steady-State Chemical Kinetics on Surface Clusters and Islands: Segregation of Reactants. *J. Phys. Chem.* 92:1538 (1988).

47. B. J. West, R. Kopelman, and K. Lindenberg. Pattern Formation in Diffusion-Limited Reactions. Proceedings of 1988 Conference on External Noise and Its Interaction with Spatial Degrees of Freedom in Nonlinear Dissipative Systems, Los Alamos, Special Issue. *J. Stat. Phys.* 54:1429 (1989).

48. E. Peacock-Lopez and J. Kelzer. One-Dimensional Reactive Systems: The Effect of Diffusion on Rapid Biomolecular Processes. *J. Chem. Phys.* 88:1997-2003 (1988).

49. L. W. Anacker and R. Kopelman. Reactions in Fractal and Small Dimensions at the Steady State: Anomalous Chemical Kinetics and Reactant Self-Ordering. *Science at the John von Neumann National Supercomputer Center, 1989,* pp. 29-36. Consortium for Scientific Computing, Princeton, NJ (1990).

50. R. Luther and F. Weigert. *Z. Phys. Chem.* 53:385 (1905).

51. K. Lieberman, S. Harush, A. Lewis, and R. Kopelman. A Light Source Smaller than the Optical Wavelength. *Science* 247:59-61 (1990).

52. A Lewis, M. Isaacson, A. Harootunian, and A. Muray. Development of a 500 Å Spatial Resolution Light Microscope. I. Light is Efficiently Transmitted Through λ/16 Diameter Apertures. *Ultramicroscopy* 13:227-232 (1984).

53. E. Betzig, A. Harootunian, A. Lewis, and M. Isaacson. Near-Field Diffraction by a Slit: Implications for Superresolution Microscopy. *Applied Optics* 25:1890 (1986).

54. K. T. Brown and D. G. Flaming. New Microelectrode Techniques for Intracellular Work in Small Cells. *Neuroscience 2* (1977).

55. N. A. McDonald. Electric and Magnetic Coupling through Small Apertures in Shield Walls of Any Thickness. *IEEE Trans. Microwave Theory Tech.* MTT-20, 689 (1972).

56. M. Pope and E. Swenberg. *Electronic Processes in Organic Crystals*, Oxford University Press, New York (1982).

57. H. Kuhn. Classical Aspects of Energy Transfer in Molecular System. *J. Chem. Phys.* 53:101 (1970).

Discussion

Ward: How important is the thickness of the sample considering the penetration depth of your probing particles?

Kopelman: Thank you for asking that question. You see, in these other methods, like electron microscopy, the electrons just go in so far because they are stopped. We have a combination of two techniques. In one we are still using evanescent photons to go into it. With photons there is no limit, really, and the limit is practical; you can go as far in as your desired resolution. The resolution and the penetration distance will be the same, so if I am happy with 500-angstrom resolution, I can go in 500 angstroms, and so on. So this is one case where one can go in deep. With the exciton, this is a much shorter-range type of interaction so this is more like a surface technique which is limited to 50 or 100 angstroms. There are some disadvantages and some advantages because you look only at what is happening at that level.

Christophorou: Are you limited by the type of excitons you can use?

Kopelman: If I understand correctly what you mean by "type," then the idea is to use all types, at least all types of *Frenkel excitons*, and that means singlet excitons over the whole spectrum, and eventually triplet excitons as well. Each one, again, has its advantage. The singlet exciton would have a range of 50 angstroms, but the triplet exciton in crude numbers will have a range like 5 angstroms. So again, by using different excitons you can choose your range of resolution. Theoretically, with triplet excitons, one should have a resolution down to a fraction of an angstrom, because if you look at the relevant matrix elements, this method has a better resolution than tunneling electron microscopy.

Christophorou: May I ask one more question? What is the role of micro breakdown in the STM microscope? Is it really a serious disadvantage? You mentioned breakdown as being a very important disadvantage of that technique. Is it so?

Kopelman: Oh, that's a good question. Definitely. When you have these fields, there are millions of volts per centimeter or more. With electrical breakdowns, people usually notice the big breakdowns, but you can have equivalent breakdowns on a smaller scale. So, you may really have serious deformations, and I don't think that this has been studied. As a matter of fact, this would be a whole new field of study, using STM to see what damage it causes, what modification it does to the samples. This may be interesting, but it shows you that you don't want to use STM as a non-perturbative method.

Kasha: Raoul, I don't know if you know the (Seleney) method for wide-angle interference for the study of electromagnetic radiation mechanism. When I think about it; and think about your small dimensions, I begin to think you will run into it. Here is what it is. It was published long ago, the theory of it, if you take an elementary light source behind an aperture which is one-tenth the wave length of light, the light coming normal and at a wide angle are in different phase relationship, and for each mutipole there is a different interference pattern. What this suggests to me is that when you get really down to these submolecular dimensions, you will start to get interference patterns which will give false images, because you get light-dark fringes in the (Seleny) method. Therefore, it looks to me like that could be a limitation on your ultimate result.

Kopelman: No, I don't think so. Interference fringes form on a scale on the order of a fraction of a wavelength, and you see them if you go far enough away from the subwavelength light source. The whole idea is to come so close that you never see interference fringes. I should have shown you a bunch of pictures you get when you do the scanning where I can show you that when you scan very close to the hole you see a clear image. If you go further away you start seeing the interferences you mentioned. [See Lewis, et al., *Optical Microscopy for Biology*, B. Herman and K. Jacobson, eds., p. 615, Wiley-Liss, New York (1990).]

Kasha: But you see, you gave us a demonstration of a hole and a spot, and I admire your quickness in covering the hole, but the exposure of the spot, even this morning, was one-tenth the size of your hole. You are very close to the hole, and now you are going to have light coming at somewhat different angles to your detector. That is when I think you might get into trouble, when your features are smaller than your aperture.

Kopelman: The spot was of the same size as the hole, and it has to be at a distance not larger than the aperture. This is based not only on theory, but the model has worked with microwaves, the work I mentioned. There are a lot of beautiful examples of what you can do with microwaves (but with no practical use) because they employ waves in the centimeter range. They can see and resolve a millimeter or a fraction of a millimeter, which is totally useless for our purposes but proves my point. So, for a subwave-length hole, as long as your distance from this hole is not larger than the size of the hole, you just don't get any interference effects. The evanescent photon comes out in a directed way and only later spreads and gives you the interference phenomena. There has been quite a bit of work on this. [See above.]

Weinstein: I think the reason you have been able to cover the hole so quickly is that you have an overview of both the hole and the top. But it is hard for me to see how you will find one exciton on this large, sticky end. With a 10^{-4} probability in a genome, how will you scan to find it? And see it?

Kopelman: There are two parts to this. Say I have a fluorophore, and I'm trying to find this fluorophore. With STM, this is one of the biggest problems because people scan for weeks to find the spot of interest, but we can use, first of all, ordinary microscopy, then near-field microscopy, so we can kind of zoom in stages to a certain place which is marked. Now, if I have a single fluorophore only, I have a problem, but if I have a number of fluorophores which are in a certain part of this sample, I am going to zoom, first of all, into this part of the sample where the DNA has accepted my fluorophore. I have come down to a fraction of a micron already; then, with near-field microscopy, I can zoom in further to about a few hundred angstroms, and then I have to fine-tune with my exciton scanner.

Weinstein: Even with fluorescent microscopy, you are not going to be able to find one fluorophore on the sticky end. How are you going to know where it is?

Kopelman: I am not going to find one fluorophore; I am going to find the area of the sample where there are fluorophores. However, in case of a single one, then the approach is the other way around. Here the idea is that we should study the end where I am going to try to create the damage. I am going to bring the damage to the image, i.e., to the place that I am imaging, rather than the other way around.

Wood: Perhaps an extension of Harel's question; one of the principal thrusts of the department's human genome program right now is to develop advanced technology for improving sequencing capability. I wondered if you would care to reflect on these approaches that you have been talking about in terms of their potential for rapid inexpensive sequencing capability.

Kopelman: I happen to have a transparency of the way that I am *not* doing it. But it is an approximation. What you can do, *in situ*, is attach a lot of fluorophores specific to each one of the DNA bases and then try to scan along the chain and see those fluorophores. There is a problem with resolution because the fluorophores are usually not small enough to prevent overlapping, so while this could give you some approximate answers, I only mention it to illustrate that, in principle, one could follow along a chain and look at the fluorophores. What we have developed is a different approach which is based on a most sensitive way of utilizing those excitonic interactions. We use the external heavy-atom effect discovered by Mike Kasha in the 1950s; then, rather than having the fluorophore on your DNA, you have a specialized fluorophore at the tip, i.e., one chosen working fluorophore, and now you want to change its luminescence color by the presence of a certain base of the DNA. One performs a mercury atom substitution (or another heavy atom) on one kind of base, say all A bases, which have mercury atoms substituted into them. The substituted bases are fed into the PCR duplication process. Then you will be able to sequence this chain with the location of the A bases, and then you feed another one which has the mercury atoms in the T base, and so on. So what we have suggested is a scheme that will do this, *in situ*, for one base at a time and then combine the information.

Ward: This technique could give you base sequences; could it be used for probing the proximity of one gene to another by using fluorescent probes for the specific genes?

Kopelman: When I described the sequencing method, it was performed on a DNA sitting on a matrix, not a "live" one. This way we just do sequencing. But if you want to have less information, if you just want to find where your other sticky point is at a certain distance; this is what I was addressing before. At this point you don't necessarily want a resolution of three and a half angstroms. You may be happy with a resolution of 10 angstroms or even 30 angstroms.

Kasha: I have lost track of Michael Beer, but he was at Johns Hopkins 15 or 20 years ago. His idea was to sequence nucleic acid bases by optical electron microscopy, using tagged bases and putting them in as markers. Now, that would be interesting to you because he had 12-foot-long photographs showing these electronmicrographs. It would be interesting because that offers perturbation right at specific bases, and you would be able to use your trick on those.

Kopelman: I appreciate that. I also collaborate with John Langmore, who is an electron microscopist, and it turns out that his Ph.D. project was on the same problem, i.e., to substitute DNA bases with heavy atoms. He has worked out this whole path of mercury substitutions. The only problem in his Ph.D. project was (what I guess everybody knows), that when you try to go to such small resolution the damage due to electrons becomes intolerable. There is no such damage with excitons.

Varma: We should respond to Harel's question. I do not think that it was answered, so let me put it another way. If you have a single base pair change of DNA, it would be a very difficult matter to see it. If you have a DNA already sitting there and you are focused on it, zoomed on it, and then expose it to either radiation or chemical events, you can see the base pair change. Is that correct?

Kopelman: Yes. But let me also point out that people using STM are going blindly for weeks to find specific places with no guidance whatsoever. We have an approach for zooming-in in controlled steps.

Weinstein: This also alarms me. John's question had to do with placing chromosomes and finding their relative positions. I believe that this will require fluorescence microscopy. With high-resolution, time-resolved fluorescence microscopy, you can have a labeled ligand and identify which part of the molecule exhibits abundant fluorescence so that you can see it. This kind of approach is followed by Josef Eisinger's laboratory in my department. It is also looking at a non-fixed preparation, and affords information and control of the dynamics as well as of the spatial relationships in the preparation.

Kopelman: We stain the chromatin (with a given color), and we zoom in onto it. We have a special fluorophore that is going to a special location with a different "color," i.e., a specific energy transfer interaction.

Weinstein: It seems that when one considers the actual magnitudes and numbers involved, there remain orders of magnitude for scanning; and scanning with the excitons is likely to be tedious.

Kopelman: We did some calculations too. Also, there is no problem when the damage occurs where you already are. [See above.]

Varma: Dr. Kopelman, the proposal you presented here is futuristic. When will we see it in practice?

Kopleman: What we have figured out is that, on a level of resolution of 300-500 angstroms, there are a lot of interesting problems, i.e., chromatin structure and dynamics. To do 300-angstrom or 500-angstrom resolutional work will take a couple of years (we hope to do it in a year, but things always take a bit longer). Now, should we want to go to a single base, this is more like five years or maybe longer (I am trying to be conservative).

Summary

Summary

Matesh Varma: *Introduction*

In my opening remarks, I indicated to you how I view the interlinking of physical and chemical processes in understanding the biological damage from the first principles. Indeed, the process is quite complex, and to understand all the processes and their interactions is clearly a very difficult undertaking. Nevertheless, our program in radiological and chemical physics can and should provide the basic information needed to elucidate the mechanisms of biological action. We should couple this fundamental information with well-justified and scientifically supported assumptions to further the goal of defining the pathways by which initial interactions at the molecular level induced by environmental pollutants such as radiation or chemicals may progress towards carcinogenesis in humans. When the radiological and chemical physics program was established by the Atomic Energy Commission some 35 years ago, very little was known about the physical and chemical processes in biological systems. A great deal of progress has been made since; and in 1974, the published proceedings of a conference held at Airlie, Virginia, in 1972 documented the progress in this field up to that time. This book became an important reference for scientists working in the field of radiation biology.

In recent years, with the advent of new and more sophisticated techniques in molecular biology, a new opportunity and challenge has been created for our program to provide the necessary physical and chemical processes and steps to interpret and understand the mechanisms of biological action at the molecular level. We have been moving toward the goal of understanding the physical and chemical mechanisms in condensed phase for those types of materials that are biologically relevant. Also, emphasis has been placed on using this fundamental information to help interpret the biological data at the molecular level. This conference at Woods Hole will thus provide documentation of the progress that has been made in the radiological and chemical physics program since 1974, and, through scientific discussions among the participants, should provide information for future research areas and priorities. The Airlie House proceedings was not intended to be used as a planning document, but was designed to be read by people working in the field, and to stimulate people to work in this field and contribute to the understanding of radiation effects. I feel that the proceedings of this Woods Hole conference, in addition to documenting the progress in the field and providing research directions, should also be used as one of the important documents for planning purposes.

We spent many hours and worked hard to plan the agenda. We invited the best experts in various fields to stimulate open and frank scientific discussions and thus achieve the goal of the conference. I think that the scientific presentations and discussions that have been presented, when published, will provide excellent documentation. At this meeting, however, we have also been rather provocative and have critically discussed the broad areas of research that should have higher priority over other areas. For example, how much more do we need to know in the area of physical process in biological systems? Can we proceed to the identification of chemical processes? Are we doing research in physics and chemistry just to elucidate physical and chemical phenomena and not paying enough attention to the types of physics and chemistry that are relevant to biological effects? As I am sure you all know, a principal objective of the Office of Health and Environmental Research (OHER) is to understand the biological effects of ionizing radiation. The radiological and chemical physics program is a major component to achieving that objective. I trust that we have clearly identified broad areas of research in the radiological and chemical physics program that will enhance these objectives of OHER.

Eric Hall: *Biology of Carcinogenesis*

We were each invited here because of our own particular microsphere of competence, but it soon became apparent after we arrived that we were supposed to interact and build bridges; this is something that many of us are not particularly skilled or practiced at doing. The international brotherhood of meeting summarizers demands that the summary start with a story, but a story with a moral--one that is relevant and makes a point. I am going to tell a very brief story that is set in a house of ill repute. The doorbell rings, the madame goes to the door, and there is a man standing at the door without any arms or legs. The madame looks at him and says, "What do you want?" The man replies, "I'm here for the night—I want the works." The madame says, "Well, what do you think you can do with the things that you have?" Then the man answers, "I rang the doorbell, didn't I?"

There is a moral in this story. We may not think we are very skilled at interactions, but we must at least try. I think that at this meeting, we at least rang the doorbell.

Of four brief summaries, I have been allocated biology. Here are priority research items as I see them at the present time.

First, there is an ongoing effort in several labs to characterize and sequence dominant-acting oncogenes activated by radiation.

Second, the obvious next step would be to compare dominant-acting oncogenes activated by different types of radiation such as X-rays, neutrons, and alpha particles. Is there an LET effect; is there a signature of high LET radiation?

Third, since the activated dominant genes are mainly associated only with leukemias and lymphomas, we want also to identify suppressor genes deleted and/or inactivated by radiation since this is the mechanism more often associated with solid cancers. That is a more difficult task for radiation.

Fourth, it is increasingly obvious that experimenters need some help and guidance in terms of hypotheses, and we need a closer interaction with the modelers to get models for radiation-induced mutations and transformation. I think that our horizons were broadened a bit by having Suresh Moolgavkar come here and talk about the models for carcinogenesis, which involve many steps. We need a big effort on models to give some guidance to the experimenters who otherwise are generating lots and lots of random data, without a central theme and with no hypothesis to test.

Fifth, carcinogenesis is a balance between the induction of the transforming event and cell killing; why then is the RBE for transformation higher than the RBE for cell killing? We don't understand this phenomenon, and we need to model it.

Sixth, we need someone to go away and do the calculations, as Marco Zaider has promised to do, to link the physics to the biology and to ask the question, "Is it, indeed, feasible for a modest dose of radiation to produce only one transforming gene in the entire genome, i.e., change one out of 300,000 pieces of DNA? Is this a reasonable event to attribute to radiation? We need to make this link, and to verify what target size is likely to be altered by a given dose of radiation.

Seventh, the new parts to this program fit into the category of identifying the mechanisms by which radiation activates the dominant oncogene or deletes a suppressor gene. We have to learn about the molecular structure of DNA and about DNA bending, as well as the generation and transduction of signals. I think Harel Weinstein began to educate us a little about this today; it is an important addition to our thinking and our program.

Eighth, then, we must go back to modeling again. I think that we would like to see some modeling of oncogenic transformation, as an *in vitro* surrogate for carcinogenesis, as a function of both radiation dose and of LET. With the revised dosimetry of Hiroshima and Nagasaki, we now have no human data on the RBE for carcinogenesis.

We didn't talk about it at this meeting since we were restricted by the time available; we focused on what was most immediately relevant to the aims of OHER; i.e., we focused on carcinogenesis primarily, and mutations secondarily. We must not forget that there is a big effort in a number of labs around the country and the world to identify, clone, and sequence radiation repair genes, and this is being done for other DNA-damaging agents. But it has turned out to be particularly difficult in the case of ionizing radiation. This needs to be done, and clearly should be supported. It should be our **ninth** aim.

Then, there is the important business of radiation-induced mutations. There was no one quite brave enough to get up and say that one of the problems is that none of the mutation assay systems we've got is very good. Every one of them suffers from a weakness. None is ideal, and we need to develop better quantitative mutation assays in mammalian cells; this is the **tenth** and last aim I shall include in my summary. It is an urgent priority. There is an obvious need to characterize mutants as a function of LET. John Ward thinks we need it as a function of dose as well; we need to investigate the balance of point mutations versus deletions as a function of dose.

These are the highest priority items that I would identify in my little corner of the vineyard, namely that of the biological effects of radiation. So now it is "farewell" to Woods Hole, and thank you, Bill Glass and Matt Varma, for organizing this meeting, and to our editor and transcriber for trying to make sense out of all of these discussions.

Herwig Paretzke: *Physics and Radiation Biology*

I think that physicists came into this game because most of us work in radiation protection, so we have to be concerned about carcinogenesis by low doses of radiation. The epidemiological data have large error bars. We have solid data at relatively low doses in certain areas, but we just don't know how this extrapolates to doses of concern to radiation protection.

The first contribution that radiation physics can make to the field of radiation biology and to carcinogenesis is perhaps in this area of models for extrapolation. The extrapolation of dose with respect to dose rate, and to radiation quality, requires basic knowledge. We need to know the effects of low doses of radiation. In most practical cases, the rate of application of the dose is much smaller than the region where we can hope to have reliable data, and we don't have much data on radiation qualities other than photons, except for alpha particles in the lungs of human beings. So we try to do calculations to help quantify the carcinogenic risks of low doses of radiation. There is a second, important aspect, the area of instrumentation. We have to design instruments to measure those quantities of radiation fields that are relevant to the associated human health risks.

What do we need to proceed along the lines where the physicist can contribute in this field? I would like to start with something which we physicists need from other fields: From biology and chemistry, we urgently need the identification of sensitive structures in the cell. That is an urgent need; otherwise, we don't know for which set of molecules or types of structures to do our calculations. That is a big question that we on the physics side are asking of our colleagues working in the other fields. Help us to identify sensitive targets and sensitive changes in these targets, and to make, first of all, an approximation how they might be related to events occurring during the energy absorption.

What physicists can provide are the cross sections or the probabilities for energy absorption, energy transfer, and transport within the molecule, for example, or by the slowing down of the electrons. I would like to use this word "transport" for both for intramolecular energy movement and for the movement of kinetic energy of charged species. This physics provides only the information about *where* things happen, but we need to know *what happens*. So we have to get more information about the decay of excited states into new chemical species. That will allow us to construct a track of effects left behind, not the dynamics for the production of excited or ionized species.

Very closely linked to this is the work, exemplified by Chatterjee's talk, which shows a very nice and almost smooth transition to chemistry in this respect. But we should design models, I think, because of their simplicity and their ability to take available knowledge into consideration; they ought to be Monte Carlo models because we now have the capabilities to deal with such open calculations. We don't have to use assumptions which were necessary a few years ago when computer capabilities were restricted. We really should try to have models that can deal with everything which is considered by the expert as solid knowledge. And the model should go, more or less, from the physics of energy absorption to observable biological or chemical effects. We need more complex models, and the wisest way to do this, for the time being, will most likely be with Monte Carlo models.

Now, we need to know the reaction of these new chemical species with the sensitive steps identified by our friends working in biology, and we have to do this for all different doses and dose rates as mentioned above, and for different radiation qualities. This analysis and these three terms help us to understand something about the mechanisms which go on at the molecular scale, which probably will not otherwise be accessible to research.

We finally have to solve problems in developing such models with optimum use of new super-computer capabilities, and this calls for optimum hardware and software design. We should never forget that all these computers which are coming to the market with the most cost-effective computing capabilities were designed or originated because elementary particle physicists and nuclear physicists could no longer afford to pay for adequate computing time. They simply went into their laboratories and designed their own computers. They made major steps forward, in certain areas of scientific computing, by one to two orders of magnitude. Computers with many parallel processors came along because of scientists who needed the power but did not have adequate computational time on serial machines. I think we in this field now have the capability to use these high-speed computing and graphic machines in an optimal way. But we have to do it ourselves and in such a way that we formulate the questions adequately and understand the results which are produced in order for us to test our working hypothesis. At this meeting, we are seeing very nice pictures about this.

That is what I wanted to say, and I think that it is more or less what we need to do in physics in the next five or ten years. Thank you.

Clemens von Sonntag: *Radiation Chemistry of DNA*

The very early effects, the energy depositions, are studied by the radiation physicist. Radiation chemistry deals with the short period after the radiation energy is absorbed, i.e., when radicals are formed and decay. This period is over after about a few seconds. The non-radical, often enzymatic, reactions start to play a role and the damage manifests itself: the early and late effects in radiobiology. Thus, radiation chemistry has to bridge the gap between radiation physics and radiation biology. Unfortunately, this gap is still rather wide.

Over the past 20 years, practically only the indirect effect has been studied, i.e., the reactions of OH radicals, solvated electrons, and hydrogen atoms with DNA. These studies were supported by investigating in more detail DNA constituents as well as tailored model systems. Thus, considerable progress has been made in this field, and we now understand quite well some of the free-radical chemistry of DNA. Although the high molecular weight of DNA is a very strong obstacle in detecting lesions, at least one quarter of the damage is now well characterized. Eventually, we would like to be able to present a complete picture.

In contrast, there is very little information on the direct effect, i.e., when the ionizing radiation is absorbed by the DNA itself. Only recently the research on the chemistry of DNA base radical cations has been started with nucleosides as models. Thus, in this area we are only at the very beginning.

Much of our future research will have to center on a better understanding of the mechanism of double-strand break formation. Here, computer modeling, which is now on the way to becoming an excellent tool in the hand of the theoreticians, will play an important role. Eventually such calculations will decide to what extent multiply damaged sites or radical transfer processes dominate these important events. For the latter we also require a better understanding of the dynamics of DNA radicals. An intensive interaction between radiation physicists and radiation chemists will be necessary to solve these problems.

A transition to radiobiology is the question of the modification of DNA damage by glutathione and the action of repair enzymes on various damaged DNA sites. Again, this will be an interdisciplinary project.

Stanley Curtis: *Modeling of Effects*

There is a long and productive history of modeling the biological effects of radiation, and new ideas relating to molecular biology have been emerging and have been incorporated into models in recent years. Because of the large amount of experimental cell survival data available, models for cell killing have been more prevalent than for other end points. However, as future emphasis is placed on other end points that will be more relevant to the carcinogenic process, it should not be forgotten that a complete theory of radiation carcinogenesis must include the effect of cell killing, particularly for high LET radiation, since killing will affect the number of cells at risk.

Three mechanisms for understanding the shoulder on the survival curve appear to dominate current thinking: i) interaction of lesions, ii) non-linear yield of initial lesions, and iii) saturation of repair processes. From the discussions at this workshop, the evidence appears to favor the interaction of lesions hypothesis, and in particular, the misjoining of double-strand breaks.

As far as growth points in model development are concerned, there is much room for expansion of current models to include physical, chemical, and molecular processes in a more comprehensive manner. We have had several fruitful discussions along these lines. The processes that were mentioned include charged-particle track structure within appropriate cellular milieu, indirect (e.g., OH radical) and direct (e.g., DNA cation) effects, and the clustering of DNA double-strand breaks, as well as DNA base changes, point mutations, frame shifts, and so on. Finally, we need to develop new models and/or theories for end points of cell transformation all the way to carcinogenesis, as well as for molecular end points such as double-strand breaks of different types ("blunt" and "sticky" ends), gene activation and inactivation, and chromosomal aberrations that are precursors. Other processes, such as cell proliferation as mentioned by Dr. Moolgavkar, must also be incorporated into the theoretical modeling.

There is a very exciting future here. The synthesis of all relevant information into a coherent picture of radiation carcinogenesis for the variety of cancers suspected of being radiogenic in humans is an extremely challenging task, and I am convinced it will, in the end, be a very rewarding enterprise.

Hans Menzel: *CEC Perspective*

I thank you for the invitation to come here, I think everybody knows I come from the Commission of European Communities and am responsible for a project which links up with part of this OHER program. I think that the point of real common interest is in the understanding of mechanisms, and I am very grateful that I have been able to come here. Also, in view of our future common projects, particularly the meeting on modeling to be held next year; we have already heard part of the discussions here. In two years' time there will be the eleventh symposium in microdosimetry, and I was particularly interested in gathering as many ideas and as much information as possible, so I can put together a scientific program which will in effect continue these things. I would just like to say one thing about those programs, and the DOE programs which are motivated and justified by the problems of radiation protection and the questions of low dose or dose rates. Inasmuch as the final answer will be coming from mechanistically based models, we really don't fully understand all the mechanisms. Thus, the "in-between" models, which give us the guidance and a basis for setting up risk factors, and setting up limits, are important. Some of you may know that the ICRP is coming up with new recommendations. I'm not sure that all this information is available. We could probably better present, with the help of models, a better base for such decisions as they are known. I think that we all should think of the practical applications, and I would like to ask you all to help in this project.

Robert Wood: *Reflection on the DOE/OHER Program*

Matt asked me to make a few comments and reflections on this past 15 or 20 years, and I am certainly pleased to do so; but before I get into that, let me recall Wednesday evening and the honor that you gave to Matt and me with the unexpected and very fine presentation of those plaques. At that time I tried to indicate some of the people at Headquarters who over the years were responsible for the development, guidance, and defense of this program. I want to get on the record here that I think the real people that are responsible for the quality of this program are, of course, you who have done the research through the years, and the others who are part of our radiological and chemical physics program but who aren't with us here today. In thirty years this has been an exceptionally strong program, as you know and I think the Department knows, and it is pretty clear that without that sort of performance and accomplishment, none of us would be here today. So we are certainly indebted to you. I might point out that this program, as you know, is housed within the Office of Health and Environmental Research. There are other groups within the Department who are responsible for the atomic and molecular physics research, chemistry research, and similar areas of related research. This program, indeed, can stand with the best of those other programs which have responsibilities similar to ours. We will have that plaque on the wall; indeed, it will be a reminder of the quality that has come out of this program over the years.

All right, some reflections—clearly the Airlie House meeting was predominantly physics oriented. A bit of chemistry was included, and almost a nodding acknowledgment of biology, with a challenge, as has been stated before, of accomplishing the coupling that we all felt was desirable, to use physics, chemistry, and biology in understanding the basic mechanisms of radiation interaction. Clearly, from the presentations that we have had in the last three or four days, that challenge was very successfully met. I think that fewer than half of the presentations at this meeting dealt with physics.

We have really moved forward in this area, and I was thinking about how we have accomplished this change in scope, in thrust over the years. There are two or three things that are rather significant. First of all, I look upon some of the components of our program, the core program, those groups that have been involved with the program over the past 20, in some cases 30 years. I've seen them, since the Airlie House meeting, make a very deliberate effort to specifically broaden the scope of their activities to include chemistry and to include biology within the research activity, either by bringing that sort of expertise into the program itself or by establishing strong collaborations with chemistry and biology research groups. I think particularly of the Argonne program, the Oak Ridge program, the Berkeley program, the Pacific Northwest Laboratory program, and of the university grants. Four of them specifically come to mind, the ones who have been around the longest: the Columbia program, New York University program, Mike Kasha's program at Florida State University, and Bob Katz's program at the University of Nebraska. When you look at the offsite university program within our office, we probably have 250 or 300 grants now. Over the past 30 years, we have probably handled thousands of different university grants, and of those there are probably less than a dozen who have been around for the length of time that these individuals have. Each of them has been here I think, about 30 years. Columbia started in the late 1950s with Dr. Failla, then Harald Rossi, and now Eric Hall. At New York University, Dr. Helmut Kallamn started, then, of course, Warner Brandt, and now Nick Geacintov. At Florida it was Mike Kasha at the beginning and Mike to the end. And in talking about modeling, in particular survival curves, one might want to model how collectively they probably survived 80 or 90 site visits and reviews—very, very successfully. Close behind that group is Bob Katz, who I think joined us in 1964 or 1965; he has been with us a good 25 years.

Certainly, I think it is very significant that we have managed to accomplish the broadening of the program to achieve this coupling. We deliberately try to bring in new expertise from outside the radiation community. I think that is clearly exemplified by Harel Weinstein, who is with us, and Raoul Kopelman, who is now with us. Also we have coupled with other parts of our Department's program, as with John Ward, to bring in chemistry insight and a new interpretation to bear on these programs.

Finally, I think that in the last several years there has been another event that has been significant, and Hans Menzel already mentioned it. We really have tried to strengthen our interaction with our European colleagues in the Commission of European Communities. We have Clemens von Sonntag and Herwig Paretzke here today, exemplifying and representing what has become a strong interaction. All of that has really contributed to the continued vitality and strength of this program.

So, to the future! I think that the final session this morning was really an introduction to the future. We have been talking about genomes, and we have been talking about structural biology; it is clear that, within our office, these areas in scientific research currently are receiving a very high priority. The director of our office, Dr. Galas, is extremely enthusiastic about the opportunities that will be provided in the future from research in those areas; he is going to give them a lot of personal emphasis. I think that our challenge is to continue to try to relate the issues of radiation interaction, at the fundamental mechanistic level, with the new insights, the new understanding that can come from continued coupling with the areas of genome and structural biology research that are going to receive emphasis over the next five to ten years.

So, there you have it. That is where I think we are going to go. In closing, let me thank all of you for joining us and making this into what I am sure is going to be a landmark meeting and a successful document. Finally, let me just say to my friend Bill Glass, "Thank you, Sir, for all the effort that you've made. You arranged for these facilities, which I think are absolutely delightful, and now you have the incredibly difficult task of trying to make sense out of all this discussion and putting it together into a readable document. I know you will." So, thank you, Bill, and thank you all.

Matesh Varma: *Conclusions*

Two people really deserve recognition for putting the agenda for this meeting together; Bill Glass and Aloke Chatterjee were really involved from the beginning to see that an Airlie House type of proceedings will be published in a timely manner. We can talk about it in the coming years and say that 1990 was the year to be successful. I think that this has been a very good place to meet and spend time together discussing these kinds of problems. Bill, thank you very much. Thank you to PNL for its support, and thank you for helping me.

Participants

Leslie A. Braby
Pacific Northwest Laboratory
P8-47
900 Battelle Blvd.
Richland, WA 99352

Aloke Chatterjee
Lawrence Berkeley Laboratory
MS 29-100
1 Cyclotron Road
Berkeley, CA 94720

Loucas G. Christophorou
Oak Ridge National Laboratory
4500S, MS-6122
Bethel Valley Road
Oak Ridge, TN 37831-6122

Stanley Curtis
Lawrence Berkeley Laboratory
MS 29-100
1 Cyclotron Road
Berkeley, CA 94720

Nicholas Geacintov
Department of Chemistry
New York University
Washington Square
New York, NY 10003

William A. Glass
Pacific Northwest Laboratory
K4-13
900 Battelle Blvd.
Richland, WA 99352

Eric J. Hall
Center for Radiological Research
Coll. of Physicians and Surgeons
Columbia University
630 West 168th Street
New York, NY 10032

Mitio Inokuti
BEM, Building 203
Argonne National Laboratory
Argonne, IL 60439

Michael Kasha
Institute of Molecular Biophysics
Florida State University
Tallahassee, FL 32306

Robert Katz
364 Behlen Laboratory
University of Nebraska
Lincoln, NE 68588

Raoul Kopelman
Department of Chemistry
The University of Michigan
930 No. University
Ann Arbor, MI 48109-1055

Dr. Hanz Menzel
Division of Radiation Protection
CEC DG 12/D/4
ARTS/LUX 243
Rue de la Loi 200
B-1049 Brussels, Belgium

Suresh Moolgavkar
Fred Hutchinson Cancer Center
1124 Columbia Street, MP-665
Seattle, WA 98104

Herwig Paretzke
G.S.F. mbH München
Institut für Strahlenschutz
Ingolstädter Landstrasse 1
D-8042 Neuherberg
Germany

Rufus H. Ritchie
Health and Safety Research Div.
Oak Ridge National Laboratory
4500S, MS-6123
Oak Ridge, TN 37831-6123

Harald H. Rossi
105 Larchdale Avenue
Upper Nyack, NY 10960

Clemens von Sonntag
Max-Planck-Institut für Strahlenchemie
Stiftstrasse 34-36,
D-4330 Mülheim an der Ruhr 1,
Germany

Larry H. Toburen
Pacific Northwest Laboratory
P8-47
900 Battelle Blvd.
Richland, WA 99352

Matesh N. Varma
ER-74 E-217/GTN
Office of Health and Environmental Research
U.S. Department of Energy
Germantown, MD 20545

John F. Ward
Department of Radiology, M-010
Univ. of California, San Diego
9500 Gilman Drive
La Jolla, CA 92093

Harel Weinstein
Department of Physiology &
  Biophysics
Mail Box 1218
Mt. Sinai School of Medicine
New York, NY 10029

Robert Wood
ER-74, GTN
Office of Health and Environmental Research
U.S. Department of Energy
Germantown, MD 20545

Marco A. Zaider
Radiological Research Laboratory
Columbia University
630 West 168th Street
New York, NY 10032

Participants at the Woods Hole Workshop

Back row: Clemens von Sonntag, Nicholas Geacintov, Rufus Ritchie, Loucas Christophorou, Hans Menzel, Harel Weinstein, Harald Rossi, Eric Hall, Raoul Kopleman, and Michael Kasha.
Middle row: Stanley Curtis, John Ward, Larry Toburen, Herwig Paretzke, and Leslie Braby.
Front row: Robert Katz, Suresh Moolgavkar, Aloke Chatterjee, Matesh Varma, Bill Glass, Robert Wood, Mitio Inokuti, and Marco Zaider.

INDEX